REVERSE DESIGN

PROCEEDINGS OF THE 4TH INTERNATIONAL FASHION AND DESIGN CONGRESS (CIMODE 2018), 21–23 MAY 2018, MADRID, SPAIN

Reverse Design

A current scientific vision from the international fashion and design congress

Edited by

Ana Cristina Broega, Joana Cunha & Helder Carvalho
Department of Textile Engineering, University of Minho, Portugal

Manuel Blanco
Escuela Técnica Superior de Arquitectura de Madrid de la Universidad Politécnica de Madrid (ETSAM-UPM, Architecture School of UPM), Madrid, Spain

Guillermo García-Badell & Diana Lucía Goméz-Chacón
Centro Superior de Diseño de Moda de Madrid de la Universidad Politécnica de Madrid (CSDMM-UPM, Fashion Center of UPM), Madrid, Spain

CRC Press is an imprint of the
Taylor & Francis Group, an **informa** business

A BALKEMA BOOK

CRC Press/Balkema is an imprint of the Taylor & Francis Group, an informa business

© 2019 Taylor & Francis Group, London, UK

Typeset by V Publishing Solutions Pvt Ltd., Chennai, India

All rights reserved. No part of this publication or the information contained herein may be reproduced, stored in a retrieval system, or transmitted in any form or by any means, electronic, mechanical, by photocopying, recording or otherwise, without written prior permission from the publisher.

Although all care is taken to ensure integrity and the quality of this publication and the information herein, no responsibility is assumed by the publishers nor the author for any damage to the property or persons as a result of operation or use of this publication and/or the information contained herein.

Published by: CRC Press/Balkema
Schipholweg 107C, 2316 XC Leiden, The Netherlands
e-mail: Pub.NL@taylorandfrancis.com
www.crcpress.com – www.taylorandfrancis.com

ISBN: 978-1-138-37011-1 (Hardback)
ISBN: 978-0-429-42821-0 (eBook)

Reverse Design – Broega et al. (Eds)
© 2019 Taylor & Francis Group, London, ISBN 978-1-138-37011-1

Table of contents

Preface	xi
CIMODE 2018 committees	xiii

Fashion and communication

Performative runway in backstage liquidity *A. Rabàdan & I. Bentz*	3
From hair to cultivated leather—the use of Alexander McQueen's genetic information *B.F. Pires & R.P. Cidreira*	13
The relevance of sketches in fashion design *A. Moreira da Silva*	21
Digital banners as shop windows: Narrative constructions of online fashion brands *D.V. Leal, D. Karam Jr. & M.C.P. Rodrigues*	27
The influence of soap operas in Brazilian fashion today *T.C. Viana, A.A. Silva, A. Adverse, R. Ribeiro & A. Horta*	37
The world and the fashion "al revés" in António Variações *M. Amaro*	45
Facets of youth culture: Discursive construction of Ahlma fashion brand *B. Machado & M. Scoz*	53
Fashion films: A semiotics approach, a multimodal signification *R. Effe*	63
The importance of language on the social medias of a brand *M.H. de Carvalho & F.A.H. de Carvalho*	71
The effects of sense contained in the fashion advertising text: The Handred brand and the reiteration of genderless concept *S.F. Bona & M. Scoz*	79
Analysis of attributes in unisex and genderless clothing *B.M. Reis, R. Miguel, N.A. Jerónimo, M. Pereira & S. Azevedo*	87
Embroidery: The narrative of the ribbon of the presentation of Arthur Bispo do Rosário—confluences with contemporary visual artists *L.U. Dantas, U.S.T. Barbosa, G.M.J. Sales, H.A. Dieb & R.R. Marques*	95
Eye tracking in fashion: An overview *A.P. Faria, B. Providência & J. Cunha*	103

Fashion, identities and cultures

Semiology: Culture as a symbol system and western influences comparison before and after the independence of 1947 *L. Vignes*	113

A geography of fashion, Medellín: 1900–1950 *W. Cruz Bermeo*	121
The *Prêt-à-porter* phenomenon and the boutiques in the dowtown area of Santiago, Chile *J. Vidal Miranda & P. Álvarez Caselli*	129
Internationalism and exoticism in the image of Aline Romanones and Consuelo Crespi *D. Gennaioli*	137
Reframing old age through design: An approach to promote empowerment *I. Rojas & B. Pino*	145
Dolce and Gabbana and Louis Vuitton: A study on the effect of gender stereotyping in the 21st century fashion campaigns *L. Watson*	153
Fashion, identity and culture in the Chinese fashion system. China, inspiration for Chinese fashion designers *R. Gaddi*	161
The pleasurable dressing of Loewe: From the store of Martínez-Feduchi to the showroom of Carvajal *A. Cano Redondo & A. Martínez-Medina*	169
Signifying the ritual legends of Brazilian Afro Religions: *Candomblé* and *Umbanda* in *Teresina/Piauí*/Brazil *L.C. de M. Tavares & V.P. de S. Araújo*	177
Globalized brazilianness on the forming way of "Elementais" *G.R.R. Vieira*	183
Class tourism, working-class porn and cultural appropriation in fashion: An introduction *M. Glück*	191
Affection sewn into hats: Identity, self-esteem, women's social relations *G.P. Lenzi*	199
The lace fabric of Bahia clothing: From artisanal to industrial *A. Okasaki*	207
Fashion design and craftwork: Subjectivation policies in the contemporary scenario *A.R.V. Peroba & C. Mesquita*	215
Chinese fashion, occidental fashion: A semiotic translation *Y. Pan*	223
The fashion design as a contribution to the preservation of Nazaré culture and costumes: 7 project *S. Moreira, M. Pereira, A. Cruchinho, J.M. Lucas & R. Miguel*	231
DIY and the slow fashion movement: Sew for yourself *L. Barrocas, G. Bezerra & M.A.V. Rocha*	241
Modes and Fashions of the colonial man from *Minas Gerais*: The indumentary of the defendants of the inconfidence of *Minas Gerais* *A.F.B. Santos & A.P.X. Vilela*	247
The aesthetics in fashion design and cultural studies *T. Lobo*	255
The physiological aspects of senescence: Comfort and relationship with the clothing *M.D. Almeida, A.C. Broega & M. Moura*	263

Fashion design through sustainability and material culture: An exploratory and experimental study 271
R. Norogrando

Castelo Branco embroidery applied to clothing 279
A. Cruchinho, A.S. Marcelo, P. Peres & A. Moura

Female jeans: The aesthetic of Brazilian "popular fashion" 287
I. Braga, M.J. Abreu & M.M. Oliveira

The liberation of women in the 19th and early 20th centuries as seen through costume 295
L. Luceño

The *Minho* traditional costume as cultural heritage in fashion design 303
S. Castro, J. Cunha & C. Morais

The casualisation and homogeneity of contemporary fashion 313
F. Spry

Product design

Practices of design: Understanding processes 323
A. Rabàdan

Contemporary textile products. Knitting as a fertile design ground for experimentation with 3D technologies 331
G.M. Conti

Clothing as an architectural project 341
B. Alcoceba

Marking embroidery in colors: The creative fashion process in *AMAC*, in *São João dos Patos – MA* 349
M.S. Lima, R.G. Noronha & L.R. da Silva

Miramar embroiderers, design and territory 357
F.P. Vallejos

Strategic approach to implement sustainability in the joineries of the city of Uberlândia, Brazil 365
F. Moreira da Silva & J. Cardoso Braga

A proposed procedure to develop clothing for pregnant women 373
D.A.N. Mentone, R.H. Osava, S.H.A. Gomes, R.A. Sanches & A.Y.S. Duarte

Functional fashion focused on the needs of people with disabilities 381
L.N. Souza, S.H.A. Gomes, C.R.G. Vicentini, R.A. Sanches & A.Y.S. Duarte

Designing for emotions: Evaluation of the drooler, a toy for preschoolers 389
B. Providência, R. Brandão & P.B. Albuquerque

Creativity and industrial fashion design—reviews, analyses and connexions 399
J.A.B. Barata, R. Miguel & S. Azevedo

Procedures and guidelines for the instruction and execution of pattern making: An analysis 407
P.A.A. Spaine, D.M. Brito, L.M. Pereira, N. Pinheiro & R.R. Andrade

Anthropometry and clothing for overweight and obese children: A literature review 415
R. Campos & M.A. Carvalho

The materialization of fashion products: An experience of the textile behavior P.M. Souza, M.M. Otani, P.P. Silva & I.C. Italiano	421
Design and redesign G. Montagna & M.J. Delgado	429

Marketing and consumption

Psychographic segmentation of female fashion consumers in Portugal B. Moreira & A. Azevedo	437
Taxhion. Model taxonomy clothing and accessories e-commerce B. Alcoceba, M. García-Ergüin, R. García, D. de las Heras & M. Adsuara	447
Cultural values of minimalist fashion A.P. de Miranda, I. Domingues & N. Souza	455
Emerging platforms for fashion design entrepreneurs: An assessment C.E. Fernandes, L. Ribeiro, M.S. Silva, M. Pereira & M.J. Madeira	463
The use of online advertisement as a TXM branding methodology's tool: FARM's case study L.W. Ribeiro & L.S. Gomez	469
The influence of social network sites on digital branding M. Barreto & L. Ribeiro	479

Teaching and education

Teaching design in real contexts to preserve local identity and memory A.H. Grácio & C. Rijo	489
Immersive business simulation in footwear design education: A bridge for reality A.M. Terroso, A. Moreno, N. Amorim, H. Palmares & J. Sampaio	497
Bounding fields of knowledge: Co-designing regional pottery of the Cávado sub-region as a strategic potential of museological differentiation and product development J. Sampaio, A. Baganha, F. Gomes & A.M. Terroso	507
A case study of didactic laboratory approach in fashion design education X. Lin & A. Dell'Acqua	515
Feminine magazines and the development of the female labor force in Valparaiso: Analysis of the graphic system of the Rosita magazine (1947–1972) as a didactic strategy for the informal learning of sewing F. Gonzalez, U. Bravo & C. Ruiz	521
Bê a Bá of sewing: Our online friend L.C. de M. Tavares & J.P.C. de Brito	529
Professional education and the formation of garments factory in Brazil V. Feldman & M.S.B. de Held	537
Complex geometry and patterns: Colonizing the surrounding space through the costume M.J. Climent Mondéjar, M.P. Moreno Moreno & A. Cano Redondo	543
Project thinking in fashion design: Strategies for facilitating cognitive and metacognitive processes M.C.F. Sanches & M.A.R. Silva	551

Pattern making books: An analysis of documentary research 559
P.A.A. Spaine, D.M. Brito, L.M. Pereira, N. Pinheiro & R.R. Andrade

Interdisciplinarity: Creation of looks and techniques used by UFPE CAA students 567
A. Camargo & R. Alves

Active applied methodologies on fashion design teaching: Possibilities and challenges 575
L.E.F.S. Rebello, G.D. Marques & M. Mansur

Sustainability in fashion and design

Fashion that cares for the future 585
E. Araujo, I. Castro, I. Pinto & I. Araujo

Design, tradition and culture—past and future united to develop depressed regions 593
I. Oliveira & M.G. Guedes

Circular economy: An approach for the fashion industry 599
E. Pinheiro, S.M.B.D. Barcelos & A.C. de Francisco

Reversal logistics: Case study in the franchises O Boticário in the city of Itabuna-BA 607
E.N. Velanes

Sustainable fashion. Strategies for sustainability and new forms of value creation
in fashion and textile 615
M.A. Sbordone, L. Di Lucchio & R.A. Sanches

Generation Y's sustainability attitude-behaviour gap 623
J.P. Bernardes, F. Ferreira, A.D. Marques & M. Nogueira

Collaborative economy: Case study of new business models 631
S.R. Fernandes, J.M. Lucas, M.J. Madeira, A.I.C. Barreiros & I.D. Honório

Imbrications and distances between the creative economy and the sustainability
of small fashion brands 637
U.S.T. Barbosa, H.A. Dieb, G.M.J. Sales, L.U. Dantas & A.N. Targino

New sustainable fashion business models: "the *coworking*" case study 645
L.F. Barcellos & A.C. Broega

Sustainability: The reconfiguration of fashion, the textile industry and the role
of the designer in a new social scenario 653
M. de L. Iracet

Tucum fiber: Reflections about Amazonian biodiversity, traditional knowledge
and sustainable fashion 659
L. Pennas & J. Baruque-Ramos

Silk: Protocols for the verification of socioeconomic impacts in the production of cocoons 667
S.M.B.D. Barcelos, M.G. Guedes, E. Pinheiro & A.C. de Francisco

A perspective towards a circular fashion design system 675
B. Melo & A.C. Broega

Fashion and sustainability's valences: Exposing gaps 683
R. Puppim, C. Jordão, L.M. Arruda, D.P. Beduschi & A.C. Broega

Design and social innovation: Methodological principles for community qualification 691
J. Oenning, J.B. Garcia Jr. & J. Cunha

Author index 699

Preface

It has been an honor to organize this 4th International Fashion and Design Congress (CIMODE 2018) with the Centro Superior de Diseño de Moda de Madrid, (CSDMM-UPM), of the Universidad Politécnica de Madrid.

We would like to take this opportunity to thank the Universidad Politécnica de Madrid joint chairs, Professor Manuel Blanco and Professor Guillermo García-Badell, who had the unyielding support of Angel Cordero, Diana Lucía Gómez Chacón, Laura Luceño and Mercedes Rodríguez.

The International Fashion and Design Congress – CIMODE – is based on six major subjects: Fashion and Communication; Fashion, Identities and Cultures; Product Design; Marketing and Consumption; Teaching and Education, Sustainability of Fashion and Design.

The 4th CIMODE theme "Reverse Design" is both a title and a topic for reflection.

Based on the contributions of researchers, academics, designers and other professionals in the fields of Fashion and Design, CIMODE aims to promote the discussion of ideas and research exchange in Fashion and Design.

In this 4th edition of CIMODE 126 full papers were presented, from which 86 were selected for this book; 17 posters of scientific initiation works; and more than 200 delegates participated in the congress.

All papers submitted went through a double-blind reviewing process only possible due to the invaluable support of 90 members of the CIMODE Scientific Committee that selected the utmost research in the field of Fashion and Design.

During the three days of the congress, academics, students and professionals engaged in rich inter and trans-disciplinary debates, sharing ideas and researches in the plenary sessions, round tables and parallel sessions.

The fourteen international Fashion and Design thinkers that acted as keynote speakers, brought in reflections on the new paths for the future of Fashion and Design with a closer vision for people and community. The plenary sessions were organized to bring to the debate both the global as the local perspectives, as well as the theoretical reflection, the experimentation and practice. In the opening session Manuel Blanco (Escuela Técnica Superior de Arquitectura de Madrid de la Universidad Politécnica de Madrid) introduced the conference theme "Reverse Design", reflecting on the new challenges in teaching and practice of design, and on the importance of the integrative process in design. In the second plenary session under the theme "Fashion Communication and Experience", Jorge Lozano Hernández described his vision of the future in Fashion Design, Daniel Rabaneda (Designer at Ángel Schlesser) presented his vision of the new creative processes, from teaching in the UPM to the creative direction of Ángel Schlesser; Alberto Gomper (CSDMM) focused on his experience in design communication. The third plenary session was devoted to design strategies where Silvia Nuere Menéndez-Pidal (researcher and teacher at Universidad Politécnica de Madrid) presented her view on project methodology: a collaborative vision between architecture, fashion and design, Giovanni Maria Conti (Politecnico di Milano) discussed Cross-fertilization in design, fashion and architecture and Maria Celeste Sanches presented her view on Fashion and Design: by a systemic and transversal approach. The fourth plenary session adressed the issue of Fashion and the Museums with Igor Uria Zubizarreta (Museum Cristobal Balenciaga) discussing about Fashion and Heritage in the light of the Balenciaga Museum, Helena López del Hierro (Director of the Museo del Traje Madrid) presenting the approach of the Museo del Traje to the preservation and exhibition of Fashion and Héctor Navarro (CSDMM) addressing the diffusion of fashion as patrimony in the academic sphere.

The round table devoted to the debate on "Reverse Design" brought together the views and experiences of various speakers: Kathia Castilho presented the approach to the social innovation design of the "Periferia Inventando Moda" project; Albert Mora Viura (head Designer at Springfield) discussed the sustainability strategy and the ethical concerns of the Springfield fashion brand; Alvaro Catalán de Ocón shared his experience in the "PET Lamp" project of Social Innovation in collaborative work in Design, Paloma G. López (The Circular Project) spoke about Ecodesign and Ecosystem in "The Circular Project" and José Lima (Ideal & Cia) shared his experience of co-working on sustainability.

It is our understanding that the themes discussed during the three days of the conference, both in plenary as well as in the parallel sessions, provided a valuable basis for new thinking about the future of fashion and design, and allowed closer cooperation links between researchers and academia, opening space for new reflections.

The CIMODE 2018 Organizing Committee

CIMODE 2018 committees

EXECUTIVE COMMITTEE

President: Manuel Blanco—*Universidad Politécnica de Madrid, Spain*
Vice-President: Guillermo García-Badell—*Universidad Politécnica de Madrid, Spain*
President: Ana Cristina Broega—*Universidade do Minho, Portugal*
Vice-President: Joana Cunha—*Universidade do Minho, Portugal*

ORGANIZING COMMITTEE

Manuel Blanco, Guillermo García-Badell, Angel Cordero, Diana Lucía Gómez Chacón, Laura Luceño, Mercedes Rodríguez—*Universidad Politécnica de Madrid, Spain*
Ana Cristina Broega; Joana Cunha; Bernardo Providência; Helder Carvalho; António Dinis Marques; Celeste Sanches (ABEPEM)—*Universidade do Minho, Portugal*

GENERAL SECRETARIAT

Helder Carvalho

SCIENTIFIC COMMITTEE

President of the Scientific Committee for UPM: Guillermo García-Badell
President of the Scientific Committee for UM: Joana Cunha

MEMBERS

Alessandra Vaccari—*Università IUAV di Venezia (IT)*
Alexandra Cruchinho—*IPCB/ESART (PT)*
Ana Claudia Mei Alves de Oliveira—*PUC (SP-BR)*
Ana Cristina Broega—*DET UM (PT)*
Ana Jiménez—*UPM (ES)*
Ana Margarida Fernandes—*ESART-IPCB (PT)*
Ana Mery De Carli—*UCS (RS-BR)*
Ana Ramos—*IPCB/ESART (PT)*
André Robic—*IBModa (BR)*
Ángel Cordero—*UPM (ES)*
Anne Anicet Ruthschilling—*Uniritter (BR)*
António Dinis Marques—*UM (PT)*
Araguacy Filgueiras—*UFC (BR)*
Arturo Dell'Acqua Bellavitis—*(IT)*
Bárbara Pino—*FAAD-UDP-Santiago (CL)*
Bernardo Providência—*EAUM (PT)*

Brezo Alcoceba—*ESNE (ES)*
Carla Morais—*FAUL (PT)*
Carlos Alberto Miranda Duarte—*IADE (PT)*
Carlos Figueiredo—*FAUL (PT)*
Cátia Rijo—*ESEPL (PT)*
Cristiane Mesquita—*UAM (SP-BR)*
Cristina Figueiredo—*UTAL (PT)*
Desamparados Pardo Cuenca—*ESDValencia (ES)*
Diana Lucía Gómez-Chacón—*UPM (ES)*
Elisabeth Lorenzi Fernandez—*UNED (ES)*
Fausto Viana—*ECA-USP (SP-BR)*
Fernando Moreira da Silva—*UTAL (PT)*
Francisca Mendes—*UFC (BR)*
Gianni Montagna—*FAUL (PT)*
Giovanni Maria Conti—*POLIMI (IT)*
Guillermo García-Badell—*UPM (ES)*
Héctor Navarro—*UPM (ES)*
Hélder Carvalho—*DET UM (PT)*
Helena Pires—*DCCOM-UM (PT)*
Herbert González—*UPM (ES)*
Isabel Cristina Gouveia—*UBI (PT)*
Joana Bosak de Figueiredo—*(UFRGS-BR)*
Joana Cunha—*DET UM (PT)*
José Ferro Camacho—*IADE (PT)*
Kathia Castilho—*ABEPEM (BR)*
Laura Luceño—*UPM (ES)*
Laura Zambrini—*FADU-UBA (AR)*
Luciane Robic—*IBModa (BR)*
Luiz Salomão Ribas Gomez—*UFSC (BR)*
Madalena Pereira—*DCTT UBI (PT)*
Manon Salles—*EBA-UFRJ (BR)*
Manuel Blanco—*UPM (ES)*
Mara Rubia Sant'Anna—*UDESC (SC-BR)*
Marcelo Martins—*UFRPE (PE-BR)*
Marcia Mello—*UNIFACS (BA-BR)*
Maria Alice Vasconcelos Rocha—*UFRPE (BR)*
Maria Carolina Garcia—*UAM (SP-BR)*
Maria Celeste Sanches—*ABEPEM (BR)*
Maria Claudia Bonadio—*SENAC/SP (BR)*
Maria Cristina Volpi Nacif—*EBA/UFRJ (BR)*
Maria da Graça Guedes—*DET UM (PT)*
Maria de Fátima da S. Costa G. de Mattos—*CUML/RP (SP BR)*
Maria do Carmo Teixeira Rainho—*(Senai-Cetiqt-BR)*
Maria Eduarda Araujo Guimarães—*SENAC/SP (BR)*
Mário de Araujo—*(UM)*
Marizilda dos Santos Menezes—*UNESP (SP-BR)*
Marly Menezes—*FASM (SP-BR)*
Mercedes Rodríguez—*UPM (ES)*
Moisés Lemos Martins—*DCCOM UM (PT)*
Mônica Moura—*UNESP (SP)*
Nelson Pinheiro—*IADE-UL (PT)*
Olga Pépece—*UEM (BR)*
Patricia de Mello Souza—*UL (BR)*
Patrizia Ranzo—*UNINA2 (IT)*
Paula Trigueiros—*EAUM (PT)*

Pedro Bessa—*DCA-UA (PT)*
Rafaela Norogrando—*UBI / IPV (PT)*
Raul Cunca—*F Belas-Artes UL (PT)*
Regina Sanchez—*EACH-USP (BR)*
Renata Pitombo Cidreira—*UFBA (BR)*
Rita Ribeiro—*ICS-UM (PT)*
Rossana Gaddi—*POLIMI (IT)*
Rui Miguel—*DCTT UBI (PT)*
Rui Roda—*(PT)*
Sandra Rech—*UDESC (SC-BR)*
Silgia Costa—*EACH-USP (BR)*
Silvana Mota Ribeiro—*DCCOM UM (PT)*
Sirlene Costa—*USP (BR)*
Solange Riva Mezabarba—*UFF (SP-BR)*
Solange Wajman—*UNIP (SP-BR)*
Suzana Barreto Martins—*UEL (BR)*
Teresa Franqueira—*DCA-UA (PT)*
Tula Fyskatoris—*CPS (PUC-SP-BR)*
Valter Cardim—*IADE (PT)*
Zara Pinto Coelho—*DCCOM UM (PT)*

Fashion and communication

Performative runway in backstage liquidity

A. Rabàdan
Unisinos Design Graduate Program, Porto Alegre, RS, Brazil

I. Bentz
Unisinos, Porto Alegre, RS, Brazil

ABSTRACT: This article intends to carry out an initial articulation exercise between the object of study—the performative catwalk (as a scenic arena that acts on the tensions between body and social power), aesthetics and design. It is a dialogue between aesthetics and design for the understanding of the performative runway. Oriented by the concepts of plot and double bond, in the polyphonic proposal of M. Canevacci; besides the Liquid Concepts of the same author, which study the liquidity and fluidity of the new times. This brings with it a new approach to working with the multiple cultures, societies, diversities and realities that meet the strategic design proposals developed by E. Manzini. It seeks to understand the design processes that are involved in the design of a fashion conceptual runway, such as performative runway. This constructive process operates from the 'Backstage' to the totality of its realization, to the understanding of its social essence. The following are considered: the designer's practice, the construction of the design problem and the relevance of the ecosystem concept, among other complementary ones. This along with possibilities of expansion to other spaces of action of the actors involved in the expression of the performative runway.

1 INTRODUCTION

What directed the choice of the theme that gives name to this text was the possibility of relating elements such as Body – Fashion – Art – Costume and Walkway through the methodology of strategic design. This is all inspired by the proposals of visual communication developed by Massimo Canevacci (2001) in his anthropological perspective. The interpretation we intend to make, therefore, lies at the interface between design and anthropology, in its creative design processes and expressions of culture of visuality, respectively, capable of showing ways that lead to the discovery of the new. It is also worth mentioning that it comes from the experience of research and teaching in the field of fashion the best inputs for this theoretical-applied essay on the runway, here adjectivized as performative, given the scenic character full of plasticity and performance that characterizes it. This experience authorizes the relationship of complementarity between design, fashion and art.

The term performative runway encompasses a context of complex relationships that, inter-related, collaborate to identify future scenarios. The political and social view that is present in the act of this runway is directly related to the agents/actors that interact in the process of its construction and concrete manifestation, among which are the designer, fashion, body, clothing, performance and creation process, among others. The understanding of this event also requires the understanding of possible interferences in the social contexts and the way in which the performative runway occurs.

To perceive the runway as a confused urban territory, because it is plural and dynamic, of traffic of ideas, people and things that speak more of society than of the product exposed there, corresponds to Canevacci's thought:

> *[...] The following paradox: always being as close as possible to the interior of the complex, confused and superimposed multiples of traffics-miasmas-traffic jams that are triggered [...] moving in the territories of abstraction, to dissolve the urban empirical specificity and to highlight some trends that are typical of the current culture, locally and worldwide.* (Canevacci, 1993, p. 20)

2 PERFORMATIVE RUNWAY: EXPERIENCES, DIVERSITY AND TRANSFORMATION

The runway as a plural urban territory is like a decoder of possibilities that the fashion designer recognizes and in which constructs, actively, expressions of culture and art. The performative fashion show is considered as the whole process of construction inserted in the *Backstage* that includes a set of distinct operations, but with connection between them. They are operations of cultural, aesthetic and design expression, whose agents are culturally multiverse agents, but organized in the construction of the runway, based on the adopted methodology. By performative we understand the conditions of the actors' representation when giving life to the narratives that they intend to tell on their projects. They correspond to narrative acts that gain special symbolic expressions, metaphors that distinguish one narrative from another, and which result in allegories, true successive metaphors. This opens, therefore, a wide range of symbolic meanings anchored, in this case, in the narrative actions projected in metaprojectual form. This term is particularly dear to design, because the projectual processes are metaprocesses that give rise to the project itself. The metaprojectual level corresponds to the critical and methodological activity that the design calls itself as relevant so that it can promote transformation and innovation.

Even if it is one out of several processes that integrate to configure this performative runway, in some moments they can be thought individually as, for example, the fashion show itself.

Here it becomes necessary to realize that the runway that is being unraveled is configured by a complex set of signs and specific codes that regulate it and make it understood as a catalyst of overlapping cultural layers. Such layers must be exposed and perceived in the social sphere, for they communicate not only the designer's conflicts, but their pre-established codes and their projectual choices. When referring to Canevacci (1993) and his polyphonic gaze in the deconstruction of the city's material codes, one intends to understand also the immaterial ones that are there. This enables, by analogy, to propose a similar look at the object—the runway (here considered as part of *Backstage*).

It is this analogy that is also recognized in the runway with the presence of complex, conflicting or convergent contents that result from the signs in competition for material or immaterial expression of contemporary social realities. In the process of exposing the layers, one of which we are trying to unravel is the process of building this runway reality. In addition, it can be perceived as a living social scenario, where the pre-established processes are shaped according to the agents that are part of this scenario, this fashion ecosystem. This multiplicity of creative systems and processes organized by fashion will be called ecosystems, a term capable of marking the solidarity, interdependence and complementarity that they consecrate among themselves.

So far, one is reflecting on the articulating gaze of thought about *Backstage* – performative runway. As part of the argument, the proposals put forward by Ginger Gregg Duggan (2002) are referred to; in the sequence, the reflection of Canevacci (2001) that in a certain way transgresses the thought of Duggan. This discussion should lead to a conceptual basis capable of responding to the fluidity of thought regarding the solid object of the performative runway, in what it expresses as experiences.

Many of the experiences in the performative *Backstage*-runway set come from reports of actual experiences. In practice, they are almost ethnographic reports of the *Backstage* as part of the runway, using practical actions that can be observed, assimilated and understood to focus on one or more fashion show categories. These actions can be elaborated and result in different visual stimuli in the performative runway. In this way, the look remains on the referred the act showing, which expands in projection for the performative runway space. This field of action is where the specialized design enables the storytelling materialization. When the runway scene is highlighted, within the fashion ecosystem, it is possible to have a wider range of questionings and to begin to dialogue with the possibilities already seen in these reports and; also, of those that have not yet happened. It is when the environment counts more of the performative runway than the fashion show presented there itself.

It is considered here the living social scene, which is present in the elaboration of the performative runway where one can propose simulations of a future—transformation and innovation—as far as one is able to reach the imagination of the fashion *designer*. This happens even with pre-established processes necessary for the articulation of this hybrid moment, as it is experienced at the moment of the show.

The social scene of the performative runway is organized in collaboration and learning, essential conditions of the design action for Ezio Manzini (2017). Were it not for them, the connection between such culturally and socially differentiated forces implied in design processes would be hampered. Thus, it collaborates in the definition and the balance of the messages conveyed by the narratives; it assists in the effort to identify the common points that the performative experiences mean, so that the connecting interlacing make the narrated acts flow, regardless of the languages in which they express themselves. And it is this fluidity that constitutes the learning matter that makes knowledge permanent, without, however, keeping it inert or sterilized. There are fluids and fluidity, as defined by contemporary society, marked by entertainment, exaggeration, speed and disappearance. It is the liquid society, thus characterized by Zygmunt Bauman (2005), in the wake of dialectical materialist thought of the striking industrial and post-industrial reality.

If there is permanent movement in the liquidity of the current scenario, it is collaboratively that recovers the factor of humanity that insists on not disappearing. People collaborate, exchange, and learn to produce innovation and to generate income, whether it is real or symbolic. The idea of community (empowering community for Manzini), can refer to the set of actors, human and nonhuman, that act in common, in convergence; both for the maintenance of states of equilibrium, and for producing important shifts to generate transformations. Because by changing things, people, places and promoting constant displacement, syncretism, the mixtures of culture, of society, of people, of different 'worlds' are favored. Nothing is permanent, everything transits and flows like the signs of aesthetics and performance that travel through the ecosystems of fashion. The performative runway gives us this mixture, this overlapping of things, people and cultures, shifting and generating new meanings for already established, everyday worldviews, but also for worldviews dreamed beyond stereotypes. The agents are moved in multiple directions, the subject-actors are merged, and the liquid space is configured. The collective, erratic and organized diaspora, in its controversy, becomes liquid and hot. It is moved. Because it is possible to do this, it is possible to perceive things, people and society in different ways, performing different roles, permanently circulating without the weight of single belonging, conventional standards or external and subjugating impositions.

In Canevacci's (2005) perspective, the place (performative runway) has a political dimension, like the city in its institutionalized, but irruptive, parameters; in the space of the performative runway, the more fluid movement of the disorder is accentuated. Fashion ecosystems synthesize from this perspective, the irregular anomie, and therefore more interesting, since it cannot be controlled or delimited, being permanent in its movements and open in its symbolic significations. The fluidity of the thought and solid status of the objects are both structured in the search of the performance runway fluidity that drives the ecosystem of fashion today.

In more specific terms, some of the performative shows that happen from the *Backstage* performative runway are characterized as public and social protests. And mostly use through the image generated by the body and its cultural layers, in environments that generate ideas and messages. Among them stands out The Sewing of the Invisible by Jum Nakao (2005). The stylist developed a show in which part of the performance made by the models on the runway was a mixture of delicacy and vulnerability. The low mobility that the pieces allowed transformed the models' coming and going almost in a ritual of time and space, running counter to the speed of the clock. At the end, when all the models were lined up, they abruptly ripped the creations, all of them made of fully detailed vegetable paper with over seven hundred hours of work, causing a commotion in the viewers. This performance act presents the approach of art and fashion on the runway from the Backstage. Here, Ione Bentz and Fábio Parode, (2015) about Ronaldo Fraga, another of these professionals who work in the same way; stylist of Brazilian and social inspiration, in a universal perspective: "Sensitive to social reality, it is attentive to the processes of discrimination or social exclusion and to the events of Latin American and international society and culture." (Bentz, Parode, 2015, p. 59)

In this sense, they complement Káthia Castilho and Marcelo Martins saying that:

> *Both dress and adornment contribute to the construction of the character's identities, as well as revealing the objects and values with which she relates and how they endow certain skills, even if only imaginary ones.* (Castilho, Martins, 2008, p. 30).

Considerations of this order contribute to reaffirm the place of fashion as an expression of art and culture, which places it at higher levels in the paradigms that propose to interpret man's way of being in society, in its ecological relations with the natural environment, society and the subjects.

3 CLOTHING, COSTUMES AND SCENOGRAPHY

In the sequence, some developments that are implicated in the performative runway. Clothing is a form of visual communication of the individual. It expresses aspects of people's personality, and it happens even daily, when they dress to go to work. However, in artistic manifestations this perception ends up happening more intensely. One works as if it were a set design, because the costumes are endowed with elements and symbologies that emphasize certain traits of that character and make laymen understand what kind of feelings that character wants to evoke in certain situations.

It is fitting to connect the costume discussion with a premise of Thomas Carlyle, which is: "The rupture with the belief that the suit is something to uniquely protect from heat or cold and understand the power of communication that clothing proposes historically." (Carlyle, 1831 apud Viana, 2015, p. 66) When relating the scenic space created with Canevacci's fluid performative runway, one can reflect intensely on its communicability. The runway as a way of adding the immateriality of the fashion designer, its creation and style, makes the performance of each show become a unique presentation, which builds an original act and that will not be repeated as such.

During the act, the clothes get a more scenic *status,* or rather artistic, like a sculpture in motion, as it is being built on a living body, a kind of second skin—a protective layer that will power the designer's artistic visuality. It is as if it no longer has its utility value of clothing. One does not seek to find ergonomics or bold design in this scenic space, but rather what it says scenically.

In this sense, it is possible to perceive how it is favored the displacement of the performative catwalk for the audience, also an agent of the construction process of the narrative, in a way more current and exact than the project was intended to execute; the meanings of the show on screen are changed. It is seen a great symbiosis in which there are no boundaries between the actors and the narratives, but fluid movement that traverses the ecosystems spaces in way that is simultaneously collective and singular. It is when the performative runway of the past loses its strength and is re-signified as a great reflection of the collaborative dialogue between all the actors involved in the performance. It is when the social force of making itself visible is more present than the questioning or the reflection that the act of the performative runway itself intended to present.

Canevacci thus expresses itself:

> *In this perspective, not only the clear distinction between different genres is overcome, but it is above all useless or inconceivable a stable, uniform, solid reading: in the irre-*

ducibly labyrinthine aspect of the assembly, the threads to be followed and interconnected are infinite. (Canevacci, 2005, p. 164)

Movements like the ones described above are only possible because it is understood that reality is liquid, not static; that the plots, the societies, the niches, the human thought, the technological advances, everything is liquid, everything is something today that tomorrow may not be. It reflects this conception in the impermanence of the social game in which one can occupy different roles.

Andrea Semprini (2006) points out that such a culture—subjective and that exposes attributes of authenticity, singularity and rarity—accompanies the transformation of the subject's identity into a changing, globalized society that uses the discourses of brands to attract and raise awareness. This conception enables then the rediscovery of the subject in relation to himself, to other subjects and to the society in which he is inserted; it creates the conditions for collaborative learning actions.

The performative show on its liquid runway refers to a classic of the performative fashion on which fell the gaze of this reflection: it is the work of the fashion designer Alexander McQueen (2001). In this specific situation, the show begins with a dim light and is designed to reveal only what is necessary, slowly proposing an inversion in the relation between the performance spectacle of the runway and the moving bodies represented by the models; the relation between models and audience is reversed, because, "Liquid concepts refuse oppositional dualisms." (Canevacci, 2005, p. 159). They seek to move the immobility of words, to follow the diversity of today, present in all areas, throughout the world. Plastering a concept will only make it obsolete because it will not allow its use in understanding all realities. Because the reality itself is illusionist, since it is not unique, and the plastering is impossible. The McQueen show translates a look of the spectator who presents himself to the shows to be seen and to walk on the runway of life. In contrast, the models walk by satisfying their own superstar egos of the fashion ecosystem, for they appeared within a mirrored box from which they cannot see the audience, but only their own reflections; and the audience can see themselves and see the models boxed in their own reflection. The immaterial present in this runway goes through paths of reflection that are greater than the physical constructions of the clothes, or even the characterizations used by the models.

The strength of the materiality present in the creations of the designer can be seen in the images in the sequel presented, in which the models use vestments constructed with symbolic elements like the bandages that cover the head and extrapolate the materiality of the clothes. The act leads the viewer through other ways of reflection, with a strangeness of the real, of the normal, of what is correct, of what must be looked at, repaired, of the disease, or leads to so many other places that the culture of the individual can contribute to make the connections.

Realizing the very strong theatricality of this performance, one can observe how the designer works two sides of the same object, as if it were a staging within the other. This design strategy allows a reflection that is both liquid and immaterial, while solid and material, at the same time. By enclosing the show and hiding it, the audience shifts everything from common sense, and by putting his creations to behold in front of the mirror, the designer deconstructs the erect body and its direct relationship with the fashion show. It shifts from the act of being seen to the act of seeing itself in its deconstructed appearances.

The block of mirrors that constitutes the scenography of the show provides a cold and disconcerting atmosphere. The whole atmosphere proposed by McQueen refers to a non-place that causes strangeness and creates a conflict in the spectator of the performative runway. One can relate this moment of the parade with the Myth of Narcissus, a Greek mythology character who was the son of the god of the river, Cephissus, and the nymph Liriope. It represents a strong symbol of human vanity. Narcissus was an arrogant and proud young man. Instead of falling in love with other people who admired him, he fell in love with his own image when he saw it reflected in a lake.

The following Figures depict scenes from the McQueen conceptual show.

As if all that has already been manifested in the performance parade was not enough, he still reserves an apotheosis, with a very instigating image, that provides a rescue of everything that has been presented in parts and it is as if it were possible to sew, which is flowing and flowing between the fingers. With a glass cube mirrored in the center of the performative runway and reserved for the outcome of performance at a particular time, close to closure, the mirror cube opens, and a woman emerges from her interior outside the fashion beauty standards —naked, with the face covered by a mask, breathing through tubes and surrounded by fluttering butterflies. It is observed in this image that McQueen scenically Liriope a rupture of the aesthetic patterns of the fashion ecosystem of the time and its reflexibility in the social patterns. It is almost a prelude, an anticipation of what we live today in fashion. A rupture of the patterns is configured in an attempt of social acceptance of the different, subject also of significant social relevance.

Figure 1. Conceptual Show (Alexander McQueen Spring/Summer 2001). Source: DAZED Magazine (Allwood, Emma).

Figure 2. Conceptual Show (Alexander McQueen Spring/Summer 2001). Source: DAZED Magazine (Allwood, Emma).

Such an act opens possibilities for reflecting on the fashion ecosystem and its social influence on beauty patterns, for example, or even on the physical structure of the models present at fashion shows. "Thinking the body therefore means confronting the 'object subject' that assumes simultaneously different trajectories in which the multiplicity of meanings brings us to different glances." (Castilho & Fyskatoris, 2008, p. 64). The body cannot be thought of separately from these multiple trajectories that it lives, allowing through the body there presented an infinite construction of possibilities in the field of imagery. This occurs when the limits of form are exceeded, and one begins to identify what else they can offer of raw material to elaborate their movement, their presence in the performance act, under the eyes of several people.

Returning to Figure 3, there is in the center of the scene a person in a calm and serene pose—a woman considered out of standards for where she was (referring to the work of Joel-Peter Witkin: Sanitarium, New Mexico, 1983), but especially within the space of which she was taking part in the moment of the performance, in a great flow of ideas and impermanence. It reflected this image wrapped by beautiful and sick women (models), wrapped in a pattern employed by the fashion ecosystem, wandering bewildered with their heads wrapped in bands, in front of their own reflection.

It is a great experience provoked by the fashion designer in his parade that takes place on a performative runway that allows the encounter of new niches of exploration for the yearnings and dreams of society. His viewers expand the designer's social policy message, which further reverberates back to the self, to the individual in his or her more specific needs. Thus, one realizes the success of his adhesion strategy and the experience of the runway introduces new and creative paths to fashion.

Making a parallel with Becker (1994, Apud Dunne, 2005, p. 96) "[...] To be effective, art must exercise the capacity for detachment [...] it must displace the viewer, reader, audience by its refusal and inability to become part of the reality principle."; art serves to defuse the already established patterns, it displaces the viewer, it provokes a new look and new perceptions.

According to Anthony Dunne (2005), quoting Marcuse, art is a place, a designated imaginative space where freedom is lived. Sometimes it is a physical entity, a place, a painting on the wall, an installation on the floor, an event marked in space and/or time, a performance, a dance, a video, a movie. But it is also a psychic location, a place in the mind where it allows a recombination of experiences, a suspension of the rules that govern daily life, a denial of gravity. This "challenges the monopoly of established reality" by creating "fictional worlds," in which one can see reflected this variety of human emotions and experiences that cannot find an outlet in present reality. In this sense, the fabricated world becomes "more real than reality itself." Art presents the possibility of realization, which only a transformed society could offer." (Dunne, 2005, p. 100).

We reflect here on the role of the professional *designer* and their relationship with the expectations of the future users, co-creating tangible visions of new products or services that can arise from the performative runway and that make use of the design skills that, in this process, are even more relevant. Designers will be sought after by the instrumentalization of their capabilities and will be recognized for their unique work in addressing the challenges arising from global and systemic issues. "The survey of cultural differences: the aim of study is impetuously affirmed as coincident with one's own cultural identity and, at the same time, constantly changing." (Canevacci, 2001, p. 11).

Recently, Manzini (2009) outlined a role for the designer that offers a new perspective, based on the ancient thinking of Italian design. The author suggests that the days of visionary design are over because of what he identifies as a kind of weariness and discontent about utopian visions. Consequently, Manzini advises the use of design skills in the search for visualizing future scenarios and alternative ways for the conceptions to be presented to the public, allowing an approximation with the futures desired by the people. According to him, to reach these futures, designers should develop new strategies, directing the industry to work together with society.

Dematerialization, therefore, has different meanings, depending on what is defined in relation to: immaterial/material, invisible/visible, energy/matter, software/hardware, virtual/real;

Figure 3. Expression of movements of centrality and decentralization. Source: McQueen fashion show Spring/Summer 2001. Dazed Magazine (ALLWOOD, Emma).

but the physical can never be completely ignored. For a change in the attitude of consumption of objects it is necessary that the industry produces solutions, not amenities. The so-called "design without object" would offer, as part of a cultural movement, an alternative to abstinence from consumption; while at the same time encouraging "the forgetting of things as objects of desire and greed" (Vassão, 2010, p. 29).

4 CONSIDERATIONS

The questionings proposed in this article enable dismantling the solid and give vent to the liquidity of the objects or even to the spaces of fashion and to new paths to be explored and investigated. In particular, regarding the articulation of the fashion ecosystem present in the performance runway, since it is complex by nature, these paths can promote relevant advances in the study of fashion and its specificities. The theoretical basis of this text also presents opportunities for deepening the issues related to culture, art and design, and the development of design strategies that bring together the stylist and the designer in collaborative and learning activities. The scope of metaproject may result in the production of design methodologies for the study of the ecosystems of fashion. Such results would give consistency to sociocultural interpretations of facts of this fashion, then added by reviews and conditions of investigative replication. It is a work that is to be done.

The designers have good references of visual aesthetics, know how to conduct creative processes, seek information and can make necessary decisions in the most varied situations; but, for the most part, they are bound to materiality and to let the fluidity of their thoughts and their reflection go deeper. The intangible, what extrapolates from the limits, the hybrid, what is to come: this is the space of the performative runway. It is present physically as a construction, but essentially as a social revelation that is built in this scenic space of projection of anticipation of the future scenario. It becomes, then, an agent capable of triggering a critical reflection for the growth of the area.

In the scenario of new perspectives for the designer, the performative runway is a vast field of experimentation to produce efficient, imaginative and stimulating designs, questioning and attuned to the fluidity of contemporary society. There, where the performative runway is divided into two great performance events: what is being proposed on the runway and what happens to the spectators that produce a parade as strong and striking as the one proposed by the stylist-designer, a situation in which his self becomes the protagonist of the story and builds a performative show of unknown strangers.

REFERENCES

Allwood, E.H., 2015. If You Like McQueen Asylun, You Will Like Joel-Peter Witkin. *Dazed Magazine*, London, Aug 24, 2015. Available at <http://www.dazeddigital.com/fashion/article/26031/1/if-you-like-mcqueen-s-ss01-voss-you-ll-like-joel-peter-witkin> Accessed on: November 25, 2017.

Bauman, Z., 2005. *Identidade*. Rio de Janeiro: Zahar.

Bentz, I., Parode, F., 2015. Moda: eXPressão de arte e design. *e-Revista LOGO* – v. 4, n. 2, 2015. Available in: goo.gl/rFJBvY. Access in: November 18, 2017. p. 57–68.

Cambridge Dictionary online. Available in: <http://dictionary.cambridge.org/us/dictionary/english/backstage> Accessed on: October 16, 2017.

Canevacci, M 2001. *Antropologia da Comunicação Visual*. Rio de Janeiro: DP&A editora.

Canevacci, M 2005. *Culturas eXtremas. Mutações Juvenis nos Corpos das Metrópoles*. Rio de Janeiro: DP & A.

Canevacci, M., 1993. *A Cidade Polifonica: Ensaio sobre a antropologia da comunicação urbana*. São Paulo: Livraria Nobel SA.

Castilho, K.; Fyskatoris, T., 2008. Dobras. V. 2, nº. 4. São Paulo: set.2008.

Castilho, K.; Martins, M.M., 2008. *Discursos da Moda: semiótica, design e corpo*. 2ed. São Paulo: Anhembi Morumbi.

Duggan, G.G., 2002. O Maior Espetáculo da Terra. In: *FashionTeory. A Revista da Moda, Corpo e Cultura*, n. 2. São Paulo: AnhembiMorumbi, June, 2002.

Dunne, A., 2005. *Hertzian tales: Electronic products, aesthetic experience, and critical design*. Massachusetts: Institute of Technology.

Manzini, E. 2009. *Design de serviços na era das redes e sustentabilidade, na concepção de serviços com métodos inovadores*. Helsínquia: Universidade de Artes e Design.

Manzini, E. 2017. *Design: quando todos fazem design: uma introdução ao design para inovação social*, Editora Unisinos: São Leopoldo, Rio Grande do Sul.

Nakao, Jum. 2005. *A Costura do Invisível*. Rio de Janeiro: Ed. Senac.

Semprini, A. 2006. *A marca pós-moderna: poder e fragilidade da marca na sociedade contemporânea*. São Paulo: Estação das Letras.

Vassão, C.A., 2010. *Metadesign: ferramentas, estrátegias e ética para complexidade*. São Paulo: Blucher.

Viana, F., 2010. *Figurino Teatral e as renovações Do Século XX*. São Paulo: Estação Das Letras.

Viana, F., 2015. Revista Modapalavra E-Periódico Vol. 8, N. 15, January/July 2015 P. 81. Modapalavra E-Periódico.

From hair to cultivated leather—the use of Alexander McQueen's genetic information

B.F. Pires
EACH/USP, São Paulo, Brazil

R.P. Cidreira
UFRB, Cachoeira, Bahia, Brazil

ABSTRACT: The human body is the object of different factors that motivate the accelerated growth of techniques and instruments directed towards medical sciences, aesthetic medicine and biotechnology. It is increasingly experiencing possibilities to modify itself through interference, more or less invasive, with the purpose of assimilating it or distancing it from the current standard of beauty and is exposed to a profusion of images of itself and others, disseminated through digital and social media.

Parallel to this, increasingly scrutinizing research on the body, generated by the possibilities given by stem cells and genetic information, opens-up possibilities never before imagined. Among them is the making of objects, in this case fabrics, carrying human DNA.

This article seeks to associate these two aspects, through reflections on the Master's project developed by the designer Tina Gorjanc, who uses Alexander McQueen's DNA to make the raw material used in her collection called "Pure Human".

1 TINA GORJANC AND THE PROJECT "PURE HUMAN"

The Master's project completed in the "Fashion Design" program of Central Saint Martins of London by the Slovenian designer Tina Gorjanc, whose title is "Pure Human", proposes the creation of pieces made from leather that is cultivated from human biological material. The leather and, consequently, the garments or accessories resulting from this process, would bear the genetic information of the individual whose biological material gave origin to the raw material of the pieces.

Gorjanc's intention with this project, published in 2016, was to raise questions related to the current gaps in the law regarding the commercial use, by companies in the area of biotechnology, of human biological material and the capacity, on the part of the luxury goods market, to absorb the innovations provided by technology, regardless of the ethical issues that they may elicit.

To dress with pieces made from the skin/hide of different species of animals that therefore possess varied genetic characteristics is, from the beginning of time, a common act in our society. Among the most commonly used epithelial tissues are those of some members of the bovine family, especially oxen, goats, sheep, and rams. Of these last two, besides the leather, the hairs are used.

The ethical issues raised by the collection "Pure Human" are a result, not from the novelty of this practice, but from the novelty of the species that will provide the raw material for the making of the pieces.

Regarding the legal gaps, the author of the project puts before us two important sets of information. The first is that all biological material, when extracted from a human body, does not belong to anyone, not even the person from whom it was taken. The second is that although legally genetic information cannot be patented, materials and objects that use this information can.

Assuming that getting the DNA of an individual is an extremely easy task, the designer hypothesizes that companies in that field could get a hold of the raw biological material of celebrities, without their knowledge, to turn them into pieces made from cultivated leather. The appropriation of this material could occur, according to the example given by the designer, by bioengineering companies, who could obtain organic materials coming from medical and surgical procedures carried out on famous people or ordinary people. This, if not explained, could lead to the falsification of the product without them knowing or consenting to the transfer or sale of their organic material. Although salons are not mentioned by Gorjanc, they could potentially offer a much easier way to get organic remnants—hair and nails—belonging to a larger and more varied number of people.

In parallel, it is interesting to note that in Western culture, the action of taking possession of a small part of the body of another and transforming it into a body adornment has taken center stage in at least two moments. In the Middle Ages the relics contained parts of holy bodies or materials that had been in contact with these bodies such as clothes or instruments with which the martyrs were tortured. And in the Victorian period with the jewels of mourning and with the so-called "hair work", both made with hair. In all three cases, the motivation for the use of such pieces does not resemble the reasons that give rise to the contemporary proposal of appropriation. In the first, the motivation was related to the belief in obtaining divine protection. In the second, the proximity to the deceased beloved, in the case of mourning, and proximity to the loved one who was alive in the case of "hair work".

In the "Pure Human" project, the pieces, which would be a great novelty in the luxury goods market, would surely arouse the desire of many fans. And certainly, as always happens, they would get false reproductions. Gorjanc, does not focus on possible falsifications, but we estimate that it is impossible to discard them from the demands that would arise from the implementation of the production of leather possessing human DNA.

The desire for a product that is absolutely unprecedented due to its raw material, coupled with a high trade value that is guaranteed due to its originality and resulting brand new and expensive production process, indicates the sale potential that the pieces made with this material would have; at first, along within the consumption of luxury and, subsequently, within mass consumption. Although these factors are enough to guarantee the consumer's interest and the value of the pieces, the product proposed by the project reconfigures the idea of clothing and allows the consumer to dress with the skin of another.

Dressing with the skin of another person means dressing with the outer layer that covers and delimits a body that is not yours. As such, it means that the individual will be affected by the world, by objects, by people, and by phenomena in ways other than the usual ones. This perspective corroborates the very notion of body-image of Massimo Canevacci who understands that the skin spreads through the surrounding tissues. In this way, the skin, as a body mask, possesses and transmits an expressive range of different meanings, capable of stimulating a full scale, also expressive, of new perceptions.

Skin is an organ that protects the organism and is effective as the body's first envelope; the natural limit between the being and everything that is external to him, between the invisible and the manifested, between the most intimate and exposed materiality, has the capacity of acting as a genetic historical map of the individual.

As a map, the skin visually reports not only the material and immaterial events that affect the individual throughout his life, be it voluntarily or involuntarily, in a distinct, specific and own way, for periods that may be transitory or not, but it reports the way how this individual reacted to each of these events. Included in the voluntary records are modifications made through techniques such as tattooing and rhytidoplasty—plastic surgery performed on the face and neck areas to minimize visible signs of aging—, etc. In the involuntary ones, the results from the passage of time and circumstances that encompass the daily life of the individual, such as climate, type of work, illness, accidents, etc.

When voluntary, inscriptions can be acquired, both to distance the individual from the current standard of beauty, and to bring it closer. Those who want to move away from the standard modify the organ that delimits the body through the acquisition of colors, textures, protrusions and recesses that in no way resemble the innate. Those who want to come closer to it modify it through the homogenization of colors, textures, protrusions and recesses with the intent, in most cases, of distancing it from the innate effects caused by the natural process, inherent in all beings, of aging.

A surface essential to the preservation of life, in which material and organic properties are summed to the immaterial and symbolic, the skin, the perceptive organ of touch, is responsible for the uninterrupted relationship of exchange that the individual makes with everything that surrounds him/her. As such, it is responsible for much of the complex faculty of apprehending what is external to us.

In "Anthropology of the Senses," Le Breton refers to skin (it) not only as a mantle, casing, material frontier, composed of organic elements, but also, and in an egalitarian way, as mantle, casing, and symbolic frontier composed of immaterial elements; these are distinguished among themselves by their origins, their contents and the importance that is attributed to these contents according to the culture and the historical period in which it is found or of which we refer to it.

They are the symbolic elements that, added to the organic ones, give the skin the extension of its depth.

> *In our societies, the body draws the outline of the self, it incarnates the individual. Its epidermal frontiers are duplicated by a no less stolen symbolic frontier that distinguishes it from others and founds a personal sovereignty that no one would transcend without an approval.* (Le Breton, 2016, p. 269)

The skin is the most visible surface of the body and thus allows that all which affects the subject, either internally, or externally, is expressed and made visible to the observer. Along with the current state of overexposure of the body to an abundance of images made with powerful cameras and displayed on high-resolution screens that expose sharp and detailed expression marks, wrinkles and blemishes, the skin becomes an important point of interest for the research and creation of new products and procedures developed fields of cosmetics and aesthetic medicine.

The amount of types of products in development, together with legal resolutions that oblige the industries—that supply these fields—to adopt methods that exempt the use of animals in their research and testing processes have significantly boosted research on development of artificial skin, also called 3D. Its morphology and physiology are similar to those of natural human tissue and its use extends to the health field in the treatment of diseases such as melanoma, cervical cancer, chronic dermatological ulcers, and burns.

Among the business groups, universities and development agencies that develop and support research aimed at improving artificial skin construction are the French multinational L'Oreal—which leads this market – the North American laboratory MatTek, the Brazilian business group Boticário, the University of São Paulo – USP, and the Foundation for Research Support of the State of São Paulo – FAPESP.

Unlike one of the issues raised by Gorjanc, the skins produced to meet the aforementioned purposes are done through the legal authorization of individuals whose cutaneous tissues that discarded from plastic surgeries serve as a matrix.

Currently, the skin-in vitro process produces fragments of up to three centimeters in diameter constructed from a mixture of genetic material from several individuals. This procedure decreases individual variability and ensures a more comprehensive result; it takes between ten to thirty days and the durability of the skin produced varies between seven and ten days.

Projects with different purposes. While the fragments destined to cosmetic and aesthetic medicine, elaborated with the combination of cells originating from different bodies, remain for short periods of time inside the laboratories where they are made, they act as small supports for tests of substances that. If approved, they will be applied on the skin of several individuals. The skin resulting from the project developed by Gorjanc, made with the genetic material of a single individual, will produce pieces of clothing that are built to move freely and indefinitely on the skin of those who acquire them.

The change in scale, the possibility of moving through different environments, the durability, and especially the fact that pieces of clothing proposed by the designer possess the genetic information of a single and well-known human being awakens higher degrees of estrangement than those manifested in relation to the epidermal fragments for the aforementioned tests.

In "Pure Human", the DNA chosen by the designer to produce the leather used in her small collection, formed by a jacket, a bag, and a backpack, is that of stylist Alexander McQueen (1969–2010). He, like her, also did his masters at Central Saint Martins.

To get McQueen's DNA, Gorjanc used hair strands that he had incorporated into his first collection, presented at the end of his Master's degree in 1992, titled *Jack, the Ripper Stalks his Victims*.

2 THE SECOND SKINS OF MCQUEEN

From the beginning of his career, British designer Lee Alexander McQueen, known as *l'enfant terrible*, in the fashion world, exhibited an original vision of fashion that could sometimes surprise and even shock. In his graduation fashion show at Central Saint Martin in 1992, McQueen exhibited, as we mentioned above, a collection entitled *Jack the Ripper stalks his victims*, and in that show he used his own hair.

Jack the Ripper was the nickname used by the London police to name a criminal who, in the year 1888, at the height of the Victorian era (1837–1901), committed serial killings on the outskirts of London. In these crimes, the victims had their throats cut and their bellies opened and, in some cases, part of their bowels removed.

In an opposite direction to that of the murderer, who removed part of the victim's body, McQueen introduced into the collection part of the body of its creator, that is, of his own body, and in a movement in tune with the Victorian Era, the chosen part was his own hair.

Hair is one of the few parts of the body that is preserved after the death of the individual and in that sense constitutes a vestige of that existence, and will be a reminder of that person. It is an element of the body that can be extracted from it without altering or compromising the physical integrity and biological condition of the person.

The use of his own hair on his clothes in his debut as a stylist marks McQueen's attitude from its outset. He wanted to make clear the relationship of overlap between him, the creator, and his collections. As we know, a single hair thread holds the genetic information of an individual, and today the DNA test is a legal device to prove paternity. Our hair is us! In this way, we understand that the stylist wore strands of his own hair not only as a kind of signature, but also as a willingness to be present even after his death.

It is worth mentioning that the hair is on the border between admiration and estrangement. When we see a hair on a plate of food, for example, it makes us sick, especially if the hair is not ours. So it resembles something dirty. In contrast, for those who keep strands of hair from a loved one, they are a good reminder and synonymous with affection. However, as Celeste Olaquiaga points out (2002), works made with hair of the dead, usual in the nineteenth century, for example, are nowadays considered bizarre and morbid things.

This incessant search for differentiated raw material gave the designer an unmistakable presentation of his clothes. The use of materials such as duck feathers, mussel shells and the hair itself promoted the sedimentation of an unmistakable style by the emotional strength of their pieces. In the autumn/winter collection of 2000–2001, in the piece Coat, from the Eshu collection, hair reappears again. Now no longer in the lining or the label of the pieces, but as raw material for the making of the piece itself, but this time, the hair is not human, it is synthetic. All made of hair, the clothes provoked a shock in the audience, reiterating the myth of the power of hair and presenting them displaced. As we mentioned before, this piece brings a mix of beauty and ugliness, difficult to be dressed, but outrageously attractive to look at. In the view of the stylist, the idea was to move the silhouette a bit and for that it would be necessary to change the thinking about what we looked at.

He was a creator who knew how to explore with intelligence and with sensitivity, the relationship between body and clothing, highlighting bodily potentialities and fragilities, in an intense dialogue with history, with life in society, plunged into surprises and contradictions.

As he liked to say, the work of the fashion designer demands the transcendence of what fashion is for fashion to become. Thus, he was able to explore the extreme, always trying to work with the possibilities of the human body, exercising new forms, exhibiting their plasticity. He used clothing as a means of expression, and was able to attract the eyes of spectators and consumers for his creations, their second skins.

Using McQueen's DNA to elaborate the fabric/leather used to make the pieces of her collection, Gorjanc incorporates to the collection and the pieces, one of the recurring elements used by the stylist in his collections: the first envelope of other beings. In order for the fabric/skin/leather used in "Pure Human" to be visually recognized as legitimate, the designer added the exact reproduction of McQueen's freckles and tattoos. These, as can easily be seen on the jacket, occupy the same position as those inscribed on the body of the stylist.

The possibility that such inscriptions, acquired spontaneously or intentionally, are easily known by people who did not live with McQueen is tied to the current phenomenon that allows easy access to, and technical quality of photographs.

3 BODY-IMAGE: HYPER EXPOSURE OF ITSELF

In contemporary times we are crossed by images; they harass us all the time: billboards, television screens, computer screens, and, more recently, mobile screens surround us, instigate, disturb, console, and ultimately appear to constitute us. Communication platforms in their multiple formats, such as film, television, photography and virtual environments have stimulated the bonds of interaction among the human beings and activated our feelings. As Walter Benjamin and Marshall McLuhan have perceived, technology, and we would say, technological images, affect our senses and perceptions by modulating them and thus, shaping new habits and behaviors that are reinforced in ways of life, reorganizing our experience. Thus, at the present time, it seems imperative to take into account how these new images that constantly circulate through mobile devices, restructure our ways of seeing and acting, even altering the environment in which we are immersed. Therefore, they establish new ways of being in the world and new forms of affective bonds between us, activating our subjectivity and emotionality.

In "Life and Death of the Image" (1993), Régis Debray summons us to reflect on how we understand the images of our time, initially tracing them, emphasizing that these "three ages of the eye" are not excluding but they influence each other. For the author we would have the *logo sphere* (era of idols) between the invention of the writing until the press; the *graph sphere* (art, between the press and TV), through the photo and the cinema, and the *video sphere* (the era of the visual), period of the video and the emergence of new devices. We are inserted in this last one, where there is a constant movement and, more contemporaneously, a disturbing instantaneity.

In the *logo sphere* we have images of worship that advocate eternity. In the *graph sphere* we aspire to immortality and in the *video sphere* we see the event. We can also say that the first phase of the gaze aims at the transition from the magic to the religious, the second, in turn, promises a slide from the divine to the human and the third invests in the transition "from the individual in his individuality to the surrounding global world, or even of being into the middle "(p. 209). In this sense, we observe that a path is established that culminates in the insertion of the individual in a network in which connections and transits are incessant and therefore there is a requirement to be in evidence and to become accessible. And if we think of how this image is, not only perceived and situated, but also how it is produced, we find that, immersed in the visual, we produce through electrons without even touching them.

In addition, in this regime of the *video sphere*, the image becomes its own reference. The association, the reversibility and the game between the images, which nowadays proliferate in intensity, send us back to a dynamic of enjoyment that evokes the power of our body full of prostheses. It is a return to the sensitive, it is the rediscovered flesh. In this way, the polarization between real and simulacra vanishes, and, as already pointed out by Debray, there is dissolution of the simulacrum to the extent that there is no more an external reference to the image itself, which now it is constituted from a mathematical logic. As the author comments,

> *Bypassing the opposition between being and seeming, similar and real, the graphically computerized image no longer has to imitate an external real, since it is the real product that will have to imitate it to exist. Every ontological relationship which, since the Greeks, devalued, and at the same time dramatized our dialogue with appearances is reversed. The "re" of the representation goes through the air, at the culminating point of the long metamorphosis in which things appeared more and more like pale copies of the images. (…) Here's the look. Finally, as in itself* (Debray, 1993, p. 277).

But beyond what Debray has already observed, we also observe that, whatever that would be exhibited as a reflection, as a representation, presents itself as real. They are the images that circulate in mobile devices, on digital platforms that, by their reproducibility and their repetition, end up constituting themselves in the body, in the skin of this hyper exposed and hyper connected individual. It is the appearance that is weaving the human being!

And these new images, we must reiterate, have reached more impressive levels of definition, aided by the latest technologies. The new vision prostheses and the various mobile devices provide a broader capacity for intervention on our environment and have altered the ways of perception and relationship between individuals. Not only the proliferation, but the high definition of the images, especially of the images of ourselves, through the circulation of the photos and, particularly, of the *selfies* in mobile devices (the newest and intense extension of the individual) has paid special attention to our body, our composition of appearance and our skin.

This hyper exposure of one self has been observed by several authors, but we would like to highlight some of the contributions of Gilles Lipovetsky and Jean Serroy (2015), who, reflecting on the increasing esthetical views of the world, characterize the contemporary individual as a trans esthetic internet user. Thus, not only do they account for the phenomenon of exacerbated exposure of themselves, but also for the hyper connection that invades life today. With the Internet and social networks we see an interesting shift: instead of focusing on professional concerns, people are connected by the pleasure of exchanging and reiterating their tastes: they meet people, signal preferences and exchange images. The authors point out "Interactions are made to amuse and pass the time, express their tastes, get on the scene, produce an image of them" (p. 371). And they also reinforce that, in contemporary times, in which we are experiencing hyper-modern individualism, a new form of self-portrait has been insinuated, an extremely democratic self-portrait in which all, without exception, are exposed in increasingly sophisticated screens. We visualize women, young people, men, professionally successful people, the unemployed, the elderly, children showing up on the internet, whether on blogs or on social networks (Facebook, Instagram, Twitter) through webcam or mobile images.

> *Then, presenting the "I" online is no longer engaged in a patient, voluntary, methodical search of self, but exposing oneself to the immediacy of its experience as it is lived, without retreat, without secrecy, without shame. No longer the intimate, hidden diary, but the continuous display: this is the time of "I" transparency, exposed on Facebook* (Lipovetsky & Serroy, 2015, p. 373).

Such an attitude reveals that we are becoming increasingly vulnerable, the images that circulate value the instantaneousness of the experience and there is no longer a concern to watch over the image itself. In this sense, we are even more and more exposed. We experience and exalt a representation of self that exalts instantaneousness, direct and passing expression. For this reason, there is a frequent update of the "profiles", since they are increasingly volatile. As Lipovetsky and Serroy comment, we exalt our "self-portrait live, perpetual review, traced in the simultaneity of the instant" (p. 374).

Massimo Canevacci, who also comments on the corporeal and imagistic transformations promoted by the culture of exacerbated consumption and new technologies, states as so many other researchers, that there is nothing natural about the body, since it is inscribed with symbols, he further establishes that the skin is not the limit of that body, since it binds to the "organic" of the metropolis, and—we would add—to the webs of the digital culture network. The body, for Canevacci, is not only body by itself, but because it blends with its surroundings. "The body expanded in buildings, things-objects-commodities, images, it is what we understand here as visual fetishism" (p. 18), clarifies the author in the introduction to his book "Visual Fetishism", 2008.

Canevacci widens his gaze and focuses attention on the emergence of the internet and social networks in an article entitled "Communication between corporations and metropolises", 2009. According to the author, there is a profound change in behavior as a result of the digital culture in which "the relationship between the eye, the screen, the hand, the mouse, the brain, and the body is much more interactive than one might imagine "(2009, p. 14) adding that "my sensory co-participation is favored and the organ that is most active on this type of procedure is the eye "(2009, p. 15). Still according to Canevacci, digital culture promotes the potentiality of the gaze, in which interactivity and creativity are strongly stimulated. We are constantly compelled to look and to look at ourselves as a way of reaffirming the other and ourselves. Our existence passes through the visual. The author defends the idea of a body-scape, a body-panorama, in which the interactions settle in the interstices. Thus, we observe that with the proliferation of images, especially those that circulate on tablets or cellular devices, our body-images multiply from the different selfies that co-inhabit these virtual spaces. Our skin and our bodies, more and more visible, become the fundamental elements of the affirmation of each one of as we exist.

4 FINAL CONSIDERATIONS

Reflecting on the Gorjanc project and on the creations of Alexander McQueen allows us to think of two hypotheses that may have been the objective, albeit unconscious, of these proposals: one is a specifically poetic order and another is of aesthetic nature. We note, on the one hand, that the need of the two designers to seek and find a raw material for the construction of the pieces reveals a symptom, namely, that there is certain exhaustion from the available raw materials for the production of their clothes. It is as if the available fabrics and textures were not sufficient or even adequate for their creations; it is as if it is no longer possible to reach the desired shapes. Therefore, it is necessary to go beyond and introduce new inputs into existing fabrics, look for the desired textures in animal linings, or more radically, create their own fabrics, as Gorjanc does. On the other hand, there is an aesthetic effect from this poetic investment: we end up realizing that these actions seek, in a certain way, to restore a lost authenticity. Up against so many interferences, by technological apparatuses and facilities, we are increasingly exposed to synthetic fabrics. In a sense, we are reinforcing the "*Frankfurtian*" thinking, especially the one supported in Walter Benjamin's work, in which we are rescuing a missing aura. Covering our skin with the skin of another, that is in fact my image and likeness, would be a kind of return to that original source.

Apart from these hypotheses, the artificial confection of the epidermal tissue used in these pieces, made from another element in the body, in this case hair intervenes in ancestral precepts related to the sacred symbology that surrounds the human body. It opens unlimited possibilities for the objectification of genetic elements and information that hitherto belonged exclusively to the sphere of nature.

REFERENCES

Alexander McQueen, 2004. *Testimony – Paris Mode January*. [Online] Available at: <https://www.youtube.com/watch?v=hOA-osK6aH0>. [Accessed 10 March 2018].
Benjamin, W., s/d. *A obra de arte na era de sua reprodutibilidade técnica*. São Paulo: Brasiliense.
Canevacci, M., 2008. *Fetichismos visuais: Corpo erópticos e metrópole comunicacional*. São Paulo: Ateliê Editorial.
Canevacci, M., 2009. A comunicação entre corpos e metrópoles. *Revista Signos do Consumo*, – V.1, N.1, São Paulo: USP, p. 8–20.

Debray, R., 1993. *Curso de midiologia geral*. Petrópolis: Vozes.
Debray, R., 1993. *Vida e morte da imagem: Uma história do olhar no ocidente*. Petrópolis: Vozes.
Le Breton, D., 2016. *Antropologia dos sentidos*. Petrópolis: Vozes.
Lipovetsky, G. & Serroy, J., 2015. *A estetização do mundo: Viver na era do capitalismo artista*. São Paulo: Companhia das Letras.
McLuhan, M., s/d. *Os meios de comunicação como extensões do homem*. São Paulo: Cultrix.
Olaquiaga, C., 2002. *The artificial kingdom: on the kitsch experience*. Minneapolis: University of Minnesota Press.
Tina Gorjanc., s/d. [Online] Available at: <http://www.tinagorjanc.com/>. [Accessed 09 March 2018].

The relevance of sketches in fashion design

A. Moreira da Silva
CIAUD, Lisbon School of Architecture, Lisboa, Portugal
CITAD, Univeridade Lusiada de Lisboa, Lisboa, Portugal

ABSTRACT: There is a deep relationship between design and drawing. The creative processes of most fashion designers are based on the development of ideas through its expression in sketches. This paper, considering the relationship between design and drawing, investigates the relevance of hand-drawing in the design process, as a booster instrument when sketching the first ideas, in its development and as a critical verification of the several hypotheses. Through the study of several statements from various authors we search to investigate the permanence of drawings important role. From this theoretical approach we also intend to verify the relevance of hand-drawing teaching in the formation of the future fashion designers, despite the paradigm changes that emerge from times changes and the new assistive technologies constant progress. Most of renowned fashion designers use sketches for their creative work. Christian Dior, Yves Saint Laurent, Gianfranco Ferré, Karl Lagerfeld, Miuccia Prada, and for many others, sketching is the natural way for exploring and showing their new ideas. This a part of a larger post-doc research project that aims to achieve contribution on the study of creative design process, enhancing hand-drawing important value for fashion design students and professionals.

1 INTRODUCTION

This paper stems from a current post-doc research project motivated for the need of producing more knowledge and reflection on drawing and drawing teaching. Our main research question is about the importance of hand-drawing as a critical instrument and as an operative support within the conceptual process in design.

Considering the deep relationship between design and drawing, in this study we investigate the relevance of drawing in the design process, as a booster instrument when sketching the first ideas, in its development and as a critical verification of the several hypotheses.

The used research methodology is based on the literary review of recognized international authors on this subject and on semi-structured interviews addressed to designers, students and design teachers. Through these interviews, we intend to verify the permanence and the importance of hand-drawing in design practice in its various forms of communication, product and fashion.

2 SKETCHING IN FASHION DESIGN

There is a deep relationship between design and drawing. The creative processes of most fashion designers are based on the development of ideas through its expression in sketches.

We can consider the relevance of hand-drawing in the act of designing not only as stimulating instrument when sketching the first ideas but also as critical verification of the several hypotheses. The use of sketches not only helps a great deal in starting the ideas generating process, but it also enables high quality concepts to begin with as initial ideas. These ideas can be accessed, combined, selected and developed. Several authors have written about the importance of drawing in architecture and design methodologies. Their general conclusion points out to drawing as an essential tool as it allows ideas to flow and provides critical investigation of the several alternative solutions in creative processes.

Nigel Cross (2005) states that drawings are a key feature of the design process. At the early stages of the projects they are communications with oneself, a kind of thinking aloud, and concludes that the conceptual thinking processes are based on the development of ideas through their external expression in sketches. For Rohde (2011) sketching is the visual thinking power tool and adding sketching to the process is a great way to amplify software and hardware tools. He concludes that the use of sketching during the creative process, reveals the central role drawing plays within the development of a project and the importance of hand-drawing, particularly in an age dominated by digital media. For Rohde, sketches can work like a visual thinking tool and as a primary language for capturing thoughts, exploring ideas, and sharing those ideas.

According to Norman (2010) the recent profound changes in design processes, due to the contemporary historical context were new technologies are so important, justify an enlarged discussion on the important role played by drawing in design courses, adapted to the new paradigm. The fact of using new technologies doesn't invalidate the important role played by hand-drawing, both at the initial stage of recording the first ideas and during their subsequent development and in the critical analysis of the different hypotheses. For Hamilton (2009) the value of hand-drawing is associated with the knowledge that drawing is the essence of creativity and its measure would lie in the individual ability to continuously express and create, using self-exploration and creative thinking through the process of drawing, linking together interesting ideas and technical know-how, enhanced by assistance and interaction with new technologies.

Sketching provides a unique space that can help us to think differently, to generate a variety of ideas quickly, to explore alternatives with less risk, and to encourage constructive discussions with colleagues and clients. Drawing holds, in itself, the making visible quality of the whole mental process underlying the conceiving stage, from the first sketches of an artifact vague idea to the creation of its final form. Drawing's transcendental importance lies on the capacity it gives to fashion designers to materialize abstract conceptualizations and to create the ideational basis for fashion costumes and accessories. For a fashion designer, the sketchbook merges as a way of thinking through drawn lines as a creative process. All renowned fashion designers use sketches for their creative work. Christian Dior, Yves Saint Laurent, Gianfranco Ferré, Karl Lagerfeld, Miuccia Prada, Jimmy Choo, and for many others, are good examples.

Karl Lagerfeld uses sketches from is early times as fashion designer to the present. Half a century after Karl Lagerfeld first drew a feminine figure in a classic but renewed skirt and coat set, in this fashion sketch the woman still seems ready to saunter off the page and into the street. This is a very good example how sketches can possess a "soul", a unique expression, not possible to pass into technical drawings.

Figure 1. Young Yves Saint Laurent showing his ideas through sketches.

Figure 2. Christian Dior original sketches (1954).

Figure 3. Gianfranco Ferré sketches for Dior (1989).

Figure 4. Karl Lagerfeld's early fashion sketches dating back to the sixties.

Figure 5. Karl Lagerfeld's first sketches for his 2011 capsule collection for Macy's.

Through his sketches he communicates to his collaborators how to execute his ideas, showing the front and back of each model and adding written notes to clarify some details.

We can also perceive through these drawings from different years how Karl Lagerfeld creative spirit evolves following the times change not only in fashion style but also in the way he expresses his ideas. In a recent interview Karl Lagerfeld stated that he draws every day and that he even needs drawing for self-fulfilling.

Miuccia Prada sketches show a different expressive style, more synthetic, but still looking for the use of sketching in order to clearly communicate her creative ideas and provide details for its implementation. Sketches are able to communicate designer's self-identity and personality.

Even for shoes and other accessories, fashion designers use sketches to communicate their creative work in their own expressive style.

Figure 6.   Miuccia Prada sketches for her Great Gatsby Collection (2012).

Figure 7.   Karl Lagerfeld for Melissa (2013) and Jimmy Choo for Cinderella (2014).

Sketching allows any fashion designer to express their ideas in an individual and original way characteristic of their creativity and uniqueness, as we can observe in Figure 7, which presents different forms of the same object representation.

Drawing today serves as a primary medium for generating and communicating fashion designers own creative and conceptual musings about an artifact.

If sketching is central for conceiving, it is also central for defining how that conception is managed as it moves from its initial stages through its actual development and realization as a material form.

Drawings illustrate various aspects of the fashion designer creative process. So, the importance of hand-drawing assumes a broad sense, conferring to the act of drawing the ability to become a means of multiple resources to practice fashion design.

part of the productions of the collections. Stores located on Passeig De Gràcia and Gran Via have been chosen because they stand out as important avenues in European fashion consumption. The material was collected in a personal and direct way regarding the contact with the situation under study. We can understand this process as being of prime importance in the development of the study, as the authors Lüdke and André (1986, p. 26) affirm that "direct experience is undoubtedly the best test to verify the occurrence of a certain phenomenon. Direct observation also allows the observer to get closer to the 'subject perspective', an important target in qualitative approaches.

Finally, the third and last procedure was carefully carried out on the same date of December 2017. Images of virtual banners were collected on the websites of the brands, always being chosen by the Spanish version, in order to establish a parameter of equivalence with the photographic material from shop windows.

It is hoped that the research carried out not only externalizes the function of communicative power between consumers and the virtual store through the banners, but also demonstrates that they, explicitly or implicitly, in a comparison with the shop windows of the same fashion brand, reveal choices of language and positioning of organizations in relation to the society of digital consumption. Finally, the communication by digital banners, especially those of the online commerce of fashion, result in a complex information system that reveals much more than the simple message intended by the photographic publisher.

2 SHOP WINDOWS AS CONSTRUCTIONS OF NARRATIVES

Without making use of orality, the showcase can be understood as a scenic space that establishes communication between store and consumers. From visual perception, the message interacts and produces meaning. Consumers, in the search for identification with the concepts of the stores and feel themselves part of them, analyzes the proposal presented there, as if unconsciously looking looked in that space, comfort and meaning. Besides the look, sensations and aromas are part of this universe, proving that in isolation the product is not sold. It is understood that the attraction power of a product is not only in its functional content, but much more in its aesthetic production.

The notion of showcase goes well beyond its functional aspect, according to the author Demetresco (2007). It identifies in the showcase emotions, sensations and involvement, growing and complexifying the space in question. In addition to exhibiting products and offerings, the windows reflect contemporary time and society, pointing to changes in habit and custom. This set of representations of the market and a specific one of the cultural universe in which they are inserted meet in this environment. For Santaella & Nörth (2010), one can think of shop windows as a space for merchandise exhibition, which are "polysemic signs associated with a plurality of meanings that are not inherent in the merchandise itself, but are generated in the intricate ways of the market. Products not only denote their use and exchange values, but also connote aesthetic, emotional, and mythical meanings." (p. 35).

Many are the signs present in the setting of the space of a showcase, in which the whole of the image must establish a communication with the passersby (potential consumers). The passerby is a spectator and a consumer: he is a spectator when he only watches the representation of the showcase and is a consumer because, even if he does not buy the product exposed, he can consume an idea or a concept. According to Ratto (1999, p. 25), "[...] the behavior of the spectator is equivalent to that of a reader who, following the literary descriptions of a novel or story, imagines and 'sees' what is being narrated, as if the places and spaces in which the 'heroes' are acting were ahead of them". The viewer can buy or not the advertised product or buy another by-product that make up the ambiance. For Oliveira (1997, p. 73), "The images are perceived by the observer from their degree of pertinence. Thus, the showcase is perceived both in terms of the practical and objective relevance of the product to the passer-by, and of the relevance of the symbolic references that it can stimulate in the observer". It reinforces the ideas of Oliveira as those of Demetresco (2004, p. 23) "We know that the windows qualify

Figure 6.   Miuccia Prada sketches for her Great Gatsby Collection (2012).

Figure 7.   Karl Lagerfeld for Melissa (2013) and Jimmy Choo for Cinderella (2014).

    Sketching allows any fashion designer to express their ideas in an individual and original way characteristic of their creativity and uniqueness, as we can observe in Figure 7, which presents different forms of the same object representation.

    Drawing today serves as a primary medium for generating and communicating fashion designers own creative and conceptual musings about an artifact.

    If sketching is central for conceiving, it is also central for defining how that conception is managed as it moves from its initial stages through its actual development and realization as a material form.

    Drawings illustrate various aspects of the fashion designer creative process. So, the importance of hand-drawing assumes a broad sense, conferring to the act of drawing the ability to become a means of multiple resources to practice fashion design.

3 CONCLUSIONS

According to many authors' statements and the professional practice of many fashion designers, we can conclude, from this research initial phases, that sketches are a key factor of the creative process in fashion design. All fashion designers use sketches on their creative work. Sketching presents itself as an operative support for problem solving and critical analyses in fashion creative process. Sketching also allows any fashion designer to express their ideas in an individual and original way showing their creativity and uniqueness.

Despite the paradigm shift required by the changing times, we believe that hand-drawing will remain inseparable from designers training and professional practice, assuming an essential operating support in their activity.

In the future, design education should pass through a systematic approach to hand-drawing in order to highlight and analyze the flexibility with which it adapts itself to various purposes, fulfilling effectively a wide range of intentions, in a constant and vital adaptation to the continuous changes in the teaching and in the practice of designers.

In order to validate these assumptions, the current post-doc research can constitute a contribution for the understanding of the importance of hand-drawing's permanence as the basis of the design creative process, although the constant progress of the new technologies.

ACKNOWLEDGMENTS

The author would like to acknowledge the support given by CIAUD—Research Center in Architecture, Urbanism and Design, Lisbon School of Architecture, University of Lisbon, by CITAD – Research Center in Territory, Architecture and Design, Lisbon Lusíada University, and by FCT – Foundation for the Science and Technology, Portugal, within Project UID/AUR/04026/2013.

REFERENCES

Cross, Nigel. 2005. *Engineering Design Methods, Strategies for Design*. John Wiley & Sons Ltd. Chichester, England.
Hamilton, Peter. 2009. *Drawing with printmaking technology in a digital age*. Retrieved from: http://www.lboro.ac.uk/microsites/sota/tracey/journal/dat/images/paul_hamilton.pdf. [Accessed 10 February 2018].
http://cinemacomrapadura.com.br/colunas/moda/366244/yves-saint-laurent-estrela-nas-passarelas-e-no-cinema/ [Accessed 12 February 2018].
http://institutoriomoda.blogspot.pt/2012/12/lesprit-dior-incrivel-exposicao-que.html [Accessed 11 February 2018].
http://vogue.globo.com/moda/moda-news/noticia/2014/07/melissa-por-karl-lagerfeld-nova-colecao-chega-lojas-em-agosto.html [Accessed 12 February 2018].
http://www.lachanelphile.com/2011/07/25/karl-lagerfelds-sketches-for-macys/ [Accessed 10 February 2018].
https://geekandchic.cl/optimus/ [Accessed 11 February 2018].
https://www.google.pt/search?q=Miuccia+Prada+sketches [Accessed 11 February 2018].
https://www.hollywoodreporter.com/news/karl-lagerfelds-early-fashion-sketches-668213 [Accessed 10 February 2018].
https://www.youtube.com/watch?v=4tmLfnLQzpQ [Accessed 11 February 2018].
Norman, Donald. 2010. *Why Design Education must change*, Retrieved from: http://www.core77.com/blog/columns/why_design_education_must_change [Accessed 12 February 2018].
Rohde, Mike. 2011. *Sketching: the Visual Thinking Power Tool*. Retrieved from: http://alistapart.com/article/sketching-the-visual-thinking-power-tool [Accessed 12 February 2018].

Digital banners as shop windows: Narrative constructions of online fashion brands

D.V. Leal & D. Karam Jr.
School of Arts Science and Humanities, University of São Paulo (EACH-USP), São Paulo, Brazil

M.C.P. Rodrigues
School of Communications and Arts, University of São Paulo (ECA-USP), São Paulo, Brazil

ABSTRACT: The digital banners came up in a online Market reality as a communicative tool that propose, as well as product display, to transmit the aggregate sociocultural values, through a variety of signs. The development of this virtual communication has as bases the shop-windows and, therefore, it is analysed in this research, fashion brands with conventional stores (physical) that have started their e-commerce activities. There are made comparative analyses between the shop-windows and the digital banners of four brands: H&M, ZARA, Stradivarius and Bershka. The objective here is show that, for the setting up of those scenarios, compete a variety of signs and the positioning choices with respect to the language used by the brand sometimes differs, and sometimes don't. For the conduct of this qualitative research, there were taken pictures of the shop windows, raised bibliographic references, and analyses of images available on the brands' websites. The digital banners are, consequently, a narrative construction of the fashion brands that has as intent the sense production on the relations between the brand, the consumer, the product and the virtual.

1 INTRODUCTION

In many different areas, the dialogue with the world is in the capacity to generate information and produce signs that form some communication. For this, fashion brands are not limited to the use of established languages, they are always in search of new ways to communicate and ways to innovate in the way they represent and present themselves. It is a challenge to identify the means that enhance the communication of fashion, above all, with possibilities for the expansion and insertion of new technologies.

The components intended for communication by fashion brands are, in general, elements of visual perception present in scenarios, shop windows, mannequins, clothing, products, materials, among others. This information relates to the sensory presentation features needed to understand fashion, which traditional (physical) brands offer their interlocutors: consumers. With the noticeable movement and insertion of these marks for the online environment, a relation of displacement of the physical scenic spaces was established (and here we speak especially of the windows) for the editorials of virtual photos, more properly denominated like digital banners in the platforms. In this way, we intend to discuss the action of signs in the fashion languages in these two coexistent environments (physical and online), seeking, mainly, to understand how the production of meanings occurs from components of representation.

For the accomplishment of the research, three main procedures were adopted. The first one was the bibliographical survey, with readings on window dressing, photographic production of editorials of fashion and semiotics, mainly. The second was the exits for photographic registration of the shop windows. The images were taken in December 2017 in Barcelona, Spain. This city was chosen for the previous perception that the sites of the brands follow the stations of Europe, serving with punctuality the stores of Spain, where is located a great part of the productions of the collections. Stores located on Passeig De Gràcia and Gran Via have been chosen because they stand out as important avenues in European fashion consumption. The material was collected in a personal and direct way regarding the contact with the situation under study. We can understand this process as being of prime importance in the development of the study, as the authors Lüdke and André (1986, p. 26) affirm that "direct experience is undoubtedly the best test to verify the occurrence of a certain phenomenon. Direct observation also allows the observer to get closer to the 'subject perspective', an important target in qualitative approaches.

Finally, the third and last procedure was carefully carried out on the same date of December 2017. Images of virtual banners were collected on the websites of the brands, always being chosen by the Spanish version, in order to establish a parameter of equivalence with the photographic material from shop windows.

It is hoped that the research carried out not only externalizes the function of communicative power between consumers and the virtual store through the banners, but also demonstrates that they, explicitly or implicitly, in a comparison with the shop windows of the same fashion brand, reveal choices of language and positioning of organizations in relation to the society of digital consumption. Finally, the communication by digital banners, especially those of the online commerce of fashion, result in a complex information system that reveals much more than the simple message intended by the photographic publisher.

2 SHOP WINDOWS AS CONSTRUCTIONS OF NARRATIVES

Without making use of orality, the showcase can be understood as a scenic space that establishes communication between store and consumers. From visual perception, the message interacts and produces meaning. Consumers, in the search for identification with the concepts of the stores and feel themselves part of them, analyzes the proposal presented there, as if unconsciously looking looked in that space, comfort and meaning. Besides the look, sensations and aromas are part of this universe, proving that in isolation the product is not sold. It is understood that the attraction power of a product is not only in its functional content, but much more in its aesthetic production.

The notion of showcase goes well beyond its functional aspect, according to the author Demetresco (2007). It identifies in the showcase emotions, sensations and involvement, growing and complexifying the space in question. In addition to exhibiting products and offerings, the windows reflect contemporary time and society, pointing to changes in habit and custom. This set of representations of the market and a specific one of the cultural universe in which they are inserted meet in this environment. For Santaella & Nörth (2010), one can think of shop windows as a space for merchandise exhibition, which are "polysemic signs associated with a plurality of meanings that are not inherent in the merchandise itself, but are generated in the intricate ways of the market. Products not only denote their use and exchange values, but also connote aesthetic, emotional, and mythical meanings." (p. 35).

Many are the signs present in the setting of the space of a showcase, in which the whole of the image must establish a communication with the passersby (potential consumers). The passerby is a spectator and a consumer: he is a spectator when he only watches the representation of the showcase and is a consumer because, even if he does not buy the product exposed, he can consume an idea or a concept. According to Ratto (1999, p. 25), "[...] the behavior of the spectator is equivalent to that of a reader who, following the literary descriptions of a novel or story, imagines and 'sees' what is being narrated, as if the places and spaces in which the 'heroes' are acting were ahead of them". The viewer can buy or not the advertised product or buy another by-product that make up the ambiance. For Oliveira (1997, p. 73), "The images are perceived by the observer from their degree of pertinence. Thus, the showcase is perceived both in terms of the practical and objective relevance of the product to the passer-by, and of the relevance of the symbolic references that it can stimulate in the observer". It reinforces the ideas of Oliveira as those of Demetresco (2004, p. 23) "We know that the windows qualify the place in which they are [...]. The shop windows are a way of manifesting the social imaginary, representing, in this way, a possible way of understanding relations".

Thinking from the perspective of semiotic significance, the sign, according to Peirce (1999), is all that comes to us from reality, which we can perceive; it is every sign of reality, every mark that represents something outside of it, but of which it is part. If a showcase environment is considered to be a sign wearing a dummy, it is also considering a number of factors that help in the proliferation of meanings of this scenario: the perception it has on the citizen, the indications suggested by metonymic processes, and the conventions secured by the social and cultural market system imposed or created. To understand these concepts with the heuristic proposal means to say that the windows have a sign dimension: everything in this space and through it constitutes a set of meanings, directing towards a communicative semiotization of the marks, through the windows, and for what they project in the construction and also deconstructing consumer and societal behavior.

The semiotic analysis also shows that the signs that compose a showcase are historical marks that contemplate not only the productive and economic development, but also the technique employed by the subject who produces it and the one who receives it. For Demetresco (2001, p. 25), "The showcase is an assembly that concentrates countless areas that merge to create an image whose purpose is to generate pleasure for a few seconds". With this definition of showcase we can cite the three phenomena of Peirce's semiotic pan: firstness, secondness and thirdness. In other words, a showcase is set up to suggest, seduce and persuade. Above all, the passerby's gaze strikes the showcase in an ambiance of dreams and fantasies and leads him to buy more than an exposed product: an ambition effect is also bought.

3 COMMUNICATIVE ACTION OF DIGITAL BANNERS IN FASHION BRANDS

Through non-verbal components, portrayed in imagery, gestures, sounds and movements, fashion as communication has the purpose of exchanging messages among the participants. A composition of signs that in modern society builds an appearance based on pleasure (Sant'Anna, 2009). One can observe the communication of fashion as a triadic model, based on the semiotics of Peirce (1999). Here, communication goes beyond imagery, also contemplating people's way of thinking. That is, as they perceive, learn and remember information. Memory stores various kinds of knowledge from past experiences, making it possible to act properly in the present moment. Sternberg (2000) says that new information collected is related to memorized knowledge and then organized into new knowledge, forming more cognitive representations of what has been seen and developing the ability to process the information received.

The fashion brands, throughout their communication trajectory, gave greater emphasis to the senses of vision and hearing. Along the same lines, with the advent of online commerce (e-commerce), a new way of establishing communication between brand and consumers, digital banners: thematic images, with graphic resources, colors, formats and illuminations. With its development, the digital banner is able to go beyond the binomial vision and hearing to a set of sensations and creation of new signs, using new technologies (videos, gifs, compositions, among others). He can, "through online communication, develop sensory experiences and create emotional connections as a complement to rationality" (Duarte, 2013, p.103).

Having as a central objective the promotion of companies, brands, products and services, with the purpose of accomplishing the acquisition and consumption by the virtual consumers, the digital banners corresponds to a particular type of communicative process. This type of process entails a cultural, social and situational environment in which the fashion brand, together with the editors, seek to produce material that establishes communication with a defined audience: it is probable consumers. An online platform is used as support for the placement of this material.

3.1 *Fashion photography in digital banners*

Images have been means of expression of the human culture since the prehistoric paintings. Today, in the information age, our daily life is permeated with visual messages and for

Santaella & Nörth (2005) today the verbal code can no longer develop efficiently without the use of images. According to Roland Barthes (2005), one must understand what is the image, what its types, and its classifications, not ignoring their importance also in the past generations. Still, according to Roland Barthes (2005, p.70), "we live surrounded, imbued with images, and yet we know almost nothing of the image [...]. The image, as a sign, of the element of a communication system, has a considerable expressive value".

The representativeness of images is also applied to fashion, more specifically, to fashion editorials. According to Caldas (2004, p. 114), one of the premises of semiotics is to consider the images created by advertising as non-verbal texts, messages that can be analyzed using the same instrument applied to verbal communication. Besides the basic function of establishing relations between company and consumer and the goal of generating sales, the generation of images, applied to the fashion area, aims not only to show the products, but also to build relationships and create identity. According to Santaella & Nörth (2010, p. 107), the objective of a commercial fashion photographic image is twofold. On the one hand, "It seeks to create the situation of a communicative act in which a message is transmitted to the reader, listener or spectator, while, on the other hand, it seeks to carry out an economic act of sale, purchase and consumption of a product."

In semiotics, the world of images is divided into two areas. The first one says about images as visual representations. In this sense are the drawings, paintings, photographs and cinematographic and television images. It is the image in the real sense, it exists and it is in front of our eyes. Already in the second area is the immaterial domain of the image in our mind. That is, where the images appear as visions, models, fantasies or mental representations. One area does not exist without the other because there are no visual representations that have not arisen from images in the minds of those who produced them, nor mental images that have no origin in the concrete world of visual objects. According to Santaella & Nörth (2005) the unification of these two areas are the concepts of sign and representation.

4 RELATIONSHIP BETWEEN DIGITAL BANNERS AND SHOP WINDOWS

New forms of communication arise daily, especially through instant tools like the internet, allowing you to reach the target audience quickly and efficiently. However, to understand this new model of communication, it is important to be aware of old techniques and not less important, such as exposing products in store windows. According to Santaella and Nörth (2010, p.97), "[...] the product needs to be shown. Online media can not show the product directly: they typically present it in the form of an image". When it comes to fashion, the touch relationship between the consumer and the product is paramount and greatly increases the chances of selling. This contact is one of the purposes of setting up a shop window. Thus, communication through digital banners, understood as an adaptation of the communication through the windows, to the online medium, does not overlap and is not understood as superior to the physical spaces destined to showcases in fashion brands, and vice versa. These environments (physical and online) coexist and, when well thought out, can establish clear and broad communication with their consumers.

Semprini (2010) believes that there are three fundamental dimensions that structure a brand, and can serve as theoretical support for the coexistence of banners and shop windows: the semiotic nature of the brand, the relational nature and the evolutionary nature. In semiotics, we have to keep in mind the meaning, the meaning. The brand needs to build meaning bonds with its audience. "The semiotic power of the brand consists in knowing how to select the elements within the flow of meanings that cross the social space, organize them into a pertinent and attractive narration and propose them to their audience" (Semprini, 2010, p. 98). The relational nature is divided into two dimensions: intersubjective and contractual. In the intersubjective dimension the brand is a continuous process of exchanges and negotiations that is divided into three poles: production, reception and general context. In the evolutionary nature the fundamental characteristic is the dynamicity and the constant evolution. The brand needs to be in line with its audience, it needs to understand what is happening in society in order to evolve with it. Since "the brand is nothing more than a cultural construction, an artifact whose semiotic nature compels it to produce permanent meanings, to renew its design, to refresh its image, not to fall into oblivion" (Semprini, 2010, p. 110), it needs to be vented, it must always be in the minds of its consumers. These three dimensions justify, therefore, the communication by banners and showcases, efficient in their purposes.

5 COMPARATIVE ANALYSIS BETWEEN THE LANGUAGE USED IN SHOP WINDOWS AND BANNERS

The analysis presented includes the following examination: comparative of the communicative environment of the photographed windows with the respective digital banners collected on the official websites of the brands. The images, both those photographed and collected on the sites, were obtained exactly on the same date, in order to establish comparative parameters. In addition, it was decided to make the photos of shops that are located in one of the most important points for the world fashion trade, in the avenues Passeig De Gràcia and Gran Via in Barcelona, Spain. These stores are contemplated with the timeliness of the arrival of the pieces of the collections because of their proximity to the production sites of the brands, also located in most cities in Spain. In this way, care was taken to collect the images of the digital banners on the official websites of the brands, destined for Spain, which is also a benchmark stipulated for the research. Since some of the elements of this communicative context are common to all the brands in the analysis, it was decided to present the photos of the shop windows first and then two digital banners, mark the brand. The brands considered were: H&M, ZARA, Stradivarius and Bershka.

In order to perform the exam, it was chosen to observe aspects such as predominance of colors used, applied materials in scene and adornments, positions and attitudes of models and mannequins, thematic, lighting and other reflections that were considered interesting.

Figures 1, 2 and 3 are the records of the shop window H&M Passeig De Gràcia/Gran Via. Figures 4 and 5 are the digital banners from the official H&M website. Here, the relationship between showcase and banner is perceived by the warm colors used in both and by the scenic spaces that allude to the environments that are in tune (forests and winter weather). The plot proposed in the photo editorial has continuity in the showcase mainly by the choice of the scenic objects.

Figures 6, 7 and 8 are the records of the shop window of the ZARA store Passeig De Gràcia/Gran Via. Figures 9 and 10 are the digital banner images from the official ZARA website. In these scenarios the dark color chart and strong lighting establish the relationship between them.

Figures 1, 2, 3. Shop windows of the store H&M Passeig De Gràcia/Gran Via (personal collection), Barcelona, December 2017.

Figures 4, 5. Banners from the official website of the H&M http://www2.hm.com/es/, access on 12/03/2017.

Figures 6, 7, 8. Shop windows of the store ZARA Passeig De Gràcia/Gran Via (personal collection), Barcelona, December 2017.

Figures 9, 10. Banners from the official website of the ZARA https://www.zara.com/es/, access on 12/03/2017.

But it is the bodily attitudes of mannequins and models that make the main connection between established communications. Here it is worth mentioning that the website of the brand uses video tools in the assembly of the main banner, thus proposing the creation of new signs.

Figures 11 and 12 are the records of the shop window of the Stradivarius Passeig de Gràcia store. Figures 13 and 14 are the images of the digital banners of the official website Stradivarius. It can be said that there is a difference between the values offered and the ways of saying of the material under analysis. The discrepancy between window and banners is wide, be it by the predominance of colors, forms, products, attitudes and languages

Figures 11, 12. Shop windows of the store Stradivarius Passeig De Gràcia (personal collection), Barcelona, December 2017.

Figures 13, 14. Banners from the official website of the Stradivarius https://www.stradivarius.com/es/, access on 12/03/2017.

employed. Thinking about the segments of fashion, the signs diverge towards almost opposite paths (party and casual).

Figure 15 is the register of the shop window of the Bershka Passeig de Gràcia store. Figures 16 and 17 are the images of the digital banners of the official Bershka website. Analyzing the two proposals designed by the brand, it is possible to point out communications that differ in the product, performance of the manikins and models, materials and thematic developed. The brand combines videos, photos and gifs on the online platform, delivering dynamic and interactive communication. Already in the space of the showcase chooses for a less daring scenic space.

In these observations, it was possible to identify the elements responsible for the first impression, that is, what the receiver qualitatively notes first: colors, textures, themes, main elements and supporting scenes. After this moment the ideas ended up generating associations, similarities and, when they are called iconic, it was possible to establish comparative parameters between the proposals of each brand. It was therefore concluded that the relationship of language, communication and positioning between the windows and the digital banners, is sometimes of similarity, and, by others, of difference.

It's possible observe two main ways to start online activities by traditional (physical) fashion brands. On the one hand there are those that opt for the integration of physical and online environments, also known as Omnichannel, which is characterized by the standardization of communication and synchronization between all forms of contact with the customer:

Figure 15. Shop window of the store Bershka Passeig De Gràcia (personal collection), Barcelona, December 2017.

Figures 16, 17. Banners from the official website of the Bershka https://www.bershka.com/es/, access on 12/03/2017.

telephone exchange, e-mail, site on the desktop, mobile site and physical store. "[...] the starting point should be the consumer, who does not distinguish channels, but discriminates brands, value propositions and buying experiences" (Serrentino, 2014, p. 42). On the other hand it is believed that consumers behave in different ways depending on the environment they are in. Some characteristics of the physical environment are incompatible with the virtual environment, which justifies the proposal of communication, at times even disparate, and directed specifically to each of the scenarios. These changes in consumer behavior due to the possibility of consumer mobility highlight the increase in Cross Channel Behavior, which refers to the alternation of online and offline channels (Gerritsen et al., 2014, Trenz, 2015).

6 FINAL CONSIDERATIONS

The semiotics, general science of the signs, is one of the ways in which we can understand the interior of the messages and the meanings that they are capable of producing, because this science penetrates the structural construction of the message, aiding in its interpretation. It was possible to understand the communication by the digital banners, analyzed comparatively with the physical showcases, through the signs as elements that represent, that are related among them and that allow diverse interpretations according to the context that are inserted and the receiver of the message, since each which interprets them in a way.

In the exam with the brands H & m, ZARA, Stradivarius and Bershka, we sought to identify and analyze the semiotic elements present in the images used (photographed or collected), associating this with a unit positioning (Omnichannel) or alternation (Cross Channel Behavior) through online and offline channels (or rather, digital banners and shop windows, respectively) proposed by the brands in question.

It is therefore concluded that, the positioning decision, Omnichannel and Cross Channel Behavior significantly influence the outcome of the communication intent of a fashion brand. It can be affirmed that there are relations of similarities and differences in the relations established between window displays and digital banners of a fashion brand, either by the choice of language used or by the development of the intended communication. I try to position myself well defined by the brand, it is possible to communicate the message (whatever the environment) with clarity and effectiveness.

Through this study, we perceive that all communication is composed of symbols and signs that aid in the process of constructing the message and understanding of what is inserted in it. In addition, the role fashion plays within an image, be it physical or online, provides the construction of ideas, and when well worked, they become extremely efficient in producing meaning, making the analysis more concrete.

REFERENCES

Aakko, M. Koskennurmi-Sivonen, R. 2013. Designing Sustainable Fashion: Possibilities and Challenges.
Barthes, R.2005. *Imagem e moda.* São Paulo: Martins Fontes, v.3, p. 380.
Caldas, D. 2004. *Observatório de sinais: teoria e prática de pesquisa de tendências.* Rio de Janeiro: Ed. SENAC.
Demetresco, S. 1990. *Vitrina: teu nome é sedução.* São Paulo.
Demetresco, S. 2002. Vitrinas: entre a tentação e a sedução. *Revista Nexos,* São Paulo, v.6, n.9, p. 7–22.
Demetresco, S. 2004. *Vitrines entrevistas: merchandising visual.* São Paulo: SENAC.
Duarte, S. 2013. *Vitrinas: expondo e revendo conceitos.* Fortaleza: Editora UFC.
Gerritsen, B.H.M. Gerritsen, K. Solberg, S. Bas de Visser, P. Hoogreef, P.J.M. Janssen, K.H.M.L. Horselenberg, L. Van Dijk, R.R. Consenheim, E. 2014. Social Media Coming to the Mall: A Cross-Channel Response. *Springer International Publishing Switzerland*: D. Schaefer, p. 169–235.
Ludke, M. André, M.D.A. 1986. *Pesquisa em educação: abordagens qualitativas.* São Paulo, EPU.
Oliveira, A.C. 1997. *Vitrinas: acidentes estéticos na cotidianidade.* São Paulo: EDUC.
Oliveira, S.R. 2007. *Moda também se lê.* São Paulo: Rosari.
Peirce, C.S. 1999. *Semiótica.* São Paulo: Perspectiva.
Ratto, G. 1999. *Antitratado de Cenografia: variações sobre o mesmo tema.* São Paulo: Ed. SENAC.
Sant'Anna, M.R. 2009. *Teoria da Moda: sociedade, imagem e consumo.* São Paulo: Estação das Letras e Cores.
Santaella, L. 2002. *Semiótica aplicada.* São Paulo: Pioneira Thomson Learning.
Santaella, L. North, W. 2005. *Imagem: cognição, semiótica e mídia.* São Paulo: Iluminuras.
Santaella, L. North, W. 2010. *Estratégias semióticas da publicidade.* São Paulo: Cengage Learning.
Semprini. A. 2010. *A marca pós-moderna.* São Paulo: Estação das letras e cores.
Serrentino, A. 2014. *Varejo omnichannel.* São Paulo: Mercado e Consumo, n. 05, p. 42–44.
Sternberg, R.J. 2000. *Psicologia Cognitiva.* Porto Alegre. Artes Médicas Sul.

# The influence of soap operas in Brazilian fashion today

T.C. Viana
Universidade do Estado de Minas Gerais e Centro Universitário de Belo Horizonte, Brazil

A.A. Silva, A. Adverse & R. Ribeiro
Universidade do Estado de Minas Gerais, Brazil

A. Horta
Centro Universitário de Belo Horizonte, Brazil

ABSTRACT: This paper offers a reflection on the influence of soap operas of soap operas in Brazilian fashion. In the contemporary world, many different media vehicles are used to sell products, such as blogs, soap operas, series, the internet itself and television. Digital influencers have become important platforms through which companies can advertise their products, just as soap operas are still a means of selling the products that compose a story's scenarios and plot. Fashion is built and influenced by social and political contexts that manifest at certain time periods and are often incorporated into soap opera scenes. Soap operas have been a part of Brazilian life for a long time and they hold the power to influence consumers' shopping decisions through financed marketing. Celebrities have become influencers of consumption and through soap operas and blogs they can transform the timing and the reasoning behind many people's purchases. Brands have already taken note of this influence and are increasingly investing in this type of media.

1 INTRODUCTION

It is perceptible that fashion is built from historical events which cause variations in clothing and in the ways a society dresses and behaves. Political and social contexts are increasingly influencing fashion trends and the history of fashion itself. Palomino (2003) points out that fashion is a system that accompanies clothing and time, which integrates the simple use of everyday clothes into a larger political, social and sociological context. According to Carvalhal (2016), fashion and clothing have, throughout history, fulfilled different purposes, such as adornment, protection and legitimation. Often these purposes seek to serve people, their dreams, the construction of an identity and self-knowledge. According to Velho (2000), soap operas in Brazil have the power to influence the viewer or consumer since they can be considered the most popular programming on Brazilian television. The author also tells us that in order to understand communication in soap operas it is essential to understand communication on television.

> We understand that soap operas are media tools that reflect the current consumption trend, as well as follow the transformations of clothing and time, integrating the use of everyday clothes to a political, social and economic context. The ease of obtaining information about trends, new collections and behaviors is influenced by what is built by the soap operas' costume design. (Limeira et al., 2016, p. 1)[1]

---

1. Compreendemos que as telenovelas são ferramentas de mídias que refletem a tendência de consumo da atualidade, bem como acompanham as transformações do vestuário e do tempo, integrando o uso das roupas no cotidiano a um contexto político, social e econômico. A facilidade de obter informações sobre tendências, novas coleções e comportamentos passa a sofrer influência do que é construído pelo figurino nas novelas. (Limeira *et al.*, 2016, p. 1)

The world of celebrities is part of our daily lives and it is full of cultural meanings that are available both to the individual consumer and to marketing, which helps generate even more interest in it. Figueiredo (2012), points out to us that this is "the promise of media products: [...] to make the consumer happier and more satisfied by giving her constant and immediate access to pleasure, all supported by media devices, by scientists and speech technologists."[2] Advertising uses this interest to advertise its products: it functions as an illusorily simple means of transferring meaning. The transfer process begins when the advertiser identifies the desired cultural meanings for the product. In the language of current advertising practice, the advertiser decides what he wants the product to say. (McCraken, 2012, p. 112)

2 SOAP OPERAS AND COSTUME DESIGN

Already in cinema, fashion brought a new perspective on life to the role that women were doomed to perform that of the mother and housewife. Marylin Monroe was prominent in this break, because while her body was indeed an object of the masculine world's desire, she was also an actress whose characters rejected this traditional role for woman. (Santos, 2013)

The costume, according to Castro and Costa (2010), is the garment or the set of accessories and clothing that will be used by the artist to compose his character. The costumes of characters in movies and television are very important because, in addition to composing a character's personality and indicating changes in time and space within the story, they can also mark an era, become a fashion reference and, consequently, appear in store windows, thus becoming an object of desire and consumption for thousands of spectators.

Just as it is today, past audiences wanted to imitate their idols and, for this, they dressed similarly—although not identically—to them. As previously mentioned, fashion has always been the mirror of the great movements of society: cinema in the 1930s, television in the 20th century, and blogs today—which arguably work as the big windows of these transformations in current times.

> *In a very general sense of the costume, being an integral part of the art direction, is to create the visual of the television narrative through the costumes of its characters. So, the costume designer will develop and/or choose clothes that understand the needs of the script and the direction, respecting the general concept of what the author and the director of the story want to go through. The costume cannot be seen as detached from a larger context. It works as an important auxiliary source to understand the narrative, and often plays a prominent role due to its importance to the vision that one has of what one intends to accomplish with the television fiction program.* (Cunha, 2009, p. 32)[3]

Kantar IBOPE Media, in a survey conducted on media consumption habits by the Brazilian population: Target Group Index, found television among the vehicles with the most social influence, with a 98% range.

The soap operas appeared in Brazil almost simultaneously to television itself. The first soap opera was produced by the extinct TV Tupi in 1951. They were adapted from the radio and were already a great success in the country. The first soap operas had their scripts imported and adapted from other countries like Argentina, Mexico and Cuba. It was only in the 1970s that Brazil began to produce its own soap operas and to create a national genre. It was also around this time that Globo Television Network emerged as the largest producer of soap operas in the country. (Cunha, 2009).

---

2. A promessa dos produtos midiáticos: [...]tornar a consumidora mais feliz e satisfeita dando-lhe um acesso constante e imediato ao prazer, tudo isso amparado pelos dispositivos midiáticos, pelos cientistas e tecnólogos do discurso.
3. A publicidade funciona como instrumento de transferência de significado ilusoriamente simples. O processo de transferência começa quando o publicitário identifica os significados culturais desejados para o produto. Na linguagem da prática publicitária atual, o publicitário decide o que deseja que o produto fale. (McCraken, 2012, p. 116).

> *Television would therefore be an important channel of interaction between the world and the individual, bringing information and culture. The question of aesthetic renewals encompasses the fact that this mass media vehicle is capable of bringing new concepts of beauty and fashion to the public.* (Velho, 2000, p. 41)[4]

Soap Operas have become part of people's lives. It is common to listen to people from different social classes discussing the previous night's chapter of the soap operas, as Cunha (2009) informs us: "The parallel world of the telenovela coexists with the world of the viewer for several months, and he becomes attached to that."[5] Velho (2000) still tells us that, since television has become part of society's life, it is pertinent that companies want to use this communication vehicle as a means of advertising and merchandising their products. Kalil (1997), in her Chic style guide: A Basic Guide to Fashion and Style, encourages us to look at what artists and presenters are wearing on television to know what the season's fashion trends are.

> *By surfing through television channels, it is possible, in a minute, to know what the trends of fashion in use are, the real fashion. Announcers, newscasters, personalities, actors parade across the screen wearing the clothes of the moment. They announce fashion at all times. It's instant information. TV is not just a record. It is a great launcher of fashions, trends and fevers. A little accessory that marks a season—like the earring of the soap opera's main actress—becomes a national fever.* (Kalil, 1997, p. 232)[6]

It can be noticed that Brazilian soap operas have a very big influence on the purchasing decisions of the fashion consumer. Velho (2000) cites some striking examples of this influence, as in the soap opera Dancing Days (1979), in which the character of Sonia Braga appeared wearing Staroup jeans. These appearances were responsible for a sale jump from 40 thousand pants/month to 300 thousand pants/month (Figure 1). Another example of the strength that characters from soap operas have is Água Viva (1980). Betty Faria, who played an important role in the soap opera, commented that she was tired of a certain color. It is said that in the following days the clothing companies called the television station non-stop, asking the actress to retract her statement, since pieces of clothing with that color were no longer selling.

> *It becomes clear that, even indirectly, soap operas are capable of making a big impact on the fashion industry. A simple comment from a character may be able to change all the industry's supply and demand forecasts, requiring great skill from apparel manufacturers as well as from their suppliers and partners.* (Velho, 2000, p. 54)[7]

The soap opera actresses have become great influencers of consumption, especially those in one of Globo Network's soap opera casts, which belong to the largest communication group in Brazil. The actress Giovanna Antonelli is considered one of the most popular, and the accessories and clothes worn by her in her soap operas are the most sought after by

---

4. A televisão seria, portanto, um importante canal de interação entre o mundo e o indivíduo, trazendo-lhe informação e cultura. A questão das renovações estéticas engloba o fato desse meio de comunicação de massa ser capaz de levar a seu público novos conceitos de beleza e moda. (Velho, 2000, p. 41).
5. "O mundo paralelo da telenovela convive junto ao mundo do telespectador por vários meses, ele apega-se àquilo."
6. Numa zapeada pelos canais de televisão, é possível, em um minuto, saber quais são as tendências da moda em uso, da moda real. Locutoras, apresentadores de telejornais, personalidades, atores desfilam pela tela vestindo as roupas do momento. Denunciam moda o tempo todo. É informação instantânea. A tevê não é só registro. É uma grande lançadora de modas, ondas e febres. Um pequeno acessório que marque uma temporada—como o brinco da atriz principal da novela—vira uma febre nacional. (Kalil, 1997, p. 232).
7. Aí fica evidenciado que, mesmo de forma indireta, as novelas são capazes de surtir grande impacto sobre a indústria da moda. Um simples comentário de uma personagem pode ser capaz de alterar todas as previsões de oferta e demanda do setor, exigindo uma grande habilidade por parte dos fabricantes de roupas, assim como de seus fornecedores e parceiros. (Velho, 2000, p. 54).

Figure 1. Sonia Braga Staroup jeans brand.

Figure 2. Police Officer Helô.

consumers. When the actress played the character of police officer Helô, in the soap opera Salve Jorge (2012), she often wore pants and a silk shirt (Figure 2). The style department of the company Ima Têxtil, an important textile distribution company in Brazil, witnessed the desire of shopkeepers to buy the character's clothes. As Cunha (2009) cites in his article:

> *It is part of the media culture linked to consumer culture to propagate purchase messages. The soap opera could not, then, escape from this, and different companies see an opportunity to take advantage of such a profitable space, much like the very communication company that produces that soap opera, and it is in that sense that Globo Marcas, for example, arises.* (Cunha, 2009, p. 30)[8]

We can see that television can stimulate the sale of products and influence behaviors and lifestyles, as confirmed by Trinca (2004) in the quote below:

> *We are now aware that the Brazilian television industry is capable of stimulating yearnings, dictating behaviors, fashion and lifestyles, seeking to sell their products by associating them with the famous artists and the "wonderful" moments of a soap opera: be it an accessory or a piece of clothing worn by the actress, everything is meticulously prepared for the dreams and ambitions of viewers. In view of this, thousands of products are purchased by consumers, who wish to be able to identify with their idol who advertised a certain merchandise, without having time to reflect on all of the ideological manipulation carried out by the culture industry on TV.* (Trinca, 2004, p. 57)[9]

Often, the fashion trend seen in soap operas is the same from the runways of international and national fashion weeks. What makes a particular trend really transform into an object of desire and purchase are the ways in which these trends are passed on to the general public. That is the differential and the advantage of soap operas. The audience begins to associate a certain tendency with a character or actress of their favorite soap opera. Another important factor, related to the fashion and the media of the soap opera, is the empathy that a character does or does not awaken in viewers. Often this timing does not appear, and the trend may not be released and approved by viewers. Trinca (2004) further states:

> *Thus, in this society, fashion would offer novelty, as well as plan its own obsolescence, imposing tastes and habits of life. An example of this is found in Brazilian telenovelas: fashion lasts the period of a soap opera, but the beauty pattern of the actresses and actors continue for a longer time; everyone seems to know how to identify a beautiful woman/man.* (Trinca, 2004, p. 56)[10]

Because soap operas cover a much larger audience than fashion shows, many companies prefer to have their products associated with characters than presented in that other traditional format. Fashion shows are generally exclusive to celebrities, bloggers, opinion makers and store owners. The general public does not participate in these events, while the soap opera can reach a larger stratum of the population and consequently become more democratic.

---

8. Faz parte da cultura midiática ligada a cultura do consumo propagar mensagens de compra. A telenovela não poderia então fugir disto, e diferentes empresas enxergam uma oportunidade de aproveitar-se de um espaço tão rentável, bem como a própria empresa de comunicação que produz a telenovela, nesse sentido que surge, por exemplo, a Globo Marcas. (Cunha, 2009, p.30).
9. Percebemos atualmente que a indústria televisiva brasileira é capaz de estimular anseios, ditar comportamentos, moda e estilos de vida, buscando vender seus produtos associando-os aos artistas famosos e aos momentos "maravilhosos" de uma novela: um acessório ou uma roupa usada pela atriz principal, tudo é minuciosamente preparado para os sonhos e ambições dos telespectadores. Diante disso, milhares de produtos são adquiridos pelos consumidores, que desejam poder identificar-se com o seu ídolo que propagandeou determinada mercadoria, sem que possa ter tempo de refletir sobre toda manipulação ideológica realizada pela indústria da cultura na TV. (Trinca, 2004, p. 57).
10. Desse modo, nesta sociedade, a moda ofereceria o novo, bem como planejaria sua própria obsolescência, impondo gostos e hábitos de vida. Um exemplo disso encontramos nas telenovelas brasileiras: a moda dura o período de uma novela, mas o padrão de beleza das atrizes e atores continuam por mais tempo; todos parecem saber identificar uma mulher/homem bela (o). (Trinca, 2004, p. 56).

Futhermore, the actresses do not always have the slim, thin bodies of runway models. (Barbosa, 2012).

In addition to soap opera actresses becoming influencers of consumption, fashion blogs are also very effective in the consumer's process of choosing and buying. Often, the actresses themselves also become bloggers. Fernandes (2013) shows us that bloggers are important parts of the buying process:

> *In addition to information research, social influences are also important in the purchasing decision process, as individuals tend to attach importance to a third party's opinion. Thus, by sharing their opinions and experiences, bloggers end up creating an affinity with their followers, and thus influencing them.* (Fernandes, 2013, p. 4)[11]

Blogs, today, have a great marketing power, and for this reason, they have become a very popular space for brands and products. This space could compete with soap operas in the sale of products. Schneider and Pereira (2015, p. 257) cite: "Ad space is often the main source of income for bloggers. This tool has great marketing power, making it a popularity thermometer and a form of advertising with lower costs than traditional media."[12]

3 FINAL CONSIDERATIONS

Fashion in Brazil, which was previously very dependent on international creations and launches, is becoming more independent. Consumers have realized that due to climatic and cultural differences, it is no longer possible to copy everything that comes from Europe and the United States. The new fashion designers, faced with this change, have been struggling to create an essentially Brazilian fashion. A good example of this new paradigm are Brazilian bikinis. They have become an international benchmark.

It can also be said that the globalization of media has helped a lot in this process of consumer choice. The internet has shortened distances around the world, making it more accessible for people to follow what was or continues to be fashionable. With that, they can decide what they really want to wear. Today, consumers do not simply follow a trend or a style, they tend to create a mix of the things they like the most, what makes them feel better. The dictatorship of what is fashionable is coming to an end, as already pointed out by André Carvalhal (2016) in his newest book, Fashion with Purpose.

It is noticeable that media contributes a lot to this process. Soap operas, which are considered the most watched program by the Brazilian population, are increasingly working with financed marketing and encouraging consumers to buy the products they sell. With this, companies want and need their clothing articles to be worn by the actresses and actors of these television programs. Fashion blogs are also part of this new form of consumption. Its articulation with the market is an element that must be studied in the future. The new generation of consumers uses this medium in their consumption decision making.

The entry of streaming services that allow the user to watch movies and series has caused a decrease in the ratings of open television channels, but their reach and repercussion on television still warrants futher study, especially in Brazil, with soap operas remaining a great vehicle of popularization for the masses.

It can also perceived that it is necessary to delve deeper into this subject, since there are still questions to be answered: why are Brazilians still so connected to soap operas if there are other forms of entertainment, such as streaming services? Why are actresses still considered

---

11. Além da pesquisa de informação, as influências sociais são também importantes no processo de decisão de compra, já que os indivíduos tendem a dar importância à opinião de terceiros. Assim, os bloggers ao partilharem as suas opiniões e experiências acabam por criar afinidade com os seus seguidores, podendo desta forma influenciá-los. (Fernandes, 2013, p. 4).

12. O espaço reservado a anúncios é, muitas vezes, a principal fonte de renda dos blogueiros. Essa ferramenta possui um grande poder mercadológico, tornando-se um termômetro de popularidade e uma forma de publicidade com custos menores que as mídias tradicionais.

icons of style for so many people? How do Globo Network soap operas manage to influence the fashion market in such a profound way?

Nevertheless, it is clear that soap operas remain as great influencers of a large part of the Brazilian population. The clothes that soap opera actresses and characters wear hold equal or greater importance than many fashion shows when it comes to the establishment of a trend. Thus, soap operas can still be considered one of the best ways to sell a particular product.

ACKNOWLEDGEMENTS

The authors thank the graduate program of the University of the State of Minas Gerais (UEMG) and the University Center of Belo Horizonte (UNI-BH) for all the support received.

REFERENCES

Barbosa, A.P. 2012. *As representações da moda pela telenovela: análise comparativa das versões de tititi*. 2012. 65f Monografia Centro de Ensino Superior do Ceará, (FAC), Comunicação Social, com habilitação em Jornalismo, Fortaleza.

Carvalhal, A.L.B. 2016. *Moda com Propósito: manifesto pela grande virada*. 1 ed.: Paralela, 414 p. São Paulo.

Castro, M.S.F. & COSTA, N.C.R. 2010. Figurino – O traje de cena. *Iara – Revista de Moda, Cultura e Arte* — V.3 no1 ago – Artigo 1 São Paulo.

Cunha, V.M. 2009. *A moda na mídia: a telenovela como expoente uma análise de "viver a vida"*. Available at: <http://hdl.handle.net/10183/32945> . Accessed em: 02/01/17.

De Castro, K. & De Castro, J.L. & De Oliveira, A.N. 2015. A moda como objeto de informação: o caso do Movimento Feminista Punk Riot Grrrl. *AtoZ: novas práticas em informação e conhecimento*, v. 4, n. 1, p. 24–33, Available at: < https://revistas.ufpr.br/atoz/article/view/41762/26063 Accessed in 28/04/18.

Fernandes, B.B.B. 2013. *A influência dos blogs de moda e beleza no comportamento de compra das seguidoras*. 2013. 123f. Dissertação de Mestrado em Marketing Faculdade de Economia. Universidade do Porto, Porto.

Figueiredo, D.C. 2012. Em busca do corpo 'ideal': consumo, prazer e controle através da mídia de massa. *Revista Intercâmbio*, v. XXVI: 42–60, LAEL/PUCSP, São Paulo.

Kalil, G. 1997. *Chic: um guia básico de moda e estilo*. 11ed.: Senac, 243 p. São Paulo.

KANTAR IBOPE MEDIA. *A Jornada do consumo televisivo em diferentes plataformas*. available at: <https://www.kantaribopemedia.com/a-jornada-do-consumo-televisivo-em-diferentes-plataformas/> accessed in 02/01/17.

Limeira, M.A.B. & De Oliveira, L.A.C. & Dantas, J. & Bulhões A. 2016. O Figurino de Verdades Secretas como elemento de composição de narrativas. In: *Intercom – Sociedade Brasileira de Estudos Interdisciplinares da Comunicação XVIII Congresso de Ciências da Comunicação na Região Nordeste, 2016, Anais., 13p* Caruaru. PE.

Mccraken, G. 2012. *Cultura & Consumo II*. Mauad, 234 p. Rio de Janeiro.

Palomino, E. 2003. *A Moda*. publifolha, 2a edição 98p. São Paulo.

Santos, T. & Ferreira, M.F. 2013. *Analise da ruptura da submissão feminina no cinema da década de cinquenta, incorporadas nas personagens de Marilyn Monroe*. 13 p. available at: http://www.ufrgs.br/alcar/encontros-nacionais-1/9o-encontro-2013/artigos/gt-historia-da-midia-audiovisual-e-visual/analise-da-ruptura-da-submissao-feminina-no-cinema-da-decada-de-cinquenta-incorporadas-nas-personagens-de-marilyn-monroe. Accessed in: 28/04/18.

Schneider, T. & Pereira, L.P. 2015. As representações do consumo e da identidade nos principais blogs de moda do País. *Revista Moda Palavra e-Periódico* vol. 8, n. 15, jan./jul. UDESC, SC.

Trinca, T.P. 2004. Moda e indústria cultural: uma relação concisa. *Revista de Iniciação Científica da FFC*, v. 4, n. 3, UNESP, São Paulo.

Velho, B.A. 2000. *A Moda Brasileira e a Telenovela: um estudo exploratório*. 2000. 120f Dissertação— Universidade Federal do Rio de Janeiro, COPPEAD. Rio de Janeiro: UFRJ/COPPEAD.

The world and the fashion "al revés" in António Variações

Margarida Amaro
ICNOVA – Instituto de Comunicação da NOVA, Lisboa, Portugal

ABSTRACT: Our analysis focuses on the concept of the world "al revés" that for Lotman (1993, p. 78) is the opposite of formal vision and the image that breaks with stereotypes to establish the logic of the unpredictable. The construction of this world is achieved through the subversion of existing values and finds in fashion a form of privileged expression of novelty, transgression and eccentricity. Eccentricity is the practical expression of the dandy who perspectives the fashion, beyond the normative forms of established clothing. A figure who embodies the forms of eccentricity is António Variações: barber as a profession, composer and musician by passion, controversial personality with an extravagant image, Variações leads the transgression of the codes and permeates the cultural imaginary of the late twentieth century in Portugal.

We will analyse the world and fashion "al revés" in António Variações on a perspective of the semiotics of culture.

1 INTRODUCTION

The inverted world (World "al revés") is constructed on the dynamics of the non-dynamic. The everyday realization of this process is revealed by fashion, which introduces a dynamic principle to those spheres of life, which appear to be non-dynamic.
(Lotman, 1993, p. 79)

The world "al revés"[1] is for Lotman (1993, p. 78) the opposite of formal vision and the image that breaks with stereotypes to establish the logic of the unpredictable. According to the perspective of the semiotics of culture, this view of a world "al revés" based on the original logic of things "al revés" and contradictory, a vision opposed to everything that is predicted and perfect, to the whole pretension of immutability and eternity, needs to manifest itself with dynamic, changing, floating and active forms of expression (Bakhtin, 1965, p. 13). For Lotman "One of the most elementary methods of evading the limits of predictability is the trope (used especially frequently in the visual arts) by which two opposing objects exchange their dominant features" (1993, p. 78). However, is Mikhail Bakhtin who first uses the expression world "al revés" referring to a carnival vision of the world, when everything that is extravagant has a particular prominence, opposing to normal time and order.

Then, the construction of this world "al revés" is achieved through the subversion of existing values and the questioning of norms at a particular historical moment finds in fashion—a synchronous, closed and unmotivated system in which the essence lies in change and, for that reason, as soon as it becomes the norm, it breaks with its own regulations, abandons them and changes—a form of privileged expression.

---

1. Although the English version of Lotman's original "Culture and Explosion" translates this expression as "inverted world", we have chosen the expression "al revés" from the Spanish translation of Lotman's Russian original, respecting the option of the Organization of Congress in presenting theme, in its English version, under the theme "DESIGN AL REVÉS". It's Mikhail Bakhtin who first uses the expression world "al revés" in his book "La cultura popular en la Edad Media y el Renascimineto" (1965).

When fashion settles in the dialectic of continuity and discontinuity, tradition and innovation, it is an example of cultural self-renewal, and for Lotman (1993, p. 79), fashion becomes a metronome of cultural development which allows us to observe the constant struggle between the tendency towards stability, immobility, and the opposite orientation, towards novelty, eccentricity. Eccentricity then becomes one of the most pregnant figures, but how does it manifest itself? Introducing a strange element into the space of fashion that contradicts tradition or even morality itself is to make it pregnant, meaning, to endow it with a semiotic meaning. And it comes to mind how Lotman contrasts the eccentric with the concentric, the eccentric can only be defined through opposition: disassociating itself from the center, it is the ex center. When everything is in the center, it is in this opposing centripetal force that fashion acts and becomes a center-fugitive agent.

Eccentricity is the practical expression of the dandy in the domain of life in society: elegant, incredible, wonderful, *fashionistas*, all times have their eccentrics. Fighting against the establishment, these artists of their own lives, are more so as the norm is powerful, strong, rigorous (Bollon, 1990, Presentation). The dandy is the antithesis of the cultural reflex and the inverted face of the stereotype through irreconcilable contrasts between his image and the common image. However, this is a "positive denial" (Bakhtin, 1965, p. 334), which the dialogue proves to be, as Lotman tells us, between a dandy and a vulgar impersonator of dandyism that translates into a dialogue between a genius of fashion and his pathetic imitator (1993, p. 8). The dandy seeks reality beyond the apparent horizon of the dominant conception and seeks, beyond the normative forms of the established clothing, of the current fashion, to envision fashion "al revés." With him, it is the unceasing restlessness and the inquiry on the other side of the thought, from which he can cast new glances on the world, manners and fashions. It is a life diverted from its normal course, a transgressing life "al revés" (Bakhtin, 1965, p. 179) by the violation of the imposed canons: the aesthetic expression of the dandy translates into a playful language that destroys norms, raises conventions and articulates an alternative way of seeing the world. Therefore, individual inspiration and creative enthusiasm can emerge as an eccentricity whose creative energy crosses an entire culture. It should be noted that culture is not limited to innovation or transgression, but according to Lotman it is the only domain of knowledge where innovation takes on an explosive character and where innovation is truly creation.

Creativity, transgression, eccentricity, excess are the dominant features that characterize the figure of António Variações: barber, composer and musician by passion, controversial personality with an eccentric image, Variações leads the transgression of the codes of conduct in force and permeates the cultural imaginary of the late twentieth century in Portugal.

So how can we understand António Variações, the figure that highlights everything that is eccentric in the demanding pursuit of excellent against the mediocre, and the banal? We will see how this charismatic, extraordinary, eccentric being breaks the frontiers and becomes a reference for new ways of being, looking, thinking and creating, ultimately, for the explosion ofthe marginal culture from Portugal: "My face without borders/My road without an end/

Figure 1. Images of the world "al revés": Frères Deckherr, 1820, "Folie des Hommes ou le Monde à rebours" e John Taylor, 1647, "The world turn'd upside down".

Flag colors I don't know of/Flag is white for you" (António Variações song, "Minha cara sem fronteiras" – "My face without borders").

2 THEORETICAL NOTE

A first theoretical note is essential to talk about the fashion system. In fact, fashion is installed in the dialectic of continuity and discontinuity, tradition and innovation: culture as a complex corpus is formed by strata that develops at different speeds so that any synchronous cut shows the simultaneous presence of several states. Explosions in some layers may be combined with gradual development in others, and for this reason, the dynamics of the processes in the sphere of language and politics or of morals and fashion, demonstrate the different speeds at which these processes move. For Lotman, as we have already mentioned, fashion becomes a metronome of cultural development that allows us to observe the constant struggle between the tendency towards stability, immobility, and the opposite orientation, to novelty and eccentricity, in short everything that ties into the idea of fashion, almost the visible embodiment of unmotivated innovation. In other words, a dialectic of cultural changes contains complex dynamic laws that allow, for example, fashion to have a compositional mobility and then eccentricity becomes one of the most preeminent figures. Lotman rightly said that eccentricity is a prerequisite of fashion (1993, p. 79). So the inverted world, according to Lotman, is constructed in the space, which lies beyond the limits of the norm (which is based on the norm and which disrupts it), we encounter a whole range of possibilities: from abnormality (destruction of the norm) to a whole gamut of positive qualities situated over and above the norm (1993, p. 78). The norm disappears with the acceptance of all variants and the diverse carries all possibilities with it; the end of the unchanging universe of traditional appearance is emerging, and "fashion can to all appearances and in *abstracto* absorb any chosen content: any given form of clothing, of art, of conduct, of opinion may become fashionable." (Simmel, 1957, p. 557). However, fashion in its aptness for otherness absorbs any velleity of distinction to quench its thirst for novelty and, in this sense, fashion systematically institutionalizes the singularity (Hammen *et al*, 2017, p. 273) or, in Barthes' words, "Fashion has exterminated all considered singularity in clothing by tyrannically appropriating its institutional singularity" (1962, p. 63). Indeed, although individual achievement is one of the aspirations of our time, it is to be noted that the Western multitudes offer an insipidly homogeneous show. In fact, fashion is a collective imitation of a regular novelty, and even when it uses the expression of an individuality or, if one prefers, an eccentricity as its alibi, it is essentially a massive phenomenon. As a system, fashion assimilates eccentricity, integrates it and institutionalizes it by creating a contradiction of terms, a paradox in which it always lives. It should be noted that fashion, like language, is aimed from the outset at the social and the dandy, in his provocative solitude, is the *a contrario* proof of this (Baudrillard, 1976, p. 94).

The dandy identifies himself with an artistic posture embodied in a form of resistance to a norm. Whether it is political, social or aesthetic, this form of resistance guarantees, above all, self-affirmation regarding the masses and the preservation of his individuality. Dandyism opposes fashion in the search for difference rather than adherence to an idea or a common silhouette and now it is through detail that he will affirm his radical stance, especially outside of the fashion system. Today, eccentricity is no longer strange and radically different, and it is a kind of vaguely nostalgic coquetry. Strangeness abandons the scene of appearances and dissolves into the realm of the diverse. More discreet, hiding its game like a social chameleon.

3 ANTÓNIO VARIAÇÕES: A WAY OF BEING MORE THAN A FASHIONABLE BEING

I would willingly call him a dandy, and for that I would have a sheaf of good reasons
(Baudelaire, 1863, p. 4)

If there is a figure that intensively embodies the forms of expression of eccentricity in Portugal it is António Variações. Endowed with a critical sense and a rare aesthetic sensibility, translated into the way he dressed, António Variações becomes an icon and shocks the most conventional ones. He was a traveller and he took different influences from his tours around the world. It is through his adventure, in the 70's, particularly in London and Amsterdam, that he discovers a new world, wanting to bring to Portugal another way of living. In Amsterdam he learns how to become a barber, a profession that he carries out in Lisbon in the first unisex hairdresser in Portugal and then opens his own barbershop. In the meantime, he founds the musical group *Variações* and starts the shows that quickly attract the attention of a young audience. On the one hand, his eccentric look does not go unnoticed and, on the other, his musical style, combines several genres, from rock to fado. From clothing to hairstyles, Variações excels at an image that still remains and inspires the new generations. António Variações was a dandy: listening to his brilliant music and behind his eccentric clothes, we can discover a cold glance, the sensibility of an artist struck by the rudeness of the great social animal, seeking a transversal path that would allow him to escape the surrounding banality and, therefore, often and willingly, taking a contemplative stance. On the other hand, his contemporaries, colleagues and friends accomplice of his aesthetic adventure are not indifferent to the greatness of his presence and the charisma that emanates from him, and so they see him as follows (Gonzaga, 2006 pp. 136–137):

> *And António was... an apparition. A figure. But what was it that distinguished him so much? Perhaps the serenity with which he walked down the street and went to work in carded boots, or felt slippers, with a zipper in the middle, which was sold at the fairs for people to use at home. Perhaps it is his indescribable figure. His pants were ballooned, his moustache discoloured, his beard in another tone. In Portugal, in 1976 there was no one else who dared to dress like that. It was a free and fascinating public spectacle.*

Aware of the impact of his image and the fracture he creates in the society of the time, Variações states: "I come from a time when I was called all the names, people made way to see me pass by and, or they found it funny, or tortured me with comments." (Gonzaga, 2006, p. 13). So, Variações, is all about the "pleasure of shocking" that marks the 80 s, and the feeling of being in some way "shaking off the dust" raised by all the excitement, the incredible party that celebrated the way out of 50 years of fascism" (Gonzaga, 2006, pp. 29, 220) or "the last flicker of heroism in decadent ages" such as Baudelaire (1863, p. 11) describes the attitude of the dandy. Note that "Dandyism appears especially in those periods of transition when democracy has not yet become all-powerful" (Baudelaire, 1863, p. 11) and it is in the confusion of these times that the figure of the last of the heroes surfaces. Now, the constant confusion between theater and life emerges in António Variações, who turns life into a show: "It is natural that I do a certain enactment, but I have always dreamed of music and the shows and on the absence of it, I made the street my stage" (Gonzaga, 2006, p. 144). However, this is an explosive mixture. Ambitioning to be a musician and performing in improvised stages, Variações, the artist, theatralizes life well beyond its limits. But exactly because life is not a theater, its transformation into theater becomes laden with meaning, as we can estimate from the comments about Variações' everyday situations: "It was a mix between Circus and Zoo. No ticket was paid, but there was a queue—a monumental queue—at the door. Some commented on the spectacle, others just stared. What? Who?" (Gonzaga, 2006, p. 136).

These are the manifestations of strangeness by the public, or others with respect to the eccentricity, to the startling attitudes of the one who destroys the norms in use to create new ones. Variações transforms his existence into a work of art, alive and in the process of becoming an act of resistance to the emergence of new forms of barbarism: "there is a sort of grandeur in all follies, a driving power in every sort of excess" because "a dandy can never be a vulgar man. A dandy may be *blasé*, he may even suffer pain, but in the latter case he will keep smiling" as Baudelaire (1863, p. 10) reiterates and as the artist "he has an elegance and a life all of its own, because everything about him mirrors his intelligence and his glory" (Balzac, 1830, p. 10). So, Variações' conduct and eccentric image stems from the theatricality of his own life.

Note that in Classical Antiquity, the theater actor uses masks to separate himself from his natural appearance and through the mask he cancels his extra theatrical personality. With António Variações, on the contrary, the personality is accentuated by his own mask and by a sequential chain of strata and sense: On the scene, Variações pretends to be an actor and the actor plays the dramatic roles of Variações. Finally, Variações manifests a passion for theater also off the stage: "He filled the scene with his Leonardo da Vinci look, a crimson beret, a strapless inner sweater, the baggy pants, gulf type, and a chequered winter robe as an overcoat" (Gonzaga, 2006, p. 134). The theater and art are mixed with reality in António Variações and confer singularity to his actions. However, any eccentric act, when repeated, can go from the domain of the explosion to the sphere of the usual and become massive, as witnessed by the experience of Variações himself (Gonzaga, 2006, pp. 144–145):

I dress the same way as ten years ago. But at that time I was practically alone. Not today. There are many more people who dress like me, I no longer feel alone and I even think I have influenced a lot of guys.

So we cannot fail to recognize in these words by António Variações an identity relative to that written by Oscar Wilde for Dorian Gray (1890, p. 93):

Fashion, by which what is really fantastic becomes for a moment universal, and Dandyism, which, in its own way, is an attempt to assert the absolute modernity of beauty, had, of course, their fascination for him. His mode of dressing, and the particular styles that he affected from time to time, had their marked influence on the young exquisites... who copied him in everything that he did, and tried to reproduce the accidental charm of his graceful, though to him only half-serious, fopperies.

Combining imitation with distinction it necessarily achieves a paradox and a risk: "To inoculate all of contemporary clothing, via Fashion, with a bit of dandyism was always going to kill dandyism itself since, in its very essence, dandyism was condemned to be radical or not exist at all." (Barthes, 1962, p. 63). When Fashion becomes more vulgarized it institutionalizes itself and its role of distinction is lost. Now only the *demodée*, the one that is out of fashion, is distinctive: "Fashion is what one wears oneself. What is unfashionable is what other people wear." (Wilde, 1895, p. 138). This is the precise consciousness of António Variações and it is also there that the *very character of fashion* is located, as Simmel rightly recalls (1957, p. 547):

The very character of fashion demands that it should be exercised at one time only by a portion of the given group, the great majority being merely on the road to adopting it. As soon as an example has been universally adopted, that is, as soon as anything that was originally done only by a few has really come to be practiced by all—as is the case in certain portions of our apparel and in various forms of social conduct—we no longer speak of fashion.

In fact, an idea or custom has its maximal innovation and its maximal significance when it is still restricted to a small group. So what interests António Variações in fashion is precisely this modern notion of permanent shows and parties and the inherent aesthetic involvement, personal experience, the fashion within the individual as stated by Simmel (1957, p. 554):

With more or less intention the individual often establishes a mode of conduct or a style for himself, which by reason of the rhythm of its rise, sway, and decline becomes characterized in fashion. Young people especially often exhibit a sudden strangeness in behaviour, an unexpected, objectively unfounded interest arises and governs their whole sphere of consciousness, only to disappear in the same irrational manner. We might call this a personal fashion, which forms an analogy to social fashion.

This is the main characteristic of António Variações, his main distinctive trait, faithful to himself, does not follow trends and creates a personal image, he does not follow fashions, he imposes them, as observed by his contemporary, the Portuguese fashion designer Manuela Gonçalves, who thus recalls (Gonzaga, 2006, p. 178):

António marked, without a doubt, Lisbon's nightlife, mainly by the way of presenting himself and being… And quite eccentric, even when everyone tried to be different… The boots, the clothes, the accessories, the straps, the chains, the rings, the extraordinary coats, the shirts, the fabrics printed with plants, nothing matched, but then everything worked out because he assumed everything in a very special way and he moved very well. It could be kitsch, but it was on purpose, without a hint of bad taste. He had a great elegance, he wasn't even tall, but stood out among people. He had light. Had anything.

This is singular glow in António Variações or the exception of the artist in the words of Balzac (1830, p. 9):

…he is elegant and slovenly in turn; he dons, as he pleases, the plowman's overalls, and determines the tails worn by the man in fashion; he is not subject of laws: he imposes hem… he is always the expression of a great thought and towers over society

An excess of artistic genius beyond the limits of predictability as further described by the author of the *Treatise on elegant living* (1830, p. 10): "The artist is always great. …There are as many lives characterized by new ideas, as there are artists. With them, *fashion* must not be forced: these uncontrolled beings fashion everything as they please."

It should be noted that at the time, Leonel Moura—plastic artist and one of the promoters of the "After the Modernism" movement, became aware of the dimension of António Variações when he wrote about him with regard to his participation in one of the events promoted in January 1983 (Gonzaga, 2006, p. 221):

This guy is not included in our movement, but is much more advanced and radical than all these people who say they are very advanced and very cutting-edge. He showed up dressed in his own way, but particularly exuberant, with a chicken wire, a door handle hanging from his ear, a belt of chains, a bit on the punk scene, but in Portugal there was nothing like that. In fact, the fashion that was on exhibition was very banal, did not leave a trace. Eventually some grew, and became well known designers, but their intervention there did not leave a mark. Variações did. And probably without knowing, he was really the postmodern artist par excellence.

Considering life as a show far beyond the limits of the scene, António Variações transforms life into a show. And life is understood as pure exteriority, but requires a subject and a qualitative and concrete interiority, inseparable from the individual, that of which it is immanent, or even in the understanding of Oscar Wilde for whom "life itself was the first, the greatest of all the arts, and due to that, the other arts seemed to be but a preparation for life" (1890, p. 93). It is the possibility of interpreting any role and exhausting all the variants and, on the other hand, going beyond the limits of possibility, accomplishing the impossible.

The important thing for Variações is this aesthetic perspective, an "aesthetic of transgression" and his dandyism consisted not in the aspiration to follow fashion, but rather in the firm conviction that he had to determine it: "(My way of dressing) has to do with my freedom. I Dress like this, different and colorful, because I feel good. However, I never cared about fashion. I care about aesthetics" (Gonzaga, 2006 p. 144).

This is another indication of the multidisciplinary aesthete and the perfectionist nature of António Variações and his talent to surprise with his secrets that reveal his awareness that fashion is not restricted to mere frivolity, such a worldly lifestyle aesthetics already equated by Balzac, who in a "treaty on elegant life" seeks to analyse the invariable principles that govern the manifestation of our thought by external life and understanding fashion as a metaphysics of things (1830, p. 26):

The material life, once the object of general progress, has undergone tremendous developments. There is not a single of our needs that has not produced an encyclopedia, and our animal life is tied to the universality of human knowledge. In dictating the laws of elegance, fashion encompasses all the arts. It is as much the principle of works and of art as it is of works of craft… By welcoming, by indicating progress, it takes the lead in everything: it brings about revolutions in music, literature, drawing, and architecture.

It is from this understanding of the dimension of fashion that Variações shares. Fashion is in his way of dressing but also in his music or in the spaces that he inhabits and frequents, the fashion understood, above all, as the manifestation of a certain moral quality of the cult of taste and of a refined aesthetic sense and, for this reason, we recognize, once again, in António Variações some of the concerns that worried Dorian Gray and, like him—both— surely inspired by Balzac, Variações "desired to be something more than a mere *arbiter elegantiarum*" and according to the protagonist of Oscar Wilde (1890, pp. 93–94):

He sought to elaborate some new scheme of life that would have its reasoned philosophy and its ordered principles and find in the spiritualizing of the senses its highest realization. The worship of the senses has often, and with much justice, been decried, men feeling a natural instinct of terror about passions and sensations that seem stronger than ourselves, and that we are conscious of sharing with the less highly organized forms of existence.

The search for elements of a new spirituality, in which the dominant characteristic would be a high instinct of beauty such as that predicted by Dorian Gray. In line with this protagonist, it was evident to Variações "that the true nature of the senses had never been understood, and that they had remained savage and animal merely because the world had sought to starve them into submission or to kill them by pain, instead of aiming at making them elements of a new spirituality, of which a fine instinct for beauty was to be the dominant characteristic." (Wilde, 1890, p. 94) guided by the development of the taste for everything that is our own and that surrounds us in the exaltation of the perfection of the external and material life or the search of that refinement of an "elegant life" already observed and dissected by Balzac (1830, pp. 24–25):

…one must be still have been endowed with that indefinable faculty (perhaps the spirit of our senses!) that always prompts us to choose truly beautiful or good things, things whose unity matches our physiognomy and our fate. It is exquisite tact, the constant exercise of which alone can suddenly reveal relations, anticipate consequences, guess the place or significance of objects, words, ideas, and people.

A sensitivity thus guided by "a noble though of order and harmony, intended to give poetry to things" (Balzac, 1830 p. 25) and if according to the aphorism also deduced by Balzac, an artist lives as he wants, or... as he can, this is the *Adventure of the Senses*[2] in António Variações (*A Canção do Engate*, The hook-up Song):

Na aventura dos sentidos	In the adventure of the senses
Tu estás só e eu mais só estou	You are alone and I'm lonelier
Tu que tens o meu olhar	You who have my gaze
Tens a minhamãoaberta	You have my hand open.
À espera de se fechar	Waiting to close
Nessa tua mão deserta	*In your deserted hand*

4 CONCLUSIONS

With Antonio Variações we find a discursive procedure that Lotman identifies as the world "al revés". It is the inquiry on the other side of the thought, from which new glances can be cast on the world. Turning life into an aesthetic experience, equating the pleasures of life with the beauty of art, breaking with moral boundaries, António Variações puts his genius in life and his talent in works, in the construction of a coherent artistic body. Suppressing the distance between being and appearing to be, between the body and the spirit. From this tension between body and soul emerges a subversive individual, a multipurpose artist and a multidisciplinary aesthete with a gaze that is simultaneously mirrored and reflector. His passion

---

2. We quote a song of António Variações, *A Canção do Engate – The hook-up Song*.

for creation freezes frivolity or any deceptive sign of superficiality: he turns his clothing into an opaque wall to become invisible in his apparent visibility and when he dresses up as a gladiator he is practically naked. Exceptional, eccentric, excessive, António Variações moves with the codes of clothing and customs with originality and insolence, defining his way of life from fashion, choosing discomfort, the vertigo of movement. The world and the fashion "al revés" and just as conjectured by Baudelaire (1863, p. 11):

> *Fastidious, unbelievable, beaux, lions or dandies… all share the same characteristic of opposition and revolt; all are representatives of what is best in human pride, of that need, which is too rare in the modern generation, to combat and destroy triviality.*

Guided by the principle of general curiosity and hostile to any form of dogmatism, with his cosmopolitan thought António Variações annuls any form of provincialism. His open character, alternating toughness and softness, detachment and great sensitivity, confers a new scale to the Pop music in Portugal and his charismatic figure embodies the ideal of dandyism: in November 1983 António Variações receives the Prize for the worse-dressed of Portugal, not without questioning the jury's ability to evaluate styles, with this provocation: "The style is me!".

António Variações, is *Beyond*[3]. This solitary person afraid of solitude ends up equating the social relation by imagining the magical relations between the solitaries, the strangers who provide each other with pleasure. António Variações embodies the transgression of the taboos of which he speaks in the songs and accomplishes an ideal of the artist, as of the lover:

Tuestás livre e euestou livre	You are free and I am free.
E há umanoite para passar	And there is a night to spend
Porque não vamos unidos	Why don't we go united?
Porque não vamos ficar	Why don't we remain?
Na aventura dos sentidos	In adventure of the senses
Tu estas só e eu mais só estou	You are alone and I'm lonelier
Tu que tens o meu olhar	You who have my gaze

António Variações, *A Canção do Engate*, The hook-up Song)

REFERENCES

Bakhtin, M., 1965. *La cultura popular en la Edad Media y el Renacimiento.* Translated by J. Forcat and C. Conroy., 1990. Madrid: Alianza Editorial.

Balzac, H., 1830. *Treatise on elegant living.* Translated by N. Jeffries., 2010. Cambridge: Wakefield Press.

Barthes, R., 1962. *Dandysme and Fashion.* In Stafford, A. and Carer, M., ed. 2006. *The Language of Fashion.* London: Berg Publishers. Ch.6.

Baudelaire, C., 1863. *The Painter of Modern Life.* [Online] Available at: http://writing.upenn.edu/library/ Accessed 3 May 2018].

Baudrillard, J., 1976. *Symbolic Exchange and Death Theory, Culture & Society.* Translated by L. H. Grant., 1993. London: Sage Publications.

Bollon, P., 1990. *Morale du masque: Merveilleux, Zazous, Dandys, Punks, etc.* Paris: Seuil.

Gonzaga, M., 2006. *António Variações entre Braga e Nova Iorque.* Lisboa: Âncora Editores.

Hammen, E., & Simmenauer, B. 2017. *Les grands textes de la mode.* Paris: IFM. Regard.

Lotman, Y.M., 1993. *Culture and Explosion.* Translated by W. Clark., 2009. Berlin·New York: Mouton de Gruyter.

Simmel, G., 1957. *Fashion.* American Journal of Sociology, Vol. 62, No. 6, pp. 541–558. The University of Chicago Press. [Online] Available at: http://www.jstor.org/stable/2773129 [Accessed 3 May 2018].

Wilde, O., 1890. *The Picture of Dorian Gray.* [Online] Victoria, Canada: McPherson Library. Available at: http://contentdm.library.uvic.ca/cdm/singleitem/collection/Literary/id/2521 [Accessed 3 May 2018].

Wilde, O., 1895. *An Ideal Husband.* [Online] University of Toronto: Robarts Library. Available at: https://archive.org/stream/idealhusband00wilduoft#page/n1/mode/2up [Accessed 3 May 2018].

---

3. *Estou Além—I'm Beyond* is the title of a song of António Variações.

Facets of youth culture: Discursive construction of Ahlma fashion brand

B. Machado & M. Scoz
Universidade do Estado de Santa Catarina, Brazil

ABSTRACT: The analysis proposed by this study is based on Greimas semiotics on the discursive construction of Ahlma clothing brand. The institutional presentation of the brand is taken as an object of study to understand how the strategies of meaning production converge to dialogue with contemporary youth culture. To do so, a brief discussion about the consumption and the construction of brands is proposed and in sequence the Generative Trajectory of Meaning is adopted as a methodological apparatus. The articulation of different levels of the model allows observing the thematization of the collective, the diversity and the new, mainly from the concept of "new era" approached by the brand. This concept is configured in the narrative effort with a greater appeal to the public, since the ideal of change is a euphoric value for these consumers.

1 INTRODUCTION

Youth culture has an expressive power within the fashion system, being responsible for great revolutions in the history of clothing and the precursor in diverse changes of tastes and styles. The popularization of sportswear, the adoption of the white T-shirt as a piece of casual clothing and no longer as underwear and the diffusion of miniskirt were aesthetics introduced by young people who marked the history of fashion. The same youthful influence was observed in the rise of subcultures and other tribes, such as the beatnik generation, the mods and the punk movement.

The predominance of the young culture in dressing has more repercussion after the 50's and until today is still relevant, since not only this portion of the population represents a considerable consumer market, but also plays a key role in the adoption of behaviors and aesthetics. Within the innovation diffusion process, the trendsetters are the group of people that researchers are particularly interested in (Vejlgaard, 2008), individuals who have adopted new habits of consumption rapidly, predisposed to changes, adventurers and fearless of taking risks. This group has unique ways of expressing itself, through music, sports, cinema, literature and fashion. The specialist in sociology of trends Vejlgard (2008) highlights the recurrence of the trendsetter profile within some social circles, for example young people, since individuals at this stage of life are exploring their own identity and therefore are more receptive to trials.

Many fashion brands are addressed to the young segment, however not all of them adopt an *avant-garde* discursive strategy. Within this approach, we highlight the work developed by Ahlma, a clothing brand that has André Carvalhal as its creative director, a professional that stands out for innovative fashion initiatives in Brazil. The repercussion of the brand since its launch in June 2017 was expressive, with strong adhesion in social networks, two stores opening since then and various issues in the specialized media, as in Elle Magazine, O Globo newspaper, VIP magazine, Pequenas Empresa Grandes Negócios Magazine, among others. Within this context, the problem of the present research is: how does the Ahlma brand construct a discourse that dialogues with manifestations of contemporary youth culture?

To answer this question, the analysis of the discursive strategies used by the brand that evidence facets of the current youth culture is defined as the central objective of this article. The aim is not to find a single definition of this group's behavior, given the multiplicity of systems of signification, as well as the plurality of identifications that are increasingly fragmented today (Hall, 2014). More precisely, it intends to point out, from the institutional statement of the Ahlma brand, manifestations of behavioral traits contained in the construction of the simulacrum of its audience.

Within this context, seeking to understand the discursive mechanisms that reveal values and intentionality of this instance that permeates the product and takes place in the brand, it becomes relevant to structure, organize and specify the construction of meaning that is constituted in the relation between objects and subjects. Thus, the discursive semiotics or Greimas semiotics is adopted as a methodological approach, which focuses on the understanding of the signification processes from an object of communication defined as text. The model used to analyze the generation of meaning is the Generative Trajectory of Meaning (Barros, 2011; Castilho & Martins, 2005; Fiorin, 2014), which consists of three different levels of signification, discursive level, narrative level and fundamental level. From the Hjelmslevian approach, the notion of text is amplified beyond the verbal discourse also encompassing other communication systems such as the visual, which contemplates and enriches the observation proposed.

2 CONSUMPTION, BRANDS E IDENTITIES

The presence of brands in the fashion market is massive, both large *maisons* and magazines seek to establish distinctive features that stand them out from the competition. Linking a name, and so to say a brand to a product, had the initial purpose of guaranteeing the origin, quality, and know-how of a company or designer, as Charles Worth did. He was personally involved in every stage of his garment manufacturing and signed his creations to legitimize the origin of it (Svendsen, 2010). Worth's professional performance in the second half of the nineteenth century was a forerunner in the construction of the designer activity as it is known today, for in this period it was up to the couturiers to only execute the orders of their clientele, while Worth was able to create his own brand.

As consumerism developed through individualization and exculpation, as well as its strong hedonistic appeal, contemporary society began to commodify lifestyles, a transformation widely explored by Lipovestky (2007). From this juncture, the concept of brand extrapolates the qualification of the offer of products and expands to a social communication dimension. The representativeness of the product for the companies is modified whereas it no longer maintains isolated, but becomes part of a syntagmatic construction. Although very relevant, the product is more an integral part of a broader formation of the brand, which is composed of all the manifestations that the company adopts, both in productive and communicational terms.

The change of identifying a product with a more comprehensive approach of marketing, lifestyle and values matches with the understanding that "we do not only consume to meet existing needs: we do it probably to create an identity." (Svendsen, 2010 p. 129). This is what amplifies the assimilation of the act of consuming beyond the satisfaction of needs such as drinking, eating, dressing, and establishes itself as a communicational dynamic of social identity construction and thus, "we have become what we consume" (Miranda 2017, p. 28). Although this construction of identity is to some extent influenced by class distinctions, Lipovetsky (2007) points out that consumption is not limited to the primacy of status and that acts of purchasing translate private motivations of self-expression. In this regard the author states "I, at least partially, reveal who I am, as a singular individual, by what I buy, by the objects that populate my personal and familiar universe, by the signs that I combine" in my own way (Lipovetsky 2007 p. 44). Added to this conception, consumption is not limited to the act of purchasing, but it also manifests itself in the consumer engagement with the brands, being it through events, videos, games, social networks and other events that show the proximity of individuals with the values promoted by brands. In this sense, Semprini (2006, p. 50) defines brand as "the instance that provides a meaningful context to an experience or an imaginary that, alone, would tend to be imprecise or very abstract." This way, it is possible to understand the semiotic nature of brands, which organize narratives through the relationship of their elements with the goal of generating identification with their audience.

Consumer habits demonstrate an emotional relationship with the products, "loaded with new collective and individual meanings" (Lipovetsky 2007, p. 24), characterized as a social process that produces meaning and identities. Svendsen (2010) points out the contradiction of this effort of affirmation, in which standardized brands are used with the aim of shaping the singularity of the self. However, this is the same fashion paradox, which is capable of individualizing subjects whereas qualifies them as members of a similar aesthetic group, concomitantly expressing individualism and conformity. Understanding the expression arrangements of a brand that has acceptance of an audience reveals possible yearnings of social visibility of these individuals.

The fashion system operates through changes in style and taste. It is the valuation of the new that is at the core of its functioning, as stated by Lipovestky (1991) and Svendsen (2010). Each season—or in even shorter time cycles—different products arrive in stores. However, as the industry or market efforts seem insufficient to alter aesthetic tastes, social behavior is a key part of these transformations called fashion trends. According to Vejlgaard (2008), trendsetters are pioneers in the adoption of new consumer behaviors, which open the way in the innovation diffusion process and therefore attract the attention of trend researchers who seek to monitor their tastes and preferences, in view of their propensity to become popular. Vejlgaard (2008) argues that within some social groups there is a recurrence of the trendsetter profile, with young people being one of them. However, it is not to be concluded that every young person fits this classification, but only that due to the identity in formation and search for affirmation it is possible to find a greater number of individuals open to novelties, experimentations and changes in this moment of life. To understand manifestations of aesthetics and *avant-garde* discourses, one turns to a brand that dialogues with these consumers in order to verify their veridictory square (Ciaco, 2013), that is, the production of truths through the organizing practices of their discourse.

3 SEMIOTIC ANALYSIS: AHLMA BRAND

Launched in 2017, the carioca brand Ahlma is under the creative direction of André Carvalhal, a personality known within the Brazilian fashion community. The advertiser figured prominently as the marketing of the female brand Farm and from his professional experience he published two books: "Fashion imitates life—how to build a fashion brand", in which he approaches branding strategies, and "Fashion with Purpose—Manifesto for the great turning point", which questions production models of the garment industry and discusses how new businesses should consider pressing values for the current moment such as sustainability, fair trade, social and cultural awareness. Carvalhal was also a co-founder of Malha, a collaborative space for fashion creation, experimentation and entrepreneurship in Rio de Janeiro. Ahlma offers women's and men's clothing for young audiences. The commercialization of the products is carried out through its e-commerce and two physical stores, one inside Shopping Leblon and another in the Ahlma Academy, which, in addition to selling branded products, also offers yoga classes, courses and various lectures and a juice bar integrated to the store.

To approach the abstract and concrete dimension of contemporary brands, Semprini (2006) proposes a model of functioning and brand identity based on semiotic theory, which comprises brand as an enunciation process, "discourse production act" (FIORIN 2014, p. 55), which is divided into two different levels: enunciative instance and statement. By terminological choices, the author defines these two levels respectively as the brand Project, its vision and purpose. It is the sociocultural value that offers a "horizon of meaning" for the consumer public in which "individuals can integrate it into their life projects, their concerns, their questions, the practical context of daily life." (Semprini, 2006 p. 158), and the brand

Manifestation, which are the actions, products, policies that give concreteness to the Project. In view of the objective proposed for the present analysis, attention is given to the discursive construction that presents the Project of Ahlma. In this way, the institutional texts, both verbal and visual, presented on its website are taken as the object. The image that follows has been fully replicated as a preview.

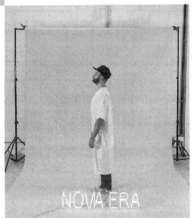

Figure 1. Institutional presentation of Ahlma brand.[1]

---

1. Translation to English from the original text can be found in attachment.

The postulates of the generative semiotics proposed by Greimas take the text as a unit of analysis. The conceptualization of the text in this approach understands it as a whole sense of meaning and its analysis aims to "describe and explain what the text says and how it does to say what it says" (Barros, 2011 p. 7). To do so, we used the theoretical-methodological apparatus known as the Generative Trajectory of Meaning (Barros 2011, Castilho & Martins, 2005; Fiorin, 2014), which examines the content plan of a text through the articulation of different levels of meaning, denominated discursive, narrative and fundamental.

3.1 *Discursive level*

The presentation of the Ahlma brand starts from the most superficial level of the text, the discursive one, in which the abstract forms of the narrative level gain concreteness, through themes and figures. It is also at this stage that discursive structures are examined from the relations established in the enunciation instance (the act of producing discourse), as well as the relations between enunciator and enunciatee. According to Fiorin (2014), these subjects are respectively the author and reader implicit in the text, projected images that leave marks on the constructed discourse. This dynamic is established because "to exercise persuasion, the enunciator uses a set of argumentative procedures, which are part of the relations between enunciator and enunciatee" (Fiorin, 2014 p. 57), that is, to whom the discourse impacts on its construction. Thus, an advertising report aimed at the young audience will have a different language from children's audience, considering the specificities of those to whom the message is addressed.

In the discursive composition of the object in question, it is perceived that the first and last image figuratively recover the theme of the collective through the photographic composition. A group of people positioned next to each other and sometimes establishing physical contact evidencing a certain degree of intimacy, since they seem comfortable with the presence and touch of the other. Although the models wear different clothes, there is coherence in the color chart of the clothes in both photos. In the first image, the participants appear wearing clothes in softer tones and in the latter, darker ones, which again creates a union between the individuals portrayed. This theme is reiterated in the verbal statements that present the brand, as in the case of the opening title "shared purpose", that, besides the choice of words, also shows the importance of this theme by the size of the font used, which suggests that a group of people shares the brand Project, as it happens in the definition "co-creation axis", that emphasizes the work performed by the team and the collaboration with other brands. Again this idea is reinforced in the phrase "A network woven together to serve dreams" and when the production process is mentioned, "Together we share processes, dreams and commitments, and the common purpose of developing a more open and responsible value and product chain," it is noticed a reiteration of such a semantic trait, by characterizing the brand through the sense of collectivity and also as a group to be integrated, expressed through word choice, the use of the first person in the plural and the visual resources.

Defining the brand as a group that acts together towards a goal builds a enunciatee that is interested in being part of this select gathering of people and when engaging with the brand shares the same ideals. Although expressed in a collective, the enunciatee is singularized because it is associated with a fashion proposal that is said more conscious, therefore different from the rest of the market. A concept also expressed by the visuality of the people photographed, although in an overall composition, its components are aesthetically diverse, both in their features and in the use of less ordinary clothing items.

Another theme identified in this text is the diversity of both gender and ethnic origin. In the statement "Beyond a brand, we are a movement whose culture allows (many) friends" ("Para além de umamarca, somos um movimento cuja cultura permite (muitos) amigxs."). In the original language, Brazilian Portuguese, it uses the "x" instead of the letter that characterizes the gender. This is a strategy employed as a form of non-binary language. This spelling is widely used in social networks as a form of neutrality in discourse and was incorporated by Ahlma. Thus, the strategy shows an inclusive approach and builds an enunciator opened to diversity, while assuming an enunciatee that does not identify with gender stereotypes.

In the images, this theme is also taken up from androgynous models, from men wearing clothes socially considered as feminine and a balance in the presence of white and black people in the photographic composition.

Recurring theme in fashion and in so many areas of consumption, the new is also incorporated into Ahlma's speech. By exploiting innovation in advertising discourses, Ciaco (2013, p. 21) states:

> *Thus, speaking about the new would be almost a requirement of these new times, anchored in a permanent process of reconstruction originated from the movements demanded by the technological, informational and communicational advances. As a result, the constancy, permanence and static character of "things" assume a strong dysphoric value in the repertoire of the immediacy of information and communication in contemporary societies.* (Ciaco 2013, p. 21)

Following this logic, the brand tackles the idea of the new figured with a greater truism in the second photo light indicator in which reads "New Era", a concept explored by Carvalhal in his second book "Fashion with Purpose—Manifesto for the great turning point". It represents an ideal of change in the production and marketing system of the fashion industry and it is revealed as the precept of the brand. The reiteration of this theme is observed in the statement "Consumption, however, sickens nature and leads the planet to its ultimate consequences. Exhaustion of resources, illegal work schemes, poor waste management. In the list of top leaders, the textile industry is among the top places—and warns: the time to change has come". Doing so, Ahlma qualifies its products positively as they represent the change and construction of this "new era" in and euphoric way.

In this premise of the new, the second picture brings the creative director of the brand into an individual portrait, wearing a piece that refers to a gown, both for the length below the knee and the color white. Barefoot also alludes to the construction of a religious or spiritual figure. Finally, in the last photo, Carvalhal is in the center of the group, this sequence of photos and their compositions personify him as the leadership of this so-called movement. Thus, it is understood that the enunciatee recognizes and probably identifies with the work done by the advertiser within the Brazilian fashion scenario.

3.2 *Narrative level*

The second level is defined by Barros (2011, p. 16) as "a spectacle that simulates the doing of the man who transforms the world." At this stage, there is an attempt to understand the narrative that, linked by enunciate of doing and of state, is structured in a canonical sequence comprising four different phases: manipulation, competence, performance and sanction.

The manipulation phase is characterized by the action of one subject over another to make him/her want and/or do something (Fiorin, 2014). It can usually be performed in four ways: temptation, intimidation, seduction and provocation, depending on how the manipulator acts on the manipulated. The syntheses of these typologies of manipulation are well expressed by Fiorin (Elements of discourse analysis, 2014, p. 30):

> *When the manipulator proposes a reward to the manipulated, that is, an object of positive value, to get him/her to do something, there is temptation. When the handler forces one to do by threats, then bullying occurs. If the manipulator leads one to do something manifesting a positive judgment on the competence of the manipulated, there is a seduction. If he impels the action, expressing a negative judgment as to the competence of the manipulated, a provocation follows.*

In the analyzed text, manipulation is identified through temptation, since the manipulator (the brand) proposes a reward to the manipulated (consumer), he can be part of the movement called "new era", become integral to the "we" created by Ahlma and have "unforgettable moments". The main object of this manipulation is to convince the manipulated to buy the clothes of the brand, its services, to develop some collaborative project or to consume its content. The verbal statement that concludes the presentation makes clear an invitation to the manipulated "From awakening to living together, and from living together to co-creating collections, services, stories and unforgettable experiences. Let's go together?" in which the brand is characterized as a precious universe to be part of, because it is for people who perceive the value of conscious consumption, and therefore select, but there is still a clear and stimulating invitation in the end.

In the competence phase, the subject is endowed with a know-how, which can be understood from the text as the knowledge of the productive and differential processes of the brand aligned with the conception of a new moment in fashion that allows the consumer to carry out the phase of the performance. Here, the change from one state to the other occurs from its engagement with the brand by the consumption of its products, services and content. The sanctioning phase, which is the validation that the performance was done, is not explicitly manifested in this text. It can be understood beyond the analyzed object, consolidating itself in the recognition of the individual by its peers as part of the group proposed by Ahlma that has more conscious consumption habits and the set of values already explored previously.

Narrative semantics deals with the valuation of objects, which are divided into modal and value objects. The first are the objects necessary for the accomplishment of the performance phase, while the second are those that are in conjunction or disjunction along a narrative path, according to the subject's goal (Fiorin, 2014). From the analysis of the presentation of the brand, the modal object within this narrative scheme can be considered as the consumption of fashion brands that, in addition to the aesthetic sense, have fairer, more innovative and sustainable bias production models, which enables the subject to be in conjunction with the object of value understood as becoming an integral part of the group of people who constructs the "new era."

3.3 *Fundamental level*

The last stage of analysis is the fundamental level, considered the simplest and most abstract of the discourse (Castilho & Martins 2005), in which elementary structures are organized to present the minimum meaning of the text expressed through a single relation of semantic opposition. Each term is qualified as euphoric (positive value) and dysphoric (negative value), these valuations are not fixed, but revealed from the internal relations of the text. According to Fiorin (2014), to put two terms in opposition they must be in the same domain so that they have something in common, as for example in the universe of aesthetics the beautiful versus ugly opposition is coherent.

Considering the articulation with the other levels of analysis, it is understood that the elemental structure of the discursive construction of the Ahlma brand occurs in the new versus old semantic opposition. By distancing itself from traditional definitions, condemning established productive processes in the fashion industry and promoting a riskier approach to the market, the brand presents itself as the construction of a possible future, embodied in this "new era" manifesto promoted by Carvalhal, that is, by introducing a break with the (old) consecrated idea of a given fashion that needs to be overcome.

Thus, the brand is repeatedly presented as the response of a new moment of fashion, characterized as a movement carried out by a group of people who act in a pioneering and daring way in dressing, in the production of clothing, in communication, in the collaborative work. This approach is strongly expressed through the verbal statement "We prefer the vitality of the new ways to the security of a single destiny, and we have already lost, for a long time, the fear of risking." From such utterances of rupture, Ahlma constructs the discourse of the new.

4 FINAL CONSIDERATIONS

From the understanding of the brand as a discursive construction that uses persuasive strategies to generate identifications with its public, it is possible to observe how Ahlma brand dialogues with its consumers, given its repercussion in the fashion market, it is assumed that these values are relevant to today's youth culture. The assumption of enunciation, between an enunciator and enunciatee, reveals relevant meaning systems to be highlighted.

The recurrence of the collective theme and the presentation of the brand as a group was verified. Verbal and visual statements evidence this stance through discursive strategies such as the photos that portray a group of people physically close to each other wearing clothes with colors that combine with each other, which brings a visual coherence. Verbal statements use the third person plural and highlight the notion of common goal to be shared.

The use of language artifacts, such as the "x" in place of the letter that characterizes the genre of the noun, thematizes the diversity of gender from a discursive strategy that implies an enunciate that shares this form of communication. The same is manifested by the people in the pictures, androgynous, white and black models, men with long hair, women with short hair. This proposal is reiterated by the more fluid dress practice, which challenges gender stereotypes from unconventional clothing choices for the male or female audience. Still, the portrayed individuals have the common profile of models, extremely young and thin. Although the brand seeks to highlight an ideal of more diverse beauty, what is found is that to some extent this still follows a canonical logic.

The "new era", a concept explored in Ahlma's discourse, summarizes the approach to the new, being constructed as a break from the traditional molds of the fashion industry. The brand criticizes the depletion of natural resources and exploitation of the productive chain, to be characterized as a response to this scenario through the reuse of raw materials, a more transparent and collaborative work. Although this movement is declared as collective, there is a leadership expressed by the figure of creative director André Carvalhal.

The identification of the brand's consumers is understood as a manifestation of a sensitized youth on the impact of their consumption habits, seeking alternative solutions that give vent to their desires. However, this awareness does not seem to be to avoid buying, but rather to choose more carefully the origin of its consumption objects. As a result, the Ahlma consumer is individualized against the remaining mass of consumers who do not share the same values, and then becomes part of the "new age" movement, which is constructed as euphoric, daring and disruptive. This idea of change is emphasized because, as discussed earlier, it is a positive appeal for individuals who identify with leading-edge behavior.

The adoption of the semiotic approach as a theoretical framework for the present article privileges the comprehension of the fashion brand as an enunciative act, which from the articulation of its persuasive strategies reveal values, habits and yearnings of the target public, and in the case of Ahlma, facets of youth culture. Finally, the limitation of this analysis, taking the issue of identity as unstable and fragmented is emphasized, it is highlighted that the possibilities of research on contemporary youth culture are not confined to a specific profile, but are manifested in their plurality.

REFERENCES

Apresentação institucional Ahlma, 2017. Sobre a Ahlma. [online] Available at: <https://ahlma.cc/p/sobre-a-ahlma > [Accessed 5 January 2018].
Barros, D.L.P., 2005. *Teoria semiótica do texto.* 4th ed. São Paulo. Ed. Ática.
Castilho, K., Martins, M.M., 2008. *Discursos da moda: semiótica, design e corpo.* São Paulo: Anhembi.
Ciaco, J.B.S., 2013. *A inovação em discursos publicitários: comunicação, semiótica e marketing.* São aulo. Estação das Letras e Cores.
Fiorin, J.L., 2005. *Elementos de análise do discurso.* 11th ed. São paulo: Contexto, 2005.
Hall, S., 2014. *A identidade cultural na pós-modernidade.* Rio de janeiro: lamparina.
Lipovetsky, G., 2009. *O império do efêmero: a moda e seu destino nas sociedades modernas.* São Paulo. Companhia das Letras.
Lipovetsky, G., 2007. *A felicidade paradoxal: ensaio sobre a sociedade de hiperconsumo.* São Paulo. Companhia das Letras.
Miranda, A.P., 2017. *Consumo de moda: a relação pessoa-objeto.* 2nd ed. São Paulo. Estação das Letras e Cores.

Raymond, M., 2010. *Tendencias: qué son, cómo identificarlas, en qué fijarnos, cómo leerlas*. Barcelona. Promopress.

Semprini, A., 2006. *A marca pós-moderna: poder e fragilidade da marca na sociedade contemporânea*. São paulo. Estação das Letras.

Svendsen, L., 2010. *Moda: uma filosofia*. Rio de Janeiro. Zahar.

Vejlgaard, H., 2008. *Anatomy of a trend*. Dinamarca. Confetti Publishing.

*ENGLISH VERSION FROM AHLMA'S INSTITUTIONAL PRESENTATION

SHARED PURPOSE

In the old world's words, we are a brand of clothing. But at the mouth of the new era, we are—nice to meet you AHLMA:) – co-creation axis.

A network made together to serve dreams. To life. And from the meeting, draw the new. We have choosen fashion as power, because we are close to it. Clothing, if not art, is legitimate desire. It is an way of expression. It is to relate to the world. It is self-knowledge. And truth be told, we all, every day, get dressed.

Consumption, however, sickens nature and leads the planet to its ultimate consequences. Exhaustion of resources, illegal work schemes, poor waste management. In the list of top leaders, the textile industry is among the top places—and warns: the time to change has come.

WHAT DO YOU WANT FROM HERE?

Through self-knowledge and the sensitive cultivation of a community of partners, producers and suppliers, we seek more fertile solutions to all that we wish to transform.

Together we share processes, dreams and commitments, and the common purpose of developing a more open and responsible value and product chain.

WE ARE OUR BIGGEST CHALLENGE

We were born to question, but we know we do not have all the answers. It's okay, after all, this is just the beginning and we are not alone. Consciousness is our guiding thread, and love for what we do, our best energy.

In addition to a brand, we are a movement whose culture allows (many) friends. We prefer the vitality of the new ways to the security of a single destiny and we have already lost, for a long time, the fear of risking.

From awakening to socializing, and from living together to the co-creation of collections, services, stories and unforgettable experiences.

Let's go together?

Fashion films: A semiotics approach, a multimodal signification

R. Effe
Instituto Politécnico do Cávado e do Ave, Portugal

ABSTRACT: We frequently come across short films that show, over a brief period of time, a whole concept of a fashion collection. These are called Fashion Films—the perfect combination of cinematographic techniques and the contents suggested by fashion, in a set of referents that communicate a product, coordinate, collection or brand. It is unnecessary, in this paper, to justify the importance of these tools as far as communicating fashion is concerned. An attempt will be made to explain some of the concerns arising when structuring and producing a Fashion Film—the discursive level and the orders of signification based on social and visual semiotics, building upon the seminal work of R. Barthes, Deleuze, Greimas and Kress. The focus of this research is not on the effective analysis of modes, but rather on furthering the inter-relational analysis ensuing. It builds upon the theories of multimodality, so as to enable the collection of data that encourage an effective brand communication and promotion.

1 INTRODUCTION

When approaching the topic of Fashion-Films, one comes across the most current projective force of brands and authors. Fashion Films are presented as a strategy that aims to communicate a concept, brand and intention within a short period of time. This tool also shows some characteristics that have the potential to fit into numerous digital applications that are justified in and by themselves, especially in the production-dissemination value ratio.

Based on the concepts of brand, gestalt structures, and technical elements of cinema, Fashion Films show a quality that offers a better and more detailed understanding of the object. The Fashion Film object, as the sum of a set of media and modes of action, looks for the discourse that is most appropriate to communicate the brand.

In order to meet their initial aims, brands commission the production of their Fashion Films to an author or team of authors who, albeit covertly, will present their own perspective, their authorial stance. This means that all authorial issues that are implemented by the creatives must be considered, in addition to the aims set by the concept of the brand. Therefore, the type of language used by another author, as well as many of the guidelines provided by the brands, will be an obvious reflection of their differentiation. From resorting to 'Fetish' models to using colours, shapes, movements and other modes of action, currently the authors of Fashion Films already make them part of their own work. That is, of their authorial work.

Social and political concerns, images of daily dynamics, and media personalities are among the tools used, which are presented on the film stage with a support that is often crucial to the artistic object. Fashion Films are more than just a digital fashion support, to become a digital object resulting from the combination of several different concepts, several discourses and diverse visual media—multimodality as a field of study for a better understanding and justification of the narrative and of the discourse of the respective brand.

2 THE STRUCTURALISM OF ROLAND BARTHES; ORDERS OF SIGNIFICATION IN THE NARRATIVE OF FASHION FILMS

The semiotic analysis of images cannot be conducted without making mention to one of its scholars. Roland Barthes, essayist, scholar and French semiotician who inscribes himself in structural semiology as a means to achieve an order of signification devotes about a decade, from the publication of his book 'Mythologies' to the publication of 'Fashion System', to the study of Semiotics.

Barthes went on exploring everything he deemed worth analysing, from 'soap powder' to 'steak and fries', from facial expressions of sex symbols to the most curious structures of thought related to anthropology, literature, cinema, plastic arts, theatre, photography and music, linguistics, psychoanalysis and semiotics. He cross-checked the knowledge of several authors and, acting as intermediary, became Roland Barthes.

The fact that he is respected by many structuralists, as a result of his modes and media of understanding the objects, accounts for one's current need to understand a whole complex system by a set of subsystems. Barthes points out that Structuralism operates according to methodologies that will enable the understanding of the whole by its parts. That is, the most effective understanding of things, of people and intentions via their smallest and most elementary structures, via the relations between them, via their signs.

> *What a piece of clothing, an automobile, a dish or an advertising image may have in common is the fact that they are signs. And in their existence of signs, they give themselves to read to the modern Man in the form of images, gestures, behaviours. These readings, however, are never innocent; they imply social, moral, and ideological values.*
> (Werneck, 2008)

It can be briefly stated that this method and this line of thought are perceived in Fashion Films. This fragmentation, this understanding of structures and their relationships, this disorganized, almost entropic, non-linear order for arriving at conclusions in the understanding of fashion also reflects on the assembly of Fashion Films. This way of telling their story, of being its true narrative, is understood to be the methodology, the script provided to the masses that it sets out to reach. Concrete cases are those that present the linearity of their narrative, fragmented, whether by the recurrent use of film planes that are distinctive among themselves, by false and absent records, or by analepses and prolepses, or even by the obvious resort to rhetorical figures resulting from the understanding of the relationships between fashion structures. There is, therefore, a need to explore all possible variables associated with object-fashion; that is, an understanding is required of space-time. The understanding of all cultural structures, to which fashion belongs, will be an added-value in the direction of a Fashion Film. Mention should be made, therefore, to the fact that Barthes had already suggested that fashion is a complete system that requires a thorough and simultaneous study of different disciplines—including, among many others, economics, history, ethnology, and linguistics.

As soon as attention is demanded by the analyst, one can understand the relationship between the structures and the result of this relationship. In a system of nearly cause-effect, based on numerous referents, the relationship between jobs and technically suitable clothes result in the Uniform being a kind of clothing that denotes the practicality of use. Barthes makes reference to this approach by citing Larmessin (Paris, 1632/94), who, in a hyperbolized and grotesque way, presented in his illustrations a series of characters dressed by a set of objects that, by direct relationship, identified the jobs. The aim is thus to emphasize the importance of signs and their relations in the creation of a character, in the creation of an Image. An example is the references made by the author in his writings, for instance, to the fact that contemporary architecture is mirrored in clothing, as is the explicit example of the chimneys of the industrial revolution that were reflected on the hats of the bourgeoisie. This is a register in which the formal transposition between creations is still evident nowadays. This is the transfer of influences, the trends. It is thus indicated that, by establishing these relationships, the promoters of images for purposes of brand dissemination have a set of crucial tools that enable a careful mass communication.

We often put aside all referents, all variables, and focus only on details. This lack of focus from all other elements sometimes takes place, disallowing the production of meaning of the image. However, the focus on only a few details will enable the construction of another meaning. Whether or not this is purposeful, it certainly deviates from the assumptions of the author of the image.

It should be underlined that this deviance from all the referents presented in the image enable the construction of a poetic image, of an image that has now been created possibly based on referents stored in memory and left by other images. Thus, this de-contextualization, this sum of parts not belonging to the same system, will allow the creation of what we will call poetic image. In compliance with the aims of the technical team, this type of imagistic input can be enabled by the most diverse techniques of image capture and display.

Barthes[1], in *Camera Lucida*, includes a legend of an image depicting two children photographed in profile on a grass field as follows: 'I put aside all knowledge, all culture... I see only the great Danton collar of the boy, the bandage on the girl's finger...', while saying that 'a detail takes hold of all my reading; it is a living mutation of my interest, a fulguration. Thanks to something that marks it, the photo ceases to be any photo' (Barthes, 2012). I still consider Image to be a text, a 'mixed fabric of different types of signs' (Joly, 2000), a system composed of a series of structural valences that, by interacting with each other, secretly proliferate in a discourse. In the case of the orders of signification of Barthes, in the specific instance of the photographic legend, the author limited himself to making a denotative, referential reading of two elements: the tape on her finger and the collar, while rejecting the connotative order, the second order that he proposed as the discourse of what is implied, his 'secret discourse.'

This example shows that we can inscribe here the intentional refusal of interpretation by many image users. There are, in fact, cases where it is impossible to interpret a second meaning of an image and, consequently, the abandonment to the second order of signification, connotation. Thus, there are some people who also refuse to interpret an image on the grounds that an image is to be lived, felt, but not interpreted. Some curious and justified cases are those of the Graphic Designer April Greiman and photographer Susan Sontag (Joly, 2000: 183). However, it is also possible that this is merely an intentional guideline, left by the creator of the image. The intentions can be varied, but all of them will have an objective that is attached to the focus aimed. A forced and intentionally produced reason for the fact that only a certain element, although inserted in a vast context, is object of interpretation, a communicative intention.

In order to mirror the narratives that are juxtaposed to the product, the format of two-dimensional communications is insufficient; however, this is a possibility in the case of the Fashion Films. Based on the numerous examples offered by the Fashion Films presented at the Fashion Films festivals, some of which are award winners, it is worth noting that, as far as this tool is concerned, it is possible to discern the various concerns that, in turn, are approached in and by the diverse disciplines. Ranging from an approach to gender issues to sustainability, values, ethics, xenophobia and racism, Fashion Films are thus inscribed in the disciplines of ecology, society, politics and economy, so as to better justify their brand concept.

3 THE POTENTIAL OF THE MOVING IMAGE

It is commonly agreed that recording of moving images enables possibilities to more compellingly present any type of narrative. It appears as a tool capable of opening a set of structures that relate to each other so that their result is more plastic.

The moving image appears as an image that incorporates a set of elements that interrelate to, act and react with each other. These images, inserted in an environment that sustains them, become agents belonging to a structure in which the dynamics among agents is evident. In this interrelationship among agents belonging to the same image, a new signification is obtained—a new signification of the agents themselves, as well as the production of the general meaning of the image.

---

1. Photograph by Lewis Hine titled: Mental Institution, New Jersey, 1924.

3.1 *Deleuze; interaction among modes*

Starting from the concepts of the fashion brand, and including the concepts explored in the collections and in the particularities of the coordinates, the entire technical team will have to develop an action panel that respects all these premises, as well as to use languages that it deems appropriate to the work in question.

Now, if every detail belonging to all the agents of an image is important; if the moving image is a set of elements that articulate and react with each other; if a Fashion Film presents an infinite number of images, care and rules will therefore have to be adopted for making meaning in communication.

In the designations made by Deleuze in his book Image in Motion, one is faced with an explanation that closes the idea of the importance of the analysis for the reading and understanding of meaning making. According to the author, '[t]here are Lulu, the lamp, the breadknife, Jack the Ripper: people who are assumed to be real with individual characters and social roles, objects with uses, real connections between these objects and these people—in short, a whole actual state of things. But there are also the brightness of the light on the knife, the blade of the knife under the light, Jack's terror and resignation, Lulu's compassionate look' (Deleuze, 2016: 159).

We believe that it is necessary to observe the relations between all the modes that operate and encompass an image, as it is thus that it will become an image.

In order to reinforce the latter statement, it is necessary to present a brief glossary about Image-Movement, Image Centre, Image-Perception, Image-Action and Image-Affection that Deleuze (2016: 317) briefly explains. Reference is made here to action and reaction, to spatial variables between constituents and modes of an image that is object of exploration via the interaction between the modes that will help determine also the modes of discourse.

To this justification, it is important to add that the exploration of the narrative, the fictionality, the creation of characters, the detailed considerations of time and space, narration, image complemented with words or other signs, historical references, and the intertextual concerns that the post-structuralist Julia Kristeva, in 'Word dialogue and novel' (1966), suggests is the interaction between two modes as a tool for new and grounded results. This, in turn, will aim to create new theories, ideas, concepts, whose goal is to determine forms, formats and that by their relations of parenthood, contrast or complement, suggest the modes of action.

This consists of establishing priorities in the choice of tools that have as main intention the manipulation of a subject by its sign structure, by a set of images rhythmically cadenced with audio schemes, voice-over narrative supports, or subtitles, among others. That is, of structuring a work that persuades, directs, guides, clarifies and informs. These are some of the contingencies that Multimodal Social Semiotics handles and whose 'meaning and form appear as an integrated whole, a sign, and the signs are always newly produced according to the interests of their producers in specific situations. In this respect, Kress states that all signs are metaphors and that metaphors, like signs, are always newly produced in specific environments for audiences with specific objectives, too' (Vieira and Trajano, 2012:258).

From the recipient, one expects the result of the manipulation via their interaction.

From the recipient, from the desiring subject (Firion, 2000), one expects the result of the articulation between modes and agents, one expects the result of the modality used in the discourse.

4 MANIPULATION OF DISCOURSE

4.1 *Greimas and the discourse of Fashion Films*

It is important to mention that the passionate sensitization of the discourse and its narrative modulation are interactive; hence, one is not justified without the other (Greimas, 1993). Therefore, passions are the property of discourse. By building upon this discursive principle, and upon the theories of the modes explored by the French Semiotics (Firion, 2000), Fashion

Films present themselves with a persuasive doing, which can be juxtaposed, both in modes of power, and in modes of knowledge.

As far as modes of power are concerned, let us consider that some of them are linked to action: temptation, which corresponds to the relationship between the manipulator and the manipulated, having as intermediary a positive object; intimidation, whose intermediary business is a threat; seduction, when the manipulator appeals by positive means; and provocation, when the manipulator provokes via negative means. It is important to recall that in this study the word manipulator does not hold in itself a negative character. These manipulations coexist with the discourses, be they political, religious, educational, ethical, commercial—that is, these manipulations coexist with the Fashion Films.

It is important to emphasize that the purpose of these manipulations is inscribed within the roots of wanting to be, must be, knowing to be and can be, modes that suggest a close relationship between the object and the subject, of the 'desirable object to the desiring subject' (Firion, 2000).

Fashion Films are discourses and present themselves to their subject within a modular structure that comprises some types of discursive manipulation entirely addressed to the subject. Therefore, the concerns of the Brand while producing and directing a Fashion Film go through the narrative dynamics; through the sequences of scenes inserted in space-time, angles and points of view; visual/audio and their close relations with space-time, the alphabet of the planes, the dynamics and movements of the cameras, the linearity of the narrative, the relation with the argument and the script that are most suitable for a discourse between object and subject.

4.2 *Discursive evidence in Fashion Films*

By resorting to differentiated discourses, let us now superficially explore some Fashion Films that are illustrative of the main points described in this article. Modal approaches are left open to serve as an interactive tool for the reader.

In their s/s season, in the film directed by Alan Masferrer in 2017, the Moncler brand presents a work that stands out with reference to the impossible constructions of the works of the Dutch graphic artist Maurits Cornelis Escher. The artist uses halftones, explores the infinity and isometries, which, in this Film, are inscribed as core valences of the Moncler s/s 2017 collection. The film was presented at the Berlin Festival Fashion Film. Metaphor, hyperbole and pleonasm are possibly the rhetorical figures that command the meaning of the work, the modality, thereby complying with the purpose of the brand.

In *https://vimeo.com/channels/alanmasferrer/204863412*

In 'Make yourself unstoppable', a film directed by Wolf & Lamm for Strellson, one can observe that, both the lyrics of the film soundtrack and its musical dynamics are the guiding principles of the narrative discourse. As it is supported by this relationship between these two main structures, sound and image, the narrative is marked by a set of concepts and themes that trample each other, without discomplementing each other.

A cover of the song 'Don't stop me now', originally by Queen, becomes the runway, the support of a rich set of images that become familiar in daily life, but which, due to their references, are unusual and gradually appear as the main character of the film as it goes through the numerous situational spaces. In a good mood, hyperbole and oxymoron are also explored, which emphasise the purpose and concept of the brand. Strellson is a brand of Men's clothing and apparel which focuses on sober lines, but which, due to its formal and conceptual richness, offers clothing and apparel for different occasions. In a narrow discursive mode of power, this Film was also presented at the Berlin FFF.

In *https://vimeo.com/182714597*

Wolfberg, the director of REVIEW BOLLYWOOD, builds upon a script, read in voice-over, that tells the story of a young man who wants to present his work in the Milan Fashion Week, but his suitcase was swapped. In a bollywood-like structure, he resorts to narrative in voice-over—narrator, music and dance—multimodality in an uncompromising discourse directed at young people's clothing. One of Berlin FFF 2016.

In *https://vimeo.com/134068100*

GEIST.XYZ shows an experimental video where textiles are presented by resorting to the use of dynamic 3D manipulations. The Film shows a clear focus on sound design, which was also awarded the Prix Ars electronica, thereby fostering the image, that flows by subscribing to the artisanal algorithmic textiles and procedural surfaces highlighted by the brand. It portrays the informative character of the textiles, in an informal and ethereal discourse, in an innovative image. Winner of the Bucharest FFF, Berlin FFF, Fashion Film Milano and Ciclope-International Festival Of Crast, all of them in 2016.

In *https://vimeo.com/168750702*

Club 99.7, Fashion Film of the Danish brand Le fix as a documentary. There is in this video a clear intention to show the honesty of the image of the brand Le Fix, which is highlighted by the personal and real account of the grandfather of the filmmaker Nina Holmgren. He is a wise poet aged 99 years and 7 months who is careful in his choices, and who states that age does not exist for those who live it lively. This is the assumption of the brand that operates in the market as multi-brand and multi-age. Truth and passion are elements that work as intrinsic motivation for the argumentative constitution of the piece.

A close relationship between the object and the subject is manifested by the honest reality presented. The Film ran in the BFFF 2016 competition.

In *https://www.youtube.com/watch?v=8WQInxgLB90*

Landmarck was the best-edited award winner at BFFF 2016. The film was directed by Romain Laurent, with changes of scenery, changes of clothing, a brand that presents itself as uneasy in the market. There are modes related to the action highlighted by the rhythms, musical cadences, the planes go through different spatial-temporal situations, showing different coordinates, such as accessories. This is a dynamics that link well the basic premises of a Multibrand company. A discourse that inscribes and proposes itself by truth and lie, ways that contribute to the interaction between object and subject.

In *https://vimeo.com/168309120*

Nowness: Bárbara Anastacio directs 'Women in Uniform', winner of the best documentary award at the Fashion Films festival in Milan. It explores the women's relationship with social life when wearing their uniform. Here the interpretative act of the senses juxtaposed to the narration is justified from a negotiation between the object and the subject. This is the use of signs indicating a profession, status, order or medium by resorting to a specific coordinate. 2017.

In *https://vimeo.com/200458309*

Ethetics episode 2 Mdingi Coutts – The film is directed in partnership with the Ethical Fashion Initiative of the International Trade Center. ETHETICS (ethics + aesthetics) embraces ethical brands that pose as holders of a significant social responsibility throughout the world. In this respect, it should be recalled the work of Amber Moelter, who directed the film for the duo Lukhanyo Mdingi & Nicholas Coutts. This work discusses the relationship of an ethical brand in a clean, expressive sense, within the sustainable premises. A genre of moving editorial of a new brand based on an ethereal explanation of appealing to the eco-conscious and the eco-curious subject by the concept, the form, and the mode.

Set in the arid African lands, the coordinates tend to respect the guidelines of a new scene in the world of fashion. 'Here we are!' A sustainable brand, an ethical brand. This multi-nominated and award-winning work is also in the official selection of the BAFTA Qualifying Aesthetica Short Film Festival 2016.

In *https://vimeo.com/236881570*

Directed and written by Sean Baker, the 11 minutes long film **Snowbird**, from the **Kenzo** brand, follows a young woman named Theo who covers a community in the desert of Slab City, California, to offer a tasting of her homemade cake. Unlike fragmented narratives left in an incomplete record, Snowbird presents itself in a so to speak complete narrative rendered by all the referents necessary for the understanding of the whole. To foster this narrative, Sean Baker finds in Slab City the ideal base for Kenzo's spring 2016 graphic collection.

The character is not superimposed on clothing, rather the opposite. This is a work that was mainly concerned with mirroring the honesty in the use of the pieces of clothing by the inhabitants of Slab City. A subtle ethnographic note. A linear discourse, a multiply-awarded short film.

In *https://vimeo.com/158305488*

4.3 *Gunther Kress, a multimodal reflection*

Some videoclips, in addition to the short films, are an example of these concerns/intentions. Beyonce's work 'Formation' was awarded, among others, the Grammy for best clip in 2016.

In this work signed by Melina Marsoukas, there is a clear drive for ecological and political problems; the former is made clear, in the video, by the scenes of a flooded city, literally expressing Hurricane Katrina; the latter is illustrated by the face of Martin Luther King Jr. in the newspaper, accompanied by the title 'The Truth'. On this subject, it is also evident the graphite that reads: 'Stop Shooting Us', a reference to the murders committed by the police, which report alleged xenophobic and racist events taking place in the United States. This is a discursive mode that infers meaning making of an ethical and political appeal to the desiring subject.

Along with these social and educational concerns, the subject Fashion is the element that ultimately claims all these issues throughout the videoclip. The main character appears in the countless scenes with several, luxurious and emblematic coordinates of Brands belonging to different authors. That's a manipulative character that at this stage delights the subject through positive media. An interrelation of signs is thus presented whose motivation is justified by the passionate sensitization of the discourse.

In 'Formation', the modes of power, from temptation to intimidation, and from provocation to seduction, emerge multimodally as social answers to the desiring subjects.

In *https://www.youtube.com/watch?v=WDZJPJV__bQ*

To this, one should add that this was justified by Gunter Kress (2010) in Multimodality. A social semiotic approach to contemporary communication: Multimodal Social Semiotics is interested in all possible forms presented as meaning.

> *As he adopts the perspective of Multimodal Social Semiotics, the author builds a theory of meaning from three perspectives. The first one is related to the production of meaning and to the categories used in all kinds of representation, communication and communication resources; the second one refers to multimodality, which deals with issues that are common to all modes and with the relationships holding between them; the third and last perspective refers to a specific mode, focusing on categories that describe forms and meanings which are appropriate to the specificities of a certain mode.* (Vieira and Trajano, 2012)

Fashion Film is, by itself, a mode that is part of Brand communication. Because of its technical characteristics and the ability to explore concepts, meaning production, with multimodal references and the specific understanding of each mode, is a crucial system, both in constructing a narrative and in determining discursive lines.

5 CONCLUSION

The use of metalinguistics for the contextualization of the Brand purposes, for the presentation of trends and values, for meaning production is effective, in the case of Fashion Films.

This tool uses a manipulative, enthusiastic discourse, a passionate sensitization to thus conquer the subject. The importance of modes in the structure of a discourse is evident for an honest communicative strategy.

This study emphasises that the discourses presented by Fashion Films express multimodal impressions that foreground, both the real testimony, and the creation of fantasies and drives. It is also my understanding that the latter are a result of the non-dissociation between reason and passion, which, in turn, allow the birth of distinct modes that enable the manipulation of the desiring subject. This explains why Fashion Films, by communicating a desirable object, open new imaginary domains by the discursive level producing meaning—new domains in the creation of fantasies and drives, while promoting the poetic image.

This research triggers the need to explore new guidelines, framed in social semiotics, which are related to multimodal principles. It is assumed that the articulation of the lines of thought encompassing the theories of Barthes, Greimas and Kress, as far as visual social semiotics is concerned, is a line worth exploring.

This study intentionally leaves the semiotic discussion open, as it expands by its resources and by other resources adjacent to it; culture, values, dynamics, social and technical structures, relationship between structures, modes, intentions of the meaning producer and agent of meanings. These buzzwords are a tool required to apply contemporary semiotics to the understanding and production of Fashion Films.

REFERENCES

Barthes, R. 2005. *Inéditos: Imagens e Moda*. São Paulo: Martins Fontes.
Barthes, R. 2017. *A Câmara Clara*. Lisboa: Edições 70.
Baudrillard, J. 1991. *A sociedade do consumo*. Lisboa: Edições 70.
Costa, L. (org.) 1982. *Teoria da Cultura de Massa*. Rio de Janeiro: Paz e Terra.
Deleuze, G. 2016. *AImagem-Movimento. Cinema I*. Lisboa: Documenta.
Fiorin, J.L. 2000. Modalização: da LínguaaoDiscurso. *Alfa: Revista de Linguística* 44: 171–192.
Fontanille, J. 2007. *Semiótica do Discurso*. São Paulo: Contexto.
Greimas, A. & Fontanille, J. 1993. *The Semiotics of Passions. Dos Estados de Coisas aos Estados de Alma*. São Paulo: Ática.
Joly, M. 2005. *A imagem e os Signos*. Lisboa: Edições 70.
Joly, M. 2012. *Introdução à Análise da Imagem*. Lisboa: Edições 70.
Kress, G. 2010. *Multimodality. A Social Semiotic Approach to Contemporary Communication*. New York: Routledge.
Kristeva, J. 1980. *Desire in language: A semiotic Approach to Literature and Art*. New York: Columbia UP.
Martin, S. 2006. *Vídeo Art*. Taschen.
Vanoye, F. 1994. *Ensaio sobre a Análise Fílmica*. Campinas, SP: Papipurs.
Vieira, J.A. &Trajano, I.S.N. 2012. KRESS, Gunther. Multimodality. A social semiotic approach to contemporary communication. New York, Routledge, 2010. *Veredas* 16(2): 257–260.
Web Links: The operation of these Web Links was checked in April 2018.
Werneck, N. 2008. Roland Barthes, A moda e as assinaturas do mundo. *IARA –Revista de Moda, Cultura e Arte* 1(1).

The importance of language on the social medias of a brand

M. Hammes de Carvalho
Universidade do Minho, Guimarães, Braga, Portugal

F.A. Hammes de Carvalho
Universidade Federal do Rio Grande, Rio Grande, RS, Brazil

ABSTRACT: Nowadays, it is common for companies, in search of new markets in the international context, to use social networks and the English language to promote their brands, using communication as one of the main marketing tools. Considering that language affects the coding, storage and retrieval of information in memory, this article proposes, from the interface between neurosciences and communication, the promotion of a critical reflection on the use of foreign languages as a conditioning aspect of communication and customer loyalty in social networks of a brand—be it Instagram, website or Facebook, in particular, pointing out the possible impact of this type of communication on the relationship with the Brazilian consumer.

1 INTRODUCTION

Nowadays, faced with the consequences of globalization processes, companies are crossing national territories and doing cross-border business. In this context, the reasons for the internationalization of brands are diverse, but the following should be highlighted: the search for growth opportunities (market diversification) and, consequently, higher profits and margins; the acquisition of new ideas about products, services and forms of negotiation; the approximation of sources of supply, the access to factors of production with lower cost or better value; facing international competition; and investing in (advantageous) relationships with foreign partners.

Following the reasoning outlined above, it is possible to say that the main concern of a company (brand) is to build a positive relationship with the globalized public, making them identify with the values of the brand. Faced with this scenario, communication is one of the main marketing tools to strengthen a company's identity. This process, according to Cavusgil, Knight and Riesenberger, was made possible by the decline of trade barriers and technological advances. According to the authors, "the double trend of globalization of markets and technological advances allows companies to be more prepared [...] to carry out marketing and global purchases" (2010, p. 23). For Lipovetsky & Serroy (2015), Brazil became a relevant actor in the design and fashion era, also facing the dynamics of internationalization.

The movement to internationalize a company implies the presentation of the brand in the face of international competition. And, because of this, brands use different strategies such as, for example, media discourse. Nowadays, the following social media are considered as high-relevance media instruments: Facebook, Instagram and website. These resources enable us to simultaneously reach a large audience, transposing geographical and temporal barriers. As technological tools, they offer conditions for the company to constantly update information and use semiotic resources, in addition to written language.

However, such media when mistakenly conducted can generate distancing to the interlocutor/client. In this case, language is fundamental, since it is the basis of every process of human interaction and, through it, it's possible to build discourses with intentionality.

For companies, the language has the objective of approximating to the client, seeking to persuade the behavior of the same. And undoubtedly, the mechanism of written language influences the decision-making process of the individual, since the deliberative process is intrinsically linked to the understanding of the message.

According to Solomon (2011), it is precisely in the language that marketers, who wish to enter in foreign markets, encounter a problem. In order to be successful and relevant to the market, some qualities are needed by the brand, which, according to Cobra (2010), are: be able to suggest something about its intangible benefits, to have an easy pronunciation, to be quickly recognized and remembered, to be inconclusive and have only one meaning for all countries and languages [emphasis added].

As a solution, the tendency of companies is to adopt expressions in languages that facilitate the transit of the brand in different countries where it is present, since national languages are not easily translatable and tend to complicate direct communication. Due to a lack of knowledge of a foreign language, advertising themes commonly move away from the original sense when translated, causing unfavorable interpretations. This difficulty of linguistic identification reaches even different countries which, despite speaking the same language, "may have communication problems due to the singular meaning of colloquial words" (Cavusgil et al., 2010, p. 109).

At the same time, adopting the partial use of a language that leads to immersion in an international market can mean a distancing from the internal market, if the language adopted is not common to the consumer. As Maingueneau (2011, p. 45) points out, "We usually manage to deal with utterances in certain foreign languages, even if we do not understand the greater meaning of their words and phrases if we can have a minimum of information about the discourse genre in which include such statements".

As a result of this difficulty, the communicative process used in the marketing tools ends up causing an interpretative limitation and negatively affecting the persuasive process of the brand. In this scenario, considering that studies in the area of neurosciences have contributed to a better understanding of the process of reading and interpretation of speeches, it is relevant to reflect on the possible impact of the use of the English Language in the virtual marketing of a brand in relation with the Brazilian consumer.

2 BRAND COMMUNICATION

Brands, especially fashion products, individualize the appearance of beings, contributing to the construction of an identity of the subjects who consume them. In Cobra's (2010) conception, brand building in the consumer's mind is intimately linked to advertising, being its purpose to create and add value to the brand, building a positive identity for it through an attractive name and, specifically, marketing strategies and communication programs. In Maingueneau's perspective (2011) with the presence of brands, non-human entities arise that are the responsibility of advertising statements.

The role of communication is essential for the brand, since, according to Dillon (2012), the brand can be understood as a distinguishing element of the company among its competitors. Thus, "from the name of the company and its products to the logo, the style of the products, the packaging and the design of promotional materials, all can be remembered from the brand" (p. 148). In addition, Solomon (2011) states that a product that brings the personality of the brand in itself contributes to detach it from competitors and inspires years of loyalty. In this perspective, the authors Agis et al. (2001) emphasize the importance of the affectivity between the consumer and the brand. According to them, this affective relationship occurs at several levels, not restricted only in products, but also addressing concepts and communication. This immaterial side of the brand becomes a decisive factor in the process of buying consumers, surpassing the price and quality of the product.

> *From one communication action to another, from one advertising to another, from one visual to another, the consumer can then more easily recognize the issuer, the brand*

that enunciates the product. All communication channels must tell the same story, without making their own procedure an end in itself, and converge to strengthen brand identity. (Sackrider et al., 2009, p. 36)

According to Cobra (2010) the objectives of the communication are: to generate desire for the product category in the various moments of fashion cycles; create brand awareness early in the fashion cycle; improve attitudes and influence future consumer intentions and facilitate the purchase. Considering the above aspects, it is noticeable the strength of advertising in brand promotion. For Maingueneau (2011, p. 212):

The name of a brand, like any proper name, is associated with a variable set of representations sedimented through time, a 'brand image' on which the company must act constantly. The evolution of this image is due in large part to the discourses that the company issues and issued about itself and its products, in particular through advertising.

The advertising message, besides presenting a message that allows the understanding of the concept of fashion and the advantages of its adoption, is more persuasive when it contains images and texts that can capture the attention of the consumer in a pleasant and disruptive way (COBRA, 2010). According to Lipovetsky & Serroy (2015), there was a reduction of the tactile relationship and an intensification of the visual relation, where the promotion of the imaginary of the consumer society directs the new economy of desire. The current consumer, through the virtual platforms, has to manage the various phases that involve consumption.

By adopting this perception, the importance of a *mídium* as a communicative device with the recipient emerges. It is extremely relevant to consider the material manifestation of discourses and their way of diffusion, since material support conditions the potential of discourse. The mode of support/transport and storage of the discursive message, therefore, of memorization is fundamental. And it was, above all with the arrival of the audiovisual media and the development of informatics, that the social role of the *mídium* was highlighted and effectively revolutionized the nature of the texts and their way of consumption (Maingueneau, 2011).

Dillon (2012), when considering that fashion communication is the way in which products are advertised and publicized, emphasizes that innovation and originality are basic elements for success in this process. And for the author, the technological advances of the 21st century allowed the internet to significantly impact the fashion industry, changing the way collections are advertised and clothing sold. Today, most designers and retailers are present on the web and websites, as well as the use of social media is a promising alternative, since online presence is the same as owning a store that never closes in a continuous traffic of information.

Thus, this type of media allows to overcome geographical and temporal barriers, reaching a globally distributed virtual public. Faced with a geopolitic that leads to the globalization of the market economy, brand, marketing, communication, fashion and product renewal have had a strong impact (Lipovetsky & Serroy, 2015).

However, since the communicative act has intentionality and is a fundamentally cooperative activity, the construction and presentation of messages must constantly predict the kind of competence that its recipient has to decipher (Maingueneau, 2011). After all, as text, requires reading, which involves perception, cognitive processes, a good knowledge of language and grammar (Eysenck & Keane, 2007).

As regards the language, which is the basis for disseminating the mark, the advertising messages have adopted the English language, since it is recognized as an official language in the field of international relations. However, language has verbal and non-verbal characteristics and the use of language is extremely subtle, being conditioned by the environment. Thus, "the concept and meaning of a word are not univocal, even though it may be well translated into another language" (Cavusgil et al., 2010, p. 109). In this sense, by disregarding the competencies of the receivers of the advertising message, the communication of fashion can introduce a problematic, negatively affecting the success in spreading the brand.

Undoubtedly, social media has been an important tool for brand disclosure in the global marketplace. As Dillon points out (2012, p. 96) "Social media is increasingly being used to increase consumer interest in a brand and build a loyal relationship through viral marketing." In an effort to "save costs," most companies believe that social media plays a crucial role in brand survival. Thus, it is common for top fashion brands to have a Facebook page and to be present in other media, such as Instagram and Twitter, since online communication allows the promotion of events, sales and trends, bringing the brand closer to the consumer.

However, even if internet use facilitates and accelerates the diffusion of a brand, global marketing, in search of an expansion in the foreign market, needs to understand how customers of other countries resemble or differentiate from those of the own country of origin of the company (Solomon, 2011).

When entering a new market, the company is subject to a new type of consumer. In addition to the characteristics such as values and ethics, it is important to emphasize and evaluate how much the language (idiom) represents in an intercultural process. Cavusgil et al. (2010) understand language as a crucial element in the formation of a culture, so much so that it requires careful treatment. Language is a crucial dimension of culture because, in addition to facilitating communication, it is a window into a person's value systems and living conditions.

> *Language differences are constant across national boundaries. Although there is a certain degree of imperialism, it is compensated by a nationalist counter-offensive. If, on one hand, globalization standardizes the superficial aspects of life that permeate the cultures of each nation, on the other, people resist these forces, insisting on their national identity and taking actions to protect it.* (Cavusgil et al., 2010, p. 36)

In this line of reasoning, the native languages of a country, as well as dialects and translation, may limit direct communication, since the lack of knowledge by managers and the company itself tends to cause uncomfortable situations. This is more susceptible when it comes to advertising, being it the element of marketing that most tends to suffer from the issue of language.

According to Silveira and Feltes (1997 apud Maingueneau, 2011, p. 75), in order to process information, an individual needs some mental effort. That is, considering that printed and/or virtual texts are distributed to a large number of readers, it is necessary to constantly predict the competence of its recipients to decipher them. The monitor of a computer connected to the internet, as well as printed magazines, offers written texts that circulate far from its origins and are accessible by unpredictable audiences. In this case, "... the enunciator does not control the reception of his utterance, being obliged to structure it in a way that optimizes his understanding and, the distance between the written text and his reader opens space for critical commentary and analysis, which conditions its interpretation" (Maingueneau, 2011, p. 79).

Solomon (2011) states that how the message is said and what is said are characteristics of the message itself that help determine its impact on the public. In addition, Squire & Kandel (2003) point out that some factors determine whether what is perceived will be remembered or not: the importance we attach to the fact, the degree to which we can organize and relate to the knowledge we had and the ease with which we can remember the material. Information is analyzed emotionally and rationally, and it generates information not only for momentary behavior, but it also creates representations for future situations.

Thus, the discourse, when constructed around a purpose and contextualized linguistic activity, consists of statements and texts. The statement designates a verbal sequence that forms a unit of complete communication within a genre of discourse: a newspaper article, an advertisement, are enunciated and are established according to the communicative orientation of their genre of discourse. The texts involve verbal sequences belonging to a genre of discourse. As an example, a commercial advertisement seeks to persuade a consumer (Maingueneau, 2011).

According to Hogan (2004), for whom persuasion can be defined as an ability to induce beliefs and values in other people, leading thought and/or actions, the advertising discourse has as its purpose an approximation with the public, in search of a partnership through well-constructed and appealing arguments with the purpose of encouraging the purchase and customer loyalty with the brand. In this way, like other discourses, the persuasive potential results from the use of language in order to guide behaviors and ideas. According to Lipovetsky & Serroy (2015), a brand aims to create emotional connections, a bond of connivance and complicity with consumers, and thus challenges the public to share a value system, creates emotional closeness, commanding by the imperative to create attention and sympathy.

In this direction, Maingueneau (2011) asserts that the advertiser seeks to capture the attention of the reader of the message. Usually advertising speech uses varied scenarios in that, in order to persuade the consumer, they need to capture the imaginary of the recipient and for this they stimulate interest through a valued scene. In this sense, the illusions are support for the extension of the mechanisms of elucidation of the message, since readers of virtual texts do not usually spend much time reading large amount of text. Usually they are colorful and attractive, collaborating for an encompassing scene, there is a stage set built especially for the context. At the same time, they use short fragments in large letters, which condense the information and attract the look. They also offer, for the reader who agrees to continue, a text with smaller letters in which some arguments are developed (Maingueneau, 2008).

Today, it is common for advertising pieces to use English as a language resource. However, despite the effort of the speech, the understanding of the meaning of a message by the reader is influenced by their existing knowledge and expectations. And as Camargo (2013) points out, consumer behavior is complex and understanding their attitudes and motivations is essential for marketing and it is in this setting that studies in the field of neurosciences can very much collaborate.

3 THE READING IN THE COMMUNICATIVE ACT

According to Sternberg (2012), semantic coding occurs in the reading process, that is, we translate sensory information into a meaningful representation that we perceive. For this, we take as a basis the understanding of the meaning of words, we search for the word stored in the memory, evoking the memory of long duration, that is, memories stored in the brain.

The reading of a text is processed in the visual cortex and passed to the language areas in the left hemisphere of the brain. The processing of visual information starts with sensory data and progresses to higher levels of cognitive processing, being operated on the basis of previous knowledge and experiences related to a given context. The fixations (series of snapshots in the read process) take longer for lesser known words. As we read, we try to retain as much information in working memory to understand what we are reading, and store the fundamental ideas in the form of a mental model. To construct a mental model it is necessary to make inferences, which depends on the understanding of words (Sternberg, 2012).

Reading a foreign language involves invocation of engram, that is, the set of memories, in the semantic case, constituted from both formal and informal experiences that people have throughout their lives. In this way, the brain searches the existing neural information networks for the interpretation of the information received (Izquierdo, 2014).

According to Sternberg (2012), sometimes coding is not possible because its meaning does not yet exist in memory, the word is not part of the lexicon (words constituting the vocabulary). Thus, taking into account that the structural characteristics of idiom affect perception and memory, different languages encompass different lexicons and distinct syntactic structures (use of subject and adjectives in sentences, for example), which interfere in the understanding of the message.

The author emphasizes that although the users of the language share the vocabulary, regional variants, denominated, dialects affect even the communicative competence between speakers of the same language, even inside the same country. In the United States, for example, it is common for national advertisers to offer telephone answering services that direct calls to the Midwest because the form of speech seems to be the best understood by different parts of the country.

According to Maingueneau (2011), interpreting a text requires communicative competence (ability to produce and interpret utterances appropriately to multiple everyday situations), linguistic competence (domain of the language in question) and encyclopedic competence (world knowledge). These competences interact in order to construct an interpretation, and a certain competence can remedy the deficiencies or the failure of another competence. And this justifies why, generally, even if we do not understand the meaning of most of their words and phrases, we are able to deal with utterances in certain foreign languages. In the absence of linguistic competence we use encyclopedic competence.

In order to highlight this problem in Brazilian market, we take as an example the figure below, made available in the Instagram of a national brand. Although the utterances emphasize the advertising intent of the text, the use of English expressions, given language limitations, has a negative impact on consumers, which can be observed in the comments that emerged.

Although the Ministry of Education advocates the teaching of the Foreign Language in the National Curricular Plans of primary and secondary education, it is noticeable that there is still an absence of linguistic competence in the scope of the English Language. In the following comment, Figure 3, it is possible to observe that the opinion of the consumers is not restricted only to the understanding of what is written, also covering the cultural and national identity question, addressing how Brazilians identify their role and reputation in the international arena.

This fragility in the language understanding interferes with the communicative act, because through the inferential process it is possible to explain how the utterances of a discourse can

Figure 1. Examples of postings on Instagram of the Brazilian brand Schutz.

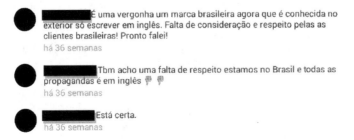

Figure 2. Reaction of consumers to the language used by the Schutz brand in their positions.

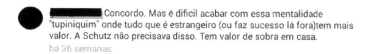

Figure 3. Commentary on the relationship between the use of the English language and the Brazilian national identity.

communicate explicit and implicit contents. However, in the processing effort, the linguistic complexity and the accessibility of the context are determining factors.

In the inferential process that is present in the search for information comprehension, discourse interlocutors try to recognize and deal with their mutual intentions and knowledge, and the presence or absence of shared information is important for the accuracy of communication (Eysenck & Keane, 2007). According to Baddeley et al. (2011), it is recalled that the words used to expose the information are already in the vocabulary of the individual, as it assists in semantic coding. Concomitantly, the difficulty in dealing with the information can generate a negative emotional stimulus in the consumer, influencing the decision to continue or not the process of a possible purchase of the product. For Damasio (2010), competent emotional stimulation is any real or remembered object or event that triggers emotion. The emotions generated, whether negative or positive, will contribute to the establishment of somatic markers that will determine, respectively, aversive behavior when individuals are exposed to the same or similar situations.

Emotional states have a neurobiological basis and play a role of communicating meanings to the subject being questioned, directing thought and generating behavior. Thus, the brain is the organ that most directly controls thoughts, emotions and motivation, including in the decision making process (Sternberg, 2012).

Considering the above, the mastery of the language by the recipient of the advertising message can be a limiting factor of the communicative act, and the communication is judged to be successful, unsuccessful, efficient or not.

4 FINAL CONSIDERATIONS

Undoubtedly, medias are means of disseminating the brand that positively interfere in the process of internationalization, since transposing geographical and temporal barriers are characteristic of these communicational devices. However, the use of the English idiom, essential for the internationalization of Brazilian brands, may negatively interfere in the intentionality effects of the advertisements and in the affective relation of the brands with the Brazilian consumers.

As discussed, advances in neurosciences have pointed out how semantic memory affects the processes of reading and interpretation before a text, including a text in a foreign language. Consumers, faced with the insufficient command of the English language in Brazil, can, by experiencing an unpleasant feeling in the process of virtual interaction, move away from the brand, impacting the consumption of their products.

In conclusion, it is important to emphasize the importance of the brand to consider how the communication techniques aimed at marketing reach consumers' feelings and memory and, consequently, direct their behavior in decision making. In the case of Brazil, another issue perceived in the comments analyzed is how the use of another language (other than Portuguese) translates the role and importance of the country in the face of the international scenario. That is, the brand's concern must go beyond the linguistic interpretation aspect, since it is evident that the use of a foreign language by a brand within its national market also affects the cultural and identity aspect of a nation.

In this sense, it is understood as important the development of research from the neuroscience triad, communication and process of dissemination of fashion brands with the intent to optimize the processes of internationalization of markets. At the same time, at a national level, Brazil being a country with a strong presence in the international fashion scene, one can see the relevance of research studies that seek to know how this subject is approached in the training of fashion professionals and what the consequences of this (lack of) knowledge as the basis in the production of advertisements before the consumer.

REFERENCES

Agis, D., Gouveia, J. & Vaz, P. 2001. *Vestindo o futuro*: macrotendências para as indústrias têxtil, vestuário e moda até 2020. Porto: Edições APIM.

Baddeley, A., Anderson, M.C. & Eysenck, M.W. 2011. *Memória*. Porto Alegre: Artmed.

Camargo, E.2013. *Neuromarketing*: a nova pesquisa de comportamento do consumidor. São Paulo: Editora Atlas.

Cavusgil, S.T., Knight, G. & Riesenberger, J.R. 2010. *Negócios internacionais*: estratégia, gestão e novas realidades. São Paulo: Pearson Prentice Hall.

Cobra, M. 2010. *Marketing & moda*. São Paulo: Editora Senac São Paulo: Cobra Editora & Marketing.

Damásio, A. 2010. *O erro de Descartes*: emoção, razão e o cérebro humano. São Paulo: Companhia das Letras.

Dillon, S. 2012. *Princípios de gestão de negócios de moda*. Tradução de Márcia Longarço. Barcelona: Editora Gustavo Gili.

Eysenck, M.W. & Keane, M.T. 2007. *Psicologia cognitiva*: um manual introdutório. Porto Alegre: Artmed.

Hogan, K. 2004. *The Psychology of Persuasion.* How to persuade others to your way of thinking. Gretna: Pelican Publishing Company.

Izquierdo, I. 2014. *Memória.* Porto Alegre: Artmed.

Lipovetsky, G. & Serroy, J. 2015. *A estetização do mundo*: viver na era do capitalismo artista. São Paulo: Companhia das Letras.

Maingueneau, D. 2008. *Cenas de enunciação*. São Paulo: Parábola Editorial.

Sackrider, F., Guidé, G. & Hervé, D. 2009. *Entre vitrinas*: distribuição e visual merchandising na moda. São Paulo, SP: Editora Senac São Paulo.

Solomon, M.R. 2011. *O comportamento do consumidor*: comprando, possuindo e sendo. Porto Alegre: Bookman.

Squire, L.R. & Kandel, E.R. 2003. *Memória*: da mente às moléculas. Porto alegre: Artmed.

Sternberg, R.J. 2012. *Psicologia cognitiva.* São Paulo: Cengage Learning.

The effects of sense contained in the fashion advertising text: The Handred brand and the reiteration of genderless concept

S.F. Bona & M. Scoz
UDESC – Universidade do Estado de Santa Catarina, Brazil

ABSTRACT: This research has the publicity of Handred Summer 2018 collection, published on Vogue-Brazil website as the object of study. From the French semiotic perspective, the elements of the expression plane in its articulations with the content plane were approached, which made possible to highlight the significant totality of the discourse in its effort to reiterate the construction of the concept of the brand. The methodology of analysis proposed by discursive semiotics was used, especially regarding the syncretic character of this type of text, which relates different homogenized languages in the plastic of the advertisement. Through the analysis, it can be concluded that the homologations between the expression and the content engendered in the discourse operated the construction of simulacra of recipients neutralized as to the gender roles predicted in the fashion system, reinforcing the position and the genderless concept proposed by the enunciator Handred in their different discursive manifestations.

1 INTRODUCTION

The present analysis covers areas of discursive semiotics, fashion and communication, having as research scope a specific advertisement used for the publication Handred brand Summer 2018 collection, marketed by Dona Santa store, on Vogue-Brazil website in November 2017. The Handred fashion brand was created in 2012 by designer André Namitala and is located in the city of Rio de Janeiro-RJ. The brand, which occupies a place outside chronological time, translates its concept through classic forms of sophisticated tailoring and at the same time easy to wear. In view of the advertising to be analyzed, it is hypothesized that the visual and verbal elements and their relationship mechanisms, contained directly in this syncretic text would allow the reiteration of the concept of the Handred brand due to the present figurative visibility and a presupposed relationship that is homologated between the expression plane and the content plane. From this, it becomes possible to elucidate the following research problem: How does Handred brand operate its verbal and visual discourse elements to reiterate the brand's identity concepts through advertising media, specifically in the discourse manifested in the advertising in question? In the face of this problem, this research aims to identify and analyze how the various visual, verbal, corporeal, and gestural elements and their mechanisms of relations present in the logical organization of the Handred brand publicity text for Handred brand Summer 2018 collection allow the construction of all meanings that reiterate the brand's speech.

This research is relevant because it scientifically analyzes this object of study and its effects of meaning through the foundations of discursive semiotics, which aims not only to describe the elements present in the text in question, but mainly to verify, through this theoretical basis, how the text says what it says through the relations of semiosis that operates. The social importance established in a fashion advertising text present in the randomness of daily life is also considered. This kind of text is the holder of a certain power of persuasion that, through a global strategy of annunciation, operates mechanisms of articulation of sense making the syncretic text a whole of meaning. To reach the objectives determined by the research, the categories proposed by the discursive semiotics and the concepts of the sociosemioticswill be used, which will enable the analysis of the object of syncretic study.

2 THEORETICAL BACKGROUND

According to French semiotics, also called discursive semiotics, a text is a unit of signification that constitutes as all of meaning. For Fiorin (2016, p. 45), "when a discourse is manifested by any expression plane, we have a text." The main interest of this research lies on the process of producing meanings in the given advertisement, aiming at demonstrating the articulation between the values proposed by it and the strategic positioning of the Handred brand, which allows the reiteration of the company's identity.

The categories of discursive semiotics are taken as the base for this analysis, considering the role of actors established in the discourse structure. Therefore, it is important to observe that the semiotic analysis process starts from the detailed description of the plastic elements of the text, seeking, in the investigation of its expression, evidence of approvals in the content plane. This measure is established by the analysis of the topology of composition, its chromatics and other classes of constituent elements of visual products. In addition, the semiotic look at syncretic objects—that is, that are constituted by the overlapping of more than one language system—considers aspects of the arrangement of the photographic image, besides the effects of coercion suffered by the text through the media where it is transmitted, as well as the existing social context. According to this perspective, one must consider that a semiotic system or the process of semiosis occurs in the relations. In analyzing syncretic texts semiotically, as pointed out by Oliveira (2009a) in the analysis of syncretic objects and considering their procedures of integrative relational functioning that makes the plastic of syncretic expression validated to the content plane, one must therefore study the compilation of all parts as significant totalities. Still, according to the author (2009a, p. 96), "apprehension is then a recognition of the association".

Plastic semiotics enables the analysis of the discourse established through the figurativeness of the elements and their systems of languages present in the expression plane and in the content plane and, mainly, in the relation of homologation of the two planes. As well as in the topological configuration of these manifestations, that allow to establish relations between certain heterogeneous units of expression. Figurativeness is the level where the enunciative acts of the complex subject of enunciation operate, this cuts parts of the world and translates them by language resources that were selected, establishing means of interaction between enunciator and enunciate. In Fiorin' words (2016, p. 45), "Discourse is a unit of the content plane, it is the level of the generative trajectory of meaning in which abstract narrative forms are lined with concrete elements." Syncretic texts are then analyzed within a general discursive theory, "being treated both in the particularity of their own materiality and in their general quality of discourses concretized in texts" (Teixeira, 2009a, p. 47). To cover the totality of the text under study, the generative trajectory of meaning will be used. As pointed out by Fiorin (2016, p. 20) "it is a succession of levels, each of which is capable of receiving an adequate description, which shows how the meaning is produced and interpreted in a process that goes from the simplest to the most complex." The generative trajectory of meaning operates on three levels, being "the depth (or fundamental), the narrative and the discursive" (Fiorin, 2016, p. 20). Also, according to the author, at each of these levels there is a certain effect of meanings that operates through a syntactic component and a semantic component.

Vogue-Brasil website, where the text was published, constantly publishes various advertising discourses related to fashion communication. Semiotics allows the analysis of the construction of meaning through a broader perspective of the communicative relations between the subjects. And, according to Oliveira (2009a, p. 100), "semiotic theory treats the subject of enunciation as installed in the syncretic object itself, in its utterance, materialized by a content manifested by an expression that holds the indications of that subject." This interaction between enunciator and enunciate composes a simulacrum of these subjects, as well as the sensitive apprehensions that work together in the meaning construction of the text. Named as interaction regimes by Landowski (1992b), these sensitive associations between the subjects of the communicative relationship, enunciator and enunciate, constitute the choice organization of the plastic assembly of the enunciation, which act in the sense of meaning, allowing different forms of semiosis. Thus, one must expect that, to the different modes of signification, some of them from the reading, others from the capture, correspond, in terms of narrativity, different regimes of interaction (Landowski, 2007a). Therefore, meaning goes beyond a notion of restricted order ending in things and begins to operate in a logic of relationship through the intelligible and sensitive apprehensions determined by a kind of agreement between these subjects.

In discursive semiotics, it is considered that there are at least three types of languages that operate the relations of articulation between the expression plane and the content plane, the symbolic, the semiotic and semi-symbiotic systems. Symbolic systems act in the relation between themes and figures, establishing a single fixed thematic interpretation for each figure, as for example, the owl represents wisdom. "The symbol is always a concrete element to convey abstract content" (Fiorin, 2016, p. 96). In semiotic systems, the languages operate in combinatorial relations established within the text itself, where the description should be separated in the expression plane and in the content plane, to achieve the relation of homologation between them. That is, the meaning is given in the relationships established between the elements of expression within the text itself. According to Floch (2009, p. 161), semi-symbolic languages are characterized not by the conformity of elements of expression and content alone but by the conformity of certain categories of these two planes. As in our cultural universe where the yes and the not correspond respectively to the movements of the head verticality and horizontality.

Considering the social context and communicative intentions in which the text is established, it will be necessary to reflect that the enunciative action aims to empower the enunciate, that is, the reader of Vogue-Brasil website, to know, to do and to want through the discourse manifested there. As Fiorin points out (2016: 75), "The ultimate purpose of every act of communication is not to inform but to persuade the other to accept what is being communicated." So, the act of communication will always be manipulative in order to make the enunciate to believe on what is being transmitted in a persuasive way. The sociosemiotics allows the construction of the current meaning in these manifestations to understand the transforming social acts situated in the intersubjective relations of the discourse. According to Landowski (2014b), the sociosemiotics is in the scope of "Saussuro-Hjelmslevian" semiotics. It is used to designate, according to the contexts, one of the specialized branches of the discipline—the one which specifically takes the social as the object—or the main theoretical lines currently offered to renew the analysis of the facts of signification in general, whatever the type of empirical domain is considered.

Based on this set of theories, this research will be underpin and the analysis of the object under study will be performed, with the purpose of understanding the structure of this syncretic discourse, the relations manifested between the elements of the expression and the content planes. As well as to understand how the result of the semiosis process allows the reiteration of the sense of identity proposed by the brand.

3 ANALYSIS

This publication, per ad format, object of this research, is constrained on Vogue-Brazil website. The advertisement is considered a textual genre that aims to promote a product or idea being served by various mass media. These texts present, in their own language, a connotative or appealing function that has the purpose of convincing the receivers of the message, in this case the readers, or possible consumers, to buy a product or an idea.

Vogue-Brazil website constantly publishes information about fashion, topics such as trends, behavior, beauty, culture and lifestyle. The fashion system can be understood, in this context, as a complex relationship between distinct codes pertinent to a specific language, which are manifested through syncretic arrangements enabling discourses with different meanings effects, as pointed out by Castilho (2004). The acceptance of rules pertinent to the language of fashion is integrated into an already elaborated narrative program, which develops through competences, such as the want-do, or even a duty-to-do extending the capacity of body meaning. Figure 1 below presents the object of study of this analysis, the publication in Vogue-Brazil (2017) website.

Figure 1. Publication announcement "Handred reaches multibrand Dona Santa".

It is from this publication that the present analysis will be developed; aiming to understand how the structure contained in the totality of this syncretic text enables to reiterate the identity of Handred brand. After the beginning of the analysis by the figurativeness present in the publication, following the presuppositions of discursive semiotics, the content manifests its materiality at the figurative level, as pointed out by Fiorin (2016, p. 41), "At the discursive level, the abstract forms of the narrative level are clothed with terms that give them concreteness". Here the manifestation of two systems, the verbal and the visual ones, can be seen and still within the system that is visually apprehensible, it is also considered the gestural and the corporal ones. In the case of syncretic texts, these systems constitute manifestations that operate in more than one language system, as pointed out by Oliveira (2009a, p. 80), "We are led to treat this kind of syncretic constitution of the expression plane by the integrator action of its parts in a single totality, since its sensitive apprehension is processed".

To deal with the reading text of the research, which is mainly due to the semantic values expressed by the figurative order of the text, that is, of its expression plane, the presupposed concepts in the syncretic plastic semiotics that establish relations interspersed with heterogeneous expressions that manifest a totality of meanings will be used. And, through these, the plastic of the syncretic expression becomes homologous to the homogeneity of the content plane (Oliveira, 2009a). According to Oliveira (2004b, p. 120), plastic semiotics must consider "the set of distinctive and pertinent traits of the material dimension (materials and techniques), chromatic (color), eidetic (forms) and topological (spaces-support)". It is about each visual formant, considering the plastic categories, which they are part of. The observation of plastic objects can select criteria that consider hierarchically or separately, or even jointly, the chromatic, eidetic and topological plastic categories (Teixeira, 2008b). First, the totality of the discourse will be observed, since, regarding the study of syncretic texts, as pointed out by Teixeira (2009a, p. 61), "the analysis always begins with the simplest and most apparent: a detailed observation, an exhaustive description". While viewing the totality of this object of study, the attention is first directed towards the dimension of the photographic image that occupies about half of the advertising text. The chromatic oppositions, black versus white, the resemblance of the two bodies composing the photography as well as the present gesture are taken into account. The title of the publication "Handred reaches multibrand Dona Santa", informs the reader the name of the brand being advertised and can be understood that it is available at a specific sales place. The subtitle "André Namitala's label has now an outlet in Recife", informs the audience that the brand has one owner, the designer of the brand, and one of its outlets, the multi-brand store Dona Santa, is in Recife.

The text begins with the phrase "Handred, genderless brand", making clear to the enunciate, in this instance, the genderless positioning of the brand, ie, without gender, genderless. The word that gives the name "handred" to the brand, translated into Portuguese as "wanted", through the relation that operates with the concept of genderless, expands the connotation of the presupposed sense, as one who is considered as not defined by the other's gaze. The genderless concept is also reiterated through the cognitive and sensitive visual elements present in the photography, from which it becomes necessary to detail its particularities later. It is in this "coming and going" between the elements expressed in the expression plane, which are verbal and visual manifestations, that allows the deepening of the apprehension of the text during its procedures of making sense that the analysis will continue.

In the scenario that composes the photographic image, the representation of an internal environment degraded by the action of time is noticed. This contrast between background and models makes visible the concepts of the old versus the new, respectively, causing rupture effects, or timelessness, a space not located at a given time. The models lie side by side in a static position, despite their apparent physical similarity, they are assumed to be a woman (left) and a man (right). It is worth mentioning, as Oliveira (2016c, 83) points out, "The synergistic communicative production also intervenes in the expression of physiognomies, kinetics, distances, postures of the body dressed in the atmosphere, with an architecture and decoration." These visibly similar body characteristics of the photographic models, who are Caucasian, have long and slightly wavy blond hair cause a neutralizing effect between the female versus the male opposition. In this sense, fashion acts as the effect of a second skin on the body dressed, as pointed out by Castilho (2004, p. 83), "Fashion is ultimately governed by continuous operations of transforming the body's opinion about the being (bio morphological)". As for gestural characteristics, the female model has hands in her pocket, hair at the front, and looks directly at the enunciate, assuming a certain gesture, judging by certain social standards, somewhat non-feminine. While the position of its lower limbs, legs and feet are placed close, which presupposes a non-male oriented gesture. Contrary to the female model, the male finds himself with hands free, body and head are slightly tilted to the side, taking for himself a certain non-masculine gesture. His lower limbs are separated, indicating a non-femininity. His gaze is not intended for the enunciate.

Both pieces of clothing worn by the models are overalls. They are similar, but not identical. As to the material qualities of the fabrics, it is noticed the fluidity through the apparent malleability presented by the fabric, smoothness and opacity. These attributes reiterate the verbal text in which we have the description contained in the speech "made in linen and other fabrics of natural fibers, ideal for the climate of the Northeast "and" In this last collection we kept the pieces focusing on the DNA of the brand with fresh, minimalist clothes...". As for the chromatic order of the clothes, both are totally monochromatic, where the model of the left dresses black color, and the model of the right dresses white. This topological arrangement of the models dressed in opposite colors causes a sense of contrariety. As pointed out by Floch (2009), the contrasting positions of elements of the same category indicate an opposite-sense effects of contrary values. In contrast, both wear shoes, also monochromatic, but opposite colors to the pieces of clothing they wear causing an effect of cohesion to the previous meaning effect, indicating the sense of opposite assimilation, conjunction, or even neutralization of opposites.

However, attention must be paid to the fact that the value of the narrative level is not identical to the concrete object manifested at the most superficial level of the generative trajectory. The value of the narrative level is the meaning that has a concrete object for the subject that comes in conjunction with it (Fiorin, 2016, p. 37). When the text is analyzed in its syncretic form and considering the publication circumscribed in Vogue-Brazil website, which is a donor of fashion skills, it is possible to elucidate a persuasive manifestation that is concretized through a certain narrativity. According to Fiorin (2016), narrativity can be considered a transformation situated between two successive and different states. This means that a minimal narrative occurs, when one has an initial state, a transformation, and a final state. Thus, in the initial state, it can be considered that the enunciate, reader of the website, is in disjunction with Handred brand and the main objective of the publication is to make it in conjunction with the brand. Still, according to Fiorin (2016, p. 29), "A complex narrative is structured in a canonical sequence, which comprises four phases: manipulation, competence, performance and sanction."

The manipulation in this ad occurs through provocation, which is, according to Fiorin (2016), when the manipulator impels the action, expressing a negative judgment regarding the competence of the manipulated. In this sense, the altering certain usual distinctions in photographic images was verified, which worked as a provocative manipulation to the ad recipients. Because, going beyond the ambiguous figurativeness introduced by the indefinite genres, the visual narrative fulfills the function of installing a imprecision, with the recipient having the role of investing in the definition of the presented genres, or being involved in the proposed value system.

Advertising does not effectively demonstrate the other discourse narrative phases that would be competence, performance, and sanction, being performed by the enunciate, but, to some extent, gives evidence through the narrativity of the ad of how they could be made or how they might happen. According to Fiorin (2016: 30), "the subject who will perform the central transformation of the narrative is endowed with a knowledge and/or power to do. Each of these elements may appear, at the most superficial level of discourse, in the most varied forms". Through the expressions in the text that reiterate the knowledge about the "Handred brand, the genderless mark of designer André Namitala is one of the New Talents of Veste Rio, and has just landed his tailoring in Recife multibrands, Dona Santa". Also a power of doing through the contact of the place indicating where the products are being marketed, at the end of the publication "Dona Santa: Professor Eduardo Wanderley Filho Street, 187, Boa Viagem, Recife. Tel. (81) 3465–3630".

The performance phase can be understood, in this case, as the attributes perceived by the apprehension of the use of product. According to Fiorin (2016, p. 31) "Performance is the stage in which the central transformation of the narrative takes place." Indications of performance are also allowed visible in the phrase "made of linen and other natural fibers fabrics, ideal for the climate of the Northeast" which indicate that the product in question is ideal to wear in warmer climates. The expressions "we kept the pieces focusing on the DNA of the brand with fresh, minimalist clothes…", indicates that the product is aligned with the semantic concepts of the brand. The last narrative phase is the sanction, where, according to Fiorin (2016), "it is noted that the performance was realized and, consequently, the recognition of the subject that operated the transformation". This phase becomes presupposed in narrative discourse through the phrases expressed by the director of the brand Juliana Santos "Handred is a new brand that is a success in Dona Santa" and "The brand is being successful for girls and boys!". Here, it is pointed out that according to this expression, Handred brand through the concept genderless, meets needs, being physical and/or subjective of both gender audience, in a process of discursive (and advertising) construction that neutralizes the opposite terms of the category /male/ and /female/.

Turning to the fundamental level, that is, to the most abstract level of the totality of meaning production, semantic categories are met, which will be the basis for the text construction. According to Fiorin (2016: 22), "A semantic category is based on a difference, on an opposition. However, in order the two terms are apprehended together, they must have something in common and it is on this common trait that a difference is established".

According to the author, "the opposite terms of a semantic category maintains a relation of contrariety among them. The terms that are in relation to reciprocal presupposition are contrary". Thus, one term gains meaning in contrariety with the other and vice versa. Therefore, to give meaning to a discourse, it must be taken as positive one of the opposite terms of the basic semantic category, which can vary its euphoric value according to the point of view, or the intention of the enunciator. Each of the elements of the text basic semantic category receives

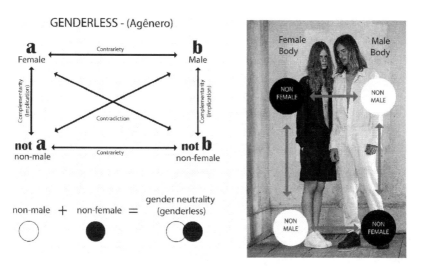

Figure 2. Semiotic Board and Chromatic and Gestural Topology. Elaborated by the authors.

the semantic qualification of euphoria or dysphoria. "The term to which the mark /euphoria/ has been applied is considered a positive value; the one given the qualification /dysphoria/ is seen as a negative value (Fiorin, 2016, p. 23). However, when dealing with the Handred brand that strongly presents "genderless" as one of its bases of discourse, it should be emphasized here that the term "genderless" itself carries the sense of negation of gender classification, that is, the absence of gender, non-gender, gender neutrality, or even that it is null and non-binary.

As for the gender classification of sexuality, female versus male, gender neutrality lies in the conjunction of its sub-contraries, that is, in the conjunction of non-feminine and non-masculine. Figure 2 exemplifies the relation of gender neutrality by the semiotic board, adapted from Fiorin (2016), and beside it the topology of chromatic order engendered in the image, that reiterates the neutrality of genera, or genderless concept.

Through the representation and relationship topologically proposed by the visual expression plane of the text, it is possible to establish a relation of homologation between expression plane and content plane of semi-symbolic order, when analyzing the relations present in the gestural and chromatic categories with the semantic category that gives meaning to the discourse. It is noticed that the chromatic topological distribution used in the construction of the photographic image reiterates the effect of contrariety. We then have the black versus white at the top, as well as the value of invested gesture, respectively black corresponding to /non-female/ and white /non-male/. That is, the relation of contrariety existing due to the opposition of the contrast of the two colors in question and the invested gesture is also the difference manifested in the basic semantic category of the sub contraries, the non-feminine and the non-masculine. However, the assimilation of gender neutrality gives meaning to the discourse reiterated in the expression plane because the two colors are in an opposite relation in the bottom composition of the image (shoes), accompanied also by the opposition of gesture, which reinforces the sense of discourse of the brand through the conjunction of these two basic semantic values. As reinforced by Fiorin (2016, p. 23) "In discourse, opposing or sub opposite terms may appear together. We will then have complex terms (compilation of a and b opposites) or neutral (conjunction of not a and not b sub opposites) (emphasis ours). Thus, the conjunction of the non-feminine and non-masculine sub opposites reiterate the sense of genderless speech of Handred brand in the text.

4 FINAL CONSIDERATIONS

Inevitably present in the daily routine, advertising texts are endowed with an internal structure whose main objective is to persuade the enunciate through discursive operations aligned with the values of the brands. In semiotic terms, they are discourses that concretize abstract values. The semiotic analysis of the advertisement used for the publication of Handred brand Summer-2018 collection allowed to ratify the hypothesis that the articulation between the heterogeneous figurative elements present in the discourse of the text allow the reiteration of the brand identity trait, or its concept, through the relationship between the expression plane and the content plane. Considering that the process of semiosis occurs in relationships, the total apprehension of the text, as we can see, occurred in the recognition of the association of these relations. When dealing with the analysis of syncretic objects, their integrative relational functioning that made the plastic of the syncretic expression valid to the content plane was considered.

However, the semantic category of the fundamental level of the text, that is, gender neutrality—or genderless—allowed to give meaning to the set of elements of the superficial level. The contrasting chromatic strategic placements of elements of the same category present in the text indicated a sense of play of contrary values allowing for approval with the content plane. In addition, the same sense is reiterated in the assumed gesture between the models, which allowed a neutralization of the differences usually adopted for gender roles in fashion advertising. Thus, the various elements of the expression plane of the visual-verbal system worked together in the operability that constitutes the syncretic text. If the first objective of any ad is to catch the enunciate's attention, the strategy of altering certain usual distinctions in photographic images was verified, which worked as a provocative manipulation to the ad recipients. Going beyond the ambiguous figurativeness introduced by the indefinite genres, the visual narrative fulfills the function of installing a dynamic of imprecision, with the recipient having the role of investing in the definition of the presented genres, or simply being involved in the proposed value system, in which one must know how to accept the indefinite genre. Thus, publicity text, besides reiterating the identity traits of the brand, operates a strategy of competence of its enunciate, taken as the subject of an interpretative performance to accept the contract of the brand, and more precisely a given moral.

It is in this way that the fashion system operates a complex relationship between distinct codes pertinent to a specific language, which are manifested through syncretic arrangements enabling discourses with different meaning effects. The syncretism present in the discourse of fashion texts requires a set of semiotic theories to perform the complete reading of the text, both intelligibly and sensibly. In fashion communication the body expands its capacity of meaning through the language that involves it.

REFERENCES

Castilho, K., 2004. *Moda e Linguagem*. 2nd ed. São Paulo: Anhembi Morumbi.
Fiorin, J.L., 2008. *Elementos de análise do discurso*. São Paulo: Contexto
Floch, J.-M. 2009. Semiótica plástica e linguagem publicitária: análise de um anúncio da campanha de lançamento do cigarro "News". In: Oliveira, A. C. and Teixeira, L. *Linguagens na comunicação: desenvolvimentos de semióticas incrética*. São Paulo: Estação das Letras e Cores. pp. 145–167.
Landowski, E., 1992a. *A sociedade refletida: ensaios de sociossemiótica*. São Paulo: PUC-SP/Educ/Pontes.
Landowski, E., 2014b. Sociossemiótica: uma teoria geral do sentido. Galáxia (São Paulo, Online), n. 27, p. 10–20, jun. 2014. http://dx.doi.org/10.1590/1982-25542014119609.
Oliveira, A.C., 2004b. *Semiótica plástica*. São Paulo: Hacker.
Oliveira, A.C., 2009a. A plástica sensível da expressão sincrética e enunciação global. In: Oliveira, A.C. and Teixeira, L. Linguagens na comunicação: desenvolvimentos de semiótica sincrética. São Paulo: Estação das Letras e Cores. pp. 79–140.
Teixeira, L., 2008b. Leitura de textos visuais: princípios metodológicos. In: Bastos, N. B. *Línguaportuguesa: lusofonia – memória e diversidade cultural*. São Paulo: EDUC. pp. 299–306.
Teixeira, L., 2009a. Para uma metodologia de análise de textos verbo-visuais. In: Oliveira, A. C. and Teixeira, L. *Linguagens na comunicação: desenvolvimentos de semiótica sincrética*. São Paulo: Estação das Letras e Cores. pp. 41–77.
Vogue Brasil Website, 2017. Handred chega a multimarcas Dona Santa. In: http://http://vogue.globo.com/moda/moda-news/noticia/2017/11/handred-chega-multimarcas-dona-santa.html. Accessed in: 15/02/2018.

Analysis of attributes in unisex and genderless clothing

B.M. Reis & R. Miguel
Faculty of Engineering, FibEnTech, University of Beira Interior, Covilhã, Portugal

N.A. Jerónimo
Faculty of Social and Human Sciences, LabCom, IFP-CFH, University of Beira Interior, Covilhã, Portugal

M. Pereira
Faculty of Engineering, FibEnTech and UNIDCOM, University of Beira Interior, Covilhã, Portugal

S. Azevedo
Faculty of Social and Human Sciences, CEFAGE-UBI, University of Beira Interior, Covilhã, Portugal

ABSTRACT: By developing the importance of fashion (a phenomenon that is significant nowadays), research in fashion, its history and the issues of gender, ascends the reflection on genderless clothing. Clothing is one of the most visible forms of consumption, and it performs a major role in the social construction of identity. (Crane 2000) In this paper we focus on the specific area of genderless clothing, having in mind the importance of gender issues nowadays. Simmel, stated in his famous 1905 essay, fashion is both a way to a sense of uniformity, as well as a form of distinction. Also, this is a proposal to explore genderless clothing and pursues to understand if it is just a trend, or it is becoming a paradigm. In this context, we pretend to analyse unisex clothing and arises the necessity of understanding the representations of men and women in garments.

1 INTRODUCTION

In this article, we intend to understand the components of unisex clothing, by regarding some social conceptions of gender. Unisex designation means that is for both sexes (woman and man), however genderless is going much more further in this issue, because gender is a social construction about the identity of each person.

This document presents a theoretical and empirical methodology, the literature review, data collection, analysis of data collection in order to present in the end an analysis, discussion and conclusions. The review aim, among the deep knowledge in these areas of study, to approach some remarkable facts in fashion history, regarding the unisex clothing; also, was made an approach in the sociological area regarding some important authors.

The aim of this work is to analyse and understand unisex and genderless collections produced and sold by fashion brands and designers. Was made a qualitative approach for understand the visible significations in the study objects, allowing a comparative analysis among the chosen collections. This goal will be attained by researching several unisex pieces produced and marketed by designers and stores such as C&A, Zara, Selfridges, Rad Hourani, J. W. Anderson, Inês Torcato and Alenjandro Gomez Palomo, by analysing the colours, the forms of pattern-making that each brand or designer used, and the form of each one is presented to public, for further named the collection: unisex, androgynous, genderless or agender.

The contribution of this work is important to clarify some aspects in this area of research, considering that our approach, never been done before in the investigation in fashion design or sociology.

2 SOME MILESTONES IN UNISEX FASHION HISTORY

Fashion was not always a gendered phenomenon, and both men and women clothed themselves with elaborate costumes until the 18th century. Costume historians argue that in elite circles, prior to the 19th century, gender distinctions in dress were not nearly as strongly marked as they have become since (Kawamura 2005).

Chanel is known by the women how challenged the fashion in her time, by wearing trousers. However, Vinken (2005) refers the overlapping of class and gender is represented by Christian Dior, when he designed the clothes for women as females, as only woman, in other words, without the detour through the masculine characteristics. Remarkably, Coco Chanel interpreted this detour through the masculine as 'naturally' feminine. Chanel herself confidently maintained that she made clothes for real women for the real life. If Chanel dressed woman as dandy, Christian Dior, with his ultra-feminized New Look, Coco Chanel believed that Christian Dior had dressed his clients up as transvestites (Vinken 2005).

Leading the way to androgyny, while androgyny might be all the rage in 2015 (Singer 2015), back in 1970 it was Rudi Gernreich who promoted the concept of unisex clothing. In his 1970 "Unisex Project", he shaved a male and female model of all body hair and had them model identical outfits, from bikinis and minidresses to trousers and Y-fronts. Gernreich's fashion concept of the future was "unisex" with the annulation of differences idea between sexes (Seeling 2000; Singer 2015).

Yves Saint Laurent in the 1960s introduced new elements in women's fashion, which have become indispensable. Particularly the women's tuxedo (smoking), which will remain forever associated with his name. He took a lot of inspiration in his men's clothes, such as Coco Chanel, a dressmaker he admired for his practical ideas, but he knew how to give the male aspect a certain eroticism (Seeling 2000).

Gender signifiers are often eschewed or overlooked in fashionable versions of South and Southeast Asian waistcloths seen on the catwalk. For many designers, the waistcloth becomes a vehicle for the promotion of unisex dressing and endorsement of an androgynous body. Jean Paul Gaultier in Spring/Summer of 1991 developed and presented the dhoti as an item for the unisex wardrobe, however in the runaway show the men shaved his head and woman's coiffure remained as signs of masculinity and femininity. In Autumn/Winter 1994–5, Gaultier developed for men and woman identical waistcloths, that both models are wearing the same style shoes and jacket with matching hairstyles. Another fashion designer it's known by the kilted male as a design source. The British Vivienne Westwood, well-known for her fascination with the British royal family, Westwood has looked in particular to the figure of King George IV for her inspirations, has we can see analysing the path that Vivienne has been doing (Bolton 2003).

Further, in our data collection we will present more important milestones regarding unisex and genderless clothing.

3 CONCEPTIONS OF GENDER IN FASHION

According to Crane (2000), variations in clothing choices are subtle indicators of how different types of societies and different positions within societies are actually experienced. Whereas in closed societies the traditional costume predominates, in open societies a sense of fashion may blossom and reign. While closed societies tend to be conservative and traditionalist, open societies are, on the contrary, ready to receive innovation and change (Barnard 2002).

Fashion implies a certain fluidity of the community's social structure, and it requires a particular type of society, like the modern world where social stratification is open and flexible. There may be differences of social position, but it must seem possible and desirable to bridge these differences. Therefore, fashion is not possible in a rigid hierarchy (Kawamura 2005).

Fashion serves two apparently contradictory purposes. On the one hand, universalises and connects, on the other hand, individualises and distinguishes. Although they constitute a logical opposition, these functions are inseparable, and they are the conditions to the social fulfilment of the fashion phenomena (Simmel 2008).

Considering Schütz (2003) awareness, that clothing and the related practices exist as banal routines and conscious expressions of esthetical ideas. To Social Phenomenology, fashion, with its codes and history, é a life-world (lebenswelt) phenomena. The intersubjective world that existed long before we were born, experimented and interpreted by others, our predecessors, as an organised world.

The connotative meanings of fashion are, thus, intersubjective constructs. Culturally assumed dichotomies expressed by clothing—austere/happy, sophisticated/simple, rich/poor, masculine/feminine—are representations negotiated by the intervening social agents, from the process of creation to the act of fruition. Fashion is a continuous negotiation of social practices and significance codes (Barnard 2002).

In fashion, understood as a cultural phenomenon and dimension of everyday life, the existing structures of creation, production and market, as well as the social institutions that surround it, would not exist without shared processes of meaning-making. In this process of construction, individuals tend to apply their repertoire of knowledge to recognize categories of meaning—a generalized typifying of reality (Schütz 2003).

Fashion meanings are reconstructed and negotiated from social representations. These have the ability to transform an unknown context into a cosy, recognisable environment, and allows the group operating as an entity where objects or events give meaning. Social representations allow that something unfamiliar obtain mask of familiarity (Moscovici 2000).

Changes in fashion produce simultaneously the novelty—and distinction—sensation, as well as the familiarity—and unity. In the contemporary individualistic society, the weakening of common codes or the loss of collective consciousness drives to informality and de-institutionalisation of practices—fashion may be relegated to a mere game of shadows (García Martínez 2012).

Among the variety of dichotomies, this paper focuses on how the erasing frontiers between masculine and feminine intertwine the forms already present in the contemporary market of fashion. And yet, the masculinization process within women's fashion and the femininization of men's fashion is not a universal absolute. This happens caused by the anthropological necessity of expressing differences between both sexes. "The lightness in looks still is a much more feminine than masculine ideal" (Lipovetsky 2015).

4 ANALYSIS OF BRANDS AND DESIGNERS FASHION COLLECTIONS

In an initial stage, for this article was made a review of the literature on some key moments in fashion, approaching some more important authors who studied and analysed the fashion considering unisex clothing.

Regarding research methodologies in Social and Human Sciences, this article is part of a kind of qualitative plan, this is, that intends to understand and describe in depth the investigated phenomenon, of emerging problems. Based on a theory founded to understand the phenomenon by the observation of the subjects, their interactions, experiences and significations, through a methodology of discovery of theory in an interactive process of data analysis (Coutinho 2011).

Thus, for the comprehension of visible meanings in the objects of study, the empirical part of this article attends as a qualitative analysis of the collected information. We selected brands and designers that offer androgynous collections, unisex, or genderless collections, thus generating our sample, from where the data were collected online, for the respective observation. The data collected corresponds to the collection of images, of their collections and garments. The selection of brands was made based on the background that the authors have about designers, brands, news and fashion trends, about androgynous and unisex fashion, since brands do not use the term "genderless" with the same frequency as the other two. The images were collected from the world wide web.

The analysis of the message content should be applicable to all forms of communication, whatever the nature of its support, valuing the exploratory attempt, increasing the propensity for discovery. Through the contents they may or may not confirm and lead to the *"enlightenment of elements of meanings capable of leading to a description of mechanisms that a priori did not have the understanding"* (Bardin 1977).

Was made a collection of data (images of fashion collections), of brands and designers that had already launched androgynous collections, unisex or without gender. After the selection was completed, the process of analysing some pieces of these collections began.

4.1 Fast fashion brands

Analysing the brand and chain of international clothing stores C&A, as well as an analysis of some of the news associated with the analysed images about the brand, it's notable that the social media made a great noise around the analysed collections. This situation is more notable in the Brazilian market. In 2016, the brand C&A released three different campaign videos. However, there is no such thing as the break of paradigm regarding garments, that is, there are no men in skirts or dresses, garments that are accredited to women. It was verified in the online site of the mark C&A (Brazilian) divides man of woman. Meanwhile, and although it does not belong well to this research and analysis, it was found that C&A has a collection of plus-size, but this is only available for women, in collections of children from 0 to 4 years there is a link that gives direction to the neutral, however, the available items were socks and robes. From 0 to 12 (ages made available by the brand site) in neutral were only white shoes, socks and robes. From the age of 10 to 16 (ages made available by the brand site) there is no neutral gender anymore, just boy and girl. Again, we do not have the interruption in the paradigm again, such as men wearing skirts or dresses.

Zara's story begins with the opening of the first store in 1975, a milestone in the company's long history. Over the years, Zara remained true to her values. Founded by Amancio Ortega and Rosalía Mera, Zara belongs to Inditex, one of the largest distribution groups in the world. In March 2016, Zara presented the collection "Ungendered", which included about 20 pieces. The colours were neutral tones (white, grey and blue), hooded coats, and jeans and t-shirts. Despite the media coverage of this minimalist collection, almost hand in hand with sportswear, their campaign images did not make an impact. The pieces of the brand's androgynous collection are not many, and in fact the pieces used by men are the same ones worn by a woman, but they are nothing but a white t-shirt, jeans and a grey sweat.

Selfridges is a department store chain in the UK founded by Harry Gordon Selfridge. In 2015, the brand inaugurated a gender-neutral space, although Selfridges herself mentioned on the website (when this initiative took place) that the gender-neutral space would also be available online in a short consultation, does not have this online space, just as the garments continue to be divided between the sexes (woman and man). In the "AGENDER" collection, which according to the brand, departs from the tradition of a definition of gender (as we know, binary), Selfridges leads its customers to believe that they can choose to buy and wear without limitations or stereotypes. However, in the campaign images from this collection analysed, the stereotype remains present. Women are sometimes dressed in skirts and clothes with an oversized modelling and fitting, whereas men, although the garments use the same type of fitting, they are always dressed in trousers or shorts, garments with clothing characteristics male gender. Also appealing to t-shirts and sweatshirts, garments that have features like straight lines of masculine attributes.

4.2 Rad Hourani

Rad Hourani soon after graduation, began working as art director. At age 23, he moved to Paris, where he studied photography and video production. He created his first "UNISEX" collection and launched the now homonymous brand in October 2007, with only 25 years. In January 2013, at the age of 30, Rad Hourani received the great honour of becoming a "guest member" of the Chambre Syndicale de la Haute Couture in Paris and, since his first

Haute Couture show in Paris in January 2013, the first designer to present a UNISEX collection Haute Couture in the history of fashion and the first Canadian to be among the greatest designers. His vision is centred on the theme of neutrality, has different perspectives addressed through the communication of painting, sculpture, photography, fantasy, sound and video. In his own vision, he is outlining a newer and freer way of living.

When making an online consultation on the designer's website, approaching the unisex clothing, the site has words like "genderless", "raceless", "ageless", "nationless" and "limitless", station specific for the launch of its garments and collections.

Regarding the images analysed, the pieces of clothing produced by the designer, it's noteworthy that, men appear in blue shoes and women in pink shoes, thus maintaining the stigma blue for boys and pink for girls.

Men are introduced to wear skirt, underneath wear trousers. Attributes of garment: the parts are of an almost oversized wider fitting, in most parts the colours are neutral (black, white and grey). One of the consequences of the fit of the pieces being of a wider character, the forms of the bodies are practically imperceptible. The modelling of the skirts, and the pants, have elastic back waistband, which makes an adjustable piece, a response or solution to the variety of existing body types. An interesting detail in the analysis of the images of Hourani's garments, was the fact that he has a kind of waistcoat, in which for woman he presents as a dress, the back of the waistcoat the way he wears a woman, is the front when he wears in a man.

4.3 J.W. Anderson

The Irish designer Jonathan Anderson established JW Anderson in 2008. At first, an elaborate collection of accessories quickly attracted attention, allowing him to appear during London Fashion Week 2008. The success of his debut collection has garnered critical acclaim and success commercial label for Anderson, whose label is now considered one of the most innovative and advanced brands in London.

The unique design aesthetic offers a modern interpretation of masculinity and femininity, creating silhouettes that provoke reflection through a conscious cross-pollination between the masculine and feminine items. This perspective led JW Anderson to present a collection of female capsules in 2010.

With images available from collections since 2011, the collections SS13, AW13, SS14 and SS17 (respectively SS – Spring/Summer and AW – Autumn/Winter) stand out in this survey. It is also worth noting the fact that the designer, despite having divided woman and man in the site, the collections for men have some feminine characteristics.

Through the analysed images of the designer, from the seasons already mentioned above, the attributes of men's garments have feminine details. Attributes such as frills, dresses, transparencies, crop-top and colours used, give evidence to the presence of the feminine gender in their collections, thus causing a breakdown of the binary system.

4.4 Alejandro Goméz Palomo

Known as POLOMO, the foundation of the brand was made in 2015, originally from Cordoba (Spain) and posteriorly has been based in London for the last five years. An analysis was made about the collections from the collections since Autumn/Winter 16, to the last Autumn/Winter 18 (Palomo 2018), it's visible a significant breakdown of paradigms in men's clothing that Palomo creates.

Ruffles, belted silhouettes, draperies, satins, we made a highlight on the type of patterns in which the fabrics are made, because they have feminine attributes. In which, the key moments of each collection already presented in runway shows, stand out the high-heeled men that are outfitted in skirts and dresses.

4.5 Inês Torcato

Inês Torcato, born in Porto, graduated in Fashion Design from ESAD Matosinhos in 2014. In 2013 he won the 2nd prize in the "*Nespresso Designers Contest*" competition. The following year he won the first prize of the *"L'Aiguille D'Or"* competition for the Atelier des Createurs and, in the same year, he presented the collection at the Bloom Space of Portugal Fashion. In 2015, she won the "Best Collection" award from the public, and the "Best Pair of Female Footwear" in the Acrobactic contest with the collection "Ícaro", which she is invited to exhibit at the Anabela Baldaque store, in Lisbon. In 2016, is one of the winners of the Bloom Contest by Portugal Fashion, ranking 2nd and wins the award Vogue awarded by the pub at the Porto Fashion Film Festival with the film "Icarus". The designer presents her own collections since October of 2016.

For the collection of Autumn/Winter 18, the designer makes an exploration and formal deconstruction of the classics. The concept of self-portrait and what this expression represents for each of us, how we interact with the other, how we share identities and what we can give from our neighbour. A collection in black, white, grey, brown and eggplant. A set of material textures with fabrics far apart from one another such as cashmere and wool in a mixture with waterproof nylons and overlapping transparencies. In an interview, for Expresso newspaper (Atayde 2018), Inês Torcato talks about her own collection: *"My collections have pieces that are genderless and others that do not, and I would wear them all, and that is what leads me to call it a self-portrait, because it is a kind of representation of what I am and I like to wear"* she explains.

However, the attributes of each garment are, no more no less than what is made in general by others, considering the neutral colours, or at least sober colours; also, the fluid silhouette of the human body is maintained through the collections that Inês Torcato developed, meaning there aren't shapes of bodies because the clothes have a very large fit.

5 DISCUSION AND CONCLUSIONS

It is inevitable to mention the fact that fashion is changeable, and just as fashion codes are changeable (Baldini 2005; Barata 2012; Amaro 2013) gender has also been continuously (re) created (Maclaran et al. 2009). Fashion changes, because the *Zeitgeist* changes, meaning the spirit of time (Flügel 1930; Baldini 2005; Vinken 2005). And genderless clothing is undoubtedly a reflection of our times, as we are going through an era in which, both socially and politically, the struggle for equality between the sexes is notable, which influences fashion designers in the development of genderless clothing.

Kawamura (2005) refers that fashion design for men attempts to extend the boundaries of acceptable forms of sexual expression for men, there is a gender division between female fashion and male fashion. Female fashion constitutes novelty and change, two important characteristics of fashion; the male population dresses conservatively in the workplace although leisure clothing seems to be gradually replacing traditional business clothing as in the 'business casual' dress code in force in many firms. Traditional male clothing styles have remained static—a characteristic which has little space in the realm of fashion.

In our data collection and the literature review that was made, it's possible to visualize that the concept referred by Kawamura, has been suffering changes, meaning that designers and worldwide brands are available to change the binary paradigm.

From the collections and coordinates presented, except for the latter, the predominance of sober colours, few adornments, almost no accessories, brings us closer to Lipovetsky (2015) in his assertion that lightness—provided precisely by the profusion of colour, accessories—a feature strongly linked to femininity remains. It seems that, in the proposals for unisex and gendered collections studied, the blurring of the differences between masculine and feminine tends significantly towards the proximity of typified representations of the characteristics associated with masculinity, probably smoothing proposals of rupture with the familiarity of the universal.

However, unlike the other collections analysed, the work developed by J. W. Anderson and Palomo shows a strong propensity to use characteristics associated with femininity in garments developed for men. Their work allows us to question whether we are dealing with the elimination of gender dichotomies or if we see the crossing of these same dichotomies to a cross gender categorization, with the representations of femininity as the standard. However, this proposal to change the meanings of gender represented by the forms and patterns typified in masculine and feminine is, in the cases studied, totally solitary.

In this way, the question is whether the main options for approaching male signification may be based on a symbolic formulation of male domination: the concept that women can (or should) be equal to men, but the opposite is not recommended.

REFERENCES

Amaro, A., 2013. *Atenuação dos Códigos de Moda na Sociedade Actual*. Universidade da Beira Interior.
Atayde, A. de, 2018. Nycole, David Catalán e Inês Torcato para mostrar que Roma pode ser conquistada num dia. *Expresso*. Available at: http://expresso.sapo.pt/cultura/moda/2018-01-30-Nycole-David-Catalan-e-Ines-Torcato-para-mostrar-que-Roma-pode-ser-conquistada-num-dia [Accessed February 26, 2018].
Baldini, M., 2005. *A INVENÇÃO DA MODA—As Teorias, Os Estilistas, A História*, Edições 70.
Barata, J.A.B., 2012. *Design de Moda e/é Comunicação O desenvolvimento de um objecto mutável*. Universidade da Beira Interior.
Bardin, L., 1977. *Análise de Conteúdo*, Edições 70.
Barnard, M., 2002. *Fashion as Comunication*, Routledge.
Bolton, A., 2003. *Men in Skirts*, London: V&A Publications.
Coutinho, C.P., 2011. *Metodologia de Investigação em Ciências Sociais e Humanas: Teoria e Prática*, Edições Almedina, S.A.
Crane, D., 2000. *Fashion and its social agendas: class, gender, and identity in clothing*, University of Chicago.
Flügel, J.C., 1930. *The Psychology of Clothes*, London: The Hogarth Press Ltd.
García Martínez, A.N., 2012. The proliferation on fashion and the decline of its code of meanings. In *Identities through Fashion—a multidisciplinar approach*. Londres: Bloomsbury Academic.
Kawamura, Y., 2005. *Fashionology: An Introduction to Fashion Studies*, Berg.
Lipovetsky, G., 2015. *Da Leveza—Para uma civilização do ligeiro*, Lisboa: Edições 70.
Maclaran, P. et al., 2009. Praxis or performance: does critical marketing have a gender blind-spot? *Journal of Marketing Management*, 25(7), pp. 713–728. Available at: http://www.informaworld.com/openurl?genre=article&doi=10.1362/026725709X471587&magic=crossref%7C%7CD404A21C5BB053405B1A640AFFD44AE3.
Moscovici, S., 2000. *Social Representations—Explorations in Social Psychology*, Cambridge: Polity Press.
Palomo, A.G., 2018. Palomo. *Site*. Available at: https://www.palomospain.com/ [Accessed February 26, 2018].
Schütz, A., 2003. *El problema de la realidad social—Escritos I*, Buenos Aires: Amorrortu.
Seeling, C., 2000. *Moda—o Século dos Estilistas*, Könemann.
Simmel, G., 2008. *Filosofia da Moda e outros escritos*, Lisboa: Texto e Grafia.
Singer, O., 2015. Rudi Gernreich: Sartorial Feminist. Available at: http://www.anothermag.com/fashion-beauty/7329/rudi-gernreich-sartorial-feminist [Accessed April 8, 2018].
Vinken, B., 2005. *Fashion Zeitgeist—Trends and Cycles in the Fashion System*, Berg.

Embroidery: The narrative of the ribbon of the presentation of Arthur Bispo do Rosário—confluences with contemporary visual artists

L.U. Dantas, U.S.T. Barbosa, G.M.J. Sales & H.A. Dieb
Centro Universitário de João Pessoa, João Pessoa, Paraíba, Brazil

R.R. Marques
Universidade Federal de Pernambuco, Recife, Pernambuco, Brazil

ABSTRACT: This paper proposes an investigation about the embroidery in the work "Manto da Apresentação" [Mantle of Presentation], done by the artist Arthur Bispo do Rosário. The Manto's narrative reports to a wide use of different kinds of elements linked to events of his life. The aim here is contemplating the embroidery as a visual language. Nowadays, it is being considered a new medium of expression, how we can check in the work of contemporary artists.

1 INTRODUCTION

This paper is the result of the master's dissertation entitled Mantle of Presentation: The ritualistic, narrative and allegoric body of Arthur Bispo do Rosário, concluded in 2016, which aimed to study the artist known as Bispo do Rosário and, in particular, one of the most significant among the others of his production, Mantle of Presentation, analyzing its aesthetic and symbolic configuration, the elements used and the relation between the Mantle and the body, understanding it as a second skin, an extension of the artist. In particular, the intent is to compare how Bispo and some contemporary visual artists, such as Leonilson and Rosana Paulino use the embroidery in their artistic works and check the approach with regard to their artistic processes on aesthetic and formal aspects, including the use and the handling of the textile materials.

At the end of the 1980s, Bispo's creations were presented in an individual exhibition and were considered an avant-garde works in the Brazilian visual arts scene. The exhibition happened in a time of great excitement, just when the contemporary Art arises in Brazil, in which artists were seeking to overreach the traditional plastic arts media (painting and sculpture), to expand their artist repertoires for a new, multiple and interdisciplinary art, who continue being experimented.

Although Bispo didn't have the artistic intent and didn't attend the discussion around the contemporary art, by using the embroidery as a very peculiar way of expression, developed, as set out Lázaro (2012, p. 21), "... a new language to perform and think about painting. His objects are true new paintings with pigments already dyed for the time of life, in frayed wires". Therefore, this artist continues using the embroidery as a proper language, empirically and inadvertent, innovated and put up with the artistic needs in the in the second half of the twentieth century.

The research presented a qualitative approach, of an applied nature, of an exploratory nature, and, as procedures, were adopted: bibliographical and documentary research, field research and open interviews with the team of the Bispo do Rosário Museum in Rio de Janeiro/Brazil.

The interest in studying the artist and his production arise from a video about his story. With it, many feelings like distress and enchantment were awakened and blended, caused by the sadness of know his condition in the face of his fabulous art work, woven with deconstruction of clothing as the basis for all his production.

Diagnosed as schizophrenic-paranoid, at age 29, Arthur Bispo do Rosário suffered an outbreak of hallucination, which completely changed the meaning of his life. After this event, he was interned for almost 50 years in the Colônia Juliano Moreira/RJ, Brazil, where he performed most of his production. With more than 800 pieces produced, the artist used discarded materials—objects and residues collected daily during his stay in the Colony—to manufacture miniatures, sculptures, assemblages and pieces embroidered on fabrics.

Bispo do Rosário said that he had received a mission to inventory the world in miniatures to present to God and, therefore, a great part of his creations was produced in his moments of delirium, guided by "sacred voices" that commanded him what to do.

Among his productions, the Presentation Mantle stands out, a vest produced by the artist during a period of approximately 30 years, ornamented profusely, with the embroidery technique, to give form to elements, names and symbols related to moments of his life.

Several factors influenced the idea of making Mantle the object of study of this research. In the first moment, there were aspects such as: its exuberance, imposture, its mystical character and the subjectivity expressed in embroidery. Consequently, the story and the discourse in which this piece was developed presenting religious, cultural and social references of the artist.

2 ARTHUR BISPO DO ROSÁRIO AND HIS WORKS

Arthur Bispo do Rosário (Figure 1, available in http://fotografia.folha.uol.com.br/galerias/18106-bispo-do-rosario, and accessed at 15 June 2014.) was born in 1909, in the small town of Japaratuba, in the interior of the State of Sergipe, Brazil. Black, son of former slaves, his parents were called Claudinho Bispo do Rosário and Blandina Francisca de Jesus (Hidalgo, 2011). According to the author, information on age is not given as certain, when other records are consulted.

At the age of 15, in 1925, Bispo left his city to join the Mariner School of Apprentices, in the state capital, Aracaju, Sergipe, and was transferred one year later to the Central Headquarters of the National Marine Corps of Villegagnon, in the state of Rio de Janeiro, also in Brazil.

There, it remained for almost ten years, during which time he traveled extensively by the seas of the Brazilian coast. When he was on board, he lived on the ships where he worked, and thus he knew "the imprisonment of the prison and the high seas" (Hidalgo, 2011). As far as is known, according to his biography, Bispo never returned to the Northeast. During this period, Bispo ventured as a boxer, participating in some tournaments.

Figure 1. Arthur Bispo do Rosário.

According to Hidalgo (2011), in 1933, because of insubordination, Bispo was dismissed from the Navy and started to work as a trolley cleaner at Excelsior Company, where he remained until 1937. During the job duties, he suffered accidents twice, which caused a large fracture in the right foot. This fact led him to leave his job and caused Bispo, represented by lawyer Humberto Leone, to file suit against the company, which resulted in compensation.

The same author tells that, after this contact with the lawyer, Bispo went to work and live in the mansion of the Leone family, where the parents lived, the brothers of Humberto Leone and a large team of domestic servants, from which Bispo was part of it. Since then he has become a faithful servant, showing himself humble and devoted to the family, in exchange for food, shelter, and security, so that he has the affection and affection of all. His room was a small room in the backyard, which became his refuge. There, his creative practice emerged with the construction of toys made with scraps.

The following year, during the week of the Christmas celebrations, Bispo was surprised by a great outbreak, an event that changed his whole life. He claimed to have received a message from seven angels communicating to him that he had been "chosen" by the Divine Father for a mission on earth, which would be "to judge the living and the dead and to recreate the world for the Day of Judgment" (Dantas, 2009, p. 30).

On December 22, 1938, led by voices he claimed to hear, Bispo left on his pilgrimage. He walked through the streets of the city of Rio de Janeiro for two days, passing by the Church of St. Joseph and ending at the Monastery of St. Benedict, announcing his mission.

After this episode, on December 24, he was intercepted by the police and taken to the Hospital of the Alleged, in Praia Vermelha, Rio de Janeiro, being diagnosed as a paranoid schizophrenic. A few days later, on January 6, 1939, he was transferred to the Juliano Moreira Colony, in Jacarepaguá, a place that housed patients considered chronic and irreversible, according to the psychiatry of the time. The doctors evaluated that Bispo suffered from "delusions of grandeur" and, according to Dantas (2009), this diagnosis was sustained from the first evaluation until the fifty years he spent in the Colony.

This time lived in the Colony was alternated with exits and returns, between the decades of 1940 and 1960. During these alternations, Bispo returned to the Leones' house in search of shelter, always being well received. There, he felt free to speak his preaching: "He spoke about God and the devil, about good and evil, heaven and hell and his mark: the luminous cross that marked his back signaling the sacred powers—and all there 'confirmed' those powers" (Dantas, 2009, p. 31). But in one of these preaching, a family member said, amid much laughter, that the mark on Bispo's back did not exist, refusing to accept him as God's envoy. In disagreement, after that fact, Bispo decided to leave the Leone house and moved to one of the disabled rooms in the lawyer's office.

There he was totally reclusive, under fasting, without contact with people and with the outside world. It was when his creations began to be born, the beginning of his "mission", the re-creation of the world and thus he built pieces such as: ships, carts, trucks made of various materials, like pieces of wood, broomsticks, etc. Over time, his mental state grew worse, until one day Humberto Leone suspected that Bispo could be seeking suicide by finding him in front of the open window of his office. Faced with this occurred, he persuaded him to reintegrate himself, on January 27, 1948.

Intern in the Colony once again, Bispo had moments of sociability, cultivating a good relationship with the employees, helping them, contributing to the cleaning and maintenance of the pavilion, and helping them to order the sick. He also had a good relationship with inmates. If there were quarrels among them, for example, the Bispo protected and defended the one he considered to be right, and thus he was rewarded with pleasure by means of objects, both inmates and their relatives, when they visited them.

Bispo was able to avoid the violent treatments that were adopted in these places during the XX century, refusing, even, to take medications. His "treatment" was, so to speak, through the creative process of his works that balanced and disciplined him by his intense dedication and involvement in fulfilling his "mission." Although for him his work was nothing but an arduous task, his work was in fact a kind of soul-treatment, a way he found to live with schizophrenia and the precarious conditions to which he was subjected, and, thus, with great virtuosity and plasticity, a huge physical and symbolic diary of its existence emerged.

Faced with this fact, much of his production was made between moments of delirium and lucidity. Bispo do Rosário said that he had been guided by "sacred voices" that commanded him – "you have already done that, have you done that? Tomorrow I want you to do this and that " (Dantas, 2009, p. 18).

The collection of Bispo do Rosário consists of a diversity of forms and styles. Their pieces were made of recycled materials from the daily disposal of the Colony, which through their creation were transformed into assemblages, miniatures made of wood coated with blue lines (Orfas), binders, banners, miss bands and pieces of clothing. The materials were as varied as possible: cardboard, wood, plastic containers, household utensils, shoes, perfume bottles, combs, etc. The pieces in fabrics were mostly made with the reuse of old sheets, and the embroidery, at the beginning of their creations, used lines that derived from the old uniforms of the Colony. With more than 800 pieces produced, it is the Mantle of Presentation the most relevant work of the production of Arthur Bispo do Rosário.

Bispo and his works came to be known to the public after a report, named Rejected Cities, held by the television program Fantástico, Rede Globo, which aired on May 18, 1980. It revealed the mistreatment and violence in the colony: filthy pavilions, infiltrations, plastered walls, restrooms, patients thrown into cells under high doses of neuroleptics and electroshocks, an affront to the dignity of the human person. It is worth noting that during this period Brazil lived the Military Dictatorship, in which the oppressive means were used in large scale with all the population that was against the political regime.

It is possible to identify, in his creations, references that relate to various experiences of the life of Bispo. With embroidery he represents, for example, the names of fighters, elements and objects belonging to the nature of boxing; the relationship with religiosity, which is present from his hometown, to the fruits of his hallucinations, and the period he spent in the Navy.

3 THE MANTLE, ITS NARRATIVES AND ITS INFLUENCES

Considered the centerpiece of the collection of Bispo do Rosário, the Presentation Mantle (Figure 2, available in http://colmeia.biz/wp-content/uploads/2012/11/Manto-da-anuncia%C3%A7%C3%A3o.jpeg, and accessed at 28 de May de 2016.) displays embroidered records of the artist's life that communicate his history. They are the graphic language that give visual form to the narrative of Bispo. Through the ritual of embroidering, making and redoing, the Mantle, little by little, over almost 30 years, was realized by the hands of the artist.

Figure 2. Presentation Mantle. Work of Arthur Bispo do Rosário.

The work, by its magnitude, provides a visual impact and ascension to something of mythical and enigmatic character, awakening meaning and enjoyment in those who see and contemplate it. It has, in its area, a profusion of embroidery, such as symbols, words, numbers, letters, among other elements.

As an imagery, the Mantle presents a visual narrative in which the embroidered elements are related to the passages of the artist's life. They are impressions of lived moments that, through line and needle, the artist wrote and subjectively wove his history, contemplating his childhood until the experience in the Colony.

As an imagery, the Mantle presents a visual narrative in which the embroidered elements are related to the passages of the artist's life. They are impressions of lived moments that, through line and needle, the artist wrote and subjectively wove his history, contemplating his childhood until the experience in the Colony. Another factor is that the arranged embroidery suggests a narrative of the facts that have permeated his life and tell his story. The embroidery activity is a recurring practice in several interior cities in the states of Northeastern Brazil, which, like Japaratuba, tradition is passed down from generation to generation, establishing itself as the predominant characteristic of each region. Being such a great representative of the cultural activities of the region.

It is perceived a peculiar interest of Bispo do Rosário by clothes. The artist constructed and decorated some robes that can be seen in his collection, as for example of the big uniforms and suitcases. In the case of the Mantle, its imposture, exuberance and allegory reveal itself as something of great power and symbolic value, an unusual garment, used in moments of celebrations and manifestations. This indicates that Bispo would like to present himself to his "God" clothed in character, worthily, as someone endowed with a certain power, his representative, or as Hidalgo (2011) said, the "king of kings". In view of this, it may be supposed that Bispo was concerned with his appearance and that he intended to characterize himself according to that moment. In fact, in carrying the Mantle, Bispo would elevate his appearance, transforming himself and elevating himself to other senses, as Castilho (2009) states, saying that a body covered by a second skin may contain several codes that will collaborate with the construction of his speech.

The Presentation Mantle is a garment adorned with the technique of embroidery to give shape to symbols and figurative elements, in a subjective and rather profuse manner. The same is the visual language of the Bispo's narrative and is related to the vestiges of his life and culture. Although in a non-linear way, it is seen in the Mantle several elements referring to reminiscences of its city, its religiosity and the feminine universe, expressed in the technique of embroidery. He becomes curious as someone of rudimentary appearance, with hands and fists, who have sometimes been used ferociously as a boxer, weave stories through delicate embroidery. It is well known, how much the embroidery by its slowness and minutia, demands so much patience and manual ability, generally attributed to a feminine universe, impregnated to the culture of the old women, named as sinhás that, through the act of embroidering they expressed their sensations and they filled their lives, among the activities of the home, as exposed Chagas (2010, p. 10):

> *The history of embroidery has been a long and long time in the history of women, bringing their brands in different spaces-times, 'basin' for a feminine 'time', with special gestures that wish to highlight, love, longing, loneliness, the necessity, the possibility, but also the exploitation to which they are subjected for centuries.*

In fact, textile handwork involving the use of threads and needles is mostly developed by women, although there are excellent works done by the hands of men. But in fact, embroidery carries the stigma of a feminized practice built socially and culturally in the history of mankind.

From this perspective, it is possible to relate the choice of Bispo using the technique of embroidery, to a feminine ancestry, certainly references of his childhood through the contact with the embroiderers, in his native city. According to Hidalgo (2011, p. 33), Japaratuba, a municipality in the State of Sergipe, Brazil, is known for its religious manifestations and popular festivities, which is still present in several municipalities of that State. The allegory of these manifestations spreads through robes, costumes, ornaments and decorations, in the festivals of Reisado, among others. They are memories of this universe that Bispo expresses his art.

The history of embroidery refers to the beginnings of mankind. It arises for the purpose of sewing the robes made with animal skins. The point used was like the currently known cross stitch. In the Middle Ages, cross stitch embroidery is now used to illustrate coats of arms and family names as a way of identifying the clothing of wealthy owners. In the Renaissance emerges in a decorative way, in addition to religious and civil robes, its use was extended to interior decoration, in tapestry, upholstery for furniture, and others. During this period, cross-stitch embroidery also contributed to women's literacy, since at that time they did not have access to school and to develop embroidery it was necessary to follow a monogram of letters. It was through this contact that they could know the letters, the digits and signs and, thus, the embroidery becomes a form of language and visual communication.

In this context, it can be said that embroidery in the Mantle, as visual language, intends to communicate something, therefore, in this perspective, a question arises: What do the embroidery on the Presentation Mantle communicate? How can one read them beyond their graphic representation, their technical and visible aspect?

In the Presentation Mantle you can see a world of possibilities. As a work of open reading, it leads to a multiplicity of interpretations. Through overlapping layers, discourses, and references, his narrative appears dense, profuse and unspeakable, in a non-linear structure of present-past-future time, so that there is no beginning, middle, and end.

His embroidery translates a narrative of time, memory, affections, feelings and anguish, references of the city of Japaratuba, his experience as a sailor, as a boxer, the contact with the Leone family, his daily life in Colônia Juliano Moreira, among other references. They are stories recounted and woven into a subjectivity of languages, referring to the historical process of Bispo do Rosário. In it are registered signs, symbols, numbers, words, texts and a diversity of figurative elements, in which several religious symbols stand out. The Mantle suggests a testimonial piece which would witness on the day of Judgment the whole history of the artist's life for the free pass and entry into the "divine" abode.

These stories woven through embroidery can be told in the most varied forms, whose origin is internal and subjective. In the field of visual arts, specifically in Brazil, embroidery appears as a medium of artistic language, approximately, from the 1970s and proves to be a new form of art expression. Therefore, it is relevant to mention some artists who have used this technique as a form of expression and who can establish relationships with the work of Bispo. After the first individual exhibition of Bispo, in 1989, his textile works served as references for the works of several other artists of the art circuit, as mentioned by Herkenhoff (2012, p. 141):

> [...] it is possible to speak of a Bispo do Rosário effect in the Brazilian art. It is not simply an extensive mapping, but a quick reference to four paradigmatic examples: Leonilson, Rossana Palazyan, Walter Goldfarb and Rosana Paulino, taken from the exhibition Registers of My Passage through the Earth, Bispo do Rosário, at the School of Arts of the Parque Lage in Rio de Janeiro, in 1989.

Considered one of the great artists of the 1980s, Leonilson excelled in seeking to integrate everyday experiences into his production, as well as most of the artists of this generation inserted into the concepts of contemporary art. Leonilson's work, especially embroidery, is extremely sensitive. In them, one can observe what is more particular in the artist, the love, the affection, the search for the other and the anguish. In the years 1992 to 1993, the year of his death, Canton (2001, p. 127) comments that the embroidery in Leonilson's work "function as synthetic self-portraits, abstract reliquaries that dematerialize together with his body."

On the experience of the act of embroidering Leonilson puts it: "it is because the business of the hand is the pleasure of giving the point, of erring, of cutting and of coming back again" (Veneroso, 2012, p. 335). The influence of Bispo's work aroused in Leonilson a universe of possibilities that was already part of his heritage, the experience of his mother, seamstress and embroiderer, and his father, a merchant of fabrics. The same author comments that the "exposition of Bispo meant for Leonilson to discover something that was latent in his memory and that has become one of the main reasons of his art" (Veneroso, 2012, p. 334). In

Leonilson's work (Figure 3, available in http://enciclopedia.itaucultural.org.br/pessoa8742/ leonilson, and accessed at 29 May 2016) there is a confluence with the work of Bispo, both for the use of embroidery and for the fusion of image and text present in his works.

Rossana Paulino, another artist who uses embroidery in her work, learned between 1996 and 1997 to manipulate sewing materials with her mother and, with these materials and embroidery, relates her work to social, ethnic and gender issues, which has as main focus the theme of the black and mainly of the black woman in the Brazilian society, sensitizing the people to the domestic violence of the woman. On the practice of sewing in Rosana's work (Figure 4, available in http://extra.globo.com/tv-e-lazer/exposicao-nos-438554.htm, accessed at: 29 May 2016), author Bamonte (2008) describes a passage from the artist's own words:

> *Lines that modify meaning, sewing new meanings, transforming a banal, ridiculous object, changing it, making it an element of violence, of repression. The thread that twists, pulls, modifies the shape of the face, producing mouths that do not scream, giving us knots in the throat. Eyes sewn, closed to the world and, especially, to the condition in the world* (Paulino, 1997, p. 114 apud Bamonte, 2008, p. 295).

Canton (2001) states that in the artistic production of the late twentieth century, female sensibility is present both in the thematic and in the choice of manipulation of materials directly related to the world of domestic craftsmanship.

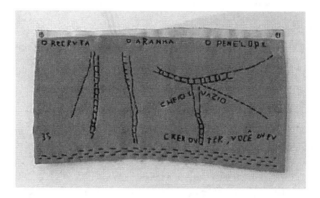

Figure 3. Work of Leonilson: The Recruit, the Spider, the Penelope. 1992 – Embroidery on felt. 18 × 35.50 cm. Photographic reproduction Fabrízio Penteado/Itaú.

Figure 4. Wall of Memory, 1994. Rosana Paulino. Screen printing on cushions, 8 × 8 × 3 cm. Private Collection.

These, among other artists in the Brazilian artistic scene, used the embroidery as a means of "language and communication". His works, as well as those of Bispo, provide a dive into the universe of intrinsic and subjective stories full of varied concepts and themes of contemporary life. In the hands of these artists, such works reveal a great deal of plastic and poetic richness and, in addition to the great aesthetic potential, carry a secular historical ancestry, since, according to Chagas (2010, p. 16).

> To embroider is to give shape, to know the way to create, to embroider is to try to recompose the history of life, is the thread of different generations to leave their marks in the spaces/times where they lived and live.

Thus, we perceive every day the rescue of the technique of embroidery by contemporary artists as a form of communication, resistance and breaking of standards, both about embroidery as a minor craft process, and art as something extremely erudite and that is only recognized from classical and traditional techniques. As we have seen, in both the work of Bispo and the most current artists, embroidery may contain narratives and subjectivities capable of achieving the true state of art.

4 FINAL CONSIDERATIONS

In view of the above, it may be suggested that Bispo's textile works echoed in the works of the renowned artists of that time, taking the technique of embroidery and the use of textile materials to another level within the visual arts. His work was built over almost 50 years and, although simultaneous with the artists who participated in the processes of change from modern to contemporary art, it was produced on the margins of this historical and aesthetic context, which makes Bispo do Rosário, even unintentionally, a avant-garde artist. And yet, it seems to reflect this state of facts, this set of questions of our contemporaneity, from which no one escapes, and to this day artists have appropriated it as a theme to develop their works.

It should be noted that embroidery is a form of language and communication that accompanies humankind from very remote times, a manual technique socially constructed in the history of humanity, as an essentially feminine, domestic and occupational practice, which gains art notoriety from consecrated artists who bring to light the true meaning behind the lines and needles. Visual narratives are constructed and reconfigured into new standards as far as techniques and materials are concerned. It is important to emphasize the importance of Bispo's work to influence a whole generation of consecrated contemporary artists and how from this new place of occupancy of the embroidery, the same goes through great value changes also in fashion, in design and even in crafts.

REFERENCES

Bamonte, J. L. B. M. 2008. A Identidade da Mulher Negra na Obra de Rosana Paulino: Considerações sobre o Retrato e a Formação da Arte Brasileira. *17° Encontro Nacional da Associação Nacional de Pesquisadores em Artes Plásticas*. Panorama da Pesquisa em Artes Visuais, Florianópolis. [Online]. Available: http://www.anpap.org.br/anais/2008/artigos/028.pdf [Accessed 25 May 2016].

Canton, K. 2001. *Novíssima Arte Brasileira: um guia de tendências*. São Paulo: Iluminuras.

Chagas, C. R. R. P. Das. 2010. *O bordado no currículo com espaço-tempo/fazer educativo*. Dissertação de Mestrado. [Online]. Available at: http://www.anped.org.br/reunioes/29ra/trabalhos/trabalho/GT23-1967--Int.pdf. [Accessed at 20 May 2016].

Castilho, K. 2009. *Moda e Linguaguem*. 2ª Edição. São Paulo: Editora Anhembi Morumbi Universidade.

Dantas, M. 2009. *Artur bispo do Rosário: a poética do delírio*. São Paulo: UNESP.

Herkenhoff, P. 2012. A Vontade de Arte e o Material Existente na Terra dos Homens. In: *Arthur Bispo do Rosário-Século XX*. Wilson Lázaro (Org.). São Paulo: Cosac Naify.

Hidalgo, L. 2011. *Arthur Bispo do Rosário: o Senhor do Labirinto*. Rio de Janeiro: Ed Rocco.

Lázaro, W. (Org.). 2012. O Artista do Presente. In: *Arthur Bispo do Rosário-Século XX*. São Paulo: Cosac Naify.

Veneroso, M. do C. F. 2012. *Caligrafias e Escrituras*. 1ª Edição, Belo Horizonte/MG, Editora: C/Arte.

Eye tracking in fashion: An overview

A.P. Faria, B. Providência & J. Cunha
University of Minho, Portugal

ABSTRACT: Advances in Affective Computing are improving the way practitioners collect and analyze data. Eye tracking technology has evolved and equipments that have emerged in the last few years are more affordable and easy to tackle. This has lead to an increasing application of eye tracking to several fields of research. Aware of the potential of this technology, researchers in Fashion are embracing it. Through a literature-based overview of the current state of research—which included peer-review journal articles between 2010 and 2018, studies were uncovered to understand how eye tracking is being applied. Results revealed that combining eye tracking with other methods makes it possible to triangulate different data sources, opening new paths for studying the processes that underlie how people explore visual stimulus. Furthermore, valuable insights are presented to help researchers and practitioners making more informed decisions when considering the use of eye tracking.

1 INTRODUCTION

Companies recognise that consumers' senses have a major role in their decisions: they help them decide which products stand out from a host of similar offerings in the marketplace (Solomon, 2015). Consequently, it is important for Fashion companies to better understand human responses to visual brand-related stimuli that are part of a brand's identity, packaging, communications and environments, by uncovering how people literally see and perceived them. Eye tracking could help answering research questions regarding this matter—it holds out the promise of another window into the mind: the semi-magical ability to know what people are looking at (Bojko, 2013).

Eye tracking—which has been applied to several fields such as human factors, cognitive psychology, marketing, human-computer interaction, among others—is a methodology that aids researchers understand visual attention (Bergstrom and Schall, 2014). It allows to detect what people see and where they look by tracking their eye movements and how long they fixate on a certain point (Barnum, 2011).

Since the beginning of the twenty-first century, eye tracking hardware and software started becoming more widely accessible and more common due to the ease of use of systems, particularly around analysis, accuracy, and mobile technology (in the form of goggles), as well as new webcam-based technology (Tullis and Albert, 2013).

Nonetheless, the set-up of an eye tracking study requires technical expertise, knowledge with respect of the planning of research studies and analysis of the obtained data sets, which may pose some challenges (Duchowski, 2007; Olk and Kappas, 2011). Usually conducted with the help of a device called an eye tracker (remote or wearable), eye tracking uses a combination of an infrared video camera and infrared light sources to track where participant is looking, based on the relative position of corneal reflection (surface of participant´s eye) and the pupil center (Bojko, 2013; Tullis and Albert, 2013).

Before conducting a study with this specialized equipment, Tullis and Albert (2013) recommend researchers to plan the study carefully, as well as taking time to explore data. Other important thing to keep in mind when acquiring or renting eye tracking technology is to check the analysis component. This latter cannot be underestimate: the vendor should tell how it will address specific questions and demonstrate that they can run tests and analysis required (Goodman, Kuniavsky and Moed, 2012).

1.1 *Present study*

The purpose of this study was to reveal how eye tracking is being applied and its contribute to research in Fashion. Thus, through a literature review, this paper will examine the existing body of Fashion which employed eye tracking technology as a method. This paper will first report research strategy: identifying databases, keywords, stating the selection criteria, among other aspects. This will be followed by the presentation of research results. Finally, the paper will conclude with a reflection about the articles and avenues for future research.

2 METHODS: SEARCH STRATEGY AND DATA COLLECTION

First, it was conducted a search in five electronic databases (b-on, Scopus, Science Direct, Web of Science and Emerald) for peer-review journal articles published between 2010 and 2018. The following keywords were used: eye tracking, fashion, apparel and clothing. Second, it was made a search in key journals of Fashion – e.g. Clothing and Textiles Research Journal; Fashion Theory; International Journal of Fashion Design, Technology and Education, in an attempt to gather more articles.

To evaluate the importance of the identified studies, it was examined the title and abstract, or the content of the article, regarding two criteria: using eye tracking technology and focusing on Fashion communication. It should be highlighted that literature reviews and articles from conference proceedings were not considered.

3 APPLICATION OF EYE TRACKING IN FASHION

From the journal articles that met the selection criterion, eight studies were considered relevant for analysis. It was verified that four studies were somehow related to traditional channels (e.g. advertising, magazines) and the other four were focused on digital media channels (e.g. websites, social media). Therefore, the results were categorized in two major sections. More detailed information of the included studies can be found in Tables 1 and 2.

3.1 *Traditional media*

Ju and Johnson (2010) used eye-tracking in order to study visual attention of young women on different types of fashion advertisements—to measure to what extent they focused their attention on models. Participants were exposed to six advertisements and were able to control their exposure to each one of them. After that, they completed a survey to measure self-

Table 1. Characteristics of included studies related with traditional media.

Author	Type of research	Variables of interest
Ju and Johnson (2010)	Survey and eye tracking	How women visually processed a fashion advertisement.
Rahulan *et al.* (2015)	Eye tracking and survey	Purchase decision behaviour of compression sportswear by Baby Boomers and Generation Y.
Amatulli *et al.* (2016)	Eye tracking and survey	How consumer recognise luxury fashion brands.
Fidelis *et al.* (2017)	Eye tracking and survey	How sexual appeal in print media influence consumers' brand recall.

Table 2. Characteristics of included studies related with digital media.

Author	Type of research	Variables of interest
Cyr and Head (2013)	Eye tracking, survey and Interview	How task framing and length of viewing time influence user website perceptions and viewing behaviour.
Ho (2014)	Eye tracking and survey	How e-consumers perceive online pictures of women's handbags.
Menon et al. (2016)	Eye tracking	Consumer attention to price of retail clothing in an f-commerce setting.
Huang (2018)	Eye tracking and survey	How the slot positioning of banner ads affected female consumers.

reports of comparison to the model, product involvement and other demographic information. They found models had great potential to influence the viewer. Thus, they argue that apparel-related product advertisements should contain human models because those who do not, may be less successful. It is noteworthy that participants' self-reports of social comparison were positively related to the time spent looking at the model and to the number of eye fixations on the model.

Rahulan et al. (2015) analysed purchase decision behaviour of compression sportswear by Baby Boomers and Generation Y. They run an eye tracking study with 40 participants. The simulation had four parts: observation of 7 different promotional images of athletes wearing compression sportswear while being active in various sporting activities; one minute viewing of three compression leggings and their packaging, with details of the products; an online purchase simulation and finally a questionnaire about products. Results showed similarities between the data collected from the eye tracker and the responses of the questionnaire regarding the attributes of sportswear advertising that most influenced them. It was found significant differences in the purchase behaviour of Baby Boomers and Generation Y for compression sportswear: Baby Boomers were concerned with protection, whereas the Generation Y respondents cared more for performance enhancement. Regarding purchase decisions, Baby Boomers were found to be confident in their choices, and spent a shorter time making purchase decisions while Generation Y were more inquisitive and took more time to make a purchase decision, attributing greater importance to information search.

Amatulli et al. (2016) aimed to discover how consumer recognise luxury fashion brands. They applied eye tracking technology to conduct the main study, which comprised two phases. First, participants observed five images, each image consisted of three total look pictures from the same brand. After observing each set of three pictures, participants were asked to identity the luxury fashion brand and to state at least three stylistic elements they believed helped them to recognise the brand. Second, participants observed five images which contained two pictures showcasing similar looks—one picture featured a luxury brand whilst the other a well-known brand that was both imitative of the first and less expensive: a fast-fashion brand. Each participant was asked to indicate which of the looks was from a luxury brand and at least three stylist elements that informed their judgment. Both eye tracking data and participants' answers indicated that luxury fashion brands are mainly recognised through accessories. Thus, these findings suggest that managers should take into account accessories for luxury brand recognition. In addition, results revealed that consumer recognition of luxury fashion brands increases when pairing these branded products with those made by fast-fashion companies.

Fidelis et al. (2017) resorted to an eye tracking experiment to understand how consumers´ perceived advertisement and how sexual appeal influence consumers´ brand recall. They asked 201 participants (men and women) to observe six advertisements: 100 participants were exposed to advertisements with sexual appeal and the other 101 observed the ones without it. Then, participants informed which brands they recalled having seen on the advertisements by filling a questionnaire. Based on the results, they considered that brands are not recalled more in advertisements with sexual appeal because consumers are distracted by models from looking at brand elements. Consequently, they advise managers to use sexual appeal in advertisements with caution: if the intention is to make consumers to recall brands name, the presence of images with sexual appeal does not have a fundamental role in advertisements.

3.2 Digital media

Cyr and Head (2013) examined task framing and length of viewing time as conditions that influence user website perceptions and viewing behaviour. They conducted an experimental questionnaire and interviews, besides eye tracking—the data collected allowed to obtain further insights into where users looked and why.

First, it was chosen a website for the basis of the experiment. Then the website was divided into three areas of hedonic interest (emotive text that describes a pleasant scene involving the garment, user comments of customers and a picture of the garment being worn on a human model expressing pleasure) and five areas of utilitarian interest (name of the website, top navigation bar focused on site functional elements, left navigation bar with contents of the website, functional attributes of garment and an order specification area). Sixty female participants joined the experiment and they were presented with two specific scenarios that aim to understand how they perceived the defined areas. After browsing the website, participants also filled a survey and answered interview questions. Results showed that images have an impact on the website, especially when viewing time is brief; utilitarian content such as adequate information is a basic requirement leading to website effectiveness, but hedonic elements are important when tasks are framed to be fun.

This study offered insights for the development of websites that aim to engage with consumers: hedonic elements such as pictures and emotive text are important and lead to user enjoyment and involvement.

Ho (2014) and Menon et al. (2016) explored how consumers perceived information online by viewing product pictures. The main data of these studies was collected with an eye tracker. In the first study it was also conducted a survey, but it was only to collect the age of the participants.

The purpose of Ho (2014) was to understand how consumers observe handbags: with this knowledge, fashion designers will know how to design and attract attention of consumers. He conducted a task free eye tracking experiment in which 33 female participants observed 74 randomly displayed pictures of handbags. Six types of areas of interest were defined, according to the general structure of the handbag for analysis. Results clarify that the first thing that attracted attention was the main body of the handbag; the handle had the greatest capacity to hold attention and the handle and strap had a stronger visual attraction than other elements. This means that websites should ensure that photos of this nature should clearly depict handles and straps and it is also necessary to add detailed pictures of this specific areas because they are areas that naturally draw the attention of consumers.

Menon et al. (2016) evaluated visual behaviour of consumers towards the price of retail clothing in a Facebook page. The study incorporated several interventions such as different price points, price visibility and the presence of a known model or a mannequin. Each of the 31 participants (males and females) was exposed to a total of 25 pictures in different combinations. The statistical analysis of the data collected through the eye tracker illustrated that participants had a significantly higher fixation on price when it is placed along with the picture than when it is placed on the right, along with company details and likes—the article did not provide figures or a description to fully understand how the price was positioned. Additionally, findings showed no significant difference in the fixation on price when retailers display clothes on a model or a mannequin. Although, males had a higher total fixation length on a page that displays a model instead of a mannequin. For retailers, the results of this study indicate that they can draw consumers' attention to price directly (through the manipulation of price position) or indirectly (through the use of a salient attribute such as a model/mannequin).

The research of Huang (2018) provides valid suggestions and contributions regarding the elements in banner ads and the positions of ad slots. The aim of the study was to examine the attention that female consumers paid to product-based personalized banner ads for online apparel retailers, and how this attention was influenced by visual content composition and slot position. The participants were instructed to browse an apparel shopping website—which include banner ads with different layouts, between three to five minutes. In addition, participants filled a questionnaire about shopping and browsing habits. The results demonstrate that central slots have significantly greater visual attention effects than those in the right sidebar, regardless of the ad content. If it is not possible to advertise in central slots, it is recommended to include models and discount information on the ad to achieve visual appeal.

4 DISCUSSION

Through the analysis of the eight articles, important factors of visual behaviour were identified. Results showed that eye tracking technology, which is starting to emerge in Fashion articles, played an important role to address specific matters related to visual attention.

Indeed, the majority of studies rely in this latter subject. Ju and Johnson (2010) recommend the use of models in advertisement because they discovered that models influence consumers. Nonetheless, Fidelis *et al.* (2017) considered that sexual appeal must be used with caution because models might unintentionally distract or divide consumers attention from brand elements.

Concerning digital media, authors focused their researches in visual details (position of ads, price, elements of websites, product pictures) – which provide insights and guidelines for designers. For instance, Cyr and Head (2013) suggest that websites should use pictures and emotive text to engage with consumers and Ho (2014) found that handles and straps are areas that attract consumers attention in a handbag.

From other perspectives, Rahulan *et al.* (2015) used eye tracking to understand purchase behaviour of two generations and Amatulli *et al.* (2016) discovered that accessories are relevant for luxury brand recognition.

Overall, studies suggested the following enlightenments:

- The lack of clarity with regards to visual stimuli could compromise the interpretation of data—vague descriptions did not allow to draw more precise conclusions;
- Four articles did not take advantage of data visualizations (i.e. heatmaps, gaze plots) that could be obtained from eye tracking software. Bojko (2013) states that when used correctly, visualizations play an important role because they help to communicate findings and can make them more persuasive and easier to understand;
- Researches used the most common eye tracking metrics for analysis: fixation duration, dwell time, number of fixations, among others. It is important to state that eye tracking has over a hundred measures that could also be of interest. For instance, the use of information about pupil response may be useful in certain situations, where the focus is on the decision making process or emotional arousal (Tullis and Albert, 2013; Weinschenk, 2015).

Researches showed results in a positive direction, indicating that practitioners and academics are aware of the potential of this technology and are exploring it to study traditional and digital communication channels.

5 CONCLUSIONS

Visual attention technology has major implications for apparel industry: it provides valuable information for marketers to study consumer behaviour that cannot be obtained from traditional means of research (Rahulan *et al.*, 2015; Fidelis *et al.*, 2017).

Fashion literature revealed that researchers are resorting to eye tracking to extract information about dwell time on print media or specific areas of an online store to provide information about how much attention it attracts—this information is useful for design purposes and to evaluate advertising campaigns (Olk and Kappas, 2011). Moreover, through the observation of eye movement, researchers are attempting to understand to which degree visual attention translates into consumer behaviour, namely, by studying how people recall and recognise brands and their purchase decision behaviour.

With regard to the method itself, researchers recognised that eye tracking reveals important information about how people see, but they know that it is not enough. Therefore, they resorted to other methods, seeking for more information, to a better understanding of visual behaviour. It was found that interviews and surveys are the most common methods used in combination with eye tracking. They allowed researchers to establish connections between data.

Thus, in our point of view, future studies should be more experimental and combine different methods—besides traditional ones, aiding researchers and improving visual measurement. Other qualitative techniques can provide more information about cognitive and affective responses of consumers (Santos *et al.*, 2015). Hence, eye tracking studies could address research questions that can go beyond the identification of areas that attract visual attention or consumer visual behaviour in different situations.

Additionally, we believe that technological devices, data mining, and social media can bring new insights to research. Thoring *et al.* (2016) argue that complementing data from one source (e.g. eye tracking) with metadata from other source (e.g. facial expression recognition) can further improve emotion measurement and help researchers understand reasons for specific emotions.

The main focus of this study was to provide a brief overview of how eye tracking is being applied in Fashion studies. However, researches in other fields could shed new light and encourage academics to try different approaches. This literature overview reinforces the words of Duchowski (2007) when he mentioned that eye tracking is a good deal of opportunity for meaningful research. We hope this paper inspires further researches.

ACKNOWLEDGEMENTS

This work is supported by FSE – European Social Fund, by national funds through FCT – Foundation for Science and Technology and POCH – Human Capital Operating Programme, within the scope of the project SFRH/BD/129900/2017.

REFERENCES

Amatulli, C. *et al.* 2016. '"Mix-and-Match" Fashion Trend and Luxury Brand Recognition: An Empirical Test Using Eye-tracking', *Fashion Theory*, 20(3), pp. 341–362. doi: 10.1080/1362704X.2015.1082294.
Barnum, C.M. 2011.*Usability testing essentials: ready, set... test!* Burlington: Elsevier.
Bergstrom, J.R. and Schall, A. J. 2014. *Eye tracking in user experience design*. Waltham: Elsevier.
Bojko, A. 2013. *Eye tracking the user experience: A practical guide to research*. Brooklyn: Rosenfeld.
Cyr, D. and Head, M. 2013. 'The impact of task framing and viewing timing on user website perceptions and viewing behavior', *International Journal of Human Computer Studies*. Elsevier, 71(12), pp. 1089–1102. doi: 10.1016/j.ijhcs.2013.08.009.
Duchowski, A. 2007. *Eye tracking methodology. Theory and practice*. Springer.
Fidelis, B.T. *et al.* 2017. 'Sexual appeal in print media advertising: effects on brand recall and fixation time', *Research Journal of Textile and Apparel*, 21(1), pp. 42–58. doi: 10.1108/RJTA-12-2016-0033.
Goodman, E., Kuniavsky, M. and Moed, A. (2012) *Observing the user experience: a practitioner's guide to user research*. 2nd edn. Waltham: Elsevier.

Ho, H.F. 2014. 'The effects of controlling visual attention to handbags for women in online shops: Evidence from eye movements', *Computers in Human Behavior*. Elsevier Ltd, 30, pp. 146–152. doi: 10.1016/j.chb.2013.08.006.

Huang, Y.T. 2018. 'The female gaze: Content composition and slot position in personalized banner ads, and how they influence visual attention in online shoppers', *Computers in Human Behavior*. Elsevier Ltd, 82, pp. 1–15. doi: 10.1016/j.chb.2017.12.038.

Ju, H.W. and Johnson, K.K.P. 2010. 'Fashion Advertisements and Young Women: Determining Visual Attention Using Eye Tracking', *Clothing and Textiles Research Journal*, 28(3), pp. 159–173. doi: 10.1177/0887302X09359935.

Menon, R. G. V. et al. 2016. 'Consumer attention to price in social commerce: Eye tracking patterns in retail clothing', *Journal of Business Research*, 69 (11), pp. 5008–5013. doi: 10.1016/j.jbusres.2016.04.072.

Olk, B. and Kappas, A. 2011. 'Eye Tracking as a Tool for Visual Research', in *The Sage Handbook of Visual Research Methods*, pp. 433–451. doi: http://dx.doi.org/10.4135/9781446268278.n23.

Rahulan, M. et al. 2015. 'Consumer behavior of generational cohorts for compression sportswear', *Journal of Fashion Marketing and Management*, 19(1), pp. 87–104. doi: https://doi.org/10.1108/MIP-05-2017-0088

Santos, R.D.O.J. dos et al. 2015. 'Eye Tracking in Neuromarketing: A Research Agenda for Marketing Studies', *International Journal of Psychological Studies*, 7(1), pp. 32–42. doi: 10.5539/ijps.v7n1p32.

Solomon, M.R. 2015. *Consumer behavior: buying, having, and being*. 11th edn. Harlow: Pearson Education.

Thoring, K. et al. 2016. 'A framework of technology—supported emotion measurement', in Celebration & Contemplation: *Proceeding of the Tenth International Conference on Design and Emotion. Amsterdam: The Design & Emotion Society*, 2016, pp. 572–576.

Tullis, T. and Albert, B. 2013. *Measuring the user experience: collecting, analyzing, and presenting usability metrics*. Waltham: Elsevier Inc.

Weinschenk, S. 2015. *100 more things every designer needs to know about people*. New Riders.

Fashion, identities and cultures

Semiology: Culture as a symbol system and western influences comparison before and after the independence of 1947

L. Vignes
GD GOENKA University, Haryana, India

ABSTRACT: This paper presents a brief review of the literature. Topics of language, evolving fashion and semiotics analysis of the behavior of the Indian society before independence, especially during the events of 1920 and 1947, and its Western influence. In 1920, the first elections in the history of the country happened in New Delhi. The year of 1947 was tough in India, with the gain of independence from the British crown, resulting in the division of India and Pakistan. Many people died during the partition, as India became a democracy. Based on this information, the first part of this work intends to investigate and analyze the images and the symbolic relation as elements of fashion in the process of signification and memory. In the second part of the article, the objective is to establish the role of semiotics as a communication dynamics, as well as the symbols, their meaning and the way the human being represents the world, giving meaning to it. The article proposes to bring a set of theoretical systems capable of assisting both the process of design projects and the inherent process of analysis of cultural pieces, which can serve as substratum for new creations.

1 INTRODUCTION

Semiotics is defined as a science that is dedicated to the study of meaning in social life. It is a study of the development of contemporary society to a society of consumption and globalization for products and services, as new problems arise with regard to groups of systems, signs, materials and cultures.

The present theme is focused on the reflection of sociocultural statutes. It defines the idea that clothing has significant meaning as a universal code of communication and interpretations. Therefore, for this study the Barthesian Semiology has been discussed, through some categories of analysis: Culture (with identity as a subcategory), and Clothing (with fashion as a subcategory). The observations resulting from this process are subjective, because according to Barthes (1980), in this case, clothing is linked to the contextual baggage of individuals added to their wardrobe.

However, we have as reference the concept of street defined by Da Matta (1997) as a place of individualization, struggle and trickery—a space where powerful relations are instituted and groups dispute geographical or symbolic territories; geographically de-territorialized by multiple identities that travel through real, virtual and imaginary spaces. The word 'street' in this context is a metaphor of the contemporary urban society and not necessarily lies in its relation with the perception of "home" (family relations, private life), but rather as an element of reference for relations, ideas and anxieties lived in the urban space.

Based on human behavior and its social relations, and reflecting the contemporary urban society, dressing the subject as a person means, through the expression of fashion, to introduce the essence that establishes the metropolis as the experience of urban living through the aesthetics of the clothing composition.

Beyond simple analogy, the interpretation of signs from different times and periods serves as the function of expression and representation of social relations between individuals, cultures, and politics. However, the style of clothing reflects the hierarchy, relations of power, status, and positions assumed and shared in the real, virtual and imaginary territories of the streets.

However, the stigma of individualization and personalization of the boundary of territories is the differentiation represented by the code of signs where it is reflected by different types and their style. The individual is worthy of his own means, wills and principles in the mass, and at the same time incorporates his autonomy into the representation that he makes of himself, through the dramatization proposed by the way he communicates to the world and how he wants to be seen, with gestures, speech and the way he dresses. He builds a persona to communicate social values or subjective aspects that he wishes to express to others.

The British influence in India had a strong influence in the behavioral changes of the society, and it has been a decisive factor in the numerous revolutions experienced within the Indian society. We can begin to enumerate the changes by the educational system introduced by Great Britain, which has brought a far-reaching impact in stimulating secondary education through the English language. Improving intellectual abilities was a unique effort by a powerful government for more than a century to educate the upper classes of society, heirs of a civilization that reacted to a religion and a very different social order. Dynamic culture was established through the use of English, used as an instrument of the society bases. English was important as a common tool for the consolidation of thinking: Sanskrit, for example, was the language of thought transmitted in India, but this knowledge was limited to a very small group. The English language reform, on the other hand, reached a national level. It is significant to know that many works that changed Hindu thoughts were written in English.

1.1 Semiology – definition

What we call globalization is defined as a set of processes of homogenization in the society, structured in the context of observation and study of language materiality (verbal and non-verbal). Researchers believe that every contemporary culture passes in some way, revealing the contents transmitted by means of communication (Santos, 2002).

Mass communication has a more concrete meaning in morphology, as the raw material with which one works is formed by collective representations (Durkheim, 1970). Durkheim believes that it is centralized beyond the sphere of individuality—Our collective actions reproduce mass communication triggers, refractory interpretations that do not cover what is published in the decipherment of its meaning.

Particularities such as anonymity, collective acceptance, dynamism, celebration, and ritual are ignited in the folklore and popular culture of each country or society, but these concepts stand out with new meaning during globalization. Like this we assume the value of tradition, but we are not influenced by valuing what is in a dynamic context.

2 THE EUROPEAN PRESENCE IN INDIA

The first to arrive in India were the Portuguese, who landed in the country in the 16th century, when establishing trading grounds along the coast of Malabar. After the Portuguese, came the Dutch, French and then the British. But it was the British who gave the cry for independence. The first British ships arrived in India in 1612. Indian products (silk, cotton, spices, etc.) were exchanged for manufactured goods from Great Britain.

Today, the traditional and the millenarian contrast in different scenarios, as for what it possesses of modern and dynamic as for the historical aspects and traditions. While facing serious socio-economic problems (hunger, poverty, overpopulation, ethnic and religious conflicts, concentration of income, disorderly urban growth, etc.), India is one of the major emerging economies with significant growth rates, which exports computer programs for companies around the world.

The role of fashion in the society serves the social structure, harmonizing the conflict between the individual impulses of each one of us. As a social individual we need the affirmation as a member of a group. – In a group we express ideas and feelings, with a language that translates into linguistics.

Within this context, fashion translates its signs through every season, always with references in the past, present and future, trying to apply an evolution in the behavioral patterns of society.

On comparison of the Image 1 (Woman in vintage Saree) and Image 2 (A bride of the British Raj), one can see clothing as a form of expression, constituting patterns with psychological, psychoanalytical motivational factors and socioeconomic status. This subjectivity reveals the behavior and the way they present themselves in society, in the same instance, proposes the inclusion and/or exclusion of people and groups in their social context.

To demonstrate this conclusion, we observe in the images the western influences in India that was not yet independent. In this context the most apparent differentiation element is the clothing that is more purposely representing the movement of fashion. However, there's a change in the way that dressing has an impact on the whole course of humanity. In this way, the society was structuring itself and showing its value through the western way of dressing.

With the Industrial Revolution, India became only a supplier of raw materials. A great example was the Indian weavings that ended up collapsing with the unfair competition of the imported British ones. Textile expert Jasleen Dhamija moved to Abbottabad, Pakistan, and provided them with chiffon and nylons and mill-made dhotis. Saris could have come from mills or local weavers.

Mahatma Gandhi was a leader who was deeply established in Hindu faith. Gandhi's religious ideas certainly had a profound influence on millions of people from all faiths, contributing to many Hindus becoming better Hindus, Muslims becoming better Muslims and

Image 1.   Woman in vintage Saree (1920).

Image 2.   A bride of the British Raj: Iris Butler on her wedding (1920).

Image 3. The March of The Salt (1930).

Image 4. The Raj & Colonial India.

Christians becoming better Christians. This reflects in this photo in his simple dress, in the white colour conveying the peace and humility of a struggle that would already have its victory.

This research is a brief review of the literature on the topics of fashion, language of signs, semiotics that deepen the analysis of society's behaviour in India, pre-independence (1920) and post-Independence (1947) to the present moment (2017). We perceive the relative change of values and customs which have influenced the designers and the great clothing market, surrounded by Western influence.

With reference from the above image we can say that a reflexive modernity exists as a broad and modifying modernization of the historical structure. Some social aspects can produce opposite effects in cultural and continental matters, such as nationalism, mass poverty,

Image 5. Lidiane Vignes wears an Indian Wedding Dress.

and religious fundamentalism of various strands, economic crises, ecological crises, possibly wars and revolutions. In short, the consequences of events like these can reflect in the sense if dynamism of conflict in a society.

Kellner (2001) says that fashion clothing offers material and models for the construction of identity. It also says that the sumptuous codes and social roles of traditional societies which immediately positions us the social class, profession and status of the person through his clothing and appearance.

The body is form of an expression that portrays the world about who we are an what are our aspirations. It works to the viewer's identity as a self-portrait, which will be translated by the interpreting part as a style, capable of expressing diverse angles of personality and to reflect extreme desires and conflicts that not always are purposeful and rational.

According to Santaella (2002) the sign is capable of provoking feelings, that is, an emotional interpretation. Icons tend to produce this kind of interpretation with more intensity. Emotional interpreters are always present in any interpretations, even when we do not realize them. The metaphorical predominance comes from a connotation, this characterization of reading the sign can be energetic, which corresponds to a physical or mental action, which wants to convey to an analogy between clothing and language.

As an example, the image above expresses elements that constitute the basis for the formation and belonging of a determined group. People do not only have an identity but a dynamics of belonging whose history reverberate a common place shared by the diverse groups that configured and still define their space of existence as individual.

Based on Indian culture we can see this phenomenon frequently, since today India applies to a new concept of globalization. In today's world, clothing has the role to create the diverse identities. Contextualizing the social relations that we are living, the language changes and quickly poses as a challenge for the changes in time and the adaptation of attitudes, beliefs, values, and desires. Yet the enhancement of these changes does not allow the old ones to lose their values, reinforcing the thesis that there are possibilities of coexistence of some.

3 SYMBOLS USED IN VARIOUS CULTURES

The symbol stands out with two interpreting objects. First, the immediate object (the symbol itself), which is how the dynamic object is represented in the sign. Then, the immediate interpreter, which is based on what the sign is capable of producing. In addition to what the symbol produces in one's mind, deriving from his nature, he can produce feelings and actions, and before interpreting a symbol we have to understand it. A dynamic interpreter investigates how symbols influence the individuals.

The second level of the dynamic interpretation is highlighted by a direct action. For example: If someone with authority delegate responsibilities to you, out of respect or fear this action will generate a dynamic energetic interpretation, structured in a concrete and real action of obedience. In this case, the action of obedience is an answer to the symbol.

We are inquisitive when we buy something that we did not intend to buy, or when we perform the appropriate behavior in traffic, corresponding to the traffic signs—According to Bense and Elizabeth (2000), we are guided by such signs. The way in which a sign stimulates its interpreter reveals the dynamic interpretation. When a sign inspires someone, this brings the influence as meaning.

The media has a lot of power over the interpreter, especially when we consider how the semiotics acts in the social behavior. Sometimes we do an immediate purchase when we see a good publicity, i.e. the signs have an impact over the actions of the interpreter.

Stylists and designers are great code interpreters who point out for product projections. In the case of clothing and accessories, they are often influenced by people on the streets or clubs, thus "they are not only indications of the emerging styles on the streets, but also costumes used by people with a special talent to manipulate the signs of clothing" (Crane, 2006).

4 STEREOTYPES

The stereotype is not related to the appearance. It is based on future plans and repeated descriptions that build articles for the public. It is very strong in area of fiction (comedies, comic books, novels, films, etc.) because they are easily recognizable and marketable. Stereotyped groups of people form different urban tribes, which are heavily targeted by the media. The society is able to understand and deal with stereotypes, but the power of the media manipulates their vision, repeating large-scale prejudices (Burton and Dimbleby, 1998).

A person's feelings, attitudes and intentions are reinforced by body language. This includes five elements: the gestures, the expression, the posture of the body, the space and proximity of the body and the touch. In addition to the gestures we have other forms of expressions, such as: whistling or gasping, guttural sounds, shouting, loud voice, tone variation, timbre, etc. We can structure these elements in the paralanguage: nonverbal signs that accompany speech. The clothes reveal a lot about the personality, situation, status and work of the people, subscribing to the identity and the groups to which they may belong. (Burton and Dimbleby, 1998).

5 WESTERN INFLUENCES ON INDIAN CULTURE

The western influences can be seen as a set of elements that are put in contrast, when we establish on individual clothing relations. Syntagmas, according to Malmberg (1974), are hierarchically organized structures and composed of discrete elements of different value. This research focuses on the syntagmas of clothing and fashion between distinct cultures of time period.

Roland Barthes (1964) discusses the elements of semiology as the form simply and coherently, the substances and the set of aspects of linguistic phenomena. Semiology sometimes leads us to a system, in which meanings are substantiated. As defined by Roland Barthes, we are faced with a system of objects that do not fulfil the role of immediate and functionally significant subsistence. Yet, it offers its own system at a certain utilitarian level. For example, the wetsuit serves to signify a situation of the aesthetic point, but also to serve to protect itself from the cold.

6 SYMBOLISM MEANING, LINGUISTICS, CODES AND CULTURE APPROPRIATION

The analysis of the importance of a culture for the development of the country in the view of manifestations and cultural expressions is totally pedagogical and formative as a factor of cultural identity. The driving factor of the time that fits in (modernity) takes us to a sphere of systems economic, administrative, educational and social systems. In this case, there is much participation in this scenario.

The various cultural symbols have always helped in identifying the oppressed among themselves for the struggle. The appropriation of what comes from the dominated class is never a spontaneous initiative of the dominant class. It is always the result of a struggle and, to some extent, the imposition of the will of the burdened sectors. With cultural appropriation it is no different.

For Fortin and Prévost (2003) "local development is first and foremost an organic process, a human phenomenon." According to Lipovetsky (2002), no theory of fashion can be restricted to the factors of economic and material life. Even if important, these phenomena in no way account for the incessant variations and excess of the fantasies that properly define fashion. That is why everything proposes us to think that it finds its force more in the social logic than in the economic dynamics.

Lipovetsky (2002) argues that we try to insert ourselves into social circles through what we wear, that concept of inclusion and social exclusion can take a broader look, from the signs left through what the society wears. After all, choosing clothes to wear is to convey a message to the world, which can either print a message from everyday life or relate a situation that is not true. The knowledge, techniques, practices and values that allow the meanings of clothes and their uses to pass between the casual and the pragmatic is what we call fashion.

7 CONCLUSION AND RECOMMENDATIONS

As a research approach to the interactions between Indian cultures with Western influences, on one hand we break with the archetype of the fashion system, on the other, it creates a nomenclature to categorize and determine the object study, differentiating it from other expressions of fashion communication existing in the large garment industry across the globe.

Fashion is a social phenomenon in postmodernity, which gradually transfers and follows the evolution of contemporary society. However, reason is fluid; it changes continuously by inserting values, meanings, attributions that constantly change, so that the associations to the term semiology in fashion within the specialized literature on the subject are diverse and not always in depth. It recognizes that the content, widely studied in this thesis, and visibly explored in the academic universe as concept, abstraction, idea, is as part of the cultural phenomena of postmodernity, a recognized concept as being hybrid, indefinite, liquid and deteriorated.

Fashion has acquired, in the last few decades, new status, depth on the subject and has established itself as a showcase of the contemporary urban society. Not because fashion brands what people wear on city streets or because it "reproduces" the movements clearly expressed by youthfulness, but rather because we have a different context and discourse that approaches art and philosophy in a symbiotic relationship with these movements.

Apart from the definitions and concepts of fashionable semiology, after months of research in the master's degree, we note that none of the definitions for this content found in theoretical and field research has dominated our intellect that has influenced the creating the collection.

Culture, as an idea, customs, and social behaviour of a particular society, is the symbolic representation of the relationship between the past and the cultural acceptance of fashion. This analogy, as fashion approaches art and philosophy "dresses", inserts important values, meanings and mass movements that take place in modern Western society, including elements of contemporary culture. In its cultural capital, fashion and trends dress us (as individuals, in postmodern cities), enclose us into a symbolic context of the culture and memories of the streets, the metropolis, the hybrid, fluent, of this mosaic of values that are translated in aesthetics and in clothing as a form of communication of ideological, philosophical discourses. And in this same aesthetic structure, the discourse on the reality lived in the disorderly and inconstant western city translates into symbolic dimensions the unfolding of our own social and cultural history.

REFERENCES

Abyoga 2017. *A Civilização dos Indus*, [Online], Available: <http://abyoga.com.br/a-civilizacao-do-indus/> [2 May 2017].
Bauman, Z. 2004. *Amor Líquido: Sovre a Fragilidade dos Laços Humanos*, Rio de Janeiro: Jorge Zahar Editor.
Barthes, R. 1977. *Elements of Semiology*, Paris: Hill and Wang.
Cavalcanti, M.L. 2006. *Carnaval Carioca: dos bastidores ao desfile*, Rio de Janeiro: EdURFJ.
Dimbleby, R & Burton, G. 2007. *More Than Words: An Introduction to Communication*. Paperback, Abingdon: Routledge.
Domingos, S.F. 1997. *Intuition: form and substance.* (Phonetics and Phonology), [Online], Available: <http://www.profala.com/arttf17.html> [2 May 2017].
Ewald, A and Soares, J. 2007.Identidade e Subjetividade numa era de Incerteza, *Estudos de Psicologia*, vol. 12, pp. 23–30.
Georgina O'hara, C. 1986. *Enciclopedia da Moda: De 1840 à década de 90*. São Paulo: Companhia das Letras.
Kenoyer, J.M. 1991.Ornament styles of the Indus valley tradition: evidence from recent excavations at Harappa, Pakistan. *Paléorient*, 79–98.
Lipovetsky, G. 1994. *The Empire of Fashion*. New Jersey: Princeton University Press.
Lóssio, R & Pereira, C. 2007 A Importância da Valorização da Cultura Popular para o Desenvolvimento Local. *Terceiro Encontro de Estudos Interdisciplinares em Cultura*, Salvador.
Nóbrega, M.R. & Figueiredo, J.G. 2008. Semiotics in fashion: an approach on non-verbal Communication through the dress of young people, [Online], Available: <http://www.intercom.org.br/papers/nacionais/2008/resumos/R3-0630-1.pdf> [2 May 2017].
Santos, M. 2009 Notas Sobre Moda, Exclusão Social e Educação, *Revista Achiote.com–Revista Eletrônica de Moda*, v. 2, n. 1.
Saussure, F. 1916. *Course in General Linguistics*. Chicago: Open Court Publishing.
Werneck, M. 2008. Roland Barthes, The Fashion And The Signature of the world. *Magazine of Fashion, Culture and Art*, v.1, n. 1, p.118.

IMAGE REFERENCES

Unknown, (1920), Woman in vintage Saree [ONLINE]. Available at: https://luxemi.files.wordpress.com/2012/08/screen-shot-2012-08-02-at-10-41-31-am.png [Accessed 5 May 2017].
Unknown, (1920), A bride of the British Raj: Iris Butler on her wedding [ONLINE]. Available at: http://i.dailymail.co.uk/i/pix/2012/07/06/article-0-13F1F7CB000005DC-648_306×595.jpg [Accessed 5 May 2017].

A geography of fashion, Medellín: 1900–1950

W. Cruz Bermeo
Universidad Pontificia Bolivariana, Medellín, Antioquia, Colombia

ABSTRACT: This article explores the development of Medellin city (Colombia) urban commerce of fashion at the first half of 20th century, analyzing the configuration of its geography of fashion. In doing so, the article plots a cartography of city places and spaces transformed by Medellin human masquerade into emblematic settings for the performance and fashion consumption. In a chronological order, the article unveils a shopping map starting at Berrío square, follows to Colombia St. ending on Junín Av. Argues that despite the historical importance of Berrío square and Colombia St. in configuring a territory for experimenting fashion as a commercial practice and cultural, Junín Av. occupies the popular imagination as the first fashionable promenade in Medellin, therefore, addresses to the role of former both places in the configuration of Medellín's fashion geography, in the context of a town undergoing a transformation: going from being a small village to a modern city.

1 MEDELLÍN'S TRANSFORMATION: A NEW CONTEXT RISING A NEW FASHION BEHAVIOR

At the beginning of the 20th century, Medellín (Antioquia)—founded in 1675—moved from an agrarian and mining economy to an industrialized economy. The coffee export and centralization of gold trade were important sources for income, in the new industrial economic panorama the textile manufacturing was one of the pioneer industries. In Colombia, the century began with the War of a Thousand Days, a civil conflict that lasted from 1899 to 1902. While other regions in the country were involved in the war Antioquia did not fully engage with, and it favored the Medellin industrial rise and its dominance in the field of modern industries. As capital of the region, wealth was concentrated to a large extent in Medellín. With the industrial dominance came an economic boom that, in addition to social mobility, brought substantial changes to the city physiognomy, due to the availability of new urban devices, the public spaces opening, and the creation of semi-private places reserved for elites.

As can be seen, Medellín was undergoing its transition towards modernity, by transforming from a small village to an industrial city. This process would take the first half of the 20th century (Botero 1996: 99). Precisely because of economic growth and social mobility, the new wealthy industrialists and their families demanded sumptuary goods with the potential to represent symbolically their new status. But it not only changed consumption practices of wealthy but also those of ordinary inhabitants, since, in addition, notions of civilization and progress were gradually widespread throughout society. Those notions were articulated by adopting fashions, attitudes and manners from industrialized countries and fashionable cities—There Paris shone with blinding effulgence for the paisas's eyes. In doing so, it was supposed where regional customs and dresses prevailed those ideals had fail to get success (Cruz 2016).

The sophisticated apparatus for imports, which paralleled the foreign trade of coffee and gold, included luxury and dry goods imports to satisfy the new demands. People with not a big budget had imitation as strategy to be in vogue or get fashionable manners, it was possible to get easily edgy fashion magazines, credit terms to buy a sewing machine or just gazing at the spectacle offered by others who you could consider stylish. With actors dressed to perform their parts were necessary to have places and spaces to stage the sartorial display and consumption spectacle. The process started with opening a first *grand magazine* at the financial and commercial core of the city: Berrío square. After that, there would be a growing opening of spaces for urban selling of fashions, configuring a real geography of fashion, emblematic streets for staging, consumption and placement of novelties.

2 THE GEOGRAPHY OF FASHION IN MEDELLÍN

In fashion studies the emergence of certain urban routes that concentrate fashion commerce and even are considered emblems of the global fashion industry capitals have been examined (Crewe 2017; Potvin 2009); such as rue Saint Honoré or 5th Avenue, in Paris and New York, respectively. London and Milan, also have been in fashion studies literature, due to the global recognition of their geographies of fashion (Breward 2003; Steele 1998). David Gilbert argues geographies of fashion are not simply collections of boutiques but places where fashion is displayed, watched, imitated and transformed by those who live and follow its game. And the rise of certain streets as fashion cores is dependent on their role as performative spaces. Thus, a geography of fashion also includes meeting places that allow citizens to stage, observe and compare each other. In its configuration, the spectacle they offer is fundamental, and the emergence of such a geography cannot be understood without reference to the intersection of numerous "cultural and economic processes bound up with the development of the modern city" (2000: 12–15). As explained above, the period 1900–1950 is for Medellín an era of urban transformation derived from economic processes, which can be understood as a base for the adoption of new fashions, attitudes and manners, and with it the emergence of a local geography of fashion.

In Medellin (1900–1950) the geography of fashion was known generically as the "Commerce". As Sofía Ospina de Navarro explains, being a witness of that time, by writing that "commerce" did not include "precisely the group of department stores of serious items, but that of the attractive fashion boutiques, with their tempting windows", and places surrounding areas such as theaters, beauty salons, ice cream parlors and record stores. Shopping was not always the main reason to go to commerce, flirting and idle walking were also great motivations. In this regard, Ospina wrote, referring to the young women: "it does not matter if you bring home only a new lipstick when you return home, if in your mind you see the image of a good-looking suitor…" As for the older people, she explains, a walk through the commerce was even more interesting because "this is the stage where the human troupe" acts, where "a variety of characters and costumes distract the eye and make people meditate on the unfolding of social plot" (1983: 17). In words of Gilbert, there the citizens of Medellín were staged, observed and compared to each other. The commerce was that place to stage the drama of seeing and being made to seen. Between 1900 and 1950, three urban spaces congregated the "commerce" made up of fashion stores, serving as the setting for elegant strutting in a village that gradually took on the face of a city: Berrío square, Colombia St. and Junín avenue.

2.1 *Berrío square, a place for the first Medellín department store*

In 1864 Charles Saffray, a French traveler, chronicled on the role of Medellin in the economic and commercial dynamics of the region. It was the collecting center of the gold coming from provinces mines. Medellin did not export, but it uses to import "large quantities of merchandise" to be distributed among neighboring towns. As for its Plaza Mayor, as Berrío square was then called, Saffray commented that the lower floors, formerly inhabited by the wealthy classes, were occupied by warehouses (Álvarez 1996: 73). This transformation in the use of space indicates the early financial and commercial orientation of Berrío square. With the concentration of market and wealth there, it was natural that the first shops dedicated to fashion commerce where opening in the heart of social and economic power.

One of those first establishments for fashion commerce was "Gran Almacén el Salón Rojo". A 1910 photograph allows to examine its exterior, it register a colonial building with two floors, the first occupied by a drugstore. In the second floor, an old colonial balcony had been transformed into a stained-glass window topped with an art nouveau cornice. A discreet sign bearing the store name also has the inscription followed: "Son of Pastor Restrepo & Co.", it was the name of a commercial company that Julio Restrepo Lalinde, owner of the drugstore and warehouse, had inherited from his father.

The opening of the Gran Almacén el Salón Rojo is estimated about 1908. That year the businessman Ricardo Olano travels to Europe in order to buy machinery and seek human capital to establish a match factory in Medellín. Doing his route went to Berlin where he made "purchases of clothing that [gave him] very good revenues". Upon returning to Medellin, he wrote in his memoirs: "I had good success with the goods I bought on this trip. Especially with finery, I sold them quickly because they were just establishing large fashion stores [...] that bought us big lots", among them "El Salón Rojo" (2004: 41-45).

When it comes to fashions it is likely this and other future stores were offering and selling German clothing as French fashions, as the Medellin urban people favoritism for French fashions would overshadow any fashion produced in Germany. In *Frutos de mi tierra* Tomás Carrasquilla warns of such favoritism by pointing out "the chic Parisian" of "the *maiceritos* [people who eat corn cakes]" and "the French Antioquia girls [...] masquerading according to the last fashion illustration" (1896, 68-67). In addition, the atmosphere of the Gran Almacén el Salón Rojo and its translated denomination of *grand magazine*, indicate an aspiration to set a space that would be experienced by costumers as a French place dedicated to fashion from France.

The Parisian *grands magasins* were the inspiration for El Salón Rojo. Emerged in the mid-nineteenth century, these commercial spaces offered a new and pleasurable shopping experience, transforming the way public used to relate with commodities. According to Wilson, for the first time you could enter the store and browse without an obligation to buy, you did not need the cumbersome haggling because prices were fixed. All goods could be touched because there was not a counter interposed between commodities and the customers. More than a simple necessity, shopping in the *grands magasins* was a playful, idle and elegant activity, therefore, they became a stage for displaying the self (2003: 146-152).

Gran Almacén Salón Rojo visual records show that it retained, at least in part, the general layout of an old warehouse. But other goods were exhibited after a Parisian *grand magasin*, allowing customers a direct contact with goods and inviting them to go through the entire space. Labeled products suggest that the fixed price strategy was embraced. The variety of goods indicates that it was governed by the "unexpected juxtaposition" of objects, under the rule that in the exhibition "instead of a hundred vases of the same size and from the same manufacturer, there would be only one sample placed next to another vase of different form" (Sennett 2011: 182), a scheme of the *grands magasins*. However, the Salón Rojo had a general appearance of small bazaar of village.

With a first space to experience fashion as an economic and cultural practice, due to new interactions fostered by a store such as the Salón Rojo, Berrío square vocation as a place for fashionable shopping and its potential as trade zone for other fashion establishments the vocation were now addressed, since, as Entwistle argues by paraphrasing to Foucault, "the layout of a building and the spatial arrangement of streets impact upon the movement of bodies, directing the individual and organizing the flow of the crowd to particular sorts of activities, relations and patterns" (2000: 229).

In the 1920s other fashion stores had been installed at Berrío square. By 1921 there was the Angelina and Leonor Molina fashion room. These businesswomen imported "special items for women and children dresses." By 1926, there was already the Canuto Toro M. Fashion Store, which had a "Fashion Atelier" run by Pastora Jaramillo de Misas. It is worth noting the role of these women in fashion business; they are not humble dressmakers or seamstresses at home, they participate as eminent businesswomen, since the commercial and financial importance of the Berrío square supposes fashion business located there were not a matter of minor figures.

The Salón Rojo had given to Berrío square its position as an elegant, elite zone where to stock up on and up to date the latest fashions. But at the beginning of the 1930s there you could find fashion stores differentiated by the purchasing power of its clients. The Gabriel Toro establishment, for example, used to advertised values such as distinction and refinement without referring to prices. In the meanwhile, La Feria de Paris, where prices apparently were something relevant for its customers, was advertised as "the best fashion store for the Medellín ladies" with the "cheapest prices" (Pérez and Jaramillo 2004: 272, 360). Modest prices were stimulating the fashion consumption, because "the cheaper things are, the more they incite consumers to change fashion, the more they force producers to create them" (Simmel 2008: 92–93).

By the 1930s the Berrío square maintained its position as a place for distinguished fashion stores, but now it was sharing its status with other surrounding streets that became specialized in the fashion trade, turning into a meeting place. Such as Colombia St. and Junín avenue.

2.2 Colombia St., a stroll du bon tone

The south side of the Berrío square is crossed by Colombia St. On this dynamic street an elegant establishment was also located around 1908, "El Buen Tono, Gran Almacén de Modas". Its name, derived from the French expression *bon ton*, alluding to the world of the refined and stylish people, therefore, suggests a store dedicated to deal with and satisfying the local social elite.

El Buen Tono visual records show an establishment with a French-style interior decoration contrasting with rusticity of the exterior space, that of an unpaved Colombia St. flanked by predominantly colonial architecture. This *grand magasin*, a department store by definition, had several galleries whose furniture scheme was repeated in each one. Between art nouveau-style wall shelves, tables and shelves, shoppers could go wandering through hats, fabrics or home decorative objects, mixing with toys and purses. These galleries did not have the spectacular exhibitions of the *grands magasins*; nevertheless, they were ruled by principle that all the store should be visible at first glance, encompassed with a single gaze. The idea according to which department stores had decadent architectural formations, in order to welcoming the crowd and retaining it to seduce it (Benjamin 2005: 75), is not evident in the visual records of El Buen Tono. Its galleries were rather the simple rooms of a colonial house modified for commercial purposes, nothing there looks grandiloquent or decadent. But maybe to the eyes of contemporaries the most grandiloquent thing in the store were goods itself. The absence of a national production of luxury goods, coupled with curiosity and waiting, made the goods itself the show offered by the store. Costumers did not need more seduction than seduction rouse by the thing itself.

El Buen Tono would turn Colombia St. into a fashionable promenade for shopping, as El Salón Rojo has done to Berrío square but unlike this square, where financial and corporate entities dominated, most of the commercial activity on Colombia St. consisted of "dry goods", that is, luxury items such as textiles, footwear, dresses, hats, jewelry, and decorative objects produced in Europe and United States. Literary and visual records illustrate the variety of stores specialized in this kind of items that were installed in Colombia St. as early as 1900–1910 decades. Due to the concentration of this branch of commerce on Colombia St. one can argues the first fashion promenade was configured there. It was the fashion street in Medellin by then. But it was not a fashion street as large as the whole Colombia St. that we know today, the fashion promenade was confined to a Colombia St. section extending from Junín to Palacé avenues.

This crowded section of Colombia St. used to bring together retailers coming from the towns of Antioquia to stock their stores, also the lady who wanted to be surprised with something new or updated in the latest fashion; and some El Buen Tono lady customer or El Salón Rojo who went to pick up orders or modify their jewelry to make a new one, since there were wholesale and retail stores for women's fashion, and other establishments providing services related to the adornment personal.

These stores were powerful advertisers in newspapers, directories, guides and commercial almanacs of the region. It comes to mention some of them and review what the stores were offering, to learn their variety of goods and services. There was the Leocadio M. Arango &

Hijos commercial company, with a jewelry shop and an apparel shop. The jewelry offered a luxurious range of products categorized as "items for presents", in which decorative objects such as glassware, mirrors and silverware entered; items for personal hygiene, such as "toothbrushes, nail brushes and for the head"; and body adornment such as "fine perfumery", combs, combs and ridiculous were other categories. It is interesting to observe how the sumptuous and variegated nature of these goods did not limit to body adornment the concept of jewel. Then there was the clothing store that, in addition to imported fashions, was specialized in fine cloths, "American footwear for ladies" and white goods. This commercial company integrated the sale of sewing machines to textiles sales. It was an ambitious commercial strategy because by increasing the number of sewing machines in homes, the potential for textile sales increased in 1906 they claimed to have sold eight thousand Wheeler & Wilson sewing machines throughout Antioquia (Silva L. 2011: 264).

El Guante Blanco store stood out on Colombia St., its motto was "to dress well, with elegance it is a sign of distinction". A few doors away from there was the La Perla jewelry, in its installations there was an optics also. On the opposite sidewalk, the Alejandro Echavarría & Hijo warehouse, founded in 1904, with its counters full of "a great assortment of foreign goods". On the same sidewalk, were Gabriel Latorre & Hno. and Eusebio A. Jaramillo, two women's fashion stores and importers of "fine American footwear" for women. These were contiguous to Sombrerería Francesa, a commercial company founded in 1857, dedicated to the exports of panama hats and importations of articles for men, especially felt and wool hats (Silva L. 2011: 318); this store, still active in 1950, occupied different places Colombia St.

Fashions for men also had their space on the vibrant Colombia St. Nevertheless, in spite of the international prestige of the men's fashion produced in London, in Medellín, stores with French names but specialized in men's fashion were dominant: Le Boulevard, owned by Luciano Villa; Le Bon Marche, by Saldarriaga Hermanos and Le Louvre, by Dionisio Ángel, are examples which ratify the domain of an imaginary where the Paris name works as metaphor for a prestigious and cutting-edge fashions.

By the early 1920s, these stores had helped to expand the geography of local fashion, concentrating on Colombia St., particularly on section from the avenues Palacé to Junín. However, other stores, such as Santander and Le Bon Marche, were installed in the same street but outside that section (Botero & Sáenz 2006: 112–326). Progressively, in other sections of Colombia St., impressive republican constructions began to be constructed, which acted as indices of a flourishing via due to its hectic commercial activity and variety of spaces for the consumption of luxury goods. For example, in the buildings on the north side, from Bolívar to Carabobo avenues, there were stores of commercial companies that included hardware, mining and electricity, giving way to furniture and tapestries, such as the American warehouse. Also stores of decorative varieties, such as the Chino-Japonés store with its offer of oriental fantasies. Or the prestigious La Marquesa jewelry store, the place to stock up on Guerlain perfumes and Shalimar facial powders. Colombia St. had a wide offer, so it did not exaggerate a 1923 propaganda album that, announcing the Europa hotel, said it was located in "the most commercial part of the city" (2006: 17). However, days of Colombia St. as a destination for fashionable shopping were being overshadowing, due to the new urban dynamics that were emerging at Junín avenue. Junín will be a crucial backdrop, installed in the popular imagination as the most important Medellin fashion promenade from early to mid-20th century.

2.3 *Junín, the avenue for fashionable people*

The "beautiful and wide Junín avenue" was "the preferred and favorite walk of the good and simple neighbors of the village", "the *rendezvous* of the most elegant Medellin's high society". The meeting used to occur in afternoon hours when the stylish people crowded Junín sidewalks to visit and admire "the sumptuous stores commerce" that arranged their merchandise in "striking and luxurious windows" (Rodríguez 2004: 44). Shopping was not the only occupation for passersby because Junín offered theaters, clubs and trendy cafes. It was a center of social life, a convenient platform for elegant strutting after watching a movie in Junín or María Victoria theaters or on the way to the Club Union, to take a break in the Astor tea room after wandering through department stores such as La Primavera or Parisina boutique, or simply after a pious retreat in Villanueva cathedral or La Candelaria church.

That image of a Junín avenue overflowing with urban life and consumption experiences began to take shape at the end of the 1920s. Previous visual records show a peaceful and residential Junín, altered only by religious processions or by car parades on a road flanked by houses in style colonial and a few in republican style, free of flashy advertising. Little has been written about the impact of opening, in 1924, Junín theater and the Europa hotel on the urban dynamics of this avenue, neither about its paving in 1925 (Betancur 1925: 200, 275) or the moving, in 1927, of the Unión Club to the house that would occupy during more than seventy years. However, not surprisingly, the massive concurrence to the different activities of these new scenarios brought a new vitality and the flourishing of new spaces for commercial and leisure activity.

While a select group of Union Club members enjoyed its exclusive rooms, a majority of citizens enjoyed the Junín theater, attending the premiere of some Chaplin film or a new addition to the Cine Colombia repertoire with the silent movie star Pola Negri, as Imperial Hotel (*Letras y Encajes*, April 1928). This new reality could contribute to transform the environment and the dynamics of the Junín avenue, transforming it into a frenetic urban stroll where different establishments for leisure, entertainment and fashion shopping would be installed.

In the decade of 1910 there were few shops in Junín, compared with those in Colombia St. The number of fashion stores began to grow in late 1920s: El Buen Gusto, Salón Blanco, Salón Francés, the dressmaker Lola Díaz Granados and several tailors (Botero & Sáenz 2006: 165–327) were among the new fashion commerce. In addition, the Ana Escobar Uribe store, whose adverts appeared in the *Letras y Encajes* magazine announced this businesswoman as an importer of the "finest and most elegant assortment of dresses, hats, shoes, socks, gloves, underwear etc." The merchant Vicente Restrepo R. had also in Junín his store Modas de París, with a section for the making of "suits and coats in all styles, according to the latest models of Paris" (*Letras y Encajes*, January 1927; July 1927). There was also the Sastrería Francesa de Posadas y Cía., announced "the largest fashion house in Antioquia" (*Letras y Encajes*, June 1927).

During the 1930s Junín is the epicenter of a luxury cosmetic arts flourishing in the city. Distinguished beauty salons add shine to nature of Junín as a fashionable urban stroll. The Beauty and Azul, Marc and Paris salons are counted. In the Beauty, owned by Ana Echavarría Vasco, permanent curls and wave combs by hand, cuts and massages could be obtained. The Blue claimed to be "a salon for elegant ladies" and its advertising showed neatly uniformed women attending to customers; hairstyles, massages and manicures were other services of the Azul. In 1938, a certain Marc puts at the disposal of society ladies a "new, spacious and modern beauty salon" that gives "to each face its accurate hairstyle"; as an added value, the Marc beauty salon had a French manicurist and a perfume shop. Then in 1941, Maruja Gómez de Londoño was pleased to announce the premiere of "her modern PARIS BEAUTY SALON", attended by herself (*Letras y Encajes*, 1938–1941). This period of blossoming of local cosmetic arts coincides with a global concern for beauty rituals that transcended reparative makeup to include facial cleansing, sports and a balanced diet as beauty methods (Seeling 2000: 170).

Nothing would be Junín as a fashion artery if its commerce were limited just to fashion stores and beauty salons. In 1930, the Astor tea room opened its doors and, more than a prosaic tea room, it would be a new space for socialization, a place to display the dressed and fashioned body. The Noel tea room also was opened. Both rooms were listed as Medellin "social centers" in a 1932 guide, with the indication not invitation was needed to enter there; on the contrary, to get access to clubs such as the Club Union, stranger people required an invitation. These tea rooms offered an experience comparable to that of sophisticated club rooms reserved for the commercial and industrial elite. Although at Junín already existed, at the beginning of the 20th century, other meeting places, these spaces were male domain; the visitor's experience was not comparable to the refined sociability of a tea room with a large female presence.

A The 5.30 o'clock tea, a traditional practice belonging to private sphere, was transferred to the new public tea rooms. This meeting had ceased to be an intimate visit of confidence where women chatted and sewed to the heat of a cup of chocolate, becoming a careful social staging whose meticulous planning included even the appearance of the maids, who, dressed in white apron and embroidered coif, announced to lady's guests the tea was already served. The 5.30 tea used to be with an excessive luxury, denounced in 1935 Sofía Ospina de Navarro in her chronicle *Los tés elegantes* [The Elegant Teas], to the extent of sensible women to economic hardships had renounced to held in their homes this traditional encounter; consequently, they had renounced also to an important part of their social relations. Others, less willing to do so, preferred to subtract apart from the already meager family budget to stay active in the social scene (*Letras y Encajes*, April 1935). So, the public tea room, being less demanding than a tea in the privacy of home, ended with the painful nervousness of hostesses careful to please their guests. As a result, for many women, going to Astor or Noel could be a practical option, coherent with modern times and the economic crisis of the 1930s.

In the 1940s, other fashion stores had joined the commercial landscape of Junín. La Primavera was not an exclusive store of fashionable clothes but rather a departmental store; it sold garments of its own brand, but also clothes produced in New York and Los Angeles. At the end of 1940s, there were two more branches of La Primavera, both at Colombia St., but fashionable clothes were exclusive to the store situated at Junín avenue. As a witness of his time, the businessman Ricardo Olano counted this branch among the "luxurious and well presented" stores that existed in Medellin in 1942 (2004: 689). Strengthening the status of Junín as the core of fashion in forties, there were also other fashion stores like Casa Christian, París-Moda, Parisina, Femina and Marilú.

Any passer-by could be captive by the visual spectacle given by a number of windows thematically arranged, either by the annual window competition promoted by the Public Improvements Society or by the voluntary impulse of merchandisers to catch attention from pedestrians. The contest of 1945 generated enough enthusiasm although there was less frenzy because "the windows live in permanent competition". Obviously, windows contest put in competition something more than fashion-related businesses. The 1945 window contest awarded with the first position to Elospina, a warehouse of metallic furniture; the photographic office of Jorge Obando, the La Primavera store and the Paris jewelry store, both located in the Junín avenue also were awarded (*Letras y Encajes*, February 1942, 1945).

The success of Junín as an avenue overflowing with urban experiences that allowed to stage the game of fashion also depended on the spectacle offered by the dressed bodies circulating there. Nothing captures better the essence of this spectacle than the popular walking portraits taken during the 1950s to unsuspecting passers-by circulated in Junín. The people in charge of these shots were photographers serving to photographic laboratories in the search for new clients, taking candid snapshots to passers by. Days after that, the portrayed would pick up their snapshots to laboratory and, in the process, they were offered copying services and photo studies (Vélez 2009:149–164). These walking portraits suspend the movement, freeze the transience of moments while leaving a record of fashion in action. In a way, the work of those photographers can be interpreted as that of the *poetae minores* who, according to Baudelaire, captured "beauty in particular, that circumstantial beauty and with certain traditionalist traits" (Baudelaire 1974: 79–124).

3 CONCLUSION

A geography of fashion is not confined to a set of streets where fashion stores and spaces for urban leisure are found. An important part of that geography is the spectacle provided by sartorial display of passers-by and the way fashion consumption is staged, both, in windows and different manifestations of the dressed self. On the other hand, the emergence of these geographies of fashion is closely linked to economic history of regions and the place such geographies occupy in local economic development. There is no need for a city to be a capital of international fashion industry, such as London or Paris, to configure spaces and places for consumption, designed to stage the drama of seeing and being seen. If such configuration corresponds to economic processes of the modern city, there capitalism triumphs and sumptuary consumption will be installed as one of its derivatives.

In Medellín case, early twentieth century is a crucial moment in the process of shaping its geography of fashion, given that it is when Medellin starts an industrialization process that will allow it to go from being a small town to be a modern city in development. This process extends until the 1950s and roots in the current city downtown and its old town. Until that period downtown will be the nerve center of commerce, leisure and sartorial deployment; and, regarding fashion consumption, the Berrío square, Colombia St. and Junín avenue will constitute the Medellín fashion geography.

Concentration of stores specializing in fashion in these three places is not simultaneous. Berrío square will be held the first department store and its role as a center for local fashion consumption will be taken by Colombia St. about the 1920s. At the end of that decade, the urban transformation of Junín would bring the opening of new entertainment places such as hotels, theaters and cafés. These places and urban dynamics developed around it, would give to Junín avenue the mythical place it has today in the popular imagination of Medellin people as "the rendezvous of the most elegant Medellín's high society". However, it should be considered that fashionable shopping took place first in the Berrío square and Colombia St. It was there where first opened all those "stores of fashions and women", where the "swarm of wealthy ladies" used to flutter, as Carrasquilla writes (2008: 679).

REFERENCES

Álvarez, V.M. 1996. Poblamiento y población en el Valle de Aburrá y Medellín, 1541–1951. In Jorge Orlando Melo (ed.) *Historia de Medellín*: 57–87. Bogotá: Suramericana de Seguros.
Baudelaire, C. 1974. *El pintor de la vida moderna*. In S. Clotas (ed.) El dandismo: 79–124. Barcelona: Anagrama.
Benjamin, W. 2005. *Libro de los pasajes*. Madrid: Akal.
Betancur, A. 1925. La Ciudad. *Medellín en el 5º cincuentenario de su fundación: pasado-presente-futuro*. Medellín: Tipografía Bedout.
Botero, F. 1996. *Medellín 1890–1950: Historia urbana y juego de intereses*. Medellín: Universidad de Antioquia.
Breward, C. 2003. *Fashion*. Oxford: Oxford University.
Carrasquilla, T. 1896. *Frutos de mi tierra*. [On line] Available at: http://bit.ly/1Siy3xZ [Last accessed: March 23, 2016].
Carrasquilla, T. 2008. *Obra completa*. Medellín: Universidad de Antioquia.
Crewe, L. 2017. *The Geographies of Fashion*. London: Bloomsbury.
Cruz, W. 2016. *Grandeza: rastros de la moda internacional en Medellín: 1890–1950*. Medellín: Universidad Pontificia Bolivariana.
Entwistle, J. 2000. *The Fashioned Body: Fashion, Dress, and Modern Social Theory*. London: Polity.
Gilbert, D. 2000. Urban Outfitting: The City and the Spaces of Fashion Culture. In S. Bruzzi & P. Church (eds) *Fashion Cultures: Theories, Explorations, and Analysis*: 7–24. London: Routledge.
Olano, R. 2004. *Memorias: 1918–1935*. Medellín: Fondo Editorial Eafit.
Oppenhaimer, H. 1935–36. *Álbum de la propaganda de la ciudad de Medellín*. Medellín: Bedout.
Ospina de Navarro, S. 1983. *Crónicas*. Medellín: Susaeta.
Pérez, L.F. & Jaramillo, E. 2004. *Medellín en 1932*. Medellín: Instituto Tecnológico Metropolitano.
Potvin, J (ed.) 2009. *The places and spaces of fashion, 1800–2007*.New York: Routledge.
Seeling, Ch. 2000. *Moda: el siglo de los diseñadores, 1900–1999*. Colonia: Könemann.
Sennett, R. 2011. *El declive del hombre público*. Barcelona: Anagrama.
Silva L., I. 2011. *Primer directorio general de la ciudad de Medellín para el año de 1906*. Medellín: Universidad Pontificia Bolivariana.
Simmel, G. 2008. *De la esencia de la cultura*. Buenos Aires: Prometeo.
Steele, V. 1998. *Paris Fashion: A Cultural History*. Oxford: Berg.
Vélez, G.M., 2009. Las historias mínimas del anónimo transeúnte. Breve reseña de un episodio urbano. *Co-herencia*, July-December: 149–164.
Wilson, E. 2003. *Adorned in Dreams: Fashion and Modernity*. New Brunswick: Rutgers University.

The *Prêt-à-porter* phenomenon and the boutiques in the dowtown area of Santiago, Chile

J. Vidal Miranda & P. Álvarez Caselli
Pontificia Universidad Católica de Chile, Santiago, Chile

ABSTRACT: This proposal analyses the first arrival of the *Prêt-à-porter* ("ready to wear") at a sector of the capital city of Chile. This new form of production brought about the installation of a net of boutiques in the eastern zone of Santiago, mainly in the district of Providencia, a phenomenon that was unknown in the national medium. The research analyses a period of two decades (1967–1897), where at first a convergence is produced between the national textile production and the creation of clothing inspired in the main international trends. This unique phenomenon was decaying as the free market economy allowed the income of textile inputs and the import of foreign clothes, which brought about the spread of this net of boutiques. It boosted the national *Prêt-à-porter* between the step from democratic governments towards a strong military dictatorship.

1 INTRODUCTION

Santiago has never been full of malls and ubiquitous retail companies. A couple of decades ago, the textile industry and the wide view of Chilean fashion was very different from the present one. In the middle of the 1960 decade and until the end of the 80 s, different agents and factors coexisted, which led to the installation of a system of fashion in Chile (Montalva, 2015, p. 71). It was inscribed in the formula of the named *Prêt-à-porter* and the meaning in a net of boutiques and retail clothing stores mainly located at the eastern area of Santiago.

The current research refers to a particular period of Chilean fashion, since 1967 until 1987, in the grip of the military dictatorship. This timeframe starts with the installation of Tai y Vog, two of the most important boutiques within the local scene of the capital city at that time. Besides, in 1967 the women's magazine *Paula* started to be edited, a mass edition with a renewed gender approach and with a liberal and open stance, that also changed the way in which fashion was perceived in Chile. Although it may seem an arbitrary date, the year 1987 was chosen as the final point of periodization of this project, considering the closure of the emblematic and still remembered clothing store "Palta", which in some way was an indicator of the declining of this system of boutiques and small enclaves of fashion and clothing in the Metropolitan area.

This work analyses this net of exclusive clothing stores from the perspective of its manufacture methods, which were mainly based on the perspective of *Prêt-à-porter*, imported from the main fashion capitals, but adapted to the local dynamics. The view of this production, created in the national territory found in the women's magazines of that time its main diffusion tool, by means of including unpublished photographs of fashion, locally produced. This allowed the introduction in the social net of the garments that the main boutiques marketed, stepping up their professionalism and impacting their working methods.

Besides, these fashion enclaves are understood as modern consumption spaces, where the acquisition of clothes was intrinsically associated to meeting and social interaction. Both aspects were stimulated by the urban configuration of the commercial landscape where they were settled, which was developed in a context of an urban regulation plan at a communal level.

The installation of a fashion system in the Chilean capital city, coincided with a series of political, economic and social changes (from democratic to totalitarian systems), which have been preferably reported from the traditional areas of history. Hence to date, only two authors have written and published this system of fashion and modern clothing of the 20th century in Chile, covering the phenomenon of boutiques with only in passing. For this reason is that the current research is mainly articulated by consulting primary sources: oral and written.

2 GENERAL BACKGROUND

The main vector that guides this work is to highlight the narrow and meaningful relation between the macro events happened in the country—corresponding to the chronological period under study—and the researched phenomenon. The origins of the Chilean *Prêt-à-porter* coincide with a series of political, economic, social and cultural changes, which deserve to be reviewed, since they impacted directly on the history of boutiques since their origins until their decline.

One of the main factors that allowed the emergence of a number of retail clothing stores in the eastern area of the capital city was the consolidation and boom of the Chilean textile industry towards the decade of the 60 s.[1] The trailers who opened their boutiques—now called entrepreneurs—counted on textile raw material of national origin with a good quality for the making of their clothes. While the commercialized garments in that period had a strong foreign influence, almost all of the models were produced with fabrics made in Chile and manufactured in local workshops.

Not only a production industry of fabrics and textile raw material existed, but also the country had a group of technicians and experts, who were able to hold the production methods of this new fashion offer. The author Pía Montalva, in her book *Morir un poco: Moda y sociedad en Chile, 1960–1976* refers to the new instances that arise during the government of Eduardo Frei Montalva, such as the technical schools of the Education Ministry, the Work Promotion Institute and Mother Centers, which allowed to professionalize the crafts related to the manufacture of clothing. The State actively participates in all of these institutions. Montalva concludes as follows: "The policies of popular promotion implemented under the government of Frei Montalva come across this area in different ways, from high fashion to *Prêt-à-porter*, fostering unsuspected work alternatives for women of the upper, middle and lower classes" (Montalva, 2015, p. 73).

Likewise, it is worth mentioning what the decade of the 60 s meant worldwide, characterized—among other events—by social effervescence, unrest and by cultural and social revolutions which meant transgression of values and customs established in the past. It is in this time, when youngsters and women turn into new historical subjects and players and find greater instances of public participation. As a corollary of what was mentioned above, the Chilean society also experimented some transformations which allowed a growing visibility for both participants and even had a way of expression in fashion and in the emergence of the national *Prêt-à-porter*. They provided fashion with new winds and turned themselves into essential characters for the development of the phenomenon of boutiques in the eastern area of the capital city, involving in modern consumption dynamics and in social interaction, unknown in the domestic environment.

3 FROM SANTIAGO DOWNTOWN AREA TO PROVIDENCIA (1967–1973)

In order to get into the phenomenon that this research involves, it is necessary to know the urban commercial expansion that Santiago suffered since the second half of the 20th century,

---

1. In the decade of 1930, after the great depression that caused a worldwide effect, the Chilean economy became protectionist and tried to reduce the dependence on foreign markets. Thus, the State's answer considering the economical risk, was to encourage the "inward development" with an industrialization policy by the substitution of imports, so called IS.

which favored a decentralization of the foundation triangle of the city. At the end of the 70 s, it was evident that there was a sub-center located in the neighborhood of Providencia.[2] This new urban space was formed as the result of the movement of the middle class in the capital city towards the eastern area, since the decade of the 40 s, which sparked the initiative of the government of Eduardo Frei Montalva to plan the layout for line 1 of the subway Metro de Santiago.

In parallel with the construction of this line, Providencia starts in 1967, preliminary studies to carry out communal regulations which will regulate the installation of trade in the area, based on the concept of *Incentive Zoning*.[3] Germán Bannen, an urban adviser in the neighborhood of Providencia, sought to combine the new private settlements with the construction of public space for pedestrians, avoiding the installation of commercial buildings "segregated from the city spaces" (Schlack, 2001, p. 20). In this new consumption space, "a place that was made for the pedestrian's time" (Bannen, 1993, p. 34), where a net of boutiques and retail stores in fashion start to set up toward the mid 70 s.

The first clothing stores that opted for Providencia—when the neighborhood was mainly residential—date from the late 50 s, starting a sort of gradual "colonization" of the commercial area of the neighborhood. This foundational projects were built mainly as family businesses and in most cases, the owners of the boutiques had, to their credit, an origin related to the fashion trade and the textile industry.

The ones who led this movement were mainly women coming from the upper classes. They imported trends from the fashion centers to the national territory and delegated the production to local workshops. It is worth mentioning that just a few of those who ventured into the installation of a boutique had previous studies related to fashion design or technical knowledge of sewing and dressmaking. Truly, the process for creating garments consisted of a copy or imitation of foreign models. For instance, Carmen Rojas—one of the founders of boutique Vog, which was marked by being an advanced store implementing, in the local area, some fashion trends that the owners observed in their trips—started to bring into Chile some garments that were designed by authors such as Mary Quant and Yves Saint Laurent, which were disassembled and assembled again, being adapted to the body of their clients (Rojas, personal interview, April 4, 2017).

Although it is endorsed that the *Prêt-à-porter* arrives in Chile in the decade of the 60 s, it sets up adapting itself to the local dynamics. Far from being a mass phenomenon, it was characterized by its exclusive offer, coming from a small number of garments of each model. As a result, the national *Prêt-à-porter* was in constant contact with High Fashion, in the sense of not stop being "custom made clothing". In general terms, boutiques showed one or two sizes per each garment, which were adapted to the clients' bodies.

The process of adapting foreign trends to the local area, was mainly given by the raw material for tailoring. All of the textile goods that were used in this first stage, came from the national textile industry. These companies designed clothing every season, which was reached by the owners of the stores by means of catalogs. It is worth mentioning that the most renowned and successful boutiques, such as Vog or Miss Paula, were closely related to textile companies, even having influence on their production, ordering special and exclusive prints for the making of garments.

After the manufacture in local clothing workshops, the owners of boutiques were the ones who marketed their products. The sale of garments was not an outsourced work, but one that demanded the owners to be in the store waiting for the arrival of clients. Such is the case of Luis Eduardo Covarrubias, the owner of boutique Wales, who had a kind of family attention (Covarrubias, personal interview, April 3, 2017). It is important to mention that

---

2. Providencia is a neighborhood in the city of Santiago, which was configured as a commercial center since the decade of the 60 s. At present it continues being one of the most important enclaves for the consumption of the author's costumes design in the Chilean capital city.
3. The Incentive Zoning corresponds to an urban regulation that consists of a bonus for greater buildability in exchange of giving some private area to public use.

boutiques not only formed sale and purchase spaces of fashion articles, but also were turned into socialization places, where the owner of the store articulated dynamics with clients and between clients. Besides they were considered experts on fashion and trends.

The acquisition of some *Prêt-à-porter* garments at boutiques was not possible for everyone. The exclusiveness of consumers was mainly given by the high prices. Moreover, in this first period when the range was not diverse and the number of boutiques and stores continued being small. However, there were other ways of accessing to modernity which meant the *Prêt-à-porter* phenomenon.

On one side, the features of the commercial landscape where the net of boutiques under study were set up, favored the outing on foot in the streets of the area. Thus, the fashion phenomenon was not only experimented by the consumers that acquired garments at the boutiques, but there is also a collective (Vaccari, 2008 p, 191); strictly a sharing, kept by all the other present agents or actants, which in some way perceived the modern offer of the *Prêt-à-porter* expressed in this innovative group of boutiques and small clothing stores, which gathered a crowd of people who toured the streets as a *flâneur*,[4] relegating the act of consumption to a second stage (Lavín, personal interview, April 3, 2017).

On the other hand, the emergence of this modern fashion offer, manufactured within the national territory, coincided with a triggering growth of the cultural industry, which—by means of various expressions—allowed the spread of the work that was carried out by boutiques and clothing stores just installed. The spread of these shops was mainly articulated around two focal points: Parades and fashion photographs published in national women's magazines. No doubt, the main publication and the one that produced a break point with previous fashion publications was *Paula*. The magazine included from its third issue, fashion photographs in the country, which meant starting with the construction of an imaginary of the Chilean fashion with productions and scenes of its own.

The last time interval of this first period of the *Prêt-à-porter*, coincides with the three-year-government of the Popular Unity, in a polarized and politically divided Chile. For some of those who participated in fashion at that time, this was a very positive stage, since the government had some interest in supporting some proposals that were aesthetically aligned with its ideology: such was the case of the so-called autochthonous fashion.[5] Another group of designers was located at an opposite side, for whom the Socialist government meant a complex phase of goods shortage and insecurity. In any case, the three years under Salvador Allende's mandate, meant—for better or worse—a period of relevant transformations which impacted the usual practices of most of the Chileans.

4 THE ARRIVAL OF NEOLIBERALISM AND DISASSEMBLING OF THE BOUTIQUES NET (1973–1987)

September 11, 1973, after the military coup, the military junta takes power. This meant that all the projects and ideologies that were in some way related to the previous government abruptly finalized or went underground. In accordance, the present research states that the unique *Prêt-à-porter* produced and marketed in Chile, which was originated in the early 60 s, concludes a stage when this political juncture takes place. The scene of boutiques and

---

4. The French term *flâneur* means walker or street. The author Charles Baundelaire used that term in his poems and afterwards Walter Benjamin turns it into an academic literary figure. With *flâneur*, Benjamin refers to the urban viewer, a typical character of modern life.

5. Within the temporary frame of this research, one of the most studied milestones has been the development of the "autochthonous fashion" It dates back to 1968, when Paula magazine publishes an interview with Marco Correa and states "a Chilean designed the Latin American fashion" (Paula, 1968, p. 64). The spokespeople of this phenomenon are 5 authors: Marco Correa, María Inés Solimano, Nelly Alarcón, Enrique Concha y Alejandro Stuven. Each of them raised on fashion, by means of different techniques and trades, a visual interpretation of the national heritage.

clothing stores does not end suddenly, but it is considerably transformed, adapting to Chile's new situation under a totalitarian regime, as a certain way of dressing at that time, could have been a reason to be identified as some of a political wing of the Popular Unity. For instance, María Inés Solimano, one of the exponents of the autochthonous fashion, suffered a depression in her sales level since the autochthonous or "hippie" fashion meant a synonymy with the leftist trends, which implied that many women of the upper class neighborhoods modified their dress code, wearing garments like "Margaret Thatcher" (Solimano, personal interview, March 23, 2017).

In addition to the censorship applied to some features related to a leftist look, the new economy model established by the military dictatorship completely transformed the view of the Chilean fashion. The measures taken meant an extreme liberalization of imports, due to the low tariffs in the income of foreign products. This new economical paradigm meant the arrival of garments of foreign designers, the chance for some authors to open themselves to new markets and, besides, it turned into the bankruptcy of all the national textile industries. Thus, it made possible that some national authors could expand their horizons and export their developments. Two of the designers who sold their products abroad were María Inés Solimano and Nelly Alarcón. Both of them manufactured their garments artisanally, the main reason why they were not able to keep pace with the production rhythm of foreign companies, which set up an unequal relationship.

In the first years, the policies for free market did not mean a complication for the exclusive national boutiques, since the products that entered the country were aimed to a mass consumption segment. However, in the late 70 s some luxury foreign brands start to settle and turn into a strong competition for local authors. Thus, the opening to the foreign market encouraged the Chilean women with greater purchasing power to prefer the innovations that the foreign brands offered. In this respect, the most affected production sector was the local textile industry, which was not able to face the competition coming from abroad, which produced the closing of a great number of these companies.

Since the mid 70 s, in the context of a dictatorship with modernizing desires, different urban sub-centers arose in the eastern area of the capital city, which were structured by agglomerations of trade and services. These places were configured according to the new needs that were demanded by their consumers. One of the great differences of these emerging commercial centers and the previous ones as well (Santiago downtown area and Providencia) was that the eccentric location meant a disadvantage, due to the absence of pedestrian public, and the recognition of a new motorized client (De Simone, 2015, p. 173).

The decline of the commercial center of Providencia, in relation to fashion consumption, is closely related to the emergence of new sub-centers in the eastern sector, where shopping habits were designed in a radically different way. The sector of Providencia Avenue, between Los Leones and Pedro de Valdivia streets provided a tour in which the pedestrian could visit fashion stores, avoiding the use of cars. The commercialization of the *Prêt-à-porter* in this space, was closely related to a new way of social interaction, allowed by the act of walking and window shopping. However, the obsolescence of this dynamism "on foot", according to Lía Fernández, dissolved this traditional way of taking advance of the public space due to the previous emergence of time compression and the mass access to cars (Fernández, personal interview, May 23, 2017).

In this way, the socialization produced in public spaces gradually went into the private area, a phenomenon that could be seen due to the closing of some stores in the capital city. Thus, in the face of the crisis and the shortage of public spaces in Santiago during the 80 s, the new commercial centers; the *mall* and the isolated stores in the exclusive sectors of the capital city, burst as gentle places, which were conditioned and safe. These conditions were the right conditions for leisure and the new consumption dynamics. The traditional concept of the walking consumer (*flâneur*), that was present in stores and shopping galleries of Santiago downtown area in the 80 s, who later moved to the commercial center of Providencia, became obsolete in front of the new purchasing options that were strengthened in the decade of the 90 s, on one side, with the *hermetic* malls, covered by retail brands and, on the other side, with exclusive and segregated stores in wide avenues, suitable for a motorized client.

Figure 1. Boutique Schock. Source: *Eva*, N° 1283, year 1969.   Figure 2. Marco Correa. Source: *Paula*, N° 26, year 1968.   Figure 3. Providencia 1987. Source: enterreno.com.

5 FINAL PREMISES

As it was previously mentioned, this research enables to recognize two phases of the *Prêt-à-porter* of the capital city. The first stage falls between the years 1967 and 1973. This period starts with the settlement of the first boutiques and clothing stores in the neighborhood of Providencia, which encouraged a sort of "colonization" of the traditional commercial sector, located in the downtown area of Santiago (the border between the "upper neighborhood" and the "lower neighborhood"). Almost all of its production depended on the raw material provided by the national textile industry; that is why, the boom of the local textile industries was one of the essential factors in the gestation of the *Prêt-à-porter*, which provided the first boutiques with raw material. Although it was a particular phenomenon created in an area of the capital city, it provided a greater expanse, given the coverage in some mass press media, such as *Paula* magazine and some others, more conservative or militant, such as magazines *Eva* and *Paloma*.

This first phase coincides with the governments of Eduardo Frei Montalva and Salvador Allende, a juncture that should be noted, since both mandates concreted measures with were closely related to the phenomenon under study. On one side, the popular promotion policies corresponding to the government of Eduardo Frei Montalva allowed the opening of work opportunities which were unknown by middle and lower class women. This triggered the emergence of a workforce that was trained to hold the boutiques production. On the other side, according to the "Chilean way to socialism" set up by the government of the Popular Unit, several cultural manifestations were supported which were aligned with a Latin Americanist aesthetics. This favored the development of the so-called autochthonous fashion. The end of the first period of the *Prêt-à-porter* coincides with the abrupt end of Allende's government. The last years of his mandate were marked by a strong political polarization and by serious economic problems, which specifically modified practices and habits related to the design of garments, manufacturing and even clothes marketing in the country.

With the violation of the democratic State in 1973, that corresponds to the second stage of the *Prêt-à-porter,* during the military dictatorship, the most immediate effects implied a transformation in the dress styles, since the features that were associated to the left wing were censored by the military government. These censure attributes could be given by women wearing long skirts, wooden clogs or jute shoes or also the long hair and beard in men, while they were indicators of the withdrawal of fashion, which shortly before seemed to move towards a more liberal attitude. Besides, the commercial sector of Providencia, which had been set up as a place for social gathering and as a cultural core, was also affected by the political disagreements of its young visitors, who on more than one occasion staged episodes of public disorder and violent riots.

Certainly, an important transformation which changed the view of the national fashion, was related to the imposition of a new political regime de-linked from the previous welfare State. The new strictly neoliberal guidelines dismantled the protectionist system that had enhanced the development of the national industry. This industrial sector was forced to compete with the foreign markets, a fact that irreparably damaged its production capacity, causing the closure of all the textile factories. This event had important consequences for the boutiques in the capital city, since they were forced to acquire imported textile goods at higher prices or cheaper ones, but of lower quality. In general terms, the decline of the Chilean textile industry, implied trouble for the stores, having no access to good manufacturing raw material. Even, for some of the most renowned stores it implied the loss of the exclusiveness of their fabrics, when the close bond with their suppliers vanished. They often designed fabrics with unique prints for particular stores.

The clothing stores and boutiques related to both stages of the *Prêt-à-porter* in the capital city,in order tovisualize their designs and commercial offer, depended upon its presence in some broadcast media. The main ways were the women's magazines of the period, being particularly significant *Paula* magazine, as the most relevant and influent in the development of the *Prêt-à-porter* during the 20 years of the time frame, which this research covers. It produced, for the first time, local fashion photographs, which enhanced the production of national stores and boutiques.

Finally, we can notice that the development of the national *Prêt-à-porter* was closely related to the purchase dynamics that followed in the commercial space of Providencia neighborhood. The design and production of garments, commercialized by boutiques and clothing stores between 1967 and 1987, was triggered by the social interaction occurred in the purchase space, since the conditions of this consumption space favored the socialization of its visitors, providing a bounded commercial route, suitable for pedestrians. This character of a walking commercial center allowed the encouragement of the urban life of the place, by means of the meeting and socialization of its pedestrians. This purchase dynamics under the paradigm of "seeing and being seen" resembles the events occurred decades before in the alleys and commercial galleries of the downtown area of Santiago, near to the Arms square. Both spaces were designed for a pedestrian consumer, prepared to social interaction.

In the late 80 s, the fashion phenomenon in Providencia suffered a decline, which was even more deepened in the following decade, with the irruption of new exclusive commercial sectors in the neighborhoods of Las Condes and Vitacura and the spread of malls in different places of the capital city. On one side, this meant the breakup of the net formed by boutiques and clothing stores in the commercial center of Providencia and, on the other side, a dissemination of the fashion consumption spots towards different sectors of the capital city. Some of the boutiques that lived their days of height in the commercial center of Providencia, moved to new neighborhoods that were designed with cars in mind, such as Nueva Costanera, Alonso de Córdova, Presidente Riesco, among others, which became consumption dynamics, very different from the ones occurred in Providencia, where the act of purchase was not encouraged by urban life and social interactions.

The above notwithstanding, a sort of resurgence of the phenomenon of boutiques and small author stores has been produced in the latest years. On one side, at the Drugstore, in the sector of Lastarria (ex Villavicencio) and in new commercial areas, such as Barrio Italia where different national brands have been settled. They commercialize their products privately or jointly in just one commercial site in a customized way. Most of these projects are led by emerging national designers, who bet on a small scale local production based on imported production materials. In these commercial sectors the fashion stores coexist with other spaces devoted to leisure, such as cafes, cinemas and stores of other items. This has favored the stay of buyers in the sector. On the other side, Nueva Costanera, Alonso de Córdova and Isidora Goyenechea have been consolidated as privileged commercial places, where some of the stores which at the time were installed in the quadrant of General Holley and Suecia, continue to exist. In these neighborhoods there are not great dynamics of social interaction in public streets, but consumers withdraw socialization to the private spaces of the stores.

REFERENCES

Álvarez, P. 2008. *Chile Marca Registrada. Historia general de las marcas comerciales y el imaginario del consumo en Chile*, Santiago, Ocho Libros Editores.

Bannen, G., 2004. El Comercio en Providencia, in *Revista CA: revista oficial del colegio de arquitectos de Chile*, N° 72, Santiago.

Cerrillo, L. 2010. *La moda moderna. Génesis de un arte nuevo*, Madrid, Siruela.

Correa S.; Figueroa C.; Jocelyn-Holt, A., Rolle C. y Vicuña. 2001, *Historia del siglo XX chileno*, Santiago, Editorial Sudamericana.

De Simone, L. 2015. *Metamall: Espacio urbano y consumo en la ciudad neoliberal chilena*, Santiago, Ril editores.

Errázuriz L y Quijada. 2012. *El Golpe estético: dictadura militar en Chile 1973–1989*, Santiago, Ocho libros.

Montalva, P. 2015. *Morir un poco. Moda y sociedad en Chile 1960–1976*, Santiago, Editorial Catalonia.

Salinas, J. 2014. *Linda regia estupenda. Historia de la moda y la mujer en Chile*, (2014), Santiago, El Mercurio, Aguilar.

Schlack, E. 2001. Producción privada de espacio público: Espacios Privados de uso público y la planificación por incentivos, in *Revista de Arquitectura de La Universidad de Chile*, N° 24, Santiago.

Vaccari, A. 2008. Reensamblar lo social: una introducción a la teoría del actor-red, in *Revista CTS*, N° 11, Buenos Aires, REDES—Centro de Estudios sobre Ciencia, Desarrollo y Educación Superior.

Vidal, J. 2017. *Moda al paso: Historia de las boutiques y pequeñas tiendas de vestuario de la zona Oriente de Santiago (1967–1887)*, Tesis para optar al Título Profesional de Diseñadora, Pontificia Universidad Católica de Chile, Santiago.

Internationalism and exoticism in the image of Aline Romanones and Consuelo Crespi

D. Gennaioli
Universidad Politécnica de Madrid, Madrid, Spain

ABSTRACT: The aim of this study is to approach from a comparative perspective the references adopted by Aline Romanones and Consuelo Crespi, Madrid and Rome Editors respectively, between 1960 and 1970 in constructing their public images. Parting from the idea of Europe ingrained in the American mind, both crafted their own image around the concepts of aristocracy, exoticism, and internationality.

1 INTRODUCTION

The 60's was the decade in which Spain tried to find a place in the new geography of international fashion. The opportunities created after World War II allowed the Italians to establish their own space in fashion publications and American stores. Whereas *Harper's Bazaar* was very receptive to the new Italian creative scene from the start, naming the journalist Irene Brin as their Rome correspondent in 1952, *Vogue* maintained a more prudent stance. Only in 1964, ten years after its rival publication, did they incorporate their own staff by hiring Aline Quintanilla and Consuelo Crespi as Madrid and Rome editors respectively.

The interest of the American market in regions outside the traditional France centered fashion world was reinforced by the increasing importance these journalists attained. Irene Brin's case is again representative, as she was able to significantly influence the molding of an Italian fashion culture to the extent of being included in the elitist international fashion intelligentsia: she was cited together with Carmel Snow, Eileen Dickinson, and Marie-Louise Bousquet in the 1955 *Harper's Bazaar UK* article "4 Minds on Fashion" (Caratozzolo, 2006). Although different in nature, the relationship of Aline Griffith and Consuelo Crespi with the *American Vogue* provides an insight into the type of self-portrayal used by each of these two editors whose personal stories seem to match perfectly. Contributors to the oldest *Vogue* edition, both had worked as New York models before joining European nobility only a year apart. But above all, both came from that bourgeois America that yearned for an European aristocratic atmosphere.

The researcher Alessandra Vaccari points out that in establishing Italy's international fame as the country of fashion, the promotion of a set of "cultural values" and both 'high' and 'low' clichés linked to the country's image had significant impact (Vaccari, 2005, p. 56). In fact, both Italy and Spain, as well as the US, had strengthened reciprocal expectations based on these elements: on the one hand the reconstruction of an *ad hoc* national identity based on a secular aristocratic past and an exotic nature emphasized to a greater or lesser degree; and on the other, the promise of progress.

How then to trigger the process of constructing a public image in which the ideals of modern America and the most characteristic values of the Spanish or Italian *genius loci* coexist? Using these premises, Aline Quintanilla and Consuelo Crespi adopted similar strategies to go from being editors to icons of style.

2 THE PERFECT SPANIARD. ALINE GRIFFITH BETWEEN MODERN AND HISTORICAL

"I hope you have a wardrobe capable of meeting the demands of Madrid's social life" (Griffith Countess of Romanones, 1988, p. 84), was the warning secret agent Top Hat gave the future Countess of Quintanilla when she first stepped into Madrid's Palace Hotel on New Year's Eve of 1943. At the time, no one could imagine that Mary Aline Griffith whose best outfit was a blue Carnegie tweed suit "with matching cape and hat" (Griffith Condesa de Romanones, 1988, p. 62) would end up representing Spanish fashion throughout the world during many years. So much so that she lost the label of "An American in Spain", title of one of her first novels (Griffith Countess of Romanones, 1980), to assume the role of the cosmopolitan ambassador of an idealized exotic Spain.

Born as Mary Aline Griffith Dexter in Pearl River (Rockland County, New York) on May 23, 1923, she studied at College of Mount Saint Vincent, a catholic school in Riverdale (Bronx, New York) graduating with a degree in literature, history and journalism. Before the beginning of World War II, during which she participated on spy missions in Europe for the OSS (Office of Strategic Services) intelligence agency under the code name "Tiger", she worked as a Manhattan model for Hattie Carnegie (Roberts, 2017). Her entry into Madrid high society was the result of her marriage to the Spanish aristocrat Luis Figueroa y Pérez de Guzmán el Bueno, Count of Quintanilla in 1947. Her contribution to the American edition of *Vogue* from the mid 1960's to the early 1980's was essential. In September 1964, the name of Aline Quintanilla, mutated to Aline Romanones as a result of the dynastic succession of her husband. In January of that same year she assumes the position of Madrid Editor, which she will occupy until December 1976.

It was a new role within the magazine that had marginalized Spanish fashion until that time. With the exception of an 1941 article titled "Drapery from Spain" in which the thin silhouette of the socialite Estrella Boissevain wrapped in a tight dress of "Balenciaga's San Sebastian house" served as a counterbalance to a Solomonic column (Vogue, 1941, p. 51), and reports on certain fashion shows from the Old Continent during 1953 and 1954, the names of designers such as Pedro Rodríguez and Manuel Pertegaz didn't appear with the same frequency as their Italian counterparts, and much less than the French. Despite these few references, Spanish haute couture seemed destined to be abandoned to a local dimension for a more contemporary narrative (Vogue, 1955). In the plan of greater international openness within the broader project of crafting a modern image of Franco's Spain, Aline Griffith emerged as the best of allies.

Already in 1955 when she first appeared in American fashion publications, the "dark, narrow-boned, American-born beauty" from Pearl River seemed to have a clear idea of the register she wanted to adopt in building her public image, a constant balance between the modern and folkloric (Vogue, 1955, p. 91). The write-up, a mere two photographs, immortalizes her image, first with arms open and the seductive gaze of Hollywood silent film stars, and then sitting on a chair wearing a cocktail dress covered with embroidery in pure Pedro Rodríguez style.

A month later she poses for *Harper's Bazaar* on a two-page spread together with Mrs. Angier Biddle Duke (Harper's Bazaar, 1955). In this case, the photographer Louise Dahl-Wolfe chooses an unusual format by placing each lady within a circular frame. Showing acumen by not only creating a reference to the romantic portrait, but also by bringing to mind the painting of the Empress Eugenie de Montijo by the painter Franz Xaver Winterhalter (The Museum of Fine Arts, Houston, EE.UU., 1854). It is not only the oval format used by the German artist that evokes the comparison, but also the similarity in posture: sitting on what seems like a low throne, wearing a flowing gown and a cape held at the waist with both hands. Unlike the consort of Napoleon III that is portrayed in profile, Aline Quintanilla is facing the camera seeming to challenge it. Interestingly, they lived in parallel conditions not only in pictorial representation but also in real life (one an American residing in Spain, the other a Spanish Empress of France), both were public figures acting in a context different to that of their origin.

The scene of the pictorial representation, which harked back to one of the most popular figures of Spanish nobility, served to reinforce her status of "outsider" in terms of geography and social standing. It must not be forgotten that if Luis Quintanilla belonged to one of the most distinguished families of the Spanish scene, Aline had lived in a bourgeois environment. As she herself recalled in her second book, her father's family had "been Maryland farmers. With the exception of my grandmother Griffith, my ancestors were North Americans from 'way back'" (Griffith Condesa de Romanones, 1988, p. 69). In this sense, who better than the unquestionable protagonist of nineteenth century European courts to personify aristocratic and high society values?

The ex-spy's adopting of Empress Eugenie's iconography should be interpreted as an undertaking of true reconstruction. Conscious of an ingrained aristocratic aspiration among the American public, Aline repeats certain attributes in the double process of defining her own public image and searching for a Spanish fashion identity.

An empire-style chair that is the only piece of furniture that accompanies Aline's debut in *Vogue*, for example, is included as a staging element in at least three portraits of the last queen of France.

In fact, even the most famous jewel of the Countess of Teba is at the heart of a unique comparison. Her legendary passion for jewelry led her to commission a tiara consisting of nine rosettes with rectangular emeralds to jeweler Eugène Fontenay. After avoiding the 1872 Christie's London auction in which the by then former Empress had to sell most of her pieces, the nine gems were inherited by the princess Victoria Eugenie of Battenberg, her god-daughter and future queen of Spain, that wore them in her 1922 official portrait. These gems where somehow related to the most recent monarchic past of Spain, given that after the Second Republic was proclaimed, the royal family had to begin a long exile in Europe. It is interesting that the most valuable piece of jewelry belonging to Aline Romanones, which was at the center of a controversial sale in 2012, was a sautoir of emeralds which could be transformed into a tiara and that had belonged to the Maharaja Ajit Sing, son of the Spanish actress Anita Delgado (Verbo, 2014).

In defining the relevant elements of her own staging, the references to protagonists of the glorious national past is mixed with the recovery of the visual repertoire of Spanish artistic tradition. When reporting for an article in the series "The Most Fashionable Women" (Rodríguez, n.d.), as she opens the doors of the Pascualete Estate (Cáceres, Extremadura), one recalls a portrait of female saint painted by Francisco de Zurbarán. The structured drooping of the coat made of raw silk and the contrast between the cape's cerulean color and the pale rose of the organza dress, colors traditionally associated with the Virgin Mary, places the photograph in the same category as authentic religious altarpieces.

Images such as these with rich iconic references, where useful in spreading a discourse of strictly nationalistic character: Aline shows herself as the most exotic of American women. Perhaps more Spanish that Spanish women themselves. "No woman today displays these classic Spanish qualities more pleasingly and dramatically than the young Condesa de Quintanilla, of Madrid", reads one of the first paragraphs of the aforementioned article, that further down, continues by clarifying: "despite her American background, looks Spanish and speaks the language without accent or inhibition" (Rodríguez, n.d.). The association between expressions of identity and national implications is often repeated both in articles about high society as in her own writings. "You look more Spanish than me!" (Griffith Condesa de Romanones, 1988, p. 103), are the words she attributes to the character Casilda Ávila, after Aline is seen wearing a Spanish mantilla and hair comb during a Holy Week religious procession in Toledo. On previous pages it was secret agent Edmundo Lasalle that reassured her by saying: "You look like a Spaniard, which is an advantage since you will fit right into the scene" (Griffith Condesa de Romanones, 1988, p. 103).

The extensive 1963 article that anticipated her later relationship with *Vogue* magazine already provided that "interesting mix" of Spanish artists, writers, and actors that where part of the inner circle of the Quintanilla family (Vogue, 1963, p. 157). In this regard, it is worth noting that Aline herself, together with the bullfighter Antonio Ordoñez, where the baptism godparents of the son of Lola Flores, the famous flamenco dancer.

Photographed by Henry Clarke wearing a classic "4 piece" suit, an outfit consisting of several pieces designed for hunting, Aline described the exciting outings organized by the matador Luis Miguel Dominguín in the hills surrounding Cordoba. It is clear that hunting, as well as attending bullfights, common entertainment for most of the Spanish nobility, represented indispensable opportunities for social exposure and were orchestrated following precise dress codes.

Perhaps the most forceful statement of the entire article is related to bullfighting: "For me it is the best of shows, and of all the most extraordinary" (Vogue, 1963, p. 161). The bullfighter Juan Belmonte also spoke of the uniqueness of the future American countess when, upon her rejecting a valuable outfit embroidered in gold that he had given her as a gift, he had to concede that: "From the lips of an American it is original; few of her compatriots understand the art of bullfighting" (Griffith Condesa de Romanones, 1988, p. 205).

Aline Quintanilla understood perfectly the fascination generated by diversity and the communicative force of extremes. To the point that while in foreign magazines she stuck unwaveringly to the values of the most distinguished Spanish elite, sometimes even bordering caricature, in the Spanish press she advocated a narrative based on the modernizing influence of her being American. The gesture of lighting a cigarette during a 1958 interview for ¡Hola! (Hello) magazine, for example, must have seemed so unusual to the journalist that it prompted her to openly ask: "Is smoking feminine? (Yale, 1958, p. 18). These "manners and gestures" were completely foreign to a woman's role under the patriarchal model of the dictatorship to the extent that it was "difficult to identify the Countess as the mother of three adorable infants" (Yale, 1958, p. 18).

Despite some exceptions, mainly in publications directed at a small circle of wealthy ladies, until the 60's for most published magazines practices such as dancing, sports, or smoking were condemned and considered transgressions not to mention unseemly in women (Di Febo, 2011). Particularly in the period immediately following the Spanish Civil War, the image of a woman wearing pants was loaded with negative connotations, being associated with the women soldiers of the recent past (Di Febo, 2011). "How can such a beautiful woman dress like a man?" (Griffith Condesa de Romanones, 1988, p. 90), a driver shouted at her on one of her first strolls in the Spanish capital.

Episodes such as appearing strategically in the middle of a large crowd of guests at the presentation of fashion designer Rodríguez's collection, cigarette in hand (Rodríguez Alfaro, 1957) or wearing jeans which in other situations would be condemned as needless provocation, could be justified if interpreted as extravagant manifestations of an eccentric *grandee* of American origin.

3 CONSUELO CRESPI AND ROME

The February 1968 American edition of *Vogue* featured a short photo-report of the reception hosted by the American embassy in Madrid, one of many that filled the back pages of the prestigious publication as of the early fifties (Vogue, 1968). As required by the finest protocol, the host, Mrs. Angier Biddle Duke, wore a Balenciaga long dress in black velvet and among the guests can be seen several marquises, some counts, and even a duke. Among the many photographs of members of the international jet set, two pictures are particularly interesting. One features Count Rodolfo Crespi and Countess Aline Romanones, the other shows Countess Consuelo Crespi and Count Luis Romanones. The symmetrical arrangement of the depicted subjects (almost a chiasmus), not only alters the "syntactic parallelism" that intended to bring together the Crespi couple in the left-hand image and the Count and Countess of Romanones on the right but is also not an accidental combination. The personal and professional circumstances of the two countesses seem to coincide as an almost perfect analogy. Like Aline, Consuelo promoted fashion from her adopted country internationally for over three decades, to the point of becoming one of the few references able to personify the "easy grace" style which in American popular imagination was the defining characteristic of the "refined and elegant" Italian woman (Vogue, 1949, p. 218).

Consuelo Pauline O'Brien O'Connor had, however, been born in Larchmont (New York) on May 31, 1928 and belonged to a bourgeois family of Irish origin. Her father had left Europe during the early years of the 20th century and had accomplished the "American dream" going from "washing bottles" in a mineral water company to becoming its president (Colacello, 2011). Having spent her childhood in Nova Scotia (Canada) and after the separation of her parents, she returns to Manhattan in 1943 and becomes a model for the agency Harry Conover together with her twin sister Gloria who would later marry insurance magnate Schiff and become an editor of *Harper's Bazaar* and *Vogue* (Pecorari, 2014). The year 1948 was decisive in the biography of Consuelo and permanently binds her to Italy when she marries Count Rodolfo Crespi and leaves New York for Rome.

The following November, she is one of four models featured in the article "Black on the Beach", a *Vogue UK* story which is couturier Simonnetta's debut in the foreign press. In an exclusive Carpri island setting, Mita Corti, sister of the designer, her husband Uberto, and the Count and Countess Crespi pose in black poplin outfits (Caratozzolo, 2008).

Her choice of only wearing clothes from Italian ateliers from that point forward, in a way becoming what was then known as *mannequin de societé*, she confirms the fashion-nobility dichotomy: one of the basic elements of the expression of Italian couture's identity in the postwar era (Gnoli, 2005 cited in Pecorari, 2014). "In many ways her own ability to wear the clothes was as effective a promotional tool as her ability to get them into the pages of *Vogue*", wrote journalist Vanessa Friedman in the *Financial Times*, a few days after her death (Friedman, 2010). In fact, it stated, the American magazine had served as a platform for many Italian designers that Consuelo had introduced into the international fashion circuit after 1963 when she was named correspondent for Italy. The new organizational chart for *American Vogue* made public the following January gave her the position of Rome Editor which she occupies until 1976.

Divided between the American and Italian editions, created in the 1960s by the transformation of the magazine *Vogue & Novità* into *Vogue Italia*, Countess Crespi had been able to blend-in to the point of becoming a mythical representation of herself: "a sort of impossible, aristocratic beauty" (Martin, 2010). Even Bettina Ballard, chronicler of European fashion, included her among the most important resources of Italian fashion. "Italian designers have the advantage of having more beautiful clothes-conscious women to inspire them than any other country", she declared in the middle pages of her autobiography, "Rome is full of pretty women like the Contessa Consuelo Crespi who like to talk clothes, and this conversation doesn't seem to bore the men at the continual round of cocktail and dinner parties that provide the elegant back-ground for their fashion parade" (Ballard, 1960, p. 254).

4 PARALLEL LIVES

The Spanish weekly art and literary magazine *Blanco y Negro* implied that in order to "make more elegant women happy" (Claraso, 1959, p. 76), *The New York Dress Institute* had not prepared just one best dressed official list for 1958. All the versions of the list, usually prepared by the journalist Eleanor Lambert, seemed to coincide, however, in a group of six names among which were included without any doubt those of the Countesses Consuelo Crespi and Aline Quintanilla. When the "distinguished Spanish lady" (Anon. 1957, p. 10) was first included in the prestigious list in 1956, Crespi had already achieved the status of socialite many years before both in American circles and Italian social gatherings.

Although with geographical and political differences due to their areas of operation, both knew how to maintain a balanced "blend" of modernism with the recovery of the local past when building their public image. For example, when Consuelo was photographed by Karen Radkai inside her luxurious Roman home (Harper's Bazaar, 1952), she did not choose the classic standing pose of Princess Domitilla Ruspoli, nor the seated position of *Donna* Laudomia del Drago, both models appearing in the same *Harper's Bazaar* article. Instead, she chose to be shown lying down, a much more impertinent posture, while wearing a dress by Simoneta Colonna di Cesarò, the most aristocratic and at the same time most international of Italian designers. "Rudi and Consuelo were like a huge burst of fresh air" said designer Federico Forquet referring to Roman nobility parties that where still too conservative during the fifties (Colacello, 2011).

If the scholar Marco Pecorari underlines how Consuelo Crespi "a benchmark for the definition of Italian aristocratic beauty" perfectly exemplifies that in fashion "national identities are not limited by the static confines of geography but are rather entities in constant redefinition" (Pecorari 2014, p. 380); it is no less true that recovering one's own roots was a definitive differentiating element relative to other local ambassadors. In both cases, as Americans, they were associated with an implicit series of values such as being practical, urbane, civilized; in a word: modern.

Aline Romanones tried to embrace these two registers in an even more radical way. Her association to the Empress Eugenie de Montijo, hunting, the world of bullfighting, and Andalucian folk dress, linked her image to the idea of what is Spanish created in 19th century France. Maintaining the late romantic clichés about Spain found in both European and American stereotypes (González de Durana, 2013), was fundamental for the effectiveness the image she created. In a certain way, the *Infanta Dress*, shown at Cristóbal Balenciaga's Paris debut, captures the same concern for offering the press an easy to transmit sense of "newcomer". So much so, that the use of the *montera* (bullfighter's hat) or the flamenco dress will be rare in the later work of the designer from Guetaría (Blume, 2014, p. 39). In those years when American magazines only attributed to French designers a creative component and considered the success of Italian fashion to be based only on superior material and tailoring craftsmanship (Caratozzolo, 2011), Spanish couturiers had to resort to exaggeration of certain clichés in order to emerge. Something that a few decades later would be considered anachronistic: when the reference to aristocratic and folk icons return to Spanish fashion, as was the case with the bull-dress of Sybilla (2000), these were already integrated in the designer's style.

5 CONCLUSIONS

This study tries to shed light on the idea that in the years following World War II, when Italy and Spain emerge as new centers of fashion, Aline Romanones and Consuelo Crespi became key figures in the process of internationalization, creating their public image in surprisingly similar ways. In fact, it is the creation of the Madrid Editor position that provides Spain with its own space within the fashion magazine world. They rekindled the underlying values shared by Europe and the United States, putting in place a strategy that combined historical reconstruction and references to an aristocratic past balanced with exotic and modern attributes.

For example, just a year after having left New York, Consuelo wrapped up the photographic sequence of noble Roman women in a *Vogue* review titled "In Rome Now". Dressed in a tight-fitting Irene Galitzine evening dress showing she was perfectly integrated into the elite nobility of the eternal city and had completed her "transformation" to become "a leading figure of high society and aristocratic Roman life" (Pecorari, 2014, p. 379).

Likewise, with the aid of her looks similar to one of Goya's *majas*, long jet-black hair and intense green olive eyes (Vogue, 1970), Aline relied on the entire visual richness of Iberian culture; even refurbishing Pascualete (in Cáceres, Extremadura), a large country estate that had belonged to her husband's family since 1232, to serve as a setting for her self-performance. Already in 1963 an article in *Vogue* magazine had established the dichotomy between her appearance of "Spanish *Madonna*" and her true origins: "was actually born in America and hasn't, as far as she knows, a drop of Spanish blood" (Vogue, 1963, p. 114). In other words, it is at this moment that designers and leading figures promoting Italian and Spanish fashion activities find in internationalism and exoticism a common language. Notwithstanding, each uses its own unique attributes in this process.

However, if for the foreign press she was the perfect Spaniard, the Spanish magazines emphasized the liberating nature of her being North American. Hence the irreverent gesture of smoking in front of the journalist interviewing her or the confession of having had to stop wearing pants once she arrived in Spain.

REFERENCES

Anon., 1957. La Condesa de Quintanilla entre "las catorce señoras mejor vestidas de 1956". *¡Hola!*, 12 January, p. 10.
Ballard, B., 1960. *In my Fashion*. New York: David McKay Co.
Blume, M., 2014. *The master of all us*. New York: Farrar, Straus and Giroux.
Caratozzolo, V., 2006. *Irene Brin. Lo stile italiano nella moda*. Venezia: Marsilio.
Caratozzolo, V., 2008. Simonetta Colonna di Cesarò. Un profilo. In: V.C. Caratozzolo, J. Clark, M.L. Frisa, ed. 2008. *Simonetta. La prima donna della moda italiana*. Venezia: Marsilio, pp. 155–188.
Caratozzolo, V., 2011. Enchanted Sandals. Italian shoes and the post-world war II international scene. In: C. Giorcelli, P. Rabinowitz, ed. 2011. *Accessorizing the body*. Minneapolis: University of Minnesota Press, pp. 220–236.
Claraso, N., 1959. La primera mujer de la lista. *Blanco y Negro*, 31 January, pp. 76–77.
Colacello, B., 2011. The age of elegance. *Vanity Fair*, [online] September. Available at: <https://www.vanityfair.com/news/2011/09/consuelo-crespi-201109> [Accessed: 29 January 2018].
Di Febo, G., 2011. Spanish women's clothing during the long post-civil war period. In: C. Giorcelli, P. Rabinowitz, ed. 2011. *Accessorizing the body*. Minneapolis: University of Minnesota Press, pp. 126–147.
Friedman, V., 2010. Magician of fashion helped shape postwar fashion. *Financial Times*, [online] 29 October. Available at: <https://www.ft.com/content/6e664f12-e38c-11df-8ad3-00144feabdc0> [Accessed 29 January 2018].
García Costova, C., 2002. La intrépida vida de Aline Griffith. *Lecturas*, 17 May, pp. 10–16.
Giorcelli, C., 2011. The cult of femininity. In: C. Giorcelli, P. Rabinowitz, ed. 2011. *Accessorizing the body*. Minneapolis: University of Minnesota Press, pp. 17–23.
González de Durana, J., 2013. Francisco de Zurbarán y Cristóbal Balenciaga: vestuario para glorias del cielo y celebridades de la tierra. In: B. Navarrete Prieto, ed. 2013. *Santas de Zurbarán: devoción y persuasión*. Valencia: Laimprenta, pp. 73–88.
Griffith Condesa de Romanones, A., 1980. *An American in Spain*. New York: Keedick Press.
Griffith Condesa de Romanones, A., 1988. *La espía que vestía de rojo*. Translated from English by E.Riambau. Barcelona: Círculo de Lectores.
Lunel, J.M., 2011. Las esmeraldas de Eugenia de Montijo. *Tendencias del mercado del arte*, 3(43), p. 71.
Martin, D., 2010. Consuelo Crespi, Aristocrat of Fashion, Dies at 82. *The New York Times*, [online] 23 October. Available at: <http://www.nytimes.com/2010/10/23/nyregion/23crespi.html> [Accessed 29 January 2018]
Morris, B., 1963. A Well-Dressed Countess Takes Fashion Role Lightly. *The New York Times*, 22 October, p. 40.
Pecorari, M., 2014. Consuelo Crespi. In M.L. Frisa, S. Tonchi, ed. 2014. *Bellissima. L'Italia dell'alta moda 1945–1968*. Milano: Electa, pp. 378–381.
Puch, J.F., 1958. Son doce las mujeres más elegantes del mundo. *Fotos*, 11 January, pp. 12–13.
Roberts, S., 2017. Model, Countess, Author, Spy: Aline Griffith is dead. *The New York Times*, [online] 15 December. Available at: <https://www.nytimes.com/2017/12/15/obituaries/aline-griffith-model-countess-author-and-spy-is-dead.html> [Accessed 29 January 2018].
Rodríguez Alfaro, J., 1957. Pedro Rodríguez anticipa la Primavera. *Fotos*, 26 January, p. 30.
Rodríguez, P., n.d. *The most fashionable women. The Young Condesa*. [print journal] Fondo Pedro Rodríguez. Sg. FD042528-05. Madrid: Biblioteca Museo del Traje.
Vaccari, A., 2005. *Wig wag. Le bandiere della moda*. Venezia: Marsilio.
Verbo, E., 2014. La condesa de Romanones: 'Estoy feliz porque Corinna lleva mi collar'. *El Mundo*, [online] 15 July. Available at: <http://www.elmundo.es/loc/2014/07/15/53c3f34 cca4741a21a8b4585.html> [Accessed 29 January 2018].
Yale, 1958. La Condesa de Quintanilla, una mujer elegante. *¡Hola!*, 7 June, p. 18–19.

Vogue y Harper's Bazaar

Beyond the Alps. 1952. *Harper's Bazaar*, 85(2886), May, pp. 108–111.
Condesa de Romanones. 1970. *Vogue*, 155(10), 1 June, pp. 102–103.
Drapery from Spain. 1941. *Vogue*, 97(4), 15 February, p. 51.
In Rome Now. 1949. *Vogue*, 113(4), 1 March, pp. 166–171, 215, 218.
Masthead. 1963. *Vogue*, 141(9), 1 May, p. 30.
Masthead. 1964. *Vogue*, 143(2), 15 January, p. 28.
Masthead. 1964. *Vogue*, 144(4), 1 September, p. 3.
Masthead. 1976. *Vogue*, 166(12), 1 December, p. 3
Masthead. 1982. *Vogue*, 172(12), 1 December, p. 18.

Spain. 1954. *Vogue*, 123(5), 15 March, pp. 112–113.

Spain. Fashion, Dramatized. 1953. *Vogue*, 121(5), 15 March, pp. 83.

Spain. Fashion-Envoy Extraordinary: The Countess of Quintanilla. 1955. *Vogue*, 125(5), 15 March, pp. 90–93.

Spanish Ladies in Yellow. 1955. *Harper's Bazaar*, 88(2921), April, pp. 176–177.

The Countess of Quintanilla. 1963. *Vogue*, 141(6), 15 March, pp. 110–117, 157, 161.

Vogue's Notebook. The party diplomacy of Ambassador and Mrs. Angier Biddle Duke, in Spain. 1968. *Vogue*, 151(3), 1 February, pp. 98–99.

Vogue's Notebook. Weekend shoot with the Guy de Rothschilds at Château de Ferrières, in France. 1968. *Vogue*, 151(3), 1 February, pp. 100–102.

Reframing old age through design: An approach to promote empowerment

I. Rojas & B. Pino
Universidad Diego Portales, Santiago, Chile

ABSTRACT: This article addresses the possibility of reframing what is understood as old age from the ongoing research "Strategic territories for older people: empowerment, inclusion and participation", by knowing new approaches which comprehend the aging process, not as a stage of deficiencies, but of potentialities and empowerment. Elderly people are a diverse age group that is growing and is as important as young people to society. From this emerging approach, concerning old age through design and its role in dealing with social matters, ageing is addressed by activating and proposing themes, considering the existing relationship between the elderly and the municipalities through the services and workshops in which they participate. This is because, when mapping or visualizing the workshops and the places in which they are imparted, we can observe trends and interests of the elderly, highlighting also their way of self-organization in the territory.

1 INTRODUCTION

Nowadays the percentage of elderly people in the world is increasing, therefore, it becomes a group as important as young people and has certain interests or particular ways of living. It is estimated that by the year 2050, the percentage of elderly people in the world will be 22% (WHO, 2018), which implies a change in the population structure.

If before it was the young people who represented the highest percentage, in the future it will be the elderly. Consequently, they are an age group that must be considered and known, since there are negative and homogenizing judgments that imply exclusion by virtue of a dominant group (Huenchuan, 2011). The focus of society is on young people and adults and not on the elderly, who are generally discriminated. Due to this situation, international efforts such as The Second World Assembly on Aging have been made (United Nations, 2002), in which different measures were proposed through the International Plan of Action on Aging to integrate the elderly into society.

From the point of view of design as an activator of conversations and socially significant initiatives that manage to give importance to what to do and how to do (Manzini, 2014), the matters of the elderly are relevant to the discipline, due to the potential for change and because it is a global concern, considering the plans and policies proposed by the United Nations and each country. Therefore, through tools and methodologies specific to design, it is possible to approach the topic from other angles and propose new areas of research or, in this case, conversation topics regarding the elderly from the emerging practices of design, that is to say, the needs of people (Sanders & Stappers, 2008).

One way to address the subject of old age and elderly people is through design. The first step is to recognize paradigms that conceive old age not only as a stage of deprivation (Huenchuan, 2011), and the second step corresponds to reframe this stage from another point of view, by understanding fashion as a behavioral way, manner or trend (Ávila & Linares, 2006), knowing the interests and forms of participation of elderly people in services, since in this way, it is possible to make issues to become relevant and their respective potentialities visible, along with valuing positive aspects of old age, considering the elderly as important as young people.

2 SOCIAL CONSTRUCTION OF OLD AGE AND EXCLUSION

The social construction of old age in Western society is mainly negative, it is even possible to talk about discrimination because of age. This, according to Huenchuan, a specialist in old age, is due to:

> "In Western society, the predominant conception about elderly people and their problems begins with the construction of old age as a stage of economic, physical and social deprivation". (Huenchuan, 1999, 2004 & 2009 quoted in Huenchuan, 2011)

When building old age only as a stage of deprivation, the qualities and potentialities of the elderly are not recognized, which determines the way in which the problems that may afflict this age group are raised, as well as promoting discrimination because of age, since it is not possible to recognize the diversity of these people or their life experience. For some authors and international organizations (Courtis, 2004 quoted in ECLAC, 2017) discrimination can be explained in part by the standard of normality by which society operates and by the lack of capacity to include those who are considered to be different, in this case the elderly. The way of conceiving old age implies that the interventions only approaches the needs of elderly people compared to other groups (Huenchuan, 2011). Therefore, if only negative qualities and deprivations are attributed to the elderly, this has an impact on the vision that society has about them and thus, on the programs and services that are intended for this age group. This situation may result in the exclusion of the elderly from different spaces and activities, just because the fact of being older.

2.1 *The exclusion of the elderly people*

In the case of Chile, a country that presents an advanced aging, being the second country with more elderly people in the region (CELADE-ECLAC, 2012 in SENAMA, 2015), there are efforts on the part of the government focused on the group of the elderly. In the Ministry of Social Development, there is the National Service for Elderly People (SENAMA), which is the institution in charge of promoting good practices. One of the main subjects addressed by this entity is the exclusion and inclusion of elderly people.

In the fourth national survey of social inclusion and exclusion of the elderly in Chile, during the year 2015, when people were asked: "Would you say that most elderly people can look after themselves or most of them cannot take care of their own?" Most of the respondents during the survey in 2015 (79.9%) answered that elderly people cannot stand on their own (Thumala et al. 2015), even though, in general, the elderly are independent (CASEN, 2015), therefore, the image of society in relation to the elderly is negative. One way to exclude elderly people is not to consider them responsible for their own aging process, which is explained by associating old age with a negative stereotype in which people are not responsible or autonomous.

On the other hand, the recommendations of the United Nations recognize the need for an inclusive society:

> "A society for all ages includes the objective that elderly people have the opportunity to continue contributing to society. To achieve this objective, it is necessary to eliminate all exclusionary or discriminatory factors against these people." (United Nations, 2002)

Recognizing that elderly people have the opportunity to continue contributing to society implies changing the perspective that the rest of the people have about them, in the first place, to be part of society, it is necessary to attribute an active role in their own aging process, recognizing their qualities and interests.

In consequence, it is necessary to develop projects that provide a positive view of the elderly, encouraging their action and autonomy. From the perspective of design, understanding the discipline centered on the needs of individuals or society, a broader approach is needed in order to achieve inclusive projects and services for the elderly (Sanders & Stappers, 2008)

3 REFRAMING OLD AGE FROM DESIGN

When evidencing the exclusion that elderly people live and the need for proposing a change in the way in which old age is conceived, it is necessary to reframe this process from an integrating point of view that allows the social construction of old age in a positive way. According to Kolko (2010), designers approach a problem in the context of a framework, but in order to analyze the problems from a new perspective, the method of reframing is necessary. In the case of elderly people, aging can be reframed, based on an approach that considers old age not only as a loss stage.

Due to the negative vision that exists regarding old age, Huenchuan (2011), through the perspective of human rights, proposes in the study material "The rights of the elderly", a distinction between the traditional approach and an emerging approach. In the first, old age is conceived as a stage of deprivation (economic, physical and social) and older people are considered subjects of benefits, therefore, the social roles assigned to them are restricted. This has implications for programs designed for the elderly, as they can perpetuate stereotypes and, consequently, lack of autonomy. On the other hand, in the emerging approach they are considered subjects of rights, autonomous and the opportunity to continue contributing to society is recognized, that is, they can play different roles. In addition to the differences mentioned, the rights approach promotes the empowerment of the elderly.

Considering that from the discipline of design it is possible to generate changes by remarking what is understood as old age, changing from a traditional paradigm to an emerging paradigm in order to show a different view of the elderly, in which they are considered autonomous subjects who can exercise different roles (Huenchuan, 2011), analyzing old age from an emerging perspective is in fact a matter of design, since, according to Sanders & Stappers (2008), the categories addressed by design are shifted from product design to design focused on the needs and purposes of people or society. This implies a change in the way of facing the problems and needs detected, since they will change what is designed, in what way it is designed and finally, who designs it. In addition to understanding design for social purposes, we can talk about design as activism, for its ability to give importance to certain themes:

> "To make design more relevant is to reconsider what 'design' issues are. Rejecting the limits we have defined for ourselves, we should instead assume that design can play a positive role in seeking answers to many different kinds of challenges." (Bell & Wakeford, 2008)

From a propositive and positive point of view, in line with the emerging paradigm regarding old age, design can play a role in the problems faced by elderly people, activating or reconsidering what is a design issue and the field of action from the design becomes more diverse. Therefore, a relevant issue is the stereotyped vision of old age, because of the negative impact it has on society, especially when compared with young people.

3.1 *Municipal services for older people: A view from design*

In the case of elderly people, there is great potential for action from the design when taking into account the services. In the case of Chile, there are different organizations and scales in which the elderly are users of services and participate. At the public level, in the Ministry of Social Development, there is SENAMA (2015), whose strategic 2014–2018 axes are: human rights, participation and decentralization.

In addition to the different programs developed by SENAMA in cooperation with different ministries and institutions, the relationship between this entity and the municipalities is fundamental, since they have the possibility of working directly with older adults in the territory. Due to this fact, the research "The work of Municipalities in Favor of Elderly People", conducted by the AMUCH (Association of Chilean Municipalities), was revised and one of the conclusions is the following:

> "(...) In most cases, municipalities are simply assigned the function of executing the decisions that this service adopts at a central level, depriving them, many times, of the possibility of generating new initiatives." (AMUCH, 2017)

The capacity of action regarding to what a municipality must execute responds to the judgements of SENAMA. However, there are some efforts on the part of municipalities, for example, in most of the municipalities in the Metropolitan Region, different workshops that respond to objectives such as participation are offered to the elderly, but the way to perform this task apparently is not absolutely directed by SENAMA, therefore, there is a field of action to explore when considering the workshops and the relationship with the municipality. According to Ferrara & Sabbat (2012) in Innovation without boundaries: ecology of innovation and municipal service design, a strategy whereby municipalities can face challenges and generate initiatives occurs when collaborating with interdisciplinary design studies.

3.2 Workshops for elderly people

In the ongoing research "Strategic Territories for Elderly People: Empowerment, Inclusion and Participation", the objective is to identify and know the actors, services and workshops in the territory that are related to the participation of elderly people. In the first place, the importance of the municipalities in the territory was determined according to the services focused on the elderly that they offer. A survey was conducted in order to gather information regarding the social programs focused on the elderly, the names of workshops and services, the places where these workshops and services were imparted and finally, the requirements. The survey was conducted in different municipalities of the Metropolitan Region through the Transparency Portal (2018) and the Transparency Law, that aims to provide the information to all people. Additionally, in some cases, information collected from the web pages of the municipalities was also used. The amount of information regarding the communes reaches the number of 34 and the communes of the metropolitan region are 52. The first result showed that, out of 36 mapped communes that had information regarding the elderly, 34 municipalities imparted quarterly, semi-annual or annual workshops at least once a week, so the elderly attended regularly. The workshops are a type of service delivered by the municipalities and are stipulated in their social programs. These workshops consist of the teaching of a technique or subject, they have a certain number of places, a teacher and, in some cases, they include materials. According to Sangiorgi (2011), the design has focused on investigating the transforming role of services as a way to conform a fairer society, therefore, services and workshops designed for elderly people have a transformative potential at a social level.

The service provided by the municipality, the role played by the elderly and the offices in which they participate, give an account of the relationship between the municipality and the elderly, that is to say, the workshops are a service present in most of the communes, responding perhaps to the particular interests of the elderly along with the municipal capacities and resources to respond to that demand (see Figure 1). The workshops and services show that there are active elderly people who are constantly developing their skills.

In a first approach, the variety of workshops is wide depending on the commune and the income. In some communes, 70 types of different workshops are imparted. This was done with the objective of analyzing the different interests of the elderly, it is also useful for the municipalities to propose new workshops and services (See Figure 1). The second step was to make visualizations with the information obtained, since, as we can see:

> "Simple visual design approaches such as mapping and visualization help us begin to understand systems and their challenges. These tools are more critical than ever, and no one else has them" (Snoad, 2012)

Figure 1. Mapping of workshops for elderly people in communes of the Metropolitan Region, Chile.

In this way, the different workshops and services are displayed, grouped into different categories corresponding to four thematic areas: Artistic and Manual Area and Physical and Recreational Activities Area, Personal Development Area, Knowledge Area.

In the results obtained and visualized through a matrix, it is observed that, in the case of the category "Artistic and Manual Area", most of the communes have some type of workshop focused on this area, even Knitting is present in all the communes that were investigated. On the other hand, in the case of the "Personal Development and Self-care Area", most of the workshops focused on these areas are concentrated in the higher income communes, which correspond to the first four on the list. Likewise, the usefulness of the matrix of workshops is diverse and depends on the interested participants, since there is the possibility of mixing workshops or areas in order to summon elderly people. For the municipalities, the mapping or cadaster of workshops is useful in order to know what initiatives are performed by other communes regarding this age group, and it is even possible to propose intercommunal meetings on the basis of a workshop or interest in particular.

3.3 Relationship between municipalities and elderly people

The interests of elderly people, expressed in the workshops in which they participate, are a first step to approach their problems and opportunities. However, it is pertinent to know the relationship between the elderly and the municipalities through the places where the workshops are imparted, since it is not the same a workshop in a municipal office as in a social office, because the second one is also used by neighbors and neighborhood organizations in order to meet, discuss and work in what they consider convenient. The organizations that usually use social offices are neighborhood councils, mother's councils, clubs for elderly people, among others.

There are two main ways to interact with elderly people, one is through the development of workshops and services in the municipalities and municipal offices and the other way corresponds to the workshops and municipal services imparted in clubs for elderly people (see Figure 2). Regarding the relationship among the municipality, the municipal office and the elderly, most of the services are provided by the municipality, in this sense, there is potential for changes or improvements provided that the elderly and civil servants of the public service are committed, according to the co-production concept:

> "(...) corresponds to an active behavior of an user that, along with the production developed by a government agency and encouraged by it, creates private and public value through results or impacts". (Alford, 2009 quoted in LIP, 2017)

Encouraging the co-production of services for elderly people, giving them a more active role, can be a matter to be investigated in order to develop projects that generate changes. An objective of public value is to generate a more positive vision regarding old age through the co-production of services and workshops. However, the main finding is that, at least

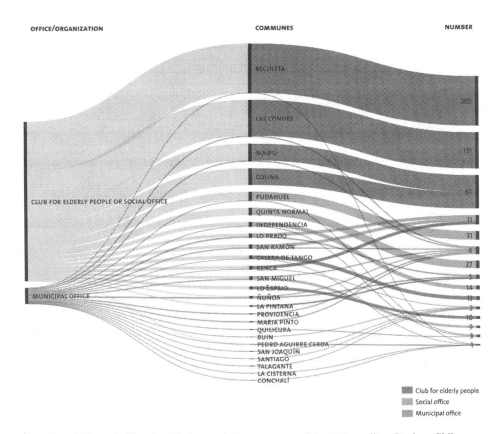

Figure 2. Clubs and offices for elderly people in communes of the Metropolitan Region, Chile.

in seven communes of twenty-five, the people participate in the workshops offered by the municipality through the clubs for elderly people. This requires a certain level of organization on the part of the elderly, since, in effect, these organizations encourage association, and this in turn enhances autonomy. This finding regarding the power that elderly people have to self-convene and associate themselves, translates into a different type of participation, since the demand for services can be closely linked to the convening ability of each club. Additionally, the clubs for elderly people generally use social offices and, in this way, they generate a link with the territory, which makes possible to organize activities outside the municipality, therefore, autonomy and empowerment are likely to be greater in the self-organization. In addition, the territorial deployment is observed in the case of clubs and social office, whose amount is higher than ten in at least eight communes, which in turn indicates the relevance of this type of organization and use of social office.

4 CONCLUSIONS

Finally, changing from a traditional approach to an emerging approach is a way to reframe old age focusing on this stage in which people can play different roles in society and empower themselves. From the perspective of design, according to the different authors mentioned, it is an important social matter, therefore, it requires certain situations to be activated to generate changes that promote a positive vision regarding elderly people.

In addition, it is essential to show what is currently happening with the elderly in relation to the municipalities, by the number of workshops they offer and co-production of services, added to the private and public value that this generates. On the other hand, when raising information regarding workshops and services and when generating categories and a visualization of these data, trends, interests and possible fields of action are observed, enhancing what already exists or proposing new topics or recombination, since this also shows what is needed when comparing the different communes of the Metropolitan Region and when crossing data such as resources or gender, other findings and topics to be investigated appear.

However, the most important finding of the ongoing research and what needs to be deepened is the relationship among municipalities, social offices, clubs for elderly people and the elderly, since in the case of the clubs, they can promote processes of empowerment and autonomy, because of the capacity of association that is required to maintain these clubs, along with the large territorial deployment that is observed in some communes that have more than 60 clubs for elderly people. Consequently, one area to continue researching is the self-generation of initiatives by the elderly to promote autonomy and empowerment, taking as starting point the social office, clubs and forms of organization. As a result, the role of the designer is to be a social actor (Manzini, 2014) that encourages the processes of empowerment and new roles for elderly people.

REFERENCES

Amuch, 2017. *El trabajo de los municipios a favor del Adulto Mayor*. [pdf] AMUCH, Available in: <http://www.amuch.cl/wp-content/uploads/2017/05/ESTUDIO-MUNICIPIOS-Y-TERCERA-EDAD.pdf> [Access 10 enero 2018].

Ávila, C. y Linares, F. 2006. Léxico y discurso de la moda. *Comunicar,* [On line] (27), pp. 35–41. Avalaible in: <http://www.redalyc.org/articulo.oa?id=15802706> [Access 15 January 2018].

Bell, B., & Wakeford, K. (Eds.), 2008. *Expanding architecture: Design as activism.* New York: MetropolisBooks.

Casen, 2015. *Adultos mayores. Síntesis de resultados*. [pdf] Ministerio de desarrollo social, Avalaible in <http://observatorio.ministeriodesarrollosocial.gob.cl/casen-multidimensional/casen/docs/CASEN_2015_Resultados_adultos_mayores.pdf> [Access 10 January2018].

Cepal, 2017. *Derechos de las personas mayores: retos para la interdependencia y autonomía.* [e-book] CEPAL, Available in: CEPAL publicaciones <https://www.cepal.org/es/publicaciones/41471-derechos-personas-mayores-retos-la-interdependencia-autonomia> [Access 30 june 2017].

Ferrara, L., & Sabat, M. (2013). Innovation without boundaries: ecology of innovation and municipal service design. In: E, Manzini & E, Staszowski (eds). 2013. *Public and collaborative: Exploring the intersection of design, social innovation and public policy*. DESIS network. pp. 105–115.

Huenchuan, L., 2011. *Los derechos de las personas mayores Materiales de estudio y divulgación Modulo 1: Hacia un cambio de paradigma sobre el envejecimiento y la vejez*. [pdf] Santiago, Chile: Naciones Unidas, Available in: <http://www.cepal.org/celade/noticias/documentosdetrabajo/2/43682/Modulo_1.pdf> [Access 30 June 2017].

Kolko, J., 2010. Abductive thinking and sensemaking: The drivers of design synthesis. *MIT's Design Issues*, 26(1), pp. 15–28.

LIP, 2017. *La co-producción del usuario en los servicios públicos Documento de Trabajo N°1*. [pdf] LIP, Disponible en: <http://lipuc.cl/wp-content/uploads/2017/07/Paper-LIP_Servicios.pdf> [Access 20 June 2017].

Manzini, E., 2014. Making things happen: Social innovation and design. *MIT's Design Issues*, 30(1), 57–66.

Portal Transparencia, 2018. *Que es transparencia*. [On line] Available in: <https://www.portaltransparencia.cl/PortalPdT/web/guest/ley-de-transparencia> [Access 10 February 2018].

Sanders, E. B. N., & Stappers, P. J., 2008. Co-creation and the new landscapes of design. *Co-design*, 4(1), pp. 5–18.

Sangiorgi, D., 2011. Transformative services and transformation design. *International Journal of Design*, 5(2), pp. 29–40.

Senama (2015). *Políticas Pública en el marco de los Derechos Humanos: promoción, prevención y difusión*. [pdf] SENAMA, Available in: <http://www.sstalcahuano.cl/file/jornada_gg/PPTUST.pdf> [Access 10 February 2018].

Snoad, N., 2012. System Design: Working with Change. *The Journal of Design Strategies, Transdisciplinary Design*. New York: Parsons The New School for Design, 5(1), pp. 55–77.

Thumala, D., Arnold, M., Massad, C., & Herrera, F., 2015. *Inclusión y exclusión social de las personas mayoresenChile*. Santiago: SENAMA-FACSO U. de Chile Ediciones, Disponibleen: <http://www.senama.gob.cl/storage/docs/Cuarta-Encuesta-Nacional-Inclusio%CC%81n-Exclusion-Social-de-las-Personas-Mayores-en-Chile–2015.pdf> [Access 12 June 2017].

United Nations, 2002. *Declaración política y Plan de Acción Internacional de Madrid sobre el Envejecimiento*. [pdf] Nueva York: Naciones Unidas, Available in: <https://social.un.org/ageing-working-group/documents/mipaa-sp.pdf> [Access 10 February 2018].

WHO, 2018. *Datos interesantes acerca del envejecimiento*. [On line] Available in: <http://www.who.int/ageing/about/facts/es/> [Access 10 June 2017].

Dolce and Gabbana and Louis Vuitton: A study on the effect of gender stereotyping in the 21st century fashion campaigns

L. Watson
University for the Creative Arts, Epsom, UK

ABSTRACT: Throughout modern and contemporary Western society, we have kept the traditional constraints of femininity and masculinity as the definition of gender identity. The thought of someone not conforming to these constraints brings an anxiety within societal views that these people cannot be considered part of our 'culture.' Using the theories of Hollander, Craik, Dyer and Veblen this papers purpose is to explore the traditional gender views alongside gender fluidity within fashion brands' Louis Vuitton and Dolce and Gabbana 2016 Spring/Summer advertisements to give an understanding towards the 'other' gender identity and to bring insight within this community of brands who differ from traditional and give a persona within the gender fluidity identity.

1 INTRODUCTION

Many young people in Western society have become open to the idea of being different within who they are and what they identify as, most brands and society as a whole have not adapted to these ideals. Fashion brands have been narrowed into the belief of gender as a sex: male and female, with collections under the same genre. An advertisement has a tendency to stereotype what is to be male and female as a way to sell an idea to the consumer of what they should be striving for. While advertisements are a way to define who the consumer would be for their brand, the simulacrum of consumer has always been a male in male clothes or a female in female clothes.

It can be argued that many women are able to delve into 'masculine' clothing without tarnishing what it is to be feminine, which might be because society has given a leeway for women to trial different paths of fashion, which is understood as being 'feminine'. Although women have a slight freedom to clothing, if they don't present the 'right' performance in men's clothes, they can be seen as too masculine and not fit into societal stereotypes.

To discuss the matter, this paper will analyse images from the advertisements of Dolce and Gabbana and Louis Vuitton Spring/Summer 2016 which have the representation of both male and female identities. This will lead us to the ideals that Louis Vuitton and Dolce and Gabbana have on how gender should be understood within their brands' aesthetic. We will be looking at the traditional femininity and masculinity of Dolce and Gabbana versus the androgyny and fluidity or gender within Louis Vuitton. The two campaigns are a "window" to understanding how gender stereotypes are performed, traditionally or not, in a broader sense.

Authors such as Veblen (2009) and Dyer (2002) have the constraining views of traditional masculinity and femininity, and these will be offset with the theories of authors Hollander and Bolin to argue how different people's views become a relying factor on how gender is seen within Western society and how brands will use these concepts as a way to choose the aesthetic they believe to provide a sense of who they are.

The Dolce and Gabbana advertisement shows how the brand uses traditional gender stereotyping within femininity and masculinity, to create a scene for the consumer to believe they will obtain if they buy their clothing they will relive this scene in day to day life. We will see for the people who don't conform to Western standards that they do not have a part in the story of Dolce and Gabbana a brand which has a specific simulacrum of consumer they are targeting.

The Louis Vuitton advertisement appears in this paper as a transition between the traditional to the fluid. The brand has a different view of gender within this advertisement with the male model, Jaden Smith, wearing a skirt and feminised fabrics within the top and leather jacket, alongside women wearing masculine silhouettes and a similar leather jacket combined with trousers printed in bold colours. Contrary to Dolce and Gabbana, the brand prefers alienating the traditional stereotypes and targeting those who don't want to conform as advocates of their brand.

Throughout the Dolce and Gabbana and Louis Vuitton advertisements, we will see how both brands stand for different viewings of gender in terms of how it will affect their consumer profiling, or who will be attracted to their brand—with this view in mind, the brands' use of gender identity is one way of differentiating what brands want to see a change in how we perceive gender what which do not.

2 TRADITIONAL FEMININITY AND MASCULINITY: DOLCE AND GABBANA

Dolce and Gabbana's first presented their collection in 1985, as part of the 'New Talent Group' in Italy. Their first collection, from 1986, was named 'Real Woman' (Segre, n.d). The brand's main descriptions focused on the 'Mediterranean style' which they source inspiration from the Sicily of Luchino Visconti's 1963 film *The Leopard*, as well as the Italian style in which they have seen the woman around them whilst growing up. The Dolce and Gabbana woman is seen as someone who is traditional, an Italian woman who is devoted to the church, strong and provocative (Segre, n.d). Dolce and Gabbana used inspiration from traditional timeless pieces from the past alongside their bright colours and prints to bring forward an innovative yet classic style within their clothing. The idea of the hyper feminine and hyper sexualisation of a woman within the 1980s coincides with the presentation of Dolce and Gabbana's collection within this era—showing that by their brand having started in that epoch, their ideas and designs have been influenced by where they once started, but also had an important part in constructing the femininity we associate with the 1980s.

Throughout the Dolce and Gabbana franchise, the main factors of traditional gender stereotypes persist—Dolce and Gabbana create a scene of which a woman should do with a purpose of entertaining the men around them, a need to be in a romantic city, specifically in *Italia,* within this advertisement and how you should be colourful, adorned in jewellery and makeup to have this lifestyle.

There is a need in Western society for women to be the ideal feminine beauty, something maintained within fashion magazines and advertisements that construct the gender identification (Ash and Wilson, 1992, p. 25) to fit with 'trends' and what society will accept—the robust, slim figured, light and delicate woman associated with the perfect attitude and poise needed to be the complete feminine (Veblen 2009, p. 95), there will then be the ideal feminine that everyone around her wants her to be. The woman is seen as useless in everything but entertaining the man, she is expensive (Dyer, 2002, p. 95) but is there for personal gain of the man who has her as their object—for both admiration and a show. This ties in with the persistent theme of men being the *alpha*, or the one who controls and owns the woman who, in return, acts perfectly and is an object to be watched. The alpha has a need for his women to be the best, which—this acts as a reflection on the man and how much of an alpha he is, and the need for the man to recognise and pursue pretty women (Wolf, 1990) in order to be the best he can be.

Figure 1 portrays the advertisement title 'Italia is Love' with red and white striped canopies and Italian words written across it, giving us the definitive location of the photographs: *Italia*. Two women are standing upon the table in brightly coloured clothing being admired by men who are sitting and standing around the table in black suits. On each side of the photograph, local women are being wooed by the men with one woman enjoying the scene whilst sitting down.

Figure 1. Dolce and Gabbana 'Italia is Love' s/s Campaign: The Show (2016). Two women standing on a table entertaining the men surrounding them: two of the locals are being wooed by the men and another local is sitting and watching "the show" outside of an Italian restaurant.

This description of the image starts to create the idea of both femininity and masculinity within the brand of Dolce and Gabbana constructing a traditional role where the woman cares about her looks and the attention she is getting, she is the one grabbing the attention of the man and pleasing him, she carries this role of being adorned in bright colours and many accessories to create the ideal feminine which people who buy Dolce and Gabbana will want to be; someone with poise who sits correctly and someone who is confident and will please the man. The theory of the leisure woman is seen as having an abstinence of productive work because of what she is wearing (Veblen, 2009, p. 105)—the outfits seen on the models are something that is clearly not for people who work in any type of labour—they are for being pretty and for pecuniary leisure.

In this photograph, all the men are dressed the same and are seated or standing in a similar position while all the women are different—dressed differently, standing differently, and displaying different ages. The men are not sexualised in this picture; yet the women are sexualised throughout, performing for the men around them and being the centre of attention in the advertisement, even though it is a Womenswear campaign. The whole point of them being there is to be watched by the men around them, hinting that the women who buy from the brand will want the same lifestyle if they want to be a "Dolce and Gabbana woman."

The 'gentleman', being the character pursued in this advertisement, is a well mannered—seeking leisure (Craik, 1994, p. 181) he would be someone of wealth that would be able to do whatever he wanted when he wanted—much like the men in Figure 1, they are being entertained by the woman around them. For men like this, clothes are a display of their strength and power rather than any desire for attention (Craik, 1994, p. 192). Men's fashion is for fit and comfort (Craik, 1994, p. 176), the suit shown in Figure 1 appearing as something that is a constant through men's dress which barely changes through time. It is timeless, something that can be worn day to day. The idea of menswear as a uniform can be seen through how each man in the photograph is wearing the same suit, because it is the epitome of men's fashion.

The stereotypes of women presented in Veblen is the woman and her children are guilty of conspicuous consumption, whereas the man ceases to consume valuable goods and focuses on producing wealth instead (Veblen, 2009, p. 67). This creates a construction not only of gender and how women are seen as flaunting and worrying about how people see them, but also the social class divide which can be seen in the fact that Dolce and Gabbana is a high-end brand, and thus catering to a specific purchase power.

To define femininity and masculinity, we have to look at the stereotypes specifically in conjunction with Western society in the present. In terms of how men and women are defined; men have always been seen in the 'important' roles where they have an authority compared to women, people are more likely to listen to men than women as men are able to provide a "dependable voiceover" (Gauntlett, 2003, p. 55). We can see that with femininity and masculinity being so inherently different together, they make the perfect match (Dyer, 2002). For masculinity, males typically have to win that title and prove they are masculine to reclaim this image (Malebranche and Donovan, 2013) this would be defined as the man being the money-bringer and the person most in control in terms of relationships and job-wise. Masculinity and femininity cannot be completely defined without one another. Femininity is seen as something that needs to be earned, with both genders being conditioned from an early age to react to specific toys, colours, and accessories. We are also taught specific routines for each gender—caring about how you look if you are a girl, playing sports if you are a boy; we are also taught how to act from an early age—boys are taught to be dominant while girls are taught to act passively, 'lady-like,' and practising things associated with fertility and nurturing (Craik, 1994, p. 42) like playing with a baby doll.

For the people that don't conform to these feminine and masculine traditions within gender, there is no leeway, especially in this campaign, to relate themselves to what they are seeing. Dolce and Gabbana have a very constricted and traditional view of gender, one in which you have to be a completely feminine female or a completely masculine male to fit their expected simulacra. There is no room for the androgynous or the fluid: it is a constricted tradition that hasn't change as of this campaign.

3 GENDER FLUIDITY: LOUIS VUITTON

Louis Vuitton opened his shop in 1854 after working as a box maker and packer since his teens. It was his trunks that became a timeless masterpiece and innovation for the designer. While most trunks had been previously made in leather, Vuitton chose his trunks to be made in grey canvas, lighter and more durable—it was more 'impervious' to water and odours. All other trunks were dome shaped—Vuitton's, however, are rectangular which would stack and be more convenient for travellers, when compared to the dome-shaped counterparts. Vuitton's innovation gives a modern embodiment to the brand throughout the time that its been established. Louis Vuitton's vision responsible for the innovation of trunks at that time became the back bone of how the brand would take on their clothing. The modifications of timeless pieces stuck through the brands philosophy where things need to be changed so that they can grow and change with how we live our lives—which is the same approach to gender ideals; they can see how important it is to adjust how things are made or seen, so that their products can spark change in society.

In opposition to the strategies presented by Dolce & Gabbana, Louis Vuitton uses role reversals to bring attention to the daring garments with bold poses and expressions. Jaden Smith—son of Will Smith—is an actor who has delved into gender fluidity throughout his teenage years, with his favourite item to wear being the skirt. He wears skirts in order to 'take the brunt' of bullying for future generations (Blair, 2016). While Jaden Smith brings forward this role reversal, he is able to create a new simulacrum of men where femininity can go alongside masculinity. The androgyny the women bring forward gives people who don't conform a way to have both masculine and feminine attributes creating their own gender identity—it is a starting point for an acceptance of who people are.

Figure 2 shows Jaden Smith with three female models. Jaden Smith keeps to the same feminine aesthetic of soft shoulders, soft face and eyes closed. The women all have a very strong poses—all in 'fight' poses, fists, strong legs bent backs, have strong facial expressions where they are either looking at the camera or slightly off. Smith is the exact opposite. The image connotes the reversal of the roles presented in the previous section; while the women

Figure 2. Louis Vuitton Series 4 s/s Campaign (2016). Three women standing in fighting poses next to Jaden Smith in a softer wider legged pose—in a city location, pavement and brick wall with elements of a fire exit in the shadow.

are strong and central to the photograph, Jaden Smith is off to the side and soft in his pose which is uncommon for a stereotypical male. Jaden Smith keeps to the fluidity, where he mixes masculine and feminine. What Louis Vuitton brings with the women are the connotations of androgyny: women in strong poses, even when wearing the Womenswear collection. While Jaden Smith is wearing the Womenswear collection to our knowledge, the Louis Vuitton menswear campaign will have a similar aesthetic in avoiding roles of gender.

Men have recently had an opening into freedoms of clothing and accessories while women have been constricted to what society wants them to conform to (Hollander, 1994, p. 22). In the present, women do have a wider choice of what they wear, whether it is seen as more masculine or more feminine. This advertisement gives a freedom to the models to reverse gender roles where Jaden Smith is able to release a more feminised pose, Smith is wearing a skirt alongside a masculinised leather jacket and softened fabrics underneath with his hair over his face, alongside the women models taking more masculine poses—this being the stern expressions, fists in front of them for the women (all women models in Figure 2) and soft shoulders, closed eyes and lifted head against strong shoulders of the man (Jaden Smith in Figure 2). This gives wider context of how gender identity is a way to bring both the personal and social side of who they are as an individual and how they are within society (Bolin, 1996)—Smith is able to show himself as a calm, emotional person through his posture that brings an acceptance for who he is because he is confident enough with his identity to be in this state.

Throughout this advertisement though, Jaden Smith is not wearing any accessories or visible makeup. Men are now more comfortable to spend money on clothing and care about what they look like, but they still aren't comfortable buying Womenswear. Even if they are, it is not widely accepted in Western society. Having someone like Jaden Smith in the campaign, who is not afraid to wear Womenswear as a model or as "himself," starts a wave of understanding gender fluidity, even if it is only a small minority. Having said that, he is still a model that has been paid to wear Louis Vuitton—he is a good match that flirts with gender fluidity in real life.

While Freud argues that a woman's masculinity soon diminishes once they have fully developed into their femininity, he also sees that a majority of men are far behind the masculine ideals of Western society (Freud, 1925). Looking into Louis Vuitton's campaign, it is possible to see a more open attitude towards people who don't conform to Western gender stereotypes. This leads to how, in many ways, the body itself does not conform to these norms (Butler, 2011): one will not always have small, delicate hands because they are born female, or strong broad shoulders when they're born male—the relation between gender techniques and biological sex is not automatic. This idea has become a way to separate (Hollander, 1994) and differentiate (Dyer, 2002, p. 24) men and women into sections that eventually made one inferior to the other. For women, with most advertisements conforming to the feminine standards we have been taught, Louis Vuitton is giving a freedom where the women can be comfortable not conforming Western society standards.

For Louis Vuitton themselves, a place is offered to those who don't conform, a representation of what they see themselves as in one way or another. This means that Vuitton doesn't want to conform to traditionally feminine and masculine ideals within their brand—bringing back the idea that the main part of Vuitton's philosophy is to innovate and adapt however it sees fit and, by bringing in someone like Jaden Smith, who is known to slip through the boundaries of gender, creates the simulacrum of who the consumer of Louis Vuitton is and how the brand is experimenting with a new idea of gender. For the people that don't conform to traditional gender stereotypes, the Louis Vuitton advertisement provides a platform where this specific group can see a representation of themselves. This may not be an exact representation of everyone but it brings a starting point for bringing in the identities of people who aren't representing a traditional gender identity.

4 CONCLUSION

There is a difference between personal identity and social identity; personal identity can only be recognised by the person themselves as everyone has an opinion of who they are, whereas social identity is formed through "sociocultural recognition" or being categorised into a gender identity society recognises (Bolin, 1996, p. 20). The advertisements analysed in this paper present a challenge for both brands to bring a consumer profile that will match who will buy their brands: Dolce and Gabbana, on the one hand is traditional, timeless, and set on a socially validated femininity and masculinity, which limits who feel they can relate to the brand, as few people can completely conform to these constraints; Louis Vuitton, on the one other hand, seem to be more open to the fluidity within gender and delves into androgyny. This is only one way of perceiving gender—the Louis Vuitton way—while one person may relate to this brand someone else may not, even if they both don't conform to feminine and masculine traditional binaries, a form of identity, still not fully validated by mainstream society.

Gender fluidity and androgyny within the Louis Vuitton advertisement, encourages the thought that the 'gender norm' we have been taught has become something so outdated that society needs to reconsider how people can relate to forming stereotypes. Although the Dolce and Gabbana traditional ideas of gender are constraining, this is still an identity some people do associate themselves with. There are some who believe they do completely fit to these standards—but saying they are the complete epitome of gender is damning for those who don't completely conform.

We have to remark that, through the way that the brands have grown and developed, a major part of ideas and ideals of gender dates back to when the brands first started to create their ethics within fashion. While Dolce and Gabbana kept to the 1980s hyper sexualised femininity, Louis Vuitton have adapted how their brand develops in a way that means a non-conforming role within gender, which fits perfectly within the innovative side that granted Vuitton their reputation.

We can see through Western society that the traditional terms of gender will be a running theme throughout history, as it is something we, as a culture, have learned through time—yet with brands starting to fulfil the needs of a fluid gender, this has started to break these traditions. People who don't conform to Western stereotypes now can see themselves associated with fashion and advertisements in a way that is not offensive and that is not limiting who they are. The identities we have as individuals come from our different make-ups and personal traits that cannot be conditioned into two categories: we can't all be the same, and being born as one sex won't automatically align with the characteristics considered "correct" for the corresponding construct of gender. That being said, advertising will always have a way to imitate real life—Advertisement gives the perfect life that people can hold only if they buy from these brands. They provide imaginary narratives people can associate with their future life, to build a place of happiness which comes associated with the purchase act. While advertisement can be misleading, it still gives a framework for people to associate themselves with. This has perhaps been true for people who do feel they conform to traditional femininity and masculinity in the past, but now there is a spiral in how people are feeling towards these traditions; there's a need for an 'update' on how brands feed gender as a whole rather than these two constrictions into the brands aesthetic.

REFERENCES

Ash, J. and Wilson, E, 1992. *Chic Thrills: A Fashion Reader*. London: Pandora.

Blair, O, 2016. *Jaden Smith explains he wears a skirt so future generations wont get bullied for not conforming to gender stereotypes* [news article online] at: http:///www.independent.co.uk/news/people/jaden-smith-skirt-gender-fluid-future-generations-bullied-not-conforming-to-gender-stereo-types-A712591.HTML (Accessed on 18 February).

Bolin, A, 1996. *Traversing gender: cultural context and gender practices.* In: Ramet, S.P (ed.) *Gender reversals and gender cultures: an introduction.* (s. l): Routledge. pp. 24–36.

Butler, J, 2011. *Gender Trouble: Feminism and the Subversion of Identity*. London: Routledge.

Craik, J, 1994. *The Face of Fashion: Cultural Studies in Fashion*. London: Routledge.

Dyer, R, 2002. *The Matter of Images: Essays on Representations*. (s. l.): Routledge.

Figure 1. Dolce, D, Gabbana, S. 2016. *Italia is Love* [Photograph] At: http://ftape.com/media/dolce-gabbana-ss16-campaign-italiaislove/ (Accessed on: 18 April).

Figure 2. Teller, J, Weber, B. 2016 *Series 4* [Photograph] At: http://www.lebook.com/creative/louis-vuitton-series-4-advertising-2016 (Accessed on: 18 April).

Freud, S, 1925. *Some Psychical Consequences of the Anatomical Distinction Between the Sexes* At: https://pdfs.semanticscholar.org/68ec/20a7889578b1e7c029a412f0a65c2d6a1d9d.pdf (Accessed on the 2 October).

Gauntlett, D, 2008. *Media, Gender and Identity: An Introduction*. Abingdon: Routledge.

Hollander, A, 1994. *Sex and Suits*. New York: Alfred A.

Malebranche, J. and Donovan, J, 2013. *Anrophilia: A Manifesto: Rejecting the Gay Identity, Reclaiming Masculinity*. (s. l.): SCB Distributors.

Roof, J, 2016. *What Gender Is, What Gender Does.* Minneapolis: University of Minnesota Press.

Segre, S, n.d. *Dolce and Gabbana brand* history [News Article Online] At: https://fashion-history.loveto-know.com/fashion-clothing-industry/fashion-designers/dolce-gabbana-brand-history (Accessed on: 18 April).

The Biography.com, 2014. *Louis Vuitton biography* [News Article Online] At: https://www.biography.com/people/louis-vuitton-17112264 (Accessed on: 18 April).

Veblen, T, 2009. *The Theory of the Leisure Class*. (s. l.): Oxford University Press.

Fashion, identity and culture in the Chinese fashion system. China, inspiration for Chinese fashion designers

R. Gaddi
Politecnico di Milano, Milan, Italy

ABSTRACT: In the long history of fashion, China has always had a place of great importance. For thousands of years it has been a synonymous of craftsmanship and quality: silk, sculptures, ceramics, the art of printing and writing. From the end of the twentieth century on, China has preferred to focus on production volumes and cost reduction, becoming a large global factory. China has opened up to the West to such an extent that it has also been influenced by its own lifestyle, to the detriment of its own identity growth. With the arrival of the big international brands on the local market, the systematic reproduction of clothes and accessories started by Chinese companies, which opened the era of a real copy of international fashion destined for the domestic market. This movement has continued up until today, despite a new phase seems to open on the horizon, leaving room for local creativity. This works tells the project of the fashion designer Chun He, who tries to reverse this trend by repositioning the territory and the Chinese tradition at the centre of the design activity.

1 INTRODUCTION – THE EVOLUTION OF CLOTHING STYLES IN CONTEMPORARY CHINA: A BRIEF HISTORY

Before the middle of the nineteenth century we cannot speak of fashion as a seasonal style, but only of customs and customs characteristic of the different dynasties: the Chinese were capable of excellent craftsmanship, creating dresses embellished with complex embroideries and decorations, but the garments they did not change at the time. Since the nineteenth century there were migrations of European merchants who spread the western tailoring in China, which hybridized with local traditions: an example was the spread of quipao, a traditional women's dress with large side slits that highlighted the beauty of oriental women. In the era of President Mao, clothing became uniform, for the military as well as for the citizens. The classic jacket, neutral in colours and shapes, would be worn by everyone, until 1990. In those same years, after the fall of Mao, begins the era of a women's fashion that, despite being stunted, begins to inspire in coarse way to Europe and to Western stylists.

Historians of fashion and costume commonly illustrate the evolution of contemporary Chinese clothing dividing four main historical periods: the late Qing dynasty, the Republican period, the Maoist period and the Reform era.

Late Qing dynasty (1644–1911), like many of its predecessors, maintained a rigid clothing code, with strong sumptuary rules for the imperial family such as for common people. The Qing mandarins were supposed to dress according to their status and social rank. This hierarchical way of dressing was not just legitimized by law. But also coded in morality: a person who dressed appropriately knew his place in society according to the Confucian concept of li, that governed all social relations and behaviours and was necessary to be followed in order to be different from barbarians. Clothing was an integral part of the imperial social order, just as described by the ancient adage "Chui yi change r tian xia zhi" ("as long as the clothing system is maintained, the social order is maintained"). Despite these strict rules, the revolution that collapsed the Imperial system for the establishment of the Republic of China, was even in the need of change of clothing style. The contact with Western states and Japan led the diffusion of different clothing styles, such as Western suits.

The Republican period (1912–1949) experienced a genuine plurality of clothing styles: more traditional changpao and magua from Qing, Western-style as suits, and new inventions as zhongshanzhuang for men and quipao for women, as hybridisation of the first two genres.

The rise to power of Mao Zedong in 1949 led to the creation of the People's Republic of China, that is historically divided into two periods: the Mao era up to the 1978 and the Reform period after 1978. This division is full of meaning even in terms of changes in clothing behaviours. The most important impact on what people wore during the Mao period was the glorification of workers and farmers, and their clothing. The qipao, for instance, was considered inappropriate, decadent, middle-class, and, furthermore, risky. The state promoted firmly sober lifestyle.

Even if no regime change occurred, the post-Mao Reform period adopted economic reform policies in China, and open-door policies with foreign countries.

The state moved from the ideological issues of Mao and entered into a renewed era of economic development, which also meant learning from the developed West. In addition to the revival of the Western suit and the—for example—traditional quipao, new styles arrived initially from Hong Kong and southern coastal areas: jeans, t-shirts, and casual and sportswear style as jumpers, jackets, blazers, wool sweaters and down jackets.

The first designer who approached the Chinese market was Pierre Cardin, who paraded in Beijing in 1979. Then, in the nineties came Dior, Gucci and Chanel. Following this success of Western names and brands (China was experiencing a new era even on Western designer catwalks and a consequent boom of fashion media) in the 1990s and 2000s even domestic luxury as brands emerged. This process continue even today, clothing styles are more independent from political control and more subject to market and popular culture influences.

This independence and changeability is essential and necessary to the functioning of fashion: it is so not strange that just in the Reform period Chinese fashion professional emerged.

1.1 *The impact of Chinese aesthetics in international fashion: The exhibition "China: Through the Looking Glass" (New York Met Museum, 2015)*

One of the most interesting fashion shows in recent years has certainly been "China: Through the Looking Glass", set up in the summer of 2015 at the Costume Institute of the Metropolitan Museum in New York. The exhibition was an exploration of the enormous impact of Chinese aesthetics on Western fashion.

A thorough research on how China has been able to feed for centuries the imagination of couturiers and fashion designers all over the world, even if, as we said, excluding the Chinese ones. The great aesthetic and formal local tradition continued to have no appeal for the Chinese creative class, which was still struggling to emerge for certain social reasons and for the strong tendency towards homologation that the regime had imposed, very difficult to erase easily.

In the halls of the Met, Western fashion has been related to Chinese traditional clothing, as well as vases, paintings, porcelains and films that have made manifest the enchantment that the Chinese imagination has always exerted on everything. The West.

> *Since the first contacts, which took place in the 16th century, the European West has been enchanted by Chinese aesthetics, also influencing fashion that, from Paul Poiret to Yves Saint Laurent, used China as a mirror to reflect different stylistic references. make fantastic pastiche of Chinese aesthetic and cultural traditions with Western forms" says Costume Institute curator Andrew Bolton.*

On show, 140 haute couture and pret-à-porter dresses by designers such as Cristobal Balenciaga, Bulgari, Alexander McQueen, Cartier, Roberto Cavalli, Coco Chanel, Christian Dior, Tom Ford, John Galliano, Jean Paul Gaultier, Valentino Marc Jacobs, Karl Lagerfeld, Jeanne

Lanvin, Ralph Lauren, Louboutin, Martin Margiela, Alexander McQueen, Givenchy, Dries van Noten, Jean Patou, Paul Poiret, Yves Saint Laurent, Paul Smith, Vivienne Westwood.

2   THE EMERGENCE OF THE CHINESE CREATIVE CLASS

To underline the condition of the local creative class, in an article written for Vogue Italia in April 2011, Franca Sozzani spoke of the almost total lack of interest in the origins and local traditions, which led to a substantial creative immobility, although full of hopes, of Chinese fashion:

When the Japanese designers like Rei Kawakubo and Yohji Yamamoto in the early '80s, or even in the '70s Kenzo and Issey Miyake or in the 90s Junya Watanabe arrived in Paris to show off their col-lections, loved or hated, had brought a new style, unusual, breaking, as I said other times, all the rules of cutting and existing shapes. It was a revolution.

Here, in the current Chinese generation, waiting for the stroke of genius of one of the emerging that is emerging now, there is no sign of radical change. For most of them the western style, a little sexy, a little minimalist, a little baroque, a little decorative is the example to follow. A bit of everything without a real choice.

Almost no one remembers their origins and traditions of their country, to change them and give life to a new way to propose themselves. They are all so fascinated by the West to look at our stylists and not look for themselves in a different source of inspiration. Yes, for now the landscape is still barely mentioned, but we must also say that it is only the beginning of a new generation. Until now they have produced for others and the idea of creating for oneself was not taken into consideration. Western stylists were the only point of reference in the fashion world.

Now they have started anyway and surely something interesting will come out. As in art, where a new artist is proposed every day and where everyone swears that the one chosen by famous galleries will be the new genius of art. We will see it there too.

But slowly, as the visionary director of Vogue Italia had imagined, things are changing.

A 2015 article published in the American magazine Forbes tells how new Chinese designers are working trying to make themselves known abroad, while the enrolment of Chinese students in fashion schools is also increasing. The reforms that have been adopted have allowed the birth of the fashion industry and therefore the emergence of creative designers and stylists, each with a personal interpretation of the traditional Chinese style.

Wang Yiyang designed a collection called "Nothing" and inspired by the twelve signs of the Chinese horoscope, earning a review of the Wall Street Journal.

Designer Laurence Xu, one of the first high-fashioned Chinese fashion designers invited to haute couture fashion shows in Paris, produces high-fashion clothes in which red and green colours are recurrent, as in the Chinese costume tradition.

Forbes also told in a long article the story of the Chinese designer Guo Pei. Despite of creating high fashion clothes for almost twenty years and has always had a great success in China, Pei has become famous only recently thanks to the showy yellow dress that singer Rihanna wore during the evening of the Met Ball 2015, the event that annually opens a fashion show at the Costume Institute of the Metropolitan Museum of Art in New York.

These examples, to which could be added brands like Emperor and Wensli, Chinese brands well known in the local market, certainly have the merit of having a strong identity linked to the territory of origin, but also the disadvantage of not being designed thinking at the international market. And in fact, it is difficult for them to make themselves known outside the national borders.

One of the main reasons is certainly a lack of attention to the stylistic demands of the international market, certainly less conspicuous and sought after in forms and manufacturing compared to the Chinese market.

3 THE ENHANCEMENT OF THE TERRITORY AND TRADITION FOR DESIGN-DRIVEN INNOVATION: THE CHUN PROJECT

An attempt to reverse this trend was led by the CHUN project, developed in 2016 by Chun He, a Chinese fashion designer based in Milan who graduated in Fashion Design at the School of Design of Politecnico di Milano.

Despite the rich, long and intense Chinese civilization and comparing it with today's global industrial culture, the Chinese designer has understood that even with the will to leave national borders, Chinese brands still do not have the strength to self-determine in an international context. Realising how today's competition between industrial forces, the main role is played by the international recognition of the brand.

Chun He has understood how the brand identity can give today a very strong push towards the promotion of the country itself. At present, China is proposing a strategy of "power" still too much based on massive production, rather than on themes such as creativity, innovation and sustainable development. The essential themes such as storytelling, product fascination are missing. Themes that create a not only visual link with the customer, but a real shopping experience capable of creating the necessary customer engagement that the fashion market requires. To allow significant customer loyalty, it is necessary to hybridize the aspirations of both parties.

Then, starting from this theme, in the first part of Chun He's work, a research was carried out on the long history of aesthetic canons and Chinese beauty, where the in-depth study of the decorative motifs of the traditional local gave the basis for the design of the line of foulard which finished products we will show later.

The second part focused on the analysis of the Chinese consumer market, on the trends of the most recent theories of the brand and on the development strategies. It analysed the most significant studies of the market (European and local) of luxury, as developed in China Today, despite the substantial urban and extra-urban areas condemned to extreme poverty, with the increase in production, consumption and internal revenues, the exponential growth of the purchasing power of its more well-off classes is being measured.

The study of the entire supply chain, from design to distribution, led to the creation of an ideal model to understand how to develop a local foulard brand that has the characteristics and specificities of a Chinese product compatible with the needs of the international market, and therefore directed to both domestic consumption and export.

As has been said, the hybridization of the European model with the Chinese one was the key project: hybridization of European aesthetic styles in terms of geometry and col-ours, combined with the most classical references of the Chinese aesthetic tradition. An-other fundamental synergy was carried out on the sales strategies (online and offline) currently used in the Chinese high-end market, to allow it to align with those used on the international market.

Figure 1. "CHUN" original brand logo.

Figure 2. Lei Shen line, inspired by the God of thunder.

Figure 3. Lei Shen, God of thunder, original inspiration.

Figure 4. Lei Shen, the God of thunder, definition of the recurring elements.

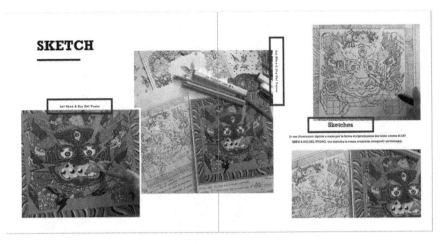

Figure 5. Lei Shen, the God of thunder, freehand sketches (preparation of digitization).

Figure 6. Lei Shen, the God of thunder, finished product and shooting.

Figure 7. Panda Line, Abstract. This line of scarves is inspired by the Chinese giant panda in the mountainous areas of the Sichuan province.

Figure 8. Panda Line, original inspiration.

Figure 9. Panda Line, preparatory sketches.

Figure 10. Panda line, finished foulard.

Figure 11. "Chun" logotype and its application on some examples of packaging.

The final product, a collection of Chinese headscarves explicitly referred to and inspired by the millenarian local culture, is a contemporary proposal and a hybridisation no longer aesthetic and stylistic, and not a real astrophile for lovers of the East, but a product finally not only "made in China", but "designed in China", where culture, territory and Chinese tradition remain at the centre of the creative process.

The project did not stop at production alone but developed a careful design of a corporate identity that also recalled design themes in graphic styles.

If the Chun brand is inspired by Chinese writing, the packaging is instead deliberately neutral and minimalist, favouring the enhancement of the product contained in it.

4 CONCLUSIONS

The Chun project demonstrates, even in the context of an evolving creative sector, how fundamental it is to recognize the intrinsic value of a territory and its unique characteristics, as a non-reproducible resource that leads to a productive and social innovation of extraordinary importance.

Today we realize how disturbing is the global scenario of a cultural scenario absorbed progressively in the economic sphere. We need a design that can reaffirm the primacy of the cultural value of the territory without renouncing the resources offered by the new economy of experience. A project capable of transmitting emotions as well as arousing respect and understanding for the rights of our time, in search of a skillful balance between values of permanence and values of change.

In the creative context of Chinese fashion and style (r)evolution, China once again becomes a source of inspiration, but this time with a new approach, the enhancement of its own specific and local culture, that does not expatriate but remains as an added value for the country.

A useful methodology for both markets: the Chinese one, which finally claims the "right to creativity", and the international one—we think about the crisis of Made in Italy, partly caused by the massive and unparalleled Chinese competition—which can finally afford to "lower the guard" to the Chinese production giant, and maybe even to fear it a little less.

ACKNOWLEDGEMENTS

For the writing of this paper, a first thought and thanks goes to the work of the fashion designer Chun He, author and soul of the "CHUN" project. We also thank all the scientific and organizational staff of this fourth edition of the Cimode International Design Congress.

REFERENCES

Bolton A. (with John Galliano, Adam Geczy, Maxwell K. Hearn, Homay King, Harold Koda, Mei Mei Rado, and Wong Kar Wai), 2015. *China: Through the Looking Glass*. New York: The Metropolitan Museum of Art.
Franca Sozzani, April 2011. E gli stilisti cinesi? E la moda in Cina? *Vogue Italia,* online magazine re-trieved from http://www.vogue.it/magazine/blog-del-direttore/2011/04/1-aprile#sthash.BEg74gFe.dpuf
Glenda Toma, December 2015, China's New wave of designers, *Forbes, online Magazine*. https://www.forbes.com/sites/glendatoma/2015/12/04/china-new-wave-designers/#711be683320f
Laurie Burkitt, November 2012, What China's Trendsetters Are Wearing, *Wall Street Journal*, online Magazine https://www.wsj.com/articles/SB10001424127887324556304578118891337284014
Racinet A., 2003. *The costume history*. Cologne: Taschen.
Williams A.G., 2015. *Fashion China*. London: Thames & Hudson.
(Curated by) Fogg M., 2013. *This is Fashion, London*: Quintessence Editions Limited.

The pleasurable dressing of Loewe: From the store of Martínez-Feduchi to the showroom of Carvajal

A. Cano Redondo
Superior School of Design of Murcia, Spain

A. Martínez-Medina
University of Alicante, Alicante, Spain

ABSTRACT: The commercial premises of the large luxury firms are, simultaneously, the point of contact with the general public and the border that separates it from a dream world, within the reach of only a minority of great purchasing power. They are also, to a certain extent, the image that the brand projects on the city, the dresses in which the business is dressed to exhibit itself in the metropolis. This work goes through Loewe's sales spaces, from its first craft workshops in the second half of the 19th century to the showroom that the firm built at the beginning of the 1960s in Barcelona. A journey that sets its sights on the relationship that these stores establish with the street, in the dialogue with the passer-by and with the city and in its evolution according to the cultural and economic conditions of each time in attention to fashion.

1 BEAUTIFUL WRAPPINGS FOR EXQUISITE WRAPPINGS

The bedroom of Lina Loos in the apartment of number 3 of Bösendorfsersatraße (Vienna, 1903) revolves around the bed. The presence of the same, in the center of the setting, generates all the space. For Xavier Monteys (2014) it is the bed that colonises the room through the prolongation of his clothing, the furniture is the one that contaminates the architecture and not the other way around. In the interiors of Loos, space, as pointed out by Beatriz Colomina (2010), is a dress that has separated so much from the body that it needs an independent support from it. Dresses that define the wrapping: architecture as a body sheath, a warm bag within which to enfold oneself.

The shops, scenographies of leisure and capital, inserted in the unfinished and undefined bases of tall buildings,[1] not only configure the image (always changing and unfinished) of

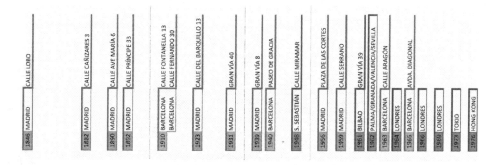

Figure 1. Schedule of the Loewe establishments. Source: Self made.

---

1. As much of the historical plots of the cities, as of the reticles of the functionalist urbanism.

the metropolis, but also showcase the modernity of each era. They are the place from which to show novelties of all kinds, from the products on sale to the technology that allows them to materialise. They are also, from this point of view, envelopes of their own selves. Clothing of the object of sale, the suits that the brand looks to launch among the public that populates the sidewalks. The stores are an object of consumption and their architectures dresses exquisitely tailored to the business, ephemeral designs that embrace the product of sale while projecting an image in accordance with the fashions of the moment. They are not covers, now, for the body. The 'Loosian' coating principle, to some extent, is inverted. The stores are not interior, but exterior. The body always remains outside and the warm bag covers, only, the object of sale.

2   MADRID, 1846–1923. THE LOEWE WORKSHOPS

The origin of the firm Loewe is well known: a small leather goods workshop located in the Madrid street, Calle del Lobo (now renamed Calle Echegaray), opened in 1846; the appearance years later of Enrique Loewe Roessberg as partner of the business and, from that moment, the growth, expansion and transformation of the small family business until its incorporation, 150 years later, to the French business conglomerate Louis Vuitton. But let us focus now on the beginnings, with the objective of taking a journey through these first spaces in which the brand was based.

Although the Calle del Lobo (near Puerta del Sol) presented a good location for the business, the increase in production forced its transfer, in 1882, to a principal on Calle Cañizares and, subsequently, to another in Calle del Ave María, looking for the commercial bustle of Magdalena and Atocha streets. The business was, in effect, focused on workshops rather than shops in their own right, although from this first transfer of the business location, an employee was exclusively responsible for serving customers (Hernández, 1995, p. 38). The workshop on Calle del Lobo first, and those of Cañizares and Ave María afterwards, were, in a certain sense, work spaces more akin to the craft tradition than to the incipient commercial retail tools, fruit of the Industrial Revolution. A shared place of production and sale (standing against the new and irreversible separation of both elements) linked, in addition, to a locality of trade guilds proximity (craft production businesses) and clientele.

In 1892 the business returns near to its first location, in the then prestigious Calle Príncipe. Although it is still a floor on the first floor, the presence of the business on the street is evident by means of two signs: one of greater dimension above the first floor that indicates the type of business ("factory of leather articles", it says), and a small canopy over the entrance to the building where the first Loewe logo is introduced, next to the street number.[2] It is at this time, in the last third of the nineteenth century, when the idea of brand as we understand it now arises (Klein, 2001, p. 23). The logo assumes the image of the brand and aspires to be globally recognised embodying the values of the company it identifies. In the words of Cristina Santamaría (1994, p. 162):

> "A name that sells is not an object of grammar but a fact of economy, a capital asset. It is, therefore, acquirable – like all goods – by means of purchase or theft. But it is, above all, consumable because it enters the market at the same time that it enters culture".

At the turn of the century, Loewe made contact with public space by opening its first stores at street level in Madrid and Barcelona. Sumptuously decorated luxury stores that reproduce in their shop windows and façades the fashions of the era. The communicative scheme, on the other hand, is the literal transfer of the façade of Calle Príncipe to the scale

---

2. Years later, Rafael Bergamín (1927, p. 261) wrote about the façade of Euskalduna showroom: *"The cover: it is the sign that shouts to the public on the street, the brand of the car that is sold inside (…). This is the cover where the name of the brand serves as a guide to the entire street pedestrian walkways, as the numbers of the portals are customarily hidden so that the one who carries a letter is obliged to play to find them".*

of the commercial façade: in Barcelona, in the Fontanella store, the description of the business coexists with the company's logo, while that in the branch of Calle Fernando, opened later, only the logo appears. A logo, however, recognisable by the use of similar typographic families, although variable in terms of format (indiscriminate use of lower or upper case).

In Madrid, the last transfer takes the business to Calle del Barquillo in 1923. Although the architectural scheme reproduces the previous ones, the terms are inverted and the store is now the main attraction, locating the workshops in the basements[3] and homes of the owners in the upper floors. The modern bureaucratic order (the study in the head and the workshops, the force of the work that obeys programs and orders, in the feet) against the artisanal order (Lipovetsky, 1997, pp. 104–105).[4]

In addition, the situation in the corner of the building allows access to the store: a staircase, modest in size, but heir to the theatricality of the Parisian department stores such as Le Bon Marché or Lafayette, is insinuated after two arches. But contact with the outside is still partial. The ground floor is somewhat elevated compared to the level of the street and the shop windows are limited to narrow windows on the façade. Although the object of sale continues to be indicated by a sign, it loses prominence in front of the sign with the Loewe logo (now made with a new typography) on the access.

The building on Calle Barquillo is certainly the evolution of the previous commercial model: now the sales space is split and overlaps that of production. It is a store that aims to be the image and representation of the brand. A foretaste of what will be the modern retail space, the one that is to come in the new Madrid, the one of the wide and luminous commercial street, expression of the new bourgeois leisure time, of the cutting-edge technology and of the luxurious products, in front to the old Madrid, the one of the narrow and dark commercial street, manifest caught in the time of work, of traditional techniques and objects of always, that of the Rastro and Ribera de Curtidores. Ramón Gómez de la Serna (1935, p. 3) would write in the following decade:

> *The new and great Madrid is that of the new cafés, the new draperies, the new shoe stores and the new pharmacies. (...) One lives in the great ways in which one reaches these constant liquidations, the noise of living, not being in gloomy corners, choosing one thing painfully. The new Madrid in that sense is already infecting the old Madrid and far from each other, the old man also tries to put more sparkling light bulbs in his stores.*

Figure 2. LEFT: Façade of the workshop of the Calle Príncipe Madrid street (1892). CEN: Façade of the shops of Calle Fernando in Barcelona (1910). RIGHT: Façade of the Calle del Barquillo store in Madrid (1923). Source: Loewe.

---

3. Workshops that are expanded in 1928 occupying the third floor to meet the increase in demand.
4. The architectural embodiment of Lipovetsky's haute couture definition (1997, p. 104) of haute couture: *"The constitution of a specialised power that exercises its own authority in the name of elegance, creative imagination and change"*.

And the new Madrid, in the 1930's, grew around the Gran Vía. The stretch whose center is located in the Plaza del Callao was colonised in the decades of the 1920 and the 1930 by many cinemas and stores, making this street the great avenue of modern leisure and business, establishing itself as a reference for a large commercial street for the rest of Spanish cities.

3   MADRID, 1931. THE STORE OF MARTÍNEZ-FEDUCHI

For Enrique Loewe Knappe, grandson of the founder, the presence of the brand in the Gran Vía, in the new and cosmopolitan Madrid, was fundamental for the promotion of the company. In number 8, in the first section, Loewe rented a store in 1936, although it was not opened until the Spanish Civil War ended (Hernández, 1995, p. 101). This store, a new representation of luxury in line with the establishment of Calle del Barquillo, was designed by the architect Francisco Ferrer Bartolomé in 1940 and enlarged in 1956 by Ferrer himself. The significance of this project in the history of Loewe is evident, however, it has now become the flagship establishment of the brand. But, previously, for a short period of time, there was another store on the Gran Vía. A smaller project, more modest, although probably much more avant-garde for its time.

In 1927 the second section of the Gran Vía in Madrid, baptised Pi iMargall, was completed. In 1931, three of Luis Martínez-Feduchi's projects were published in issue 147 of the *Arquitectura* magazine, the three located on the sidewalk of the new avenue, at the height of Callao square, and a few metres apart: in the current number 36 (before 14) was the Chrysler Seida car shop, in 44 (before 22) Rekord Records and, between both, in the now number 40, Loewe.

The photographic report of Jesús García Férriz published in *Arquitectura* consisted of two unique images: one of the façade and another of the interior. The caption read: "Exterior and interior of the Loewe store, in Calle del Barquillo". It is, in fact, an error. That shop was, as we have seen, the most well-known and representative of the company, and it is likely that the publisher would allow himself to be guided more by an intuition than by a certainty, since Martínez-Feduchi's store was in one of the ground floors of the building owned by Ramón López-Rumayor designed by the architect José Miguel de la Quadra Salcedo in 1926 and completed in 1930. The confusion, a priori anecdotal, seems to predict an unjustified future forgetfulness.

Luis Martínez-Feduchi was then, at the age of thirty, a recently titled architect who was about to build (along with Vicente Eced) a few metres from these three small shops, the Carrión building, an architectural landmark of cosmopolitan Madrid. The relationship between the skyscraper and the Loewe store is not limited to its neighborhood. The art-deco, luxurious language used in the Carrión has been previously tried, more modest and subtle, in the shop: the padding of the interior walls, the materialisation of the access door, the furniture, and even the verbal communication by means of signs and lettering. In a definitive way, the business description is displaced, almost ejected, towards the upper part of the façade, outside of the façade, while the name of Loewe forms the lintel of the door-showcase couple. It is an urban scale sign, to be perceived from the speed of the new avenue. The logo has changed again, which shows the intention of modernisation of the company, although the result seems more an essay than a step with conviction. In this sense, the contribution of Martínez-Feduchi is the substitution of a Roman typeface for another of Sans-Serif font, carrying out a compositional game with the first letters.

This stylistic typography aspect is not the only communicative question that Martínez-Feduchi puts on the table. The resolution of the showcase (a theater directly connecting with the street) is not bringing anything relevantly new in the commercial space concept, since it was even at that time a usual solution, but the new design does make a difference when compared with the narrow display cases of the façade of the Calle del Barquillo store. If there the shop was a mystery, almost inaccessible, here in Gran Víathe interior is put on display behind the glazed curtain of the shop display window.

The photograph of the interior shows us a modest space in scale, but tremendously interesting. The luxury, for the architect, are Loewe's pieces, and the store is a discrete bag that wraps them. The objects needed for the sale (light fittings and furniture) must recede to a second plane. Thus, the lighting is built between the mouldings of the ceiling, an illuminated sky that is part of the wrapping, as opposed to the lighting from superimposed fittings hung in the shop of Barquillo. On the other hand, the furniture (support of the object of sale and also for the clientele) responds to the same criterion of discretion. The objects (display cabinets and shelf units) to show objects to best effect leave out unnecessary mouldings, showing behind large glazed elements the objects for sale. It is modern furniture designed *ad hoc*, geometric lines and simple delineation, but of noble materials. On the contrary, objects for the human body (tables and chairs) respond to a criterion of industrial production. That is to say, they become prototypes for their serial production,[5] as can be seen from their resemblance to some of the series that Martínez-Feduchi designed for the Carrión building manufactured by Lledó.

However, and despite modernity in the approaches of Martínez-Feduchi, the first establishment of Loewe in the Gran Vía has gone unnoticed, not only in the architectural historiography, but in that of the company itself.[6] The glory was reserved for the aforementioned Ferrer's shop. Loewe, in the full challenge of the postwar period, ignored the move to modernity in design to embrace luxury, to show a dream world, unattainable, but hopeful. Although the luxury will be the manifestation, perhaps ostentatious, of wealth, the *"triumph of appearance over substance"* (Munari, 2011, p. 13).

The path opened by Ferrer transited to the language of closed shop windows, Roman typographies, chandeliers, the dazzling brightness of marble and recurring golden adornment. The model was exported to Barcelona in 1940 and to San Sebastián in 1948. A model that had its *raison d'être* in a sought-after homogeneity of the brand's image (the shop windows, all identical with their characteristic semicircular arch, were for years designed globally, ubiquitously exportable, by José Pérez de Rozas). The enlargement of the Gran Vía store in 1955 was, perhaps, the epitaph of a time and a society in need of change. The old clothes would have to be replaced by new suits. And Javier Carvajal, architect, and Vicente Vela, artist, will be the new tailors.

4 BARCELONA, 1963. THE CARVAJAL SHOWROOM

After the Spanish Civil War, Loewe's production had spread in small workshops. In 1950 the factory was centralised in a new building located in Madrid, in Calle Batalla del Salado 25

Figure 3. LEFT & CEN: Façade and interior of the store of Gran Vía 40 (Martínez-Feduchi, 1931). Source: *Arquitectura*. RIGHT: Façade and interior of the store of Gran Vía 8 (Francisco Ferrer, 1939). Source: Loewe.

---

5. At this time, the industry did not offer a catalog of furniture to the taste of avant-garde architects. *"Today, in the primitive situation of development, the Architect and the 'Furniture Designer' constitute almost a single personality, which has to be unfolded later..."* (Giedion, 1932, p. 87).
6. In the detailed narration of Hernández Cava in 'Tiempos de Loewe', although it is true that he names the existence of this establishment, he does not mention its authorship or provide any other relevant information. (Hernández, 1995, p. 96).

Figure 4. Interior of the showroom in the Loewe factory of the Calle Aragón (Carvajal, 1963). Source: *Cuadernos de Arquitectura*.

(Hernández, 1995, p. 123). The separation between the point of production and the point of sale is already irreversible and supposes, *de facto*, the definitive abandonment of the hierarchised model of Calle del Barquillo.

At the end of the 40's, in the United States, companies began to become aware of the importance of having a brand identity beyond the value associated with their products. Therefore, stores acquired a universal vocation, in which the sale of the concept of the offered brand is superior to that of the product itself. Enrique Loewe Knappe, aware of this need for change, places himself in the hands of Carvajal so that the firm can go from being a leather house, to becoming a fashion house (Martín and Chocarro, 2008, p. 17).

Javier Carvajal began working for Loewe at the end of the 50's, becoming a fundamental agent in the modernisation of the image of the luxury brand in Madrid. And not only carrying out projects of exemplary stores, such as the *Calle de Serrano* (Madrid, 1959) or the Hotel Presidente (Barcelona, 1965), also putting the firm in contact with artists such as Vicente Vela, the creator of the well-known logo of the four "L".

From this collaboration with artists to work together with Loewe, the relationship with the Manchego sculptor José Luis Sánchez[7] was born. His are the friezes, handles and reliefs in concrete, copper, brass or ceramics for the Loewe stores in Madrid, Barcelona, Bilbao and London (Ruiz, 2012, p. 66). Carvajal understands the collaboration with the plastic artist as an anonymous complement within his aesthetic plan for Loewe. They are not figurative works whose theme revolves around products for sale, but geometric abstractions that accompany and enrich the reading of the façades, providing a certain elitist distinction that, in some way, determines the social segment of its clientele.

Thus, the fabrics of the new Loewe costumes are the result of the 'Le Corbusierian' thought of the *synthèse des arts majeurs*, where sculpture, painting and architecture combine in a unique plastic event at the service of poetry (Calatrava, 2010, p. 22). The fabrics of the new suits are, also, transparent. Faced with the previous secrecy, now open to the street through a new concept of showcase, much more permeable than those of Ferrer-Pérez de Rozas. Now the interior is not an unattainable and hidden world, but it is glimpsed filtered by shelves or curtains, elements that are recreated in the theatrical and scenographic aspect of the sales space.

---

7. Sánchez's work is prolix in collaborations with some of the most relevant characters in the history of Spanish architecture of the 20th century, apart from Carvajal himself, such as Gutiérrez Soto, Blanco Soler, de la Sota, Fernández del Amo, Anasagasti, Corrales and Vázquez Molezún, Bonet Castellana or de la Joya (Ruiz, 2012).

However, the end of this succinct tour is not located in a store at street level, but on the first floor of the factory that Carvajal built for the company in Barcelona in 1963. If the exhibition and sale space was detached from the artisan workshops of Madrid in the second half of the 19th century, the new industrial workshop required a place of exhibition, which is currently known in the retail arena as the showroom.

The characteristic of this space, which takes us back to the starting point, is nothing other than its lack of relationship with the outside world. The upper floor of the new factory is not a sales space and, therefore, does not need to be shown to the street, even to insinuate itself. Carvajal cancels any kind of connection with the exterior, encloses the most precious of the firm, its pieces, in a hidden space, only accessible to a small elite. Carvajal's showroom is an exquisite protector, a wrapping for the bags, a dress separated from the body. As Quetglas (1980, p. 2) would point out, with regard to Lina Loos' bedroom, "*a warm bag to wrap around oneself*". Dresses, in any case, that define the wrapper in a continuum where the vertices and the edges disappear vanished between curtains and mirrors. The pleasurable dressing of Loewe.

5 CONCLUSIONS

For Simmel (1998), fashion is the tendency of the individual to imitate the way of dressing to feel integrated in the group and in society. *Haute couture*, aesthetic pleasure reserved for the most solvent social strata, is not alien to this feeling. However, it does not aspire to become a broad social phenomenon. It lives an exclusivity that, paradoxically, must be exhibited among the general public, although only a part of this society can access its products.

Loewe, paradigm of luxury firm in Spain, does not need to display its merchandise in the shop windows. However, it requires showing to position itself in the market, and it does so in a discrete, elegant and distinguished way. The commercial establishments of Loewe, exclusive fashion products, have always sought a location that brings them closer to their clientele: the court and the nobility of Madrid at the end of the nineteenth century, the Catalan bourgeoisie at the beginning of the 20th century, the cosmopolitan Madrid of Gran Vía of the 20's or the European San Sebastián of the 40's, until the expansion by the great Spanish cities from the late 50's, after the economic stabilisation of the end of the autarchic Spanish postwar period.

The stores (the costumes with which Loewe dresses for the eyes of the city) are designed more in terms of what they hide than what they show. The balance between exhibition and exclusivity is achieved by showing a luxurious façade that hides, to varying degrees, the interior of the establishment. In the workshops located in the main streets close to the Puerta del Sol (1846–1892) the contact with the street is virtually non-existent. Through the timid contact with Calle del Barquillo (1923), Loewe makes patent its exclusivity, hiding the client and barely showing the product of sale. Martínez-Feduchi (1931) situates commerce in full view of the public, relying on an avant-garde design that abandons, for a moment, the classical style of the brand, recovered years later by Ferrer (1939–1955). A model, the one of Ferrer (still in force), also based on the opacity of the establishment towards the outside. An opacity that Carvajal (1959–1965) replaced by the superposition of filters that reveal the interior to the street, with the exception of the pleasant autism of the showroom in the main one of the factory in Barcelona.

REFERENCES

Bergamín, R. 1927. Tiendas nuevas en Madrid: La tienda de automóviles Euskalduna. *Arquitectura* (99): 260–264.

Calatrava, J. 2010. Le Corbusier y la synthèse des artsmajeurs, 1945–1950. In *Arquitectura y Cultura Contemporánea*: 9–38. Madrid: Abada.

Carvajal, J. 1960. Instalación y decoración de una tienda. Madrid. *Arquitectura* (16): 32–36.

Colomina, B. 2010. *Privacidad y publicidad. La arquitectura moderna como medio de comunicación de masas*. Murcia: Cendeac, Coamu y Obs.

Diseño de Interiores. 1964. *Arquitectura* (64): 49.
Exterior e interior de la tienda Loewe. 1931. *Arquitectura* (147): 247.
Fábrica de artículos de piel para Loewe S.A., Barcelona, 1964. 1965. *Cuadernos de Arquitectura* (59): 19–21.
Fábrica Loewe en Barcelona, 1960. 1974. *Nueva Forma* (104): 10–11.
Ferrer, F. 1941. Tienda de artículos de cuero en una avenida principal de Madrid. Loewe. *Nuevas Formas* (1): 19–22.
Giedion, S. 1932. El arquitecto Marcel Breuer. *Arquitectura* (155): 82–90.
Gómez de la Serna, R. 1935. Del Madrid Viejo al Madrid Nuevo. La Ribera de Curtidores. *Madrid Turístico y Monumental* (1): 2–3.
Hernández Cava, F. 1995. Tiempos de Loewe. In: Argullol, R. (ed.) *Loewe 1836–1996*: 17–183. Madrid: Loewe.
Klein, N. 2001. *No logo, el poder de las marcas*. Barcelona: Paidós.
La nueva exposición de Loewe. 1956. *ABC* (February 2, 1956): 48.
Lipovetsky, G. 1997. *El imperio de lo efímero. La moda y su destino en las sociedades modernas*. Barcelona: Anagrama.
Martín, C. and Chocarro, C. 2008. Loewe 1960. Cuestión de estilo. In: UNAV (ed.) *Loewe años 60. Cuestión de estilo*: 15–25. Pamplona: Servicio de Publicaciones de la Universidad de Navarra.
Monteys, X. 2014. *La habitación. Más allá de la sala de estar*. Barcelona: Gustavo Gili.
Munari, B. 2011. *¿Cómo nacen los objetos? Apuntes para una metodología proyectual*. Barcelona: Gustavo Gili.
Quetglas, J. 1980. Lo placentero. *Carrer de la ciutat* (9–10): 2.
Ruiz Trilleros, M. 2012. *La escultura construida de José Luis Sánchez*. Thesis (PhD), UCM.
Santamaría, C. 1994. Las palabras del mercado. *Revista de Occidente* (162): 151–162.
Simmel, G. 1988. *Sobre la aventura. Ensayos filosóficos*. Barcelona: Península.
Tienda en Barcelona. 1965. *Arquitectura* (79): 29.
Tiendas de Loewe en Bilbao, Barcelona y Madrid. 1968. *Arquitectura* (111): 27.

Signifying the ritual legends of Brazilian Afro Religions: *Candomblé* and *Umbanda* in *Teresina/Piauí*/Brazil

L.C. de M. Tavares & V.P. de S. Araújo
Instituto Federal do Piauí—Campus Teresina Zona Sul, Piauí, Brazil

ABSTRACT: The article is part of the Master Dissertation, developed in the Postgraduate Program in Anthropology-PPGANT of the Federal University of Piauí-UFPI, entitled Ropa de santo: identity markers of religions of African matrix. It had as empirical field the Ilè Asé Oloomi Wura and Ilè Oyá Tade. It aims to show the importance and meaning of the ritual vestments for the practitioners of these services and for society in general. We used a field book, participant observation, collection of photographic and filmic images, as well as an exhibition composed of 20 dolls/parents with the robes of the sixteen most worshiped or is has in Brazil and four other families of priest, Bahian, men's clothing and feminine. We base the work with theorists such as: Cidreira (2015); Damatta (1987); Prandi (2001); Santos, (2012); Santos (2010); Souza (2007), among others. The results indicate that these robes are markers of identity for practitioners of these cults and are triggering religious intolerance and racism.

1 INTRODUCTION

Whether they are rituals or not, clothes keep much of us: our scent, our sweat, are part of our memories. Who never noticed remembering happy or sad moments and associates them with clothing or punctuates: "that day I was wearing those clothes". The researcher Cidreira says that, referring to this reception of clothing:

> And this welcoming makes them have the ability to identify an absence, hence their imaginary and affective dimension. And also their relation to memory [...]. They present traditions, memories, customs, values, as well as actualize them, insofar as they are also re-signified, reformulated. (Cidreira, 2015, p. 8).

The garments used by the children-of-saint of the African-born religions, both the ritual vestments and the vestments used in the work in the *Terreiros*, are considered sacred; deserve special care on the part of the followers of these religions. In *umbanda terreiros*, not so much; in *Candomblé*, however, there is a hierarchy involving the garments. Mother *Estela de Oxóssi* (Santos, 2010, 43) tells us that "dressing properly, with good looks and without affectation is part of the learning of the newly initiated". As citizens, teachers and researchers we were not allowed to remain indifferent to the problems that were presented to us. These students preferred to be invisible instead of facing their colleagues. We were inside these houses, we did not see anything evil in rituals and at parties, we had to somehow demystify, show the beauty and the cultural importance that exists in these religions. That even the sacrificial ceremonies of animals considered by many to be barbarous were nothing of the sort, on the contrary they were very respectful, which are sometimes forbidden, not because of an evil character but because they are in direct relation with the degree of initiation and, consequently with the physical and spiritual capacity of the individual (Santos, 2012).

It was our objective to first analyze the clothes of the women and men of the saint, we wanted to know what they represent for their users and the people with whom they live, the processes of sacralization of the same and when they represent power within the *Terreiros*, which means wearing certain clothes in different ways. Our proposal, for reasons and situations explained here, was to construct an exhibition with the most desecrated *orixás* in Brazil, four dolls dressed in a Bahian costume, a priest and two raiment clothes, one female and one male. We did not only engage in the construction of the exhibition, but continued to visit the *Terreiros* chosen as our field of work. We used the field notebook as a working tool as well as sensitive listening and participant observation, we interviewed, photographed, filmed and we propose that the part of this work where we describe the ritual garments should not only be read but analyzed.

2 DEVELOPMENT

During the research, we did a detailed study of the ritual vestments and obtained information on the theorists of the Afro Brazilian religions, besides questioning the fathers, sons and daughters-of-saint of the researched *Terreiros*. We will score each piece of clothing.

2.1 *The skirt*

In both *umbanda* and *candomblé*, the skirt is always round, resulting from the five cloths of it and a number of other skirts underneath, seven, all white or six white and one of the color of the saint of the daughter. The skirts are underneath and the first, the skirt from above, can be white or colored, always very ornate, but every detail has a meaning. For example, each tape represents a year made from the daughter. It can be made of cotton fabric, lace (synthetic or not), but always very round and beautiful. These skirts are washed with scented water and gums. As stated above, they are considered sacred.

The one we visualize in the photograph above is all made of cotton fabric, without any embellishment; just very bulky. We have simple skirts like that, but we can also have embroidered skirts, well adorned with ribbons and other garments. In the *umbanda*, the skirt embroidered and/or much worked, with many adornments may be used by any daughter. In *candomblé*, only daughters who are already many years old can wear embroidered skirts and gowns. The information about the skirt of the daughters-of-saint was shared by Profª. Drª. Joanice da Conceição, daughter of a saint when she accompanied us to *Terreiro Ilè Oyá Tade*, an empirical field of our research.

Beneath the last skirt, for the saint's daughters to feel comfortable, they usually wear loose trousers, ankle-fastened by elastic or a kind of button-like cuff, the famous cuirasses.

2.2 *The blouse (shirt and robe)*

Accompany the skirt a blouse, which can be the shirt or the robe or both at the same time. In the *umbanda*, there is no such distinction, but in *candomblé*, you can only wear robe that daughter who has already been seven years old.

Figure 1. Skirt.

Figure 2. Bata.

This piece, most of the time is white but may be of other colors, and the fabrics, most of the time, are of cotton, lace (synthetic or not), always very worked with lace, ribbons, and either embroidered in *richilieu* or other embroidery, by machine or by hand. In our wanderings through the many *Terreiros* and festivities of the people of *santo*, we saw many robes, beautiful, but the embroideries in *richilieu* are the most beautiful and showy.

2.3 *Turban*

An extremely important prop for the people of *santo*, the turban, or torso, known ritually as *ojá* is a fabric made from simple cotton to extremely elaborate embroidery in *richelieu*, as well as fabrics adorned with daisies and also with shiny threads.

Ojá is a piece of many meanings. It can be tied up by covering the ears or not. The men use with ears totally discovered, whereas women use it with ears covered, exhibiting only the part where the earring is placed. The information was given to us on 11–02–2017, by Father Maurice of *Oxum*. In the *candomblé terreiros*, this aspect is seriously observed; in *Umbanda terreiros*, this is not always a concern.

The size and width of this prop may vary, depending on the desired effect, from 1.50 m to 3.00 m in length, for a width ranging from 0.30 m to 0.50 m. Father *Maurício* assures that two can be used at the same time, depending on the arrangement that one wants to make. Mother *Stella* draws our attention to a fact that can not be overlooked: the history of every house in *axé*, the ethnicity of that house, all influencing the way children dress. In some houses in *Bahia*, the *ojá* can only be used by daughters from seven years and over, which is not the case of *Ilè Axé ÒpóAfonjá*, where, by determination of Mother *Aninha*, even *Abiyans* on special occasions must use *ojá* and have to know how to use it (Santos, 2010).

The *ojá* is a piece whose main purpose is to cover the *ori*—the head; the language, the codes, the symbolism is in the form as it is placed upon the head, which we have already described. Sousa (2007, apud Escorel 2000, p. 27), draws our attention to the utilitarian value of the torso, that is to cover and keep the hair tied, which refers to a time when these religions were practiced by black men and women who wanted to hide the "bad hair" under the turbans, the capes and the *filas*; an attitude fruit of contempt for the hair of the black people and that was already sung in verses and prose. Today we know: one of the most successful forms of domination used by the so-called white race to subject Afro-descendants.

Today, black women are proud of their little curls, a symbol of beauty; use their *ojás* in a ritual manner, to cover the *ori* with all the symbology on their turbans, to gracefully cover the head, beautifying it or even as a form of resistance. This is what we heard from a militant of the black movement in *Teresina* who said:

> *The use of the turban is an AFRONTMENT, which is the mixture of afro and confrontation, and that, nothing more than this ability to use the turban in order to demystify the exotic, to reinforce all our ancestry, all our resistance in preserve our culture, our values.* (São Terra, 2017).

Figure 3. Turban.

2.4 *Cloth of the back*

The cloth of the back, part of the clothing of the sons and daughters initiated in the cult of the *orixás*, is a piece used not only in the parties – depending on the house of *axé* – but as in other activities of the shed. Mother Stella affirms that:

> *The cloth of the back is the feminine piece of greater historical significance. Together with the torso, it is part of African clothing, survival of Terra Mater, since the skirt, the shirt and petticoats are European survivors of past centuries* (Santos, 2010, p. 61).

And it is Mother *Stella* (2010) who insists on affirming the importance of correct use of the cloth of the back, and how it should be made. According to her, it must be made of good consistency, remembering the traditional African cloths, never of fabrics as light as silk; if smooth, it should be a soft color, no bright colors; however, can be striped or squared reminding the traditional Nigerian cloths. She states that, worn on the waist or chest, is synonymous with work; is used this way at parties and other rituals; on the shoulder, when worn by women, only in non-religious activities. He concludes by stating, "If we do not hold onto these teachings, history will be lost (Santos, 2010, 60)."

Father *Mauricio* de Oxum, priest of the *Terreiro*, on one of the many afternoons of religious orientation, was telling us about the importance of the cloth of the back, which should be worn by women tied to the waist or chest; forming a loop (butterfly-shaped or tie), depending on whether the daughter's *orixá* is female or male, respectively, and that should be worn on the left shoulder only in non-ritual activities. Men wear on the right shoulder, in ritual activities or not. He also told us that, according to an *oriqui*, women use the cloth of the back to protect themselves from the attack of the Ima Mi, hence it is an indispensable and important piece for the followers of the *orixá's* cult.

The cloth of the back measures about three meters long and ninety centimeters wide. It is a liturgical play, with which the sweat of the faces of priests and/or priestesses can be wiped away; it is to cover someone who for some reason fell into a sudden trance, or to cover someone who has gone to embrace an *orixá* and is wearing a clothing of a color he (the *orixá*) does not support, like black and red for *Oxalá* and Iemanja, among many other utilities (Sousa, 2007). In *Africa*, the cloth of the back also has utility value: it is used to carry children, tied behind their backs or spread, leaving the hands of the carrier free. The cloth of the back is so significant for the people of *santo* that it is common, when arriving at another *axé*, to offer of the same; and it is usual for the father or mother to put it on the breast immediately. Contrary attitude may seem as an insult or lack of knowledge.

2.5 *Men's vests*

Men, both at parties and at other activities in the yard, always wear pajama-style pants and tunic whose length will vary with festivity and with the hierarchical degree. The longest and

Figure 4. Cloth of the coast.

special fabrics are reserved for children-of-saint who already have more than seven years of craftsmanship; the shorter ones and cotton fabrics are reserved for the simpler services of the *Terreiro* and for the beginners. The pants can be ankle-fastened and made from the most unusual fabrics.

3 CONCLUSION

The readings and discussions of the texts studied were useful, provided us the theoretical basis and the understanding of many facts that have occurred so far. We are aware of how artisanal the anthropological research is and we are in agreement with the thought of authors who say that this research is influenced by humor, phobia and everything that is human in who researches and who is researched (Da Matta, 1978), and of course, all the cultural formation of both. We have to respect our training, what we believe in, our values.

If we continue to follow this path, if we continue to try to understand African-born religions more and more, it is likely that tomorrow, when we look at this work, we will have many corrections to make. Some will be necessary for us to delve into others. Certainly, there will be new discoveries, because whenever we read an article, essay or book, they refer us to new authors and so on.

What we have seen so far is that research in religion is not an easy thing, the religious choice of a group leads this group to live in a certain way, to dress in a certain way, to feed on this or that, to make certain restrictions, to behave this or that way. Religion is embedded in the way of living of these groups, it ties and organizes everything. We must be aware of the importance of the same for the social group. How important it is for each element individually and for everything. We can spare no detail, especially those that are part of the visual composition of the followers.

The awareness that clothing is extremely symbolic, that clothing is our second skin, according to the so-called *Hundertwasse*, although for many it may be considered futile it is only less important than our epidermis. It informs us a lot about us—what we think, what we feel, what we want others to think, what group we belong to, our historical, geographical and social localization and so on.

We went to *terreiros*, talked with sons and daughters inside and outside the houses of *axé*. We went to other events of the people of Santo. We are aware of the social and anthropological importance of them, we are also aware that this work, in order to bear fruit, must be increased. And much more than that, as a researcher, I was the person who was most benefited by this study; today I am more aware of the value of the black people in the construction of this country and of the enormous social debt that this nation has towards Afro-descendants.

The answers to our questions, our observations during our visits to the houses of *axé*, the observations made in the presentations of the exhibitions, the results of the many conversations with the people of the *terreiro* and with the visitors of the exhibition led us to intuit that the ritualistic vestments and the props that make up the look of the saint-people identify them and, more than that, these robes are triggering religious intolerance and racism. We will close our work telling a story occurred with a mother of saint and that was passed on to us by Father *Rondinelle* de *Oxum*. The priest, the State Coordinator of *CENARAB*-National Center for Afro-Brazilian Resistance and Afro-Brazilian Affairs, tells that, in the groups of *Terreiro's* dialogues, undertaken by the institution in 2016, a mother of saints reported the following:

> *She was in the terreiro, participating along with her children of religious activities and stepped inappropriately, twisting her foot. She was taken by one of the children to a care unit, she was in great pain. Upon arriving, she informed the employee who answered her about the incident. She was brave, round skirt, robe, turban, and many guides. When she was taken care of by the doctor, the first thing he asked, before she even asked what she was feeling, was to remove the turban and the guides. She was outraged at the disrespect and told the doctor: Dr. My problem is not in the head, much less in the neck, my problem is in the foot* (Father Rondinelli of Oxum, n.d.).

The testimony we have seen above shows very clearly the religious intolerance to which religions of African origin are subjected and, more than that, the garments used in these cults unleash this intolerance.

REFERENCES

Brito, J.P.C. 2016. Bata. (Source: *L'Hosana Tavares* personal archive).
Brito, J.P.C. 2016. Cloth of the coast. (Source: *L'Hosana Tavares* personal archive).
Brito, J.P.C. 2016. Skirt. (Source: *L'Hosana Tavares* personal archive).
Brito, J.P.C. 2016. Turban. (Source: *L'Hosana Tavares* personal archive).
Cidreira, R.P. 2015. The robes of the Good Death. *Cruz das Almas*/Ba: *UFRB*.
Damatta, R. 1987. Relativizing: an introduction to social anthropology. *Rio de Janeiro*: Rocco Publishing House.
Prandi, R. 2001. Mythology of the *Orixás*. São Paulo: Companhia das Letras.
Santos, J.E. 2012. The *Nagô* and death: *Pàde, Àsésè* and the Cult *Égumin Bahia*; translated by the Federal University of *Bahia*. 14th ed. *Petrópolis: Vozes*.
Santos. M.S.A. 2010. My time is now. *Salvador:* Legislative Assembly of the State of *Bahia*.
Souza, P.R. 2007. *Axós and ilequês*: rite, myth and aesthetics of *candomblé*. São Paulo: USP.

Credit: *João Paulo Brito*-2016.

Globalized brazilianness on the forming way of "Elementais"

G.R.R. Vieira
UFBA, Salvador, Bahia, Brazil

ABSTRACT: The purpose is to reflect the style composition and assertion in the collections of the female clothing brand from Bahia, Elementais, by articulating its creations elements from the diverse cultures which compose the hybrid Brazilian imaginary, from the region Northeast and the state of Bahia. In the dialogue between local culture and fashion design, the analysis is devoted to the comprehension of fashions that can handle this multiple environment that set Bahia in plural, constituted in its essence by miscegenation of the three races: black (Bantu and Sudanese mainly), white (Portuguese mostly) and indigenous. But a Bahia connected to the world. This connection is submerged on its formation and permanent natural condition of Colony (economical, social and cultural dependence); nevertheless a connection in movement, configurator until nowadays and enlarged by the countless digital network, that tear down and also build borders and ties in a quotidian marked by universality.

1 INTRODUCTION

There is more than 25 years since the fashion production female brand Elementais has acted upon the region of Bahia state in Brasil, besides investing currently on its national expansion through the system of franchising[1] and virtual commerce, in order to throw oneself into this wider and more demanding market. It is yet overcoming the barriers that polarize Brasil with a depreciative image of Bahia which was built and supported along history by the power center São Paulo, Rio de Janeiro and Minas Gerais, as Risério (1993, p. 117) points out.

Therefore, Elementais has put faith in a new dynamics that requires a constant professionalization on the production, linking innovation, singularity, productive excellence, strategic planning without abandoning its unique forming way, however (PAREYSON, 1993). According to Pareyson, *in Esthetics: Formativity Theory* (1993, p. 59), forming means doing, but it means doing meanwhile inventing the way of doing. From the consciousness emphasized by the author that there is artisticity in every area of human activity, fashion is taken as one of the realms in which the formative process "includes gestation, incubation, birth, growth, maturation" (Cidreira, 2013, p. 103).

As a consequence, immersed in such a diverse socio and anthropological context, fashion produced in Bahia comes across in a knowledge that "emerges from a colorful palette of interculturalities" (Miguez, 2002, p. 46). Débora Leitão (2009, p. 146), considers the "imaginary triangulation among Rio de Janeiro, Amazônia and Bahia" as the symbolic geography of the Brazilian nation. The Brazilian type must be thought, she says, as a heir of Brazilian nation ethnic and cultural diversity. It would be through this heterogeneity that Brazil could build its otherness before the other. So it matters, from the style composition of Elementais, to understand formativity process that characterize the Brazilian design, on the other hand, there is the clearness to be about objects produced in specifics historical cultural technological circumstances, as it is highlighted by Ana Beatriz Simon Factum (2009, p. 58).

---

1. Business model that allows to use a well-stablished, tested and padronized brand, testada e padronizada, reducing management risks and problems. It is based on the use lisence sale for using a business with a brand, know-how, routines and product/services developed by a franchiser (SEBRAE, 2017).

2 ELEMENTAIS: MAKING ITSELF KNOWN BEYOND THE NORTHEASTERN BORDERS

The investments in technology made by Elementais are constant, through modern machinery to a bigger production capacity, in addition to explore more and more the dexterity of workmanship and use of smart fabric on its collections—among the most important are satin of technological silk, dull crepe, dull satin and *chiffon*. With the unification of the factory and office in a only and plane shed, the company earned space to amplify and horizontalize its production. When the factory was placed at Rio Vermelho (Lucaia's neighborhood unit, in Salvador, Bahia's state capital city), the cut used to stay in a vertical space, which made logistic for distribution (transporting fabric rolls for example) to be difficult. In this section, all the factory cuts are made, including hiring extra workers according to the necessity.

The style manager of Elementais, Karina Facó[2], emphasizes on her speech the relevance of progress in the textile industry and the investments of Elementais for a productive efficiency that serves, mainly, the demands of the contemporary woman:

> *The pure silk does not fit so much in the today's world. I have to have an organization to do the ironing to go out, the moment I pack my luggage and travel. So in the modern world of the woman who takes its role on society, it does not fit. I have to dig for a bit of technology to give me comfort so I can enter and leave at any moment without having that master structure, of dry-washing because it can not get wet. In this moment, I need the technological fabric and the textile industry is very important. In the past, when I spoke about technological fabric, I remember that thing of uniforms. Industry has grown too much so it is able to produce a synthetic fabric that looks like natural. I have millions of polyesters that you can not be sure if it is silk or synthetic. I work with synthetic, artificial (viscose) and natural (cotton and linen), which has a lot of this leisure woman, of this contemporary woman's lifestyle.*

Elementais has released itself on the franchising system in 2001, aiming specially a business model favorable to its joining on national fashion environment, when they started an expansion plan for the brand. According to the owner partner and communications director of Elementais, Emilia Medauar, in a article "DNA Baiano" signed by the journalist Ronaldo Jacobina in October 18th 2015, at the beginning the idea was to keep the model of street store. In 2005 otherwise when they made the project, they decided to try on the shopping mall, to understand how the brand would behave in another environment. Consequently there was the inevitable increase and acceleration of production. The style manager[3] stands out some production strategies for production, above all:

> *We have reduced the time to finish the collection. With the recessive economy in terms of credit, we have reduced it to decrease the operational costs to give the final clients the discount. If I receive with more time, I have a longer deadline to pay the tickets. This generates an administrative cost, because it is not always you have the cash to pay off everything. So we used to pass along the product. Nowadays, I can get a better selling price because my administrative cost gets better.*

Two large collections are developed annually and monthly collections are made that reach 22 unpublished models per week and between 80 and 110 new models per month[4]. From the development of the datasheet to the product in retail, it is approximately 70 days for the production. In parallel, there are also the capsule collections that meet the specific demands especially related to the calendar of popular festivals and commemorative dates—among the most significant to the brand are Mother's Day, Brazilian Valentines Day and Saint Jhon. For the June's typical parties in the Brazilian Northeast (Santo Antônio, São João and São

---

2. In an interview given to this study's author in February 9th, 2018.
3. In an interview given to this study's author in March 13th, 2017.
4. These informations were provided by the style manager Karina Facó, in an interview given to this study's author in March 13th, 2017.

Pedro), there is an increase in the investments for the development of products for the season in order to meet the customers' expectations and consequently reach a higher percentage of profitability. To increase and cope with production, the company also works with the faction system—outsourcing services with the hiring of sewing workshops. In the faction section, the distribution of cut pieces for outsourced sewing workshops is made, the workshops assemble the pieces and send them back to the Elementals. The service orders are delivered to the teams every 15 days approximately, says Karina Facó[5]:

> *Everything that is cut by the company comes to the faction sector that will do all the distribution logistics for the workshops that work exclusively for Elementais. We cut everything and after the cut, it comes here. There are about 60 workshops, usually. We work with collections of two months. They are always busy with a work order. They deliver a dynamic, and we deliver another cut for them. On average, they spend 15 to 20 days producing a work order.*

The brand—which started as a t-shirt manufacturing company in 1992, inside a shed in the Rio Vermelho neighborhood of Salvador—currently occupies an exclusive manufacturing space within the Bahia Têxtil condominium (Uruguay neighborhood, in the Lower City Area of Salvador). The image of a pine tree was chosen by the founding partners of the brand Fernando Albuquerque and Maria do Carmo Queiroz. "The pine tree is a tree that does not bow, even with heat or a storm, it keeps standing," according to Karina Facó. *Elementais* was born inspired by the elements of nature, in tune with a mystical universe, which the partner Fernando believes to exist. In addition to the production of clothing and accessories growth aiming at its target audience (especially women aged 35–45) – there was a need to update store spaces, making them more comfortable and attractive. The style manager[6] underscores the growth of Elementais in its 25-year market:

> *At this February [2017], we are celebrating 25 years of existence. For a brand, in our country's economy, this is a very big glory, because even with all uncertainty, we last for so long. Right here in the state of Bahia, we really have a very great reputation, a very good acknowledgment. Our goal is to achieve this in the whole of Brazil. It is hard work, we still have not been able to gain a greater breath to be very well in other states away from Bahia.*

With an annual income of between R$ 13 and 14 million[7], the network operates today with 14 units, including street stores, shopping malls and e-commerce. Among the physical stores, four are owned and the others are managed by franchisees. The only franchise outside the state today is located in the Aracaju capital, in the northeastern state of Sergipe. According to Karina Facó[8], franchising trials were also carried out in the Southeast—São Paulo, Rio de Janeiro, Minas Gerais—and in Ceará, but they were not successful:

> *Today, thank God, we are much better than we were in 2015, when this thing started [crisis][9] in Brazil, we were much more bloated. Some franchises have not gone ahead because of the crisis and this notoriety thing out there is not easy when you do not have large investments in advertising. Our vision is to expand greatly and to be noticed as an important brand in the Southeast mainly. It is very difficult. We even do market research, analyze the entrepreneur, but it's union; you want it so much, there are huge*

---

5. In an interview given to this study's author in March 13th, 2017.
6. In an interview given to this study's author in February 9th, 2017.
7. These data were given by Karina Facó in an interview given to this study's author in March 13th, 2017.
8. In an interview given to this study's author in February 9th, 2017.
9. Since 2015, Brazil has reached a considerable economic crisis, with direct impacts on the Gross Domestic Product (GDP). The number of unemployment increased and consequently affected income and consumption. The impacts of the crisis on the world economy were significant; however, the moment of political uncertainties in the country—especially after the impeachment of President Dilma Roussef, and the presidency being assumed in 2016 by Michel Temer—also interferes directly and hinders a short-term recovery.

Figure 1. Warehousing Sector. Source: Gina Reis. Photos taken by the author of this study on March 13th, 2017.

> *promises in the process, you do not want to continue in business. I believe that the lack of success is perhaps linked to the fact we have not had the possibility of doing mega advertising structure, this might have been a big part of the failure.*

Apart from the adversities to compose a structure that handles the required apparatus for a greater national penetration, Elementais, as well as other brands and designers from Bahia, faces its own obstacles of the productive chain in the state. Figure 1 below shows the Warehouse sector, responsible for receiving all the primary material used in the production, including the outsourcing material (bags, shoes, jeanswear, knitting and costume jewelry) made by the Bahia brand. From left to right, it is possible to observe the supply of fabrics and sewing supplies. All the sewing supplies come from the South and Southeast of the country. The brand's managing partner, EmíliaMedauar[10], states in the article "I bought in Bahia!", signed by journalist Luís Fernando Lisboa, for B + magazine, in December 2014:

> *It is not easy to produce fashion here. We face daily the lack of primary materials, sewing supplies, specialized labor. It is necessary to bring all the material from outside and to train the teams that often have few options of work.*

3 FOR A BOSSA, A PERFUME IMPREGNATED WITH PLURALITY

Therefore, it is essential to emphasize the evidence to the continuity of the barrier that polarizes Brazil, signaling the subjection that the Northeast and other Brazilian regions maintain in relation to the Southeast. Agreeing with Risério's (1993, 117) reflections, such obstacles can be perceived as effects supported by a depreciative image of Bahia built by this same power center—São Paulo, Rio de Janeiro and Minas Gerais—and which is mainly reproduced regionally in the face of Bahians' own resistance to consume products made in Bahia.

It is not by chance that the challenge of Elementais and other names of fashion design currently made in Bahia—among them, Gefferson Vila Nova, Aládio Marques, Jeferson Ribeiro, Vitorino Campos, Márcia Ganem, Carol Barreto—is permeated by a kind rancidity that the Northeast and, more specifically, Bahia—as "Motherland"—became impregnated. Besides disputing a slice of such a complex and inconstant market, the designers and brands originally from Northeast and Bahia have one more mission that is charged at all times. They need to prove that their creations are particular (in the frequent game of fashion between imitation and differentiation), done with excellence (breaking the taboo that *baianada* is "a poor service, a mess") and above all, they are not caricatured creations. In her speech, Karina Facó[11] points out such a concern that still persists in today's local fashion design:

---

10. Revista B+, edition nº 27, p. 84 | December 2014/January 2015.
11. In an interview given to this study's author in March 13th, 2017.

People still look at the textile composition on the "label", but this has demystified, it has declined more, customers are accepting it more because they are understanding the need to use a technological thing. But there are still those people who think "oh, polyester", polyester refers to the oxford uniform of old days, but that has changed significantly. I believe that our work focused on this has helped a lot of customer awareness specially here in Bahia. In the Southeast, people have had this culture for some time. In the Northeast there is still this prejudice thing.

Thus, an album[12] of stereotypes for Bahia was created. However, when thinking about stereotypes, something negative is normally thought; as what must be taken in account on this album about Bahia is the way that highlights this mythological dimension seized in the national imagination, through which some are the cultural traits are expanded "as an excessive measure that unilaterally shifts as the boundaries of wholeness," as Simmel states in Caricature (2009: 198):

As soon as a form is achieved, the course of life leads it beyond that, it transforms the objective balance into the being of things by the excessive intensification of a trace, the unilateral displacement of one or the other boundary. In fact, based on a more intimate law of our spiritual movements, the excess seems to align with our essence.

For Simmel (2009, p. 199), not all excess is caricature, since "the premise of any caricature is the unity of the personality, that which after having split up into the plurality of qualities, movements and lived experiences, is presented as its determinate proportion". For a better understanding, the author (2009, p. 200) makes an interesting reference between the caricature cordial or humorous and the caricature of the artistic accentuation of character:

The humorous caricature does not fully realize the concept, it is half way, to say it so, when it only triggers the momentary process of equilibrium which, however, as a promise of reintegration into the totality, it is still perceptible behind the current disproportion. (...) The excess of the singular trait breaks the unity of the Self and it proclaims that this deformation is its durable form, which is, so to speak, its normality, or even better, only in this way does it deform it. This precisely differentiates the caricature from the artistic accentuation of character. If the playwright or the sculptor expresses a character trait or an affection in an intense and absolute way that the actual experience does not present, they must make us feel the universal existential greatness of the personality to which they belong at the same time. Just as such presupposition this unilateral intensification is not disproportionate (...). It presents here a deep meaning of what we call stylization.

In this way, by the hands of creators, fashion shows itself as an efficient "magnifying lens" (using Simmel's expression) for the recognition of this unity of personality that would be the caricature. The formative attempts that fail to achieve such a recognizable synthesis—which would be stylization—are marked by disproportion, so they are half way and they do not achieve success in attempts to shape their way of forming. Within this ambience that claims elements of a disproportionate mythical localism at every moment, Karina Facó[13]—when questioned about the way in which the Elementals places their signature, their way of forming from their experiences with the cultures of Bahia—rushes to say "this Bahian root thing does not have much to do with it. We are spectators of the world fashion scenario in our own way". However, soon after, the style manager[14]—who is from Recife city in Pernambuco, a northeastern state—adds, emphasizing:

---

12. Album: among the ancient Romans, board or blank panel where commemorative phrases, edicts of praetors, postures, announcements etc. were transcribed and exposed for public reading.
13. In an interview given to this study's author in February 9th, 2017.
14. In an interview given to this study's author in February 9th, 2017.

Figure 2. Valorization of handmade and stamping in brand promotion. Source: Elementais.

> *What we carry in the baggage of the local culture is to have something special, to make a surprise, to do the handmade thing in the finishing, the bias, in an internal customization, but our scenario is focused on the world.*

The valorization of the handmade and the stamping in the brand promotion signalize to the differentiation from the color, the exuberant nature, the joy and the seduction. In the caption of the post published in the official Instagram of Elementais (Figure 2), there is still a concern to highlight the client's unique/exclusive experience: "Floral Dress + Embroidery to be the most beautiful of the weekend".

Images published in the official profile of the brand in the social network Instagram on February 20, 2017 and January 20, 2017, from left to right respectively.

The style manager's speech plots this recurring concern of Northeastern and Bahian stylists and brands to be continually seen and recognized exclusively from stereotypes "too fragile for a harmonious enlargement, but not strong enough to completely reject such expansion" (Simmel, 2009, p. 201). Engaged in the mission of being desired and consumed by Brazilian women from all points of an extensive country, in the first moment Elementais affirms it does not want to see its image associated to the cultural markers proper to a representative conception of Bahia. At the same time, the brand uncovers in an unpretentious way (see Karina Facó's[15] declaration below) her bossa, her way of forming that is in constant dialogue with the various Bahias inscribed in the territory of Bahia, although not deliberately assumed:

> *Bahia is a very important cultural center of significant importance in our Brazil. I bring this root in art. We always want to personalize, bring the art to our clothes, or in a ribbon that I will put on a label, or in a Swarovski that I will customize my size, or in a sewing supply. So, this artistic cultural side I bring in a cool way. Without getting that yokel thing, it is a more international thing, but it does not lose the root of originality. It is the bossa, I would say that it is the perfume of our collections, it is that little flavoring that you put in the end that makes your dish to shine and to have movement, to have color, to have will, to have its own brightness.*

The "perfume" pointed out by the style manager in her speech still appears literally (Vieira, 2017). The brand has developed a specific aroma to perfume its own stores and franchises. When entering an Elementals unit, the customer is surrounded by a characteristic scent, which is gentle but present and it leads her immediately into the brand universe. The fragrance was

---

15. In an interview given to this study's author in February 9th, 2017.

Figure 3. Elementais Aromatizing and image published on Instagram. Source: Elementais Website http://www.elementais.com.br.

Figure 4. Frontage of the store Elementals and publicity piece of the Villa Tropical Campaign | Summer 2017. Source: Photos taken by the author of this study on September 20, 2016, August 20, 2016 and August 4, 2016, from left to right respectively.

created exclusively for Elementais and, as a response to this identification among customers, the aromatizing (see Figure 3 above) is now available for sale both in the physical stores and in the store virtual and it can be found in sachet and vaporizer.

Above all such "perfume" can be perceived from the details of the physical store decoration (see Figure 4) – one of the brand's main business cards for its customers—that explores tropicalelements—palms, greens, fruits, bright and solar colors, sea, beach. In Figure 4, it is possible to recognize such characteristics beyond the name of the collection Summer 2017, named Villa Tropical. The shops still invest in a fluid architectural project, set in a cozy atmosphere, with plenty of wood in natural tones and very white, which are always in harmony with adornments, many made of natural materials, referential of a Brazilianness poetics. As an example (Figure 4), the scene consists of painted pineapple pottery, hay basket and palm tree in the center as a hanger. There is a detail with the natural palm tree that illustrates the Visual Merchandising of the store and still appears in the plots made on the interior walls to the decoration of the environment.

When molding its forming way, thus Elementals reinforces its interest in the interrelations between the regional, national and international, triggering fragments of the Bahian collective memories for their creations, which even pass through the African[16] identity markers; but also through the native tropical[17] imaginary, that is exuberant, colorful, from Brazilian

---

16. We agree here with Antonio Risério, on his book *Caymmi: uma utopia de lugar* (1993), to unveil as a mythical tripod of Bahia its historical antiquity, cultural originality and natural and urban beauties.
17. Exuberant, colorful, cheerful and Brazilian imaginary, constructed from elements alluding to a Brazilian outlined from countless historical, sociocultural interlacings—fauna, rich flora (tropical motifs), artistic and architectural influences of blacks, whites and Indians—where the origin point is always confusing, especially when you think of Bahia.

beaches, or even by the European influences so outstanding. Despite all these confluences, the Bahian brand of women's clothing has been able to print a unique signature that is recognized beyond regional borders, based essentially on one of the most traditional stylistic techniques in the construction and communication of the symbolic life of peoples—Textile Printing. The colors, vivacity, joy, fluidity, movement, nature, religious syncretism and the mixture appear as distinctive elements in the collections of the brand emphasized, mainly, by the performance of their stamped fabrics and reaffirmed in their communication.

REFERENCES

Cidreira, R.P., 2013. As formas da Moda: Comportamento, estilo e artisticidade. São Paulo: Annablume.
Elementais, 2017. Elementais. [Online] Available at: <http://www.elementais.com.br/>. [Acessed 14 March 2017].
Factum, A.B.S., 2009. Joalheria escrava baiana: a construção histórica do design de joias brasileiro. Tese de doutorado. Faculdade de Arquitetura e Urbanismo—Universidade de São Paulo.
Jacobina, R., 2015. DNA baiano. Revista Muito, A Tarde, 283 (18 de outubro), p. 23.
Leitão, D.K., 2009. O Brasil é uma paisagem: moda, nação, identidades e outras invenções. Iara—Revista de Moda, Cultura e Arte, [Online]. Available at: <http://www1.sp.senac.br/hotsites/blogs/revistaiara/wp-content/uploads/2015/01/07_IARA_vol2_n2_Artigo.pdf>. [Acessed 16 April 2018].
Lisboa, L.F, 2015. Comprei na Bahia. Revista B+, Edição 27, p. 84.
Miguez, P., 2002. A organização da cultura na "Cidade da Bahia". Tese de doutorado. Faculdade de Comunicação—Universidade Federal da Bahia.
Pareyson, L., 1993. Estética: Teoria da formatividade. Traduzido do italiano por Ephraim Ferreira Alves. Petrópolis: Vozes.
Risério, A., 1993. Caymmi: uma utopia de lugar. São Paulo: Perspectiva.
Sebrae, 2017. Como funciona o sistema de franquias. [Online] Available at: <https://www.sebrae.com.br/sites/PortalSebrae/sebraeaz/como-funciona-o-sistema-de-franquias,46cf39407feb3410VgnVCM1000003b74010aRCRD>. [Accessed 11 March 2017].
Simmel, G., 2008. Filosofia da moda e outros escritos. Translation from German to Portuguese Artur Morão. Lisboa: Texto & Grafia.
Simmel, G., 2009. A caricatura. Bakhtiniana—Revista de Estudos do Discurso, [Online]. São Paulo, v.1, n.2, p. 197–204, 2º sem. 2009. Available at: <https://revistas.pucsp.br/index.php/bakhtiniana/article/view/3021/1952>. [Accessed 15 February 2018].
Vieira, G. R. R., 2017. Um diálogo entre a cultura local e o design de moda. Dissertação de mestrado. Universidade Federal da Bahia.

Class tourism, working-class porn and cultural appropriation in fashion: An introduction

M. Glück
CSDMM, Universidad Politécnica de Madrid, Madrid, Spain

ABSTRACT: Fashion constantly (re-)fabricates styles as potential identity markers. A style close to an imagined 'original' identity feels more authentic and gets, therefore, validated. *Authenticity* is often staged and socially constructed by personal and collective meanings of *nostalgia*. In the absence of the 'original', the fashion system holds the copy as the authentic. A copy can acknowledge its cultural references or it can result from *cultural appropriation* (often finding its source in minority groups and subcultures). The appropriation of nonconforming styles temporarily validates those trends and brings relevancy—but also a commodification of the divergent styles and a de contextualisation of their original meaning. In this essay, I will discuss some of the terminology used in the context of cultural (mis-)appropriation in fashion with a focus on their relation to authenticity and nostalgia: 'class tourism in fashion', 'working-class porn', 'cultural appropriation of subcultures' and 'appropriation of cultural heritage'.

1 COPY, TREND AND AUTHENTICITY

In 1989, while creating the fashion documentary film *Notebook on Cities and Clothes* (commissioned by the Centre Georges Pompidou in Paris), Wim Wenders started to explore questions of identity, fashion and the notion of the original as obsolete. In the opening monologue, he reveal show in the digital age with ever-faster changing images everything is a copy and all distinctions are arbitrary. If so, what does fashion have to do with identity? And how do agents and institutions in the field of fashion produce various designs and styles when the original is absent? If there is no original, the copy validates itself through expressions of authenticity. This points at the obfuscated connection of a copy-item with the real. The authentic mimics the essence of the original by defining its capacity of being real through personal and collective expressions of nostalgia. How does the copy become a trend? And how does the copy claim its authenticity? Isa trend's origin closer to the original? Or is the validated copy more authentic and, thus, more real?

Fashion functions as a cultural system. It simultaneously allows, on the one hand, the introduction of styles and meanings of opposition and disruption, and on the other, it dictates oppressive and predominant values of social conformity. From a poststructuralist perspective, the existence of these countervailing values provides the consumer with a narrative to develop critical and personalised interpretations of those meanings while constructing a social identity. The critical identities resulting from the decommodification of the cultural experiences allow manifestations of subculture styles and expressions of non-conformist fashion (Thompson and Haytko, 1997).

Fashion's organic composition implies a flow of multidirectional influences. Several theories have tried to analyze those fashion movements and trend flows. First, the *Trickle down theory*—shared by Bourdieu and Simmel—explains that the origin of trends is found within the higher sphere of a hierarchic social structure and emanates from top to bottom to those who ultimately assimilate and copy the trends as acts of class distinction and class emulation. This theory fails to account for the significance of mass fashion as well as other processes of cultural production and consumption indifferent to class differentiation (Rocamora, 2002);

second, the *Trickle across theory* claims fashion trends move horizontally between groups of similar social levels (Robinson 1958; King 1963); third, the *Trickle up* or *Bubble up theory*, according to which trends and style innovations are created within the lower income groups, subcultures, and minorities. These trends engage the upper-income groups who absorb and appropriate them to ultimately—after a period characterised by intensive publicity (hype) and exclusivity—trickle down again as fast fashion styles to the high streets; there is a forthcoming new theory called *Boiling up* which accounts for the irruption of new figures such as the new consumer-creator, co-design, DIY, indiepreneur, customer-made, and crowd-sourced models that will shake up the flow of fashion trends in the close future (McMahon, 2012).

2 CULTURAL APPROPRIATION IN FASHION

Since the so-called *Bubble up theory* implies that a trends' origin is found within the lower income groups, minorities and subcultures, the action of appropriating those styles—while leaving aside the original meaning given to them—is a form of cultural appropriation. There has been an array of concepts and a, perhaps, misused terminology to define the effect of cultural appropriation in fashion. A classification of those concepts helps us to frame the discussion on cultural appropriation in fashion, authenticity, and 'original' identity. Appropriation is conceptually equivalent to an illegitimate act of borrowing or taking unconsented possession or property of an item of symbolic, economic, social or other intangible value. The historical approach to appropriation was lacking the negative connotation inherent in today's perception of the term. Appropriation was understood as the human's—creative— capacity of transforming or developing new uses of an item for a particular purpose (Young, 2010).The appropriation of culture—so-called cultural appropriation or in a more descriptive way *cultural misappropriation*—comprises a dominant culture's adoption of elements, ideas, images, art, and styles originally created by a minority culture. Cultural appropriation is generally understood as an expression of colonialism as opposed to the ways of *cultural borrowing* and *cross-fertilization* which imply mutual exchange. Fashion industry's ambiguous stance between appreciative imitation and blatant exploitation of minority groups and subcultures' styles have induced several debates around the actions and consequences of cultural appropriation. Through the process of appropriation, non-conformist fashion may be subjected to what is called *fashion defusion*—a risk of decontextualisation of a trend's origin and meaning. Some of the most frequently used concepts discussing cultural appropriation in the field of fashion are: *class tourism* in fashion, *working-class porn* in fashion or, more generally speaking, *cultural appropriation* in fashion. Despite the commonly unspecified use of all the above-mentioned concepts when describing general acts of cultural appropriation in fashion, I propose to define these terms separately in relation to whom they are referencing (those subjected to appropriation); under which distinct circumstances are they more descriptive; as well as introducing theories and paradoxes. Hence, I classify the terminology to adapt it to a more specific context in relation to the more general subtext of (cultural) appropriation in fashion.

2.1 *Class tourism in fashion*

Class tourism is usually defined as the act of obtaining contentedness on making use of customs, behaviour, aesthetic styles or any other quality referring to the lifestyle of lower social class members. Class tourism concept has its roots in the so-called *Slum tourism* or *Slumming*, a type of tourism originated in the slums of 19th-century London and New York that implied wealthier people visiting impoverished neighbourhoods such as Whitechapel and Shoreditch to see how they lived. It turned poverty into something that could be experienced—and then escaped from—to get back to the comfort of their lives. The novelist George Orwell experienced himself slumming in Paris and London by living a life of poverty for a period of time. Poverty he could have left at any moment and that he enjoyed and glamorised as seen in his *Down and Out in Paris and London* book. Orwell believed that experiencing a life in poverty meant experiencing the *real*, the *authentic*. In order to have an authentic experience (closer to the desired real), the voyeuristic original tendency turns into a fetishising performance. Class tourism in fashion accounts for the act of designing or consuming fashion that is inspired on, or referencing, or directly taken from the style of lower economic social groups, a style/trend that can be dropped at any time since its consumption is an actual transitory choice with no binding consequences. Class tourism in fashion is a performative act of the imagined authentic (aesthetic) identity and projects certain lower economic groups' style signifiers as 'desirable'. Behind these acts of appropriation—that first appear as a colonialist manifestation of authority and exploitation—, there is perhaps an unsubtle attempt to materialise an identity that accounts for real. It is a faithful copy of what is thought to be an original version or an original shape or an original idea. Those copies are based on a style that is considered to be more real and thus relevant. Designers Demna Gvasalia (*Vetements* and *Balenciaga*) and Gosha Rubchinskiy have been leading the fashion trends and market in the recent years by representing a combination of elements that bring the consumer closer to this yearned authenticity. Both designers come from the former Soviet Union and have brought to catwalks of Paris and other Western fashion hubs a 'post-Soviet aesthetic' that may be called *new realism* (Sebastian, 2017). This aesthetic shows youth from Eastern Europe as models (another reference to authenticity: *real people,* not models) wearing fashion that mimics styles actually worn by youth in the run-down outskirts of post-Soviet cities. These aesthetics sum up a mix of scarcity, creativity, simplicity, functionality, and a sort of realness. D. Gvasalia and G. Rubchinskiy grew up in former Soviet countries which provides their fashion discourse with a particular validation within this post-Soviet nostalgia. This nostalgia communicates a desire for the real, for an original that is lost and that can only be reached through authenticity, as a faithful copy to the presumed original, yet remote enough from the mainstream.

What does it mean when fashion's trendiest styles describe a desire to look like *common people*? Normcore was a precedent for the trends on *new realism* since it already described a style that looked for the ordinary as a statement. Though normcore does not explicitly imply class tourism, it opened the door to the more recent demand of real-ness or authentic styles that would fit in the sphere of class tourism in fashion. Considering that the hipster consumer is not driven by the classic notion of luxury when it comes to purchasing and consuming habits but rather by irony, camp and insider humour (Rubkin, 2016), normcore already raised the question whether these trends are just an expression of irony and integrity associated with hipsterism, or an actual perspective on the wider consumer's drive for authenticity (Collins, 2014). Consumers of fashion brands like *Vetements* and *Gosha Rubchinskiy* are not only driven by the mentioned desire for authenticity to overcome the unattainable *real,* but they are also attracted by the idea of affiliation to a group, a collective that holds a meaning, a sort of collective identity. In order to maintain the relevance and validation of their fashion discourse (against other contemporary styles or trends), both designers have developed strategies of *logo appropriation* (another expression of class tourism). This sort of appropriation accounts for the use of classic working-class fashion typography to use it in their high-end and luxury logos (e.g. use of *Champion* font on *Vetements* sweatshirts) or the direct appropriation of logos or emblems distinctly related to lower income groups (e.g. *Vetements* 'DHL' – shirt spring/summer collection of 2016). When *Gosha Rubchinskiy* appropriates the flag logo of *Tommy Hilfiger*, in the same manner kids with low economic resources do in Russia (a copy of the copy), we wonder if that is a plain act of exploitation or an act of irony against the corporate greed in fashion (Rubkin, 2016). More recently, logo appropriation has given place to *logo collaboration*, a differently shaped contractual relationship between the validated fashion insiders and traditional lower-middle class and working class brands. On the one hand, we find that *Vetements* initial *DHL* logo appropriation has turned into a profitable collaboration between both companies (or *Gosha Rubchinskiy*'s collaboration with brands like *Fila*). And on the other hand, with a different and not-appropriative fashion discourse, designer brand *Telfar* has collaborated with *White Castle* creating unisex uniforms for the fast-food chain's workers. Telfar Clemens has a reputation as an independent designer, perhaps being identified as a member of a minority or subculture himself, yet being validated by the art world (he designed uniforms for the 9th Berlin Biennale for Contemporary Art) and the fashion system (winner of the 2017 *CFDA/Vogue Fashion Fund* award).

'I want to live like common people, I want to do whatever common people do, I want to sleep with common people (…). Laugh along even though they're laughing at you. And the stupid things that you do. Because you think that poor is cool.' (Song: *Common People*. The Pulp. 1995).

Despite Pulp's song *Common People* openly criticizes class tourism, journalist B. Ellen argues that the song is not about the girl (class tourist) but about the actual people who are suffering the conditions of precariousness. By extension, it raises the question whether some designers may have a genuine intention to bring to notice the social situation of underrepresented fashion styles, their meanings and the collectives they signify. Actions of appropriation, whether ironic, political or merely exploitative, have given place to new relations and translations in fashion. In this sense, *Hypebeast* journalist T. Armstrong declares pride instead of offense towards this shift of attitude from the wealthier groups of society—that have gone from feeling mistrust to fetishisation. He claims there should be a sense of 'pride' on the fact that again new identities and styles are being created within the (British) working class and that the exchange should not be understood as hypocritical and an offensive appropriation but as an acknowledgement of those creative new styles. Styles that do not need the fashion system's validation in the first place, which by extension means they are not victims. This is an expression of the subculture heritage's relevancy that otherwise would be disregarded (Armstrong, 2016).

2.2 *Working-class porn in fashion*

Porn is understood as the act of 'emphasizing the sensuous or sensational aspects of a non-sexual subject and stimulating a compulsive interest' (Oxford Dictionary, 2018) in an audience. If we extend this concept to the working-class as the objectified subject in the field of fashion, *working-class porn* describes the appropriation and exploitation by the fashion elites of items belonging to the working-class attire, uniforms or clothing used in the means of (hand-)work. Fashion industry's appropriation of work wear implies sexualisation and banalisation of working-class utilitarian styles by using these work/labour aesthetics to produce similarly looking styles (copies) in the high-end and luxury sector. Jeans are one of the most notorious results of appropriation of a working-class utilitarian garment in fashion (originally produced as workwear, denim pants were used by workers in the mines due to their resistance). Denim pants or jeans have a history of being appropriated by the luxury sector, but also by subcultures, the youth, or the working-class themselves. Some examples of working-class porn in fashion are the introduction by *Nordstrom* of the Barracuda Straight Leg Jeans: a pair of jeans that was artificially 'mud-stained' giving the appearance of being the jeans of a hard hand-worker; and the also cynical production of jeans or denim pants that appear heavily used, an effect (usually achieved by an extensive use of an item) that can only be attained in short time (and without being used) by a dusting process that is highly toxic to the lungs of the workers that manipulate them. The fashion industry has a tendency to ignore its workers and their conditions but then paradoxically produces clothes and styles that emulate used work wear. Re-thinking Bourdieu's outdated concept of the working-class taste for necessity and emulation, we may recognize how fashion's style markers find an attraction for those taste-for-necessity items—due to their potential authenticity and familiarity with a 'real'.

2.3 *Cultural appropriation in fashion*

In regard to *cultural appropriation in fashion*, I will discuss two concepts: cultural appropriation of *subcultures* and appropriation affecting *cultural heritage*.

2.3.1 *Appropriation of subcultures*

The notion of *subculture* implies a group of people which differentiates from the larger prevailing culture surrounding them. These groups share values and conventions and are always non-conforming or dissenting, tending to oppose mainstream culture. Subcultures find common means of stylistic expression within the group e.g. the riot grrrls, skinheads, punks, hippies, goths, hip hop, black block, rastafaris, zoot-suiters, and mods. These subcultures exhibit negative attitudes towards capitalist work ethics and class affiliations; they have a tight association to their own territory (e.g. 'the hood'); display stylistic overemphasis and openly reject massification (Gelder, 2007). Often subcultures are correlated with negative traits due to their rejection of the dominant expressions of culture in their attempt to find their own identity and meaning. Ethnicity, race, class, and gender or a mix of them can be physical distinctions of subcultures (Hebdige, 1979). Appropriation of subcultures' styles is a reiterated act within the fashion industry of incorporating trends originated within dissenting social groups. Those subcultural stylistic expressions hold an attractive innovative intensity due to their affiliation with a small group. Fashion's tendency to admire those disrupting elements ends ultimately with the commodification of the style: The production of a luxury version followed by a wave of mass-produced versions lead to a loss of meaning of the original identity. Perhaps the rejection itself (in order to maintain those original meanings of disruption) of fashion industry's appropriation of subculture styles, equates those subculture stylistic approaches to the avant-garde and modernity as opposed to fast fashion and mass fashion (mass culture). Subcultures frequently develop their style identities by personalising other existing items and styles, yet this process differs from the cultural appropriation since that *personalisation means decommodification* whereas *appropriation implies commodification*. Polhemus identified two processes in the street style, *dressing up* and *dressing down*. Dressing down accounts for the tendency, usually of the middle class, to make a stylistic descent in the social class in an attempt to reach towards authenticity (e.g. hippies, beatniks); dressing up would illustrate the opposite intention (e.g. zooties, mods, rappers). Once fashion has commodified these fetishised expressions, we observe a meaningless use of former counterculture signifiers. In 2016, Joe Corré (the son of Vivienne Westwood and Malcolm McLaren) burnt down his £5 million worth punk collection in protest for the lack of meaning of 'punk' nowadays. 'Punk is a McDonald brand', he said. It was a punk action against the commodification of punk.

2.3.2 *Appropriation of cultural heritage*

Tangible and intangible cultural heritage is often subjected to acts of misappropriation in art and fashion. Heritage is understood differently depending on the cultural contexts it is framed. Following the definition given by the *Intellectual Property Issues in Cultural Heritage Project*, we understand cultural heritage as including elements of a collective past that remain meaningful to a culture today, connected to a shared memory, or linked to a collective identity. Misappropriations occur when there is a one-sided process and one entity is benefiting from another group's culture without acknowledging it, without permission and without giving anything in return. These actions of misappropriation do not only imply disrespect but often reinforce stereotypes underlining racism and bigotry as well as acts that result in a diminished value of the original items, designs or styles. Typical examples of appropriation of cultural heritage in fashion are those related to the exploitation and commodification of patterns and textiles developed by native communities, without their consent, without acknowledging their origin and without sharing the resulting benefits or profits; or the appropriation of traditional garments of certain cultures, as with the recent accusation towards *ZARA* of appropriating the 'lungi' (a traditional inexpensive garment from South East Asia, worn around the waist with a knot) and re-naming it as 'skirt'; or the use of dreadlocks on mainly white female models as seen in *Marc Jacobs*' spring/summer runway collection of 2017. Outmoded discourses which may call 'straightened hair-styles' imitations of Western norms by POC communities may find their reverse analysis throughout the era of cultural modernity, in which 'it is the white people who have been doing a great deal of the imitating while black people have done much of the innovating' (Gelder and Thornton, 1997).

3 CONCLUSION

In the discussion on cultural appropriation and fashion, I define commonly used concepts such as *class tourism* in fashion, *working-class porn* in fashion, *appropriation of subcultures* and *appropriation of cultural heritage* in fashion, as well as the significance of *authenticity* and the *original*. In the absence of a 'real' or 'original' identity in fashion, we witness the ubiquitous influence of *nostalgia*. Authenticity is often staged and socially constructed by the subjective reality of the cultural experience. The copy the fashion system (re-)fabricates aims to be authentic by deliberately referencing an (imagined) real as a cultural source. This copy is frequently validated as authentic, resulting in negative implications for the original style signifier.

When discussing the designer as a protagonist of appropriation, it raises questions such as whether the designer's actions are simply irresponsible acts of colonialist ignorance; whether the designer belongs originally to a subculture or an underrepresented group (making the action of appropriation an act of *return* to a more meaningful and authentic style); whether the appropriation documents the designer's genuine interest in the origin of those non-conforming styles (with irony as an expression of resistance); or whether the designer could manifest a trend towards the so-called political dressing (opening the question: can fashion be transformative?).On the other hand, the consumer as an active appropriator adapts, personalises and assimilates styles and trends, but is ultimately yearning for the construction of an 'original' identity. Since there is no more original, then there must be a validated copy, an item holding the character of authenticity. Perhaps the designer and the consumer are experiencing multiple identities in order to find the unattainable *real* in a society governed by nostalgia, the loss and the perception of authenticity. These multiple identities are materialised through expressive experiences (like those categorised as *class tourism*) entering the sphere of *consumption of experiences*, which implies lesser consumption of traditional sumptuous items and a higher consumption of the experiences themselves. If so, fashion could be understood at times (as in these performative and appropriative expressions) as an *experience*—and consumption of fashion as consumption of experiences: '*The fashion designer is not only seen as an expert in creating attractive and current garments but as someone with expertise in staging and creating experiences of newness*'(Petersen, Mackinney-Valentin and Riegels-Melchior, 2016).

The fashion system transforms a DIY punk style into a luxury product or a shabby looking workwear into a costly and desirable commodity. Consumers are buying an expensive copy-version of an inexpensive product (e.g. a *DHL*-shirt). In this search for the *authentic*, the consumer accepts the high costs of the presumed authentic style—almost real, nearly original. This could explain the paradoxical coexistence of a *snob demand* based on the desire of unusual, limited and expensive items (fashion's validated copy), with the almost identical, costless items on which they are based (e.g. *Vetements*' 'DHL' – shirt or *Balenciaga*'s 'Ikea' – bag as 'copies'; and the inexpensive originals as 'references'). These paradoxes work because the copy which fashion industry creates and validates, claims itself as authentic. '*When the real is no longer what it used to be, nostalgia assumes its full meaning*', as Baudrillard asserts. In other words: Nostalgia for the real or nostalgia for the authentic.

REFERENCES

Armstrong, T. 2016. Why Fashion's Reappropriation of British Working Class Culture Isn't a Bad Thing. [Online] HYPEBEAST. Available at: <https://hypebeast.com/2016/1/fashion-british-working-class> [Accessed 17 October 2017].

Aronowsky Cronenberg, A. 2017. Demna Gvasalia on Authenticity, Irony and Resistance. [Online]. OP-Ed. Business of Fashion. Available at: <https://www.businessoffashion.com/articles/people/demnag-vasalia-on-authenticity-irony-and-resistance> [Accessed 20 December 2017].

Baudrillard, J. 1981. *Simulacres et Simulation*. France: Editions Galilée.

Blumer, H., 1969. Fashion: From Class Differentiation to Collective Selection. *The Sociological Quarterly* Vol. 10, Iss. 3.

Bourdieu, P., 1984. *Distinction*. London: Routledge.

Brown, M.F. 2005. Heritage Trouble: Recent Work on the Protection of Intangible Cultural Property. *International Journal of Cultural Property*. Cambridge University Press, 12(1), pp. 40–61. doi: 10.1017/S0940739105050010.

Bubble-Up Effects of Subculture Fashion. [Online]. Scribd. Available at: <https://www.scribd.com/document/236268737/Bubble-Up-Effects-of-Subculture-Fashion> [Accessed 25 January 2017].

Choufan, L. 2018. Logomania? Blame the Hipsters. [Online]. OP-Ed. Bussines of Fashion. Available at: <https://www.businessoffashion.com/articles/opinion/op-ed-logomania-blame-the-hipsters> [Accessed 10 February 2018].

Cocker, J.B., Doyle, C., Banks, N. & Russel Senior, Mckey, S. P. 1995. *Common People.* [lyric] London: Universal Music Publishing Group.

Collins, J. 2014. Normcore: when being ordinary becomes a fashion statement. [Online] The Conversation. Available at: <https://theconversation.com/normcore-when-being-ordinary-becomes-a-fashion-statement-31608> [Accessed 27 January 2018].

Craik, J. 1994. *The face of fashion: cultural studies in fashion.* London: Routledge.

Ellen, B. 2015. Common People: Jarvis Cocker's song is not all about the girl. The Guardian. Opinion. [Online]. Available at: <https://www.theguardian.com/commentisfree/2015/may/10/common-people-pulp-cocker-greek-girl> [Accessed 20 January 2018].

Gelder, K. & Thornton, S. 1997 (eds.). *The Subcultures Reader.* London and New York: Routledge.

Gelder, K. 2007. *Subcultures. Cultural histories and social practice.* London: Routledge.

Hebdige, D. 1979 *Subculture: The Meaning of Style.* London: Methuen and Co.

Intellectual Property Issues in Cultural Heritage Project, 2015. Think Before You Appropriate. Things to know and questions to ask in order to avoid misappropriating Indigenous cultural heritage. Simon Fraser University: Vancouver.

Jenss, H. (ed.) 2016. *Fashion Studies. Research Methods, Sites and Practices.* London: Bloomsbury.

Kinmont, B. 2012. Streetstyle: The Bubble Up Theory. Fashionunstiched. Contemporary Fashion blog, [blog] 10 September. Available at: <https://fashionunstiched.wordpres.com/?s = The + Bubble + Up + Theory> [Accessed 23 December 2017].

McMahon, K. 2012. Pass the scissors: consumers close in on the fashion industry. In Shanbhag, Vinod (Ed.) *Fashion Beyond Borders 14th Annual Conference Proceedings.* Pearl Academy, Jaipur, India, pp. 260–270.

Notebook on Cities and Clothes, 1989. [Documentary Film]. Directed by Wim Wenders. Paris: commissioned by the Pompidou Center. Distributed by Axiom Films.

Orwell, G. 1933. *Down and Out in Paris and London.* London: Victor Gollanz.

Oxford dictionary. 2018. [Online]. Available at: https://en.oxforddictionaries.com/definition/porn [Accessed 21 Feb. 2018].

Petersen, T.B., Mackinney-Valentin, M. & Riegels Melchior, M. 2016. Fashion Thinking. *Fashion Practice*, 8:1, 4.

Polhemus, T. 2010. *Streetstyle.* PYMCA.

Rabkin, E. 2016. The Rise of The Fashion Hipster. [Online]. OP-Ed. Bussines of Fashion. Available at: https://www. businessoffashion.com/articles/opinion/op-ed-the-rise-of-the-fashion-hipster [Accessed 15 July 2016].

Rocamora, A. & Smelik A.(eds.) 2016. *Thinking Through Fashion: A Guide to Key Theorists.* London: I.B. Tauris. ISBN 978 1780767338HB.

Rocamora, A. 2002. Fields of Fashion. Critical insights into Bourdieu's sociology of culture. *Journal of Consumer Culture*, vol. 2, 3: pp. 341–362.

Sebastian, J.E. 2017. The Cynical Realism of Demna, Gosha and Lotta. [Online]. OP-Ed. Business of Fashion. Available at: https:<//www.businessoffashion.com/articles/opinion/gosha-rubchinskiy-demna-gvasalia-lotta-volkova-vetements-pseudo-cultural-fad> [Accessed 30 June 2017].

Shand, P. 2002. *Scenes from the Colonial Catwalk: Cultural Appropriation, Intellectual Property Rights, and Fashion.* Dissertation. University of Auckland.

Smelik, A. 2011. The Performance of Authenticity. In Address. *Journal of Fashion Writing and Criticism.* September Issue, vol 1, num 1: 76–82 ISSN 9772044 927001.

The Gale Group, 2005. Theories of Fashion. Encyclopedia of Clothing and Fashion. Available at: <http://www.encyclopedia.com/fashion/encyclopedias-almanacs-transcripts-and-maps/fashion-theories> [Accessed 16 January 2018].

Thompson, C.J. & Haytko, D.L. 1997. Speaking of Fashion: Consumers' Uses of Fashion Discourses and the Appropriation of Countervailing Cultural Meanings. *Journal of Consumer Research*, Volume 24, Issue 1, Pages 15–42.

Young, J.O. 2010. *Cultural appropriation and the arts.* Blackwell Publishing, 2010, 168 pp, ISBN 978-1-4443-3271-1.

Affection sewn into hats: Identity, self-esteem, women's social relations

G.P. Lenzi
Universidad de Salamanca, Salamanca, Castilla y León, Spain

ABSTRACT: The hat, a clothing element in gradual disuse nowadays, was applied to a group of women with breast cancer. We thought that its use might contribute to the maintenance of these women's identity and self-esteem while they were hairless, which is an effect of chemotherapy. More than only wearing this accessory, they could make the hats they should wear or, in case of physical frailty, women already cured of this disease, but still under post-cancer treatment, were their colleagues' hands at hatmaking. Besides understanding in which way the hat could be capable of helping their self-esteem and identity, we aimed at understanding how that technique could be a means of getting these women together and in which way that social relation could benefit them. We found out that the hat had created a new narrative in this group's identity context, besides becoming a tool for social proximity.

1 INTRODUCTION

The hat, as a clothing element, is almost in disuse nowadays, and, together with it, also the techniques of handmade millinery are in gradual disappearance, as mentioned by Longoni (2003). Also according to the same author, this accessory may be found nowadays in groups where its symbolic weight is seen as representative, as well as in restricted groups or niches. Of a protective nature, however also of extremely symbolic content, the hat carries, along its history, the capability of guarding the identity of those who wear it (Vanni 2004), as well as of being responsible for guiding the thought and ideas when worn on the individual's head (Jung & von Franz 1968, Machado de Assis 1884).

So stated, we aimed at introducing the hat—with its innate capability of identity protector and guardian—in a group of women with breast cancer, who felt the need for the use of an accessory, especially during the period when they were under the chemotherapy treatment, that is, when they were hairless. By confronting breast cancer, these women have to deal with several types of fear and anguish, such as the fear of death and issues related to the financial situation (for example, to pay for the treatment). However, the loss of identity referentials, together with aesthetic factors, such as the loss of the breast and of hair, occupies a position at the top of the list of their main worries (Silva 2008).

Recognizing the factor mentioned by Silva (2008) as a relevant problem for this group, we felt the need to understand how the hat could be able to help the maintenance of these women's self-esteem and identity during the chemotherapy treatment. We then hypothesized that already-existing elements commonly worn by this group, such as scarves and even wigs, are stigmatizing symbols (Goffman 1988) of this disease. Besides the use of the hat as an identity element and its possible support for self-esteem, the millinery technique was revisited, employed and developed in the group where the hats were created and made. We also aimed at understanding how that technique could be employed as a means of getting these women and their peers together and in which way the social relation developed in the group could benefit them. In this sense, we thought that there is some mutual benefit, in relation to the accessory as well as to the craft which retains the technique. Both—the hat and the millinery technique—are in gradual disappearance nowadays even for this group of women who intend to restore their self-esteem and identity during the delicate phase in which they find themselves; thus, particularly, the social and cultural context to which they belong should be considered.

To achieve the objectives proposed herein, this paper follows in terms of structure a sequence which includes, firstly, in section 2, the field description, which involves a brief description of time and space circumstances as well as of subjects, so that the contexts of their experience could be understood. Following this, in the same section is the methodological framework of the research, which was set out with the help of authors such as Aguiar (2015), Kottak (2013), Taylor & Bogdan (1987), Angrosino (2009), Bastide (1972), Foster (1974), Geertz (1992, 2014). Later, in section 3, the results and discussion are presented, sprawled in two subsections. The first is called "When the hat guards self-esteem and identity throughout breast cancer" and gives an account of the accessory as a support element to identity and self-esteem during and after the chemotherapy treatment process. The second—"Women's social relations through the hat and the millinery technique"—aims at understanding the social relations established among the subjects, through handmade millinery and the use of the hat. In order to establish a dialogue with the results from the empirical study, we got advice from authors such as Goffman (1988, 2014), Castilho & Martins (2005), Camargo (2000), Santana (2014), Stallynbass (2012), among others. Lastly, we laid out the final comments, which validate the objectives.

2 METHODOLOGY

2.1 Field description

This research was carried out in the town of Blumenau, located in Santa Catarina State, South Region of Brazil, between the years of 2015 and 2017, period during which the millinery technique was applied and taught to women suffering from breast cancer. Besides the social relation developed in the group and the learning of a new craft, the women had the opportunity of covering their heads, which were hairless as a result of the chemotherapy treatment. Because of this, the setting chosen for carrying out the research was Blumenau's Women's Cancer Network, a nonprofit organization whose goal is to prevent gynaecological and breast cancer and support mastectomized patients (Rede Feminina de Combate ao Câncer de Blumenau 2015).

At first, we intended to invite only patients under the chemotherapy treatment and, because of that, with no hair, so that each one could make their own accessory. However, we noticed that, right from the pre-field approach, due to the physical impairment experienced by them in this period, it would be unfeasible. Because of that, we decided that the already-cured patients who were not undergoing chemotherapy anymore, but that came to the organization because of some post-cancer treatment, could be the creative hands behind the hats which would be given later to those colleagues under treatment and who had no hair. Fourteen women enrolled in the millinery classes. Among them, 3 made the accessory for themselves, because, although they were still under the chemotherapy treatment, they considered themselves physically stable in order to participate.

In the second fortnight of February 2016, we launched the research, with the development of the project called by the very participants the Hatmaking Workshop. Some volunteers from the institution could be found among them, who would also help with the entity's organizational issues, as well as with the patients enrolled. The lessons were held weekly and had, in the beginning, the duration of one hour and a half. After a few weeks, the women expressed the intention of staying one more hour and so the lessons were two hours and a half long. The workshop came to an end in mid-July 2016, and, as a result, 23 hats were made and given to the wearers in an event promoted at the organization, so that, more than just receiving the hats, integration among them could be possible.

Following up on the research, between November 2016 and January 2017, we interviewed, with a set of semi-structured questions, 10 women who were wearers of these hats, being among them the 3 who, besides being wearers, also made them. The criterion for the selection of those women interviewed was based on each one's physical conditions, since many of them, in this period, were fragile because of the disease and/or the treatment they were undergoing.

2.2 *Methodological framework*

It is a research of anthropological and qualitative nature, with the use of ethnography for data collection, during which we took the role of a participant observer, with indirect and semi-structured interviews during the different moments of the participation in the field. Participant observation, to Aguiar (2015), is a more effective way of noticing what can and can not be done in order to get to the comprehension of a culture; it was the compass which guided the research, having in mind the delicate environment that we entered and to which we dedicated a considerable amount of time using observation as a method of subject approach (Kottak 2013, Taylor & Bogdan 1987). Together with participant observation, indirect interviews were conducted, which occurred randomly during the participation in the group and were of a strong nature (Angrosino 2009), so that the comprehension of the symbolic system behind the intense socialization in the field could be possible. The observation and interview moments were supported by the use of a field journal, as well as of audio recordings and photography and, in a few specific moments, even video. In order that these data collection tools could be used in accordance with the due ethics, the research was submitted to the Ethics in Human Research Committee, which approved it. Thus, every participant signed the free will informed consent form, which aims at their protection and integrity, accepting, thus, their own participation.

In regard to the Hatmaking Workshop mentioned in the previous subsection, since it was the application of a craft to a group of women with breast cancer, it is regarded as being applied anthropology or, as it is was referred to by Bastide (1972), an action research. Therefore, we got advice from Foster (1974), who, from his first writings, says that, when the anthropologist applies theoretical, factual and methodological knowledge in programs designed for the solution of current social, economic and technological dilemmas, he/she is doing applied anthropology. Finally, the interpretation of the data collected was done by means of thick description, in accordance with what is proposed by post-modern anthropologist Geertz (1992, 2014), since this description, besides portraying every detail, is capable of capturing the significant intentions involved in the observed conduct (Velasco & Díaz 2009), which allows a more legitimate and authentic exposition of the experiences in the field.

3 RESULTS AND DISCUSSION

3.1 *When the hat guards self-esteem and identity throughout breast cancer*

Débora started participating in the Hatmaking Workshop after it had begun. She was one of those participants who, besides making a hat in order to be donated to a colleague, also made use of one. On her head, in that moment, tiny hair strands which looked like plumes were popping out, and nothing could be more symbolic than this occurrence: she was reborn and, like a baby bird that has just come out of the eggshell, practiced her first and new experiments with the look and curiosity of those who have just arrived in this world. The Hatmaking Workshop, to her, was in this new life of hers. After entering the room, one of the earliest Débora's reports was about how traumatic had been the loss of her hair. To her, hair loss was worse than the loss of the breast, due to the exposure of that part of body: "*So, losing the hair was very much [sigh and pause], it's that when it's only the breast you don't feel so much, ah [short pause], it's something people don't notice, so much so that after the surgery I soon went back to work and also went to parties. It was a party at my sister's, her daughter's birthday. People saw me like that with all that hair, so they didn't see me weakened, it seemed that it had been a surgery to remove an ingrown nail, let's say, huh? You know, I was as if unconcerned, huh?*

Because no one noticed. Then I thought: 'My God, the hair is going to fall out, what do I do?' Then I started thinking and getting desperate: 'What am I going to do without my hair?'" The fear and the uncertainties caused by hair loss expressed by Débora are also seen in the studies by Camargo (2000), who pointed out that hair loss is one of the most stressing factors during this process, and, many times, seen by women as more shocking than the very discovery of the disease, as she reports: "However, there are collateral effects of the therapy, which have psychological consequences (physical and aesthetic) such as hair loss, in such a way that, for the woman, knowing it beforehand does not reduce the devastating impact of its occurrence on her. Some women say that this loss is more stressing than the breast removal surgery, in part due to it being a visible indicator of the disease, but also because it is disfiguring."

The same way as Débora did, Maristela pointed out that hair loss was, among all the misfortunes which come along with cancer, the one which had distressed her more. Maristela confirmed that she preferred losing the other breast to losing the hair. According to her, besides people labeling as "sick" the woman who had no hair, they stigmatize her socially. According to Goffman (1988), even self-recognition factors are considerably affected in this moment. Thus, Maristela dealt only little face to face with her new appearance. She avoided looking at herself in the mirror and never showed the bald head to her husband or to friends and family. After being asked about the reason for such a behavior, she gave a straightforward answer: *"I am not that person, I can't see myself in her."* Not only seeing herself in this way, but also showing up socially with her new appearance was the reason for Maristela's great fear, since she moved out of a social representation which made her comfortable in front of the social group to which she belonged and that, in that moment, stigmatized her in some way. In this sense, Gonzaléz Rey (2006) says that social representation ensures, in a certain way, the stability through which the individual recognizes himself/herself and is recognized; therefore, one finds appropriate to participate in the group of which one is part, particularly considering the sociocultural context in which one lives, where the hair has a relevant meaning, especially for women (Santana 2014).

Débora's and Maristela's short accounts show that traumatic hair loss was as indication of the need to have some support to soften the anguish experienced with this disease. By recounting these experiences, they had, in their reports, an introductory means of speaking about the relevance of wearing the hat in that period, a hat which was not only an adornment, but also a physical and social protective element for hairless heads. Under this perspective, after a week wearing the hat, Débora came to the Workshop bringing news to the group: *"Girls, you can't imagine what happened to me this week! For the first time since I lost my hair, people didn't look at me with pity, as if I were sick, people looked at me with awe and said: 'Wow, how stylish you are with a hat!' People didn't look at me anymore with those eyes full of pity, but instead saw me as a stylish woman, in awe."* Débora told that, by making use of a scarf, people looked at her with compassion, since wearing a scarf indicated that, in fact, she was suffering from the disease. To understand this account, we refer to Goffman (1988), who mentions "symbols of stigma, that is, signs which are especially effective to arouse the attention to a degrading identity discrepancy which breaks what, on the other hand, could be a coherent global picture, with a consequent decrease in our appreciation of the individual."

Tânia, who was of the most timid participants at the Hatmaking Workshop, also made the accessory that she should wear, since she was under treatment and, because of that, had no hair. Although she did not talk with the same intensity as the other participants, it was possible to notice, in the sublime lift of the corners of the lips, that it was a pleasure to be there, as well as by the gentle way she passed the hands over the felt with which she would make the hat. From the first encounter, she wore a long wig which hid the bald spot, though gave away her disease. After finishing the accessory quickly, due to the anxiety to wear it, Tânia came the following week without a wig and wore only the hat, naturally, as if she had always had it on her head. On the other hand, she was seen by the group as another woman. With this newer version of herself, she started to talk more and expose her feelings. Then she told the group that, after breaking the barrier of appearing without a wig socially, she also felt more confident to express what she experienced. She also said that she had felt an increase in her self-esteem, not only because of aesthetic factors, although she had felt she was prettier, but also because she felt self-confident about her self-image before society (Goffman 2014). Wearing the hat created a new social meaning, and she started to feel integrated. Castilho & Martins (2005) emphasize that it is natural for the subject this need for resignifying and rebuilding oneself by means of devices which connect one with values adequate to one's culture and to the social environment in which one lives.

In accordance with what was mentioned by the authors, Andiara's accounts show the capability of the hat of putting together these women and the cultural and social values in which they were inserted. Andiara, who was given a hat to wear while she was hairless, reported that the adornment was very useful in the process of maintenance of her self-esteem and identity throughout the disease, since people noticed it and commented on it around her when it was worn: "*The hat helps a lot as a matter of fact, so the hat makes us more elegant, it's different, isn't it? So it helps a lot, since it not common to see people wearing a hat for daily affairs, then it catches attention, but it is good attention, because people make compliments. Every time I was wearing a hat, when someone who saw it, or if someone saw a picture, they used to say: 'Oh, how nice, how beautiful!' 'Oh, cool, where did you get if from?' So it is very nice to wear a hat in this phase, you feel prettier with it and this helps a lot.*"

Clothes have the capability to disguise, hide or show, reshaping the body by means of the plastic compositions present in them, as expressed by Castilho & Martins (2005). Therefore, the hat worn by the women participating in the Hatmaking Workshop, although it disguised the bald spot in some way, did not play the role of hiding or only covering the hairless head, but, instead, unveiled the woman who was wearing it. Sandra, who wore two hats made at the Hatmaking Workshop, such her identification with the object was, told that it not only had caused the others not to notice her disease, but even she, when she walked in the street, forgot she had no hair: "*I think that people don't look for hair under the hat, you know? I always say that, with the hat on, they want to see if you are fashionable and if you don't have a problem, you know? So, you can feel prettier with the hat, since the looks are different. Until they notice you have cancer, some time has already passed [laughs]. However with a scarf and with the bald spot they know immediately, don't they? I think that the hat adds a distinct charm. It has a distinct charm! That's good, isn't it? Just look at the compliments you get ...*"

3.2 *Women's social relations through the hat and the millinery technique*

Leonida, who made a hat to be donated to her colleagues, told that her initiative to participate in the Hatmaking Workshop was because of the challenge: "*I had never had a way with these things, you know ... but then I thought: 'Why not?' Then I enrolled and put myself a challenge. I had always wished to learn these things, such a craft. Then the opportunity was there, huh.*" To Leonida, besides the challenge given to her, what stimulated her throughout the development of the project were the social relations with her colleagues. "*While the mouth was talking, for we talked a lot, didn't we? [laughs] The hands kept working and we didn't even notice time passing. When you did you had already embroidered a lot ... the hand almost went about alone then! At the end, we talked more than worked, it looked like a lot of parrots,*" she told us while she was laughing and reliving the moments.

As expressed by Leonida in her pragmatic account, several Greek myths refer to women and to their relationships with textile artifacts. Besides this, the results coming from this relation have laid, at its roots, the communication and the ties among women, as it can be observed in Ovid's poem (1994) Procne and Philomela: through texture, Philomela, a young woman raped by their brother-in-law, who cut out her tongue so that she could not tell her sister Procne about what happened, was able to narrate what she had experienced in a tapestry woven by herself. Besides depicting the symbol of communication, the tapestry expressed the bond that it promotes with the female world woven into the thread of the texture. Mythologies serve as clues of these relationships between the female world and the artifact and of their communication. Certainly, objects are loaded with information and narratives, but also spoken words, present in groups of women who make some craft, are commonly found in similar situations even nowadays or from a not so distant past. The very craft of weaving straw for the hat factory, present in the region of Florence, Italy, until last century, was usually seen in the hands of women who, while they were weaving the material, chatted with colleagues and neighbors. They met in groups, in the backyards, chatting and weaving. Together, they not only guaranteed support for their family's subsistence, but raised their children, who sat by their feet while they were working and who also learned how to weave from an early age (Bellini 2007).

The goal of the women from Blumenau's Women's Cancer Network at enrolling in the Workshop was based on the initial purpose of getting to know this craft, however with the intrinsic goal of being able to build a network of friendships with women who were their peers. Gorete told that it was the organization's physical therapist who had told her about the Workshop which would teach how to make hats and that that, in that very moment, aroused her curiosity. "*I do want!*" Gorete said it at the same instant to the physical therapist. According to Gorete, more that the desire of learning the technique, which was very strong, she thought that, in that group, she could also create friendship ties, since she came from another town and missed this type of contact: "*Women like these things, don't they? They like talking, chatting and also like this fashion stuff and craftmaking. Even more because it would be a group of women which had also gone through cancer, like me. I needed this type of contact, you know.*" Like Gorete's words, the reports and the accounts given by the group moved usually around the craft being learned and the social relations originated from it, seeming almost impossible the separation of both.

To the women who were given a hat to be worn, every accessory brought along the narratives of this group of women which had so much in common with their histories. "*As a caress to the soul*" was how Sandra viewed the hat received. To her, every trimming embroidered in it was a word of love and friendship spoken by the woman who made it. She told that, for this reason, she found it difficult to choose one accessory, because she looked at each one of the hats admiring, surely, their aesthetic beauty, but thinking also about the love placed in them by the "*artisan*". She felt as if, by taking a hat for herself, she was touching the words of affection from someone about whom she knew so little and yet so much. "*It seemed as if those hats talked, you know? When I saw all of them that day, huh, I felt like I should even be silent to be able to listen to what every one had to say … I felt that that voice, it was spoken by someone that understood the pains which I had also felt [wet eyes]. And, you know … it looks like something crazy, but I could feel all the affection they put in them and, you know, only a woman who has lived what you have lived knows how it is,*" Gorete expressed herself this way.

By all means, these hats brought together with them the accounts of those women who made them, through the possibility of social relations during the application of the technique or through the symbolic communication present in every stitch, and even through their use, in which the wearers could also imprint their narratives, being this an innate characteristic of costumes, because of their capability of embodying and of being transformed by the one that makes them as well as by the one that wears them (Stallynbass 2012). Because they were women who were also their peers, they knew, by their own experience, what they had experienced, their fears and anguish, particularly regarding the issues related to the identity losses which stigmatized them. They knew likewise the relevance of the use of the accessory. Hats, on their turn, carried along related narratives, both of those who had made them and of those who wore them. In this sense, we noticed the connection among the social relations developed by these women, fruit of millinery and reflected in the object.

Besides that, we felt, at first, the surprise of the receivers of the hats donated to them in regard to the dedication devoted by the colleague who made them. In this regard, Maristela expressed herself thus: "*It's a caress donated, like this, from the bottom of the heart, which fills you there with joy, fills you with smiles, huh, gives you new horizons to think about, to imagine things you didn't imagine before. Then being given the hat was more or less that, feeling loved by someone by whom you didn't expect to be loved. At the time you even get scared, because in today's world we are not used to this in everyday life, and when I got there to receive the hat I felt loved by a person whom I not even knew, huh. To feel loved by someone whom you know is different, but to feel loved by someone there whom you don't know, it does scare you in today's world [laughs].*"

The fact of feeling loved and supported by the colleagues whom they met at the Hatmaking Workshop made the object take another value, that is, even not seeing each other so much frequently, seeing the object and feeling surrounded by it was enough to feel the presence of all of them. In this sense, Mara said that, when she felt alone, it was enough to look at the hat, wherever it was, to feel supported and accompanied by the woman who had made it.

Still under the same perspective, Tânia told that she had felt inspired by the women she had met at the project: "*They are out there, you know ... They are well and cured! Some for 10, 5, 3 years and that told me that if they got cured, I too could be cured. So this meeting of ours was inspiring to me, huh, gave me strength and, it was all because of the hat, let us say that.*"

Thus, having social ties with women who are alive and cured of cancer, whose main fear is of death, made the women still under treatment feel encouraged to face the disease. Under this perspective, Feather & Wainstock (1989 in Biffi & Mamede 2004) reiterate that this is the main role of social support: the favoring of psychological, physical and emotional resources which make the individual capable of facing the difficulties found along the way.

4 FINAL COMMENTS

The loss of hair makes the woman with cancer face, in a decisive way, the fact that she is sick. Besides the personal confrontation that she herself experiences, stigmatizing social issues also come up to the surface. Before a woman, now sick, this is the way she starts being socially represented, and the looks of compassion coming from the others make her question her own identity. In this regard, the use of the hat made at the Hatmaking Workshop was able to create a new narrative, which allowed the women with breast cancer to leave the interface of a sick person and to start guarding their own personal image embodiments. Such a right as that of guarding one's identity was fundamental to self-acceptance and also to help self-esteem.

The support coming from the women's social relations was based on the bond promoted by millinery during hatmaking, and was also reflected in the accessory. The fact that the object creators and makers were women who had had breast cancer made this network of relations become a space of acceptance and understanding. In Ovid's poem (2009), Philomela communicated with her sister through the tapestry woven by herself. In an analogous manner, to these women, the hat was a way of saying to the women to whom they made the hats that they were not alone, that they understood them and that they could count on them. In the same way, this was also felt by the wearers of the hat, who saw in it not only a new way of resignifying themselves, but also of feeling backed for this. Exchange and communication were fully present in the group, through chatting, or when they were making the accessories together, or in the events where they all met, or imprinted in the hat and in the narratives inspired by it. This support perceived inside the group and present in the accessory was also seen by those women as responsible for the increase in self-esteem, since that they felt accepted and strengthened at the Hatmaking Workshop so that they could face the disease.

The benefits obtained were noticed by the women with breast cancer, but also through the very technique of handmade millinery, which, nowadays, is little practiced, as well as by the object, which discovered a singular space so that it can be found again in the streets nowadays. The win-win relationship developed between the group and the practice was responsible for arousing new interest for the application of the same technique to these women, who see fashion and clothing elements from a discourse of sociability at the service of what is human. Therefore, if this application takes into consideration these women's perspective, it will be possible to lay out the continuity of the Hatmaking Workshop project inside the same organization because of the positive results obtained.

REFERENCES

Aguiar, E.P. 2015. "Observación Participante: una introducción". Revista San Gregorio, *Metodología de* la Investigación (1): 80–89.

Angrosino, M. 2009. *Etnografia e observação participante*. [Coleção Pesquisa Qualitativa]. Porto Alegre, Brasil: Bookman.

Bastide, R. 1972. *Antropología aplicada*. Buenos Aires: Amorrortu.

Bellini, M.E.T. 2007. *La manifattura della paglia nel Novecento:* da Signa e dalla Toscana nel mondo. Firenze: Polistampa.

Biffi, R.G. & Mamede, M.V. 2004. Suporte social na reabilitação da mulher mastectomizada: o papel do parceiro sexual. *Revista Escola de Enfermagem da USP*, São Paulo, Brasil 38(3): 262–269.

Camargo, T.C. 2000. *O ex-sistir feminino enfrentando a quimioterapia para o câncer de mama*: um estudo de enfermagem na ótica de Martin Heidegger. [Doctoral Thesis]. Universidade Federal do Rio de Janeiro, Rio de Janeiro, Brasil.

Castilho, K. & Martins, M.M. 2005. *Discursos da Moda:* semiótica, design, corpo. São Paulo, Brasil: Anhembi Morumbi.

Foster, G.M. 1974. *Antropología aplicada*. México: Fondo de Cultura Económica.

Geertz, C. 1992. *La interpretación de las culturas*.14ed. Barcelona: Gedisa.

Geertz, C. 2014. *O saber local*: novos ensaios em antropologia interpretativa Petrópolis, Brasil: Vozes.

Goffman, E. 1988. *Estigma*: notas sobre a manipulação da identidade deteriorada. Rio de Janeiro, Brasil: LTC.

Goffman, E. 2014. *A representação do eu na vida cotidiana*. Petrópolis, Brasil: Vozes.

González Rey, F.L. 2006. As representações sociais como produção subjetiva: seu impacto na hipertensão e no câncer. *Psicologia: teoria e prática*, São Paulo, Brasil 8(2).

Jung, C.G. & von Franz, M.L. (eds). 1968. *Man and his symbols*. New York: Random House LLC.

Kottak, C. P. 2013. *Um espelho para a humanidade*: uma introdução à antropologia cultural. Porto Alegre, Brasil: AMGH.

Longoni, G.M. 2003. *L'ereditá dei Cappellai – memoria, mito e realtá di una avventura del lavoro*: a cura di G.M. Longoni. Cinisello Balsamo: Silvana.

Machado de Assis, J.M. 1884. *Histórias sem data*. [Capítulo 7]. Rio de Janeiro, Brasil: Garnier

Ovid. 1994. *Metamorphoses* [Translated by Frank Justus Miller]. London: Loeb Classical Library.

Rede Feminina de Combate ao Câncer de Blumenau. 2015. [online], Blumenau, Brasil. Available at: <http://www.redefemininaccblu.com.br> [Accessed on June 12th, 2015].

Santana, B. 2014. Mulher, cabelo e mídia. *Communicare Dossiê Feminismo, Revista semestral do Centro Interdisciplinar de Pesquisa*, Faculdade Cásper Líbero 14(1).

Silva, L.C. 2008, abril/junho. Câncer de mama e sofrimento psicológico: aspectos relacionados ao feminino. *Psicologia em Estudo* [online], Maringá, Brasil, 13(2): 231–237. Available at: <http://www.scielo.br/pdf/pe/v13n2/a05v13n2.pdf> [Accessed on June 12th, 2015].

Stallybrass, P. 2012. *O casaco de Marx*: roupas, memórias, dor. 4ed. Belo Horizonte, Brasil: Autêntica.

Taylor, S.J. & Bogdan, R. 1987. *Introducción a los métodos cualitativos de investigación*. 2ed. Barcelona: Ediciones Paidós.

Vanni, M. (org). 2004. *Identità e diversità. Il cappello e la creatività*. Siena: Carlo Cambi Editore.

Velasco, H. & Díaz de Rada Á. 2009. *La lógica de la investigación etnográfica – un modelo de trabajo para etnógrafos de escuela*. Madri: Trotta.

The lace fabric of Bahia clothing: From artisanal to industrial

A. Okasaki
UNISO and ESAMC, Sorocaba, Brazil

ABSTRACT: This article deals with lace fabrics in the range of the traditional dresses used by Bahians. Reflected on the origin of the Bahian costumes, the research aims to define the meanings and motives for the use of these lace fabrics. This production begins in a bibliographical reference to define the composition of the clothes of the Bahians and the history of laced fabrics in this garment, during the Brazilian slave period. Finally, the symbolic functions of these fabrics are discussed in relation to Bahians clothes up to the present.

1 INTRODUCTION

This research deals with the use of lace fabric, in the typical Bahian costume, after the second half of the 16th century. From this theme, the question that led to the research arose: What are the reasons for the use of lace in the traditional clothing of Bahians today?

The most obvious hypothesis is that there would be reasons of cultural historical context, derived even from the period of Brazilian slavery.

Some of the justifications for the importance of this research is the recognition of the suitability of Bahian dresses as cultural heritage, as happened with the craft of making *acarajé*. With this cultural protection, elements such as the clothes of the Bahians, can obtain protection by the cultural patrimony and be historically perpetuated. In addition, exploring the theme of lace in Bahian costumes is important to demonstrate that the motivations for this choice were not arbitrary or a fashion trend. There were cultural historical motivations related to the resistance and survival of slaves in Brazil. For this reason, recognition of this struggle through clothing is elemental.

1.1 *Objectives*

To show the evolution of the use of lace fabric by the Bahian from the arrival of the African slaves to Brazil. To identify the meaning and motivations for the use of lace fabrics in the Bahian dresses.

2 METHODOLOGY

This article deals with a descriptive research on the dresses of Bahians. More precisely, on the use of lace fabrics in these dresses. The approach is qualitative, focusing on the meanings of lace on this outfit. The research is based on references and bibliographic images. Determining the investigation, these works are centered in the second half of the sixteenth century, when the first slaves contributed in Brazil, and only in Bahia, the traditional place of these costumes.

3 OCCURRENCE OF BAHIANS COSTUME

Clothing is a trademark of Afro-Brazilian culture. It emerged between the interaction of African culture (religion, social structure, ethnicities, etc.) and Portuguese-Brazilian culture, after the sixteenth century, in northeastern Brazil. It is possible to verify the environment in which the Bahian costumes appeared in Figure 1. The interaction between the African slaves and the Portuguese was essential in the creation of the costumes seen on Debret's canvas (Fig. 1) with lace. One might note the cotton loom and the contact of raw and white fabric with strong colors. These colors marked, among other characteristics, the ethnic groups from which the blacks descended, and their orishas of devotion. The situation of more than three centuries of slavery, as well as cultural shocks, have also influenced this costume (Monteiro, Ferreira, 2005).

In Bahia, the ethnic groups with most influence in culture and clothing, due to the number of slaves who arrived and the resistance to acculturation, were the *Banto, Daomeans (Jeje), Nagô-Yoruba* and *Kêto* ethnic groups (Verger 2002).

It is true that the traditional African dresses were modified after the arrival of the slaves in Brazil. The influence of the European settlers, the Brazilian raw materials and the conditions of survival of the blacks modified the clothing. The clothes of the slaves slowly influenced and came under the influence of the clothes of the slaves 'masters. European settlers also took French and English fashion with costumes worn as hereditary assets, such as silk fabrics, brocades, velvet, taffeta and lace. The influence of the English fashion brought heavy fabrics for female garments, such as Bulky skirts, preceded by the petticoat (skirt used under the dress) with long pleats and tight bodice with square or circular neckline. However, the differences in costumes between the ladies and the slaves inside the houses were smaller in everyday life. It happened because the status and power that the ladies' clothes should show when they were in the eyes of society were diminished by the need for thermal comfort. So, at home, the ladies walked barefoot, or in slippers, in shirts of transparent cambric, with wide collars that slipped down over their shoulders, leaving their breasts exposed. On the other hand, the slaves closest to the masters dressed differently from the slaves who worked on plantations, mills or in tasks with little contact with their masters (Martini 2007).

It should be noted that the current Bahian costume; with many white laces, volume on the skirts, necklaces and turbans, as in Figure 2, derived from the Creole costume. That occurred because the clothes of the slaves who arrived from Africa were simpler in comparison with

Figure 1. Painting Praça do Palácio [Palace Square], 1835/1839 by Jean-Baptiste Debret (Ribeiro, 2010).

Figure 2. Bahian selling acarajé, Ana Bela dos Santos (Frazão, 2009).

the customs of the Creoles (black people born in Brazil). The Creoles had a better social position and could work in domestic functions or as trading function slaves, i.e., slaves who sold products in the street (Monteiro, Ferreira 2005).

The Creoles served as a trophy for the master, who could show how the slaves he supported were exotic. For this purpose, the slaves who were closest to the masters, such as "milkmaids", maidens, pages, nannies, cooks, laundresses, coachmen, cupbearers and messengers, in short, all those who were seen with their masters should present a wealth appearance. Thus, these slaves were often seen wearing clothes borrowed from their masters, clothes that were no longer worn or wearing uniforms. When the master showed a well-treated slave, it was like showing a material asset, synonymous of power (Resende 2011).

4 LACE HISTORY

The origin of lace is unknown, however, the craft techniques for the fabrication of these fabrics gained prominence in the West as of the sixteenth century, after the Crusades. There are stories indicating that lace was brought to Europe from the East, along with various spices and other high-value exotic products to the European market. Another factor that corroborates this theory is the similarity of the typical figures in lace, with arabesques. In the fifteenth century, the artisanship elaborated from lace weaving became a symbol of wealth adapted by fashion. Catherine de Medici was one of the influencers in the insertion of the lace in the royal costumes, taking the fabric from Venice to the French court, in 1547. This increased lace imports in such a proportion, that due to the financial crisis of the court, the king decreed the prohibition of lace usage in various costumes. This decree did not last long, and a domestic French production started in the city of Le Puy (Pezzolo 2007).

Certain European regions were outstanding in promoting different types and methods of knitting lace. In Italy, especially in Venice, needle laces were highly developed. As seen in Figure 3, it has an almost imperceptible background, making a closer fabric, less transparent and heavier. On the other side, in the city of Flanders, Italy, and Antwerp, Belgium, the improved technique was bobbin lace. The bobbin lace is lighter and has a much larger area without a design (only with a background). In addition to Italy and Belgium, there was a large production of lace in France in the fifteenth century. Therefore, these three countries request the title of creators of this fabric (Martini 2007).

In Brazil, laces arrived in the sixteenth century, at the time of the first convents and catholic collecting. Women should devote themselves to household affairs and handcrafts, such as lace altarpieces for altars, to prepare them for marriages (no two uses), and to seek mundane thoughts. Almost in the nineteenth century, with the Portuguese royal family in Brazil, the colony got more contact with the use of lace in the costumes and not only in the decoration of houses and churches. The use that used to be restricted to the clergy and nobility, in silver and gold trimmings, began to be used also in clothing, in ruffles, finishes and details of dresses, hats, gloves, caps, umbrellas, shawls, among other pieces. Models of French costumes also appeared in magazines and imported books, influencing the clothing of the rich ladies of the colony. The presence of bobbin lace in Brazil is striking on the coast, alike the city of

Figure 3. Needle lace on the left and bobbin lace on the right (Casa.com.br, 2013).

reference in this technique which is also coastal, Florianópolis. This is because in many cases these women wove with yarns used on men fishing nets, in view of the greater thickness of the yarns to weave the lace, besides the fact that coast cities were the entering doors of laces, brought by the Portuguese (Silva 2014).

The lace production expanded after 1809, when Englishman John Healthcoat invented the first lacemaking machine. In the nineteenth century, Leavers machines expanded industrial production of lace in the United States, Germany, Switzerland and Austria. Currently, the industrial production of lace exceeds the French artisanal, disseminating a cheaper product for use in popular costumes (Pezzolo 2007). For this reason, it is possible to verify the presence of lace in the daily dress, in addition to festive costumes.

5 BAHIANS CLOTHING

Among the different Afro-Brazilian dresses, one of the best known is that of Bahians. Unmistakably dress, the Bahian dress was initially used in Brazil by the slaves closest to housewives and who performed tasks inside the big house. Currently, it is a popular costume in candomblé fields and by sellers of *acarajé*, besides being present in the wings of the carnival with remarkable models. The suit draws attention through the volume, where it is unknown if the Bahians are fat or if the layers of the skirts are bulky. This volume is due largely to laces. The corpulence of skirts is a historical feature of fashion, which simulates a larger female hip. For a long time, this quality was an attraction in women. This is because larger hips meant good bone formation for the birth of many children, in addition to becoming a sex symbol. Because it is a lightweight fabric, lace was also used in the skirts of the Bahian costumes, to produce this exceptional volume.

In addition, the lace present in the Bahian dress had a sexual delight to the white masters, due to the transparency of the fabric. The French anthropologist Pierre Verger had already perceived this sensuality through lace, in the photos and researches carried out with several Bahians, between 1946, 1960 and 1970. In a report from 1718, examined by Verger, we have the record of opulence and luxury in the costumes of the slaves (Martini 2007).

Compared the creoles costume to the dresses of the ladies, it is possible to verify in Figure 4 that there was a moral religious pressure by the Catholic Church on the way of women dressing in the Brazilian colonial period. A very popular writing of St. Paul at the time was: "As for women, let them have decent clothes, and adorn themselves with pudency and modesty; no braids, no gold objects, no pearls, no sumptuous clothing [...]" (Silva 2014).

On the other hand, the clothes of the slaves who were forced to work with prostitution were so opulent and sensual that the Church itself rejected certain transparent fabrics, as it was in the preaching of Frei Manuel Calado in the seventeenth century (Silva 2014).

However, it is necessary to identify the composition of the Bahian costume. The Bahians are a worldwide-recognized cultural symbol, e. g., traditional in the sale of *acarajé* in Bahia, candomblé areas and Brazilian carnival. Despite having only the business of selling *acarajé* as immaterial patrimony, IPHAN indicates that the clothing is the characteristic element of the Bahians:

Figure 4. Costumes in São Paulo (Silva 2014).

> *The first and marked identification of the acarajé Bahian starts by the dress with an abundant and complex assembly of cloths. Turban, fabrics in different formats, textures and techniques of formation according to social, religious, ethnic intent, among others; petticoats, several strands with interwoven fabrics and tip, something like a second skirt, skirt, in most cases with five meters of wheel, various fabrics, with ribbons, ties, among other details in the whole; T-shirt, usually flanged at the bust, robe up and in thinner fabric; a shawl for the back of different usages, weaving of loom, other industrialized cloths, rectangular, of visual approach to those of the parts of Africa* (IPHAN 2004).

The dress described by IPHAN (2004) and seen in Figure 5, is the contemporary costume of the Bahians. In Figure 5 it is possible to see each piece daily used by the Bahians. In this costume, we have the tradition of modeling, knick-knacks and colors, allied to industrialized fabrics and manual made shoes (Monteiro, Ferreira 2005).

5.1 *Ceremonial dress*

This clothing is a cultural element of the African aesthetic of the Afro-Brazilian religion of Candomblé. It is necessary to understand that since the slavery period of Brazil, for the culture of Candomblé, there was no art for art. This was because there was no difference in daily life, in religious, in deities and in nature. Thus, the whole aesthetics of clothing has significance with the beliefs of life and religious up to the present.

In candomblé, the dress is modified according to each cult. Each costume has a ceremonial significance. The lace is found more frequently in the cults named *Kêtu* and *Angola*, in which the skirts are bulky. In addition to the ceremonies, the *orixás*, *voduns* and *inquices* also determine the models, fabrics and colors of the costumes. From these cults, the Bahian dress was expanded for its uses in *Maracatu* in Recife, for the *Cortejos* and dances of São Gonçalo, the *Congadas* in Sergipe, for the various Brazilian samba schools, among other popular festivals (IPHAN 2004).

In Figure 6, it is possible to see the Bahian dresses in a religious ceremony. In Debret's painting, named Black youth on the way to church for baptism, it is possible to note the presence of more elaborate fabrics, with laces on the edges, in addition to imitation jewelry (necklaces and accessories). Debret portrayed the use of the Bahian dress delivered to the catholic ceremony, but the most common use was within Afro-Brazilian religions, such as candomblé. However, on a day-to-day basis, during the period of slavery in Brazil, Bahian laces and dresses had little use, and were used for special occasions, such as the religious ones. That is, the clothes of the religious parties gained space in other celebrations. It is important

Figure 5. Composition of Bahian dress by Wasth Rodrigues, 1891–1957 (Rebello, 2009).

Figure 6. Black youngsters on the way to church for baptism, Jean Baptiste Debret, 1834–1839 (Martini, 2007).

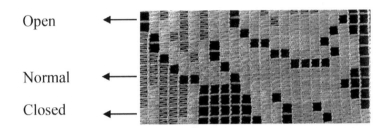

Figure 7. Jacquard Shade (Sanches, 2011).

to note that this dress, although popular, is a party dress, for unique events. It is true that the sale of *acarajé* has become customary, but still remains within a ceremony, the cooking in this case. The analysis of this dress should focus on the fact that this is a costume for rituals and for parties.

6 ARTISANAL AND INDUSTRIAL LACE FABRIC

6.1 *Point fabrics*

Lace can be manually or industrially woven. The textile sector that makes industrial lace is the point fabric.

In the warp mesh, there are currently two looms that dominate the market (although there are other types): Kettenstuhl and Raschel. Raschel machines produce smooth and embroidered fabrics. The designs can be simple, very similar to handicraft items, or even elaborated, with jacquard tones of greater or lesser coverage factor at one point, as shown in Figure 7 (Romero 1994, Sanches 2011).

6.2 *Lace*

The lace is an open-knit fabric that forms the background on which the design is superimposed. The structure of the lower part of the tissue will determine both the coverage factor of the fabric and the fall, and will interfere in the design of the lace. Even with more closed bottom structures, the lace still has the main characteristic of being a transparent fabric. Transparency and motifs (usually floral and arabesques) have guided the use of lace in women's clothing. It is true that many collars and cuffs of sixteenth-century men's clothing wore lace, but their usage was more exploited in dresses, scarves, skirts, and women's blouses. The delicacy of the lace fabric also comes from the fiber used. In handmade lace, the commonly used yarns are linen, silk or cotton, while in the warp industry the fast Raschel Jacquard machines accept multibar and silk, cotton, synthetic fibers, such as elastane, cellophane, silver and even gold (Pezzolo 2007).

In Figure 8 it is possible to analyze a lace fabric fabricated by the Raschel Jacquard warp loom. With a hexagonal bottom, the design has several shades, with higher or lower coverage

Figure 8. Lace canvas—industrial Raschel (Huaxing, 2013).

factor. This promotes a more defined design, with gradient, smooth curves and designs that are more elaborate. Using nylon thread, the fabric acquires greater elasticity for its most diverse uses. Despite being a Chinese fabric, orders take up to ten days to arrive in Brazil, which makes it a very competitive fabric in the market. The production of fabrics with Raschel machines adapts to the demand of the buyer, varying the segments, the background structure and the drawings. The Raschel machine that produced the fabric on Figure 8 has the capacity to produce 4572 meters per month (155 centimeters wide). This production is low if we compare it to machines with circular and rectilinear meshes; however, it is bigger when compared to handmade lace (Huaxing 2013). This led most of the lace fabrics, which are currently used by the Bahians, to be industrialized and no longer handmade.

Although there are a variety of products that Raschel machines can produce, handmade lace is more valued because of the uniqueness of lace with unique motifs and workmanship.

7 ANALYSIS OF RESULTS

We observed that the Bahians dresses marked, as of the nineteenth century, the differences between white, black, slave, free and emancipated women. However, it is possible to say that there was an imposition so that, through the clothes, the slaves resembled their women masters. For such reason, the use of lace is so striking. It was a fabric rich in details and elaborated Being a fabric abundant in detail and elaborate weaving. Because of that, the slaves who could use it were seen as the guardians of a better social standing. Thus, the use of lace continued to be a symbol of imitation of prestige during the period of slavery in Brazil. Imitation of prestige is the term given by anthropology to refer to the copying of habits, beliefs, customs, appearance and also clothing to approach a higher figure.

Lace transparency gives this type of mesh a certain sensuality by revealing what is behind the fabric; often the body of the one who is dressing. It should be noted that the wearing of expensive clothes, with quantities of lace, jewelry and transparencies by slaves in the eighteenth and nineteenth centuries is due, in part, to the imprisonment of the white ladies in their houses. Because of religion and husbands (and fathers), women rarely went out to the streets, so decorating her maids was a way to channel that luxury and sensuality. However, this sensuality of transparencies and lace designs was not free as it also symbolized the sexual service to which slaves were imposed. The slaves had to fulfill their masters' wishes. This way, the opulence and richness of the clothes played against the blacks, transforming them also into sexual slaves. It is important to emphasize that this eroticism came from looking at the European Catholics. This is because, for many African ethnicities, breasts' nudity, for example, was not erotic. At the moment, the clothes of the Bahians do not have a so strong sexual allusion, since lace has different layers of skirts and lining in the blouses. This is due to Catholic apostolic interference in this garment, incorporating its values and morals.

The use of lace in the dresses of the Bahians expanded in relation to the period of its emergence, also due to economic factors. The lace fabrics were backed by the high industrial production, as well as the higher purchasing power of the Bahians. With the Raschel and Leavers machines, the lace mesh has greatly increased production. With the more affordable and inexpensive Chinese fabrics, lace has become more widely used in dresses. In festive costumes, lace is already in the everyday clothes of the iconic Bahians.

8 CONCLUSION

The lace that marks the dress of the contemporary Bahians refers to its Afro-Brazilian origin, to that of the dresses of the Creole slaves. This fabric is present in the dressing of Bahians due to the imitation of prestige, the eroticization of the masters, the channeling of the sexual and aesthetic freedom of the masters and the Afro-Brazilian religion of candomblé. These are the main historical reasons identified. However, as important as historical factors are the cultural factors that have made the use of laces remain in this process up to date. The tradition of local culture, the religious issues of rituals and cultural preservation, through the aesthetics of the dress, brought the dress of the Bahian girl up to the present.

The industrialization of lace fabric was possible as of the development of the knit fabric processes, with the warp knit and the Raschel machines. With this machinery, the production increased, maintaining the delicate characteristics of the lace. Thus, the fabric began to have a lower cost, making it more accessible to the Bahians in general.

Therefore, lace brings a whole set of cultural historical meanings to Bahian clothing and there is a need to document and preserve it, so that the use of this fabric does not seem random or a fashionable characteristic of a given period.

REFERENCES

Casa.com.br, 2013. O incrível universo das rendas. *Revista Bons Fluídos*. São Paulo: Grupo Abril. Available at: http://casa.abril.com.br/materia/o-incrivel-universo-das-rendas [Accessed 18 November 2017].

Echaniz, I., 2006. *Paixão e Glória – História da Companhia de Jesus em corpo e alma – tomo 2: Verão (1581–1687)*. São Paulo: Loyola.

Frazão, H., 2009. O estilo casual dos soteropolitanos. *UOL Mulher. Moda*. 08 set. Available at: http://mulher.uol.com.br/moda/album/salvador_album.htm#fotoNav=18 [Accessed 15 December 2017].

Huaxing, 2013. Chinês tecido renda raschel. *Alibaba.com*. Available at: http://portuguese.alibaba.com/product-gs-img/chinese-raschel-lace-fabric–475062820.html [Accessed 21 November 2017].

IPHAN, 2009. Ofício das baianas de acarajé. *Dossiê IPHAN 6*. Brasília: IPHAN. Available at: http://portal.iphan.gov.br/portal/baixaFcdAnexo.do?id=919 [Accessed 06 November 2013].

Martini, G.T., 2007. *Baianas do acarajé: a uniformização do típico em uma tradução culinária afrobrasileira*. Brasília: Tese (Doutorado em Antropologia Social) – Departamento de Antropologia, Universidade de Brasília. Available at: http://repositorio.unb.br/bitstream/10482/1302/1/Tese_2007_GerlaineMartini.pdf [Accessed 06 November 2017].

Monteiro, J.; Ferreira, L.G., 2005. As roupas de crioula no século XIX e o traje de beca na contemporaneidade: símbolos de identidade e memória. *Mneme*, Natal: UFRN. Available at: http://www.periodicos.ufrn.br/ojs/index.php/mneme/article/viewFile/329/302 [Accessed 05 November 2017].

Pezzolo, D.B., 2007. *Tecidos: histórias, tramas, tipos e usos*. São Paulo: Editora SENAC.

Rebello, C., 2009. O traje típico da baiana. *África: saberes e práticas*. 27 out. Available at: http://africasaberesepraticas.blogspot.com.br/2009/10/o-traje-tipico-da-baiana.html [Accessed 15 November 2017].

Resende, L., 2017. *Cultura Popular Brasileira e Folclore*. Available at: http://www.lendorelendogabi.com/folclore/cultura_popular_e_folclore_pag1c.htm [Accessed 22 November 2017].

Ribeiro, M.G., 2010. A arte no século XIX: Um estudo da peculiar obra artística no Brasil do pintor francês J.B. 19 & 20, Rio de Janeiro: IFCS. Available at: http://www.dezenovevinte.net/obras/jbd_monike.htm#_edn12 [Accessed 15 November 2017].

Romero, L.L., 1994. *Malharias*. Rio de Janeiro: BNDES. Available at: http://www.bndes.gov.br/SiteBNDES/export/sites/default/bndes_pt/Galerias/Arquivos/conhecimento/relato/malha.pdf [Accessed 29 November 2017].

Sanches, R., 2011. *Malharia Urdume: malharia I, Bacharelado em Têxtil e Moda*. São Paulo: Universidade de São Paulo, 08–25 nov. 87 slides. Notas de Aula. Apresentação Digital.

Silva, S.R. de A., 2014. O figurino da colônia: uma análise da sociedade colonial brasileira a partir da indumentária. *Revista História*. Rio de Janeiro: UER J. Available at: < http://www.revistahistoria.com.br/index.php/Revista_Historia/issue/viewIssue/8/7> [Accessed 22 April 2018].

Verger, V., 2002. *Fluxo e refluxo do tráfico de escravos entre o Golfo do Benin e a Bahia de Todos os Santos: dos séculos XVII a XIX*. Salvador: Corrupio.

Fashion design and craftwork: Subjectivation policies in the contemporary scenario

A.R.V. Peroba & C. Mesquita
Universidade Anhembi Morumbi, São Paulo, Brazil

ABSTRACT: This paper presents a frame of the doctoral research on PPG Design UAM-SP Fashion design and craftwork: subjectivation processes and resistance in the context of the IWC, and deals with the contemporary scenario in which sectors of the economy and culture come closer to present some relationships between Design and Craftwork in the Integrated World Capitalism (IWC) context. Key concepts from the theories of modeling of contemporary subjectivity and product homogenization (Guattari & Rolnik) are used as tools. From this critical apparatus, we have tried to approach the the fashion designer performance with artisan groups in the creative-productive processes, as well as to raise some of the socioeconomic forces and cultural resistance possibilities in the field of craftwork.

1 INTRODUCTION

The research that generated this article explores the so-called process of globalization or Integrated World Capitalism (IWC), terms coined by Guattari (1981) to approach capitalism in contemporary times. This approach is here to understand transformations in the dynamics of culture and in the formation of the concepts of creative industry and culture economy in Brazil.

In this context, the research considers in particular the analysis of the modes of subjectivation that permeate the policies of promotion of the artisan activity. This approach includes how globalization involves people and places, either by directing the actions by the dominant social actors to hegemonic interests or by provoking the emergence of possible escape lines as of small craft communities.

The hypothesis considered is that fashion design can act as a transforming agent of the social context through the work of training groups of artisans. These new craftspeople, in turn, would be able to act by creating small cracks in the homogenizing power block, engendering artisanal techniques to design processes in order to imply desire and subjectivity, thus becoming vectors of micropolitics of desire, economy, culture, etc. Such small open spaces, in turn, would carry the possibility of breaking with the block, configuring themselves as nuclei of resistance. The micropolitics of desire and their subjectification fields, fundamentally political and micropolitical, are instruments of our cartography (Guattari, 2007).

The aim of the research is to articulate a critical perspective of understanding the relationships between fashion design and craftwork—Whenever the terms in the text refers to the field, capital letter is used; when it refers to the action, it will be in small letter—, especially regarding the configuration of their creative-productive processes within the scope of the Creative Economy (CE) policies in Brazil. As far as artisan production is concerned, it can be seen that, besides the socioeconomic feature, the relationships between design and craftwork can contribute to future political actions through the construction of critical thinking.

According to Guattari & Rolnik (2013), the IWC can be understood as an update of the capitalist model, since it focuses on the structures of production and collective labor forces, changing social relations by directing consumer desires and the formatting of behavior patterns.

The methodology of the study was divided into five stages: exploratory, documentary, case study, data analysis and construction of a critical perspective based on the cross-referencing of the information collected, resulting in a cartography of the studied population and the micro political and economic movements that permeate the productive and creative processes in artisanal practice. In this paper, we will present a general outline of the paths covered by the ongoing research.

2 THE CONTEMPORARY CONTEXT: PROCESSES OF SUBJECTIVATION, CULTURE AND CRAFTWORK

Brazil seeks to chart its own path to cultural issues in the 21st century, when Design and Craftwork have struggled to guarantee proper spaces and rights. The activities related to Design—project-design, creation, craft objects, utilities or fashion products—gain new attention, appearing in the discussions established in public and private spheres as essential for the positioning of the country in the international market. This new scenario increases the responsibility of its actors and the need for considerations regarding the interrelationships with the design activity.

According to Guattari & Rolnik (2013, p. 21), the concept of contemporary culture was established as a way of imposing values through the "subjectivity's power taken", that is, culture has become the "collective force of social control" (Idem, p. 26) and its manifestations constitute, in the model in operation, the form by which subjectivation is controlled. However, making craftsmanship would, in a way, escape this kind of subjectivity production, since it would (until then) be a production sector outside the control of contemporary capitalism (Guattari, 1981).

Hence, artisan know-how can be taken as an instrument of both power and resistance, which is confused in contemporary agencies (Deleuze & Guattari, 1999). While the appreciation of the artisanal aspect of fashion products is a global macro trend, artisan groups struggle to stay in the market of cultural industry, the so-called Creative Economy. This cartography helps to investigate the set of forces articulated in this encounter between Craftwork and Design, in an attempt to deal with the complexity of the processes of subjectivation that take place there.

Subjectivity, according to Guattari & Rolnik (2013), would be a continuous production that occurs from the encounters between beings (people, society, nature, objects, events), in contrast with a centrality produced within the individual, i.e., subjectivity "results from a cross-referencing of collective determinations" (Idem, p. 43); for the authors, subjectivity is always a composed phenomenon.

> *The production of subjectivity by the IWC is serialized, normalized, centered around an image, a subjective consensus referred to and overcoded by a transcendental law. This scrutiny of subjectivity is what allows it to propagate itself at the level of production and consumption of social relations in all media (intellectual, agrarian, industrial, etc.) and in all parts of the planet* (Guattari & Rolnik, 2013, p. 48).

Therefore, subjectivity is understood as a process in which the subjects (or individual subjectivities) are produced as a result of the constant interaction between the social context and people. Thus, the subject is made as he comes in contact with others, with the world, with political, ethical, moral, and economic positions, with nature and the differences and conflicts that occur in the most diverse dimensions. From all those encounters, he alters his actions and creates, settles, becomes singular, becomes man, even if temporarily (Rolnik, 2003; Mansano, 2009).

Capitalist profit, in Guattari & Rolnik's (2013) conception, would have been, since the second half of the 20th century, concentrated on the production of subjective power. Power is present in the most diverse social relations, presenting itself as a clash between those who take and exercise it and those who accept it. The exercise of power removes from the other his power of realization, because by mastering the fruits of actions (work, creation, etc.), they belong to the dominator. The existence of actions that escape the dominant production model can be configured as resistance when it carries creative potencialities in itself (Deleuze, 1988). In this sense, resistance movements can occur and put the dominant powers in question, that is, presenting possibilities to modify the relations of production and, by extension, to imply changes in social relations.

The artisanal productive mode is recognized for representing cultural richness and carrying this potential of invention; like Design, it is part of the so-called creative sectors (MinC, 2012). In some cases, the Design/Craftwork encounter can produce what is being understood as resistance or creation. In as many others, it only reproduces the logic of the dominant power.

Contemporaneity is marked by attributes related to fragmentation, speed and immediacy, and especially to the context of consumption in which the individualistic desire is one of the instruments of subjectivation policies. In this field, the production of social relations in general serves an economy of desire (Guattari & Rolnik, 2013). By desire the authors understand "all forms of will to live, will to create, will to love, will to invent another society, another perception of the world, other systems of values" (Guattari & Rolnik, 2013, p. 261).

The strategies of this economy in the social field can illustrate what Guattari calls micropolitics. However, micro and macro politics refer to individual issues as well asto social or large-scale problems, since they coexist in the most varied systems. The difference is that the macropolitics refers to the molar segments, whereas the micro political one to the molecular flows (Deleuze & Guattari, 1999). Micro political forces engender a process involving desire and subjectivity, according to Rolnik (2011).

It is possible to understand that two of the main vectors of the subjectivation policies—culture and economy—have approached in the contemporaneity, generating a field called the economy of culture, which in turn will reconfigure the variables of this context. In contemporary subjectivation policies, by aligning the economic agenda with culture—perceived as an economic asset—the latter has become a priority issue for the State, shifting to the center of the world stage the creative and productive processes of the scope of design and craftsmanship.

Hence, CE has its concept related to creative goods and services and would be based on a tripod of creativity, the symbolic and economy. Creativity, according to this organization, would be the most expressive factor for the production of goods and services (Howkins, 2001). Thus, culture and its agents and producers are driven to assume other roles in the current stage of capitalism, which, in addition to operating in a logic of concentration, seeks to attract and condense the communities and groups that used to produce and survive apart from the system. Therefore, in its various dimensions, the performance of culture in social and personal organization produces subjectivities aimed at reproducing hegemonic modes.

The emergence of the cultural industry and the mass media—and of the mass culture itself—is fostered by the phenomenon of industrialization in the 18th century. For Teixeira Coelho (1993, p. 10), "It is this [industrialization], through the changes that it produces in the mode of production and in the form of human labor, that determines a particular type of industry (the cultural one) and of culture (the mass one), implanting in one and the other the same principles in force in the economic production in general".

The cultural industry, even today, is understood as the one that produces and distributes the most diverse cultural products. This industry would usually work by extracting some elements—such as spontaneity, rusticity, plastic aspects or which would trace back to traditions—from the so-called "popular" products, by adding to them a quality standard aimed at reproducibility, within a planning for adapting to the "consumption of the masses". In such engineering, it would direct and create its own consumer market, representing the main instrument of social control.

In this context, modeling systems are understood as a set of norms (instructions, codes) that generate cultural models as from their structure. Therefore, as Guattari & Rolnik (2013, p. 36) state, "the production of the IWC is not only that of representation, but that of a modeling which refers to behaviors, sensitivity, perception, memory, social relations", with standards of values and procedures imposed by contemporary society.

The phenomenon of mundialization is understood as a multidimensional, complex and comprehensive process, with several aspects: economic (capitalist system), political (neoliberal ideology) and cultural (modeling). Therefore, the integration process is based on the accumulation dynamics of capitalism, which, through collective equipment and the media, produces a new type of subjectivity, the capitalistic one (Guattari & Rolnik, 2013). Yet, as Guattari (1992, p. 171) warns, "This division of the mundialization productive forces and capitalistic powers is not absolutely synonymous with a homogenization of the market, quite the opposite." To continue to exist and attract the public, the market needs differentiations, innovations, singularities, special values, creation. From this point of view, the capital would function in a complementary way to culture as a concept of equivalence: the capital dealing with economic subjection, and the culture with subjective subjection of society (Guatarri & Rolnik, 2013).

3 FASHION DESIGN AND CRAFTWORK: MATCHES AND MISMATCHES

In this paper, it is considered that it is possible to expand the understanding of design and craft products from an approach on the process used in the production of such objects, which constitute—beyond their utilities and functions—expressions of both the individuals and the communities that produce, carry and use them.

However, there is no interest in categorizing or delimiting the manifestations of one or the other, since none of their production processes, circulation channels, market-nominated value, consumers, attributed uses and meanings, among other elements, serve as a parameter to delimit their conceptions. In the perspective of this study, what arouses interest is what such productions have in common.

For Guattari (1996, p. 181), the artisanal mode of production operates as a "collective system of memorization of data and knowledge" and its artifacts are understood as "technical (low technology) and aesthetical material devices." On the other hand, the fashion design products, besides incorporating both low and high technologies, would also incorporate science, aesthetics and marketing functions. In other words, while the fashion product is composed by a stream of stages and operational sequences that start from socioeconomic-cultural analyzes of the contemporary world and depend on the target public and available technologies, investment, weather conditions and market, among other variables, craftsmanship results from the constant exercise of generating forms in the dialogue between ideas and practice. One of the main characteristics of the artisanal productive method is the action of the artisan in all phases of the process, from the collection or creation of raw material to the contact with the final customer in the commercialization. Therefore, more time to elaborate, experiment, mature and modify the product is what characterizes craftsmanship, besides the skill, the ability of the artisan.

Craftwork and design are living systems of human production and creation. The increasing rapprochement between Design and Craftwork, between practice and theory, between the craftsman and the user is evident in the contemporary society. This movement occurs in the sense of reconstituting the connection that was broken with the advent of fragmented industrial processes and the scientific separation between body and mind. Thus, understanding the concepts that involve Craftwork and Design, as well as the mechanisms of acting in one or another field, is important not to show differences or make separations, but to recognize similarities, exchanges and confluence points which are so rich in information.

For Guattari & Rolnik (2013, p. 55), what characterizes the "experience of a subject group" in its processes of singularization is its ability to grasp the elements of the situation and generate for itself its own practical and theoretical references, without staying in a position of dependence on global power. In this way, artisanal making is understood as a process of singularization, while the design line that shares knowledge and enables the exchange of information, methods and techniques acts as a trigger in the process of aggregation of groups with the purpose of expressing their creative potential.

The designer, in attempting to activate the "will to care for the quality of individual and collective existence" in his "creative power", may stimulate the "creation of new modes of subjectivation, new modes of existence, new types of society" (Rolnik, 1995, p. 13). Hence, the design process can act as a facilitator to creative and collaborative work, as the production of subjectivity presents a dimension of procedural creativity that can extend possibilities of existence when put into practice.

4 FINAL CONSIDERATIONS

As this is an ongoing study, the considerations presented here aim to bring up a scenario contextualisation of the case study in question. The research was based on a survey in Brazil, especially in the northeast region, of several projects that work with artisan communities and fashion designers, to then focus on the work of the Brazilian designer Ronaldo Fraga with the communities of Nossa Senhora Penha and Jacarapé, both in João Pessoa, Paraíba (Fraga, 2012).

Craftwork is a skill connected to the process of man's creation, and it is possible to consider that when the capitalist productive system needs the injection of creative triggers, it turns to the arts and crafts. In general, craft objects emerge in the world market as niches of novelty with characteristics desired by increasingly large portions of consumers: naturalness, sustainability, originality, distinction, solidarity and fair valuation are some of the characteristics that add value to them.

In 2012, the Municipality of João Pessoa (PMJP), through João Pessoa Artisan (JPA), along with the Brazilian Service of Support to Micro and Small Businesses (Sebrae) and the Federal Institute of Education, Science and Technology of Paraíba (IFPB) devised a social inclusion project to serve the inhabitants of the two communities around the João Pessoa Convention Center and contribute to improving the economic and social situation of that locality.

The first action was carried out in 2014 aiming at: "Potentializing craftwork actions by investing in qualification and opening spaces for the commercialization of craftworks produced in the city of João Pessoa – PB" (Sereias, s.d.). The target women of the project were found to have no knowledge about techniques, nor did they perform artisanal productive activities. Training was carried out in two more stages, with Ronaldo Fraga's presence in the last one, in order to promote the identity and improvement of the products made. After three months of training, the bio-jewels developed by a group of 17 women were presented in April 2015 in the designer's fashion show held at São Paulo Fashion Week (SPFW – Spring/summer 2015/16).

Work in the community of Penha was perceived as a space where processes that help us problematize and discuss micropolitics of the dynamics of the artisan group, which is the focus of this research, occur. It is possible to perceive that the ambience created and worked between the group and the designer has allowed processes of singularization and ruptures of the original models to emerge, which can be understood as points of resistance to the capitalist model.

The partial result of the investigation leads us to understand that some of the projects that combine design and crafts can act as catalysts for the agglutination between manual and project-design work, setting up as empowering agents of artisan communities previously modeled in the scope of the Creative Economy and in the context of the Integrated World Capitalism. Fashion Design, by enabling a rearrangement with Craftwork through actions promoted in certain communities, can become a scenario for singular processes which are not only new products, but also affirmations of possibilities and encounters that produce different modes of subjectivity production that subvert domains and unfold in social transformations. The fashion designer can act precisely by allowing the processes of singularization to occur, differing from the reproduction of models and working to ensure micropolitics compiled from other references and praxis.

Guattari & Rolnik (2013, p. 66) point out that "revolutionary micro-processes" can arise from the relationship of an individual with an agent that awakens "an entirely new process of perception and sensitivity". In the case of the community of Penha, observation and data collection have allowed us to point out that women were affected and stimulated by the qualification process of artisanal making, expanding the development of their perceptions and connections with themselves and with their surroundings. Learning the techniques, developing skills and engaging with the production process for a fashion collection have given rise to new visions and connected them, allowing for the recognition of similarities and acceptance of differences. Thus, "the three levels of the molecular revolution: the infra-personal level, the way in which the social relation is lived and the presence of the relations of political forces" were put into operation (Ibidem). In addition, the perception of the surroundings—family, craft partners, locality, community, fashion system, international market and media—was expanded, operating the construction of new reality outlines. It is worth mentioning that neighbors who did not know each other started to interact every day and to have a common goal. Throughout the process, they left the community together and traveled out of the state for the first time in their lives to participate in an event that launched the fruit of the group's work on the largest fashion launch platform in Latin America.

As from that work, the Sereias da Penha Artisan Association was created, a group that effectively started to work together. One of the interviewees has stated that this is a way out of "the routine, household, husband, children" in order to become productive, with the possibility of managing small expenses without needing to ask for husband permission, that is, they gained a certain financial autonomy.

Moreover, the process of working with the fashion designer resulted not only as a trigger for the women's creative force and productive potential, but also as an extension of the connections between craftwork and design. According to statements made by the members of the group, before the project fashion was what was seen on TV and soap operas. They had no knowledge about the fashion designer or the work he performed. As from the course and the interaction with Ronaldo Fraga, they realized and came to believe in the value of local materials such as fish scales and skin, beginning a process of establishing new cultural values and a new way for the community of Nossa Senhora da Penha to assert itself in its peculiarities, with international reverberations. As an outcome of the project with the Sereias da Penha, at the end of 2017 the city of João Pessoa joined the Unesco Creative Cities Network (UCCN), with special emphasis on craftwork (João Pessoa, 2017). The Sereias were produced and chosen as an example of a successful production group within the scope of the CE. They have begun to represent the growth in the economic potential of craftwork. Still, their actions establish powerful creative lines that can form networks of resistance.

In summary, the process involved the formation of a group that became a reference, the opening of a space for interaction, the scheduling of periods of shared creation and production, the configuration of a context of collective aspirations. It is possible to consider that as from the work done along with the designer, the new artisans appropriated the process itself and the outcomes illustrate the micropolitics of resistance, which, even though originally configured to meet a particular power relation, expressed differences and produced singularities. Obtaining a critical view in a group that shared the experience and investment in their own subjectivity functions as a strategy of resistance in the modus operandi of the IWC.

REFERENCES

Brasil, (2012). *Plano da Secretaria da Economia Criativa: políticas, diretrizes e ações, 2011–2014.* [pdf] Brasília: Ministério da Cultura. 156 p. Avaiable at: <http://www.cultura.gov.br/documents/10913/636523/PLANO+DA+SECRETARIA+DA+ECONOMIA+CRIATIVA/81dd57b6-e43b-43ec-93cf-2a29be1dd071> [Accessed 18/10/2016].

Deleuze, G., (1988). *Deleuze R de resistência.* In: O Abecedário de Gilles Deleuze. [pdf] Transcrição de entrevista concedida à Claire Parnet. Diretor: Pierre-André Boutang. Elenco: Gilles Deleuze e Claire

Parnet. Gênero: Documentário. VHS, 459 min. Paris: Éditions Montparnasse. Avaiable at: <http://stoa.usp.br/prodsubjeduc/files/262/1015/Abecedario+G.+Deleuze.pdf> [Accessed 05/08/2016].

Deleuze, G. & Guattari, F., (1999). *Mil Platôs:* Volume 3. São Paulo: Editora 34.

Fraga, R. (2012). *Blog.* Avaiable at: <http://ronaldofraga.com/blog/> [Accessed 14/02/2016].

Guattari, F., (1981). *O capitalismo mundial integrado e a revolução molecular.* [pdf] Translated from French by S. Rolnik, São Paulo: Brasiliense. Avaiable at: <https://praticasautogestionarias.files.wordpress.com/2013/06/o-capitalismo-mundial-integrado-e-a-revoluc3a7c3a3o-molecular.pdf> [Accessed 12/10/2016]

Guattari, F., (1992). *Caosmose: um novo paradigma estético.* São Paulo: Ed. 34. Coleção TRANS, Translated from English by A. L. Oliveira, e L. C. Leão.

Guattari, F., (1996). *Da produção de Subjetividade.* In: Imagem Maquínica: A Era das Tecnologias do Virtual. Parente, A. (org.) Rio de Janeiro: Editora 34.

Guattari, F., (2007). *Microfísica do poder e micropolíticas dos desejos.* [online] In: Queiroz, A.; Cruz, N. V. (org.) Foucault Hoje?. Rio de Janeiro: 7 Letras. Avaiable at: <https://territoriosdefilosofia.wordpress.com/2014/12/16/1985-microfisica-dos-poderes-e-micropolitica-dos-desejos-felix-guattari/> [Accessed 24/08/2015].

Guattari, F. & Rolnik, S., (2013). *Micropolítica: cartografias do desejo.* 12ª ed. Petrópolis: Vozes.

João Pessoa (Gov.), (2017). *Projeto Sereias da Penha impulsiona Capital a integrar Rede Mundial de Cidades Criativas da Unesco.* Avaiable at: <http://www.joaopessoa.pb.gov.br/projeto-sereias-da-penha-impulsiona-capital-integrar-rede-mundial-de-cidades-criativas-da-unesco/> [Accessed 20/12/2017].

Mansano, S.R.V., (2009). *Sujeito, subjetividade e modos de subjetivação na contemporaneidade.* São Paulo: UNESP. In: VIII Encontro Científico da Pós-Graduação em Psicologia: Modos de subjetivação no contemporâneo, realizado no dia 25 de novembro de 2009, na UNESP – Assis. Revista de Psicologia da UNESP, 8(2): 110–117.

MinC, Ministério da Cultura, (2012). *Plano da secretaria da economia criativa: políticas, diretrizes e ações 2011 a 2014.* [pdf] 2ª Ed. Brasília: Ministério da Cultura. Avaiable at: <http://www.cultura.gov.br/politicas5/-/asset_publisher/WORBGxCla6bB/content/plano-da-secretaria-da-economia-criativa-politicas-diretrizes-e-acoes-2011-a-2014-636560/10913> [Accessed 22/01/2015].

Oliveira, J.M. de; Araujo, B.C. de; e Silva, L.V., (2013). *Panorama da Economia Criativa no Brasil.* [online] In: TD 1880, Texto para discussão, Brasília: Rio de Janeiro: Ipea (Instituto de Pesquisa Econômica Aplicada), outubro de 2013. Website <http://www.ipea.gov.br/portal/index.php?option=com_content&view=article&id=20292&catid=337> [Accessed 22/01/2016].

Rolnik, S., (1995). *À sombra da cidadania: alteridade, homem da ética e reinvenção da democracia.* Em M.C.R. Magalhães (Org.), Na sombra da cidade São Paulo: Escuta: 141–170.

Rolnik, S., (2003). *Cadernos de Subjetividade: O reencantamento do concreto.* São Paulo: Editora Hucitec; EDUC.

Rolnik, S., (2011). The Geopolitics of Pimping. [pdf] Holmes, B. (trad.) In: *Critique of Creativity: Precarity, Subjectivity and Resistance in the 'Creative Industries'.* Gerald Raunig, Gene Ray & Ulf Wuggenig (eds). London: MayFlyBooks: 23–39. Website <http://mayflybooks.org/?p=212> [Accessed 24/10/2017].

Sereias, s.d. *O projeto.* Website <https://sereiasdapenha.wordpress.com/o-projeto/> [Accessed 14/02/2016].

Teixeira C., J., (1993). *O que é Indústria Cultural.* Coleção Primeiros Passos, n 35. São Paulo: Brasiliense.

Chinese fashion, occidental fashion: A semiotic translation

Yingqiao Pan
University Complutense of Madrid, Madrid, Spain

ABSTRACT: Fashion is always semiotic, as said by Yuri Lotman (1999, p. 114). The insertion of fashion is also a continuous process of transformation of the no-signifier in significant. This paper tries to analyze the mutual influence between the occidental and the oriental culture in fashion from a semiotic perspective. In the 17th Century, the Chinese Culture inspired the *Chinoiserie* in Europe. In the twenties of the 20th Century, due to the occidental influence, the *Qipao* was born in China. Later, the Chinese signs began to inspire Western fashion shows. Within the Chinese signs, this paper has chosen the white and blue Porcelain as an inspiration for Western designers. The influence between East and West is unstoppable.

1 INTRODUCTION

According to Yve Zimmermann (1998, p. 111), the "design" and the "designate", both share the same root: sign (from Latin *signa-signum* = sign, mark or insignia). The design is the intention to bring the object to its sign. According to Saussure (1916), a sign is the connection established between a signifier and a signified. Semiotics is the science that studies the signs from this connection. The design of clothing can be communicative by a non-verbal form. But within this communication, a code system that allows us to translate should be established. Umberto Eco (1972) indicated us that if in its origins fashion was something functional, fashion soon lost this functionality to become a communicative value. The dress is expressive, and there are dress codes that speak with more clarity than words.

Fashion is always semiotic (Lotman, 1999). If we consider that a given culture is a limited semiosphere, fashion innovation comes from the periphery or the border of this semiosphere (Lotman, 1996). The process of the insertion of fashion in a culture begins when a certain element that is not a signifier when it is outside, passes the border, and becomes significant. This process talks about the exchange between the inside and the outside of a cultural system. From this perspective, the present article tries to analyze phenomenon of fashion exchange between China and the West from a semiotic perspective.

1.1 *The fantasy of these occidentals to the orient*

In May 2015, the Metropolitan Museum of New York organized the exhibition "Through the look glass" that lasted three months. It wanted to show the influence of the Asian country in the design of fashion from antiquity to the present day. The Chinese name of this exhibition is "Jing (mirror) Hua (flowers) Shui (water) Yue" (moon), which means "flowers in the mirror and moon in the water". This translation, along with its English title, shows the imagination and the fantasy of the western designers when they look up to the Chinese culture. The exhibition was divided in two parts. The first part dealt with the historical stages: Manchú[1] in the Qing Dynasty (1644–1912), Qipao in the Republic of China (1912–1949), and the Mao Costume in the times of Mao Zedong. The second part was called "the empire of signs" and included different elements as silk, calligraphy, blue-white porcelain, perfumes, etc.

---

1. An ethnic minority of China that lives in the northeast.

When we see an exhibition like this, the word *chinoiserie* comes to our minds. The *chinoiserie* entered Europe approximately in the last quarter of the seventeenth century and its boom occurred in the mid-eighteenth century, when it was assimilated by the Rococo. It was mainly characterized by the use of Chinese designs, asymmetry, capricious changes in size, the use of lacquered materials and abundant decoration. From the Renaissance to the 18th century, Western designers tried to imitate the technical sophistication of Chinese ceramics, achieving partial success. The imitation of Chinese designs began in the late seventeenth century, and was incorporated in the production of porcelain, usually in tea sets, and became very popular during the Rococo *chinoiserie* (around 1740–1770). We can see that Chinese culture certainly left an important influence on Rococo. Through this cultural exchange, Chinese signs entered the Western semiosphere, becoming significant in terms of of its exoticism. For instance, the upper classes used it to differentiate themselves from the lower classes (Simmel, 2014). But the exchange between cultures is never one-sided. Two centuries later, Western culture left a great impact on Chinese fashion. The birth of the Qipao dress can be understood as a mixture between the East and the West. In the hall that the MET dedicated to Qipao, there were some designed by Dior and Tom Ford that showed, in a certain way, Western's adoration of Qipao.

1.2 *The occidental influence on the Qipao*

The Qipao is a Chinese dress that was born in the 20 s of the last century. It is a tube dress, with a closed neck and side openings, that was originated during the Qing Dynasty (1644–1911), dominated by Manchu, a group of farmers and ranchers from the North who wore loose and dark uniforms in one piece, with the same pattern for men and women, and that did not allow to show any part of the body.

According to Simmel (2014), "When a time is more nervous, the more quickly the fashion changes, since one of their essential supports, the thirst for exciting new things, goes hand in hand with the depression of the nervous energies." The fall of imperial rule began with the Western invasion. After that episode, China became a semi-feudal and semi-colonial state. The Qing dynasty lost its domination after the bourgeois revolution of Xinhai, which led to the establishment of the Republic of China, which had democratic ideals. For a time, the traditional dress and the western suit coexisted.

From 1911, China left feudalism. After a series of revolutions, China underwent great changes in the scientific and political field. Strictly speaking, Chinese fashion coincides with

Figure 1. Dress of Dior inspired by Qipao.

Figure 2. Dress of Tom Ford inspired by Qipao.

Figure 3. Qipao.

the beginning of modern Chinese society at the beginning of the 20th century, so it is still a very short journey. The appearance of Qipao is a phenomenon whose cultural meaning is deeper than a simple dress. It is a mixture of tradition and modern style, and its narration links the colonial, the feminine, the Western and the Oriental. Qipao represents the ambiguity of the essence of modern Chinese culture itself, and that is why its "limits" are fuzzy.

According to Zhangailing, a Chinese writer: "After a series of revolutions in science and democracy, women were very interested in Western culture, and though they fought for their rights, those who continued to dominate society were men. Since then, the women showed their anger and wanted to do something to show their desires. At that time, the clothes that the men wore were called Cheongsam, so dressing Cheongsam was one of the attitudes that best showed their political position".

In the beginning, these clothes were worn only by the students when they attended protest marches in favor of a democratic change in China. Later, women of high class and celebrities began to imitate the students. In this sense, Cheongsam began to be a sign of education and modernity. This was an antecedent of the appearing of Qipao.

Simmel points out that social change has a great impact on fashion. For instance, as the absolute monarchy is at its end, and the level of democratization increases, the scope affected

Figure 4. Coexisted the traditional dress and the western suit.

by the fashion gradually widens. The rhythm of changes in fashion accelerates faster and faster, while the presence of each specific form of fashion is shorter. Undoubtedly, social progress favors the development of fashion, since the imitation of the lower classes and the innovation of the upper classes are represented more quickly.

Said this, we can conclude that the Qipao is not just a dress, but a mixture between the inside and the outside, tradition and modernity. China's modern fashion, and especially the popularity of the Qipao dress, was under the influence of the new ideological values such as independence, national identity, the illustration of women's rights, Western aesthetics, etc. In a stricter sense, we can consider that Qipao culture is like a system of signs, and that the space where this dress exists and evolves is its own semiosphere. The term Semiosphere, based upon Vernadski's biosphere, was defined by Lotman as a "semiotic space, outside of which the existence of semiosis is impossible" (Lotman, 1996).

1.3 *The oriental influence in porcelain dress*

In the second part of "China Through the look glass" exhibition, showed a vase of blue and white porcelain, and next to it, there were two printed dresses with white and blue drawings, as shown in Figure 5. The one on the left was designed by Roberto Cavali, the one on the right was designed by Alexander Mcqueen.

The blue-white porcelain had appeared long before the Tang Dynasty (618–907), but it reached its peak during the Yuan Dynasty (1271–1368). At that time, a great amount of blue-white porcelain was being produced, some of them real masterpieces. Jingdezhen (**景德镇**), the small town that invented the classy white-blue porcelain during the Yuan Dynasty, became the center of production. With the arrival of the Ming Dynasty (1368–1644), the establishment of the imperial ceramic oven in Jingdezhen promoted the maturity of the art, and a large number of valuable pieces were produced during this period. Currently, blue-white porcelain is the most representative among Chinese porcelains.

From the middle of the 20th century, blue-white porcelain was a source of inspiration for Western designers. In 1952, Christian Dior became the pioneer who designed a blue-white porcelain dress. Later, Chanel designed the *Porcelain Doll* in 1984. Following them, Valentino (1986, 2013), Roberto Cavalli (2005, 2013), Dior (2009), Tory Burch (2013), Carolina Herrera (2013), etc., designed many high fashion series of blue white porcelain printed dresses on their fashion shows.

Why had they chosen the blue-white porcelain? Or rather, why the blue-white porcelain was highlighted compared to other Chinese signs in the fashion world?

Figure 5. Cheongsam.

Figure 6. The porcelain dress of Roberto Cavalli (2005) and that of Alexander Mcqueen (2011).

Figure 7. From left to right: Roberto Cavalli (2005), Rodarte (2011), Chanel (1984), Guopei (2010).

Roland Barthes, in his work *System of fashion* (1978), invented the matrix of signifier OSV, which helps us to analyze the dress codes, in which, O is object, S is support, V means variant.

Let's have a look at the following table. The semiosphere 1 (S1) implies the field of ceramic art, whose object is a vase, its variant the white and blue drawings, and the support can be any part of the vase. The set of codes gives us the image of a Chinese porcelain.

227

On the other hand, the fashion design represents the semiosphere 2 (S2). The object in S2 would be a dress, the support can be any part of the dress (the neck, the sleeves, the waist), and the variant is what varies and gives a meaning to the support. According to Lotman, there is a border that separates two cultures, a non-closed border, "a set of points that belong at the same time to the interior and exterior space", a bilingual mechanism that acts as a filter for the translation between the external and internal texts.

According to this table, we see that the object and the support have to be a physical material. Within the semiosphere of fashion, the vase is not taken into account. Then, the only thing that is allowed to pass the border is the variant.

The designers moved the variant of a signifer to another signifier from a different semiosphere, and here fashion emerges. In our case study, the white and blue drawing as a variant of the Chinese porcelain passes the border of the semiosphere of fashion, and after the trans-

Figure 8. Dior in 1952.

Figure 9. Dior in 2009.

Table 1. Analysis of *OSV*.

Semiosphere 1	Semiosphere 2
Signifer 1	Signifer 2
Object: vase	Object: dress
Support: body	Support: waist
Variant: white and blue drawings	Variant: white and blue drawings

fer, it mixes with the object (dress) creating a new meaning, a new sense. As Lotman (1999) would say, fashion always comes from the outside, and the insertion of fashion is a continuous process of transformation from the non-signifier to the significant.

Lars Svendsen (2010) said that designers try to approach art, and that the reason behind it is that art is a sign that has a lot of symbolic value. In other words, if someone wants to add symbolic value to his or her designs, one effective way is to connect it with something that has artistic value. And why has art so much value? In the 30 s, Nikolai Trubetzkoy coined the term "markedness". It referred to the non-dominant element of two opposed elements. Within a cultural society, there is always a dominant tendency surrounded by sub cultural tendency, precultural tendency, etc, which are marked or non-dominant tendencies. Art is marked and should not be reachable by a dominant society, because it would lose its value when it is available for a majority. Artists always go further, because they want to be marked and accepted. Among all Chinese artistic signs, the adoption of porcelain by Western Culture has been the most successful due to the combination of white and blue. In the sixteenth century, the *Delft Blue* manufactured in Holland was influenced by Chinese porcelain. The combination of white and blue not only satisfies the taste of the Chinese but reaches the standard of world aesthetics.

If we extend the view to the globalized world, Chinese culture is looked as exotic in the Western world, and like art, exoticism is also marked. Therefore, Chinese art plays a double marked for Westerners but with a relatively small trend of markedness. So, it reaches a symbolic value of great depth. That can be one reason for designers to appreciate Chinese art.

2 CONCLUSION

Fashion innovation always comes from outside or from the border that separates one culture from another. Thanks to social changes, the texts of the periphery can vary a lot. As a proof of this, the Qipao, which initially was a cloth that resisted changes when it was under the regime of a traditional culture (Qing Dynasty), was subjected to great modifications after the Western influence, generating therefore new dynamics around this type of dress. Thanks to the entry of Western texts in China, the Qipao culture was enriched, andin this regard, its limits have expanded, generating new meanings and dress codes. The culture of the Qipao is, above all, a cultural mixture between the western and the oriental fashion and develops in a process between explosion and gradualism (Lotman, 1999), as Lotman would say. The Qipao is not a simple dress, it has been a great witness to the Chinese social changes for almost a century.

Today, under the great impact of the western fashion's development and globalization, the West, on one hand, has left a certain influence on Chinese fashion, but on the other hand, it has not stopped being inspired by Chinese art. Western fashion can be enriched with texts from outside. According to Roland Barthes' matrix, the variant of a signifier in a dress is what gives a new meaning to a dress. In our case of study, the emergence of the porcelain dress design belongs to the transfer of the porcelain variant to the dress. Porcelain is one of the best known Chinese artistic signs abroad, and not only has an artistic value, but also a symbolic one. That's why designers have never stopped looking for new combinations between porcelain and fashion. From the *chinoiserie* until today, the time of globalization, the problem of fashion has been a social phenomenon where the power of appearances is manifested in an increasingly powerful way, playing the role of a metronome of cultural development, as Lotman would say.

REFERENCES

Barthes, Roland 1978. *Sistema de la moda*, Editorial Gustavo Gilli, Barcelona.
Eco, Umberto, *et al.* 1972. *Psicología del vestir*, Lumen, Barcelona.
Liu Yu (2011), 中国旗袍文化, Shanghai People's Fine Arts Publishing House, Shanghai.
Lotman, Yuri M. 1996. *La semiosfera* I, Cátedra, Madrid.
Lotman, Yuri M. 1999. *Cultura y explosión. Lo previsible y lo imprevisible en los procesos de cambio social*, Editorial Gedisa, Barcelona.
Lotman, Yuri M. 2011. "La moda es siempre semiótica" en Revista de Occidente, Número 366, noviembre 2011, Madrid.
Lozano, Jorge, ed. 2015. *Moda. El poder de las apariencias*, Casimiro Libros, Madrid.
Simmel, Georg 1923. *Filosofía de la moda*, Casimiro libros, Madrid 2014.
Song Yin 2015, Forum of semiotics of University of Sichuan. [Online] (Updated Dicember 2015) http://www.semiotics.net.cn/index.php/view/index/news/4948
Sun Peidong 2008, "论西美尔的时尚观", Journal of Northwest Normal University Social Science, Shanghai.
Svendsen Lars 2010, *Fashion: a philosophy*, traducido al chino por Manze Li, Universidad de Pekin.
Zhen Wendong 2006 符号域的空间结构—洛特曼文化符号学研究视角 Journal of PLA University of Foreign Languages.
Zhu Jirong 2016. 解读「中國鏡花水月」展. Contemporary Art and Media Culture, Ta wán.
Zimmermann, Yves 1988. *Del Diseño*, Editorial Gustavo Gili, Barcelona.

The fashion design as a contribution to the preservation of Nazaré culture and costumes: 7 project

Sueli Moreira
Department of Textiles, University of Beira Interior, Covilhã, Portugal

Madalena Pereira
Department of Textiles, University of Beira Interior, Covilhã, Portugal
FibEnTech, Covilhã, Portugal
UNIDCOM-IADE-UE, Lisbon, Portugal

Alexandra Cruchinho
ESART, Polytechnic Institute of Castelo Branco, Castelo Branco, Portugal

José M. Lucas & Rui Miguel
Department of Textiles, University of Beira Interior, Covilhã, Portugal
FibEnTech, Covilhã, Portugal

ABSTRACT: The aim of this paper is to present a project that aims to contribute to the dissemination and affirmation of the ethnographic culture and roots of Nazaré, in a Fashion Design contemporary approach inspired by the concept of Nazaré's traditional costume of its cultural, traditional and environmental values, in particular the beaches and waves of Nazaré. In addition, it also intends to contribute to the implementation of these values in local tourism and create value for the local and national economy. Based on design methodologies adapted from several authors such as Baxter, Munari, among others, the Fashion Design project was developed to contribute to the dissemination and affirmation of the ethnographic roots of Nazaré. After the development of the various design phases, prototypes were tested and validated. It is concluded that fashion design can contribute to the valorisation and dissemination of cultural, traditional and environmental values and, in view of the increase in tourism in Nazaré, it could be a contribution to the creation of added value in the region.

1 INTRODUCTION

Nazaré is an important point with regard to the use of the traditional dress, in which many citizens continue to wear it in their daily life, mainly in festive times, and being the beach of Nazaré a major touristic point of sale in the last years, it is considered an opportunity for the creation of value in the different areas. The S. Miguel Fort received 400,000 tourists in four years of regular opening to the public, according to Nazaré City Hall. In 2015, the second year of the opening of the monument to the public, there were 80 thousand visits, in 2016 there were 121,374 entries and in 2017, there were 174 thousand visits (Publituris, 2018).

According to the Expresso, the growth in the last years of the Nazaré region is due to this characteristic:

> "We were in January 2013 when a photo of giant wave surfer Garrett McNamara, a tiny spot in a monstrous wave over 30 meters and the Nazaré lighthouse in the foreground, covered the English newspaper The Times. Nazaré already knew the American surfer McNamara and him by this time have recognised the potential of the region. But it was from then on that the world came to know the two, with the image becoming viral on the internet. We perceive in Turismo of Portugal the power of that photograph. We

went to see the surfer and we went to Nazaré to talk to him. We saw that he truly loved Portugal and we paid him to promote our country, which at the time gave some controversy, because he was a foreigner", recalls Cotrim de Figueiredo (Expresso, June 17, 2017).

This confirms a growth of potential consumers of products associated with Nazaré tourism in the different areas (beach, surf, culture and religion), knowing however the need for a more in-depth study in this area. During the research period and in an informal way, it was found that tourists and shopkeepers would be open to added value products, associated with the ethnography and culture of the region. This justifies the opportunity to study and develop new projects.

1.1 *Nazaré*

Nazaré, a fishing village located on the Atlantic coast, a little more than a hundred kilometres north of Lisbon, is from its origin in the 19th century, a community marked by the presence of tourism. (Trindade and Penteado, 2001). Nazaré was naturally shaped between high contours of the Sítio sea cliff, a place that in the sixteenth century was still bathed by the ocean and the interior sea of Pederneira. However, with the sharp decline of the sea during the seventeenth and eighteenth centuries, the whole covered area of sand of the continental shelf along the rocky masses became land, being increasingly occupied, first by boats and fishing equipment, and then by people and families carrying their belongings (Granada, 1996).

The giant waves of Nazaré find the ideal conditions to be formed due to a big canyon, called the Great Canyon of Nazaré. It is relevant to explain the failure of the Nazaré Canyon to understand how these perfect and sometimes so great waves form (Município da Nazaré, 2015).

According to the Portuguese Geographic Institute (2013) the Nazaré Canyon is a submarine gorge of tectonic origin located off the coast of Nazaré, linked to the Nazaré/Pombal fault.

The Nazaré Canyon contributes to a riche wildlife with species of commercial interest because it generates abundance on the surface, abundant in nutrients and plankton. It has been the target of several studies by the Portuguese Navy and other institutions, both national and foreign, with the aim and the argument of expansion of the Exclusive Economic Zone (EEZ). Recently the researchers found, for example, a shark at 3,600 meters deep and several coral colonies.

The Círio of Nazaré is a procession in devotion to Our Lady of Nazaré and is celebrated on September 8 in the village of Nazaré. Known from 1793, the term "Círio" has origin in the Latin word "Cereus", that means big candle. It was from these processions to the Círio that the Sítio of Nazaré was developed. At the time, so many people arrived that housing was not enough and because of this people used to live boats and collected equipment outside the tents to house the various families. They replaced the hardwoods and thatched roofs with walls. The stones that were removed from the beach, were structured with clay and finished with tiles. (Granada, 1996).

According to Cortesão (1966), "Sítio became a centre of diffusion of fashion: and we remember that the set of skirts, demarcating the hips, swinging and tangling like a bell that proclaims feminine exuberance, is a reflection of balloon skirts, that ladies wearing ornamented clothes at the end of the 7th century were strutting there, as they come out from the seats and carriages". The fortress is built, called Real House of Our Lady of Nazaré in the construction of the palace and remodelling of the old house of the chaplains, transforming it into a hospital. In 1736 the bullring and the theatre were built and later, in 1818 the wall was built.

According to Pedro (1942) quoted by Granada (1996) with the French invasions in November 1807 valuable objects of the churches were destroyed and following the end of the invasion the population rebuilt them. In 1898 was installed the headquarters of the county of Nazaré, in the House of Alberto Carvalho Remígio, located in the street of the Customs.

Besides this there was also an expansion and a mentality open to the future, in the way of a constant tourism. The popular Saints and festivals of the Site have always been celebrated with their own and very joyful characteristics, but when there are processions, caution and prudence are predominant factors.

Nazaré is perhaps one of the few places where folklore is part of everyday life. Women wear even many skirts, not seven as the traditional, but they will be five or six, lace pleated, embroidered aprons, rings on almost every toe, threads around the neck (with crosses or the portrait of the man on a medallion), earrings on the ears, wool covers that trace in front and hold behind the back (Vieira, 1997).

Nazaré is also known for the belief of Our Lady of Nazaré and the miracle of D. Fuas Roupinho, which is one of several reasons Nazaré religious tourism. This version of the sanctuary's past was first told at the end of the 16th century by the chronicler Frei Bernardo de Brito, monk Bernardo de Alcobaça. The account was based on the donation letter of the Sítio by D. Fuas Roupinho, which the chronicler would have discovered in his Monastery and would publish in the work Lusitana Monarchy (Caneco, 1999).

1.2 *Ethnography*

As for ethnography, Nazaré has the following cultural elements evidenced: (1) Costume that represents part of the significant identity of this locality. There are two types of women's clothing, the party dress and the work outfit. Both are full of history, myths, legends and traditions, and are still worn today. For this reason, the traditional costume of Nazaré is one of the strongest inspirations in the collection, since this translates visibly that the collection is about Nazaré and from no other location. (2) Art Xávega: it was the first form of fishing carried out in Nazaré and the community does not stop keeping it alive until the present days (Silva, 1970 and Vasconcelos,1975).

Craftsmanship: Most of the handicrafts that are marketed in Nazaré are related to the sea and its connections to the land. The most typical and commercialized pieces are the dolls (representing a Nazaré woman dressed traditionally) and the regional boats (small boats of wood or ceramics of different colours and dimensions) (Figures 1–2). In addition to these two pieces, there are also puppets dressed in the typical fisherman costume, paper or wooden paintings with different scenarios, sailor knots and boats with half-built hulls (CMN, 2014).

1.3 *Tourism*

Nazaré continues to record a growing number of tourists. Something that the autarchy justifies with the already known giant wave surfaced by Garrett McNamara and the international promotion that has been made (Vieira, 2015). But Nazaré is from its origin in the XVIII century, a land marked by the presence of tourism. It was for centuries the largest sanctuary

Figures 1 and 2. Nazaré women dressed in typical costume in their daily life; Representative boats ofxàvega art. Source: Nazaré City Council.

of Marian devotion in Portugal, the Sanctuary of Our Lady of Nazaré played an important role in the development of the new settlement, the Praia of Nazaré. The stunning natural landscape, the exoticism of the fishing community and the beach have become important elements for tourism, allowing Nazaré to continue to be an important destination, even when the apparitions of Fatima came to take the leading role at the beginning 20th century (Trindade and Penteado, 2001).

According to Trindade and Penteado (2001) in the 19th century, Nazaré bathing beach is supported by fishing and tourism. Since recent years, surfing and body boarding have become a new attraction in Nazaré. The "giant wave" (tow-in) and the international record established by Garrett McNamara when surfing a wave of 30 meters are factors of notoriety for Nazaré, with the consequent attraction of practitioners and followers of this sport. (CMN, 2014).

According to the news of the official website of the Portuguese television RTP "the study, made under the Gulbenkian Oceans initiative, concluded that "estimated economic impact of the giant waves is 10.4 million euros" spent in Nazaré by domestic and foreign tourists between the years of 2011 and 2014. According to Vieira (2015), the brand "Praia do Norte", which sells products linked to the marketing of surf and big waves in Nazaré, reached sales in the order of 30 thousand euros and at a very short term has a potential to reach 100 to 150 thousand euros of annual sales". This economic dynamism is confirmed by local merchants, such as the sellers of dried fruits in Sítio da Nazaré.

1.4 Traditional costume

The Nazaré women are the image that the language of love is introduced very often in the costumes of Portugal when, covered by their cloaks, they awaited the arrival of their men (husband and children). Showing colour in clothing always supposes the demonstration of feelings of life as a sense of communication. Thus, the variety of coloured skirts that the women of Nazaré wear also correspond to an act of seduction and amorous evocation (Ormonde et al., 2003).

The work costume (Figure 7) consists of a shirt, a shawl, a simple apron, tartan socks below the knee, three skirts at the most and a pocket. (CMN, 2014). The party dress (Figure 8) consists of a lace shirt with different patterns, seven white and coloured skirts with flannel florals, an embroidered satin apron, a scarf on the head under the hat with a pompom on the side, a black cape and another inside, some black lacquer and various gold accessories (necklaces and earrings). The seven skirts may essentially represent the sets of the curl, the seven waves, the seven virtues, the seven days of the week, the seven colours of the rainbow, among others (CMN, 2014). The key garment pieces of the female costume are the typical blouse, the bodice, the rounded skirt below the knees with several layers and some pleats,

Figures 3 and 4. Nazaré costume and party dress. Source: Nazaré City Council.

apron, cape, scarf and hat. The men usually walk barefoot and when they are on shoes they wear black leather clogs with white or black painted wood sole, or they also wear braid slippers (Figures 3–4).

2 PROJECT

2.1 *Fashion design project*

The concept of "7 project" is intended to be a fashion based on cultural roots and, therefore, with an affective and enduring attachment to the individual. The new age of consumption modifies the merely functional character of the design since the created products are concerned to generate some form of meaning. Hypermodern design (emotional and affective) proposes emotion through sensuous products centered on the imaginary and the emotions of the consumer. Lipovetsky (2015) argues that the mentalities and the way of creating and communicating these products are undergoing changes, since luxury no longer means excess or ostentation, but identity when identifying the profile of its consumer. A collection inspired by a tradition and culture such as Nazaré, ends up communicating to the consumer what it represents, such as Nuno Gama did when he created garments inspired by the scarves of the lovers and the history of Portugal, and Dolce & Gabbana, which uses ethnographic inspirations constantly in their collections, as for example in 2014 with a Byzantine theme.

The development of a project in the area of Fashion Design is based on a methodology, since in this way it is possible to establish a coherent connection between the entire process of idealization, construction and the post-project. The methodology that best suited this project was that of Munari but some key points of other authors were also added, taking into account the opinion of Lupton (2012) in which each designer should develop his own method.

Problem: The traditional clothing fell into disuse. How to preserve the culture of the Nazaré costume so that it can create wealth for a region with growth potential in tourism, combining tradition with the contemporary for the market?

Opportunity: Look for an opportunity, in this case the growth of tourism in Nazaré, which makes it a business opportunity to create clothing on and for this region.

After defining the problem and identifying opportunities, the various components of the problem, data collection and data analysis, such as the history and ethnography of Nazaré are analysed; the market; trends, colours, silhouettes, lines and typical pieces of Nazaré. The creativity component is followed by drawing sketches of inspirational images and Sketchbook, ending this phase in the final illustrations among other points based on the design methodologies studied (such as Bonsiepe (1992), Munari (1981) and Baxter (2005) and Jones (2005)) as the definition of materials and technologies available and appropriate to the project followed by experiments, models with prototype development according to the choices of the experiments and their verification and final solution.

The name given to the collection comes from the fact that the number seven is present in almost everything, seven are the skirts that the Nazaré women wear, seven are the children they usually have, seven times they count on the launch of fishing boats overboard. Based on this, the collection has seven coordinates, seven materials, seven attachments, seven vivid colors, seven buttons on shirts, and many more secrets around number seven are present in this original creation.

The ambience panel (Figure 5) thus illustrates a woman with representations around her inspirations such as the giant waves, the traditional costume of Nazaré, embedded in an old frame, showing the inspiration for what is traditional and the ethnographic roots of the region, in the fishing nets found in the bottom of the frame, and on the top of the young woman's head is a boat, representative of Xávega art, still used today in Nazaré as a form of fishing.

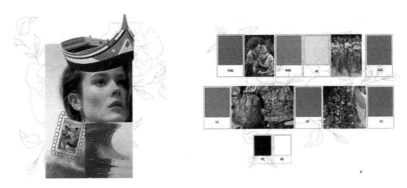

Figure 5. 7 Project, result of the creativity phase. Source: Author.

Figure 6. Illustrations of 7 Project, result of the creativity phase. Source: Author.

With regard to the color panel (Figure 6), this one represents seven main colors, which are colors already used in the traditional costume of Nazaré and two secondary ones, white and black. The collection is very colorful, and the color is one of the elements that most draws attention to the garment pieces, these were used carefully so that the coordinates between them became harmonious.

The materials and supplies panel contain seven textile articles and seven articles of materials used in the collection. The materials were chosen according to what is worn in traditional costume of Nazaré, but with the introduction of new materials related materials that also apply to sportswear and casual. Each material is characterized, and one example can see in Table 1.

The patterns have been developed with inspiration in traditional costumes such as floral patterns and tartan patterns; or the inspiration in surfing with the great Canyon and Xávega art in the typical boats of Nazaré.

Lastly the final illustrations were developed and completed (Figure 6). The techniques as technologies used for its application were the embroidery and the fingerprint, available in the University of Beira Interior/Textile Department workshops. After their experimentation, the technical sheets were developed, and the prototypes were made.

For a young female target audience, for a medium segment, the products developed have as direct competitor the Praia do Norte brand.

After the elaboration of the prototypes, the final coordinates were developed with the respective patterns and embroidery applied, as can be observed with the example presented here of the result of the elaboration of the communication images (Figures 7 and 8).

Table 1. Fabrics properties.

Assignment of the garment	Sample of fabric/price	Features	
Scarves		Type of faric/Knitting	Fabric
		Designation	Cotton satin
		Composition	100% CO
		color	white
		Width	150 cm
		Weight/m	162 g/m
		Weight/m^2	108 g/m^2
		Torsion	S
		Special Properties	Wash fabric

Figure 7. 7 Project, result of the communication phase. Source: Author.

Figure 8. 7 Project, result of the communication phase with elements of culture and ethnographic roots of Nazaré. Source: Author.

3 CONCLUSIONS

This project aimed to contribute to the dissemination and affirmation of the culture and ethnographic roots of Nazaré, namely the fact that a tradition of Nazaré has fallen into disuse, the costume, and at the same time seize an opportunity in the market with the growth of tourism of the region of Nazareth. In this way and utilizing the knowledge and skills acquired during the academic training, Bachelor and Master in Fashion Design at the University of Beira Interior, it was intended to contribute to the dissemination and affirmation of the cultures and ethnographic roots of Nazaré through the hypermodern Fashion Design. Project 7 is made up of clothing and accessories from a collection inspired by the culture and ethnographic roots of Nazaré, considering the region and its market as well as the increase of tourism in the region.

Assmann (2008) analyzed something similar to the costume of Nazaré in the case of the custom of using the Kimono in Japan, in which its sales are constantly high. Being a traditional piece especially for parties, the kimono happened to be a national garment that is being constantly reinvented. This has become a symbol of the transmission of an individual attitude towards social conventions and national identity.

At the same time, the Sueli More brand was created inspired by the typical costume of Nazaré and nowadays, more specifically in surf waves with an enterprising attitude and with the objective of taking advantage in the future of the continuity of this project for its implementation in the market and commercialization, using the value of intellectual property as a tool that contributes to the exclusivity of some characteristics of the project. At the end was developed an on-site fashion production and simultaneously product testing against the consumers that appeared on the site and addressed with that goal. It was concluded that the adhesion and the opinion were quite positive, questioning some of them where the product having the characteristic of identifying the culture of the region was available to be acquired.

This project contributed to apply to a fashion design project the culture and the ethnographic roots of Nazaré that are tending towards its disappearance, using current concepts, methodologies and techniques. One of the next phases to be developed will be the industrialization of the product and its commercialization in a segment directed to national and international consumers that visits the region. It is also intended to develop a business plan to submit to the contest of ideas and thus acquire a minimum capital to start the project in the market.

REFERENCES

Assmann, S. 2008. Between Tradition and Innovation: The Reinvention of the Kimono in Japanese Consumer Culture. *Fashion Theory* 19:3, pages 331–361.
Barthes, R. 1983. *The fashion system*. London. University of California Press, Lda.
Baxter, M. 2005. *Projeto de produto: guia prático para o desenvolvimento de novos produtos*. São Paulo. EdgardBlücher Lda.
Bonsiepe, G. 1992. *Teoria e prática do design industrial: elementos para um manual crítico*. Lisboa.Centro Português do Design.
Caneco, J.1999. *Nazaré—tradição e história*. Nazaré.Câmara municipal da Nazaré. at: http://www.cm-nazare.pt/
Cortesão, J. 1966. *Portugal: a terra e o Homem*. Realizações Artis, Lda. Lisboa. Imp.Nac. casa moeda.
Cunha, P. e Gouveia, M. 2015. *The Nazaré Coast, The Submarine Canyon And The Giant Waves – a synthesis*. MARE—Marine and Environmental Sciences Centre. University of Coimbra, Faculty of Sciences and Technology. Department of Earth Sciences. Coimbra.
Expresso, 2017. *Turismo. O segredo do nosso sucesso*. Sociedade, 17.06. Access in January 16, 2018. Em: http://expresso.sapo.pt/sociedade/2017–06–17-Turismo.-O-segredo-do-nosso-sucesso#gs.HaQBjv0
Feroldi, S. e Anselmo, T.K. 2011. *Metodologia projetual, um diferencial no desenvolvimento de produtos de moda*. Universidade do Vale do Itajaí. 1–3 pp. Anais do 7° Coloquio de Moda. 4ª Edição Internacional, Maringá/Paraná.
Gomes, L. (2010). *Criatividade: Projeto, Desenho e Produto*. Ed. sCHDs. Santa Maria.

Granada, J. 1996. *Nazareth: pederneira, sítio, praia para a história da terra e da gente*. Nazaré.Gráfica da Batalha.
Jones, Sue Jenkyn. 2005. *Fashion design: manual do estilista*. Tradução de Iara Biderman. CosacNaify. São Paulo.
Lipovetsky G. e Serroy, J. 2015. *A estetização do mundo: viver na era do capitalismo artista*. Companhia das Letras. São Paulo.
Lupton, E. 2012. *Graphic Design Thinking*. Editorial Gustavo Gili, Lda. Barcelona
Munari, B. 1981. *Das coisas nascem coisas*. Edições 70. Porto.
Município da Nazaré. 2015. *Plano estratégico de desenvolvimento humano*. Consultado em Janeiro de 2018 em: http://www.cm-nazare.pt/sites/default/files/documentos/
Ormonde, H., Nunes, C., Anastácio, D., Sancho, E., Vieira, J., Isaac, A. e Natas, A. 2003. *O traje do litoralportuguês*. Câmara municipal da Nazaré. Museu etnográfico e arqueológico Dr. Joaquim Manso. Nazaré.
Publituris. 2018. *Ondas da Nazaré aumentam visitantes no ascensor e forte de S. Miguel*. Access in January 24, 2018. Em: https://www.publituris.pt/2018/01/23/ondas-da-nazare-aumentam-visitantes-no—ascensor-forte-s-miguel/
Silva, A. 1970. *O trajo da Nazaré*. Editorial Astónia. Lisboa.
Silva, G. 2005. *Design 3D em tecelagem jacquard como ferramenta para a conceção de novos produtos: aplicação em acessórios de moda*. Dissertação de mestrado em Design e Marketing, opção Têxtil. Universidade do Minho. Guimarães.
Trindade, José e Penteado, Pedro. 2001 "*A Nazaré e os seus Pescadores: entre as representações sociais e novas leituras histórico-antropológicas*", in Oceanos, 47/48, pp. 80–92.
Vasconcelos, J. 1975. *Etnografia portuguesa*. Volume VI. INCM. Lisboa.
Vieira, Inês. 2015. *O desenvolvimento da Nazaré através do Marketing Territorial e do Branding*. Dissertação de Mestrado. Faculdade de Economia da Universidade de Coimbra.
Vieira, Vera, Rui Miguel, Madalena Pereira, Liliana Ribeiro, José Lucas (2014). Rainwear Collection Development for Inspired Lifestyle Brand of Surf Culture, *Proceedings of 2º International Fashion and Design Congress*, Polytechnic of Milan. Milan.
Warkander, P. 2014 "NO PANSIES!!": Exploring the Concept of "Style" through EthnographicFieldwork. *Fashion Theory*.18:3, 233–250.

DIY and the slow fashion movement: Sew for yourself

L. Barrocas
Universidade Federal de Pernambuco, Recife, Brazil

G. Bezerra
Universidade Federal do Ceará, Fortaleza, Brazil
Universidade do Minho, Guimarães, Portugal

M.A.V. Rocha
Universidade Federal Rural de Pernambuco, Recife, Brazil

ABSTRACT: The fashion industry has been currently dominated by fast fashion: a model system that streamlines the production stages, turning out new products and low quality constantly in the market. Reacting to this industrial culture, we have the rise of slow fashion movement, created with the aim of making production and consumption in a more conscious activities. Drawing a parallel, in the same way that the slow fashion is against the current regulations of the fashion industry, the DIY (Do It Yourself) is somehow here for the rejection of the options market, with practitioners developing projects of artefacts that are closer to their own desire. The purpose of this article is to analyse these two contemporary movements with the intention to find ways to unite them in the act of sewing for anyone.

1 INTRODUCTION

Since the beginning of the industrial revolution, companies have offered facilities at the time of consumption. Being the industry in its current phase, the mass market, we find lots of products easily at our glance, replacing simple domestic activities, like to cook, for frozen meals, no more needing to waste "preparing" time.

But if nowadays almost everything is ready on store shelves, why some people insist on produce or modify the artefacts? Some authors (Atkinson, 2006; Ferris, 2013) claim that despite the difference of motivations, the sense of autonomy and customization with the artefact is what makes the practice generally grouped around the acronym DIY (do-it-yourself) be made until today. To Atkinson (2006), the DIY can be considered since painting your own house doors, to reform and build a furniture, through the transformation of existing products.

The phenomenon of massification of industry, with the constant supply of new products also finds space in the fashion system, being this phenomenon called fast fashion (Calliope, 2015). It is characterized also by the faster consumption, since the search is by offering the lowest price and, consequently, dismissing the quality (Fletcher, 2007). However, some sectors of society no longer admit this production schema, which also requires a huge production of clothes every week and damages the worker inside the factories with small wages and controversial working conditions. In the discussion of alternatives, the thought and the practice of slow fashion (Fletcher, 2007) comes on the agenda, as this movement values for the attention to the life cycle of products and the quality in the production stages.

This article is intended to discuss the Slow Fashion movement and DIY culture, present today in our contemporary culture through the practice of garment sewing for the user him/herself. Although Fletcher considers that it is possible to patch and customize your clothes, nothing was said about the practice of sewing as a whole, passing from the choice of the model, cutting and sewing apparel. The questions for this article would then be: how is it possible to point approaches between the Slow Fashion and DIY culture? How sewing practitioners connect these two points? Therefore, we will be analysing theoretically the two streams that redefine contemporary practices of wear, combined with the practice of sewing.

2 THE SLOW FASHION MOVEMENT

Today the major international fashion magazines such as Zara, Forever 21, H&M and the Brazilian one, Riachuelo, works in the model called fast fashion: a quick production of goods, where new products are dumped in the stores each week (Fletcher, 2007). In addition to the speed of production, the voracious consumption is stimulated, with the interest of making more and more money. To achieve the maximum profit, the production stages are shortened, reducing the life cycle of the clothes. For Pinto and Souza (2015), the fast fashion is the most successful production model in recent years, combining the ability to predict trends with the agility in production.

Calliope (2015) demonstrates how you can decrease production steps for achieving a low selling price through a line of products of fast fashion. Starting with low quality materials used to produce low quality clothes, this chain results in cheap clothing and along with the constant motion of the fashion trends, the same pieces are discarded in a short time. Fletcher already warns that despite this model be very lucrative due to the speed in the production, somehow someone is paying for this, either the worker or the environment, highlighting the contradictions of this model.

In this devastating production scenario, especially when you have a glance of behind the scenes, Kate Fletcher (2007) and other researchers tried to conceptualize a new way to work with fashion, borrowing the concept of slow food. Founded by Carlo Petrini in 1986, the slow food movement encourages people to seek pleasure in food in a conscious way. Calling it by the name slow fashion, this movement is about a consumer awareness search, in the design and production of clothing, making it somewhat more sustainable (Clark, 2008). The "slogan think globally, act locally" has become one of the mottoes of the slow movement (Clark, 2008), encouraging the local resistance, defying the global enforcement of fast fashion.

The word slow, when associated with the word fashion, means a slowing of the process and an appreciation of each step in the production of a garment, turning the attentions to the quality (Fletcher, 2007). It is a move to give more value to details, care and long-term thinking (Silva and Busarello, 2016). In a comparison with the offer of fast fashion stores, purchase a slow product becomes a matter of choice and information. The product will be more expensive, but with more value and durability. Calliope (2015) explains that the slow fashion consumer opts for versatility of the wardrobe, keeping true to his/her style, what causes a feeling of purchasing well-made and durable items. Fletcher (2007) explains that this consumer awareness allows entrepreneurs to have more time to organize their applications, and to invest in their long-term workers, generating a richer interaction between the designer and the producer, the producer and the garment, the garment and the user. Fletcher (2007) and Calliope (2015) already predict that the user him/herself, as a way to align the slow fashion movement, perform minor repairs and recycled, thus prolonging the life cycle of products.

In short, Clark (2008, p. 3) presents specific points of the slow fashion movement to help question the global order: (1) questioning the hierarchy of the designer, producer and consumer; questioning the notion of fashion being everything that is new; (2) challenging the interlacing of fashion with the image; (3) introducing the fashion as choice and (4) fostering the collaborative work, especially among women. All these points represent new ways to think about fashion, making it somewhat more human. Moreira (2016) mentions that inside the slow movement slow design is associated with slow fashion, where the design focus is on the project and on the benefits to people involved in the production and use of these artefacts. The project becomes more holistic, trying to reflect on the entire life cycle of the artefacts and the impact it may have. Moreira (2016) says that slow design is emphatic about the democratisation of design, aiming to achieve a greater range of interested. So, slow fashion becomes a promising democratic process regarding the approach with the design, since it aims to make relations closer and valued inside the fashion chain.

3 THE DO-IT-YOURSELF

The do-it-yourself (DIY), it is as old as the industrial revolution, since from the beginning of the urbanization of the cities, people needed to perform minor repairs on their homes, or produce artefacts they could not buy, starting from raw materials like wood (Atkinson, 2006; Kuznetsov and Paulos, 2010). To this day, the practice of doing on their own has more or less space in people's lives, due to the availability of financial resources of the population and the popularization of the practice in the media. Currently, it would be more natural if there was no concern of doing things, due to the availability of artefacts, but Atkinson (2006) raises evidence that DIY is in expansion, including the practice of sewing their own clothes.

Ferris (2013) explains that one of the main reasons for an increase of craft projects is that it deals with one of the many contradictions of current world: we acquired artefacts from the market, but we modified it, so we have a sense of autonomy, embodiment and work, acquiring the value offered by the industry. Shiner (2012) explains that in the proposed craft, what is implied is that the person who does it be the same person who designs. Campbell (2005) elucidates that point, explaining that this attitude is somewhat against the division of labour, because joining the project and the execution in a single person, a division that contrasts on one side with the "inalienable, humane, authentic and creative work, on the one hand, and purely mechanical, unfulfilling and alienating labour, on the other".

According to Atkinson (2006) this DIY's premise serves as an antithesis of the design process, making it more democratic in an amateur way, bringing the artefact to near the end user, overcoming the stigma of manual work cited by Campbell (2005). Taking the definition of Cardoso (2013) for the design, as an area that, at the same time, acts of conformation of materiality and also in the informational field, influencing the perception of experiences, we see that, in fact, the design can be practised in the amateur level as an essential tool to which we can relate better with the surrounding artefacts.

Here we realized a first approximation of the DIY with slow fashion, as both speak of democracy and of the designer's approach with the end user, more radically present in DIY, as the designer and the user are often the same person. As much as there is DIY projects semi-ready on the market where the industry already has also found a space, the user can still modify the final result (Shiner, 2012). Cardoso (2013) comments that the design should still turn to the craftsmanship, understood by the author as a "high degree of attention to detail and care in execution, from a peculiar sense of pride at work, the pleasure of doing well done", notions that Fletcher (2007) have already explained that it was possible to achieve with the slowing process.

Deepening in the understanding of DIY, Atkinson (2006) proposes a division of intentions with the practice, division that he explains not being accurate, changing categories as the user needs. There is a (1)DIY proactive, which consists in activities with a high creative input, resulting from the combination of raw materials with pleasure or gain; the (2) DIY reactive, that would be a hobby, mediated by kits or ready-made templates, practised in leisure hours, and may contain a financial gain; the (3) DIY essential which is housekeeping activities, practised there or not being able to hire professional help, and finally, the (4) DIY lifestyle, where the activities are made for personal consumption choices and work. Once again we found a possible dialogue between the slow fashion and DIY when we realize that there is a criticism of consumer options available. If the option is to make a chair, the producer probably will buy wood from suppliers near their home, handling the materials chosen to result in an artefact that has a bigger emotional attachment than a chair purchased in a store. The awareness of the choice when purchasing the materials for the projects can be part of the activity, as well as part of slow fashion, where the use of local resources is an alternative against the standardisation of products (Clark, 2008).

4 POSSIBLE INTERSECTIONS IN THE ACT OF SEWING

We saw that some DIY principles are similar to those of slow fashion so that they appear to be siblings in movement, in response to the massification of the products. The concern of the human engagement with the materials that exists in the craft and DIY practices (Shiner, 2012) can be understood as one of the reasons for the slow fashion, as Clark (2008) focus on the material aspects of the garment. However, Clark (2008) does not make it clear if the sewing of clothes activity is an aspect of slow fashion, as it requires some tools more expensive than a pair of needles and a more specific knowledge, not causing the same interest that knitting has today. This issue has already been presented by Bain (2016), indicating the lack of academic interest on domestic sewing when the knitting was the subject of several studies.

Despite an apparent resistance of the academy and the non-inclusion of sewing in the slow movement, there is evidence that the global interest in sewing is alive, with the production of reality shows (The Greatest British Sewing Bee-GB; A Caixade Costura-BR) and the great number of Youtube channels and blogs on the internet about the subject.

Pereira and Nogueira, in 2013, did a study on how the practice of ordering outfits for a seamstress who works in the home environment is aligned with the principles of slow fashion. They claim that the detail and the deadline of the clothes that will be tailored along with the social aspect of the dressmaker studio is one of the factors of consciousness in the slow movement preaches. Having to wait a month for clothes to be ready to use causes the project to be well-thought-out and "liked". So if the attention and detail in clothing construction, along with an appreciation of the social aspects important to the slow, the DIY your own clothes can be considered an important part of the movement, even if buying a sewing machine is a little expensive at first.

Going through the process of sewing something for yourself involves a lot of attention and affection from the choice of the fashion pattern, to the finishing, because one of the intentions to sew for themselves is to make an outfit that will last and be worth all this effort. For Walnes (2014), the pleasure of stitching involves many issues, including knowing that nobody in the world has the same outfit that you, the experience of making something unique with your own hands, and wearing clothes that fit well on your body and make you feel good.

The seamstresses who keep blogs on the internet consider themselves as part of the slow fashion movement, with a hashtag created especially to celebrate the movement among the makers: the #slow fashion October as during the entire month of October they are encouraged to post creations or aspects of their life that have to do with the movement. The American magazine Seam work (2016), focusing in the amateur seamstresses, has a story about this event that happens every year, where is explained that it does not meet just who sews, but who knits, mends and consumes conscious fashion, raising the issues of waste, disposal and exploitation. Powell, the author of this text, indicates that in addition to seeking manufacturers and designers with ethical origins, sewing and mending clothes are aligned with the principles of the slow movement which also includes exchange clothes and limit the number of trips to the stores.

However, it's not just the act of sewing for oneself that automatically turns the seamstress into a slow fashion practitioner because even using the sewing machine involves the process of deceleration and awareness of what is being done. You should not buy fabrics that will not be used quickly or pick a project that produces a lot of waste. It is also unnecessary to sew garments which are not part of the utility wardrobe, as it encourages the movement. You need to have the same consciousness that slow fashion brands are using, taking the same principles to the domestic scale.

The Brazilian vlogger Karina Belarmino on her Youtube Channel discusses the conscious option of producing their own clothes in response to the imposition of the trends found in stores. As she states in her videos, she teaches how to sew clothes that looks good in her and are in her style, since she should not do something that she will not use later. To her, consumer awareness is also choosing fabrics of good quality, keeping few trims at home. Managing a wardrobe that also has fast-fashion items, Karina says that she would like to sew all her clothes, serving as a real-world example of how to adopt the slow fashion gradually day by day.

5 FINAL CONSIDERATIONS

We observed in this article an alignment of these two movements present in society, the DIY and slow fashion, with the rejection of the mass products found on store shelves. If one can make the act of consuming a conscious act full of value, others can redefine the movement as an act of production for him/herself. What we need is do more with what is offered making trends a choice, not an imposition.

The act of sewing suggests a bridge between these two movements that involves both emotional and creative aspects of DIY as productive and consumer aspects of slow fashion. To sew, even if it is stitching by hands, you need patience and planning, designing the garment according to the user's needs, taking into account the time available to carry out such activity. Overall, sewing can be considered an act of leisure as it breaks the acceleration that the whole society is currently involved, adding more layers to this meaningful activity. Tying the seam in two movements, this activity can be viewed as searching for a personal style at your own pace.

Bain (2016) says that studying the stitching involves a lot of meanings that can be understood in a political, social and cultural way, in addition to the design and creation of artefacts. Although the movement slow fashion is more comprehensive and encompass all an industry, as well as the DIY, we can democratize the practices, bringing to the domestic sphere and so turn consumers also as producers. For future researches, the question would be how the practitioners see themselves inside the two perspectives, if they claim they are inside one or both movements, as well as investigates their vision of other specific reasons the act of sewing are involved in, from this old and tradition activity to its current context, revealing some contradictions of the contemporary society.

REFERENCES

Atkinson, P. 2006. Do it yourself: democracy and design. *Journal of Design History*, 19 (1): 1–10.
Bain, J., 2016. "Darn right i'm a feminist…sew what?" the politics of contemporary home dressmaking: sewing, slow fashion and feminism. *Women's Studies International Forum* 54.
Calíope, T. 2015. Moda e Sustentabilidade: uma relação contraditória? Um ensaio sob a perspectiva do ciclo de vida. In: A*nais—Engema. Encontro Internacionalsobre Gestão Empresarial e MeioAmbiente.*
Campbell, C. 2005. The Craft Consumer: Culture, craft and consumption in a postmodern society. *Journal of Consumer Culture* 5: 23–42.
Cardoso, R. 2013. *Design para um Mundo Complexo*. [ebookversion] São Paulo: Cosac Naify.
Clark, H. 2008. SLOW + FASHION—an Oxymoron—or a Promise for the Future…? *Fashion Theory*, 12 (4): 427–446.
Ferris, M. 2013. Making Futures. The Return of Craft in a Post-Global Sustainably Aware Society. In: *Making Futures Journal* 3.
Fletcher, K. 2007. *Slow fashion* [online] Available at: https://theecologist.org/2007/jun/01/slow-fashion [Accessed 24 February 2018].
Kuznetsov, S. & Paulos, E. 2010. Rise of the Expert Amateur: DIY Projects, Communities, and Cultures Proceedings. Iceland: *NordiCHI 2010*.
Moreira, A. 2016.*Gesto, Significado e Superfície: um estudo no contexto do Slow Fashion*. Trabalho de conclusão de curso. Porto Alegre: UFRGS.
Pereira, D.R. & Nogueira, M.F. 2013. Moda sob Medidauma Perspectiva do Slow Fashion. *9º Colóquio de Moda*. Fortaleza.
Pinto, A. & Sousa, C.S.M. de. 2015. Roupas Feitas de Roupas. *Iniciação – Revista de Iniciação Científica, Tecnológica e Artística* 5 (3). São Paulo: Centro Universitário Senac.
Powell, L.A. 2016. Slow Fashion OCTOBER. In: *Seamwork magazine* [online] Available at: https://www.seamwork.com/issues/2016/10/slow-fashion-october [Accessed 24 February 2018].
Shiner, L. 2012. "Blurred Boundaries"? Rethinking the Concept of Craft and its Relation to Art and Design. *Philosophy Compass* 4 (7): 230–244.
Silva, S.P. & Bussarello, R.I. 2016. Fast Fashion e Slow Fashion. O Processo Criativo na Contemporaneidade. *Revista Estética* 12.
Walnes, T. 2014. *Love at First Stitch: demystifying dressmaking*. London: Quadrille Publishing.

Modes and Fashions of the colonial man from *Minas Gerais*: The indumentary of the defendants of the inconfidence of *Minas Gerais*

A.F.B. Santos
Universidade FUMEC, Belo Horizonte, Brazil

A.P.X. Vilela
Universidade Federal de Minas Gerais—UFMG, Belo Horizonte, Brazil

ABSTRACT: This article has as its central theme the recovery, identification and recording, through illustrations, of the dress code during the colonial period of Minas Gerais, referring to how it related to the objects adopted from the inconfidents as illustrated in the *Autos de Devassa*. To do so, a survey of the historical importance of textile material in the mining lands as well as the value of dressing among the colonial population of Minas Gerais was conducted. In addition, we sought to present the view of different authors regarding inconfidence and how they dealt with the material conditions of the inconfidents by presenting the textile genres as significant contributions to their inventories. Finally, we illustrated the difficulty of creating images from sparse materials and records and the way in which historical and literary references become the object of image translation.

1 THE TEXTILE MATERIAL: FROM TRADITION TO INTERDICTION

In spite of the availability of important documents related to the textile traditions, nothing is known about the type of fabric used by the primitive man from Minas Gerais that transited through these lands, whether utilitarian textile weaves or fabric intended for some type of garment. Likewise, no concrete reference has been identified for any textile activity, whether woven products or objects and tools related to the art of weaving.

The materials used in clothing, that is textile fibres of organic origin, are fragile and deteriorate easily when subjected to the conditions of a tropical climate. This was the main factor that affected the deterioration of these fabrics, making it impossible to preserve specimens as records of the customs related to the dress code of this primitive population. This explains why there is a gap that has always existed in the history of Brazilian culture, as far as the prehistoric man's costume is concerned. The few references that have been made regarding the customs and traditions of the Minas Gerais man, who lived here a thousand years ago, such as during the colonial period, are not easily justifiable. It is known that when the Portuguese arrived here, the knowledge and use of the fibres of wild cotton was common among our American ancestors, but cultivating the plant, obtaining a little more from it than was freely offered by nature, and transforming the fibre into a product of greater value were ideas brought and implemented by the colonizer, during the early decades of the 1500s.

According to Frei Vicente do Salvador, cotton was already being worked on by the people from Santa Cruz and "it was done with less work than there is made of linen and wool, because under the cotton can the spinner be harvesting and spinning, nor are there any paints with which it dies" (Salvador, 1982, p. 76). In the same work, he describes the use of the cotton thread as a commodity of exchange with the European man. Fibre was shipped for trade along with loads of chickens, parrots, honey, and wax, and the Indian received knives, sickle or combs (Salvador, 1982) in exchange, a fact that confirms the importance of the cotton produced there at that time.

From the second half of the 18th century, the occupation of the production of textile pieces was recorded in Minas Gerais. In these regions which were distant from the coast, due to transportation difficulties, the weaving tradition became important, and was aimed at meeting the needs of the local population. Here the fabrics were produced for domestic use and for the clothing of slaves, and eventually they were fabricated to meet the needs of the wealthiest classes of the population. After the last two decades of the century, following the gold mining decay, the cotton plume sustained the economy of the mining province.

Were it not for Queen Maria I of Portugal's ordinance in 1785, which demanded the closing of all existing manufacturers in Brazil and prohibited the making of fine fabrics in wool, linen, and cotton (Fonseca, 1984), textile production in Minas Gerais would possibly have had another outcome, since the weaving of cotton was already expanding. The queen's document only authorized the production of coarse cotton fabrics, intended for the clothing of slaves, the production of bales, and the packaging of products. The purpose of the charter was to stimulate Portuguese production and favour trade with England, which at the time was selling cotton to Brazil through Portugal. This prohibition resulted in the importation of fabrics by Brazilians with the intention of producing sophisticated pieces for the nobles, meeting the needs of the religious brotherhoods for dressing and religious vestments, and using in the furniture and building walls of the most affluent. Thus, until the revocation of the charter by King João VI in 1808, velvets, damask, lavender silks, and other textile genres used in Brazil were imported from Europe through trade with Portugal with the aim of providing the colony with the necessary infrastructure for the future headquarters of the Portuguese monarchy. If during this period the fabrics were already valued in their place of origin, in Brazilian lands they represented products of extreme luxury and were traded at very high prices by the merchants who brought them to Brazil by ship.

These fabrics of European origin not only served to produce costumes for wealthy people, but the silk drawn with their floral motifs inspired the artists of Minas Gerais because they accommodated the floral compositions designed to decorate the rococo interior of the most important and sacred places of the churches and also the exquisite spaces of the noble houses of Minas Gerais. In spite of the dictates of the Queen's license, textile activity which was considered illegal in the colony, continued to be developed clandestinely in a multitude of looms spread throughout Brazil, highlighting the production of the state of Minas Gerais. Here the linen and wool fabrics were produced under the guise of cotton production. This production met the needs of most of the population and over the following century it spread to other provinces especially Rio de Janeiro where the Court had already been established since 1808.

Although the production of wool fabrics was common in Brazil and Minas Gerais state, cotton fibre was the basis of almost everything that was produced in the state until the late 19th century. After the states of Maranhão and Pernambuco, Minas Gerais had the best cotton fibre and from it originated the fabric that was identified in the 19th century as "Pano de Minas". The dress of the free and poor population of colonial Minas Gerais, more specifically in the provincial capital, resulted in "an instrument of social differentiation and maintenance of social status in a glorifying society of appearance" (Januário, 2004, p. 01). Carla Maria de Carvalho Almeida in her doctoral thesis emphasizes that the inhabitants of Vila Rica were happy to invest in objects that guaranteed the appearance of fruit of the aristocratic ideal present in the region, since it had in its territory a village that was the seat of the captaincy of Minas Gerais (Almeida, 2001).

Four years after the signing of the charter by Queen Maria I in 1789, and in the middle of the gold cycle, a rebellion against the captaincy of Minas Gerais was initiated, due to extortion through taxes by the Crown on the gold extracted in the territory of Minas Gerais among other reasons. Inconfidence or conjuration was one of the most important social movements in the history of Brazil and because it represented the struggle of the Brazilian people for freedom against the oppression of the Portuguese government in the colonial period. Brazil, still a Portuguese colony, suffered from political abuse and high taxation and duties. In addition, the metropolis had enacted a series of laws such as the aforementioned law that prohibited the operation of manufacturing industries in national territory, which hampered the industrial and commercial development of the country.

Thus, one of the testaments of the inconfidents in 1789, quoted in the testimony of Dr Domingos Vidal Barbosa and published in the *Autos de Devassa da Inconfidência Mineira*, was related to the democratization of costumes and the use of satins and silks. The testament begins with speculation about Brazil's self-sufficiency if it were independent of Portugal and states that "Dr. Claudio, Canon Luís Vieira, and Judge Gonzaga had already made the laws to govern themselves, in which it was ordered that all the plebeian man could wear satins, that the diamonds would be frank" (AUTOS DE DEVASSA, 1976, p. 214). From this quotation, it is possible to conclude that the commoners desired freedom of the use of costumes and that the inconfidents intended to abolish the sumptuary laws.

Upon the arrival of fabrics of Spanish origin to Brazil, Frei Vicente do Salvador states that Flemish fleas dispatched and chartered in Lisbon, Porto, and Viana with farms of their land and Portuguese merchants to bring sugar (Salvador, 1982, p. 292).

It is known that cloths intended for clothing, interior decoration and ecclesiastical use were considered to be precious commodities and along with other goods, they only arrived in Minas Gerais on pack animals after being disembarked from the ships in the coastal ports. At the time, the ships took in sugar, cotton and mainly shipments of redwood, dyeing wood widely used in Europe as a dye in textile manufacture, for exchange.

Frei Vicente do Salvador describes the arrival of ships that landed on the Brazilian coast. "Among them was one to the captaincy of the Holy Spirit and asked the captain of her to the superior of the house of the priests of the Company... a load of brazil wood... and the following year would make it search again and would bring them the payment in ornaments for the church or whatever they wanted" (Salvador, 1982, p. 292–293).

About the arrival of fabrics in the collections of the churches of Minas Gerais, the book Inventory of the Factory of the Cathedral of Mariana, from the period between 1749 and 1904 reveal a considerable number of liturgical pieces in its collection. These inventories recorded more than two entries of textile objects such as sacred vestments, complete pontifical ornaments, inner vestments, and altar vestments. It should be noted that the majority of the tiled fabrics, damask, velvets, and brocades among other genres reached the collections of the churches in ready-made pieces which generally composed the complete sets of liturgical textile implements.[1]

2 TEXTILES AND GARMENTS OF APPAREL MARKETED IN COLONIAL VILA RICA

The studies of the researcher Alexandra Maria Pereira (Pereira, 2013) on the trade of a commercial establishment of Vila Rica in the 18th century present important references on the categories of products marketed in the colonial region. The author analyses in detail the establishment of the property of Mr. Jorge Pinto de Azevedo and identifies a wide variety of textile and other materials related to the production of clothing, such as clothes, and accessories among the various products sold there. The document identifies tissue as the most consumed genre in the store: approximately 65% of the total marketed genres and twenty times more than household utensils. Fabrics including the garments, threads, and lace came in second and third were the articles of clothing. This information is sufficient to prove the concern and effort to appear noble in the colonial lands of Minas Gerais, and the quest for good appearance, a conflict between being and opinion.

The papers of the commercial establishment of Mr. Jorge Pinto de Azevedo uncovered the variety of the commercialized fabrics: a total of 58 textile categories. Despite the much diversified demand, the *baeta*, the cloth, the velvet, and the nobility were the most sought after by the Minas Gerais consumers. The amount of each fabric and its value made it possible to verify the differences between genera for common use and those intended for luxury. Fabrics such as nobility, a kind of silk corrugation common in the silks drawn from the Spanish 18th century costumes, and even the *galacê,* and the chambray of India were of great value, even in small quantities, which supposes to treat the most luxurious textile genres among those commercialized. The *baeta* was the best-selling textile genre both in quantity and value. One cubit[1] of the fabric was sold on average for 800 *reis*. The *baeta* was a genus of thick fabric made of cotton or wool. It was used extensively in the making of pieces of clothing such as the rudder, the type of sweater or jackets, veils, and coarser pieces used by slaves, women and, children of lower classes as protection against the cold. It was a very versatile fabric that was not only used for garments, but also for interior decoration including in linens, upholsteries, and bed and table pieces. As regarded colours, they came in different shades of blue, green, red, yellow, orange, peach, clove, pinion, sulphur, vinegar, gold, silver, rose, lead and black. Among the shades of blue, the most sought-after colour of the *baeta*, the store offered ferrette, celestial, dark, clear, and simple blue. The greens were also sold in varied tones, sea, dark, the colour of parakeet, the colour of cane and simple green.

The cloth was another very common term among the categories of fabrics marketed in the commercial establishment in Vila Rica, second in value and fourth in quantity. According to the Dictionary of Authorities of the Royal Spanish Academy the term *paño* is related to a generic name for a textile product and is used as a synonym for fabric (REAL SPANISH DICTIONARY ACADEMY OF AUTHORITIES, 2016). Considering the values and quantities of the genera identified as cloth traded in the store in Vila Rica, possibly fabric of varied characteristics or distinct genres, it is believed to be the cloth which was already at the time a generic denomination for the fabric. Marco Aurélio Drumond raises this possibility, from the term cloth to have been used to denominate all the fabric not identified by the seller or inventory taker (Drumond, 2018), which is believed to be the correct one.

Velvet was the third most consumed textile in the commercial establishment in Vila Rica. The fabric was offered in wool, silk or cotton. Like the cloth, it embodied varied characteristics and could be beaded, carved, or worked. The most common colours were black, best seller, crimson, blue, green, cinnamon, fire, lead, and grey. Velvet was a noble fabric and it traded at 3,000 *reis* a cubit.

The most sophisticated fabric marketed in the store was the *Galacê*, or *Glasé* in Spanish and *Glacé* in French (D'ávila Corona *et al.*, 2004). This textile genre was produced with silk thread, usually in taffeta ligament because it was the most suitable for the fine clothes of the nobles. The fabric was widely marketed in Vila Rica and most commonly in combinations of gold and crimson, gold and fire colour, silver with green, silver with blue, or silver with yellow. The *Galacé* was a high-cost fabric; one cubit of the genre was sold for 8,500 *reis* and was present in the attire of the highest classes of the 18th century Ouro Preto society.

Another category of products sold in Vila Rica and also used for the noblest garments was the *gros of tours*. This was a fabric of French origin, quite common in the Spanish textile workshops during the 17th and 18th century and constantly used in dressing the nobles of the time. This fabric was a basis for several categories of washes, including the characteristic French and Spanish *chamalote*. The *gros of tours* is a genus produced from silk and characterized by a variation of the ligament, with a double passage of weft and with a corrugated effect. A cove of tours was sold for 300 *reis* at the Vila Rica store. The *gros of tours* is the same nobility or the common fabric of the 18th century television industry (Dávila Corona *et al.*, 2004).

Other fabrics such as damask, taffeta, drogue, silk crepe, bretagne, the bombazine or velvet cord, the fustian, the cheetah, and the morim produced from the cotton fibre, were among the textile articles traded along with various other species and therefore common fabrics in the dressing of the sophisticated man of the Colonial Vila Rica.

---

1. The cubit was a measure of length used by several ancient civilizations and based on the length of the forearm: from the tip of the middle finger to the elbow. It is not known when this measure came into use. The cubit used by the Babylonian Egyptians was an average of 50 cm, while that used by the Hebrews and Romans was an average of 45 cm.

Trade in good quality textiles in Minas Gerais was not restricted only to the surroundings of the capital of the province; the same variety of noble fabrics was also traded in commercial establishments in the Region of *Rio das Velhas*. According to Marco Aurélio Drumond's studies, the three stores owned by Mathias de Crasto Porto, the *Vila de Sabará, Roça Grande,* and *São Romão* sold varieties of taffeta, damask, nobreza, crepe, velvet, and chamalote among others (Drumond, 20018) in response to the idea of luxury and pomp that was prevalent in the colonial society of Minas Gerais. In the beginning of the following century, Auguste de Saint-Hilaire described the way women dressed in Villa Rica. According to the naturalist, these women were worthy of criticism from a Frenchman from Paris, and he was amazed not to see "so great a distance from the coast, a more sensible difference still between the ways of these ladies and those of the European" (Saint-Hilaire, 2000, p. 74).

3 THE INCONFIDENCE AND THE INCONFIDENTS IN THE VIEW OF THE AUTHORS AND ACTORS

The Inconfidence of Minas Gerais, a movement that occurred in the Brazilian state in the late decades of the 18th century, was the theme for several authors, at different times, reporting from different perspectives. Woven into a weave of double strands, the movement and its characters were widely explored by poets, artists, and historians. On the one hand, it triggers a fictional and poetic bias and on the other, the historical aspect is triggered. Among the figures of the inconfidents, the targets of an infinite amount of studies and essays, the image of the poet Tomás Antônio Gonzaga appears.

The character is present in such literary works as *Dirceu de Marilia*, Joaquim Norberto, *Romanceiro da Inconfidência*, Cecília Meireles (Meireles, 2008), *Gonzaga e a conjuração de Tiradentes*, Antônio Gonçalves Teixeira e Souza (Souza, 2017), *Gonzaga ou a revolução de Minas* (Alves, 1956), and two more recent works, Antônio Barreto's *Barca dos Amantes* (Barreto, 1996), and *Boca de Chafariz* by Rui Mourão (Mourão, 1991). The poems that express his love for his fiancée, Maria Doroteia Joaquina de Seixas Brandão, *Marília de Dirceu* (Gonzaga, 2006), published in Lisbon in 1792, and the *Cartas Chilenas* (1744–1810) (Gonzaga, 2006), one of the most emblematic satirical works of this period, composed of poems that circulated in the city of Vila Rica in the context of Inconfidence of Minas Gerais. For a long time, this work was considered anonymous.

The movement of the conjuration observed and analysed from a distance was recorded in the work of Englishman Kenneth Maxwell, *A Devassa da Devassa – A Inconfidência Mineira: Brasil e Portugal 1750–1808* (Maxwell, 1978). The book, published in England in 1973, presents the foreign look on the subject and proposes to present a revision to the official version of the movement. The author presents his hypotheses supported by documents and reasoning, reporting a more logical version of the events that informed the relationship between Brazil and Portugal, with a critical approach to historical facts.

Other authors and actors of the mining conspiracy have written important documents about the movement and these works are now available for interpretation and research. In 1955, Augusto de Lima Junior writes, at the request of the Governor of the State of Minas Gerais, Dr. Juscelino Kubitschek de Oliveira, *Pequena Historia da Inconfidência de Minas Gerais* (Júnior, 1975). In 1989, José Lino Grünewald organized the book The *Os Poetas da Inconfidência* (Grünewald, 1989), a book released for the commemorations of the bicentenary of the Inconfidence of Minas Gerais. This book contains the most significant aspects of the works of the inconfident poets, Alvarenga Peixoto, Cláudio Manuel da Costa, and Tomás Antônio Gonzaga, including a selection of the Chilean Letters which satirize, through allegorical verses, the political situation of the end of the 18th century. The work illustrates the historical moment in which the production of these poets took place, presenting in general lines, the Arcadianism.

Two more recent publications: *O Manto de Penélope* (Furtado, 2002) by João Pinto Furtado and *Os fios e os Bordados* (Oliveira, 2012) by Ilca Vieira de Oliveira, have dealt with the theme in a more realistic way. The first deals with the historiography of the inconfidence presenting a vision of a more humane character to the figure of Tiradentes, separating the memory of the myth, and showing a more reliable version of the history of the mining movement. The author refers very closely to the material conditions and structures of the possessions of the inconfidents, transcribing and analysing the goods sequestered by the Portuguese crown during the period between 1789 and 1792. The second book, *Os Fios e os Bordados* presents a detailed study of the traditional images of the poet Tomáz Antonio Gonzaga present in the Brazilian literary fiction.

Among a great number of publications on the subject, we highlight the document *Autos de Devassa da Inconfidência Mineira* (AUTOS DE DEVASSA, 1982), as the reference upon which this research is based. The eleven volumes of the publication deal with the proceedings of the legal process brought by the Portuguese crown against the ensign Joaquim José da Silva Xavier, the Tiradentes and other participants in the conjuration movement. The main purpose of the document is to investigate crime and treason, provided for in the Philippine ordinances and Portuguese legislation in force at the time.

4 THE MATERIAL CONDITIONS AND STRUCTURES OF THE POSSESSIONS OF THE INCONFIDENTS—THE CASE OF THE KIDNAPPING

Among the eleven volumes of the *Autos de Devassa da Inconfidência Mineira*, is an issue supervised by Heculano Gomes Mathias and dated 1982, the sixth volume in its entirety dedicated to the publication of the *Translados dos Autos de Sequestros* executed in assets of the defendants of the *Conjuração Mineira* of 1789–92, where he points out evaluations and assessments of the same, in compliance with the orders of the *Judges of the Devassas* opened in Rio de Janeiro and Vila Rica. The publication presents data related to the assets of the actors, the personal credits and debts of the victims, the nature of their movable and immovable assets, and a list of the collection of household and personal use of clothing, including white clothes, precious stones, jewellery and ornaments, and various assessments which formed the basic elements that supplied data and information for this research.

Contrary to the evaluation of several authors who identify the movement of inconfidence as a plot by scholars, one can clearly conclude from the possessions and wealth presented in the case, that many of the inconfidents were men of few words or using an expression of the time, rustic. The pieces of clothing and other items listed among each asset display various categories.

Of all the defendants of the inconfidence only José Alvares Maciel and José de Resende Costa did not have their property exposed to seizure. According to Rodolfo Garcia's explanation in the publication of the *Autos de Devassa da Inconfidência da Biblioteca Nacional* of 1937, "because they were family-children and lived under the patria power of their parents, the chief captain and their colonel and their names" (AUTOS DE DEVASSA, 1982, p. 15).

The inventories and deposits for the confiscation of assets of the defendants of the Conjuration of the Minas Gerais began in May 1789 at the home of Domingos Fernandes da Cruz, the ensign Joaquim José da Silva Xavier, or Tiradentes, in the city of Rio de Janeiro, where he had taken refuge and was arrested on May 10, 1789. There were seized an infinite amount of movable goods, silver and gold objects, bed and tablecloths, clothing and personal items, furniture, household utensils, ornaments, books, tools, and credit receipts. In April of 1792, Domingos Fernandes da Cruz was acquitted in the judgment issued in the records of the inconfidence coming from Minas Gerais regarding the return of his assets.

5 ILLUSTRATED INVENTORY CONFISCATION

The rare iconographic records about the dressing of the Minas Gerais region of colonial Brazil of the 18th century, as well as the dressing of society at the time, inform the work of recovering pieces and costumes by means of illustration based mainly in written documents, reports and texts of diverse natures. From these references, illustrations can be approximated based on images of paintings and engravings of the time as well as pieces preserved in museums, and clothing collections mainly from Europe, where there are a greater number of collections that comprise costumes and pieces of the 18th century.

Figure 1. Illustration of Marília de Dirceu dressed in the wedding dress from the poem "Dirceu bordando" and Tomás Antônio Gonzaga dressed in pieces described in the Autos de Devassa.

However much one tries to work within a historical and documentary context to create images through the interpretation of texts and approximation by similarity, it inevitably becomes an act of fictional construction. Moreover, even the cataloguing of clothing items requires a certain level of interpretation. Some terms did not designate at the time what they have now come to designate. In addition, there are names of parts and fabrics whose comprehension of meaning requires searching in specific documents. This factor made the historical, documentary, literary, and poetic texts interesting to the research.

If the relation of goods of the inconfidents can help to infer anything regarding their habits, customs, and social position, it is possible to risk using these goods to reveal a shadow of who these characters were. In a society in which the game of appearances guides social relations, the goods that are revealed, and even those that are not, it helps to compose the scenography and costumes of the great social staging and the roles played therein. In this way, both the civil subject character, Tomás Antônio Gonzaga, and the Arcadian Dirceu, in love with Marília, who in Lira XXXIV borders the dress of the beloved are of interest here.

Along this line of reasoning, two illustrations are presented in this article, one by Tomás Antônio Gonzaga wearing pieces identified in his relation of goods, and the other by Marília, the Arcadian name of Maria Doroteia Joaquina de Seixas, object of the love of the inconfident dressed in the unfinished dress in the poem "Dirceu Bordando" (Oliveira, 2012), by Bueno de Rivera.

The clothing collection of the inconfidents thus has more than textile pieces, pieces of clothing and pieces of accessories in the list of goods sequestered contained in the *Autos de Devassa*, to it are poetic objects and fictional constructions that the Devassa cannot seize. They are objects of fabrics made up of text.

6 FINAL CONSIDERATIONS

Despite the importance of the textile material and the knowledge of the colonial population's way of dressing in the history of Minas Gerais, there is still a lot to be researched about the subject. The studies or bibliographies on the history of Brazilian clothing are full of gaps and most of the time they illustrate misinformation already consecrated by society. Likewise, the literature on fashion, clothing, and textile materials and techniques deals almost exclusively with the dressing and customs of civilizations in Europe and Asia, without citing the customs, traditions and ways of living of the American people. Therefore, researching and identifying the clothing worn by the defendants of the mining conspiracy is of fundamental importance for the culture, the history of clothing, and Brazilian history in general.

There is a need to know the dress of the public man from the colonial period of Minas Gerais, contributing to a better understanding of the clothing and customs of the time, filling gaps and updating this subject with more accurate and truthful information. The images resulting from this research are characterized by dissemination and promotion of the culture of Minas Gerais in general, as an important means of recording and documenting the clothes of the unconfessed defendant, in order to fill this gap.

REFERENCES

Almeida, Carla Maria de carvalho. Homens ricos e homens bons: produção e hierarquização em Minas Gerais, 1750–1822. Tese de doutorado defendida na Universidade Federal Fluminense, Niterói, 2001.
Alves, Castro. *Gonzaga ou a revolução de minas*. Coleção Dramaturgia Brasileira. Rio de Janeiro: Editora Progresso, 1956.
Autos de Devassa da Inconfidência Mineira. Câmara dos Deputados do Governo do Estado de Minas Gerais. Belo Horizonte: Imprensa Oficial de Minas Gerais, 2a. Edição, Volume 6, 1982.
Barreto, Antônio. *A barca dos amantes*. Belo Horizonte: Editora Lê, 1996.
Dávila Corona, Rosa Maria; Duran Pujol, Monserrat; García Fernández, Máximo. *Diccionario histórico de telas y tejidos*. Salamanca: Junta de Castilla, Conserjería de Cultura y Turismo, 2004.
Drumond, Marco Aurélio. *Indumentária e Cultura Material: Produção, comércio e usos na Comarca do Rio das Velhas (1711–1750)*. Dissertação de mestrado apresentada na Faculdade de Filosofia e Ciências Humanas da Universidade Federal de Minas Gerais. Belo Horizonte, 2008.
Fonseca, Maria Cecilia Londres. *Tecelagem Manual no Triangulo Mineiro*. Brasília: IPHAN – 1984.
Furtado, João Pinto. *O manto de Penélope. História, mito e memória da Inconfidência Mineira de 1788–9*. São Paulo: Companhia das Letras, 2002.
Gonzaga, Tomáz Antonio. *Cartas Chilenas*. Curitiba: Educart, 2006.
Gonzaga, Tomáz Antonio. *Marilia de Dirceu*. São Paulo: Editora Escala, 2006.
Grünewald, José Lino. Os Poetas da Inconfidência. Rio de Janeiro: Nova Fronteira, 1989.
Januário, Erlaine Aparecida. A roupa como instrumento de distinção social dos segmentos sociais de pobres livres e libertos de Vila Rica (1789–1807) <http://www.cedeplar.ufmg.br/diamantina2004/textos/D04A075. PDF> acesso em 17.05.2017.
Junior, Augusto de Lima. Pequena Historia da Inconfidência de Minas Gerais. Belo Horizonte: Imprensa Oficial, 1955.
Maxwell, Kenneth. A Devassa da Devassa—A Inconfidência Mineira: Brasil e Portugal 1750–1808. São Paulo: Editora Paz e Terra, 1978.
Meireles, Cecília. Romanceiro da Inconfidência. Rio de Janeiro: Coleção Folha Grandes Escritores Brasileiros, 2008.
Mourão, Rui. Boca de Chafariz. Belo Horizonte: Villa Rica Editoras Reunidas, 1991.
Oliveira, Ilca Vieira de. Os Fios e os Bordados. Imagens de Gonzaga na ficção literária brasileira. Belo Horizonte: Editora UFMG, 2012.
Pereira, Alexandra Maria. Das Minas à Corte, de caixeiro viajante a contratador: Jorgje Pinto de Azevedo. Atividade mercantil e negócios na primeira metade do século XVIII. Tese de doutorado apresentada na Faculdade de Filosofia, Letras e Ciências Humanas da Universidade de São Paulo, 2013.
Pereira, Alexandra Maria. Um Mercador de Vila Rica: atividade mercantil na sociedade do ouro (1737–1738). Dissertação de Mestrado apresentada no Instituto de Ciências Humanas da Universidade Federal de Juiz de Fora, Juiz de Fora, 2008.
Real Academia Espanhola—Dicionário de Autoridades. <http://ntlle.rae.es/ntlle/SrvltGUIMenuNtlle?cmd=Lema&sec=1.1.0.0.0.> acesso em 06.06.2016.
Saint-Hilaire, Auguste. Viagem pelas províncias do Rio de Janeiro e Minas Gerais. Belo Horizonte: Itatiaia, 2000.
Salvador, Frei Vicente do. Historia do Brasil: 1500–1627. Editora Itatiaia, Belo Horizonte: 1982.
Santos, Antônio Fernando Batista dos. *Los tejidos labrados de España del siglo XVIII y las sedas imitadas del arte rococó en Minas Gerais (Brasil): análisis formal e analogías*. Tese de doutorado apresentada na Escola de Belas Artes da Universidade Politécnica de Valencia, Valencia, Espanha, 2009.
Souza, Antônio Gonçalves Teixeira e Souza. Gonzaga e a conjuração de Tiradentes. <http://www.caminhosdoromance.iel.unicamp.br/biblioteca/0221/Gonzaga.pdf> acesso em 10 junho de 2017.

The aesthetics in fashion design and cultural studies

T. Lobo

UNIDCOM/IADE – Universidade Europeia, Lisboa, Portugal

ABSTRACT: In this paper, we will explore the notion that beauty—an emergent value of human attention and emotional attraction—can help fashion designers plan and craft products that offer a rewarding, memorable encounter. Within fashion design studies, aesthetics continues to be one of the least-examined areas that deserve serious attention, and the research methods and interpretive techniques used by cultural studies can serve as appropriate models. The paper offers an overview of how cultural studies research may inform a specific aspect of fashion design, namely aesthetics. It also suggests that formalist evaluations of aesthetics and meaning, though valuable, are incomplete and need to be supplemented by other interpretive strategies that include social and cultural concerns. Engagement with the discourses outside the discipline can only enrich the discourse within, providing us with better tools for understanding the extent of the impact design has on the everyday lives of people.

1 INTRODUCTION

Aesthetics in design has been a neglected area of research, even though there has been some attention given to understanding the aesthetic qualities of the non-functional, "emotional" factors in design. (Norman, Donald A., 2004; Jordan, Patrick, 2000) Attempts to establish a scientific discourse for design have instead placed emphasis on analyzing and prescribing the *methodology* in designing (as in the practice-based framework of Design Methods) (Lawson, Bryan, 2005; Cross, Nigel, 2006); or the impact of *culture* and social processes on the making and consumption of design (as in studies of design history and the material culture of design, where matters of aesthetics are often consciously set aside due to an ideological struggle with the pervading notion of "good design" and its prescriptive aesthetics of outer beauty leading to moral improvement) (Forty, Adrian 2005; Attfield, Judy, 2000); or the issue of *meaning* in design—that is, how "form follows meaning"—and how design, on a semantic basis, makes sense in different contexts (e.g. contexts of use, language, life cycle, and ecology) (Verganti, Roberto, 2009; Buchanan, R., and V. Margolin, V., 1995). These positions have left out any analytical consideration of aesthetics. However, raising the issue of aesthetics in design is crucial, and not doing so leads to diffuse and sometimes unqualified discussions.

This paper, will establish a conceptual framework for discussing, theorizing, analyzing, and practically addressing aesthetics in design. It presents mainly the theory of phenomenology but also touch upon various aspects of the tradition of aesthetic theory in European philosophy. The paper aim is, however, not to use a philosophical, conceptual discourse to establish the "true" meaning of the word "aesthetic" to define it once and for all. The history of the concept itself has led in many directions—it was coined by Alexander Baumgarten in *Aesthetica* (1750–58) to describe a philosophical discipline that investigates the "lower" sensual aspects of human experience as opposed to the higher realm of logics. This led to the debate on taste and value judgment of beauty and the sublime in Kant's *Critique of Judgment* (1790), which preceded the close link between the work of art (Kant, Immanuel,1978: pp. 204–208) and the philosophy of aesthetics from Schelling's Romantic-idealistic celebration of the work of art in *Philosophy of Art* (1802) to Adorno's Modern-critical investigations of the communicative means and utopian potential of art in *Aesthetic Theory* (1970). (Adorno, Theodor W., 1998)

The paper's aim is to a new understanding of aesthetics in design will not go through the traditional discussions of art as a medium of aesthetic appreciation and communication, as these risks reducing design to a matter and medium of artistic aspiration. A design object can be the result of purely artistic and autonomous self-expression, but it often has a wider context. In relation to design methodology, it will be more justified to speak of design as a meeting point of multiple interests (those of a client, designer, and manufacturer) and as a complex negotiation between "problem formulation" and "solution generation." (Cross, Nigel, 2006) From a point of view of cultural analysis, design is a practice of innovation and change, not to be separated from the culturally circumscribed patterns of consumption. Further, an appropriation of design by the aesthetics of art, implying a view of design as art, may hamper an understanding of the unique complexity of almost every design object or solution: that design is not the expression of a lone artist, but the result of commercial and societal processes (Forty, Adrian 2005) and, at best, of an ambition to grasp the potential power of giving shape to our environments in innovative and progressive ways that are appropriate to human needs.

Aesthetics is no longer the exclusive domain of art but applies to our immediate, sensuous experience of the world.

To demonstrate these points, it will be examined two examples: the avant-garde fashion designs by the Japanese fashion designer, Issey Miyake and various Guinea textiles used by the African fashion designers.

2 FORM AND SENSUOUS EXPERIENCE

Evaluating aesthetics in fashion design is mainly a matter of grasping its sensuous qualities, or, rather, design's distinctive appeals to the senses. This does not mean that assessing aesthetic qualities in fashion design exhausts all the different properties that design encompasses (for example, functionality and sustainability). (Kyndrup, Morten, 2008) The purpose here is to explore how form and appearance can be qualified as means of a type of aesthetic communication that challenges experience, and to discuss the role of form as a challenge to our understanding of things.

These issues of form, experience, and understanding in design can be situated within two powerful frameworks. First, in recent years there has been a tendency to try to loosen the connection between art and aesthetic theory, and, to revisit Baumgarten's original idea of applying aesthetics to sensual matter (in Old Greek, *aisthetá*, "that which can be sensed"). This movement from Works of art to general sensuous experience and, further, to questions concerning how reality is arranged and perceived aesthetically, is the topic of a new era of aesthetic theory that has been unfolding since the 1990's in works by philosophers Richard Schusterman, (Schusterman, Richard, 1992) Martin Seel, (Schusterman, Richard, 1992) and Gernot Böhme. Tellingly, the title of one of Böhme's recent works features the Greek root of the word aesthetics: *Aisthetik*. (Seel, Martin, 2000).

Second, this bias of recent aesthetic theory can be seen in the contextualization of phenomenology as a philosophy that addresses the fundamental premise of the importance of experience and the basic conditions of experience. The term "phenomenology" was coined by the philosopher Edmund Husserl based on Old Greek etymology as the doctrine (*logos*) of that which shows itself (*phainomenon*). The point is that phenomenology, as a theory of experience, can address certain aspects of aesthetics related to sensuous appearance and experience. In the following, the paper will use the theory of the French phenomenologist Maurice Merleau-Ponty to discuss various modes of sensual qualities in design. In an important essay, "L'entrelacs—Le chiasme," (Merleau-Ponty, Maurice, 1964) Merleau-Ponty introduces two kinds of interlaced structures in experience to which it will be referred in the following discussion of two important aspects of aesthetics in design.

2.1 *An aesthetics of sensual relation*

Merleau-Ponty's first structure takes its departure in immediate and concrete experience. Here, Merleau-Ponty follows a basic assumption in phenomenology: That experience is a matter for a concrete and specific subject whose consciousness is incarnated in a body that is in a concrete world of things and intersubjective relations.

Reversely, the "world" is only ever a matter for a bodily-incarnated subject. Merleau-Ponty criticizes the traditional dichotomy of subject and object. Further, in a sort of deconstructive gesture he attempts to reverse the dichotomy to show that it has a common foundation in a figure of continuity that he calls the *flesh* chair. "He speaks of density of the flesh" ("l'epaisseur de chair") as a means of communication between the viewer and the thing.

For Merleau-Ponty: "The body participates in the order of things and likewise the world is universal flesh." (Merleau-Ponty, Maurice, 1964) Experience, in Merleau-Ponty's phenomenology, is an ongoing exchange between subject and object that takes place in the common material of "chair".

Almost as an explication of Merleau-Ponty's notion of "chair," the German philosopher Gernot Böhme has developed a powerful concept of ambience, *Atmosphere*, to analyze how things, situations, and surroundings appeal to us. The point is that ambience can only evolve if there is an experiencing subject. However, it is not an inherent part of the subject but rather objective as the result of an effect evoked by a specific constellation of things. (Böhme, Gernot, 1995) Thus, to Böhme the concept of ambience becomes the main designator for the conditions of perception, the "primary object for perception": (Böhme, Gernot, 1995) and even though they are not characteristics of the objects, they are obviously produced through the characteristics and interplay of objects. That is, ambiences are something *between* subject and object. They are not something relational; they are the relation itself .... it is the ambience the first reality of perception out of which subject and object can be separated. (Böhme, Gernot, 1995).

In this context, three aspects of Böhme's theory are particularly important:

- First, as a theory of sensuous experience and relation, to Böhme the main concern of aesthetics is how ambience Works and constitutes a specific relation between subject and object: "For aesthetics, the ambiences are therefore the first and essential reality. They are the perceptible co-existence of subject and object." (Böhme, Gernot, 1995).
- In Böhme's perspective, there might be a "real reality" behind the operations of ambience, but what is important for aesthetics is the "reality of appearance" which puts an emphasis on how (perception of) "reality" is mediated through ambience, on the effect of surface and form, and on the value of staging meaning. (Böhme, Gernot, 1995).
- Second, ambience is experienced and expresses itself as a coherent unit. Instead separating the various aspects of sensuous experience (i.e., sight, hearing, scent, etc.) and asking how one sense can evoke effects in another, ambience functions as the perceptual background upon which things and surroundings present themselves, and where one may look for sensuous differentiation. In this context, Böhme discusses the traditional aesthetic concept of synaesthesia and especially the power of colour. (Böhme, Gernot, 1995).
- And, third, ambience is not only something to be experienced but also something to be *made*, or manipulated. Böhme speaks of "aesthetic work," the intention of giving things, surroundings, and people certain qualities that let them appear as something special with a power of appeal to be perceived in a certain (controlled) way. (Böhme, Gernot, 1995) In this context, he mentions creative areas such as stage work, commercials, art, architecture, and design as examples. This notion of aesthetic work is clearly linked to today's prevalent concept of experience economy (Hagman, George, 1994; Schulze, Gerhard, 2005) and to the way in which our surroundings—especially with the help of design—is "aesthetically calculated," where the artefacts in question are conceived with a high degree of "aestheticity," construed to be perceived "aesthetically." Attfield, Judy (2007).

3 DESIGN AS A STRUCTURE OF APPEARANCE

The strength of Merleau-Ponty's phenomenological and Böhme's aesthetic-philosophical framework is that they conceptualize the relation of sensual experience between subjective apprehension and objective appearance. Merleau-Ponty thus follows the phenomenological dogma of reducing the world of phenomena to abstract to investigate the basic structure of experience. Böhme, on the other hand, through the notion of ambience, seeks to conceptualize the importance of the *specific* world we encounter, but in the end, he too remains in the realm of abstract speculation through his main philosophical interest in issues of, for example, the notion of perception. (Böhme, Gernot, 1995) In dealing with an increasingly designed and aesthetically staged world, we need more precise concepts to discuss the structure of appearance. In relation to this, in a philosophical, cultural, and material context, design is important as a major means of structuring the appearance and the surface that signifies "world" in our perception and cognition.

As an example of a kind of design that creates an ambience and thus stages a certain kind of relation between body and garment, Pleats Please clothing is innovative in its process: the clothes are first cut and sewn together from fabric that is nearly three times larger than the finished item of clothing, then sandwiched between sheets of paper and hand fed into a pleats machine. From tube dresses to cardigans, skirts, shirts, or elastic-waisted pants: the clothes emerge with permanent pleats. This industrial process allows both texture and form to be created at the same time. Vertical pleating is used to create different effects and architectural shapes. Pleats Please clothes are very functional and practical; they store easily, travel well, require no ironing, can be machine-washed, and dry within hours. Shapes are simple, and the colours and prints diverse (a set of basic colours is available each season, plus seasonal colours and prints). The clothes' simple beauty, comfort, lightness, and ease of care have changed the way many women all over the world dress. (Kitamura, Midori, 2012) Fashion design and textiles often evokes a high aesthetic effect of ambience because it can create an encapsulating and highly calculated setting. This is certainly the case in Issey Miyake's Pleats Please (Figure 1), (Figure 2), (Figure 3), and (Figure 4). Miyake's projects show design at its extreme, rethinking and remodelling our conception and perception of the structure of the garment (English, Bonnie, 2011) Miyake Since 1971 has been paving the way in design and technology, experimenting and expanding the relationship between the human body and clothing. Combining pleats with design, Issey Miyake's Women has been at the forefront of innovation opening new horizons for clothing for women. In the words of Kazuko Koike, "Issey Miyake brings the Japanese fashion designer's story to life through floating wire bodies and rainbow colour pleats. [...] Good ideas aren't born overnight and, as Miyake's chronology proves, creative genius stems from lifelong cultivation". (Koike, Kazuko, 2016)—a key word for Seel—of the world, in the sense that the world is given to us as "a momentary and simultaneous abundance of appearance", but also intensifies the appearance of the pure present that is otherwise inaccessible to ordinary perception". (Seel, Martin, 2000) Thus, to Seel, aesthetic perception is a matter of looking in a certain intent way that involves attention for the play of appearances. The focus is still on the given objects, which are simply seen in another way, that is, with an enhanced sense of the presence of the situation. (Seel, Martin, 2000) The argument in this context is that Miyake fashion s design points reflectively to itself and urges a kind of "aesthetic perception", apparently "wanting" to be perceived with an enhanced sense of presence, of being in exactly this ambience, here and now, and achieving exactly this through "designerly" and sensuous means such as colour, materials, and form. By combining these means into a whole, one can create not only fashion design but also a reflective aesthetic that questions how aesthetic is perceived.

3.1 *An aesthetics of communicative self-reflection*

Merleau-Ponty's second interlaced structure is also bound to concrete experience but should do with the way in which every concrete, visible manifestation carries with it an invisible idea or meaning.

He speaks of a bond "of the flesh and the idea, of the visible and the inner brace that the visible makes "manifest and hides," meaning that the idea is not the contrary of the sensual but instead it double and its depth." (Merleau-Ponty, Maurice, 1964) An additional point is that the idea, though always a part of the sensual, cannot reach the surface of direct manifestation; instead it operates as a "transparence behind the sensible." (Merleau-Ponty, Maurice, 1964) It is about this structure that the paper investigates the context of aesthetics and design. In the same way that the sensuous relation of an appealing objects and a sensitive subject can be called aesthetic. That Merleau-Ponty's notion of incarnated ideas can be applied to design is obvious: every piece of

Figures 1 and 2. Miyake's "Pleats Please" dresses.

Figure 3. Miyake's "Pleats Please".

Figure 4. Miyake's collection.

design contains an idea, a dimension of immateriality; vice versa, design is only conceivable as something concretely manifested—when speaking of immaterial design, Merleau-Ponty's structure of interlaced meaning indicates that it is nothing without some sort of physical manifestation. This could be considered a matter of communication, that is, specifically, how the relation of manifestation/idea displays itself in design. Whereas the question up till now has been how design establishes a sensuous relation with a perceiving and experiencing subject, the question now relates to the object itself, asking how the object in its sensual being points to a level of idea content or meaning, which, in a complex process of displacement, it simultaneously contains and conceals. (Verganti, Roberto, 2009) This aesthetic operation could be considered in two ways:

- First, it unfolds through the sensual being of an object, which links it to the aesthetics of the sensual relation.
- Second, the relation of physical manifestation and idea, which can be direct and problematic, has also been a topic of modern, art-based aesthetic theory. The question has been how the work of art is constituted through a specific "form" that (un)reveals its meaning and/or resists understanding. (Adorno, Theodor W., 1998; Bubner, Rüdiger, 1989; Bohrer, Karl-Heinz, 1998; Jakobson's, Roman, 1960) In the following, it will be focused this aspect under the heading of *aesthetic coding*, which examines how an object can not only attract attention and appeal to the senses (as in the sensual relation) but also be constituted in a way where it, in establishing a specific relation of physical manifestation/idea, demands or even commands a specific order of alignment or mode of understanding. Jakobson speaks of a self-reflective "poetic function," which in focusing on the act of communication itself could be activated within language, thus proposing "poetic language" to have a dominance of poetic function. (Norman, A., 2002) Thus, we can speak of objects with a high degree of "aestheticity," that is, with an implicit, communicative construction those points in this direction.

4 THE CONCEPT OF ADDED QUALITY IN AESTHETIC OBJECTS

How aesthetic objects contain something "more" has been a central topic of modern, art-based theory, from Schelling to Adorno. The ability to articulate this aspect has been one of the major benefits of this kind of theory and is far from obsolete today, although it may at one time have been too narrowly focused on art. (Menke, Christoph, 1999) Besides, it holds considerable potential for criticism of the operations and contexts of aesthetic phenomena—something that has been sorely neglected by the aesthetic theory directly related to design. Thus, in his influential *Aesthetic Theory*, Adorno discusses art as a medium that paradoxically is inevitably bound to the reality of the given, while at the same time having the potential to transcend the given. (Adorno, Theodor W., 1998) He says that "phantasy" cannot be "that cheap ability to escape being in proposing a non-being as if it existed"; instead it can transform "what the works of art always absorbed from being, into constellations, through which they become the other of being, is it also only through the specific negation of being". (Adorno, Theodor W., 1998).

Martin Seel claims that art's ability is to "bring forward otherwise unrepresentable circumstances". Art, in his view, has to do with: "…ways of human commitment in the real or the unreal, in conditions of the world in the past, the present, or the future. Objects of art are medium for an experience that takes place as a process of an understanding that isn't oriented towards a result of an understood … Understanding art is more about an otherwise impossible meeting with otherwise impossible possibilities of perceiving ourselves." (Seel, Martin, 2000).

As garments of everyday life, it may perhaps be difficult to see design in this context of an aesthetic negation of reality and proposals of new models of understanding. Still, though, it is worth asking designed objects the difficult question concerning how they define a relation to reality in the relation of physical manifestation/idea, and how they are mediums for meeting the world in new and/or reflective ways where new kinds of experience and of experiencing are evoked. In the case of Miyake, the conceptual framework of inquiring about the aesthetics of communicative structures can lead to different levels of questions.

Miyake's design is a provocative response to a climate of increasing and pervasive cultural conformity with little room for alternative ways of living. In this broad ideological context, Miyake's design, roughly speaking, proposes a new model for life. Second, we can observe how Miyake's design proposes new orders of experiencing and meeting the world. Miyake's design contains a strong and ideologically biased idea of using garments differently but only expresses this idea through a physical manifestation. In short, His design tries to lead us, "afford" us, (Curtis, William J. R., 1994) to live in new ways that could hardly be imagined *before* the realization and presentation of the design. In this sense, his design also encompasses a dimension of *performatively* implying an irreversibility of a "before" and "after"—the way we think of and experience design can never be quite the same again. Thus, it performs the new kind of being that it states on an ideological level. In and through its physical manifestation, Miyake's design not only suggests an idea of using garments differently, it fundamentally challenges our very understanding of design.

5 WORKING WITH AESTHETICS IN DESIGN

The paper argues that aesthetics in design is a matter of how design relates to meaning. It is not enough to ask *what* the meaning of a specific design is on a conceptual level (the "idea"), we must also ask how it performs or reflects this meaning in its physical form, and how it relates to the kind of self-reflective "aesthetic function" where it displays a surplus of meaning.

In this way, discussing aesthetics in design is a way of consciously focusing on dimensions of meaning in design, but also, on behalf of the designers, on the construction of meaning. How can a surplus of meaning be invested in design, and how can it be reflected in an actual piece of design?

Miyake points to one possible direction in allowing the basic idea to be so pervasive and effective in his design that it not only stands behind the sensual relation of creating a garment but also produces a surplus of meaning on an ideological level of a different way of using them.

As with Miyake, the idea pervades and determines the design, d in both cases there is an almost perfect integration of idea and physical manifestation—the idea is only relevant in so far as it is "put to work," and the physical expression of form has hardly any relevance without an idea or meaning content. In my view, this is a hallmark of aesthetics in design. But where Miyake's design reflectively points to the fact that there is idea operating in and through the design (clear in the way her design, appealing directly and aggressively to the senses, performs the utopian idea of a different way of life), sometimes in other fashion designers the idea is a more subtle, pure form experiment. This structure of investigating how an idea can be reflected in the design and how it can create a surplus of meaning (that is, the overall aesthetic question of how design relates to meaning on a general level) can not only be described in design, it can also be used more actively (by designers) as a tool of reflection in the design process.

In relating these two aspects of design as aesthetics of communicative self-reflection, where the x-axis represents the relation to the "aesthetic function," (that is, the degree of surplus of meaning in relation to functional qualities) and the y-axis represents the reflection of the idea, it is possible to see how different kinds of design communicate differently aesthetically. This coordinate system encompasses different modes of aesthetics:

6 FRAMEWORK FOR CONCEIVING AESTHETICS IN DESIGN AS THE FORMULATION AND CONSTRUCTION OF MEANING

Way of reflecting the idea

"Functionality" is not opposed to "aesthetics" as such but according to the two axes has its own kind of aesthetics with a non-surplus in the appearance of the sensuous relation. Designs in this category include the purely functional design of everyday objects that may also reflect the idea content in different ways. At one end of the spectrum there is *anonymous* design, where we simply see through the inherent idea; at the other end of the spectrum there is the kind of *functional* design that displays its idea in a way that only reflects that there *is* an idea but which also, through this mechanism, often explains itself in a process of "natural mapping." (Norman, A., 2002) Likewise, there can be (as described in the cases of Miyake) different modes of aesthetics linked to a great surplus of meaning and appearance at one end of the spectrum there is the purely conceptual design, which does not, however, entirely circumscribe the modality of Miyake's highly sensuous experiments, but which is prevalent when the conceptual aspect is formulated on the ideological level. The other end of the spectrum is where most "life style" design is found, a type of design that uses a high degree of outer appearance with a surplus of appeal to the users, or rather consumers, and where it is not important that the underlying idea is reflectively stated. Other fashion designers the design could be more experimental than "life style", but he operates with the same approach of indirectly putting the idea to work. The experimental focus of the design of other designers could *challenge* the relation of idea and physical manifestation so that the idea does not take over but has the status of Merleau-Ponty's inner structure, manifesting and hiding itself at the same time. In Other designer, the aesthetics in design could express as an ongoing dialogue of outer appearance, constantly hiding and revealing its meaning content.

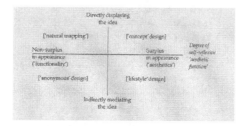

Figure 5. Framework.

7 CONCLUSIONS

The theoretical framework proposed here can be used in analyses and discussions of aesthetics in design, but it can also inform designers who need to deal practically with the challenges of the aesthetic in design. The aspects of aesthetics in design that are put forward in this paper—design as a structure of sensual appearance, and design as an act of communication that may contain an aesthetic coding that lets an idea or content of meaning be physically manifested and reflected in different ways—can lead to a more theoretically focused inclusion of aesthetic matters in the process of designing.

Thus, indicating how the questions raised in this paper can be turned into a series of aesthetic challenges for designers, that will conclude it. The first issue is the challenge to work consciously and strategically with the *sensuous* impact of design, that is, to draw specific attention to the nature and function of the sensual when designing. In this way, the concept of "ambience" can become an important addition to the toolbox of design methodology. Further, we may consider how an object can be designed to urge a kind of "enhanced perception." This does not, however, necessarily mean that design needs to flash and mark itself as "design;" it can also be accomplished in the anonymous design of everyday objects through more subtle aesthetics and a more discreet appearance. However, it may prove productive to challenge the aim and scope of design and its means of creating an entire universe of sensuality, as demonstrated in the case of Miyake's design, where the power and importance of a sensual relation are achieved through designedly means. In sum, these instruments can be used as an aesthetic challenge to the conventional way of conceiving design and how it is created, thus facilitating the overall development of designedly and practical means of addressing aesthetics in design.

REFERENCES

Adorno, Theodor W. 1998. *Aesthetic Theory* (pp. 122–258). Minneapolis: University of Minnesota Press. Attfield, Judy. 2000. *Wild Things* (pp. 125–28). Oxford: Berg, pp. 125–28.
Böhme, Gernot. 1995. *Atmosphäre. Essays zur neuen Ästhetik* (p. 33). Frankfurt a.M.: Suhrkamp.
Bohrer, Karl-Heinz. 1998. *Die Grenzen des Ästhetischen* (pp. 87–89). München: Carl Hanser Verlag.
Bubner, Rüdiger. 1989. *The Innovations of Idealism*, (pp. 124–125). Cambridge University Press.
Buchanan, R., and V. Margolin, V. 1995. The Proposition that 'Design is Making Sense (of Things)', in *the Idea of Design*, eds. London: MIT Press, pp. 156–184.
Cross, Nigel. 2006. *Designerly Ways of Knowing* (p. 34). London: Springer Verlag.
English, Bonnie. 2011. *Japanese Fashion Designers: The Work and Influence of Issey Miyake, Yohji Yamamoto and Rei Kawakubo*. New York: Bloomsbury Academic.
Forty, Adrian. 2005. *Objects of Desire* (p. 45). London: Thames and Hudson.
Gallimard. Norman, A. 2002. *The Design of Everyday Things* (pp. 9–13). N. Y.: Basic Books.
Hagman, George. 1994. Aesthetic experience: beauty, creativity, and the search for the ideal (pp. 29–42). N. Y.: Editions Rodopi B.V.
Jakobson's, Roman. 1960. Closing Statement: Linguistics and Poetics". *In Style in Language* (pp. 350–77). ed. Thomas A. Sebeok, Cambridge: MIT Press.
Jordan, Patrick. 2000. *Designing Pleasurable Products* (p. 55). London: Taylor & Francis.
Kant, Immanuel, 1978. *The Critique of Judgement* (pp. 204–208). N. Y.: Oxford University Press.
Kitamura, Midori. 2012. Pleats Please: Issey Miyake. New York: Taschen, Mul edition.
Kyndrup, Morten. 2008. *Aesthetics and border lines: 'design' as a liminal case* (p. 106). Copenhagen: Gyldendal.
Lawson, Bryan. 2005. *How Designers Think* (pp. 203–205). Oxford: Architectural Press.
Menke, Christoph. 1999. *The Sovereignty of Art: Aesthetic Negativity in Adorno and Derrida* (pp. 241–24). Massachusetts: The MIT Press.
Merleau-Ponty, Maurice. 1964. *Le visible et l'invisible* (pp. 170–201). Paris: Éditions
Norman, Donald A. 2004. *Emotional Design* (p. 9.). N. Y.: Basic Books.
Schulze, Gerhard. 2005. Die Erlebnisgesellschaft: Kultursoziologie der Gegenwart (p. 67). Frankfurt: a.M. Campus.
Schusterman, Richard 1992. *Pragmatist Aesthetics: Living Beauty, Rethinking Art* (p. 46). Oxford: Blackwell.
Seel, Martin. 2000. *Aesthetics of Appearing* (pp. 75–76). Stanford: Stanford University.
Verganti, Roberto. 2009. *Design Driven Innovation* (p. 241). Harvard Business Press.

The physiological aspects of senescence: Comfort and relationship with the clothing

M.D. Almeida
University Sagrado Coração, Brazil
PPG Design (UNESP), Brazil

A.C. Broega
University of Minho, Portugal

M. Moura
PPG Design (UNESP), Brazil

ABSTRACT: The clothing for the elderly is still a product to be questioned as to its levels of comfort and satisfaction regarding the physiological and ergonomic needs of the user. Thus, this study intends to raise discussions about the natural aspects of aging as well as how the clothing must correspond satisfactorily to the needs of this public in terms of comfort.

1 INTRODUCTION

In Brazil, the elderly person is not considered a predominant consumer by the fashion industry. This might happen, on one hand, because this clothing market is not attractive, and, on the other hand, because these items of clothing do not have the commercial value desired by the fashion industry, which raises the question about the levels of comfort offered by such products. The contemporary elderly person has undergone significant changes, for example, an extension of life expectancy, which leads to the increase of the elderly public. The tendency is that there will be an expressive increasing amount of wearers with peculiar demands that are specific to this age group and very different from the other people.

Thus, this study aims to understand the physical and mechanical needs of the contemporary elderly, as well as to evaluate the comfort levels of the elderly clothing. The hypothesis is that although there are studies and research addressing the textile area, there is a research lag in the clothing sector, which demonstrates the need for innovation and for the presentation of products that have a significant value for the elderly public, fulfilling their needs beyond the physiological ones.

Therefore, firstly, we will approach the characteristics of the contemporary elderly public in an analytical way; secondly, we will discuss the definition of comfort and its parameters; and finally, we will discriminate the relationship between fashion design and the ergonomic aspects of comfort.

2 THE CONTEMPORARY ELDERLY

The contemporary reality regarding the aging of the Brazilian population has led the reflection on the future of the elderly since we are facing a new social formation, structural (body) and psychological, which is very different from the elderly portrait of the past decades. Becoming an elderly person is no longer a chronological aspect since we cannot consider someone an elderly exclusively because of the age. According to Moragas (2010) and Okuma (1998), the chronological age is no longer enough to determine the senescence, which is an

> *Important data, but not determining—other personal and environmental conditions are added to determine the general state of an individual* (Moragas, 2010, p. 20).

Moragas (2010) adds that what matters is the quality of the time lived, the experiences, and the local environment that embraces them. Thus, there are other parameters to be considered.

> *Aging is also experienced "from the inside", through physical manifestations* (Caradec, 2011:25).

The physiological matters concerning the elderly body are under-discussed when it comes to clothing development. Although there are doubts, in some areas, about which factors can be considered normal of the aging and which are pathological, it is correct to affirm that aging isn't and shouldn't be considered a contracted illness, but a crucial chronological process. First of all, to differentiate pathology from natural aging, it is convenient to associate the absence of acute illness with the quality of life (Moragas, 2010). Caradec adds (Caradec, 2011:25):

> *Firstly, the organic body register refers to health and physical abilities. In this register, it is expressed, on one hand, the statement of preserved physical skills and the absence of illnesses and, on the other hand, the debilitation of those skills and even functional limitations and illnesses, which are considered to be signs of aging.*

Aging is the phase of life in which significant physical, cognitive, and social changes occur; however, this still is a natural process. In the past decades, it would be normal for the elderly to wait for death. But nowadays, because of the increase of life expectancy, the valorization of the well-being and quality of life requires more attention, as well as motivational elements, and an emphasis on social actions and more inclusive public policy.

As they are more active, the "new elderly" present new needs, mainly regarding the satisfaction with their body changes since the organic modifications are perceptible, for example, in the skin. The skin is the target of the cosmetic industry, which offers product lines that claim to enhance the skin aspect and others that try to delay aging. However, there are other aspects to be addressed (Figure 1), such as a considerable loss of muscle tone, stooped posture, weakening of the bone, loss of balance, walking pattern changes, and slower movements (Netto, 2004).

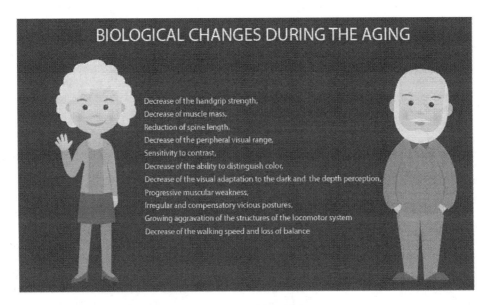

Figure 1. Biological changes during the aging (Almeida, 2016).

Although the elderly are more active and perform many activities nowadays, there are biological factors regarding their body health that emphasize the need to observe everything that surrounds their relation with the environment because, in this phase of life, the individual is more susceptible to disease sequels and effects that can be cumulative (Moragas, 2010). The elderly aren't so physically strong and may have their health harmed by a series of factors. However, there are other determinants, such as environmental ones, which may influence their health in a positive or negative way.

Caradec (2011) points out that there are few researchers who study the elderly body. The author believes there are two questionings. The first concerns the devaluation of the body, especially the female one. The terms used by Goldenberg (2012) are "invisible" and "unclassified". This is how society treats the elderly, who aren't taken into consideration; however, at this moment, there isn't a classification because the creation of such names to the contemporary elderly is no longer convenient.

The aesthetic standards of the eternal youth raise the pressure to avoid aging. Thus, although the elderly in their sixties seem to be more vigorous, the stereotype of the tired elderly is passed to the following age range. The second questioning concerns the phenomenological point of view, in other words, the physical experience of aging and the interpretation of the aging signs. The author mentions that physical aging is composed of three registers: organic body, appearance, and energy,

> *which express concerns about health, beauty, and shape, respectively.* (Caradec, 2011:25).

When we think about the body considering its registers, we think about its maintenance and preservation. The science of comfort becomes the determining factor for the analysis of all the parameters to measure the elderly comfort levels.

3 COMFORT

Despite of being easily felt, comfort is a subjective parameter, and to reach its concept can be even more complex. This debate has already been raised by authors such as Lueder (1983) and Noyes (2011), among others; however it is based on the study of Van Der Linden, Guimarães, and Tabasnik (2005) that we can find fundaments to reach a concise definition (Van Der Linden, Guimarães and Tabasnik, 2005:2):

> *[...] comfort is a mental status which occurs in the absence of uncomfortable feelings. This model was accepted by other researchers, assuming the existence of a continuous that varies from the extreme comfort, with high levels of comfort; going through an indifference state (neutral point); to the most extreme discomfort, with many levels of discomfort.*

Comfort is correlated to a process of mental perception that is communicated by the senses. According to Broega and Silva (2010), it is more easily identified when the feeling of discomfort is described negatively, that is, we easily notice the absence of comfort. To measure the aspects of comfort, Van Der Linden, Guimarães, and Tabaski (2005), list three dimensions: the physiological, psychological, and physical. However, considering the clothing area that we will address in this study, we will consider the four aspects of comfort (Table 1) raised by Broega and Silva (2010). The factor included by the authors is one more point that can satisfy the needs of the clothing and individual relationship.

Based on these four topics, it is possible to analyze the objects in order to verify how much it can be comfortable for an individual. The idea of harmony indicates that all the dimensions must be fulfilled (Van Der Linder, Guimarães, and Tabasnik, 2005:2), since they act as a combination of several factors that make an object acceptable or comfortable in its several forms.

3.1 *Comfort and clothing*

The clothing is something that is constantly in contact with the skin of the individual who wears it and it acts as a "second skin", as observed by Martins (2008), in which the epidermis inhabits and creates a very close relation. Thus, the clothing must provide satisfaction and well-being to the individual. The author affirms that a clothing project must fulfill all needs, capacities, and limitations related to mobility, age range and the activity performed.

The clothing is a protection of the body, which offers favorable conditions for the physical performance and must assure in its textile layers the circumstances for the organism survival (Broega and Silva, 2010). The authors also affirm that designers give relevance only to the visual and aesthetic fields, neglecting other senses such as smell, hearing and touch, which are very important to ensure a bigger interaction in all dimensions of comfort of the clothing and the individual (Table 2).

Thus, clothing must correspond to all the comfort levels, ensuring the usability of the object. To evaluate all these parameters might be a complex and exhaustive task, but it is relevant to prevent complicating effects in the usage of the clothing, such as causing damage to physical and emotional health. Therefore, the comfort evaluation must have scientific basis, with less subjective parameters. Another way is to find a way to make them more quantifiable, since nowadays its evaluation is made by an empirical and hedonic way. (Broega e Silva, 2010:4):

> *All these variables contribute to the high evaluation complexity and quantification of the comfort of the clothing, which until nowadays, has been rated by customers, manufacturers, and fabric producers in an empiric way, by the "touch" and by the "well-being when dressed". It is about an evaluation that is completely subjective, based on sensations and experiences, without any scientific basis.*

Comfort is one of the most important aspects of the clothing as it is close to the body. The answer of discomfort is often an instant one; however, the evaluation of the object must be conducted accurately and concisely. This is a reality that is not observed in the manufacturing industries, but that is becoming more and more necessary, especially with more inclusive products. There is a demand for clothing products that fulfill the wearer needs, especially

Table 1. Dimensions of comfort (Broega and Silva, 2010).

Comfort	Description
Thermo-physiological	It translates a thermal and moisture state to the surface of the skin, which involves the transfer of heat and steam.
Sensory of "touch"	The set of several neural sensations, when a surface comes into direct contact with the skin.
Ergonomic	The ability an object presents to allow the freedom of movement of the body.
Psycho-aesthetic	A subjective perception of aesthetic assessment, based on vision, touch, hearing, and smell, which contribute to the overall well-being of the wearer.

Table 2. Answer of the clothing to comfort (Vianna and Quaresma, 2015).

Comfort	Clothing
Thermo-physiological	Relation with the sensations of heat and cold, demanding fabrics that provide the well-being through their natural or technological characteristics, such as the threads of irregular surfaces that facilitate the circulation of air.
Sensory of "Touch"	It is connected with the fabric softness that is a result of its composition.
Ergonomic	It includes all sensations that are connected to the shaping and sewing of the clothing.
Psycho-aesthetic	Textile type material, the modeling of the clothing on the body, the choice of colors, and the smell of clothes worn.

the elderly, and satisfy the needs of performing daily tasks that are vital, for example, simply dressing a jacket without aid. The elderly public, as observed in the first topic, shows a change in its lifestyle, but such fact seems to be invisible to society. In this regard, the present study aims to elucidate arrangements to be taken in the product design to the elderly people to give them more autonomy and quality of life.

4 THE INCLUSIVE CLOTHING DESIGN AND THE COMFORT OF THE ELDERLY

In the field of design, there are studies that worry about the elderly in projects of environments, packaging, and graphic design; however, in the fashion design, there aren't parameters that fully and efficiently meet this population.

The fashion designs are recognized by presenting advertising speeches with lively aesthetic standards. Machado (2005) states the senescence condition does not fit the current stereotype in the media.

It is true that there is no formal prescription to address all the current challenges (Cardoso, 2012) regarding the product design requirements, especially for this group of wearers (the elderly), whose health and a social condition can be very diverse, according to different lifestyles and economic power. Thus, noticing their needs and limitations is a challenge. It is time to consider the opinion that these wearers make of the products that are directed to them.

Aging is inevitable; therefore, it is necessary to prepare products that fulfill the demands of this public, who is increasingly high in economic power and has the right to current products with a fashion design that addresses their specific bodily needs (physical and physiological) and sociological needs, and that does not include them in a stigmatized group, outside the media universe. It becomes necessary for markets (and the means of production) to observe their behavior and to understand that their requirements have changed.

It is important to elucidate the point of view of Krippendorff (2000), who defends another parameter for the design, which is to be centered in the humans, in their present needs. This author states that we have been through a phase in which the design system was object-oriented (discussions on form and functionality were raised) and strongly supported by the industrial era, but that we are currently undergoing a phase of changes and have to take consciousness of the meaning of something when it changes in accordance with our way of living. Therefore, we no longer react to the physical qualities of things, but to what they mean to us (Krippendorff 2000: 89).

Thus, we ask about the design that proposes the inclusion of the elderly: is it being inclusive or exclusive? What if it is necessary to change things to make our lifestyle easier? We believe the design in all its aspects has presented practical and viable solutions, for example, the case of kitchenware, which initially tends to develop ways to include another group of individuals, the elderly, but end up universalizing more the product because it makes the daily routine easier.

However, when we try to observe this in the clothing products, the speech may be different, the concept of inclusion in fashion can take the notion of exclusive products, often stigmatized. Regarding the elderly, it is not just about embracing the idea of physical needs, they lose their aesthetic appeal and fail to be fashion objects. Ideally, objects should continue to stimulate individuals, and the project should be implemented to make the product effective for all wearers.

A piece of news published in Folha de São Paulo (2016) reported that many elderly people have great difficulty finding out clothing that suits their needs: pieces either look too youngster or "super-elder". However, nowadays, the elderly are more demanding, they do not accept objects that do not fit in their universe.

The absence of more inclusive (but not stigmatizing) products in the clothing sector makes us rethink the current design and marketing behavior since we cannot remain stuck in face of social changes and new paradigms—we must (re) design their language and discourse, be continuously critical, not accepting the impossible or dubious assumptions (Krippendorff, 2000). Thus, an inclusive design must respond in a more direct way to the needs felt and expressed by the population, bringing the community closer to the decision-making capacity and reinforcing their sense of responsibility (Simões and Bispo, 2006: 46).

The Researchers Simões and Bispo (2006) claim that social inclusion originates more balanced societies, which provide a better lifestyle—not only for the disabled and the elderly but to society as a whole.

Putting communities, instead of individual wearers in the center of the concerns of the design opens the way to something totally different (Krippendorff 2000, p. 92).

Therefore, inclusion in the clothing for the elderly is not only about appropriate measures for the development of a product; it is the design of clothing thinking about physical diversity or age group. However, in the situation of unpreparedness in which companies find themselves, they end up not seeing the current scenario, leaving the elderly in the margin, just making adjustments of what already exists. The redesign practiced is not enough, a new attitude of innovation is necessary.

Thus, questions of the physiology of the elderly should be taken into consideration, since the levels of comfort are correlated with the clothing, as previously discussed. Based on the authors Vianna and Quaresma (2015), Broega and Silva (2010), we developed a table (Table 3), with certain points to be met in the development of a clothing product for the elderly wearer.

Thus, with these questions to be raised in the clothing product design, the object gains significance to be more inclusive, is not just a redesign, because, when the wearer needs (namely the elderly) are known. the relation with the object of clothing becomes more trustworthy, bringing welfare and, consequently, more comfort, which involves aesthetic aspects, but also formal and functional ones through modeling and proper fabric, suitable to a mature public (Vianna and Quaresma, 2015). Completing these questions (Broega and Silva, 2010:4).

The body-clothing interactions (thermal and mechanical) play very important roles in determining the comfort state of the wearer, as well as the external environments (physical, social and cultural)

Table 3. Comfort of the clothing for the elderly.

Comfort levels	Requirements to be fulfilled
Thermo-physiological	It is related to thermal sensations (of heat and/or cold) to the skin's surface with the textile material, which allows an appropriate humidity control. This derives from the porosity of the textile structures, the way the fabrics provide well-being through its natural or technological characteristics. For example, fabrics that are manufactured from the threads of irregular surfaces that ease air circulation.
	The combination of porous textile surfaces made of natural fibers (or a mixture of natural and synthetic ones) may favor thermal comfort.
Sensorial "touch"	The sensorial comfort is related to the fabric softness which is the result of its composition. The skin of the elderly is drier, so the fabric pleasant "touch" is fundamental for avoiding skin irritation and providing a sensation of comfort. Softer textile materials that have a "pleasing touch" can be achieved with special trims and/or with a functional mixture of natural/synthetic fibers, as well as with other solutions more oriented to the fabric structure.
	There is a decrease in the tactile sensitivity and in the grip strength; therefore, the use of trims with more noticeable textures, less abrasive, and with an unusual use can offer significant changes.

(Continued)

Table 3. (Continued).

Comfort levels	Requirements to be fulfilled
Ergonomic	The ergonomic comfort is related to the modeling and manufacturing of the clothing, and it must provide an adequacy to the new configuration of the elderly body in order to offer well-being and freedom of movements. It can be conditioned (or optimized) by thermal and tactile factors of the textile materials that constitute clothing.
	With the slowness in walking, the loss of muscle tone, and the movement limitation, clothing must be thought in terms of guaranteeing an increasing amplitude of movement, through looseness, modeling allied with the right choice of textile materials. Clothing items must ensure greater comfort.
	However, modeling must also pay attention to the posture impairment, to body axis levels such as knees and elbows, in which there is a tendency to flexing, and consequently a tendency to the excess of textile materials in those regions that can be configured in probable harmful effects.
Pyscho-aesthetic	The object appearance is a crucial factor so the wearer can feel included; therefore, the adequacy of the language of the individuals with the surroundings in which they are inserted is fundamental. The elderly will feel comfortable and safe with the pyscho-aesthetic comfort, being able to express their own personality, individuality, and style. They will recognize the kind of textile material desired, the adequacy of the clothing item on the body (to dress well), the option for colors that they like to see themselves, and also the feeling of safety transmitted bt the exemption of unpleasant smells in the clothes (a sensation that can be undetected by the lack of olfactory sensitivity).

5 CONCLUSION

The relationship between clothing and comfort is a vitally important factor in product design; however, this is true only in the textile area. There is still few references to comfort in the design area of the clothing product, and this is a gap that still has issues to be researched, such as the aspects of comfort parameters for the development of more inclusive clothing to the contemporary elderly. There is also a need to understand and analyze the universe of the elderly. The lack of knowledge of the comfort of the elderly clothing demonstrates how fashion design needs to improve its products, make them appropriate for an audience that has a series of physical and physiological urgencies, as well as other pertinent psychological needs, which are not met maybe due to a lack of preparation in the area. The most worrying situation is the disregard of the fashion industry for this "new market niche"; leaving a growing elderly population, who is enjoying an increasing economic power, to the edge of fashion, at least as far as to the comfort-fashion satisfaction binomial.

REFERENCES

Broega, A. C.; Silva, M. E. C., 2010. O conforto total do vestuário: design para os cincos sentidos. *V Encuentro Latinoamericano de Diseño*. Buenos Aires.

Caradec, V., 2011.Sexagenários e octogenários diante do envelhecimento do corpo. In: Goldenberg, M. *Corpo, envelhecimento e felicidade*. Rio de Janeiro: Civilização Brasileira, p. 387.

Cardoso, R., 2012. *Design para um mundo complexo*. São Paulo: Cosac Naify.

Folha De São Paulo., 2016. *Idosos reclamam da falta de produtos adequados*. Disponível em: <http://www1.folha.uol.com.br/fsp/mercado/183649-falta-produto-adequado-a-3-idade-diz-consumidor.shtml> [Acessado em 09 de junho de 2016].

Goldenberg, M. (2011). *Corpo, Envelhecimento e Felicidade*, 07–19. Rio de Janeiro (RJ): Civilização Brasileira.

Krippendorff, K., 2000. Design centrado no ser humano: uma necessidade cultural. *Estudos em design*, Rio de Janeiro, v. 8, n. 3, p. 87–98, Setembro—dezembro.

Lueder, R. K., 1983. Seat comfort: A Review of the Construct in the Office Environment. *Human Factors*, v. 25, n. 6, p. 701–711.

Machado, M. C., 2005. *Análise ergonômica em uma instituição geriátrica*: estudo de caso. 2005. Dissertação (Mestrado em Engenharia de Produção) – Universidade Federal de Santa Catarina, Florianópolis.

Martins, S. B., 2006. Ergonomia e usabilidade: princípios para o projeto de produtos de moda e vestuário. *Congresso Brasileiro de Ergonomia*, ABERGO. Curitiba.

Martins, S. B., 2008. Ergonomia e moda: repensando a segunda pele. In: Pires, D. B. *Design de moda*: olhares diversos. São Paulo: Estação das Letras e Cores, p. 319–336.

Moragas, R. M., 2010. *Gerontologia Social*: envelhecimento e qualidade de vida. São Paulo: Paulinas.

Netto, F. L. D. M., 2004. Aspectos biológicos e fisiológicos do envelhecimento humano e suas implicações na saúde do idoso. *Pensar a prática*, Goiânia, v. 7, n. 1.

Noyes, J., 2001. *Designing for humans*. Hove (East Sussex): Psychology Press.

Okuma, S. S., 1998. *O idoso e a atividade física*. Campinas: Papirus.

Simões, J. F.; Bispo, R., 2006. Design inclusivo: acessibilidade e usabilidade em produtos, serviços e ambientes, Lisboa: Centro Português de Design.

Van Der Linden, J. C. D. S.; Guimarães, L. B. D. M.; Tabasnik, R., 2005. Conforto e desconforto são constructos opostos? *3º Congresso Internacional de Pesquisa em Design*. Rio de Janeiro: [s.n.]. p. 1–8.

Vianna, C.; Quaresma, M., 2015. Ergonomia: Conforto têxtil no vestuário do idoso. *15º Ergodesign/ Usihc*. Recife: [s.n.], p. 1–9.

Fashion design through sustainability and material culture: An exploratory and experimental study

Rafaela Norogrando
ID+ Research Institution for Design, Media and Culture, Portugal
FibEnTech-UBI, Portugal

ABSTRACT: The present article was based on projects developed within autarchic professional spheres and consists of an exploratory and experimental study of fashion design. We aim to demonstrate the potential of using material culture not only as a visual reference, but also to attribute value and create product differentiation.

Contemporary fashion market is permeated with brand massification in a globalized and pasteurized logic. In this sense, this work intends to present the strategic difference between a design approach that integrates material culture and/or its agents and an approach that perceives cultural heritage merely as a methodological phase of the creation process.

We conclude by presenting a study based on a design methodology that upcycled a Portuguese cultural identity into a fashion product—which was integrated into the heritage collection of the National Museum of Costume (Portugal).

1 SUSTAINABILITY, EMOTION AND MATERIAL CULTURE

The concept of sustainability is intrinsic to the concept of design when seen only as a way to solve a problem and contribute to people's quality to life. In this sense, it is possible to find and combine several theorists who, in general or with specific perspectives, have explored design theory through a more holistic, innovative and responsible approach, such as Victor M. Manon, Ezio Manzini, Michael Braungart, William McDonough, Victor Margolin, Alistair Fuad-Luke, among others.

When considering fashion design, or even fashion as a social phenomenon, we can find many approaches regarding its history or theory. But it is only in the beginning of the 21st century—and especially after its first decade—that we notice an effective development and registration of practices. In this period, some theoretical works brought about new possibilities that went along with unsustainable actions and were more suited to serve the causes of a fervent capitalist world and its globalized gears (Black, 2008; Fletcher, 2008; Blanchard, 2008).

With the disasters that took place in Bangladesh in 2013 and the surfacing of situations of slavery in the fashion industry, new approaches have gained strength and visibility and have broadened the critical sense about the ethical commitment of each agent of the system—from designer to consumer. For designers, choice awareness has been reinforced, given that much of what involves the production chain and processes can be reviewed to better consider ethical relations between workers and options for decreasing waste. In addition, human creativity, which is heavily involved in the training of these professionals, is considered an important economic resource—as seen by the reflections of Richard Florida since 2002 (The rise of the creative class). For consumers, their relationship with the concept of sustainability permeates individual choices and changes in habits, which require not only an empathy for the cause, but also a conscious, critical and emotional commitment.

To meet all these events and reflections, the Fashion Revolution movement that began in 2013 has been rapidly globalizing. Throughout the year and in several regions of the world, it promotes events for the assessment of values and attitudes on behalf of everyone involved. Its mission statement says that "it wants to unite people and organizations to work together to radically change the way our clothes are supplied, produced and consumed, so that our clothes are made in a safe, clean and fair way" (Fashion Revolution website). 02/26/2018).

In a truly iconic manner, the exhibition Redefining, Redesigning Fashion: Designs for Sustainability, held at the Goldstein Museum of Design from January to May 2013, graphically highlighted the five themes that structured the exhibition. The themes were conceived through curatorial brainstorming and debates with a select group of designers justifying their products/projects (DeLong, Heinemann, Reiley, 2014). The majority of themes were related to design methodologies (choice of material, processes, techniques, etc.) and the articulation of the production chain (for example, local suppliers). However, the icon depicting a heart demonstrates the emotional and human component in the bond that exists between person and object. That is, from a perspective of extending the life cycle of a product, emotional relations can be a key component to add value and credibility to the sustainability of the object, since its disposal is delayed. Moreover, its life cycle is revised (restarted) in function of its implicit emotional appeal. Thus, emotion is one of the catalyst elements, bringing about feelings that, when evoked, can perpetuate convictions and stimulate actions.

In this sense, companies are increasingly being advised about the need to review their values and complement their commercial activities with less short-term profit-oriented principles and more long-term prospects—with attitudes that aim to consolidate an ethical stance (Eckschmidt 2017). This situation is reflected in the behavior of consumers that want more information on what they buy, or even to buy less. Thus, against the notion of fast-fashion that disqualifies clothing—and our relationship with it—as "soulless and homogeneous" products (Fimmano in OVNE 2017), an inverse order is obtained: a more inclusive, identitary and consequently emotional order, with a recovered sense of caring for our second skin.

2 FASHION DESIGN

The use of references for the design of new products is a major part of design methodology. In fashion, the appreciation of images and objects from an earlier period or from other cultural manifestations is a common practice that provides inspiration and helps renew contemporary perspectives. Nowadays, accounts of designers' collections are widespread (Salles, 2015), providing us with a good record of the recent past[1]. This reflexive and critical approach to past material culture or ethnic cultural manifestations can also be expanded to post-fashion situations, for example, when narrated in curatorial moments in museums or other exhibitions (Norogrando 2015).

When considering the Portuguese context and the development of objects and fashion collections, we can say that there has been a steady growth in the recording of conscious actions in favor of an appreciation of local culture in recent years. It is important to mention that a new age for Portuguese fashion has developed since 1974, with greater freedom, democratization and visibility, as well as the consolidation of new authors and brands in national and international contexts (Duarte, 2004).

We shall now present some works that were inspired by the material culture of certain regions of Portugal, also taking into consideration the reflections on the fashion design final product. In other words, local cultural artifacts and manifestations are not only references for the creative process, but also raw material for the creation of fashion coordinates in a reinterpretation of these artifacts and manifestations for the contemporary market. Some of the examples brought here were carried out through partnerships with local authorities and intended to promote local development, economic growth and the perpetuation of immaterial

---

1. As, for example, the Mariano Fortuny artifact collection belonging to the Museo Del Traje (Madrid) collection.

culture by promoting the knowledge of techniques and the survival and appreciation of these techniques and the people who practice them.

2.1 *Portuguese heritage in fashion design*

Among a multitude of cultural manifestations that the Portuguese territory provides, much remains to be explored in the fashion industry. Following are some very pertinent actions, sometimes strategically planned or else randomly disseminated. We also draw attention to the monitoring and registration of actions by researchers, providing a scientific perspective to help mobilize a better appreciation of cultures and peoples.

In the Portugal, one of the strongest examples is the so-called "Valentine's handkerchief" with its peculiar embroidery, or only written words, having been explored in several products: a planned initiative that overcame copyright issues due to its popularization. In fashion, they were particularly notable in the collections of Nuno Baltazar, Anabela Baldaque, Filipe Faísca and others (Norogrando and Broega 2012).

To exemplify cases that involve actions that go beyond the design process (the inspirational/conceptual phase—or stamping), we can mention two cases in which the local public authority also assumed some responsibility towards the material and immaterial culture of its territory, dignifying actions, people and economies.

The first case is the Guarda Municipal Council project, supported by fashion designer Alexandra Moura. Through her fashion shows in the 2016 Fashion Weeks of Lisbon and Porto, the designer shed light on the huge potential and creative possibilities of the "*cobertor de papa*", by using it in her autumn/winter collection (Cruchinho, Peres, Moura 2016 and Jornal Público 2016). A local action subsequently took place at the Municipal Theater, with the exhibition of pieces by the stylist and students of the School of Applied Arts of Castelo in the "*Cobertor de papa é moda*" fashion show (images from the Terras da Beira Newspaper, 2016). This action can be considered a design process due to the choice of material, which is used to fabricate the so-called "*cobertor de papa*" (sheep's wool manufactured with a specific method). Even before this movement to rescue material and immaterial cultural heritage, the "*cobertor de papa*" had already appeared on the catwalks ten years earlier, in 2007, with Filipe Faísca also promoting the regional heritage, and again in 2018 at ModaLisboa autumn/winter 2017/2018 (Diário de Notícias, 2018).

With regards to the *lérias* embroidery, the Fundão Municipal Chamber and the Póvoa de Atalaia and Atalaia do Campo Parish Council joined to organize the "*Lérias* – Art of the Lines" Fair and the "Traditions of Beira Interior in Fashion" contest that takes place since 2015. This competition including young students of fashion design has the support of the University of Beira Interior and the Polytechnic Institute of Castelo Branco, and results in a fashion show of selected works (according to information contained in the Fundão City Hall website).

Fortunately, many other cases could be presented and we hope that actions in this direction will be increasingly explored (not only in Portugal), since they promote the growth of local identity in its conception of value, as stated in the heart of the goals of immaterial cultural heritage promoted by UNESCO (Pelegrine and Funari 2008).

3 FASHION DESIGN PROJECT + ARTIFACT OF MATERIAL CULTURE

As an attempt to verify artistic and design feasibility, an exploratory and experimental study was carried out aiming to transpose a piece of Portuguese material culture into a fashion design object for contemporary users. Since the purpose of the study was to build upon an existing out-of-use artifact of material culture, this implied on finding exclusive solutions to face the reality of the situation concerning the design and unique character of the product.

Having this in mind, the purpose of this fashion design project was not directed toward mass production of the fashion coordinate, but primarily for upcycling a cultural identity through a theoretical and practical study. As a result, we found that the process could lead to low-scale production and, even in the case of a collection, each product would be the result of appropriate adjustments to the raw material used.

3.1 *A product to be worn: Target audience and use situation*

The project was intended for use in real situations. It was made for a (Portuguese) girl in the age group of 2 to 3 years, to be used in ceremonial occasions in the spring/summer in Germany. The choice for this target audience was, at first, random, based on the opportunity presented. However, when it comes to clothing, size involves investment in material and labor and a project for this type of public would provide a more effective and detailed action, given the proportions to be worked. Moreover, as evidenced by almost a decade of professional experience in the fashion industry, objects for children and infants instill memories and emotions that bring about more empathy from different publics, which could reinforce the possibility of greater acceptance of the design proposal of this study.

3.2 *"Raw material culture": The artifact as inspiration and textile support*

As raw material and source of research and inspiration, we selected an accessory of Vianese women's clothing, characteristic of the northern region of Portugal. This choice was based on the fact that this dress, with emphasis on the red *Lavradeira* costume, is an important icon of Portuguese identity and imagery[2], according to the study presented in "*Uma imagem da Nação–Traje à Vianesa*" (Medeiros, Pereira, Botelho, 2009).

Given the complexity of the costume in question, we decided to use the apron, since it presents characteristics of the chromatic and textile visual language of the cultural heritage under consideration. We also decided to escape from elements that were already overly explored, such as the pouch, or more commercialized, such as the scarf printed with flowers and adorned with fringes.

As a conceptual prerequisite, the apron should not be made exclusively for the project, which could be a possibility, or with definitions of size and other details made by companies dedicated to this—such as the Folklore Store, among others[3].

Figure 1. *Left:* Vianese Costume displayed in the Costume Museum of Viana do Castelo. Source: Olhar Viana do Castelo (2018); *Right*: Typical costume apron purchased for the project.

---

2. I do not intend to discuss here Portuguese image(s) and identity(ies), which is a much deeper topic than the extent of the Lavradeira costume. I consider here the historical records of this relationship, according to the references presented, and the caricatural dimension and collective memory of the icon.
3. https://www.lojadofolclore.pt

Having in mind a conceptual stance in favor or recovering material culture in association with concepts and processes of sustainability, the raw material had to be second-hand. That is, it had to be a traditional costume apron no longer in use. With that in mind, we chose an online platform as search and purchase tool, where the transaction could be made directly. In OLX.pt, it was possible to scroll through a vast collection from which to choose. We purchased a *lavradeira* apron in the color red, which, according to the seller's indications, was over 40 years old.

This search and purchase option was made intentionally in order to provide the fashion design product development process with an element of the contemporary practices of communication and marketing. The objects are seen through a screen, the transactions are online and the product delivery is made by couriers or delivery companies, without intermediaries between the consumer and the original product.

3.3 *Upcycling of a cultural identity: Adaptations of the concept to the practice of design*

At first, we opted for a classic language of children's dresses with straps and a wide skirt. For our conceptual objectives, one of the difficulties in the adaptation of the "raw material culture" to the project was to fulfil the idea of maintaining the memory of the artifact without losing its characteristics. This can be better verified in the back of the hem of the dress, where the original finishing was maintained with the application of a new intervention (Figure 2). In the chest of the dress, in order to adjust the old apron to the correct measures of the object (a children's dress), no cuts were made, but the structure itself was explored and the dimensions reduced by pleating.

In order to compose the other parts of the dress, since only the front of the apron was designed, we sought to maintain the color red, but with a different fabric that had textures that differed from the apron in textile design, albeit still dialoguing. The option was for vertical lines and, in order to emphasize this graphic choice, we explored the surface design with the incorporation of other elements that reflected part of the complex textile work of the apron. To soften the ethnographic character of the product, we opted for a different color composition, with more emphasis on red and pink (with different saturations and luminosity). We also made more use of dark blue to soften the green so strongly present in the composition of the *Lavradeira* apron. Orange (applied to the lining and to the ends of the straps at the back of the dress) was used as a transition point between red and yellow—heavily used in the apron's surface/textile design).

The white fabric lining, stitched together with invisible seams, was also deliberately chosen. Instead of being smooth, it refers to finishings made on other pieces of clothing. It also received a finishing of white lace tape in the hem of the dress (as can be seen in Figure 3 *Right*).

For the straps of the dress, which retained the original color of the ribbon tying the apron to the waist, a pin was applied to adjust the height of the handles as the child grows or for donation in a perpetual collection of stories about the artifact, the object and the dress.

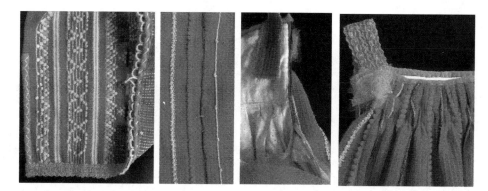

Figure 2. "Tati-Viana" dress details.

Figure 3. *Left*: Red Bustle, Yohji Yamamoto, 1986. Photo: Nick Knight. *Right*: side of the "Tati-Viana" dress.

The pin was adorned with red and orange tulle so as to maintain a certain monochromatic character in the upper part and create a meeting point (through the color orange, since it has yellow in its composition) with the green stripe available only in the back sides. They are only there as a link between the front of the dress (richly adorned and colored, including yellow and green) and the back (predominantly red).

3.3.1 *Assessment of a design as a function of the outcome of clothing design*
It is important to remember that the product design process predicts the verification of the prototype before the final phases. In the case of our study, it was necessary to assess the initial idea already in the final stage of preparation, due to the trimming the fabric provided in a "vanishing" motion, which constrained all fluidity and relaxation needed for a project intended for children and also impelled the work towards a conservative version of party dresses for girls, which did not contribute to the purpose of the project. This result was only possible at an advanced stage of the production process with the connection of the side of the dress of the old apron (comprising the front) and the back.

As an alternative, the use of the apron as the front was assumed more consciously and clearly, with the sides stitched together only by long ribbons and more tulle frills introduced in the lining in an inspirational reintepretation of the look produced by Yohji Yamamoto, punctuating visually deep contrast zones.

3.4 *Global project: The dress + the act of dressing*

Although we usually could consider the project complete with the completion of the dress, the same cannot be said for the public, given that the act of dressing requires a clothing composition.

In this sense, and taking into consideration the cultural origin of the chosen artifact, the inspirational source for the overall composition was also extended to the whole Vianese *Lavradeira* costume, including all accessories and their materials and characteristic expressions.

However, making choices that were similar to the inspirational outfit would result in a caricatural dress, semiotically bringing about the idea of a costume, which was not the intention of the project. For this reason, the chromatic weight given to the lower accessories, such as sock and shoe, was softened. Socks remained white and were designed not to cover the entire leg, softening a possible direct interpretation. The shoes chosen were not single color and texture (usually black clogs for the *Lavradeira*). Instead, we opted for a bicolor design with leather in crude lace fabric and golden toecap. This choice of color was a reference to the value of the Vianese costume accessories, with many gold chains and filigrams.

Figure 4. Images of the photographic edition of the "Tati-Viana" global dress project: costume composition with children's dress produced with artifact of Portuguese material culture. Photos: Vava Etchepare (June 2016/ Siegburg).

Conversely, for the upper part of the body we chose to use more saturation in a monochromatic composition to distance it from the visual idea of the traditional Vianese costume composition, which uses a white shirt. Therefore, the dress was composed with a pink shirt—darker than the apron, in cotton fabric in a circular mesh; and a red bolero with a metal blend of straight-stitched long-sleeved knits and frills on the shoulders (worn over the undershirt and under the dress). This piece of clothing was also purchased in a second-hand store (brechó/second life store), specialized in children's articles, in order to reinforce the concept of the full composition of the project.

The same tapes used in the dress were used in a simplified language to produce the headdress. No excessive production was made in order to escape the "doll-like" patterns of the children's figure and to maintain, even in a ceremony suit, an element of relaxation—natural of the age of the public in question.

4 FINAL CONSIDERATIONS

Based on the reflections and practices here presented, we were able to demonstrate the pertinence of approaches that aim to rescue material culture through fashion design. There may be different reasons behind this type of action, with many different levels of intention. However, a strategic action—going beyond the research and creation phase in the design process—can contribute in a broader way for sustainability and to provide added value to the product through ideas that involve cultural heritage, people and creativity.

We believe that our study can be explored in a professional and academic context, given that it provides a practical approach to the valuation of cultural artifacts, a stimulus toward creativity in the face of the challenge of upcycling and building consciousness concerning designer performance in the context of the fashion industry. In addition, the continuity of this type of laboratory practice/workshop by students of fashion design, scientists and/or professionals may provide possibilities for research and documentation of technical and aesthetic knowledge of material cultures through design. In this sense, the presented study tended to open a new strand of work to the *Xi-Coração Litle Project* (Norogrando 2016), started in 2016 and with 3 editions carried out by the author's undergraduate students.

The dress/composition "Tati-Viana" was integrated into the collection of the National Costume Museum (Lisbon, Portugal), and is therefore a culture object now accessible to researchers and the general public.

ACKNOWLEDGMENTS

To Uwe and Carmela for the opportunity, to Ana Norogrando for making this project viable, to Mafalda Grigoletto Goellner for participating and contributing in the process and Vava Etchepare for conceiving the photo editorial with the little Francesca.

REFERENCES

Black, Sandra, 2008. Eco-chic. The fashion paradox. London: Black Dog publishing.
Blanchard, Tamsin, 2008. Green Is the New Black: How to Change the World with Style. London: Hodder & Stoughton.
Câmara Municipal do Fundão. 2017. Concurso "Tradições da Beira Interior na Moda – 2017" Available at https://www.cm-fundao.pt/municipionews/concurso_tradicoes_beira_interior [Accessed 26 February 2018].
Cruchinho, Alexandra & Peres, Paula & Moura, Alexandra, 2016. *Cobertor de Papa* applied to Fashion Design. Global Fashion Conference. Available at http://gfc-conference.eu/wp-content/uploads/2017/01/CRUCHINHO_ET_AL_Cobertor_de_Papa_applied_to_Fashion_Design.pdf [Accessed 26 February 2018].
DeLong, Marilyn & Heinemann, Barbara & Reiley, Kathryn. 2014. Redefining, Redesigning Fashion: Designs for Sustainability. Fashion Practice: The Journal of Design. 6. 10.2752/175693814X13916 967095073. Available at https://www.researchgate.net/publication/262580097_Redefining_Redesigning_Fashion_Designs_for_Sustainability [Accessed 26 February 2018].
Diário de Notícias, 2017. ModaLisboa prenuncia estação fria cheia de cor, 12/03/2017. Available at https://www.dn.pt/sociedade/interior/modalisboa-prenuncia-estacao-fria-cheia-de-cor–5719447.html [Accessed 26 february 2018].
Duarte, Cristina, 2004. O que é moda. Lisboa: Quimera.
Eckschmidt, Thomas. 2017. Palestra no Fashion Revolution UBI—II Jornadas de Moda. [vídeo em fase de edição, publicação prevista para 2018].
Fletcher, Kate, 2008. Sustainable Fashion and Textiles: Design Journeys. London: Earthscan.
Jornal Público, 2016. Jornal Público, 5/3/2016. Available at http://lifestyle.publico.pt/noticias/358856_cobertor-tipico-da-regiao-da-guarda-chega-a-moda-pela-mao-de-alexandra-moura [Accessed 26 February 2018].
Jornal Terras da Beira, 17/7/2016. Available at https://www.facebook.com/195678817128420/photos/pcb.1355675924462031/1355674437795513/?type=3&theater [Accessed 26 February 2018].
Medeiros, António; Pereira, Benjamim; Botelho, e João Alpuim, 2009. Uma Imagem da Nação: Traje à Vianesa. Viana do Castelo: Câmara Municipal de Viana do Castelo.
Norogrando, Rafaela & Broega, Ana Cristina, 2012. O lenço dos namorados, Revista dObra[s], vol 5, nº 12. São Paulo: Estação das Letras e Cores. ISBN 9771982031115 Available at https://dobras.emnuvens.com.br/dobras/article/view/85/85 [Accessed 26 February 2018].
Norogrando, Rafaela, 2015. Narrativas patrimoniais sobre moda: análise das temáticas expositivas e das escolhas museográficas, Revista dObra[s], vol 8, nº 8. São Paulo: Estação das Letras e Cores. ISBN 9771982031115 Available at https://dobras.emnuvens.com.br/dobras/article/view/109 [Accessed 26 February 2018].
Norogrando, Rafaela, 2016. Xi-Coração Little Project. Available at: http://littledressproject.wixsite.com/xi-coracao.
Olhar Viana do Castelo. 2018. Traje à Vianesa: Certificado para garantir autenticidade e preservação das tradições. Available at: https://olharvianadocastelo.blogspot.pt/2017/02/traje-vianesa-certificado-para-garantir.html [Accessed in 26/02/2018].
OVNE. 2017. OVNE Entrevista, Philip Fimmano: "Produtos sem alma não atraem mais as pessoas" 7/04/2017. Available at https://medium.com/ovne/produtos-sem-alma-não-atraem-as-pessoas-3cf60219fb7f [Accessed 26 February 2018].
Pelegrini, Sandra C. A.; Funari, Pedro P., 2008. O que é patrimônio cultural imaterial. São Paulo: Brasiliense.
Salles, Manon, 2015. O colecionista Walter Rodrigues. In Márcia Merlo (org.), Memórias e museus. São Paulo: Estação das Letras e Cores. ISBN 9788568552117.

Castelo Branco embroidery applied to clothing

A. Cruchinho, A.S. Marcelo & P. Peres
ESART/IPCB—CIPEC, Portugal

A. Moura
ESART/IPCB, Portugal

ABSTRACT: The main question of this research is: will it be possible to develop garments with the application of Castelo Branco embroidery? It is around this question that research objectives have been defined. Castelo Branco embroidery is an identifier of the culture of a region in the interior of Portugal and presents itself as a reference of albicastrense cultural heritage and traditionally it was used mainly in quilts. The wealth of this patrimonial element lies in the raw material—silk and linen—but also in the vast diversity of elements and motifs that together, control a narrative that may vary according to the composition chosen. This richness is also represented in the vast palette of colors and the diversity of stitches that constitute it. In terms of methodology, a bibliographical review about Castelo Branco embroidery was first developed. Secondly, a briefing was prepared to present to Fashion Designers. The new products presented should reflect the use of an element of culture and the tradition of a region in a contemporary approach where Design differentiates and presents itself as a differentiating element, important for added value and sustainability.

1 INTRODUCTION

The main objective of this research is to develop products in the Fashion Design field, applying Castelo Branco embroidery, an important element of culture and identity of the region of Castelo Branco.

The richness of this heritage element lies in the raw material—silk and linen—but also in the diversity of elements and motifs that, together, build a narrative that may change according to the chosen composition. This richness is also represented on the diversity of colors and stitches.

Castelo Branco embroidery was named after this city because it was in Castelo Branco that its production was resumed in the beginning of the twentieth century. Quilts or kilts are the master pieces of this embroidery style which, in Portuguese territory, is considered "a obra prima da arte de bordar" (Ormonde, 2008, p. 16), "as mais ricas, vistosas e de mais complexa feitura" (Pinto, 2004, p. 15).

Athough Castelo Branco embroidery has always been associated with quilts, gradually finding a place in panels and smaller decoration artifacts, in this project we tried to integrate it in fashion.

The challenge was launched to Luís Buchinho and Alexandra Moura, two Portuguese fashion designers. The designers should be acquainted with the type of embroidery and its applications and develop a proposal for a piece of cloth integrating it. The embroidery should be made by professional embroiderers and the final product would be later presented in a fashion show.

The methodology for the implementation of this project was the following: in the first place, after the identification of two fashion designers that would develop the proposal, to bring them to Castelo Branco, so that they could collect all the information they needed to develop their proposal. The fashion designers were invited to visit Francisco Tavares Proença Júnior Museum, where they could find some copies of embroidered quilts and familiarize themselves with the various embroidery motifs and different compositions.

The Silk Museum, was another place to visit. In this museum, designers could learn about the history and origin of silk, production methods, features and applications, followed by a visit to the Interpretation Center of Castelo Branco Embroidery, where embroiderers are the main characters. Here fashion designers would have a closer contact with Castelo Branco embroidery, knowing stitches, types of linen used in quilts, patterns, colors, etc.

After the visit to collect information, we've tried to answer the main question of this research, with the development of the creative process made by the designers. They should take into account the characteristics of the embroidery, its application, target public and available market, as well as costs involved in the production of the final product.

As Galliano mentioned, quoted by Seivewright (2008, p. 7) "A investigação criativa é o segredo e o truque que realça todo o design original." (Creative research is the secret and the trick to enhance any original design).

Each fashion designer has different characteristics and ways to express his/her creativity. In this article we chose the project presented by fashion designer Alexandra Moura. The choice fell on her proposal too, because it was the one that presented a product closer to embroidery identification, not only at the level of stitches chosen, but also motifs and the raw material used in its design – linen.

Finally, some conclusions will be presented about the research carried out and the implementation of the practical project.

2 CASTELO BRANCO EMBROIDERY IN ALBICASTRENSE CULTURE

Research carried out in this area stitces to the year of 1891 as the date of the first written reference that associates this type of embroidery to the city of Castelo Branco, in an article published in Correio da Beira newspaper (eg Pires, 2008, p. 11); "elemento patrimonial definidor da identidade cultural (*de Castelo Branco*)" (Rechena, 2007, p. 7), (after a century it has continued to assert itself as the patrimony element defining Castelo Branco cultural identity).

However, the history of the quilts nowadays denominated Castelo Branco quilts dates back, at least, to the XVIII century; some studies stitch to the possibility of its production and commercialization having already begun in the XVII century. However, the lack of information and written documentation makes it impossible to establish a more accurate date for which the motifs used in this embroidery are the main resources to be considered (eg Pinto, 1993; Melo, 2008; Pires, 2008). In its genesis there are markedly Eastern influences (China, India, Persia,...) and European ones, that allowed the creation of sumptuous objects not confined to the quilts function (IMC/Francisco Tavares Proença Júnior Museum, 2007; Pires, 2008); Vila Tejero, 2008).

The largest production of Castelo Banco embroidery took place during the seventeenth and eighteenth centuries, in workshops that worked under the orders of professional designers or flossers (IMC/Francisco Tavares Proença Júnior Museum, 2007, p. 11; Ormonde, 2008, p. 11). During the nineteenth century the production decreased considerably (Pinto, 1993); the abandomn or lack of interest in this technique even led to the use of some Castelo Branco quilts as lining to other quilts, such as damask ones (Pinto, 1993, p. 11). In this period there was a generalization of the production in non-workshop contexts, although there is no official confirmation that it was only domestic production (IMC/Museu Francisco Tavares Proença Júnior, 2007, p. 11).

In the 1940s and 1950s, in the historical context of *Estado Novo*, some workshops units were created in Castelo Branco, of which the *Oficina da Mocidade Portuguesa Feminina* stands out, with an emphasis on handmade production embroidery and the *Casa Mãe de Elísio José de Sousa*, which already combined modern technology with traditional embroidery techniques (IMC/Museu Francisco Tavares Proença Júnior, 2007, p. 12). In this local regionalism period, Sales Viana, Luís Chaves and José Ribeiro Cardoso led a movement of identity appropriation of the embroidery from the region of Castelo Branco "o processo de apropriação identitária do Bordado pela região e pela cidade de Castelo Branco, (*o que*) levou inclusive a incluir nesta designação, muitas peças que a investigação comprova hoje não o serem" (IMC/Museu Francisco Tavares Proença Júnior, 2007, p. 12). One of the most widely circulated versions, which served the political-economic objectives of *Estado Novo* advocated the designation of engagement quilts which were placed on the bridal bed and then stored in trunks. Thus, the folklore, traditionalist and popular dimension of an icon materialized as an invented tradition and adapted itself to taste changes (Melo, 2008, p. 146). On the other hand, the recognition of the symbolic value of these Castelo Branco embroideries was felt with their integration in the curricula of technical schools, in 1952 (Melo, 2008, p. 156).

In the decade of 70, the Revolution of the 25th April 1974 caused the extinction of "Mocidade Portuguesa Feminina"; in 1976 the Regional Embroidery School ("Oficina Escola de Bordados Regionais") was created in Francisco Tavares Proença Júnior Museum (IMC/Francisco Tavares Proença Júnior Museum, 2007, p. 12), reinforcing its role as a promoter of Castelo Branco embroidery (Melo, 2008, p. 157).

In the twenty-first century Castelo Branco embroidery was consolidated as a live cultural patrimony, breaking with atavisms and blockages from Estado Novo period "património cultural vivo, rompendo com regionalismos atavismos e bloqueios práticos que vinham do período do Estado Novo" (Melo, 2008, p. 158). The publication in 2007 of the Brochure "*Castelo Branco Embroidery—Technical Specifications Manual*", by the Institute of Museums and Conservation/Francisco Tavares Proença Júnior Museum, demarcated the district of Castelo Branco as the production area of this embroidery. On August 24, 2016, a positive opinion was given to the registration of traditional embroidery production in Castelo Branco by the Advisory Committee to the Certification of Traditional Craft Productions, following the request made by the municipality of Castelo Branco. On July 25, 2017, the Embroidery Interpretation Center of Castelo Branco was opened and in a near future an application of this embroidery to Intangible Heritage of Mankind is planned.

In terms of this embroidery certification, in an extract of the Anex from Despacho nº 15559/2016, December 27, there are three main categories, defined from the existing ones in the Francisco Tavares Proença Júnior Museum: classical embroidery, those pieces that faithfully reproduce the embroidery produced in the 17th, 18th and 19th centuries; classical recreation embroidery, those pieces that are inspired by the old embroidery but introduced some change; contemporary creation embroidery, those pieces that open the way to innovation, either going deeper at the level of design, structures and colors (when proposed by artists or professionals from arts and design fields) or in terms of application of drawings on different applications and with diversified functions (IMC/Museu Francisco Tavares Proença Júnior, 2007, p. 97).

For the appreciation of Castelo Branco embroidery contributes the use, since its origin, of two natural raw materials: linen, in taffeta fabric, and silk, in twisted thread. "The linen cloth was generally made up of several cloths to the desired width" (Pinto, 1993, p. 13). This fabric presents shade variations from darker, in historical and classical pieces, to lighter ones, in classical recreation pieces and contemporary creation, and is also dyed blue or brown, in some parts; the silk is presented in both natural and dyed tones (IMC/Museu Francisco Tavares Proença Júnior, 2007, p. 15).

Castelo Branco recovered silk production through an institution—APPACDM (Associação Portuguesa de Pais e Amigos do Cidadão Deficiente Mental) which supplies extra quality and wild silk "produces several (...) kilograms of extra-quality silk and (...) of wild silk" (IMC/Museu Francisco Tavares Proença Júnior, 2007, p. 18).

The production of a piece of Castelo Branco embroidery requires a set of steps, which are described in detail in the Technical Specifications Manual, and which are summarized below: (1) compose or select a drawing, (2) decalate the design to the linen fabric with the help of chemical paper and a round tip object, cut and polish the fabric, (3) fasten the piece to the frame (a large piece can be embroidered by several people—up to six people, (4) embroider the garment from the left to the right, and from the bottom to the top, with silk thread, (5) detach the garment from the frame, sew the garnets, apply lining and suspending garters, if applicable (IMC/Francisco Tavares Proença Júnior Museum, 2007, pp. 25–26).

For Pinto (1993) the distinctive characteristics of this embroidery are, on the one hand, the decorative grammar and, on the other, the technique employed—preferably the Castelo Branco stitch. In the Technical Specifications Manual of this embroidery, other authors' opinions add to the specific characteristics of this embroidery, the (intentional) presence of a drawing and the privileged use of linen and silk as raw materials (IMC/Francisco Tavares Proença Júnior Museum, 2007, p. 11).

The great diversity and richness of motifs used in this embroidery manual, in the opinion of several researchers, result from influences that extend in space and time, namely "the European drawing and engraving from the XVII and XVIII centuries, Portuguese tiles, Indian textiles and Cinese textiles and china ware" (IMC/Francisco Tavares Proença Júnior Museum, 2007, p. 81). Thus, they are currently grouped into five typologies: vegetal motifs, anthropomorphic motifs, zoomorphic motifs, mythological and specific symbolism motifs and inanimate motifs (IMC/Museu Francisco Tavares Proença Júnior, 2007, p. 81). The group of vegetal motifs, most often represented in a very stylized way, include, among many others, cloves, pomegranates, artichokes, forget-me-nots and acorns. In some cases the stylization makes their identification difficult and leads to disagreement among researchers. In the group of anthropomorphic motifs, stylization is also visible; both male and female elements are represented separately or in pairs (male/female and female/female), placed mainly in the center of the quilts. In the group of zoomorphic motifs, a strong stylization prevails, as in the previously referenced groups; the bird is undoubtedly the most represented zoomorphic motif, and deserves differentiated adjectives ranging from domestic to exotic; horses, dogs, reptiles and winged animals appear more sporadically. In the group of mythological motifs and specific symbology, the Castelo Branco Embroidery Technical Specifications Manual only considers the two-headed eagle; this motif is also represented in more or less stylized forms and sometimes "with the heart pierced by an arrow, constituting the most bizarre form of representation" (IMC/Museu Francisco Tavares Proença Júnior, 2007, p. 84). In the group of inanimate motifs some elements which stand out are vases, palms, shells, ribbons and ties as well as other elements "reminiscent of the architectural adornments of both the Gothic and the Baroque" (IMC/Museu Francisco Tavares Proença Júnior, 2007, p. 84).

Regarding the structure of Castelo Branco quilts, that is to say, the way in which the individual motifs are grouped in the area of the quilt, Pinto (2004) establishes two great decorative groups: in one of these groups the structure develops around a central medallion which fits in a simple field or in a field with a bar all around; in the other group the structure is organized around the dominant module which can be a Life Tree or a repetition module (p. 26).

A more diversified proposal is presented in both the Castelo Branco Embroidery Technical Specifications Manual and the Anex from Despacho nº 15559/2016, of December 27, which defines six typologies: central medallion quilts, defined by the intersection of the bisectors of the corners, the most common and identifiable structure of these textile objects; meandering quilts, bar and corners quilts, life tree quilts, five-senses quilts and tile quilts (IMC/Francisco Tavares Proença Júnior Museum, 2007, p. 84; Despacho (extract) nº 15559/2016, from 27 December).

Based on the collection of quilts from Francisco Tavares Proença Júnior Museum a survey of stitches used in this embroidery was carried out by the Regional Embroidery School, finding a total of 48 stitches. This inventory highlights the mandatory application of Castelo Branco stitch in the trunk of the trees being also dominant in some historical examplars. It has as peculiarity the fact that only the right of the embroidery is filled with the silk thread, which allows to save raw material. Another aspect to consider is its technical flexibility, having six variants of this stitch, namely color variant (stitch with equal stripes with two contrasting colors), variant in the orientation of the silk thread (launched vertically, horizontally or following the shape of the motive), variant in the orientation of prisons. (IMC/Museu Francisco Tavares Proença Júnior, 2007, p. 29)

Ormonde (2008) reminds us that embroidery is more studied as an artisanal production and that, because it is carried out mainly by women, it is considered subsidiary. But he adds the that "the other no less important component is embroidery as the bearer and producer of meanings and values thus participating in the symbolic sphere" (pp. 10–11). In this perspective, the Castelo Branco Embroidery Technical Specifications Manual presents the symbolism associated with the most frequently used elements, but recalls that "symbolic language adapts itself, loses itself, recovers itself and metamorphoses itself" (IMC/Museu Francisco Tavares Proença Júnior, 2007, p. 92).

Concerning the use and meaning of symbols, Vila Tejero (2008, p. 211) defends that:

The elements that make up the decoration of an object, especially if it is old and mostly if its origin is distant, are endowed with a symbolism that approximates them to almost magical or protective objects, when in most cases these elements are no more than a decorative repertoire that repeats itself over and over again, adapting to numerous positions.

In what concerns Castelo Branco embroidery, this author defends that "it has been written that they are like a garden full of symbols" but rather than symbols maybe we should say that they are inhabited by elements of the collective imagination of a people that has inherited and assumed naturally what has been part of its great History (Vila Tejero, 2008, p. 164).

However, Vila Tejero also says that "undoubtedly, the Castelo Branco embroidery is much more than a technical and formal description. It is a small universe that (...) always surprises us" (p. 198); the author ads that each quilt "seems to be endowed with a life of its own and determined to show us its uniqueness, and it is very rare that it does not succeed" (p. 198).

In her argument she says that usually vegetalist, naturalistic or even fantastic elements are used as structures or schemes to attach or highlight other elements that may contain a symbolic function. In this way, and when these symbols, by the repetition of their representation, become known "they keep repeating themselves for centuries, even when their initial symbolic charge has been forgotten" (Vila Tejero, 2008, p. 211).

As for Castelo Branco embroidery, this author argues that "it has been written that they are like a garden full of symbols" but that instead of symbols maybe it should be said that "They are inhabited by elements of the collective imagination of a people that has inherited and assumed naturally what has been part of its great History" (Vila Tejero, 2008, p. 164). However Vila Tejero also emphasizes that "Undoubtedly, the Castelo Branco embroidery is much more than a technical and formal description. It is a small universe that (...) always surprises us" (p. 198); the author ads that each quilt "seems to be endowed with a life of its own and determined to show us its uniqueness, and it is very rare that it does not succeed" (p. 198).

3 CASTELO BRANCO EMBROIDERY APPLIED TO CLOTHING – ALEXANDRA MOURA

During the visit to several places, cited above, fashion designer Alexandra Moura had contact with images of several quilts from the seventeenth century, that influenced her inspiration.

Themes and symbology mentioned in some bibliography consulted by the designer led her to the choice of an oriental quilt from the 17th century—Adam and Eve, belonging to Alberto Sampaio Museum in Guimarães, whose theme and its deteriorated appearance served as a starting point for her decision (Fig. 1).

Pereira (2008) "(...) as the representation of masculine and feminine, as the essence of life of which Adam and Eve, the primordial pair, flanking the tree of wisdom where the serpent is curled, represented in the center of a quilt... are the paradigm" (p. 82). This was also the paradigm of Alexandra Moura, until inserting these elements in a garment where the absence of gender is well marked.

The designer began by choosing the silhouette of the garment to be created. Several elements influenced her choice because the identity of the embroidery is associated with the identity of the designer and the taste for details that are associated with the concept of her collection developed at the time (collection AW 16), which has to do with the absence of gender and where the silhouette although referring to something feminine comes from the early eighteenth century men's attire. The image of the jacket, the coat, men's underwear (frills, ties), romanticism of the eighteenth century. The application of the ties on the sleeves, in addition to referring to male romantic traits in, also shows versatility of the piece itself.

Figure 1. Oriental quilt, XVII century(?) – Alberto Sampaio Museum.

Figure 2. Alexandra Moura's coat.

The developed piece, Jacket, reveals all that inspiration and combines ties, frills and a very versatile silhouette (Figure 2). It also allows for the deconstruction of the piece by allowing the sleeves to be removed, leaving the piece in the image of a long vest.

The most rustic linen was the base chosen for the coat and the choice lies in the quest to make a metaphor for the pure state of the plant woven with the delicacy and subtlety of the silk that embroiders it. Also the contrast of the opaque linen with the brilliance of the silk are elements that help to reinforce the concept that seeks to return to the past while projecting itself into an aesthetic of the future. The use of contrasts and the return to the origins in the search for inspiration is something very contemporary and current.

The chromatic palette chosen by the fashion designer seeks the natural state, the raw. The pursuit of linen in its original color in a blend with silk whose brilliance helps in relieving the nuances of a monochromatic palette. The texture of the embroidery stands out where the brilliance of the silk helps to emphasize different shades in the same thread.

The choice of stitches and motifs is also justified by the search to reproduce as much as possible the theme and composition of the quilt used as inspiration. Added to these motifs is a repetition of circles, some filled in, others just contoured, much as a reinforcement of the designer's own identity. The circle is associated with a geometric symbology with affection, love and, at the same time, represents a new breath from the embroidery itself.

Although it is a piece inspired by an extremely traditional and cultural product, with deep roots and historical references from the albicastrense tradition, we are able to visualize and launch the piece into the future. It is a contemporary, conceptual piece that, at the same time, is transverse in terms of time. It is a piece of garment time, that does not follow any trend but adapts to any season or occasion.

Alexandra Moura, presented a proposal that includes Castelo Branco embroidery in all the components that define it: the base—linen; the choice of the theme, stitches and motifs so easily identified by the Castelo Branco embroidery.

4 REVELATION AND PROJECTION OF CASTELO BRANCO EMBROIDERY

Communication strategies in fashion are a topic that holds special interest due to the impact of fashion in today's European and world economy. According to Díez Vial (2003), the fashion industry is characterized by being one of the main sectors of wealth creation in the European Union. Through different communication techniques, fashion companies, following the example of companies from other sectors, seek to achieve an advantageous position in the current context of the *"Digital Economy"*. In this sense, it is urgent to think about the impact and effectiveness of information and communication tools which, integrated in a marketing communication strategy, are especially important in challenges that companies face in the *"Globalization Era" (Hines and Bruce, 2017)*.

Krohling Kunsch (2003, 2007) argues that there are two key areas of communication management: public relations and marketing,

> *The first one would cover, by its theoretical essence, institutional communication, internal communication and administrative communication. Marketing would respond for all marketing communication* (Krohling Kunsch, 2007, p. 13).

In this sense, the author emphasizes that

> *Communicative actions need to be guided by a philosophy and an integrated communication policy that take into account the demands, interests and demands of strategic audiences and society. That is, there must be a total integration between internal communication, institutional communication and business communication to search for and achieve effectiveness, efficiency and organizational effectiveness, for the benefit of the public and of society as a whole and not only of the company alone.* (Krohling Kunsch, 2007, p. 13).

It is important to define an integrated communication strategy to promote not only in national, but also international context, Castelo Branco embroidery, through several initiatives such as the embroidery project which resulted from the cooperation between the Portuguese fashion designer Alexandra Moura and de Castelo Branco municipality.

Among the planned communication actions, the presentation of the piece by the designer Alexandra Moura at the Portuguese Embassy in London (Fig. 3) was highlighted. This presentation took place under the London Fashion Week initiative and was supported by the Portugal Fashion Association and the National Association of Young Entrepreneurs.

In articulation with this action, fashion designer Alexandra Moura also presented her collection in a fashion show and showroom. In this way, in a space where potential buyers,

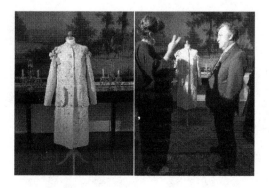

Figure 3. Coat displayed at the Portuguese Embassy in London.

international press and other professionals of the fashion industry are interspersed, and as a result of an integrated communication strategy, which includes other actions, Castelo Branco Embroidery wins international projection in the fashion field, a privileged environment for the valorization of this icon of the Portuguese cultural heritage.

5 CONCLUSIONS

As Ormonde (2008, p. 25) states

The work of the designer plays an increasingly important role in the creation of consumer objects, including the projects of recovery of these traditions. This role or intervention may become problematic when it comes to preserving emblems of national identity, but responds effectively to a contemporary tendency to blend tradition with innovation. In a way, it is with the category of objects that result, in which the old and the new are mixed, the local and the global, in which territorial relations are stifled and in which cosmopolitanism is celebrated made of pluralities from the places and from the practices associated with it, that new identities and new circles of consumption of embroidery are being built today, whether this is a massed piece or whether it is an antiquity.

The challenge of applying Castelo Branco embroidery in garments design led to the search for knowledge of the history and productive method of this cultural product. Fundamental elements resulting from research such as color, shapes, silhouettes, textures, details, will define the choices and influence all the aesthetics and narrative.

The designer seeks his inspiration through research and reveals his own sensitivity and visual aesthetics in the development of unique pieces that unify the initial concept.

It is possible to reinterpret Castelo Branco embroidery by giving it a new approach associated with product design. As a result on the one hand, we are capable of reinforcing the identity and culture of a region, on the other hand, it can boost the development of production for these products by increasing their demand.

REFERENCES

Despacho nº 15559/2016, 27 december)—approves the inclusion of traditional production Bordado de Castelo Branco in Registo Nacional de Produções Artesanais Tradicionais Certificadas.

IMC/Museu Francisco Tavares Proença Júnior (Coord.) 2007. *Bordado de Castelo Branco—Caderno de Especificações Técnicas*. Castelo Branco: Parceria Ex-Libris.

Melo, D. 2008. A 'Cidade-Símbolo' de Castelo Branco e o Bordar da Identidade Beirã (séc. XX). In IMC/Museu de Francisco Tavares Proença Júnior e ADRACES (Coord.), *Colchas de Castelo Branco—Percursos Por Terra e Mar*. pp. 65–89, Castelo Branco: Ex-Libris.

Ormonde, H. 2008. Bordar no quotidiano e no imaginário. In IMC/Museu de Francisco Tavares Proença Júnior e ADRACES (Coord.), *Colchas de Castelo Branco—Percursos Por Terra e Mar*. pp. 8–31, Castelo Branco: Ex-Libris.

Pinto, C. 1993. *Colchas de Castelo Branco*. Lisboa: Fundo Vip e Instituto Português de Museus.

Pinto, C. 2004. O Bordado de Castelo Branco, In Pinto, H. (Org.). *Artes Tradicionais de Portugal*. pp. 23–37, Lisboa: Fundação Calouste Gulbenkian.

Pires, A. 2008. Bordado de Castelo Branco. Elementos para uma geografia. In IMC/Museu de Francisco Tavares Proença Júnior e ADRACES (Coord.), *Colchas de Castelo Branco—Percursos Por Terra e Mar*. pp. 292–397, Castelo Branco: Ex-Libris.

Rechena, A. 2007. Prefácio, In AAVV. *Bordado de Castelo Branco—Caderno de Especificações Técnicas*. p. 7, CasteloBranco: Parceria Ex-Libris.

Vila Tejero, M. 2008. La universalidad del bordado através de los Bordado de Castelo Branco. In IMC/Museu de Francisco Tavares Proença Júnior e ADRACES (Coord.), *Colchas de Castelo Branco—Percursos Por Terra e Mar*. pp. 162–239, Castelo Branco: Ex-Libris.

Female jeans: The aesthetic of Brazilian "popular fashion"

I. Braga
Federal University of Piauí, Teresina, Brazil

M.J. Abreu
University of Minho, Guimarães, Portugal

M.M. Oliveira
University of Minho, Braga, Portugal

ABSTRACT: The Brazilian "popular universe" in the last decade has drawn attention to its expansion and an aesthetic expression of "popular wear" (Braga & Abreu, 2016). The present paper presents part of a literature review, the PhD study, about the aesthetic constitution expressed in clothing and fashion, with the main focus being the Brazilian publications on the aesthetic constitution of the Brazilian peripheral classes. Therefore, the objective of this article is to know the scientific areas and the research methods applied to study the aesthetics of the "popular feminine jeans". With the accomplishment of this investigation, it is possible to affirm that the publications dealing with the Brazilian "popular classes" are developed from three main methods: ethnographic, discourse analysis and symbolic analysis. It was concluded that although the "popular classes" have formed a segment of great opportunities for successful negotiations and creative potential, there was a lack of fashion and semiotic studies on the aesthetics of female "popular jeans". On that account, the importance of conducting fashion investigations that seek to understand the aesthetic expression of the Brazilian "popular classes" and their consumption behaviors.

1 INTRODUCTION

To understand the construction of the aesthetic expression of Brazilian "popular fashion", fashion created, produced, marketed and consumed by the "popular classes", specifically the female "popular jeans", it is necessary to know the studies developed on the aesthetic constitution, with clothing being the main element of research. It is questioned how the clothing has been investigated according to the substrate of use of this representation and which research methods are applied to the understanding of the meanings of the components of such elaborations.

The direction of this research for the universe of jeans happens because of its historical trajectory, initially emerging as work clothes (Michels, 2005), definitively settled in the market with the repercussion of the blue jeans of Levi Strauss in the fashion market (Araújo & Broega, 2014) and, since then, it crosses the most diverse spaces and is currently one of the most consumed pieces in informal "popular markets" in Brazil (Braga & Abreu, 2016).

Therefore, the primary objective of this paper is to know how the aesthetic expression of Brazilian "popular jeans" has been presented in Brazilian scientific publications, the research areas, and methodological procedures.

For this purpose, the present text carried out a bibliographical research and a research of scientific publications. It was initiated by the collection of an author who investigates the aesthetic compositions of clothing and fashion, followed by a research in Brazilian publications that have as a theme the fashion of the Brazilian "popular peripheral class", with emphasis on female jeans offered in Brazilian "popular markets" and its visual compositions.

The relevance of this study lies in bringing the "popular peripheral fashion" to the academic discussion of fashion, and pointing, as a research opportunity, the semiotic study of "popular fashion" and, especially, the "popular female jeans".

2 THE AESTHETIC EXPRESSION OF CLOTHES AND FASHION

The subject related to the act of aesthetic composition of clothing is researched by different areas of investigation: history (Laver, 1999), anthropology (Veblen, 2007); sociology (Simmel, 1895), philosophy (Lipovetsky, 1989); psychology (Flügel, 1966); semiotic (Mesquita, 1997; Barthes, 1999; Castilho & Martins, 2005).

These studies applied different methods, such as descriptions, observations, analyses and interpretations of clothes; of its elements of the constitution and its interaction with the body and the environment, to elaborate concepts and theories about the Fashion theme. Among the areas that investigated the aesthetic composition of clothing, garments, and fashion, the emphasis is the area of semiotics.

About the methods used by semiotics, Mesquita (1997) claims that semiotics investigates clothing and fashion through the description of the physical elements of clothing, not only as an aesthetic tool but also as a set of meanings that construct a system of nonverbal language.

Semiotics began to investigate fashion only from the 1950s, through Roland Barthes, the first semiologist to inaugurate fashion research as a sign of society, even declaring that clothing is, in its entirety, an imaginary field (Sant'Anna, 2007).

Barthes (1999), in "The Fashion System", developed an investigation based on the structuralist theory of Saussure, in which he claims that clothing has three aspects: the real piece (produced), represented (written) and used. Semiotics is the study that inaugurates a line of thought that seeks to reflect on the expressive dimension of the dressing act, on the one hand, and fashion as a social phenomenon, on the other hand. Thus, notwithstanding its limitations, it can be said that this approach initiates a relevant reflection on the anthropological, cultural and symbolic value of clothing, accessories associated with them and fashion as the organizing system of these elements.

In Brazil, fashion studies from the perspective of semiotics became widespread in the academic field of fashion, based on the investigations of Castilho & Martins (2005), with the publication of the book "Discursos da Moda: semiótica, design, corpo", in which the authors start from the conception of the body as a channel of materialization of thought, existence, environment and the other. They point out that through the descriptive analysis of the human body constitutions in the course of human development, in different historical periods and the various methods of manipulation, it is possible to understand the cultural values inscribed in the bodies.

The investigations on the semiotics applied to fashion published in Brazil emerge from the theoretical perspective of the constitution of clothing and the act of dressing the body, not only as an aesthetic element but as a tool of signification (Nacif, 2007).

Nacif (2007) based himself on the theoretical conception of authors such as RUFFIÉ, BARTHES and BOURDIEU, with the aim of understanding clothing as an interface between the human body, the natural and the cultural environments, its practical and symbolic aspects in which these are inseparable, resulting from the cultural elaboration from which they abstract language and the creation of objects.

Nacif (2007) still considers clothing as a form of expression, a visual language that, at the same time, refers to the individual and to the society that produced it. He uses the analysis of photographs applying the historical-semiotic methodology in the study of the vestimentary forms. Thus, he concludes that the physical forms and characteristics of clothes reveal economic conditions and technological knowledge, modes of production, systems of thought, social organization and symbolic representations of society and individuals, also adding that the history of clothing forms takes into account the diverse representations of the human body, in time, in space and within the various social layers.

Therefore, it is understood the importance of the investigative look at the aesthetic expressions printed in the act of dressing, as part of a conception of the re-signification of bodies, that communicate their cultural, natural and symbolic aspects, and relevance to observe the context of the forms and symbols used in each different space, time and social class.

Taking as pertinent to observe the processes of composing the dress of contemporary social groups, in the different layers of society, the following section presents publications made from studies that focus on the "popular universe" and the image of the "popular classes" in Brazil.

3 THE POPULAR AESTHETICS IN THE ACADEMIC VIEW

The behaviors and aesthetic expressions of the Brazilian "popular classes", namely the social classes of low incomes, only appeared as subject of scientific investigations in the last 10 years, most frequently from the year 2012 (Mauro & Trindade, 2012; Rahde & Tietzmann, 2012; Torreglossa & Jesus, 2012; Trotta, 2013). The researchers on aesthetics or "popular" expressions appeared, initially, as a collection of themes for fashion designers.

This is corroborated by Portela & Brandão (2010), who report that the investigations about the languages of the peripheral "popular classes", denominated as cultural production, appear as elements of decoding and "re-signification" by famous designers.

Portela & Brandão (2010), based on the research of discourses of the fashion communicators in blogs, television channels, newspapers and magazines, verified that the cultural production of significance of Brazilian peripheral fashion, when mentioned, is homogenized to the standard model of the peripheral fashions of the cities of Rio de Janeiro and São Paulo, disregarding the North and Northeast regions of Brazil. These regions, when approached by creators, designers, and journalists, are folkloric and/or misrepresented and aren't considered as investigation objects by fashion authors, because they are classified as signs of "low culture" values and tacky.

The misrepresentation of the image of the Brazilian "popular class" is most commonly found in the image of Carmen Miranda. Guedes & Teixeira (2009) affirm that the image of Carmen Miranda is a representation of the Brazilian people of the 1940s, in the sense of highlighting the aesthetics of the actress as an identity characterization of the Brazilian popular. Through the analysis of the elements of composition, costumes, and behavior of the characters represented by Carmen Miranda in Hollywood movies, they identify expressions of cultural and geographical meanings that characterize the Brazilian popular aesthetic, stereotyping the Brazilian people for color and carnivalization.

This stereotyped image of the Brazilian popular, the Carmen Miranda's symbolism, was used in the "reconstruction of the Brazilian identity", invented by the Estado Novo in 1937, and disseminated massively by the media, becoming a potential influencer of the attitudes and customs of the Brazilian and North American populations (Bonadio & Guimarães, 2010; Raslan, 2014).

The use of the "popular" image in the media continues to be caricatured, misrepresented, re-signified, being dissipated mainly through television programs. In the last years, stood out the investigations, specifically, the representations of the working classes in soap operas and comedy programs such as "Cheias de Charme (2012)", "Avenida Brasil (2012)" and "A Diarista (2004–2007)".

Rahde & Tietzmann (2012) stated that with the economic growth of the country and the new socioeconomic profile of the population, the television in Brazil keeps talking to all audiences and focusing mainly on the bigger social class (Class C). The focus is on reporting the symbology of valorization of the "popular" classes expressed by the soap opera "Avenida Brasil (2012)", describing scenarios and costumes, translating printed codes into the construction of the images of the plot.

Thereby, they mention the concern of TV Rede Globo (television channel) in the characterization of the aesthetic expressions typical of the popular class. Because they understood that the segment deserves particular attention, since it represents 68% of the Brazilian population, and with their economic ascent, it has also increased their self-esteem, representativeness, and visibility.

Lastly, they concluded that the "popular class" is increasingly reinventing itself aesthetically and that experiences in their neighborhoods and communities strengthen their identities. Such strengthening explains why the "popular class", even when ascends economically, doesn't want to move from the periphery to the elite neighborhoods of the big cities. They want to remain in their communities, with their principles, values and consumption habits.

Because of the contents of the soap opera "Cheias de Charme", Torreglossa & Jesus (2012) reveal that the concern of the TV programs of TV Globo consists in exalting the new emerging consumers and the power of influence and dissemination of content and peripheral languages in other social strata. The soap opera "Cheias de Charme", the diffusion of electricforró by the personage of the singer Chayene, represented by the actress Cláudia Abreu, and the dress of the personages "Empreguetes", Rosário, Penha and Aparecida (Cida), respectively represented by the actresses Leandra Leal, Taís Araújo and Isabelle Drummond.

Torreglossa & Jesus (2012), through the studies about the fans and the contents reception, conclude that Rede Globo television channel, by means of the contents of the soap opera "Cheias de Charme", and the creation of tools such as plots, site, backstage images, spaces of interaction, controls and establishes the phenomenon of the "soap opera fan", maintaining the audience and the consumption of contents and products.

Trotta (2013) emphasizes that because of the strong power of influence and diffusion, these soap operas are investigated in different aspects as a reflection of the symbols of the corporal aesthetics of the housemaids.

Mauro & Trindade (2012) point up that the "popular" universe comes to be identified, the housemaids are transformed into themes of soap operas, and so, what was formerly considered subaltern, the "Empreguetes" style, the peripheral exuberant creativity with decorations, glitter, and color, becomes a fashion trend.

The symbolic aspects of the "popular dress" are studied by Bessa (2013), where he presented a report on the critics' community of television and fashion, which censures and disqualifies the image of the singers Joelma and Gaby Amarantos (singers of eletricforró). Singers have a representative visual composition and exuberant aesthetic with glitter and color, being singled out as the singers of the working classes.

Still, by means of theoretical surveys and analysis of combinations of aesthetic elements, Bessa (2013) stated that the emerging and instant Brazilian social celebrities have as visual characteristic to wear clothes with brightness, colors, props combined with pieces of trends, and in the search to compose the clothes with elements of luxury, exaggeratedly, come to be pointed out as tacky. Even the tacky, disapproved by fashion critics, is idolized by the fans, creating a consumer market of this style. To generate positive movements in the market, the tacky style becomes mediatized, and thus, it becomes kitsch.

In the scope of research on the context of the tacky signs that have become kitschy, is the publication of Jacinto (2006) that discusses the products manufactured by "popular groups" from the periphery of the city of Rio de Janeiro, supported by the media.

Jacinto (2006) analyzes the combination of texts with the expressions of the corporal movements dressed, from a semiotic evaluation of the clothes as a "speech" producer of sense. Thus, he identifies that the Daspu women's t-shirts have a cluster of signs, they are fashion products imbued with language that expresses speech, tells something to someone and articulates an interweaving of meanings. The shirt-pamphlet is a media medium and has a strong communicative appeal.

Among the elements and characterizations of the "popular dress" mode, in addition to the t-shirts, another important piece in the aggregation of symbolic value, and that emerged as work clothing, is the jeans.

4 JEANS: SYMBOLIC VALUE AGGREGATION DEVICE

Among the pieces of clothing evaluated as having a high capacity of aggregation of symbolic value, besides the T-shirts, are the jeans.

Almeida & Emídio (2012), through the combination of bibliographical research and field research, identify this capacity of symbolic value aggregation through a survey of the different codes attributed to jeans since its emergence as working class clothing in the coal mines of the United States of America, moving through the artistic environments, political activists, motorcycle gangs, being considered opposition clothing, passing through the genre of the youth spread by the cinema of the 1950s. It had the adhesion of the middle class and the urban tribes, was torn, adorned according to the codes of identification of the street groups. Therefore, it was part of the collection of designer labels, remodeling, adding design and attaching labels, and, finally, elevated to the plateau of representation of power, status, luxury, sophistication, and seduction.

After collecting the symbology of jeans, Almeida & Emídio (2012) interviewed Brazilian consumers and found that jeans are the most used garments for women of different ages and that for every age there is a perceived image to use. Finally, they identified that the image associated with the piece is related to the type of denim, modeling, washes, and finishes applied to the piece.

Altafim, Campos & Catelão (2010), in the semiotic analysis of the advertising campaign of Levi's brand, concerning the languages attributed to the jeans, concluded that the models and the meanings of the jeans were changing, not only according to the fashion of the time, but also according to the lifestyles and their ways of wearing the jeans.

Mizrahi (2011), through the anthropological perspective, reports on the dress of the women who attend Funks dances in the suburbs, in the south zone, in the city. The article is "Brazilian jeans: materiality, body, and seduction at Rio de Janeiro's Funk Ball" from Rio de Janeiro. The author describes the type of material to provide jeans more glued to the body and the modeling in order to give a look of raising the "butt", emphasizing the rounded forms of the female bodies and, thus, characterizing them as the model of the Brazilian jeans.

Based on the studies described above, Bruno (2016) analyzes the consumption behavior of jeans of the "popular fair" of the city of Goiás, called the Hippie Fair of Goiás, and identifies, from the consumer's perspective, that the purchase of the models of jeans varies according to the occasion. She says that when consumers are looking for jeans to go to parties, they look for pieces without much detail, but that emphasize body shapes, elastane pants, and modeling that lift the "butt." When looking for jeans to wear on a day-to-day basis in work and/or study environments, they look for looser, comfortable pieces that have embellishments like rhinestones, glitter, embroidery, pearls, and applications of income.

Braga (2015) enters the universe of jeans having as a field of investigation the parties of forró of the city of Fortaleza, Ceará, specifically in upper-middle-class clubs that are frequented by people from peripheral popular classes.

Through the research of observation of the forms of speech, consumption, and dress, when describing the clothes worn by the "forrozeiros", she observes that jeans are the typical garments of partygoers in the forró. She identifies that the "forrozeiras" distinguish themselves, regarding the class to which they belong, through the models of the jeans that they wear. The "fangs" who wear the most fair-to-the-body jeans and bright-glitter props belong to the popular classes, living in the suburbs, while the middle-class "forrozeira" wears basic jeans with a simple wash and carries a label of a luxury brand.

Braga (2015) concludes that jeans are pieces that make all "forrozeiros" uninformed and, at the same time, distinct. What differs is the added value that is perceived through the type of raw material used, the specificity of the denim, the models of the pieces, the washes and the amount of props applied to the jeans, being that each element applied to the piece conveys different values, status, and meanings.

5 CONCLUSION

The aesthetic composition of the bodies, imprinted in the elaboration of clothing, in clothing and fashion over time, has been investigated by different scientific aspects such as history, anthropology, sociology, philosophy, psychology, semiotics, in the composition and overlap of artifacts and props and the impact of these elements on the environment. These investigative perspectives aim to understand the interaction between different models of aesthetics of bodies, human behaviors, and social relations.

Although semiotics is indicated as an efficient method in the analysis and interpretation of the values, signs, codes, and expressions present in the act of composing the dress, when carrying out this research it was not possible to find semiotic studies that deal with the analysis of aesthetic expressions of "popular fashion", nor of the "popular female jeans".

The investigations on the aesthetics of the Brazilian "popular class" are still very recent, and the most significant number of studies have emerged since it appeared at the Brazilian television. Through the soap operas, they presented popular representations, even if stereotyped, highlighting the exuberance in the combination of models, the richness of the props, with colorful prints, embroidery, glitter, and sequins. Such publications, listed here, are basically reports of how the image of the body's aesthetic of the peripheral population has been treated by the media. This subject has been examined explicitly by scholars of the science of communication such as Portela & Brandão (2010), Mauro & Trindade (2012), Rahde & Tietzmann (2012), Torreglossa & Jesus (2012) and Trotta (2013).

However, it was possible to find Bessa's (2013) research, a fashion study that surveys the discourses of fashion and media critics, and a brief description of the dress of the Brazilian "popular class".

The reports on jeans are commonly approached in the aspect of the image printed in advertising campaigns, as in the publications of Almeida & Emídio (2012) and Altafim, Campos & Catelão (2010). Being that the latter make a semiotic analysis of jeans printed in commercials. As for the studies specifically related to the compositions of the elements of jeans, no semiotic investigations were found on the subject.

Specifically for "popular jeans", anthropology has focused on consumer behaviors, dress, and the cultural symbology of wearing jeans in different peripheral social settings (Mizrahi, 2011, Braga, 2015, Bruno, 2016).

However, it has been observed that although it is still little investigated by fashion researchers, the "popular" mode of dressing, especially the Brazilian "popular jeans", has specificities in the choice of forms and aesthetic composition. Nevertheless, it was identified that the eyes of fashion scholars recognize the existence of an aesthetic of the segment, but they still disregard the importance of the movement and treat it as inferior, distant, and construct disturbed readings of the aesthetic expression of "popular" peripheral consumers.

Because of this study, we confirm the lack of semiotic studies on the aesthetic constitution of "popular fashion" and "popular women's jeans", and it is essential to carry out fashion investigations that seek to understand the processes of elaboration of dresses of the "popular classes" and their consumption behaviors. As warned by Rahde & Tietzmann (2012), the "popular classes" represent 68% of the Brazilian population, a segment of great profitable trade opportunities, in addition to having strong creative potential.

REFERENCES

Almeida, A. & Emídio, L., 2012. A Evolução da Calça Jeans e do Comportamento do Consumidor: uma reflexão como parâmetro para a concepção do produto. *Projetica, [online]* 3(2). Available at: <http://www.uel.br/revistas/uel/index.php/projetica/article/view/14124> [Accessed 28 May 2015].

Altafim, E., Campos, M. & Catelão, E. de M., 2010. Comunicação, moda e semiótica: pressupostos para o estudo da história do jeans em campanhas publicitárias. In: *coloquiomoda.com.br. [online]* Available at: <http://coloquiomoda.hospedagemdesites.ws/anais/anais/6-Coloquio-de-Moda_2010/71393_Comunicacao,_moda_e_semiotica.pdf> [Accessed 25 May 2015].

Araújo, R. & Broega, A.C., 2014. Jeans: ícone de moda e cultura material. In: *2o CIMODE: 2o International Fashion and Design Congress.* Milan, pp.1402–1412.

Baldini, M., 2006. *A invenção da moda: as teorias, os estilistas, a história.* Lisboa: 70 Arte & Comunicação.

Barthes, R., 1999. *Sistema da moda.* Lisboa.

Bessa, R., 2013. O brega, o luxo e o kitsch-em busca da fama e status na cena brasileira. *coloquiomoda.com.br.* [online] Available at: <http://www.coloquiomoda.com.br/anais/anais/9-Coloquio-de-Moda_2013/

ARTIGOS-DE-GT/Artigo-GT-Moda-Midia-e-Estilos-de-Vida/O-brega-o-luxo-e-o-kitsch-em-busca-da-fama-e-status-na-cena-brasileira.pdf> [Accessed 25 Apr. 2015].

Bonadio, M.C. & Guimarães, M.E.A., 2010. Alceu Penna e a construção de um estilo Brasileiro: modas e figurinos. *Horizontes Antropológicos, [online]* 16(33), pp.145–175. Available at: <http://www.scielo.br/scielo.php?pid=S0104-71832010000100009&script=sci_arttext> [Accessed 19 Feb. 2018].

Braga, I. & Abreu, M.J., 2016. The jeans in the popular Brazilian panorama. In: *TIWC 2016*. Poznan, pp. 534–540.

Braga, R., 2015. Eu era feio, agora tenho carro: encenações e práticas de consumo em clubes de forró de Fortaleza. [online] *Universidade Federal do Rio Grande do Sul*. Available at: <http://www.lume.ufrgs.br/handle/10183/109704> [Accessed 4 May 2015].

Bruno, S.S., 2016. O suporte para pensar o consumo—Uma etnografia na Feira Hippie de Goiânia. [online] *Universidade Federal de Goiás*. Available at: <http://repositorio.bc.ufg.br/tede/bitstream/tede/7546/5/Dissertação—Bruno Souza Silvestre – 2016.pdf> [Accessed 15 Feb. 2018].

Castilho, K. & Martins, M., 2005. *Discursos da Moda: semiótica, design, corpo*. São Paulo: Editora Anhembi Morumbi,.

Flügel, J., 1966. *A psicologia das roupas*. São Paulo: Editora Mestre Jou.

Guedes, R., Texeira, E. & Silva, H., 2009. Moda na década de 40–A performace de Carmen Miranda. In: *V Enecult. [online]* Salvador—BA. Available at: <http://www.cult.ufba.br/enecult2009/19553.pdf> [Accessed 27 May 2015].

Jacinto, A.P., 2006. Daspu apresenta coleção em São Paulo. *[online] Terra.com.br*. Available at: <http://moda.terra.com.br/galerias/0,,OI28012-EI1119-FI329194,00.html> [Accessed 27 Apr. 2015].

Laver, J., 1999. *A roupa e a moda: uma história concisa*. São Paulo: Companhia das Letras.

Lipovetsky, G., 1989. *O império do efêmero*. São Paulo: Companhia das Letras.

Mauro, R. & Trindade, E., 2012. Telenovela e discurso como mudança social na análise da personagem Maria da Penha em Cheias de Charme. *Em Questão, [online]* 18(2), pp. 196–182. Available at: <http://www.seer.ufrgs.br/EmQuestao/article/view/33380/0> [Accessed 26 Apr. 2015].

Menta, G., 2012. Carmen Miranda em Hollywood: vestindo formas, cores e culturas. [online] *tede.utp.br. Universidade de Tuiuti Paraná*. Available at: <http://tede.utp.br/tde_arquivos/2/TDE-2013-05-24T083739Z-381/Publico/CARMEN MIRANDA.pdf> [Accessed 27 May 2015].

Mesquita, R., 1997. Comunicação não-verbal: relevância na atuação profissional. *Revista Paulista de Educação Física.*, [online] 11(2), pp.155–163. Available at: <http://citrus.uspnet.usp.br/eef/uploads/arquivo/v11 n2 artigo7.pdf> [Accessed 26 Apr. 2013].

Michels, M., 2005. *L'Evoluzione del workwear in denim e le origini nei marchi americani*. In: Jeans! Le origini, il mito americano, il made in Italy, 68th ed. Prato: M&Mmaschietto&ditore, pp. 55–84.

Mizrahi, M., 2011. Brazilian jeans: materiality, body and seduction at a Rio de Janeiro's Funk Ball. In: *Global Denim*. pp. 103–126.

Nacif, M., 2007. O vestuário como princípio de leitura do mundo. In: *XXIV Simpósio Nacional de História. [online]* Rio de Janeiro/RJ, p.10. Available at: <http://snh2007.anpuh.org/resources/content/anais/Maria Cristina V Nacif.pdf> [Accessed 7 May 2015].

Portela, A. & Brandão, L., 2010. Um Quilt entre carne e concreto. *Iara, [online]* 3(1). Available at: <http://www1.sp.senac.br/hotsites/blogs/revistaiara/wp-content/uploads/2015/01/IARA_vol3_n1_Completa_2010.pdf#page=25> [Accessed 27 May 2015].

Rahde, M. & Tietzmann, R., 2012. Avenida Brasil: o popular como pós-modernismo televisivo. *Estudos em comunicação, [online]* 12. Available at: <http://www.ec.ubi.pt/ec/12/pdf/EC12-2012Dez–16.pdf> [Accessed 27 May 2015].

Raslan, E., 2014. Disseram que voltei americanizada: a construção da imagem de Carmen Miranda pelos meios de comunicação. [online] *Pontifícia Universidade Católica do Rio Grande do Sul*. Available at: <http://repositorio.pucrs.br:8080/dspace/handle/10923/6645> [Accessed 27 May 2015].

Sant'Anna, M., 2007. *Teoria da moda: sociedade, imagem e consumo*. Barueri, SP: Estação da Letras.

Simmel, G., 1895. *Philosophie de la mode*. [online] Editions Allia. Available at: <https://scholar.google.com/scholar?q=Georg+Simmel%2C+Philosophie+de+la+mode&btnG=&hl=pt-BR&as_sdt=0%2C5>.

Torreglossa, S. & Jesus, A., 2012. Estudo sobre fãs na telenovela brasileira e sua representação modelar em 'Cheias de charme'. *In: Intercom—Anais do XXXV Congresso Brasileiro de Ciências da Comunicação. [online]* Fortaleza,CE. Available at: <http://www.intercom.org.br/sis/2012/resumos/R7-1435-1.pdf> [Accessed 27 May 2015].

Trotta, F. da C., 2013. Entre o Borralho e o Divino: a emergência musical da 'periferia'. *Galaxia, [online]* 26, pp. 161–173. Available at: <http://www.scielo.br/pdf/gal/v13n26/v13n26a13.pdf> [Accessed 27 May 2015].

Veblen, T., 2007. *The Theory of the Leisure Class*. Oxford University Press. Pennvania: The Pennsylvania State University.

The liberation of women in the 19th and early 20th centuries as seen through costume

L. Luceño
CSDMM – UPM, Madrid, Spain

ABSTRACT: Costume expresses and embodies social changes. Today's feminist and gender equality movements began in the 19th century and found an echo in the fashion proposals of the time. Women's aspiration to comfortable and functional clothing, like that of men, was materialized in new garments. In other cases, the dressmakers and artists of the time transformed the prevailing silhouette, freeing women from the costume that oppressed them physically, psychologically and socially. Some of the proposals were ahead of their time and were adopted only by a minority of women. In other cases, liberating costumes were reserved for the interior of houses and gardens. Avant-garde garments created a debate and a far-reaching meditation on the artificiality and lack of functionality of the conventional female costume of the 19th and early 20th centuries.

1 INTRODUCTION

Costume is a non-verbal language. Throughout history, clothing has expressed social status, economic power, occupation, age and—above all—gender. Even in the early stages, characterized by draped garments for both sexes, there were gender differences in the way they were worn: length of garments varied, and women were usually more covered than men. In Western culture, women wore long skirts and gowns that restricted their freedom of movement, virtually since Mesopotamia to the twentieth century. Their bodies were compressed and molded by stays, bodices and corsets for four hundred years. Ladies dressed in bulky, heavy gear to hollow out their skirts from the 16th through to the 19th century. In the meantime, men's clothing evolved and adapted to men's need to ride, fight in battle, and work. The women of the 19th century were not independent, not even in their attire. We sometimes forget that one of the worst inconveniences for women when dressing was that they needed help adjusting their bodices and fastening the many buttons on the back of their garments.

In bourgeois society in the 19th century the social roles of men and women remained unchanged, yet critical voices began to be raised in favour of gender equality, women's suffrage and the liberation of women. This entailed a transformation of dress, which oppressed them physically and socially. Regarding the dominant fashion of the time, François Boucher said: "As for the few attempts at change, some flowed ideologically from the struggle for gender equality through the feminist movement, which developed from 1848 to 1914." (Boucher, 2009, p. 373). Although some of the elements mentioned only became established and widespread at the beginning of the 20th century, it was the 19th century that marked the break with old social paradigms. The global changes and the new mindsets of modern men and women in the 19th and early 20th centuries made room for the gradual emancipation of women. The effect of this transformation on costume translated into new silhouettes and new garments that liberated the female body. The 19th and early 20th centuries were a propitious period for fashion: many inventions crystallized and a new era, modernity, was born.

2 HISTORICAL, ECONOMIC AND SOCIAL CONTEXT

At the dawn of the 20th and 19th centuries, costume was no longer simply handcrafted but mechanized in France, thanks to the invention of "Thimonnier, who, in 1830, brought his invention to Paris: the sewing machine". (Latour, 1961, p. 83). In industrial manufacturing, "the first garments made with template patterns did not appear until 1870" (Latour, 1961, p. 83), encroaching on the bespoke tailoring business. Artificial and synthetic textiles also emerged in this period: At the Exposition Universelle of 1889, Chardonnet presented a synthetic form of silk and a spinning machine (Deslandres, 1998, p. 60) and "In Great Britain, a team of chemists invented viscose at the end of the 19th century" (Deslandres, 1998, p. 60). Dyeing of textiles also developed exponentially in the 19th century: "In 1859 in Lyon, Verguin discovered magenta red or fuschine. 1860saw the invention of aniline blue, and 1863 was the year of aniline black, produced by a process of oxidation. When Perkin created mauveine in 1856, there were only three solid colors among natural products" (Deslandres, 1998, p. 68). So this was a time of great scientific and technological discoveries that were applied to costume and fashion.

At this same stage, great changes arose in the trades of seamstresses, tailors and fashion designers. The Briton Charles Frederick Worth, regarded as the originator of haute couture, opened his fashion house with his partner Bobergh in 1858. Haute couture brought renewed social status to the occupations of seamstress and tailor, who created exclusive bespoke models reflecting their creative genius, emulating artists, and not at the bidding of customers. In the same context, "in 1869 a merchant named Stockmann became a specialist in busts and standing mannequins for seamstresses, tailors and dressmakers; they were stuffed with cardboard and built 'according to human anatomy.'" (Deslandres, 1998, p. 94). French fashion schools emerged at the same time: in 1841 Alexis Lavigne launched the first dressmaking courses, the forerunners of the later Esmod school (Esmod (2016) *Historique de l'école*. Available at: http://www.esmod.com/fr/content/historique-de-lecole [retreived 10-04-2018]), considered the oldest in the world, with its exclusive and patented techniques. Lavigne is credited with inventing the tape measure. In 1864, Elisa Lemonnier, considered the founder of women's vocational training in France, set up a school of sewing arts and crafts on Rue Duperré to train women in the fashion profession (École Duperré (2018) *Histoire de L'école Duperré*. Available at: https://duperre.org/lecole-duperre/histoire-de-lecole-duperre [retreived 10-04-2018]). In the beginning of the 20th, the first photographs were published in fashion magazines. This conveyed new proposals with heightened realism, communicating and popularising new trends.

3 THE NOVELTY OF DEPARTMENT STORES

The first, incipient democratization of fashion took place with the new mercantile concept of the department store. According to François Baudot (2008), department stores played an important role in changing customers' tastes, and hence the prevailing aesthetic: together with the decorated, complex garments of the great designers, the stores introduced (for everyday use) more sober and practical clothing, although simplicity did not necessarily gainsay elegance. Haute couture customers gradually became accustomed to this new paradigm clothing, and then sought it out among the great couturiers. Baudot (2008, p. 33) cites as the leading Parisian department stores: "the Bon Marché, the Printemps and others". According to François Baudot (2008, p. 33), although a true democratization of fashion was yet to arise, there was already

> the future influence of "the street": sober and often well-made garments offered by department store counters familiarised many women with the concept of a garment that is not yet bold enough to be termed "practical" but is already clearly less ornamental.

The new commercial phenomenon of department stores began in Paris in the mid-19th century, as did *haute couture*: *Le Bon Marché* opened in 1852, *Printemps* in 1865. Though *Harrods* opened 1835, its major overhaul took place in 1861. In these new establishments one could buy ready-made fashion, available to women from all walks of life, in an elegant setting.

4 WOMEN'S ACCESS TO SECONDARY AND HIGHER EDUCATION

Women had traditionally been excluded from secondary and university education. In the United States, since the mid-19th century some women sought greater political and social equality with men. These new women were being educated, wanted the vote, and demanded new garments that would allow freer movement and less constriction. According to Deslandres and Müller (1986), women began to have regular access to secondary and university education in France at the end of the 19th century. In 1880 high schools for young women were created on the initiative of the parliamentary deputy Camille Sée. Fashion illustrations of the time echo this change: women getting on a train, visiting exhibitions or travelling to Swiss lakes. Their clothes are not reformist, but it is clear that at the end of the 19th century women's lifestyles began to change significantly. At the beginning of the 20th century, images were published of suffragettes under arrest, or dressed in their university robes. This was a symbol of their acquired status and of their equality of university training with men. Women's access to university education had a decisive impact on their self-esteem and subsequent empowerment.

5 THE SUFFRAGETTE MOVEMENT

The first feminist proposal on dress emerged in the midst of American women's struggle for the vote. In May 1851, the American women's rights campaigner Amelia Bloomer used her feminist periodical *The Lily* to promote a loose reformist garment consisting of a short crinoline skirt over baggy Oriental-style trousers. According to the designer, educator and researcher Diana Fernández (Fernández, Diana. (2012), writing on her blog, this "rational" clothing was actually invented by the activist Elizabeth Smith Miller, although it ended up bearing the name of its most prominent advocate: *"the Bloomer Costume"* or *"bloomers"*. These women campaigned for women's suffrage and sought to change their clothing; costume was to reflect a new social and political role for Western women. Fernandez (Fernandez, Diana. 2012) reports that Amelia's periodical addressed women directly:

> *To you, housewives, we say: unbutton your clothes and let your garments hang loose on your bodies. Then, breathe freely, fill yourselves with air as much as you can, and fasten your clothes at that moment. Then crop those wavy skirts at the height of your knees and put on some baggy pants fastened at your ankles.*

Although the "reform dress" was often mocked by the public of the time, "bloomers" began to be used for cycling or sea bathing. These were sport-related uses, but the costume could be considered an early minor victory. To Bloomer's contemporaries, the trousers she advocated seemed disturbingly to erase gender difference in dress. This outfit, with baggy, ankle-length Turkish-style trousers, made the majority of the population uncomfortable. Marnie Fogg (2014) indicates that the idea of trousers did not seem appropriate for women. The intention to add a skirt, the female garment par excellence, was insisted upon, and the trouser-skirt was allowed: "the debate on clothing reform reintroduced the idea of the trouser skirt in the 1880s and 1890s" (Fogg, 2014, p. 193). The trouser-skirt, along with the tailored suit, became two crucial forms of attire in women's liberation.

6 CLOTHING FOR SPORT AND OUTDOOR

Sportswear opened the debate on the functionality of clothing, a concept that did not exist in the dictates of *haute couture*. With the development of means of transport, the pace of life accelerated. Clothes had to be functional to adapt to new changes. At the same time, sport became increasingly popular, including among women. In the 19th century, women did not want to give up the bicycle as a means of transport. In the United States, sport and outdoor activities were touted as beneficial for health. Women sought out clothes that allowed them to move freely, and were therefore simple, sober and functional. By the end of the 19th century, some women were already mountain climbers, tennis players, skaters and cyclists. The garments required for sport made for an antagonistic aesthetic to *haute couture*, and aided the gradual acceptance of sober, light, flexible, unadorned and purely functional clothing. These features powerfully permeated fashion in English-speaking countries, while Parisian fashion became defined as the attire for salons and city avenues.

7 ANGLOMANIA

In the late 19th and early 20th centuries, Britain was a major economic and political power, following the proclamation of Queen Victoria as Empress of India in 1877 and the occupation of Egypt in 1882. Throughout the history of costume, there is no example of a nation that, while exercising great power economically, politically and culturally, failed to impose its fashion. This was the case with the British Empire. Anglomania spread around Europe and the taste for the English style became a trend that gradually liberated female dress, which until then had followed the artificial and ornate French fashion emerging from the recently created *haute couture* salon. Sobriety and the search for comfort in dress arose from the spirit of fashion in the English-speaking sphere, which was closer to nature and the open air. At the time, cartoons were published of people "before and after" their trip to England: before, they dressed in French women's fashion, ornate, colorful and with large volumes; on their return from London, they were to be seen in sober and functional garments, in neutral colors. François Baudot (2008), from his perspective as a Frenchman, comments on the influence of Anglomania on French fashion in the late 19th century: "a new word, impregnated with barbarism and English chic, became increasingly common parlance: 'sport.'" (Baudot, 2008, p. 33) English-speaking culture seemed more permeable to feminist demands coming to London from the United States than the Mediterranean culture.

8 MASCULINIZATION OF WOMEN'S CLOTHING

The Redfern company, originally English but based in Paris, was responsible for creating the women's tailor-made suit after 1885. It is thought to have been designed by the dressmaker's son, Charles Poynter Redfern, for the then Princess of Wales. At the end of the 19th century, the house brought out the *trotteur* or "walking suit." The skirt, which initially reached down to the ground, eased movement and was combined with an American-style jacket inspired by the cut of the equivalent male garment. Women started wearing these clothes on the coast when on holiday, but ended up wearing them in the mornings, too, to go shopping and move around the city. These two garments are clearly designed for functionality, but still had an inside corset. Upper-class women began to wear this clothing, which was perceived as functional, modern and liberating. Morales (1947, p. 122) adds:

> However, the trotteur or tailleur did not at the time fully satisfy the ambitions of haute couture or the demands of its illustrious customers. They were perhaps the simplification of elegance, which is what is needed, but not the apotheosis of elegance, which is what the exclusive clientele yearned for.

However, this suit has remained current to this day, with few modifications, as is the case with the men's tailored suit, for its versatility, comfort and "modernity".

9 THE ARTISTIC DRESS OF THE PRE-RAPHAELITES AND THE ARTS & CRAFTS MOVEMENT

In the 19th century, some artists proposed new women's clothing to dress the women in their social circle: their muses and family relations. These intellectuals ahead of their time sought to express the new social role they believed women should play, and their desire for liberation with respect to traditional dress. François Boucher (Boucher, 2009, p. 375) claims that at the end of the 19th century

> *costume was far more subject to the influence of the decorative arts than of the fine arts, such as painting. The abundance and truculence of the "upholstery" style that emerged in interior decoration was adopted by fashion between 1870 and 1895. It only took new colors or color combinations from the avant garde art of Manet or Gauguin.*

However, several artistic styles saw costume as one of their aesthetic concerns. A new form of dress was proposed for a modern woman who wants to be freed from tight corsets and oversized skirt hollowers. For Marnie Fogg (Fogg, 2014, p. 184):

> *In the mid-19th century, supporters of an alternative form of dress known as "artistic" challenged the defenseless femininity of the "angel of the house", constrained by bulky, increasingly voluminous skirts and tightly fastened corsets. The new style was especially associated with loose garments and natural pleats, worn by the live models and lovers of Pre-Raphaelite painters during the 1860s.*

The Pre-Raphaelites drew inspiration from medieval art, reviving the idea of the wide and fluid Gothic robe, which does not draw the garment in at the waist and does without the artifice of hoops or tight corsets. Fogg (2014. p. 184) states that the costume proposed by the Pre-Raphaelites is composed of

> *a waisted bodice but no corset, a simple and pronounced round neck and sleeves with tails up to the neck (floor). This garment was worn over a skirt drawn up into a high waist, which fell into folds around the feet.*

As to the lifetime of this artistic and literary dress, Marnie Fogg (Fogg, 2014) comments that it fell into disuse in the 1870s but was revived by the Arts and Crafts movement in the 1880s. Fogg (Fogg, 2014) states that: "Morris eschewed heavily printed Victorian fabrics" (Fogg, 2014, p. 184), seeking a measure of sobriety and simplicity in the material. William Morris abhorred cheap industrial fabrics and reintroduced handlooms and woodblock printing. For Morris: "any ornament was limited to hand embroidery of folk motifs" (Fogg, 2014, p. 184). This therefore meant removing the excess of ornamentation and coloring of fabrics used in the prevailing fashion of the time, marked by a luxurious and eye-catching *haute couture*. One of the striking features of the artistic dresses of the Arts and Crafts movement is that they have folds in different fabrics, with no pressing. The artistic costume of this movement introduced neutral colors and a new chromatic sobriety into women's clothing that eventually shaped future modern fashion. According to Marnie Fogg (Fogg, 2014, p. 185), in this period the *avant garde* in dress was finally adopted by mainstream fashion:

> *the cult of beauty permeated every aspect of the aesthete's life. When the formerly marginal avant garde attire was absorbed into conventional dress, people began to enjoy what had been a minority fashion in the past.*

Along with the trouser-skirt and the tailored suit, a new type of dress appeared in the 19th century, closer to conventional dress but with some genuinely distinctive characteristics.

10 THE TEA-GOWN AND THE LIBERTY DRESS

In London, the simplification of the women's costume of the time prompted the creation of a new type of dress called "tea-gown". Marnie Fogg (2014) explains that the tea-gown is a garment that can be worn at home or in the garden in the late afternoon. The dress could be put on without help, and was clearly more comfortable, less stiff, lighter in weight, and could sometimes be worn without a corset. The Liberty dress is in some sense a "signature" tea gown. Fogg (2014, p. 189) explains:

> *The new garment was made by Liberty's Artistic and Historic Costume Studio under the direction of E. W. Godwin, the designer who headed the fashion department from 1884. A follower of the aesthetic movement, he created women's garments inspired by classical, medieval and Renaissance clothing.*

According to Marnie Fogg (Fogg, 2014), the dress bears the name of the young Arthur Lasenby Liberty. Although the suit was not designed by him, in 1875 Arthur had persuaded his employer to open an Oriental department in the store known as Farmer and Roger's Great Shawl Emporium. Liberty eventually established his own business. In 1884, he opened an "artistic and historical costume" department, directed by the architect E.W. Godwin, where he intended to offer alternative garments to Parisian fashion under the label Liberty & Co.

For the Victoria and Albert Museum (V&A, 2017) the Liberty suit is a "reform" dress related to the Arts and Crafts movement in an effort to free women from the corset. Marnie Fogg (Fogg, 2014, p. 189) speaks of a: "Liberty dress in dark green silk velvet, reflecting the interest of the time in 15th century fashion and the favourite chromatic range of aesthetes, the "greenery yallery", or green-yellow." Yet the Liberty dress commented on by Fogg (2014) was designed to be worn with an inner corset. There was no single type of the artistic "reform" costume, but they all shared various features. Art Nouveau proposed its particular vision of this trend.

11 ART NOUVEAU AND "REFORM" DRESS

In the late 19th century and into the 20th century, reform garments were of a piece with a total art concept that included architecture, furniture and clothing. In this conceptual setting, the Belgian architect Henry van de Velde exhibited a dress, designed by him, at a German exhibition of reformist fashion. The author remarked that it was a dress to be worn when entertaining guests: it would be worn at home, and not on the street, where the prevailing fashion was different. The velvet dress was sober and printed with organic Art Nouveau motifs. According to Max von Boehn (Von Boehn, 1945, p. 191):

> *Establishing a new architecture and a new art therefore required the emergence of a new style of clothing that would harmonize with this framework. Henry van de Velde would have considered it an offense for his wife to wear a "haute couture dress" within the interiors he himself had designed.*

The so-called artistic, aesthetic or reform fashions, which proliferated in several western countries since the mid-19th century, did not prevail. They were generally used at home, to welcome and entertain guests, and only among a narrow group of women: "a minimum percentage of freethinkers and aesthetes who embraced the principles of the Arts and Crafts movement (...) aspired to universal Empire lines" (Watson, 2004. p. 18) However, these sporadic proposals brought about a decisive change in the way of seeing the fashion of the time. They succeeded in raising a public debate on the health implications and comfort of clothing, challenging exaggerated and artificial shapes, which *Vogue* magazine's editor, Josephine Redding, dubbed "humps" (Watson, 2004, p. 18). In the work *Fashion: History and Styles* (2012, p. 236) the claim is advanced that all these initiatives, despite reflecting only an isolated minority, had common elements that allowed female costume to evolve:

> *Despite the diverse range of solutions, the designs had common elements. The corset, considered the root of all evil, was widely discarded; dresses were waisted high and not too tightly; the voluminous sleeves and skirts allowed greater freedom of movement.*

12 CONCLUSIONS

The gradual empowerment of women in the 19th and early 20th centuries was driven by several factors. This was a time of major social change and technological progress, affording a glimpse of the foundations of the current system and framework of fashion. Women's desire for liberation was expressed in new fashions—more comfortable and functional, less oppressive and artificial—springing from the great economic, social and artistic changes that shaped new mindsets during this period. Modern fashion reflected the desire for liberation of women who could now go to university, play sports, and travel alone—and therefore demands to vote and participate in society on an equal footing with men. The first feminist movements, landing in Britain from New Zealand, Australia and the United States, and then spreading from Britain and the Nordic countries to the rest of Europe, stirred up the outlooks of the men and women of the time. As is the case today, revolutions and social transgressions are heralded by visible signs, embodied in costume, our personal image. In the 19th century, the first proposals for the liberation of traditional women's clothing emerged; they became entrenched at the beginning of the 20th century. A woman's standard wardrobe was made up of long dresses that hung down to her feet: chiseled by corsets, hollowed out with baleen hoops and richly decorated. These clothes were neither comfortable nor functional, and prevented free movement of the body. In this period, the role of department stores and fashion magazines became crucial in the democratization of fashion tastes and consumption—and therefore, of society. The adaptation of dress to sport, which requires freedom of movement and functionality, was another key aspect in the transformation of costume, along with the incipient masculinization of female attire through the tailoring suit. The Pre-Raphaelite, Art and Crafts and Art Nouveau movements also proposed new garments, whose common feature was that they no longer cleaved to a woman's body like a glove. These artists sought new silhouettes in the shapes of ancient costumes or in the "Empire" cut, to banish the tortuous hourglass figure. As a rule, the new silhouettes were straighter, looser and freer. The patterns and colours of the fabrics also underwent a shift towards less eye-catching motifs and more neutral shades. The emancipated woman of the 19th and early 20th centuries could ride her bicycle with bloomers, run to the department store sales in a tailored suit and entertain guests in her garden while wearing a tea-gown. Fortunately, neither fashion, nor art, nor feminists were insensitive to the discrimination suffered by women in the domain of costume. During the 19th and early 20th centuries, the foundations of modern women's clothing were forged.

REFERENCES

Baudot, F. 2008. *La moda del siglo XX.* Barcelona: Gustavo Gili.
Blackman, C. 2012. *100 años de moda.* Barcelona: Art Blume.
Boucher, F. 2009. *Historia del Traje en occidente. Desde los orígenes hasta la actualidad.* Barcelona: Gustavo Gili.
Deslandres, Y. and Müller, F. 1986. *Histoire de la mode au XXe siècle.* Paris: Somogy
Deslandres, Y. 1998. *El traje imagen del hombre.* Barcelona: Tusquets
Diseño, Historia y Teoría del Traje y la moda, Cine, Teatro, 27 de abril. Available at: https://vestuarioescenico.wordpress.com/2012/04/27/el-vestido-reformista-y-el-origen-del-termino-bloomer/ [retrieved 20-01-2018]
École Duperré. 2018. *Histoire de L'école Duperré.* Available at: https://duperre.org/lecole-duperre/histoire-de-lecole-duperre [retrieved 10-04-2018].
Erner, G. 2014. *Sociología de las tendencias.* Barcelona: Gustavo Gili.
Esmod. 2016. *Historique de l'école.* Available at: http://www.esmod.com/fr/content/historique-de-lecole [retrieved 10-04-2018]
Fernández, Diana. 2012. El vestido reformista y el origen del término "Bloomer", *Vestuario Escénico.*
Fogg, M. 2014. *Moda. Toda la historia.* Barcelona: Art Blume.
Lehnert, G. 2000. *Historia de la moda del siglo XX.* Barcelona: Colonia, Könemann
Moda. Historia y Estilos. 2012. Deleatur: Dorling Kindersley
Morales, M.L. 1947. *La moda. El traje y las costumbres en la primera mitad del siglo XX.* (Tomo noveno- siglo XX- 1900–1920). Barcelona: Salvat Editores.

Toussaint—Samat, M. 1994. *Historia técnica y moral del vestido*. Madrid: Alianza.
Victoria and Albert Museum 2017. *Dress, Liberty & Co. Ltd.* Available at: http://collections.vam.ac.uk/item/O108865/dress-liberty-co-ltd/ [Accessed 20-10-2017].
Von Boehn, M. 1945. *La moda. Historia del traje en Europa. Desde los orígenes del cristianismo hasta nuestros días.* (Tomo octavo—Siglos XIX y XX – 1879–1914). Barcelona: Salvat Editores.
Watson, L. 2004. *Siglo XX moda.* Barcelona: Edilupa.

A# The *Minho* traditional costume as cultural heritage in fashion design

S. Castro
Department of Textiles, University of Beira Interior, Covilhã, Portugal

J. Cunha
Department of Textiles, University of Minho, Guimarães, Portugal

C. Morais
Faculty of Philosophy and Social Sciences, Universidade Católica Portuguesa, Braga, Portugal

ABSTRACT: For this research we propose an analysis of traditional national costumes, in the sense of perceiving the connection with cultural value in an anthropological and historical approach. The Minho's traditional costume emerges as a reference of the local communities, of the region, with its own identity of valorisation and conservation, of the history, experiences and memories of its people. The concept of cultural heritage is increasingly widespread in the area of fashion, in the sense that it presents itself as a singular characteristic of a textile product, between communities and nations, in a trickle-up pattern. In order to understand the different scientific approaches of the interrelationship between fashion and tradition culture we have analysed some cases of designers who recreate the costume as a characterizing and diffusing element of culture, history and identity in a new vision about fashion design. Thus, we realize that concepts such as intelligent fashion design allows us to create a cooperation between traditional and recreated products, in order to promote the bridge with the Minho costume, enhancing it.

1 INTRODUCTION

Traditional costumes evoke regional culture, with a deep burden added to what was once a period costume. The different forms and ways of dressing express moments of everyday life and of an entire community.

The origin of things is born with the cultural and historical legacies of places and people that aim for the conservation and preservation, of what has been and can be,

> [c]ulture is defined (…) as a learned meaning system that consists of patterns of traditions, beliefs, values, norms, meanings and symbols that are passed from one generation to the next and are share to varying degrees of interacting members of a community» (Ting-Toomey e Chung, 2005, p. 28).

In order to present the different variations and offerings of traditional costumes, the investigation focuses the traditional national costume. In addition to the ways of life and the realities of day-to-day dressing, the religious, social, and economic character transform the perception of seeing fashion from and in local communities to general societal entities in a sense of inclusive design.

The focus on tradition and legacies of traditional costume enrich local communities with a sense of belonging, with their own concepts and values,

> integrated approach in which nature meets culture, the past meets the present, (…) contributes to the development of societies and the building of peace. By virtue of its multifarious origins and the various influences that have shaped it throughout

history, cultural heritage takes different tangible and intangible forms, all of which are invaluable for cultural diversity as the wellspring of wealth and creativity (http://www.unesco.org/en/culturaldiversity/heritage/ cit. in Csapó, 2012, p. 210).

As such, it is necessary for society to contribute to its understanding with creations and value proposals, in the fashion world, which in recent years have been noticed in emerging communities. With more and more adherence to fashion creation, it is necessary and urgent to promote products with an affective value, in order to appeal to our heritage, calling into question the ephemerality of consumption and making the process and the final product into emotional durable objects.

2 CASE ANALYSIS: THE TRADITIONAL COSTUME

The traditional costume goes back to historical periods of a given nation and a particular community, that is,

may be defined as the ensemble of garments, jewellery, and accessories rooted in the past that is worn by an identifiable group of people. Though slight changes over time in colour, form, and material are acknowledged, the assemblage seems to be handed down unchanged from the past (...) (Encyclopaedia of Clothing and Fashion, 2005).

We make an analysis of four traditional costumes - *Kimono, Vyshyvanka, Hanbok and Saree* (Figure 1) –, being the most expressive towards our object of study as a historical and contemporary concept, in order to understand the relation between the cultural heritage and appropriation by fashion design.

Therefore, in many cultures, traditional costume takes on a preponderant role in current fashion. The *Kimono* went through different chronological periods, as

born in the Heian period (794–1192). Straight cuts of fabric were sewn together to create a garment that fit every sort of body shape. It was easy to wear and infinitely adaptable. By the Edo period (1603–1868) it had evolved into a unisex outer garment called kosode (Green, 2017).

With Samurai being the first elite to use *Kimonos*, it eventually became a personal expression of wealth, coupled with new Western techniques of production and colour. Silk fabric was produced with higher quality and prices became more accessible to all, however the woman wore the costume in her domestic and outdoor life while the man only wore it in the domestic sphere. From the nineteenth century, with the introduction of chemicals, *Kimonos* gain vibrant colours that predate until the mid-twentieth century. Currently the *Kimono* is

Figure 1. Traditional female costumes of the world: *Kimono, Vyshyvanka, Hanbok* and *Saree*.

used mainly by older people, geishas, actors and servants of the restoration and in formal occasions that have the tea as the principle character. The reality of the *Kimono* of second hand is an attraction for the youngest who propose combinations with other pieces of clothing (Victoria and Albert Museum, n.d.).

For the second case, we analyse the *Vyshyvanka*, traditional Ukrainian male and female shirts that are presented as elements of denomination among the national decorative arts. They are known for their embroideries with vivid colours and ornaments of a sacred character, improved by women during the sixteen to seventeen centuries. The creative development was such that it allowed the creation of a style of its own, in which the cultural heritage was passed down from mother to daughter by generations. Embroideries are different from region to region and with their own embroidery geometries and techniques; there are six basic colours with a symbolic value and their own meaning—black, red, white, blue, green and yellow—with geometric, floral and zoomorphic symbols. The geometric motifs include e the cross, the circle, the rectangle, the wavy line, the six-pointed star, the diamond and the spiral; as for floral symbols they use oak, millefolia, poppy, grapes, lily and violet; in the zoomorphic elements the most used are the rooster, the peacock and the cuckoo with a strong mythical demarcation of the Ukrainian culture (Proud of Ukraine, s.d). It is still today, used either the complete costume, as far as play activities in Ukraine, and shirts are implemented in society, to this day for daily use. The *Hanbok*, Korea's traditional costume, is characterized by colourful clothing, worn as a formal attire, day-to-day clothing and holiday clothing. Its shape is due to the adoption of the Confucian style code—robes of Buddhist monks, called *Rakusu* –, the Hanbok allows an easy asy movement and is composed of four pieces: jeogori (coat), baji (pants, only for men), chima (skirt) and pho (outer cloak). The Korean costume design is analogous to the eaves of Korean traditional houses, with delicate lines and angles, and is ruled by the chromatic symbolism of ying and yang, with white representing metal, red the fire, blue the wood, black the water and yellow the earth. The chromatic range and the fabric type reveal the social status—lower classes with white or light attire and upper classes with gaudy colours: children wore bright colours; single women wore yellow coat and red skirt, while matriarchs wore green and red, respectively; women with children opted for navy blue. As for iconography they were embroidered with floral and zoomorphic themes, such as peonies, lotus, pomegranates, dragons, phoenixes, bats, cranes and tigers. *Hanbok* is seen as a Korean cultural legacy, exquisite, for its historical and artistic value (Ladner, 2017).

Saree is considered one of the oldest forms of clothing with more than 5,000 years, as the ancient writings (Vedas) mention the use of it by the Indus Valley Civilization, circa 3000 BC-. A single piece of cloth about 4.5 to 8 meters long, cannot be pierced with a needle for it will become impure, as the Hindu culture defends. In India it is assumed that the *Saree* evolves from 3 pieces, without seams: lower part, chest band and a worn part over the shoulders or head; as well as other coordinates will seek inspiration from the *Saree*, such as *lehengas* (ankle skirt), *ghagras* (long skirt with embroidery, mirrors and bells) and *cholis* (short-sleeved bodice). Traditional Indian attire allows warmth in the winter and refreshes in the summer, hence the choice of many women, especially from South Asia, given that variations in the drapery of the piece depend on the type of fabric used, cotton or silk, and weaving methods applied; there are more than eighty variations to drape the *Saree* and styles vary, regionally. *Saree* is now completely rooted in Indian culture, as it has always been (Nambir, 2016).

In all cultures, here presented, traditional costumes are elements of historical and cultural characterization, but they are also the reflection of a former fashion. However, the current fashion will draw from the past and its references the potential of traditional costumes as elements that characterize the history and culture of a community, in order to offer a reinterpretation that reflects this cultural heritage in new recreations by designers.

The need to allow fashion industry to pursue the tradition and culture of a nation becomes an asset for the continuity and persistence of concepts, symbolisms and iconographies, meaningful and that are spread throughout the world, for knowledge of the heritage of each location. In this sense, we make a relationship and analysis between fashion industry, culture and tradition in the following point, as we also analyse the ability of recreate to preserve, with works of designers, that take to the fashion shows their recreations.

3 CULTURE, TRADITION AND FASHION

The cultural identity assumed by the fashion industry is increasingly developed, in the sense that multiculturalism allows the tradition to be experienced in a unique way. Culture is settled from our existence, with a rational and strategic basis; it is in the culture that we review our experiences and references. Foucault (1996) argues that "fundamental codes of culture to which we belong, those who govern this language, its perceptive patterns, the hierarchy of its practices. These codes set the empirical content that participants of this culture can access "(Foucault, 1996, pp. 10–11, authors translation)".

The essence of culture in the world of fashion industry is characterized by the variety of techniques and materials that can be combined to meet the needs of the market, where Design plays a key role in its dissemination and transformation. The continuity of the use of traditional costume among regional communities, from generation to generation, demonstrates the collective acceptance of it, in the sense that it aims to remain in the society of consumption, opposing the ephemerality of clothing, the current trend, where fashion assumes the continuous development of original products. The course of the market, today, is moving too fast, from mass consumption we moved to fast consumption. In this sense, we need to create products aimed at durability, in order to achieve the emotional aspects of both the traditional offering, as the mass consumption, to appeal to our cultural heritage by means of affective products. As Ferro (1987) assumes,

> *fashion (...) even came to exceed the mere act of communicating, passing from medium to an end, the naked body itself needs customs to be able to shine, and it is the dress that makes women to be wanted, since "there are women who are too modest to dress themselves to better prepare the apotheosis of their naked bodies"* (Ferro, 1987, p. 32, authors translation).

In Portugal, as in many other countries, social differences were always noticed by dress, with the upper bourgeoisie imposing a fashion culture, based on its aesthetic needs and values, which was followed by the working class. In this research our target is the *Minho* traditional costume, linked to a mentality of traditions, with strong influence of what is produced in fashion. As Mattoso (1998) argues,

> *Portugal did not originate in an ethnic formation, but in a political and administrative reality. (...). The Portuguese State has added to itself a series of territorial areas with few ties to each other, with marked cultural differences and very different living conditions* (Mattoso, 1998, p. 67, authors translation),

however, the traditional regional costume gains importance by the identity of each region and the construction of a national consciousness; evokes an affective memory marked by pieces of clothing adapted to different situations of daily life, in view of social and cultural needs.

The role of Design today is to assume the cultural and operational dimension so that with tradition one can innovate, in a conscious and strategic sense in face of the qualitative changes of the life of our days.

3.1 *The Minho traditional costume*

The analysis of the traditional regional dress allows us to approach the Design in interdisciplinary terms: in the reinterpretation of the current tendencies. According to Albino *et al.* (2011), design is

> *a communication interface between past and future—is to propose new products that are not only seen as experiences' mediators, but are especially capable of promoting quality of life, fostering more intense experiences of places, acting as emotional relationships with the space through the indexing of the body, and anxious to resolve concerns of a symbolic nature* (Albino et al., 2011, p. 99).

In order to know the traditions we should not only focus on the past and habits of that time but we must know the people and their ancestral values, the techniques to be able to create, innovate and also to associate tradition and our cultural heritage with the dimensions of fashion design and to our identity. Thus, the *Minho* traditional costume emerges as an expression of various mindsets that have created groups and reflect a culture. For a long time the rural society, far from the influences of the city, wore dresses of past generations, with slow alterations, greater decorative detail and of choice of fabrics, and with formal style,

> *each one wears the same type of clothing as everyone else since generations ago, but each clothing is different: women do not spare work to add an embroidery or a lace in a way to differentiate themselves and became more beautiful* (Henri Merendas cit. in Esteves e Barreto, n.d., authors translation).

The *Minho* traditional female costume (Figure 2), which is the object of this paper, has general characteristics intended for ceremonial events, used on designated days, or weekends and according to Joaquim Pais de Brito (1996), the "minhota[1] dressed with the Minho traditional costume [also referred to as Vianese costume], composes a portrait in which the Minho and the country itself are reviewed" (Brito, 1996).

When we think of Portuguese women, the image that comes to mind is that of the *Minhota*. This costume consists of a scarf with a red background, printed with floral and plant elements, and cornucopias; a white shirt embroidered in blue—in the wrists, front and shoulders -; a vest, deeply embroidered with several colours of Baroque influence, that works like a corset with several functions: it defines the forms of the body, it allows a better movement of the arms and the border follows the line of the diaphragm, facilitating the breathing. The skirt is round and marked by a wide bar, embroidered with the same elements as the vest. The apron in pleats is decorated with pulls that recreate a magnificent "garden" in relief, with three base colours: black, red and yellow. The pocket in the shape of a heart, has the function of transporting money and the handkerchief. On her legs, the *Minhota* wears white lace socks and black slippers, embroidered with floral, vegetable and geometric motifs (Esteves and Barreto, n.d.).

For this research we analyse, in general, the *Minho's* costume for it is a manifestation that indicates the difference between the rural and the urban environment, with own marks referenced by the social hierarchy.

3.2 *Fashion as a challenge for the Minho costume*

Given the current state of many world economies, the fashion industry faces considerable challenges. It is necessary to create competitive market niches to obtain higher margins for the products, and consequently to have greater profitability,

Figure 2. *Minho's* traditional female costume.

---

1. Minhota—Portuguese term used to designate the women from the Minho region in the north of Portugal.

> *[f]inding your market profitability is an effective way of analysing your business' financial health. By understanding how forces influence parts of a given market, you can begin to make important decisions about starting a new business or ensuring your existing venture remains competitive in your chosen area* (Maguire, n.d., p.n.p.).

Modern design grows by products and services associated with a set of methods applied to a social aspect, that is, "the design process has once been identified, or a market opportunity has been detected" (Cunha and Broega, 2009, 864), and when we combine it with cross-disciplinary knowledge and disciplines, it becomes a holistic phenomenon.

The reinterpretation of cultural heritage and our legacy of traditions emerges in the textile sector as an added value both in the maintenance of Portuguese roots throughout the world and in the importance of creating more lasting and affective products that allow to reach and revive memories and historical events. The designer must then use and recover tools so that he can respond to the needs and expectations of the consumer; and should also take fashion design as a structured, tested and reflected activity, combined with ancestral know-how in complementarity with current techniques and materials.

4 CASE STUDIES: REINTERPRETATION AND RECREATION OF TRADITIONAL COSTUMES

More and more designers apply reinvention and creativity inspired by the traditional costumes of their nations; the demand and recreation of the traditional legacy is now a reality in the world of the fashion industry. Therefore, in this research, we propose a comparative approach between two costumes mentioned above: *Vyshyvanka* (Ukrania) and *Hanbok* (Korea).

Vita Kin is a designer and photographer who recreates the traditional Ukrainian shirt to create dresses, blouses and overalls. She creates and produces pieces in fine linen, with embroidery. Her inspiration is also based on the Aztec culture; Portuguese, through cotton and blankets and Greek with pompoms. The colors are of Guatemalan inspiration. The first step towards the recreation of Vita Kin is through the hand of two artisans in the production of the "virgin" dress. Only later, embroideries are placed where Vita Kin "adds a contemporary twist; a midi-length line, a saucy to-the-thigh split, bigger sleeves, bolder embroidery and a hint of the Seventies". Vita Kin assumes that:

> *[f]or me, it was always important to try to preserve the authentic feel of the hand embroidery, so that it would look as if it was made by the old masters. Because I know*

Figure 3. Recreation of the Ucranian traditional costume by the designer Vita Kin.
Source: https://www.vogue.com.

Table 1. Comparative approach between the traditional Ukrainian costume and the reinterpretation by the Ukrainian designer, Vita Kin.

Traditional costume: *Vyshvanka*	Reinterpretation/Recreation by the designer Vita Kin
Brief description: – Shirts embroidered with bright colors; – Ornaments of a sacred character; – Cultural heritage passed down in generation; – Different embroidery and costumes, by region. Traditional costume elements – Embroidered shirt and skirt; – Apron with small pattern; – Wreath on the head—*vinok*	Brief description: Vita Kin is a designer and photographer who recreates the traditional Ukrainian shirt to create dresses, blouses and overalls. *For me, it was always important to try to preserve the authentic feel of the hand embroidery, so that it would look like it was made by the old masters.* *Because I know exactly how I want my things to look, I broke all the rules about modern embroidery machi-ne technology. (Vita Kin cit. in Fair-a-porter, 2015)*
Fabrics: cotton and wool	Pieces made of fine linen, with Aztec inspired embroidery, Portuguese cotton and blankets and Greek pompons. It takes two craftsmen, for two days, to make a dress.
Colors and meaning: • black—land and family livelihood • red—passion, happiness, love and joy • white—purity and holiness • blue—water, sky and inner peace • green—spring and youth • yellow—stars, moon and separation	Guatemalan inspired colors
Processes: – By hand or machine; – Handmade embroidery.	With inspiration in embroidery, *[s]he adds a contemporary twists; a midi-length line a saucy to-the-thigh split, bigger sleeves, bolder em broidery and a hint of the Seventies (Fair-a-porter, 2015).*

Iconography
Geometric elements:
  Cross: sun and light
	Circle: solar signs
	  Rectangle/Square: Fertility and Earth
	  Wavy line: water
		Hexagon: amulet
		  Spiral: secret knowledge
		Floral elements:
		  Oak: sacred tree, protection of diseases
		  Millelet: symbol of the Ukrainian nation
		  Poppy: ancestors and eternity
			Grapes: fertility
			  Lily: mystery of life
				Violet: youth
				Zoomorphic elements:
				  Rooster: happiness, male health, wealth and fire
				  Peacock: family happiness
				  Cuckoo: mythological figure, linked to the other world
					Innovation and cultural heritage Vita Kin is the ultimate exponent of the return of the Ukrainian traditional heritage; brings traditions into the modern context. She matches old to modern patterns, programmatically, tested for several weeks. The assembly of sleeves, collars and cuffs is handmade according to methods previously used. (Fair-a-porter, 2015).

exactly how I want my things to look, I broke all the rules about modern embroidery machine technology (Vita Kin cit. in Fair-a-porter, 2015).

Vita Kin makes matching patterns from old to modern, programmatically, tested for several weeks. The assembly of the sleeves, collars and cuffs is done by hand, according to methods used in the past (Fair-a-porter, 2015). Vita Kin is the ultimate exponent of the return of the Ukrainian traditional heritage; placing the traditions at the service of all and available to the current context.

Leesle Hwang finds inspiration in the *Hanbok* and there is a reinterpretation in terms of cultural heritage that goes from traditional costume to the recreated one, with an intrinsic connection between the past and the present. Leesle Hwang published a book "I go to Hongdae in hanbok", where she reports all her experiences over 8 years. In pursuit of tradition and its roots, Leesle Hwang uses the same colours of traditional costume in the contemporary proposals she presents to the public. Using latest generation Information Technologies, she reproduces and applies ancestral techniques.

In this recreation, the designer appropriates the composition of the traditional costume, in order to make it contemporary, in light of today's offer. For this, the designer applies methods of composition, patterns, materials and details full of skill, linked to its cultural heritage. Using classic elements with wearable designs, transforms the traditional patterns and structures in simple materials, although seeking the tradition of fabrics and fibers, as in the case of working with cotton and linens. The designer assumes that the fashion she creates has its value in overcoming the regional and cultural level:

> *[t]he person in Leesle can be anyone from any country. Leesle reaches forward to create clothes that bring out one and only personality, clothes that give an exhilarating experience, and clothes that you want to have in envy. Leesle is dreaming of becoming a global fashion with a new perspective of looking at tradition* (Leesle Hwang cit. in Leesle, n.d.)

In the same way that these designers work the traditional costume and manage to transport the local cultural heritage into a contemporary language, the *Minho* costume has the same potential to be analyzed and worked in this sense of the creative intervention. Particularly for its strong presence in the daily life of Minho, it presents itself as an important reference for the development of less ephemeral fashion products, in a perspective of greater durability. In addition to the great variety of the visual grammar, from the forms, to the materials, passing through the colors, the costume has a series of elements that can be worked in the sense of a recognized and historical cultural presence, so that we can preserve this iconography, unique and singular.

Figure 4. Recreation of the Korean traditional costume by the designer Leesle Hwang.
Source: https://www.kooding.com/.

Table 2. Comparative approach between traditional Korean dress and reinterpretation by Korean designer Leesle Hwang.

Traditional costume: *Hanbok*	Reinterpretation/Recreation by designer Leesle Hwang

Brief description:
– Colorful clothing, Confucian style;
– Pieces allow fluidity of movement;
– Thin top and wide bottom;
– Ceremonies and special occasions.
Traditional costume elements
  – *Jeogori*—coat;
  – *Chima*—skirt;
  – *Pho*—outer cloak.
	Brief description:
	Leesle Hwang produces her design based on the traditional style of Hanbok, with an intrinsic connection between the past and the present.
	The person in LEESLE can be anyone from any country. LEESLE reaches forward to create clothes that bring out one and only personality, clothes that give an exhilarating experience, and clothes that you want to have in envy. LEESLE is dreaming of becoming a global fashion with a new perspective of looking at tradition. (LEESLE, n.d.)

Fabrics: linen, ramie and cotton. Fabric coloring made with natural dyes.	Classic motifs with wearable designs that make traditional patterns and structures in simple materials with cotton and linen
Colors and meaning: Ying & Yang and the 5 Elements • white—metal • red—fire • blue—wood • black—water • yellow—earth	In pursuit of tradition and its roots, Leesle Hwang uses the same colors of traditional costume in the contemporary proposals he presents to the public.

Processes:
– sewing techniques and straight-line fabrics, linear and flat woven fabric.
Iconography
  the patterns on the bottom of the skirt were only allowed to the royal family
Floral elements:
  Peonies: desire for honor and wealth
  Lotus: hope for the nobility
  Pomegranates: children's desire
Zoomorphic elements:
  Dragons: representation intended for the King
  Phénix: representation for the Queen
  Bats: children
  Grous: civil Officials
  Tigers: military officers
	Using state-of-the-art Information Technologies, the designer reproduces and applies ancestral techniques.
	Innovation and cultural heritage LEESLE is a fashion brand that has its value in overcoming the regional and cultural level.
	Leesle Hwang published a book "I go to Hongdae in hanbok", where he recounts all his experiences over 8 years.
	Methods of composition, patterns, materials and details full of skill, linked to their cultural heritage are the basis of LEESLE (http://leesle.kr/about2/).

5 CONCLUSIONS

It is increasingly necessary to produce from the inside out, in order to respond to consumer expectations, with personalized effect, in order to promote and value the cultural richness of popular tradition, with reinvention and reconcile with the artistic and cultural communities. With this new contemporary language, products take on a greater value by themselves, revealing the need to recover traditions in fashion design. There are already many designers who invest in a perspective of fashion show, in a reinterpretation mode adapted to our societies and people, where they propose and create spectacle in the fashion industry, being seen and recognized all over the world, from Paris to New York.

On the other hand, the evolution in the development of materials, specifically of intelligent textiles, opens doors for the development of potentially innovative products, with a focus on sustainability, as opposed to fast-fashion.

ACKNOWLEDGMENTS

This research is supported by FEDER funding in the Operational Competitiveness Factors Program—COMPETE and by national funds through the FCT—Foundation for Science and Technology –, under the project POCI-01-0145-FEDER-007136 and UID/CTM/00264.

REFERENCES

Albino, C. et al, 2011. Reinterpretation of tradition values in Minho territory: handcraft—a Reading key. *Strategic Design Research Journal*, 4 (2), p. 99–105.

Brito, J.P., 1996. *Retrato de Aldeia com Espelho: Ensaio sobre Rio de Onor.* Lisboa: Publicações D. Quixote.

Csapó, J. (2012). The Role and Importance of Cultural Tourism in Modern Tourism Industry [Online]. In Kasimoglu, M. & Aydin, H. (eds.). *Business, Management and Economics: Strategies for Tourism Industry—Micro and Macro Perspectives*, Chapter 10. ISBN 978–953–51–0566–4. Available at:: https://cdn.intechopen.com/pdfs-wm/35715.pdf (Accessed: 17 February 2018).

Cunha, J. & Broega, A.C. 2009. Designing Multifunctional Textile Fashion Products. *AUTEX 2009 World Textile Conference*. Turkey: Izmir. Available at: http://repositorium.sdum.uminho.pt/handle/1822/19207 (Accessed: 17 February 2018).

Esteves, L.M. & Barreto, A.C., n.d. *Aspectos do Traje Popular nos arredores de Braga na mudança do século (XIX-XX)*. [Online]. Available at: http://www.folclore-online.com/trajos/regioes/minho/aspectos-traje-popular-arredores-braga2.html#.WqWtSOjFLIV (Accessed: 20 February 2018).

Fair-a-porter. 2015. Stories—Vita Kin: When traditions become fashionable. Available at: http://fairaporter.com/vita-kin/. (Accessed: 17 February 2018).

Ferro, A., 1987. *Teoria da Indiferença. Em Obras de António Ferro*, 1.º Volume. Lisboa: Editorial Verbo.

Green, C. 2017. The Surprising History of the Kimono. *JSTOR Daily*. Available at: https://daily.jstor.org (Accessed: 20 February 2018).

Johnston, L. (2015). *Encyclopedia of Clothing and Fashion.*

Ladner, M. 2017. Hanbok: An Introduction to South Korea's National Dress. [Online]. In Home: Asia—South Korea. Disponível em: https://theculturetrip.com/asia/south-korea/articles/hanbok-an-introduction-to-south-koreas-national-dress/ [Accessed: 20 February 2018].

Leesle, n.d. Available at: http://leesle.kr/about2/. (Accessed: 17 February 2018).

Maguire, A., s.d. *How to determine your market profitability.* [Em linha]. Quiick Books: Business planning. Disponível em: https://quickbooks.intuit.com/r/business-planning/determine-market-profitability/ [Accessed: 20 February 2018].

Mattoso, J., 1998. *A Identidade Nacional*, Lisboa: Gradiva.

Nambiar, S., 2016. *A Brief History of India's Traditional Saree.* [Online]. Disponível em: https://theculturetrip.com/asia/india/articles/a-brief-history-of-indias-traditional-saree/ [Accessed: 120 February 2018].

Proud of Ukraine, s.d. *Ukrainian embroidery: history, regional features, colors and patterns.* [Online]. Disponível em: http://proudofukraine.com/ukrainian-embroidery-history-regional-features-colors-and-patterns/ [Accessed: 17 February 2018].

Ting-Toomey, S e Chung, L., 2005. *Understanding Intercultural Communication.* New York: Oxford University Press.

Victoria and Albert Museum, n.d. Content—Articles: A History of the Kimono. [Online]. Disponível em: http://www.vam.ac.uk/content/articles/h/a-history-of-the-kimono/ [Accessed: 17 February 2018].

The casualisation and homogeneity of contemporary fashion

F. Spry
University for the Creative Arts, Epsom, UK

ABSTRACT: Fashion is a key indicator of social conditions on a global scale. As fashion has lost the power it once held in terms of being indicative of status, consumers no longer wish to stand out and portray individuality through their clothing. In fact, anxieties to fit in within a local and global context have produced an international homogeneity of dress. The purpose of this paper is to explore the occurrence of clothing casualisation and homogenisation within the 21st century. I analysed the cities of London, New York and Tokyo as case studies and have utilised street photographs capturing inhabitants of each city in order to compare them. The key theorists that have aided my observations include Hill (2005) and Lipovetsky (1994). Hill discusses the lack of taste in clothing within the 21st century and the environmental impacts on dress. Lipovetsky explores the loss of interest towards clothing and the subsequent homogeneity thereof. These theories have allowed me to examine and identify how and why the casualisation of clothing has occurred within the 21st century and how this has led to homogeneity in dress on a global scale.

1 INTRODUCTION

It can be argued that today's society has more freedom and ability to express and identify through dress than ever before. With hundreds of shops, both on the high street and online, consumers have millions of options to choose from. Accessing fashion has become easier than ever before; technological developments and the increase in large global brands has led to the mass production and distribution of clothing at an unprecedented rate. It is now possible to order garments and receive them within the same day without even having to get out of bed. It is not a requirement to have a large amount of disposable income as it is now possible to buy many garments with limited funds due to cheap production costs. Likewise, access to media is also far greater and consumers can see hundreds of clothing options and trends in magazines, on the television, on blogs, websites, videos and of course social media. Advertisements also inform consumers as to exactly what kind of look and lifestyle they can buy into.

However, despite having easier access to fashion and being able to consume more readily, urban inhabitants, in an increasingly urbanised world, are continuing to look more homogenous by the year. With similar shapes, cuts and colours, it seems there has been a severe decline in the ability, or indeed desire, to dress ourselves in a way in which we feel individual. Our aspiration to fit in has led to a continuously homogenous fashion scene. However, it can also be argued that it is not just our eagerness to belong that has led to the homogenisation of dress but the fact that we all feel the same way and thus express ourselves through fashion with increasing similarity. Hill, after observing people on Oxford Street in London, described them as being "indistinguishable by their clothing." (Hill, 2005, p.67). The fact that consumers dress so similarly has been normalised. When actually scrutinising other's dress it soon becomes apparent that Hill's observation is not an anomaly but, in fact, when studying the clothing of inhabitants of most cities in the world, particularly those where globalisation is the strongest, the homogeneity in dress will be more noticeable.

The fact that modern urban dwellers have far more choice than ever before when it comes to clothing consumption has led to many turning their backs on fashion. Even if an avid follower, the changes made in terms of style from one season, even year, to the next, will probably be very minor. The vast choice that consumers have allows the freedom to choose whatever style desired. However, it is far easier to decide what to buy and wear by looking at those around us. This takes away the stress and confusion that may come with making clothing choices and alleviates any anxieties about not fitting in if the wrong choice is made. By wearing what everyone else is wearing, you are sure to be safe. Fashion has always been about looking at others; inspiration used to come from the royals, then from other prominent figures such as first ladies, actresses, singers, then from celebrities and influencers and, now, from each other.

To explore the impact of globalisation and the influence of our environments on fashion, I have used the case studies of London, New York and Tokyo. I have used photographs taken in the streets of each of these cities with inhabitants going about their daily business as the subjects. These urban scenes have allowed me to analyse examples of what inhabitants wear in each city and then compare them with one another and consider the similarities that are made apparent in each. I chose to study London as Hill already discusses the streets of London in his paper (2005) and I wished to explore this example further. New York allowed me to examine a similarly Western city which I know has much influence upon not just the world of fashion but on the world itself and all forms of culture. I also chose Tokyo as it is not a Western city, but it is one that has been influenced by Western culture, even though its own historic culture is largely differential. The impact of globalisation upon a city such as Tokyo becomes apparent through the use of comparing these photographs. It is through these secondary ethnography comparisons that I have been able to explore the loss of distinction in dress in each of these cities. The choice of these cities, and subsequently of these photographs, has been intrinsic in my observations of how what is happening globally affects what we wear. The way they influence and have been influenced by one another has led me to count them as an emblematic object that reflects a global scenario.

I have also utilised the work of theorists Hill (2005) and Lipovetsky (1994) amongst others to support my analysis. Hill explores the link between urban environments and how they have an effect on the type of clothing we wear and how we feel about fashion in general. He also suggests that an increasingly urbanised lifestyle has led us to lose passion when it comes to dressing ourselves as we lack inspiration due to simple architecture and design along with busy working lifestyles. Finally, Hill explores the relation between our increased access and choice of fashion and the decrease in our involvement to make it our own. Lipovetsky analyses the homogenisation of fashion and how a shift from being constricted to certain clothing due to sumptuary laws, to having free choice of clothing, has created a disinterest in dress having removed the importance it once held. These theories have allowed me to develop the present work, as they investigate fashion in the 21st century and how history, along with our local environment affects us. They have also allowed me to realise how events happening in a global context have influenced how fashion has become both more casual and more homogenous.

2 THE RELATIONS BETWEEN CASUAL WEAR AND IDENTITY

The relatively recent development of casual wear has been perhaps the largest factor leading to the decline of individualism through dress. Sumptuary laws, regulated by the government, once visibly separated people into classes by enforcing what they could and could not buy and wear. However, from the start of the eighteenth century these were not renewed, and people could wear and buy whatever dress they desired within their economic capacity (Chrisman-Campbell, 2011). Without these laws and the obvious distinction between social status, clothing began to lose the power it once held. As the rules relaxed, so too did the clothing. As a result, clothing worn today is predominantly comfortable, practical and simple, with people no longer expressing their perceived worth. This is due, in part, to the transformation of class citizens and their lifestyles since this era. From peasants and servants, working in rags or uniforms, to having more fulfilling careers, working for themselves, rather than directly "serving" someone else. This new class that fits into neither slave or servile, nor that of royalty or aristocracy has changed the way we view dress. Today, it can be said that your "everyday consumer" would be a member of the middle class which developed during this transformation and is an extension of the 19th century bourgeoisie. Present day consumers are able to work for themselves and have enough disposable income to afford garments other than the ones that they work in, thus they are able to define themselves as more than just their occupation. This has led to a certain nonchalance when it comes to the clothes that we choose to wear and has indeed led to a general casualisation in fashion.

Choosing to purchase more comfortable, casual clothing has led to many wearing very similar garments, as Lipovetsky expressed: "Everybody looks like everybody else" (Lipovetsky, 1994, p. 123). Following the eradication of sumptuary laws and the loss of regard for the expression of status, coupled with fast fashion and the ability to manufacture clothing quickly and cheaply gave consumers the opportunity to wear whatever clothing style they wanted. Subsequently, practically nothing was off-limits and the aspirational urge to have something unattainable was lost and as a result so was the importance and sentiment people held towards dress. Lipovetsky argues "clothing no longer arouses the interest or passion it used to elicit" (Lipovetsky, 1994, p.120), which is apparent through the way certain groups of individuals dress today; there is often no devotion or excitement contributed to the acts of consuming or dressing. As a result, a true sense of individualism or identity is not being displayed. The decline in needing, or indeed wanting, to distinguish oneself through dress is evident.

Casual wear, according to Collins English Dictionary: "*noun:* informal articles of clothing or footwear" (2014) is clothing, this definition suggests, as not being suitable for a formal occasion. Yet, casual wear is so common in our society today that many offices have "dress-down Fridays" or "casual Fridays". The office is an environment where casual wear would, and perhaps should, otherwise be deemed inappropriate. Originating in Hawaii as "Aloha Fridays," a day in the week to wear a Hawaiian shirt to work, the idea of a casual wear day soon spread to the rest of the United States and then the world (Clark, 2012). To wear a more casual look to work became something that companies could do cheaply to allow their workers to relax (Clark, 2012). It is, conceivably, the wish of employers to keep employees happy; to allow workers to be more at ease and dress as themselves. However, by dressing as themselves, it is all too common that they dress just like everyone else. Dressing up in the office is an important part of our culture and society. We "dress up" to show our seriousness, respect and obligation to our occupation. The clothing we wear not only portrays to others what kind of outlook we have, or what kind of individual we are but also has the power to depict a persona. By dressing down on a Friday, confusingly both our work and home personas and outlooks are merged. A more relaxed individual is portrayed but in a formal setting in which the clothing is not always appropriate.

It is interesting to consider office workers within the context of their work environment. This will be the place that they spend most of their time and it is probable that their colleagues will, whether they like it or not, be the people they spend the most time with on a daily basis. Average actual weekly hours of work for full-time workers in the United Kingdom was 37.2 hours per week in September to November in 2017 (Office for National Statistics, 2018). This means that over five occupational days, workers are spending, on average, 7.44 hours working per day. This is similar to the USA which, in 2016, had average weekly working hours of 34.40 (The Organisation for Economic Co-operation and Development, 2017). This presents an average of 6.88 hours working per day. Japan's was slightly lower in 2016 at 33.25 (The Organisation for Economic Co-operation and Development, 2017). However, this still presents an average of 6.65 hours working per day. This data shows that workers within these countries spend much of their time during each day in their occupation. Within an office environment, many of the employees will possess jobs that are similar in terms of intellectual ability and economic status to the others that they work alongside. Living in the same location, with similar incomes, workers will often have home lives that aren't that dissimilar. To study office workers and their lives in this way brings to light the similarities that one colleague would hold with another and suggests that the idea that they would wear similar clothing outside of work or when "casual Fridays" come around as not incomprehensible.

The similarities of those around us is something that can be studied on a global scale. Technological advancements have allowed most people to access technology such as smart phones and the internet. This access allows the spread of information and has influenced how we all experience the world around us. If we are all exposed to similar influences, such as particular advertisements which are prominent on different social media platforms, then this is going to alter how we perceive and experience certain things. If we are all viewing the same advertisements, campaigns and information then it is unsurprising that our perceptions and experiences of clothing begin to become homogenous. After all, our experiences play a large part in shaping how we choose to portray ourselves. However, it can still be argued that we have input into our own experiences both on and off social media. In fact, social media allows us the opportunity to portray ourselves to the world as an individual far more than ever before.

When it comes to casual wear we must consider what it is that makes people feel more at ease in this attire. After all, employers use it as a tool to relax employees so there is definitely something about this particular type of clothing that makes wearers calm. The looser cuts and more relaxed styles are clearly important factors but why is it that so many individuals simultaneously feel at ease wearing the same thing? The very anxiety to belong within a certain group, be it our work colleagues, our neighbours, our friends and the people we follow on social media has led to the homogenisation of clothing through similarities held with others.

With no overarching rules telling consumers what they can and cannot buy and wear "the meaning of clothing flattens out, it empties, it fades." (Hill, 2005, p. 73). Many then turn to advertisements and magazines for guidance on what to wear and continue to buy clothing every season in the new colour, even if the cut is the same. Lynch and Strauss argue that consuming the same products: "reduces anxiety and improves self-image…resulting in higher levels of individual security." (Lynch and Strauss, 2007, p. 30). Our instinctual need to belong in a group, as humans, has led us to feel secure when we have the same products as those around us and this is what drives consumer culture. Consider the neighbourhoods where each residence has the same extension built from it, the same style or make of car parked outside it and the same breed of dog playing in the garden. These neighbourhoods allow us to see how our anxieties to fit in within a society impact consumption. Clothing is something that can always be seen and immediately marks us as someone who belongs within fashion, or indeed society, or someone who does not. By continuing to consume in a way that effectively imitates those around us, whether subconsciously or not, individuals can always be in fashion, and therefore blend in. Barthes argues that: "Fashion is health, it is a moral code of which the unfashionable is nothing but illness or perversion." (Barthes, 2006, p. 68). To pursue fashion is not only due to a deep rooted will to survive within society but it is due to our perseverance in following what is good, healthy and right. The motivation of following fashion, for many, is as a means to achieve not just inclusiveness but overall happiness.

3 CASUAL WEAR AND HOMOGENEITY IN A GLOBAL CONTEXT

The ability to see what others are doing, wearing, watching, eating and consuming across the world instantly has made it smaller. Globalisation and the continual development of technology has allowed for the spread of cultures, chiefly the dominant Western culture, across the world. Globalisation via mass media has led to an often-unavoidable suppression of locality leading to our global cities losing the traits and marks which define and separate them. This is apparent through the clothing worn in these cities and the similarities that are evident when studying them.

Figures 1, 2 and 3 all depict crowds of people walking the streets of major cities in the world: London, New York and Tokyo. These images have been collated as they depict comparable scenes and the similarities in dress of the subjects are obvious. Predominantly, individuals wear jeans and a dark, short jacket and silhouettes are reasonably slim line throughout. The colours remain practical; blocks of dark colours feature the most. The images have a gap of four years between them; Fig. 2 was taken in 2012, Fig. 1 in 2014 and Fig. 3 in 2016. It is compelling to note that although fashion is so fast paced and constantly changing, that the overall style of dress has remained largely the same across this time scale. Through these images we can deduce that even though small details may change, dress has and will remain predominantly the same within this era. Fashion, as an industry, must generate variation

Figure 1. Shoppers walk to stores on Oxford street, (2014) Ben Pruchnie http://www.gettyimages.co.uk/photos/ben-pruchnie-oxford-street?excludenudity=true&mediatype=photography&phrase=ben%20pruchnie%20oxford%20 street&sort=mostpopular (Accessed on 18/11/17).

Figure 2. Crowd of people walk across street in midtown Manhattan (2012) blvdone https://www.shutterstock.com/video/clip-2271371-stock-footage-new-york-circa-april-crowd-of-people-walk-across-street-in-midtown-manhattan.html (Accessed on 18/11/17).

but consumers are not looking for dramatic changes, so only small details are adapted from season to season. Fashion journalist Jamie Wolf argues "Fashion is based on change…which change, exactly, is unimportant." (Wolf, Cited in Steele, 1997, p.152). This links back to the idea of consumer culture and the need to buy what everyone else is buying. Our sense of security is heightened when we have made the same purchases as those around us. For this reason, if fashion changes slightly and the plain tank top that was in last season, is in this season but with a frill on it, then that is why it becomes a trend. It is, undeniably, the same product but a frill has been added and voila, it is a new style and we buy into it because everyone else does. The same overall style of casual wear persistently remains.

    The consistencies of dress in these images are apparent and it would seem that jeans, jumper, jacket and trainers have, to some extent, become a uniform throughout the years. Craik refers to the homogeneity of dress as "everyday fashion" (Craik, 1994, p.204). This indicates the casualisation of clothing as fashionable due to most people choosing to wear it. The popularity of casual wear has granted it its place in fashion. However, modern dress could be described as more of a social uniform, a term which is certainly less amiable than Craik's "everyday fashion" (Craik, 1994, p.204). A uniform is something that someone is required to wear, usually for their occupation. It can be argued that, socially, casual wear is

Figure 3. People walking in slow motion in the streets of Tokyo Japan crowded pedestrians commuters walking in the city (2016) Zanuck https://www.videoblocks.com/video/tokyo-japan-circa-november-2016-people-walking-in-slow-motion-in-the-streets-of-tokyo-japan-crowded-pedestrians-commuters-walking-in-the-city-documentary-editorial-image-rtbtvhjkxiz4ogfeu/ (Accessed on 18/11/17).

a requirement. As discussed before, those that are not purchasing and wearing the same as everyone else have lower self-image and security and are excluded as they do not fit in. In order to find belonging then, one must be wearing what is in fashion and casual wear seems to be the persistent trend in the 21st century.

The cities in the figures are geographically thousands of miles apart, yet the people that live and work within them all seem to dress alike. When considering the geographic distances and the rich cultural histories that all these cities possess it is disconcerting to acknowledge that the similarities in dress are so apparent. London, for example has incredibly distinct cultural history which has both extended into, and been determined, by clothing. The diversity of clothing and ideas have been vast, for example the many subcultures and their identifying features. These groups of people, such as the punks in the 1970's, used dress to distinguish themselves from others in society (Vannini & Williams, 2009). New York has also been the centre of many cultural events, particularly those involving music such as jazz in the 1940's and disco in the 1970's (Edmondson, 2013). These styles of music had a massive amount of influence on what people wore in the city. Compare this to the traditional Eastern values and costume that were both so intrinsic to places like Tokyo, where kimonos and other customary garments are now often only worn for special occasions. We must also not forget the relatively recent street styles of the 1990's and early 2000's where Harajuku style, gothic Lolita's and bright anime hair could all be seen in the streets of Tokyo (Egli, 2015). This leads us to ask the question: how have these places lost their distinction in dress?

The development of mass media and rapid globalisation has closed the gap between countries and cultures and created a kind of homogeneity which has not been seen before. Giroux refers to this as a "world without borders" in which "differences" become unified (Giroux, 1993–4, p. 18). Gibson argues that the development in technologies in the eighteenth century pushed Western society from rural to urban and allowed the spread of mass media (Gibson, 2011, p.40). Gibson goes on to argue that this, along with political events, such as the French Revolution in 1789, created modern capitalism and as society moved away from aristocracy, consumer culture as we know it today began to emerge and spread across the world (Gibson, 2011, p. 40). The spread of globalisation, continuing Western dominance and consumer culture has led to the lack of distinction between cultures, and indeed individuals, through dress. Maynard refers to this as "global attire" (Maynard, 2004, p. 33), and is an accurate description of the unification of dress across the world.

Western fashion becoming the dominating style is a reflection on attitudes from Western cultures in terms of what is regarded as normal and appropriate, or healthy and good, as stated by Barthes (2006, p. 68). Historically, there has been a lack of regard from Western countries in terms of the importance of other cultures, their traditions and what they choose to wear: on the contrary, cultural manifestations from places outside the Western empires were considered culturally inferior, or even "savage" (Lévi-Strauss, 2012, p.12). As technology, mass media and social media have developed over the years we can see how these attitudes have been influential. London and New York are cities that have always held cultural inspiration and influence. Casualisation has had a massive effect on fashion around the world and it has now become stylish to blend in as Egli discusses in his article on the depletion of subcultures in Tokyo: "…the continued popularity of Normcore in Japan—a unisex fashion trend characterised by unpretentious, average-looking clothing—means standing out from the crowd isn't necessarily the 'in' thing anymore." (Egli, 2015). Even casual wear has sub categories and subcultures and from once loud, bold, colourful styles, in Tokyo, fashion is now about blending in, rather than standing out.

It can be argued that urbanisation has played a critical role in how international dress has become homogenous and lacking in individuality. London, for example, is built up, crowded and often lacking in creative stimuli. Public transport systems, particularly the tube, forces users to be pressed up against one another and generally endure the discomfort and boredom that comes alongside this. When speaking of London, Hill argues "It is not the type of environment that encourages people to dress in a more distinctive or individually spectacular way." (Hill, 2005, p. 70). Our clothing reflects the environment that we are in, most people moving about the city are commuting; moving quickly around a built up, crowded area. As a result, the clothing is practical, functional and plain. This can be seen in all the figures, particularly Fig. 1 where the linear architecture and plain colours of the city are clearly reflected in what people are wearing. Our need to blend in does not just stop at blending in with one another but with the urban landscape around us. The unification of architecture directly influences the clothing that we wear. Cities around the world have become more and more similar, not just culturally but physically. Urban architecture has developed in a way similar to fashion; effected by globalisation and the trend of modernistic simplicity, it has become increasingly unvarying from one city to the next. Finch argues that in modern architecture there has been: "… an emerging condition in which local tradition and identity comes under threat." (Finch, 2012). The way in which modern architecture has evolved has been due to the effect of globalisation and a dismissal of distinctive cultural features. Thus, the homogenisation of architecture has not only been influenced by globalisation, like fashion, but has also, itself, influenced fashion.

4 CONCLUSION

It can be argued that dress does help subjects to identify themselves as it does allow them to fit in in a local and global context. The world has become smaller, cultural distinctions lesser and perhaps we all feel like, as Hill described: "distinct individuals in the same way" (Hill, 2005, p. 75). The dominance of Western culture has led us all to believe that we are only our true selves in casual wear. This would then be indicative of the kind of individuals we have become: lost in a world of consumer culture.

We all go to different shops, buy different garments and style them in a way in which we believe portrays our identity and expresses our freedom to wear what we want. However, when we all have the same image of our cultural identity, this idea of our own individuality is diminished, as, without becoming aware of it, our own sense of identity through dress is more than likely to be the same as that of millions of others. It is apparent that dress is more a reflection of society than it is of the individual. As Cunningham and Lab argue "[clothes] have the potential to reveal a great deal about society." (Cunningham & Lab, 1991, p. 2). Fashion reveals our technological advances as it is influenced by mass media and globalisation. It also reveals the kind of work we do and the environments in which we live, as well as revealing our desiderata to belong.

In conclusion, it is clear that dress has become largely unified across the globe. Our anxieties to fit in have led us to draw influence from one another and follow trends that are made of small alterations to the persistent overarching trend of casual wear. We also have a desire to blend in with our surroundings and urban architecture has a large impact on how we choose to dress ourselves. Like fashion, modern architecture is influenced by the global movement of homogenisation through globalisation and as with most forms of popular culture, is a result of events or activities within an environment; in this case it is global. The way in which we live in and around these buildings also has an effect on the way we dress and has led to more practical clothing that does not mark us apart from one another. Likewise, the people that we spend our time with; possessing similar, fast paced lifestyles, influence the clothing that we buy and wear. Again, it is within our nature to feel more confident and secure when we know that we have what others have. It is also common to have the same interests and opinions to those that we spend a lot of time with and have similar lifestyles to, and this is where the homogenisation really becomes apparent, as, globally, city dweller's lives are becoming more and more similar. Finally, as fashion, and culture as a whole, have become less diverse, our individuality has followed suit. It is true that our sense of identity through dress is probably akin to that of the person next to us.

REFERENCES

Barthes, R. 2006. *The Language of Fashion*. UK: Berg, 68.
Chrisman-Campbell, K. 2011. *From Baroque Elegance to the French Revolution: 1700–1790*. New York: Berg, 37.
Clark, K. 2012. *Dress code: The History of 'Business Casual'*. Podcast accessed from https://www.marketplace.org/2012/08/17/business/workplace-culture/dress-code-history-business-casual (06/12/17).
Craik, J. 1994. *The Face of Fashion: Cultural Studies in Fashion*. UK: Routledge, 204.
Cunningham, P & Lab, S. 1991. *Dress and Popular Culture*. Ohio: Bowling Green State University Popular Press, 2.
Edmondson, J. 2013. *Music in American Life: An Encyclopedia of the Songs, Styles, Stars, and Stories that Shaped our Culture*. USA: Greenwood.
Egli, J. 2015. *What the hell has happened to Tokyo's fashion subcultures?* Online article accessed from http://www.dazeddigital.com/fashion/article/28687/1/what-the-hell-has-happened-to-tokyo-s-fashion-subcultures (accessed on 13/02/18).
Finch, P. 2012. *Architecture has globalised, but design can still be different*. Online article accessed from https://www.architectsjournal.co.uk/opinion/architecture-has-globalised-but-design-can-still-be-different/8636642.article (accessed on 13/02/18).
Forsyth, M. 2014. *Collins English Dictionary*. UK: Collins.
Gibson, P.C. 2011. *Fashion and Celebrity Culture*. London and New York: Berg, 40.
Giroux, H. 1993–4. *Consuming Social Change: The "United Colors of Benetton"*, Minnesota: University of Minnesota Press, 18.
Hill, A. 2005. *People Dress so Badly Nowadays: Fashion and Late Modernity*. New York: Berg.
Keedwell, P. 2017. *Headspace: The Psychology of City Living*. UK: Aurum Press, 1.
Lévi-Strauss, C. 2012. *Race and History*. San Bernadino: Ulan Press.
Lipovetsky, G. 1994. *The Empire of Fashion: Dressing Modern Democracy*. Princeton: Princeton University Press.
Lynch, A & Strauss, M, D. 2007. *Changing Fashion: A Critical Introduction to Trend Analysis and Meaning,* New York: Berg, 30.
Mayndard, M. 2004. *Dress and Globalisation*. Manchester: Manchester University Press, 33.
Office for National Statistics. 2018. *Average actual weekly hours of work for full-time workers (seasonally adjusted)* Webpage accessed from https://www.ons.gov.uk/employmentandlabourmarket/peopleinwork/earningsandworkinghours/timeseries/ybuy/lms (accessed on 13/02/18).
Organisation for Economic Co-operation and Development, 2007. Webpage accessed from https://www.oecd-ilibrary.org/employment/data/hours-worked/average-annual-hours-actually-worked_data-00303-en (accessed on 20/04/18).
Vannini, P & Williams, J. 2009. *Authenticity in Culture, Self and Society*. United Kingdom: Taylor and Francis Ltd.
Wolf, J. 1997. Within *Fifty Years of Fashion: New Look to Now,* Steele, V. Connecticut, Yale University Press, 152.

Product design

Practices of design: Understanding processes

A. Rabàdan
Unisinos Design Graduate Program, Porto Alegre, RS, Brazil

ABSTRACT: The proposal of this article is to revisit a creative practice of the author, articulating with concepts of strategic design that were not initially used to construct the final object—clothing for the discipline of alternative material. The intention is to dialogue with some authors who seek to understand the practice of design projects that start with a design problem. The interfaces of this creative process are present in the work of alternative material that was developed at the undergraduate level, in which to give vent to all the author's yearnings and thoughts, it was necessary to have a perspective that was broader than simply building clothes. During the processes to solve the problems present in the activity, there was a redirection, a resumption of the initial problem, where decisions were made, and some variables were corrected, helping to improve the process, in the search for a solution of the design problem.

1 INTRODUCTION

The process developed in the search to find a decision of the 'problem territory' needs to be malleable to the point of adapting to the demand of needs that change according to the progress of the 'problem solution'. This thinking also takes place in the construction of clothing and the creative processes developed by fashion designers that result in the materialization of their creations. As can be seen throughout the text, these processes are most often constructed according to the references that each designer is using; incorporating some elements of its reference to the clothing or the collection being created. In this article we sought to clarify this discussion a bit with some design authors such as Kees Dorst (2003), Herbert Simon (1991), and Donald Schön (2000).

This article deals with the relationship between fashion design and its creative processes from the body as a support for the creations. This theme was present in the project to be revisited. During the discipline of alternative material in the University of Caxias do Sul (Universidade de Caxias do Sul, UCS), the work was done directly with the probabilities, testing the hypotheses of rationality that could be associated to the method and possible transforming agents of the process. According to Simon (1991), in artificial theory, there is no optimal method, but the possible method to test the theory. This is because the method can be changed according to the progress of the project, to improve what is being designed and with that interfering directly in the final product.

When faced with the construction of a project, what is found is the creation of a service or even a product that is nothing more than the encounter with the artificial, which has its origin in a human action, different from natural. Natural objects and the way they work have been studied historically by scientific disciplines. For example, a tree in bloom belongs to this world of the natural. In the first instance, engineering schools are responsible for the teaching concerning the artificial, man-made to satisfy some of the social needs of desire, such as the creation of beams, walls and furniture from the wood extracted from the tree. This is only a simple way of understanding the difference between the natural and the artificial—one comes from nature and the other is created by man.

In the case of the 'Alternative Material' discipline, the study goes in the direction of artificiality and with a very close proximity to engineering, as it is part of a fashion course formed more than 20 years ago, with its matrix derived from architecture. Even if this is not the focus of argumentation at this point, it is important to understand the origin of the process being investigated.

2 THE PROCESS

The process carried out in the discipline uses unconventional materials to construct a product, with a more conceptual focus, in the ways of generating ideas attracting other views of the clothing presented there. It is important to situate the body as capital, territory of investigation and for its appropriation. The body can be understood as the raw material of the project that meets the reflexive creation needs proposed in the 'Alternative Materials' discipline. Such proposal also refers to an author of strategic design called Flaviano Celaschi, who quotes in his text:

> *Ezio Manzini dedicated his essay "Artefatti" (1990) to the evolution of dualistic conception between "natural" and "artificial", seeking to define the characteristic of a new superindustrial landscape (Maldonado, 1987) in which these two worlds tend to integrate as destiny. Manzini uses the term "geology of the artificial" (p.15) to identify the context surrounding contemporary man, to which all that precedes is certain, has always existed, even if some traits seem older than others, such as the ones from the terrain on which we are walking* (Celaschi, 2016, p.59).

Celaschi further adds that "the history of man and that of the artificial are practically coincident" and that "the artificial is deeply human, for good and evil, [...] for man, to produce the artificial is an activity absolutely natural" (Celaschi, 2016, p.59) This dichotomy is still present in the process of creating the clothing constructed of Alternative Materials, prepared by fashion designers whom, without having the whole understanding of this process or even of the duality, operate in this space.

There is also the intention to look at this academic process more carefully, seeking to raise observable data in the processes present in many design courses, to which academics are projecting something they do not understand fully, often due to weakness of teaching. The more organic constructive processes present in this creation, which will be dissected throughout this article, show that the form of association and understanding of the proposal made in class interferes with the aesthetic result presented in all the pieces of the Alternative Materials course.

In the first step of the protocol study, Dorst & Cross (2001) state that the empirical basis of the research consisted of protocol studies of nine experienced industrial designers working on a small design mission in a laboratory setting. The field of industrial design is particularly interesting for the study of creative design because it calls for the new, integrated to solutions to complex and multidisciplinary problems. Being able to measure a process of creativity present in the design action in a relatively reliable way.

> *Apparently, they are much more in agreement (in an admittedly intuitive way) about recognizing the creativity of a design than the inconclusive discussions about the definition of creativity would suggest. For our purposes, the results suggest that it is reasonable to claim that creative design can be assessed dependably in this manner* (Dorst & Cross, 2001, p.426).

In this way we seek in the project of alternative material the path of articulation that may show the purpose of the creation made. Thus, we tried incessantly to understand how the processes that resulted in the final product were constructed. In this research we used the protocols belonging to Dorst & CROSS (2001) to complement the research already present in the current process. The chosen cut gives the exact meaning of this research, which favors the analysis of the hybrid path between the study universes present in fashion design when materializing a creation. In other words, creating solutions to the design problems that emerged was a gradual process.

The initial solution would be the construction of clothing in alternative materials with the lowest possible cost that could be shown in the University and with a development process that presented the conceptual domain of the designer in his work. This domain is seen when the product seeks to question the practices of the fashion ecosystem and finds other ways to articulate its processes. In this case, the activity was in accordance with the performance practices employed by the art discipline.

The production of clothing that meets the needs presented works the co-evolution concept of Drost, where the problem lies in the result. This generates other problems directly related to the solution of the previous problem, which was the creation of clothing with alternative materials to be shown in a more conceptual runway. This generates a non-tangible product to all who attended the show. This type of product involuntarily works with the flow of information that the object can generate. This conceptual construction protocol allows the approach to the center of the design problem according to Dorst, where the problem and the object coexist.

The second step presented by Dorst is the experimental procedure, making it clear that for the experiment to have the expected result it requires a special condition and a right way by which the information was provided to the designers. "All the necessary information was prepared in advance on information sheets, with one specific topic on each sheet. Topics included interviews with the client, technical information about materials and production techniques, [...]" (Dorst & Cross, 2001, p.426). The Alternative Materials show in the project of the semester in question was christened Velô, in honor of the Brazilian artist Caetano Veloso.[1]

As of the delimitation of the subject, an intensive research on the artist's life and discography began, seeking the most interesting face to be worked in the form of clothing. It was found in the relationship between siblings, Caetano and Bethânia, more than a consanguineal link, the fusion of artistic souls. Thus, it was perceived that the singer and his sister are faces of the same soul, face and crown, completing in the origin and the continuity of the Brazilian musical discourse. By detecting Eastern influences in his work, the research was redirected to the figures of the Indian gods, and Shiva was understood as the God of the artists. Shiva is one of the 3 main gods of Hinduism, the divine Hindu trinity. He is known to be benevolent, kind and to bring good news. It is called 'The Destroyer' or 'The Transformer'. He builds the new.

Through clothing, one can at the moment make the visual communication of the individual that will perform, because it transmits aspects of people's personalities, and this happens every day when one gets dressed to go to work. However, in artistic manifestations this perception ends up happening more intensely, as if it were a set design, because the costumes are endowed with elements and symbology that emphasize certain traits of that character and make laymen understand what kind of feelings that character wants to evoke in certain situations.

We then resumed the first construction attempt for the alternative materials show taken it to the clothing/costume concepts that were applied to the project. We then found the name that would give rise to this creation: "Faces de Minha 'alma" (Faces of my Soul), which presented beyond what was requested by the discipline, some questionings that were already manifested in the professional process of the author. We then sought to integrate these questionings into the project. There was no interest in a process that would result in clothing that was directly linked to a pre-established theme where creative thinking could not be expressed or exercised. A differentiated investigation was begun in the search for signs that would aid in the clothing development. We then started to relate fabrics, colors, tastes and intentions, all

---

1. Brazilian singer, musician, composer, producer, political activist, born in Bahia in 1942, who built a musical work of great cultural value for the Country and the world. Caetano has been considered one of the most influential Brazilian people in the past 5 decades. He is one of the founders and leader of the Tropicalismo movement, which revolutionized Brazilian music in the 60 s and 70 s. (Author Summary)

without a greater plastic concern with the form or with what would be the product resulting from the research.

The path to construct this process was at the discretion of the available materials of the designer and their perceptions of which would be the most suitable choices for the solution of the design problem, with the aid of two guidelines from the professor of the discipline. The purpose was to translate the artist's universe into a piece of clothing in the best and most creative way possible. To translate all feelings into the material and the clothes and to transpose them later to the runway.

Dorst & Cross (2001) put the reader in the face of a creative impasse: that of reliably measuring this design process where the design disciplines propose for the designers themselves to execute it. Even contemplating each of the links in this creative process, we are always resuming methodological questions of creative construction in the discipline. Dorst himself clarifies, however, that some recent empirical studies that are descriptive of the creative event have begun to shed more light on this mysterious (and often confusing) aspect of the project.

It should be noted that this article is far from wanting with this investigation to turn this into the study that enables the total clarification of the question. The intent is to raise possible empirical paths in the process already carried out by the designer. At the same time, as Dorst & Cross (2001) put it, the more independent design creativity studies are made, the better one can develop a better understanding of how creative the designer proceeds in their decisions. The increasing number of project protocol studies tend to be constructed as studies of normal project activity, with no specific intent to seek creativity.

In this case of the Alternative Materials discipline, it is important to always consider the body's spice, or rather the body that communicates through performance, where a more performative thinking allows one to work with what is not categorized. "Performance is what has not been named, which lacks a tradition, even a recent one, which has no place in the institutions. A kind of matrix of all the arts" (Glusberg, 2006, p.07).

The idea of associating the performance with the proposed show allows a communication dialogue with the references of the theme and those used until then by the fashion designer. It was designed to happen on the alternative material runway, an empty white walkway, where women would also enter in white and bring with them water, recalling the cleaning of the stairs of Senhor do Bom Fim ceremony, along with incense and offerings. The integrative path that enables the sum of all elements came from the performance language. "The performance uses a sum language: music, dance, poetry, video, avant-garde theater, ritual. [...] In performance, the important thing is to present, to formalize the ritual" (Cohen, 2009, p.50).

In this proposed ritual of staging, the models enter: a man and a woman, united by the clothes all made in papier-mâché. They have faces painted in white, white contact lenses, oriental ornaments, a few body tattoos, and rehearse steps of an Indian dance. It is worth remembering that the initial objective proposed for the discipline was to build a piece of clothing with Alternative Materials, where performance could be a part of, but not the goal. This reminder may be associated with Dorst's thinking when he explains the deviation from the original goal:

> *Studying creative design is seen as problematic because there can be no guarantee that a creative 'event' will occur during a design process and because of the difficulty of identifying a solution idea as 'creative'. As always, creativity can be found in each design project—if not in the apparent form of a distinct creative event and then as the evolution of a single solution, possessing some degree of creativity* (Dorst & Cross, 2001, p.426).

Some ideas emerged during the research, more precisely in the album called 'Livro' (Caetano, 1997), bringing the idea of using the raw material for the construction of books: paper. Such a proposal to work using the cultural baggage of the academic is the same as that used by Dorst in the protocols, "[...] feasible in the time available and within the researchers' sphere of knowledge" (Dorst & Cross 2001, p.426). The task at UCS was to construct an outfit with Alternative Materials that would allow for an extension of concepts related to the material and its origin and which remain present in the practice of fashion design in fashion courses.

The process of creative design is immersed in the practical proposal, where transforming and (de)conceptualizing is inherent to the process investigated here. The product that will give the three-dimensionality of the creation comes from artificial origin, built by man's need to record their stories: paper, which has in its raw material the tree of natural origin. The idea was to transform the sheets of paper used in papier-mâché and thus to build the clothes. Besides having a low cost, this would enable the recycling of the material. Paper refers to simplicity and leads to endless possibilities. If one thinks of the universe that exists on a blank paper, it can be said that it is purity, the infinite; an idea that approaches the experimental procedure of Dorst.

The project was increasingly redirecting itself and moving away from its original purpose, which was the construction of a costume that would use the unconventional material for its existence; leaving the construction aspects in the background and prioritizing the performance. This remoteness has resulted in a very high burden of investigation that should be directed towards creative solutions with low investment value. For this reason, the creation process was resumed in an attempt to find an alternative so that the project could gain life in a leaner way, but without completely abandoning the fusion of concepts by performance. As Dorst presents in the text The Problem of Design Problems (2003), it should have been foreseen the part of the problems that can occur and cannot be avoided, but one must reserve time for the design of indeterminate problems:

> *They are partly determined by 'hard' (unalterable) needs, requirements and intentions. The designer will have to reserve time in the early part of his design process to unearth these 'hard facts' by information gathering and analysis and live with these specifications. This information can be seen as a necessary input at the start of the process design, and this type of interaction can be well described and modeled within the rational problem-solving paradigm* (Dorst, 2003, p.3).

Simon (1991) says that design can be seen as a 'rational process of problems'; in the same sense the project is understood as a rational process of search, which at no time uses the intuition to solve the design problem. This shows that space is what defines the totality of the problem and is what should be researched to solve it.

Using problem solving theory, it is shown what possible ways people or artificial systems should come up with the solution. This theory can be captured by the four propositions: – the gross characteristics of the human information system which do not cooperate in the resolution; – such characteristics only help to determine the problem space; – the solution of tasks is determined by the environment and its possibilities; – and, the structure determines the programs that can be used to solve the problem. The difficulties that can arise from the application in the design in defining the problems generates the 'mal-structured problems'.

It was possible to restructure the project that initially had a badly structured design problem with a new beginning, starting from the same initial problem: the construction of a piece of clothing with Alternative Materials that allows other looks towards the creation that is not only a commercial product of design. The creative process was rescued and reconfigured in another concept, with other creative paths that took the project to the field of the cuisine from the Brazilian state of Bahia, where the honoree was born. The restructuring foresaw new ways of working on the problem presented. The last stage of the project was titled 'Comendo Caetano' (eating Caetano). After different approaches and ideas, it was made the decision to work the relationship with food, which is always present as a result of human culture. The performance would occur through a Bahian recipe of a fish codenamed Caetano Veloso. The fish was carefully prepared in palm oil, using a very specific symbology of cooking and peppering with a much more involving situation. As a starting point, the song 'Vamos Comer Caetano' (1998) (let's eat Caetano) by Gaucho artist Adriana Calcanhotto, which talks about the moment when Caetano Veloso was stripped as a spectator of the theater play "Bacantes", directed by Zé Celso, from São Paulo.

The act of describing the process occurred in the discipline allows one to look at the connections made and thus seek a delimitation, even if provisional for the project stages. It also must find the evolutionary path for the construction of the alternative material clothing, working with the eye of Schön (2000); where the path ends up being rationalized to become executable, and only then solving the initial design problem in this project.

Donald Schön (2000) describes the project and the constructive steps involved in a reflective practice of the problem. This constructivist project comes from a reaction to the problem, the approach of the faults perceived by Schön in the conventional methodology. The beginning of this methodology involves the feeling caused by the paradigm of the technical rationality in the agents (involved in this practice) that could solve the problem. Such inquiry has been wrongly directed within the discipline and reflects on the process so far referenced, where the human being and its limitations are being poorly developed or at least not fully ascertained during the creative process. The incessant search for the solution does not allow only the act of verifying the totalities of the problem if not by the look of the solution.

> *He believes that the design component of the professions is underestimated, and that the nature of human design is misunderstood. He has shown that in the training programmer of professional schools that recognize design as a core activity, design knowledge is defined in terms of generalities about design processes and declarative knowledge needed to solve design problems. No attention is paid to the structure of design problems and the crucial problem of linking process and problem in a concrete design situation* (Dorst, 2003, p.4).

The new and final stage of the project: 'Comendo Caetano' (eating Caetano) is the problem's solution through the look of the result, less process-oriented and fully connected with the solution. For the Alternative Materials show, the universe of Caetano Veloso was brought to the clothing, with elements that allowed an amplitude of concepts and at the same time were pertinent to a more singular thought about the work of the proposed theme. Of course, the most important thing was to work with something that was handy and did not require financial investments to build. The design of the clothing was developed simultaneously with the possible raw material and execution solutions.

It was made the decision to reformulate one of the costumes created before the fashion course for a children's play by Caio Fernando Abreu called 'A comunidade do arco-íris' (The Rainbow Community). This has directed the creation and the raw material that would be used from now on: used sheets of X-ray examinations, for they were already present in the costumes. Everything from that moment on had to talk with this material; both the staging proposal and the soundtrack composition that fed the performance act and the language of expression.

It is noteworthy that the creative format of this work is a result of the unforeseen process, bringing it closer to Schön's theory, which puts the prior knowledge of the agent as vital to solve the problem, and this step is essential for action during resolution. He acknowledges, however, that this implicit 'knowing in action' is difficult to describe and convey to students. What can be thought and taught is the explicit reflection that guides the development of habits of knowing-in-action, to this he calls reflection-in-action.

In all, there were four stages of work: – a photographic essay with a stuffed fish, an allusion to the cook who offers her best tidbits to her customers, which should be projected at the back of the runway during the reading of the recipe; – the creation and recording of the voice over with the fish recipe to give start to the performance on the runway; – the clothing construction; and finally—the choice of a model that not only paraded, but became the delicacy of the recipe.

Proud of his spice, the model would offer people their best dish: Caetano Veloso, the fish. This photographic essay was used both for the publicity in the newspaper as for the scenery at the back of the runway. Meanwhile, the voice over would sharpen the senses and the curiosity of the audience, with a recipe of preparation of a special fish dish, recorded in a sensual way, directing the viewer to eat the fish called Caetano. The moment the model began to parade, the background image would be replaced by another, now without the fish, which would be offered to viewers on the runway.

In a somewhat disordered way, the creation was born next to the process of production of the photographic essay, as if the two perspectives of the subject were linked to each other. It was still necessary to define the clothes and the plastic form to express this hunger, as well as to find the suitable person to satisfy it. Research began on male models; the criteria were essentially aesthetic, since the model chosen would be literally served to the viewers. The preference was given to the exotic features, someone distant from the conviviality of the city and unknown to the eyes of the people.

The clothing was reworked on top of an already existing costume of mermaid tail, which had its materiality made with the moulage techniques in cotton knit. The mesh formed an entire tail lined with X-ray plates, manually cut in a circular shape and subsequently embroidered overlapping the entire tail, leaving no mesh base structure of the tail. The model had his back oil-lined with urucum seeds and several crystal necklaces that reminded the umbanda guides.

'Comendo Caetano' was not just a show: it took on the category of performance, with a well-directed narrative with the use of all possible enchantment for the moment of presentation at the University of Caxias do Sul (UCS), using other senses beyond touch and hearing.

3 FINAL CONSIDERATIONS

When reflecting on the proposed project in the discipline of alternative materials of UCS, it is perceived that the language was built inherent to the knowledge of the study techniques and processes. Its process was completely focused on the solution and not on the size of the problem presented in the activity; with all the glances in need of expression, in the solution of the problem and how could be better understood the creation, after its execution. Revisiting fragments of this design practice, one notices that the practice of creative construction processes is present in the activities of academic design. However, it is lost in relation to what may be the process or even the understanding of the proposed initial problem of investigating nonconventional materials for the construction of a conceptual clothing/wardrobe that could be shown at the end of the discipline's semester. The relationship between the designer and his problem-solving process is now better understood, even though it is difficult to measure it fully assertively.

The study presented here is much more a process of conducting ideas and thoughts about what has been done and can express through the fusion of academic processes and practices than a rigorous conceptual analysis, employing the authors of the strategic designer. It is believed that content – form and theory – practice were not originally connected in the transformation resulting from these knowledges. We perceive creation as an inherent action to the human being. Thus, it is understood that the act of creating encompasses the understanding, which is composed of the acts of relating, ordering, configuring and signifying the available agents to solve the design problems.

It is necessary to realize that during the research there was a reformulation in the agents of the problem with a considerable redetermination of clothing and knowledge acquired and consolidated throughout this activity. The creative processes in each time and space increase their possibilities of innovation thanks to the flexibility of the making during the process employed by the fashion designer in their projects. In this specific case, with a closer look at conceptual artistic doing.

The apprehension in writing these final considerations is that they would mistakenly consider that when something is concluded the subject is closed, as a passing and non-returning step. Here, one has the certainty that it has been opened a path of discussion and learning in which to continue working.

With a constant look at the possibilities raised here, which belong to the path of artistic expression, the designer seeks to create forms of clothing, to contextualize and reflect on the specific actions of this creative process and the result obtained.

REFERENCES

Celashi, F., 2016. O Corpo Como Matéria-Prima Do Projeto. *In*: *Cadernos De Estudos Avançados Em Design*. Ed. UEMG: Belo Horizonte, pp.57–71.

Cohen, R., 2009. *Performance Como Linguagem*. 2ª Ed. São Paulo: Perspectiva.
Dorst, K., 2003. The Problem of Design Problems. In: *Design Thinking Research Symposium*. Sydney: Sydney University of Technology.
Dorst, K.; CROSS, N., 2001. *Creativity in The Design Process: Co-Evolution of Problem–Solution*. Eindhoven: Delft University of Technology.
Glusberg, J., 2006. *A Arte da Performance*. São Paulo: Perspectiva.
Schön, D.A. 2000. *Educando o Profissional Reflexivo*. Porto Alegre: Artmed.
Simon, H.A. 1991. *As ciências do artificial*. Coimbra: Armênio Amado.

Contemporary textile products. Knitting as a fertile design ground for experimentation with 3D technologies

G.M. Conti
Department of Design, Politecnico di Milano, Italy

ABSTRACT: This paper sets out to provide an overview of the current state of the art in design for the creation of industrial objects that take advantage of the latest innovations regarding the electronic 3D knitting processes. The analysis stems from a desire to illustrate the great importance of fabrics for technical applications and of design for industrial textile products. Specifically, it focuses on the evolution of two-dimensional and three-dimensional fabrics as well as on three-dimensional knitting or smart knitting technologies. The issue is explored in relation to the difference between 3D knitting and 3D weaving technologies with the aim of demonstrating the broad scope of application in different areas of design, no longer pertaining to fashion and clothing alone. The paper will conclude by showing that fabrics, especially those produced by smart knitting, hold great promise and advantages not only in the field of clothing but also in the field of industrial design in terms of both performance and production processes.

1 INTRODUCTION

When the materials and technologies used in design are examined, it may be noted that textiles are frequently taken into consideration only for projects in the furnishing or clothing sectors; a rather marginal role considering the wide range of product markets and sectors that exist today.

Only in certain projects do textiles play a more functional role; indeed, they exhibit great potential and characteristics which other materials (such as polymers, metals, etc.) do not and cannot have. Nevertheless, the first great obstacle encountered in designing with textiles is that studying and designing their behaviour and predicting their performance often turns out to be a costly, complex process. This is because currently available technologies are all geared towards a traditional use of textiles, while few firms have the capacity to invest in order to attempt to modify existing technologies and making them more competitive with a view to broadening the market for textiles by incorporating them into various industrial sectors.

The search for textiles with superior technical capacities has led to the creation of three-dimensional textiles; this 3D alternative is carving out a space for itself on the design scene as it tends to use new materials and seeks to advance the use of textiles in hitherto unexplored fields. This search has produced many positive outcomes in spite of the fact that it is difficult to invest in these new manufacturing technologies.

The textile field, too, is therefore seeking to evolve and, given its characteristics, is perhaps the area of research which, more than any other, lends itself to bridging the gap between the world of aesthetics and that of functionality.

Aesthetics and functionality should not necessarily be considered separate from each other; on the contrary, the greatest promise in the world of textiles lies precisely where it can push in this twofold direction. On today's contemporary textile scene, the drive towards innovation comes thanks to three-dimensional textiles which are expanding their range of applications, require greater attention and offer designers more options and choice.

1.1 *Technical textiles*

The sector represented by technical textiles is the one which shows the most rapid growth within the more general industrial textile sector, accounting for around 19% of total global consumption of textile fibres used in all fields (Byrne 1997 in *Handbook of Technical Textiles,* Horrock 2000).

The challenge posed by technical textiles consists in the need not only to understand and apply the principles of textile science and technology in order to provide primarily technological solutions but also and above all to investigate, in parallel, the areas in which they will be applied. For example, in order to apply geotextile fabrics, it will be necessary to have knowledge in the civil engineering field, or manufacturers of technical medical textiles will have to talk to the doctors and nurses who will be the end users of the finished product. This approach, which is typical of design and defined as cross-fertilisation between areas of scientific knowledge[1], will intensify the design challenge and allow designers to experiment more.

In *Textile Terms and Definitions* (published by the Textile Institute[2]), technical textiles are defined as "technical materials and products manufactured primarily for their technical and performance properties rather than their aesthetic or decorative characteristics."

This description, however, leaves much open to interpretation, especially in view of the rapid growth which modern textile products that combine decoration with functionality are experiencing.

For many years, the term *industrial textiles* was used to define any textile products designed to be applied in the various fields of clothing, furniture, furnishing and interior design. Nowadays this would seem to be no longer adequate, especially if one considers the development of the textile products themselves. Today we are dealing with *performance textiles, functional textiles, engineered textiles, high-tech textiles*, which are often used in different contexts. For example, performance textiles are often related to the world of sportswear, despite the fact that they have and are related to a precise meaning[3].

1.2 *3D textiles*

It is in the area of technical textiles that three-dimensional textiles are being developed for the first time. Textiles have also entered the field of composite materials, or—simply—composites, that is, materials consisting of at least two distinct phases: a continuous phase, called the *matrix*, and a dispersed phase, called the *reinforcement*. The result is the combination of the characteristics of the two materials to obtain a new one which has superior properties to those of each of the two initial components. Various kinds of 3D textiles exist. The first is based on the technology with which the textile has been created and therefore on its interior structure (Khokhar classification), while the second is based on the forms that can be obtained from the various three-dimensional textiles, as set out in Table 1.

Table 1. Structures of 3D textiles and textile architectures.

Structure	Architecture	Form
Solid	Multilayer Non-Interlacing Interlacing	Compound structure, with regular or tapered geometry
Hollow	Multilayer	Uneven surfaces, tunnels on different levels in different directions
Shell	Single-layer Multilayer	Spherical shells or "open box" shells
Nodal	Multilayer Non-Interlacing Interlacing	Tubular nodes and solid nodes

---

1. Conti G.M., *Cross Fertilization: un approccio al progetto per la moda*, Mondadori Università, ebook, Milan, 2012.
2. The Textile Institute is a professional body founded in 1910, Its headquarters are in Manchester, UK. It is concerned with textile fibres, clothing and footwear and has an academic research and development department.
3. It should be remembered that these different terms may also take on different nuances of meaning when translated into different languages.

Three-dimensional textiles are defined as those which have a third dimension in the thickness direction. Indeed, in 3D textiles, the thickness or dimension along the z axis can be considered substantial in relation to the longitudinal and transversal dimension on the x and y axes respectively (Badawi, 2007).[4]

Given a definition of the term, it is thus possible to classify the different types of 3D textiles according to their interior structure. Based on Khokhar's classification (1998), five categories of three-dimensional textiles can be identified.

- *Interlaced 3D fabric*, consisting of the conventional 2D weaving process to interlace the two sets of yarns which are orthogonal to each other (warp and weft) with the addition of a third set of yarns which act as a "binder" and which extend through the thickness (in the Z direction).
- *2.5D fabric (polar fleece)*, consisting of the conventional 2D weaving process, using three sets of yarns (back warp, fleece warp and fleece weft) to create the polar fleece fabric.
- *Non-interlaced 3D fabric*, consisting of the conventional 2D weaving process with three sets of yarns to produce a non-interlaced fabric with yarns in the directions of the weft and warp and through the thickness.
- *Fully interlaced 3D fabric*. The process of 3D shuttle weaving designed to interlace three sets of yarns which are orthogonal to each other. The shed of the weave is present both vertically and horizontally. This produces a fully interlaced three-dimensional fabric in which all three sets of yarns are bound orthogonally to each other, which is possible with a machine designed specifically for 3D weaving.
- *Non-woven fabric*. A three-dimensional fabric which creates thickness as it connects three sets of orthogonal yarns but without weaving, knitting or braiding processes. The fabric is held together by a special, usually mechanical bonding process.

These fabrics have a very broad field of application; they have made their main appearance in the worlds of fashion and furniture yet are present in many other sectors, from the automotive industry to sportswear, all the way to more technical applications such as inserts in objects with various functions or as preforms for composites.

We are familiar with some of these materials and use them daily: for instance, fleece, commonly used for sweaters, jackets and blankets, has the characteristic of having its own "open" third dimension, making it extremely soft and warm to the touch.

Whole garments can be used as medical garments: the technology is often used in the manufacture of gloves, knee-length stockings, knee supports and compression stockings. They are custom designed for users with specific difficulties in certain points of the body, and being seamless they ensure the necessary comfort for the success of the treatment or sporting performance.

2 WEAVING PROCESSES

Until now we have discussed textiles, their variety and the various fields of application in which they are used. Now we will examine the technological processes involved in the weaving of these textiles.

Various technologies exist today to produce textiles which differ greatly from each other and are each suitable for well-defined applications. Thus we have orthogonal fabrics, also termed shuttle-woven fabrics; warp-knitted fabrics, weft-knitted fabrics or in tulle, cohesive or non-woven fabrics, fabrics obtained with hybrid technologies, that is, with the combination

---

4. Other definitions exist beside the one cited here. For example: "single fabric systems, the constituent yarns of which are supposedly disposed in three mutually perpendicular plane relationships." (Behera & Mishra, 2008).
"A structure that has a substantial dimension in the thickness direction formed by superimposed layers of fabric or yarns." (Chen, 2011).

of different technologies. The number of types of fabrics that can be obtained is therefore greater than ever, to a large degree because the combinations of yarns are more varied than ever: indeed, synthetic fibres lend themselves to being blended with natural fibres, thereby giving the fabric specific effects of resistance, toughness or sheen, as well as the virtue of specific technical characteristics.

2.1 Technical characteristics of knitted garments and textiles

In order to create an orthogonal or shuttle-woven fabric, two fundamental elements are necessary. These are, specifically, the warp (or chain), consisting of parallel threads and the weft, made up of threads arranged transversely to the weft. It is clear that textiles obtained using different methods are distinguished from each other above all in terms of extensibility.

Unlike shuttle-woven fabrics, knitted fabrics do not have a warp and weft but a single thread which passes through the needles creating small linked rings, the so-called "stitches" or "wales".

There are two technologies for their production, which can be distinguished according to the configuration of the basic knitting unit: weft knitting technology and warp knitting technology.

To make a comparison with hand knitting, knitting needles produce warp knitting, whereas crochet work is an example of weft knitting.

The production of knitted fabrics takes place on looms and machines which, by working essentially in different ways, give them their specific features for their end use.

The indispensable element for their production is the knitting needle which enables the formation of curvilinear knitting patterns consisting of stitches which extend in a transverse or vertical direction.

The needles fitted to knitting machines may be of different types depending on how the machine works. There is a range of wearable products that come under the product category of knitwear, from knitted underwear to outerwear, from corsetry to hosiery. What mainly distinguishes knitwear production is the machines used as well as the type of finished garment, and assembly of the knitted fabrics. The basic distinction is between cut and sewn knitwear and full-fashioned knitwear.

Several machines and looms are used to manufacture knitwear fabrics; the position within the textile product of the various kinds of knitwear is characterised by the types of machines and looms used. The distinction between machines and looms is determined by how the needles work. On machines the needles move individually and are able to pick up the yarn *(maglie raccolte)*. On looms the needles work cumulatively and need to be fed the yarn *(maglie gettate)*. The stitches are formed with the yarn which passes horizontally over the needles, giving rise to the term weft knitting; the stitches are formed one after the other with the passage of the carriage with the yarn. For weft knitting, in contrast, the yarn works longitudinally, the number of yarns arranged is the same as the number of needles which form the stitches simultaneously.

The basic classification of knitted fabrics that can be produced on the various types of machines available (on flat machines and looms and on circular machines and looms) is the following:

Plain knit or simple jersey fabrics obtained on a single needle bed, that is, with a single series of needle;

Ribbed knit or double jersey fabrics obtained with two series of needles which cross each other while working together with the passage of the carriage and will thus be offset from each other;

Interlock fabrics obtained with two series of facing needles which work alternately (the needles are selected by means of the long-butt and short-butt needle selector cams);

Links-links fabrics with a series of double-ended needles (machines set up for this type of knitting are in fact called links-links machines) that move on two adjacent needle beds on the same plane: the weave has alternating rows of face and back stitches.

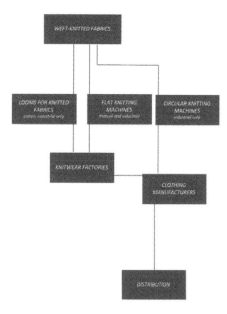

Figure 1. Diagram of supply chain for knitted textiles.

2.2 *3D knitting and its potential*

Looking at available technologies is essential if we are to have the right tools for the project. Understanding what a given machine can and cannot do and understanding the limitations and resources of a certain technology is the necessary step towards an informed design process.

While the processes of manufacturing and assembly of knitted garments on cotton looms and flat machines are performed entirely by specialised firms called knitwear factories, fabrics produced on circular machines, incorrectly termed jersey fabrics by those in the industry, can be assembled both by knitwear factories or by following the process of assembling woven fabrics within clothing manufacturing firms. A large textile sector specialising in the production of knitted fabrics for cutting and sewing exists; we find these fabrics in shops in the form of garments ranging from the simplest of t-shirts to evening wear. This type of fabric figures strongly in the collections of all luxury brands as it is much cheaper compared with full-fashioned or ready-made knitted garments.

Naturally this comparison is made while taking into consideration garments that are equivalent in terms of weight, fabric and fibre used. What drastically inflates the price of full-fashioned knitwear, as it is called in the industry, is the assembly process. All of the shaped sections of fabric are assembled by hand with the linking machine.

Shaped knitwear assembled with the linking machine is termed full-fashioned knitted outerwear, and is high-quality knitwear requiring great skill to manufacture precisely because of the manual dexterity and time required for its assembly. As a result, the most recent technological development in the knitwear area has been that of the full-fashion knitted garment. Complete garments with a total absence of stitching are manufactured on flat knitting machines specifically to overcome the obstacle of the high assembly cost.

The two main manufacturers of industrial electronic flat knitting machines, Shima Seiki (Japan) and Stoll (Germany), which have historically shared 90% of the global market for flat knitting machines almost equally between themselves, have developed machines that produce an entire knitted garment from the edge to the back and from the front to the sleeves which comes off the machine already assembled and completely stitchless and with reductions where the garment must tighten in order to adhere to the body in an optimal fashion.

335

Each machine operates at a given fineness[5], that is, its needles can work yarns with diameters, or yarn counts, of greater or lesser thickness. It is important to understand that if a machine can work an extremely fine thread, it will not be able to do so with thicker threads, and vice versa. Solutions can be found, however, if the fabric requires the use of thread of different thicknesses, by passing several thin threads into one single larger needle. The choice of the fineness of the machine that is to be used determine how thick or fine the garment produced is. Furthermore, different types of machines exist, enabling the creation of garments with different characteristics.

By analysing 3D preforms and their development it is possible to set up the machine so that it creates three-dimensional preforms which, once they have come off the machine, can be closed with one or few stitches. This process is extremely useful, for example, in the creation of coverings or linings. Yet it is also useful when, instead of creating a whole garment, only a few connection points are analysed. The figure below shows a garment devised by Stoll which brings together the various technologies that can be obtained from its machines. This balaclava represents an excellent example of optimisation of technologies on a product.

The structures that can be obtained are as many and varied as the processes that enable them to be achieved. A general classification allows at least three of them to be identified:

- manufacture of fabrics with high thickness by overlaying several layers of yarns;
- total interlacing of the fibres, in both the vertical and horizontal direction, by means of a specific shed opening system;
- creation of shaped textiles (Hu, 2008).

Nevertheless, this does not give a clear overview of the configurations that it is possible to obtain. A classification which differentiates between the available manufacturing technologies provides a clearer picture. Based on such a classification, at least six product categories can be distinguished:

- 2D WOVEN 2D FABRICS.

Fabrics obtained using traditional two-dimensional weaving technology from the weaving of two sets of yarns, warp and weft. Three-dimensional effects are achieved through pleating or weft control technologies. In the former case, permanent, more or less regular folds are formed on the surface of the fabric, while weft control makes it possible to obtain not only three-dimensional patterns on the surface but also shaped pieces.

Figure 2. Weave-in and plating applications have boosted the development potential of performance knits even further, resulting in the Performance + collection, thanks to Stoll techonology.

---

5. The number of needles on a single needle bed contained within a specific unit of length. This measurement of fineness is required in order to understand a priori how thick or fine the knitted garment will be. The yarn count is therefore closely related to the type of machine that is intended to be used.

- **2D WOVEN 2.5D FABRICS.**

Fabrics consisting of three sets of yarns, one of which is arranged in the direction of the thickness. Known as pile or double wall fabrics, they are manufactured on traditional looms using selective yarn tensioning techniques and face-to-face weaving.

- **2D WOVEN 3D FABRICS.**

More commonly known as multilayer fabrics, these are obtained using a traditional loom to which a third set of yarns is added to act as a "binder" or the weave is structured in such a way that one single set of weft yarns is able to interweave and interconnect several levels of warp yarns.

- **2D WOVEN NON-INTERLACED 3D FABRICS.**

These are also multilayer fabrics. However, in this case the sets of yarns, assembled on a traditional loom with a flat configuration are not interlaced with each other.

- **3D WOVEN NON-INTERLACED 3D FABRICS.**

Some looms for three-dimensional weaving enable 3D geometrical structures to be created without interlacing the three sets of yarns involved in the manufacture. An example is polar weaving.

- **3D WOVEN 3D FABRICS.**

Fabrics in which three sets of yarns arranged orthogonally are completely interlaced by means of a multidirectional shed opening mechanism. this process requires the use of specific machinery and is termed true 3D weaving.

3 DIFFERENCES BETWEEN 3D KNITTING AND 3D WEAVING

3D knitting and 3D weaving are two of the technologies most widely used to give textiles three-dimensional characteristics.

However, they differ from each other in fundamental ways, which means that one technology may be more suitable than another for certain purposes.

With 3D knitting it is possible to obtain specific forms. The fabric comes off the machine already formed with the edging already finished (full-fashioned whole garments). With 3D weaving, in contrast, shaping is only possible after the fabric has come off the loom through cutting and hemming.

With regard to the precision of designs, weaving allows for much more precise, well-defined colours and shapes when compared with knitting.

Digital knitting machines make it possible to alternate the various techniques in a single machine while 2D weaving looms do not also allow 3D structures to be obtained.

By their structural nature, knitted garments are intrinsically elastic while woven garments have a rigid structure. This is why 3D weaving is used for the creation of preforms for composites: the interlocking of warp and weft, in combination with fibres such as glass and

Figure 3. Textile as concrete reinforcement.

Figure 4. Textile for carbon fibre composite.

carbon fibre, mean that it becomes structural and that it can even substitute reinforcing elements in metal.

3D knitting, on the other hand, is excellent for coverings and clothing, as its structure is able to adapt to the contours of the body, while the patterned effect peculiar to knitted fabrics is able to simulate the rigidity of weaving.

4 SMART KNITTING. A SHOE DESIGN CASE STUDY

Smart knitting is one of the most innovative manufacturing processes for the creation of more contemporary sports shoes. This type of production process has met with great success among users as the shoes it creates are lighter and more breathable, in addition to displaying the effect of the stitches on the upper part of the shoe. The interaction between looks and functionality is described in the following manner on the Adidas website regarding the Adidas Adizero shoe: "using the latest design tools with our new seamless engineering technology, Adidas has created this first-of-its-kind running shoe, the Adizero Primeknit. While sport shoes are usually made from lots of separate pieces, this breakthrough method digitally knits the entire upper in just one.

"Knitting fused yarn allows us to fine-tune the exact amount of flexibility and support needed in every part of the shoe. This means lightweight comfort that wraps seamlessly around your foot, whilst fewer materials produce less waste."

Compared with the steps in the manufacturing process for a traditional shoe, there is a great saving: the great advantage of this technology lies in precisely in the fact that a ready-made upper is obtained in one piece as opposed to needing to sew several pieces together. The steps involved in the manufacture of a shoe using the smart knitting process are the following:

- design
- testing on a sample
- fitting
- choice of yarn
- programming the machine
- closure of upper (sewing)
- fitting to the sole

The production process chain is much shorter and above all free from the defects can occur during manual stitching. This type of apparel, on the other hand, is perfectly reproducible[6].

Figure 5. Detail of the shoe obtained using a manual machine.

Figure 6. Detail of shoe obtained using an electronic machine.

---

6. The only defects which might be encountered are external temperature and machine operation. If the machine is too hot, the needles, being made of metal, could dilate, giving rise to slight differences between the pieces created.

What is interesting about the process is that it makes it possible to join the various parts necessary for the upper into a single piece. What was done with the cut-and-sew method through the selection of different fabrics is done in smart knitting not only by choosing different yarns and fibres but also by being able to combine them with each other with structures chosen on an ad hoc basis. This is possible as the machine enables every point, every pixel of fabric to be programmed individually.

5 FUTURE INNOVATIONS AND APPLICATIONS

Recent advances in materials science and the textile industry are lending textiles a new role that makes them potentially suitable for the use of new technologies based on the integration of several disciplines.

The plus point in the textile industry's favour will be the use of technological platforms based on multidisciplinarity, which will lead to a new way of conceiving the use of textiles in the future. For example, in the clothing textile field, the aim is to create garments that behave like a "second skin"; in other words, a shield that is protective and functional at the same time. Currently many research centres are working on developments based on innovative processes and materials, such as the process of plasma treatment, which can modify fabrics on the surface without altering their intrinsic characteristics, thus obtaining additional functionality such as water repellency, hydrophilicity or adhesiveness, or enhancing their antistatic performance, sheen, permeability, biocompatibility or tactility.

Another sector undergoing development is nanotechnologies, which operate at the molecular level, combining principles of chemistry and physics with elements of science and information engineering. The term "nanoproducts" applies to structures with dimensions of less than one hundred nanometres, that is, 800 times smaller than the diameter of a human hair, with totally different properties from those of the same materials but of greater dimensions. Applications and studies currently underway in the textile field regard the creation of materials which have a chameleon effect, are antibacterial, protect against ultraviolet light, are flame retardant, are antistatic, protect against chemicals or are *self-cleaning*.

Finally, and perhaps the most advanced research frontier, regards clothing which "communicates"; that is, clothing equipped with micro embedded computerised systems which fully integrates with mobile phones, computers, remote health checks, music devices, and so on.

6 SOME POSSIBLE CONCLUDING REMARKS

A veritable clothing revolution is taking place and is leading to a new way of dressing, of living daily life, of interpreting materials in a more functional, less superficial way. In this new context, Italian creativity has been able to prove itself by seeking out new approach to textile design, one in which yarns and fibres are combined with different structures and functionalities in order to obtain the necessary performance characteristics.

Currently around 60% of textile products manufactured globally use fibres which only fifty or sixty years ago were not yet on the market, while according to some analysts 30% of the products that will be sold in fifty years from now have not even been conceived of yet. According to a survey conducted among researchers all around the world, products that will arrive on the market over the next few decades are expected to feature self-repairing materials, garments equipped with digital devices, smart nanomaterials, and so on.

It is therefore clear what the areas of growth and development are for the industry in the near future.

Producing innovation requires trying out new directions that others have not taken yet, exploring new production methods and anticipating changes that are still to come.

The textile sector is proving increasingly to be a research area capable of becoming an ideal arena of experimentation for technology, aesthetics and advanced functionalities. It therefore becomes essential that we completely transform our design approach, no longer separating design from technical knowledge and engineering in the true sense with a view to final product innovation.

REFERENCES

Briggs-Goode A., Townsend K., (eds.), *Textile Design: Principles, Advances and Applications*, Woodhead Publishing Limited, Philadelphia, 2011.

Chen X., Waterton Taylor L., Tsai L., *An overview on fabrication of three-dimensional woven textile preforms for composites*, Textile Research Journal, January 26, 2011.

Conti G., Gaddi R., Motta M., *3D technology and industrial design to breathe new life into product design. Is this the future of fashion and textiles?* International Conference on Design (CODE 2016), Gurgaon, India.

Conti G.M., *Cross Fertilization: un approccio al progetto per la moda*, Mondadori Università, ebook, Milan, 2012.

Conti G.M., Mello de Souza P., *Cross Fertilization: Um Direcionamento Para A Inovação*, 12º Congresso Brasileiro de Pesquisa e Desenvolvimento em Design, Belo Horizonte, Brazil, 2016.

Conti G.M., *Sul filo dell'innovazione. Progetti e visioni tra moda, tessuti, filati* in (exhibition catalogue), various authors, *Textile Vivant. Percorsi, esperienze e ricerche del textile design*, Editoriale Silvana Spa, 2014.

D'Ercole M., Rosa G., *Industria tessile*, Enciclopedia italiana Treccani, Appendix V, 1995.

Del Curto B., Marano C., *Materiali per il design, introduzione ai materiali e alle loro proprietà*, Ambrosiana, 2012.

Durante, V., *Sportsystem, the Fashion Performance*. Caerano San Marco; Danilo Zanetti Editore, 2004.

Ferrara M., Lucibello S., *Design follows materials*, Alinea, Florence 2009.

Fiorani, E., *I panorami del contemporaneo*, Milan, Lupetti, 2005.

Frassine R., Soldati M.G., Rubertelli M., *Textile design. Materiali e tecnologie*, Franco Angeli, Milan, 2015.

Horrocks A.R., Anand S.C., *Handbook of Technical Textiles*, The Textile Institute, Woodhead Publishing Limited, England, 2000.

Redi G., *Tessuto tra le due e le tre dimensioni*, Politecnico di Milano, 2014.

Ricchetti M., *Moda. Neomateriali nell'economia circolare*, Edizioni Ambiente, Milan, 2017.

San Martin, M., *Materiali innovativi per la moda*, Modena, Logos 2010.

Shukla A., *3D Fabric and Its Application in Clothing*, Indian Institute of Technology, Delhi, 2013.

Tempesti A., *Dallo sport alla moda: nuovi trend style*, in Forma e Materia, Maggioli Editore, 2012.

Unal P.G., *3D Woven Fabrics*, Namuk Kemal University Department of Textile Engineering, Turkey, 2012.

Clothing as an architectural project

B. Alcoceba
ESNE – Escuela Universitaria de Diseño Innovación y Tecnología, Madrid, Spain

ABSTRACT: This paper highlights the broad character of architecture, an attribute normally associated with the construction of buildings whose essence encompasses the design logic found in many other traditional and emerging creative disciplines. Departing from the consideration of clothing as the architecture closest to the human body, the central theme of this research is the manipulation of this interface, while describing and analysing the challenge of covering a three-dimensional body from a two-dimensional perspective. The plain surface, conceived as the main format of human apparel, has been adapted through the centuries to its topographical irregularity by means of draping, piercing, fragmentation, mapping, parameterisation and interaction, being thus transformed into an increasingly more complex and perfect artificial skin. Furthermore, its design involves the consideration of variables such as size and scale, function and form, structure, material and construction, technique and tools. In short, clothing is an individual habitational architecture, a bodily limit that acts as a connection between the inner and outer self, the inherent and the foreign, the you and the I; a simultaneously specific and abstract filter; an interface in which clothing is the container and the body its content.

1 INTRODUCTION

The body, as the sum of the parts making up the organism, is a mutable entity. Human beings have attempted to give continuity to its ever-changing condition through different actions of an architectural nature. Based on the observation of natural processes, individuals artificially describe themselves, transforming their innate reality into a distorted version of itself.

By adding, subtracting or modifying, the skin, as the last natural layer, became a primordial canvas for plastic manipulation in order to guarantee human existence and to control individual and collective identities.

The experimental evolution of these initial endeavours made it possible to supplant the natural skin with a composed reinterpretation of itself, namely, an exempt and removable skin with which to project provisionally a different self. The constant use of this removable and interchangeable 'prosthesis' led to the transformation of the naked body into a clothed one, in a social milieu in which nakedness had ceased to be the natural human state.

This paper is thus based on the principal assumption that since its advent clothing, as the first and foremost artificial covering, has been a project of an architectural nature incorporating in its design process the concepts of size and scale, function and form, structure, material and construction, and technique and tools.

In order to draw conclusions about each one of the aforementioned aspects, it is first necessary to conduct an analysis of the strategic decisions and design tactics developed throughout the history of the production of this bodily architecture. Hence, the focus is placed here on the different design paths that human beings have trodden so as to satisfy their physical and mental need to create their own anthropometamorphosis (Bulwer, 1653).

2 METHODOLOGY

The architectural treatment of the body surface spans from the dawn of mankind to the global present. Since this period is as long as human existence itself, this study has been performed using the interpretative-historical method (Groat and Wang, 2002: 136).

When deciding on the design methodology, the contributions of the French archaeologist and historian Léon Heuzey (Heuzey, 1922) and the German ethnographer Karl Max Tilke (Tilke, 1922, 1990), who after observing the dearth of information on the architectural analysis of ancient costume at the beginning of the twentieth century, performed comprehensive research in this respect, have been particularly enriching. This interest was also shared by the architect Bernard Rudofsky (Rudofsky, 1947, 1971), the historian Rudolf Broby-Johansen (Broby-Johansen, 1968) and the researcher Dorothy Burnham (Burnham, 1973), who in the middle of the twentieth century approached the history of costume from a more critical and comparative analytical perspective. Their works take us back to a spatial-time-body-architectural dimension of clothing far-removed from conventional observation, insofar as for all of them, as well as for the Argentinean Andrea Saltzman (Saltzman, 2004), what is involved is an architectural project.

Analysing the state of the question, it is François Boucher (Boucher, 2004), who has put forward the methodological approach that has had the greatest bearing on the object of study here. According to this author, 'the manifold creations of costume, stripped of all accessory elements, can be reduced to five archetypes' (Boucher, 2004: 12), which combined can in turn give rise to other different ones:

- 'Draped costume, obtained by wrapping a skin or piece of material round the body.'
- 'Slip-on costume, made from one piece of skin or cloth, pierced with a hole for the head and worn hanging from the shoulders.'
- 'Closed sewn costume, made of several widths of light stuff, fashioned round the body and fitted with sleeves.'
- 'Open sewn costume, made of several widths of material assembled lengthwise, worn over other garments and crossed in front.'
- 'Sheath costume, fitting closely to the body and limbs, particularly to the legs.'

What can be gathered from this classification is that the increasingly greater morphological complexity of costume over time is directly related to that of the design decisions made regarding the fabric surface. This two-dimensional surface is taken as the main architectural dimensional unit; it is human beings who produce and constructively transform it in order to adapt it to the three-dimensional reality of their bodies.

Boucher's proposal has been redefined, broadened and updated in order to be able to draw conclusions about the architectural variables of clothing as bodily architecture. Understanding costume as a deployable architectural project, there are many different creative and constructive operations that can be employed to undertake it.

For a proper understanding of all the possible transformations, it has been considered indispensable to apply a method based on design formulas[1], the following being those that define the archityphes of the main classification (Alcoceba, 2015: 114):

1. Unitary surface: surface + draping + body = draped garment

The surface is draped, establishing a direct link to the body through movement. The fold is a resource of spatial construction, giving rise to a developable ephemeral garment which was usually held in place with different types of fastenings, on whose reversibility its design versatility depended.

Figure 1. Design formula for the draped garment.

---

1. What is understood by this is the process as a formula.

2. Pierced surface: surface + piercing + body = slip-on garment

The surface is pierced to establish a stable anchor point for the body. Taking into account the position of each limb, the surface perimeter is cut to ensure an adequate adjustment through its folds.

3. Fragmented surface: surface + fragmentation + seams + body = sewn garment

The surface is fragmented to define the body morphologically, thus allowing it to adjust to the segmented reality of its topography. The rectangle is divided into smaller ones whose size is determined by that of the different limbs. The Euclidean surface is broken, the seams being the main focus of an innovative articulation between planes that allows for its full reconstitution as a three-dimensional covering.

4. Outlined surface: design + pattern + surface + outline + seams + body = designed garment

The surface is outlined to reproduce the body morphologically, the curve correcting the orthogonal matrix (Puerta, 2006). The garment is not now remarkable for the number of pieces (Arnold, 1985) needed to make it, but for its formal quality thanks to the advent of a geometry engendering complex designs either of an anthropomorphic or spatial character. The pattern thus constitutes a tool that facilitates its design. Waste of material becomes a global concern (Albayceta, 1580).

5. Parameterised surface: design + surface + parameterisation + construction + body = digital garment

The surface is parameterised by becoming involved in the design and construction of garments in the digital environment. Consequently, more efficient design tactics have emerged, including the concepts of system, modulation and algorithm. The number of techniques used to join planes or three-dimensional blocks have increased thanks to the development of an interdisciplinary approach.

6. Interactive surface: design + surface + post-production + body = augmented garment

The surface interacts in order to transform the garment into a screen for projecting augmented realities. Its function is not now only to protect or identify, but it also begins to serve human beings in alternative ways, becoming a complex nomadic habitat.

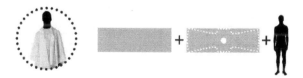

Figure 2. Design formula for the slip-on garment.

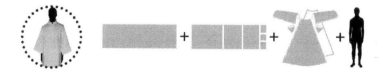

Figure 3. Design formula for the sewn garment.

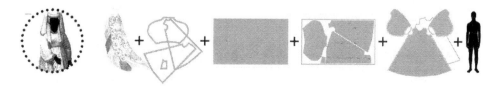

Figure 4. Design formula for the designed garment.

Figure 5. Design formula for the digital garment.

Figure 6. Design formula for the augmented garment.

In this study, it has been noted that the evolution of clothing design has not been linear, inasmuch as there have been many overlaps. Thus, a rhizomatic proposal is put forward here in which each historical morphology has been associated with design systems developed in any other space-time.

In turn, each resulting garment can be analysed according to the concept of variant. In his book *The Fashion System*, Roland Barthes provides an interesting inventory of variants through which the metamorphic capacity of artificial skin can be enhanced. As an 'independent and irreducible substance' (Barthes, 2003: 105), the variant becomes an intangible entity that determines the final result of the three-dimensional surface.

Variants (Barthes, 2003: 105–151):

– Variants of configuration (form, fit and movement).
– Variants of substance (weight, suppleness, relief and transparence).
– Variants of measurement (length, width, volume and size).
– Variants of continuity (division, mobility, closure, attachment and flection).
– Variants of position (horizontal, vertical, transversal and orientation/right, left, high and low).
– Variants of distribution (addition, multiplication and balance).
– Variants of connection (emergence, association and regulation).
– Variant of variants (degree).

Depending on the historical period, each one of them modifies '*la première et la plus intime des créations de l'art*'[2] (Heuzey, 1922: 232) according to the architecture of the moment.

3 RESULTS

Clothing design can be regarded as the first and most basic architectural contribution of humans to their bodily existence by adapting both their physical and emotional needs, both variable in time and space, at an individual and collective level.

Despite the complexity of its design analysis, this study has recorded specific archetypical processes with which to define the strategic decisions made and tactics deployed for producing clothing through the ages. In light of the above, a number of conclusions can be reached in relation to the variables that corroborate its status as an architectural project.

As regards size, it can be claimed that clothing is a deployable three-dimensional architecture configured using two-dimensional segments. However, this assertion has been called into question in the contemporary age. The traditional use of fabric, as a pliable orthogonal surface, has been substituted by a concept closer to that of 'proto-clothing' in which dimensional versatility emerges from the direct creation of reconfigurable three-dimensional segments.

---

2. 'The first and most intimate creations of art' (own translation).

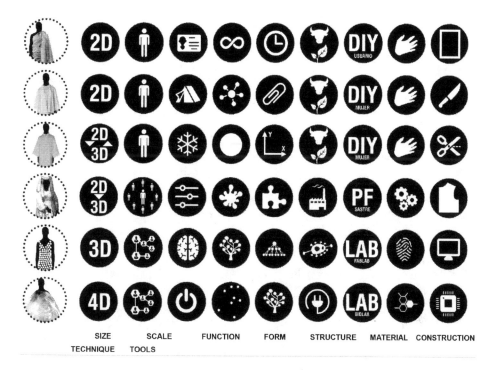

Figure 7. Overview of the relationship between the architectural variables of clothing and its design methodology.

The artificial skin becomes a three-dimensional fabric, a topography whose fourth dimension is linked to time.

Clothing is an individual bodily architecture whose components are scaled independently to create an outfit adapted to the size of the user.

Despite its intrinsic personal condition, clothing can be reinterpreted as a limit capable of encompassing several bodily units, thus allowing us to explore its impact on spatial and social behaviour. The materialisation of community-wide clothing has fostered the perception of a shared reality, amplifying the concept of fullness to promote not only a traversable volume, but also a new emotional state.

To satisfy the desire to self-construct a removable identity has always been the basic function of clothing. The need for plastic intervention is greater than the physiological needs of the body, until artificially covered it forgets to adapt to the medium on its own accord. It is only then that a climatic relationship of dependence emerges which converts the garment into a covering with habitational characteristics.

The body has been faced with a simultaneously exogenous and endogenous reality, increasingly more adverse, two-dimensionally intellectualising its relationship with its covering until its current transformation into an interactive membrane that has transported us from the physical environment to augmented reality. This transmutation to a bionic prosthesis challenges the traditionally passive character of costume as a fixed and immutable reality.

Notwithstanding this functional development, modesty continues to be the psychological reason that induces the vast majority of human beings to get dressed every day. Since clothing is a social imposition, natural nakedness is interpreted as an uncomfortable situation that is also embarrassing for others.

Body volume is the formal departure point for artificial skin and there are many clothing morphologies—which combine to create hybrids—for covering it. At present, all imaginable configurations are being designed, from the reproduction of anthropomorphism to a total lack of definition.

The interest in morphology is progressively shifting towards morphogenesis by converting biomimetics into one of the new human challenges.

The body is the structural basis necessary for clothing, insofar as this lacks three-dimensionality and meaning without its presence. Despite its symbiotic status, over the ages many self-supporting initiatives have been explored which have redefined body topography defying the force of gravity.

Nowadays, modular conceptualisation has been established as an efficient structure that optimises the creation of three-dimensional coverings, taking the repetition of equal units as a design criterion for generating multiple solutions. Customisation is conceived on the basis of maximum standardisation and its transition to parameterisation involves an increasingly greater approximation to the structure of a living organism.

Skin suppleness, transpiration and tautness, as models to be imitated, limit the material possibilities of clothing. The concept of fabric was established as the closest artificial substitute until, in addition to natural fibres and their artificial versions, the recent introduction of elements belonging to other fields, such as architecture, electronics, biology and medicine. This has led to the reinvention of a non-textile garment whose production is gradually shifting from the FabLab to the BioLab.

The construction of garments is directly linked to design. When design and constructional logic are correlated, this fosters the sustainability of the process, thus avoiding energy waste, which has been, and still is, one of the major concerns of the clothing industry.

Women were once the main designers of body architecture, but they were then relegated to second place by men. Although this historical displacement still holds, clothes making has again become a domestic activity for women thanks to the DIY concept and its open-source digital derivate.

Thus, it can be held that the evolution of clothing is the result of a process of exponential technologisation. Craft and mechanised and automated systems have now succumbed to digitisation to continue developing through the fusion between hardware and software.

Computational couture is today's clothing. The garment as a definitive product has lost its appeal versus a mutable covering organised as a mapping of technological superimposition designed from a sensitive rather than visible perspective.

Technique and art combine to pursue a sole object that evolves over time along two complementary paths: efficiency and aesthetics; reality and imagination.

Innovation in tools allows the manufacturing process to be perfected, increasing design opportunities since each tool is linked to a constructive action. Virtual tools are now replacing real ones and mathematical tools, formal ones, diversifying design methods more than ever before.

The uniqueness of the pattern as a preconceived idea has been replaced by the algorithm as a prescribed set of instructions.

The demands of the body are relentless. Clothing design is on the threshold of the greatest conceptual transition in its history. To the constant quest for the mutable identity and adaptive habitability must be added the desire to achieve customised features of a totally ethereal nature. Versus the physical garment, what is now pursued is a halo of energy.

Augmented reality is promoting the abandonment of artificial skin as an added layer to be substituted by a skin manipulated by means of the integration of nanotechnology.

This epidermal recycling allows the natural skin—the most perfect layer—to return as an interface capable of modifying its functional, spatial and formal attributes, on the basis of conscious, subconscious or unconscious orders. The epidermis, like that of a technified chameleon, is thus transformed into a phenomenological membrane that expresses the physiological and psychological traits of its owner.

In a future regression to its origins, the body will reconnect with its nakedness through technological reinterpretation. What is wearable nowadays will cease to be a removable garment to become a prosthesis immersed in its own body, allowing future urban nomads to experience an authentic anthropometamorphosis.

REFERENCES

Albayceta, J. 1580. *Geometría y trazas pertenecientes al oficio de los sastres donde se contiene el modo y orden de cortar todo género de vestidos Efpañoles y algunos Eftrangeros, facandolos de qualquier ancharia de tela, por la vara de Aragón y explicada efta con todas las de eftos Reynos y las medidas que veian en otras Provincias Eftrangeras*. Madrid: Casa de Guillermo Drouy.

Alcoceba, B. 2015. *Piel artificial. Metamorfosis arquitectónica del cuerpo a través de la superficie*. Thesis (PhD). Universidad Politécnica de Madrid.

Arnold, J. 1985. *Patterns of fashion: The cut and construction of clothes for men and women c1560–1620*. New York: Drama.

Barthes, R. 1983. *The Fashion System*. New York: Hill and Wang.

Boucher, F. 2004. *A History of Costume in the West*. London: Thames & Hudson.

Broby-Johansen, R. 1968. *Body and Clothes. An Illustrated History of Costume*. London: Faber & Faber.

Bulwer, J. 1653. *Anthropometamorphosis: Man Transform'd*. London.

Burnham, D. 1973. *Cut my cote*. Toronto: Royal Ontario Museum.

Groat, L. & Wang, D. 2002. *Architectural Research Methods*. New York: John Wiley & Sons.

Heuzey, L. 1922. *Histoire du costume antique d'après des études sur le modèle vivant*. París: É. Champion.

Puerta, R. 2006. *La segunda piel. Historia del traje en Espana*. Valencia: Generalitat valenciana.

Rudofsky, B. 1947. *Are clothes modern? An essay on contemporary apparel*. Chicago: Paul Theobald.

Rudofsky, B. 1971. *The Unfashionable Human Body*. New York: Academy Editions.

Saltzman, A. 2004. *El cuerpo diseñado: Sobre la forma en el proyecto de la vestimenta*. Buenos Aires: Paidós.

Tilke, M y Hamilton, L. 1922. *Oriental Costumes. Their Designs and Colors*. New York: Brentano.

Tilke, M. 1922. *Studien zu der entwicklungs Geschichte Orientalischen Kostüms*. Berlin: Verlag von Ernst Wasmuth A-G.

Tilke, M. 1990. *Costume Patterns and Designs*. Wigston: Magna Books.

Marking embroidery in colors: The creative fashion process in *AMAC*, in *São João dos Patos – MA*

M.S. Lima
Federal Institute of Maranhão, São João dos Patos, Maranhão, Brazil

R.G. Noronha
Federal University of Maranhão, São Luis, Maranhão, Brazil

L.R. da Silva
Federal Institute of Piauí, Teresina, Piauí, Brazil

ABSTRACT: This article discusses the relationship between innovation and creativity in fashion, in the Creative Needle Women's Association – *AMAC*, in *São João dos Patos—Maranhão*, specifically in the "Marking Embroidered in Colors"), project which occurred between 2009 and 2011. This article aims to understand the project in all of its extension, and in particular in the event that made the *patoense* society see the craft of embroidery in a new light. The research is characterized by being a case study, with the focus group technique. These relationships that have occurred all along the way are understood on the basis of theorists, but also perceived through the discourses collected and confronted in the field. And it is in this approach that links speech, the discourse, to that which we refer to as "something", be it object or material, contextual and daily, that we notice the relationship among the actors involved in the research, and observe and confront what theory and practice say, and we come to conclusion that with the respect and trust among those involved we were able to create a free, democratic and relaxed environment, and thus let the collective creative potential flourish and aggregate value to the research subjects' know-how.

1 INTRODUCTION

In the area of embroidery, a technique that especially links the two subjects of this research—artisans and designers –, placing the fabric in the frame is what represents the beginning of the creative process and delimits the area to be worked on. Similarly, at the start of this article, we deemed it necessary to also present its delimitations.

The relations between fashion and crafts have always been present since the formation and construction of societies, and, even today, they must be discussed and questioned from a wide range of points of view, including that of innovation.

The reflections proposed here are part of a broader work that consists of analyzing the performance of the consulting in fashion design in the handicrafts in São João dos Patos—Maranhão. This paper presents partial results of the dissertation still in preparation.

In this article, in particular, we will get to know the Marcando Bordado em Cores project—henceforth referred to by its translated name, "Marking Embroideries in Color", for the sake of clarity –, which took place in the period between 2009 to 2011, in the Associação de Mulheres da Agulha Criativa ("Creative Needle Women's Association"), AMAC, and culminated in an event that made the Patoense society see the craft of embroidery in a new light.

To better understand this process, we will present important concepts about design, fashion, craftsmanship, creativity and the creative process, to assist us in our later analyzes.

Design, according to what Thackara (2008) tells us, is an innovative and creative practice with the potential to transform societies and contribute to social development. As such, it is required of the designer, in addition to their designing capacity, a deeper understanding of societal matters in their field of activity so that they may make associations between elements, codes, and concepts of multiple meanings, all of which originating from diverse behavioral models. Designers, therefore, find themselves in a context of complexity, as pointed out by Cardoso (2016).

The theme of craftsmanship is talked about by Lima (2010) in its symbolic dimension, considering a tripod formed by the artifact, as a product of the human act of making; the artisan, as the axis of artisanal existence in the objects they create and their memories and references of place; and the consumer, the one who seeks not only the object in its materiality but also stories of places and people that inhabit these objects.

In defining what it is to be an artisan, the embroiderer of AMAC, in an interview with the author in 2016, tells us:

> *It means to build. Build life day by day, in the context of how craft – it expresses what you are. When you sit down to make a piece craft work, you get that piece ready and you look at that piece, right? You see you put what was inside you there... your story, right? Your moment! Because in that moment you organize the colors, you choose the motives you'll work on, the details of that motif, it expresses how you are! So for me, that's what being an artisan is about.*

São João dos Patos is a city located in the Sertão Maranhense, located 570 km away from the capital, São Luís, and with 24,928 inhabitants (IBGE, 2013). Among its peculiarities is the craft work that makes it known as the "embroidery capital."

The craft of embroidery common to many women of the region is what unites them in association. AMAC, belonging to this context, contributes directly and indirectly to the local social growth.

In order to better understand the relationship between fashion and crafts, it is necessary to define the meaning of the term "fashion". We consider fashion, according to Calanca (2008), as a social phenomenon of the cyclical change of customs and habits, choices and tastes, collectively validated and made almost obligatory. Fashion, according to Erner (2005) does not follow chance, but it instead is the result of a collective choice.

Fashion is the product of a collective choice, but it uses trends to make itself concrete. These trends, as Erner (2005) tells us, have the goal of mass domination, since they do not apply only to clothing, but to all industrial and cultural sectors.

Creativity enters this process as something interdisciplinary from the outset, from the moment the artisans think about what they will make, going through the trends presented by the consultants, or, oftentimes, transforming something they see in magazines or on the internet, in order to put a hint of their identity in their products.

And this contact of realities so different and, at the same time, so close, requires the understanding and explanation of the dynamics of social relations. For that, we use a methodology that applies the focus group technique. From the registration, transcription and categorization of the data obtained with this tool, we present hereafter our analyzes, in the intent of making the creative process visible, in addition to the approximations and limits found in the speeches of those involved. Thus, we contribute to the problematization of the role of fashion design for social innovation in artisanal communities, and we raise continuous and increasingly in-depth research, critically addressing the performance of this professional.

2 CREATIVITY AND CREATIVE PROCESS

According to Morais (2011), creativity is a subject of interest in several areas of knowledge, and because of the immense number of concepts, defining it becomes a challenge, since some discussions demonstrate that it is related to the generation of original ideas or products in a given context, with the potential to solve daily problems. On the other hand, as affirmed by Morais; Almeida (2016), there are illusory notions that creativity is associated with the characteristics of individuals as potential creators.

According to the concepts above, creativity belongs to any academic or personal context, and is not restricted to the artistic one. We understand that the creative process is related to social construction, and is not a divine enlightenment reserved for a few chosen by nature (Bachert et al, 2011).

Ostrower (2007) presents creativity as an inherent potential to mankind, and the realization of that potential, one of its needs, continues, in stating that creating is, basically, shaping. It's being able to give form to something new. The act of creating therefore encompasses the capacity to understand; and this, in turn, comprehends that of relating, ordering, configuring and giving meaning.

We understand that creativity is something that is linked to doing, and, for Zanluchi apud Smirnof (2005), it is not a game of free imagination, but an activity that requires great work, and that all stages of creation require an organized and systematic work.

We realize, therefore, that the cultural aspect contributes to the knowledge of reality and to imagining its fantasies. However, regardless of the cultural context, man remains a conscious and sensitive being. Culture represents the social development of man and his way of living with people.

Alencar (1996) shares these ideas when he presents the creativity as related to the thought processes that are associated with imagination, insight, invention, intuition, inspiration, enlightenment and originality.

There is a consensus that creativity and innovation are related but different processes. The former is a process of emergence of the new idea. The latter, on the other hand, represents the process of application, introduction and operationalization of the new idea in the organization.

For Bono (1994) creativity within organizations plays a fundamental role in unlocking the potential of professionals, motivating people to take an interest in what they are doing and promoting proactive thinking. It is also important in finding more efficient and effective ways of doing things, getting information, solving and avoiding problems, as well as achieving objectives, quality, cost reduction, continuous improvement and innovation.

About innovating, Bonsiepe (2010) presents an innovation-stimulating approach, in which he defends the artisans' autonomy in order to improve their subsistence conditions, which are almost always precarious. Such an approach requires the effective participation of the creators. In this case, the result of this approach may be extremely important for the country, promoting what we call social innovation.

The development of new and better concepts can contribute to better understand what we see in fieldwork. The investment in creativity and innovation is important, because through these tools we can collaborate with processes that aim to develop and ascribe new meanings to a research object, as is the case of this work, which aims to understand the relationship between different knowledges and ways of doing things in a project which has the mission of aggregating value to products made through the creative process of those involved.

3 METHODOLOGY

According to Yin (2005), the use of the case study is adequate when one intends to investigate the how and why of a set of contemporary events. The author argues that the case study is an empirical investigation that allows the study of a contemporary phenomenon within its real-life context, especially when the boundaries between phenomenon and context are not clearly defined. In this sense, we will use some techniques to understand the creative process in the Marking Embroideries in Color project, which took place in the AMAC, from 2009 to 2011.

We used the focus group—FG, because we understand, like Debus (1997), an intentionality where there is at least one point of similarity between the participants. We chose, in this study, for the composition of the groups, the criterion of sharing the same craft, one consisting of embroiderers and the other formed by designers and consultants who work with crafts

Figure 1. Focus groups with artisans, designers and consultants.

and culture. This will favor the reports of experiences, needs, values and beliefs among those who interact with the theme in focus.

We performed this technique in two moments: first with designers/consultants, in the Casa de Nhozinho ("Nhozinho's House"), located in the historical center of São Luis, where we structured the meeting in blocks, in order to understand the arrival and stay of the consultants in the association, the creative process of fashion in the projects and confection of the products, as well as the insertion of the products in the market.

The other FG session took place in SJP (São João dos Patos), with the embroiderers at AMAC, in which we understood, through the speeches of the artisans, how they deal with the trends of fashion brought by the consultants, the inspirations they seek on the internet in general, and their own creations.

It should be noted that the FG technique allows the revelation of the meanings that express the point of view of those who were researched. In this sense, we opportunize the unveiling of the singularities present in the cultural complexity of the context.

It is worth mentioning that the FG technique applied methodologically in this study led us to realize that this tool contains implicit procedures with rules, norms, values and meanings of an ethical nature, such as respect, dignity and commitment. Our intention is that the discourses presented in this meeting allow for the categorization of the data grouped and analyzed by affinities, when they will then compose themes related to the research.

We would like to inform that the data collected were more comprehensive, because these techniques help us in data for a larger research. However, we also focus on the specific part that deals with the Marking Embroideries in Color project, and from this information we transcribe the actors' discourses and categorize them so that they may be better analyzed.

4 MARKING EMBROIDERY IN COLOR PROJECT: THE DEBATE

The idea of the Marking Embroideries in Color project came from SEBRAE through handicraft projects it developed in the region. *São João dos Patos* was one of the places chosen because of its main attraction; which was the traditional cross-stitch embroidery.

The project aimed to bring fashion trends with the objective of aggregating value to what the embroiderers already did in the association. The initial intent would be to provide courses, workshops about new products and new techniques along with the artisans, with the intention of, together, creating a collection of products that was being built along the way through the creative potential of the agents involved.

In this sense, we align the concepts of Vascocelos (2007) with the objectives of the project, when he states that the act of creating encompasses the capacity to understand, relate, configure, order, apply meaning and, above all, re-apply meaning.

The consultant and designer of SEBRAE began with a research on the history of the city and the association, visiting various embroiderers, the history of embroidery of the city, always questioning the origin of everything. He did this research in the beginning and, from then on, together with the artisans, decided to develop traditional, but without losing the particularities of that work. In fact, the greater intention would be to find the essence of embroiderers of the well-known "embroidery capital."

In this sense, we must mention Calanca (2008), which brings fashion in this reminiscence of social events such as a change of customs and habits, choices and collective tastes.

As the embroiderers were open to innovation, they decided to give a unique touch to the products they made, making use of creativity to aggregate value to the pieces they make the most, such as toilet towels, placemats, table centers, among others. In the face of the research done by the designers with the informants of the city, the exchange of experiences between the consultants and the embroiderers, the idea of making colored pieces came up, since they always only embroidered in raw white or beige fabrics.

The challenge now would be to embroider the opposite, the fabrics would now have more colors, tone on tone and even the black fabric was suggested to be embroidered on with white threads. The iconography would be based on what had been researched, and the color palette was also intervened on.

In this sense, it should be emphasized that fashion trends are present in one's everyday life in a clear and objective manner, in the form of windows, billboards, and other means of communication.

In this way, not only the designer, but also the craftsman uses information related to fashion and trends to develop their products and aggregate value to traditional crafts.

Here we allude to Alencar (1996), who conceptualizes creativity as an intuitive process using the knowledge acquired by not only by designers and artisans, but also by the informants who contribute important information to the process. And in this interrelated web a collective cultural repertoire is built, and value is aggregated to the project and the products.

Two months were spent experimenting with the tonalities in the fabric. During that time, creativity flourished with the intention of elaborating something that had the essence of AMAC, but with a new approach. In the eyes of the embroiderers involved in the project, this is called innovation:

> *I'm doing work I've always done, even as a teenager... I never thought of doing it differently, because that's how I learned it. Now this light has showed up to us. The work is the same, but the possibilities are endless, because we have in our hands some materials like engravings and cloths embroidered by ladies who started this technique here in the city. For me, this is something innovative.*

With the artisan's speech, we refer to what Bonsiepe (2010) brings us a innovation-encouraging approach, in which he defends the artisans' autonomy in order to improve their subsistence conditions. In this sense, he presents the effective participation of producers, showing the importance to the local community and what we call social sustainability.

SEBRAE provided the project with all the necessary raw material for the production of these new products, so that it would not compromise the resources of the association for this experimental phase, as the artisan states:

During the period that we were doing these tests, we didn't spend any of our stuff we have here, neither line nor fabric... so it was good... it gave us more freedom to create, because I believe that if we were to test with the material that we have, we'd be afraid to make mistakes and even would give up on making anything, and it could even stop our creativity from showing, which we only noticed we have from the moment we got to work.

Here we understand what Zanluchi apud Smirnof (2005) tells us about how creativity is linked to doing, and Bono (1994) brings creativity in the sense of unlocking the potential of professionals, motivating people to take an interest in what they are doing and promoting proactive thinking so they can find more efficient and effective ways of doing things, getting information, and solving and avoiding problems.

The for the execution of the project was going by and the work team composed of the artisans, designer/consultant, marketing manager was testing and adjusting within the possibilities established timeframe, since they had a set time for the execution of the project and an expected date for the presentation of the products to society.

In addition to innovation in the product, it was suggested, through observation, that some products could have part of the raw material reduced, thus lowering the final cost of the product and facilitating a subsequent sale, as the artisan tells us:

> *The designer did a job for us to reduce the raw material, like crochet, because our traditional work is larger, with more crochet. Then in this new product, he made it smaller to reduce costs make us earn a little more, because we work hard and, in the context when he was adding work hours, the time it takes for everything, we earn little for how much we work. So, he reduced the work to get a fair price. Small details make a big difference. It also makes a big difference in our views, working time, so we can get pieces ready faster, to get that money faster.*

The artisan explains that, actually, the designer had greatly reduced the crochet used in the pieces: "[The designer] placed it in a more delicate and differentiated way. The traditional ones took almost a whole ball of thread, and the other takes only a third. And when we're selling, we see the difference. We can sell for a good price", she said. For Bono (1994) cost reduction and continuous improvement is seen as an integral part of creative thinking, which stimulates and encourages innovation.

The final exhibition of the products had a photoshoot of the embroiderers with the products they had made be taken in some attractive points of the city. The creative process of fashion occurred as a collaboration (Bueno; Padovani (2015)), in which the combination of the cultural repertoire of those involved in the process brought new visions and perspectives, in order to better solve some problems and, in this case, accomplish the exhibition and presenting of the products made throughout the project.

Throughout the event site there were elements and information about the research for the guests. A slide show was screened, which was a surprise for the embroiderers. The video contained the process, the results and, in the end, emphasized the creative process of the women behind this work. In this sense, local authorities, Patoense society and the artisans' relatives were able to recognize and, many of them, get to know the traditional know-how of these women who embroider the history of São João do Patos.

These relationships throughout the course of the project are understood based on the theorists, but they are also perceived through the discourses collected and confronted in the field. And it is in this approach that links speech, the discourse, to that which we refer to as "something", be it object or material, contextual and daily, that we notice the relationship among the actors involved in the research, and observe and confront what theory and practice say.

And so, Foucault (1997) states that discourses go far beyond speech, far beyond the linguistic system. And it is in this broader sense, in the construction of the knowledge, practices, institutions, actions and reactions that we oppose the discussion of creativity defined in common sense as a result of the divine enlightenment reserved for the few chosen by nature, and we have presented many examples in this project that affirms in each discourse that creativity is a collective and social construction and that fashion is the product of a collective choice.

5 FINAL CONSIDERATIONS

According to the theoretical foundation raised in the points regarding innovation and creativity in fashion, we conclude that creativity arises from the need of the individual or the group of which they are part.

The exchange of information and experiences among artisans and designers, for example, provides learning to the individual who also shares within their group, leveling them and stimulating work in a cooperative way. The groups are always seeking skills which, in addition to improving their product, provides greater insight and knowledge of their enterprise, also exploring their creative potential.

Within this perspective, we have the clarity that through these discourses we seek above all to explain the actions developed throughout the research process, as well as to present what we observed in the field.

It is important to point out that this work is open and is in continuous construction, with the contributions of other researchers to become as close as possible to good research being essential.

In this way, we note that innovation and creativity in fashion arises from a reinterpretation of the world, of new forms, or even of existing forms that assume other meanings in the face of society, which evolves and creates new values and concepts.

In this sense, fashion appropriates the most varied cultural and social elements to create trends and launch new products in the market. Therefore, fashion also appropriates popular elements, such as handicrafts that can be present in different ways.

REFERENCES

Alencar, E.S. 1996. A gerência da Criatividade. São Paulo: Makron Books.
Bachert, C.M.D'Antona; Damasceno, Y.S.L., Nakato, T. de C., Wechsler, S.M. 2017. Criatividade: desafios ao conceito. In: *Congresso Internacional de Criatividade e inovação*, Manaus. Anais... Manaus: UFMA, 2011. p. 8–28. Disponível em:. Acesso em: 28 mar.
Bono, E. 1994. *Criatividade levada a sério: como gerar ideias produtivas através do pensamento lateral*. São Paulo: Pioneira.
Bonsiepe, G. 2010. Design, cultura e sociedade. São Paulo: Blucher.
Bueno, J.; Padovani, S. 2015. A importância do processo de design na criação e desenvolvimento de materiais didáticos digitais. *7th CIDI 7th Information Design International Conference*, v. 2, n. 2010, p. 243–258.
Calanca, D. 2008. *História social da Moda*. São Paulo: SENAC.
Debus M. 1997. *Manual para excelencia en la investigación 2. mediante grupos focales*. Washington (USA): Academy for Educational Development.
Erner, G. 2005. *Vítimas da moda: como a criamos, por que a seguimos*. São Paulo: Senac.
Foucault, M. 1997. *A arqueologia do saber*. Rio de Janeiro: Forense Universitária.
Instituto Brasileiro de Geografia e Estatística. 2013. *Coordenação de População e Indicadores Sociais. Perfil dos Municípios Brasileiros: Cultura*. Rio de Janeiro: IBGE.
Kintzinger, J, Barbour RS. 1999. *Introduction: the challenge 4. and promise of focus groups. In: Kitzinger J, Barbour RS, organizadores. Developing focus group research: politics, theory and practice*. London (UK): Sage.
Lima, R.G. 2010. *Objetos: percursos e escritas culturais*. São Paulo.
Morais, M. de F.; Almeida, L.S. 2016. *Percepções sobre criatividade: Estudo com estudantes do Ensino Superior*. v. 29, n. 2, p. 141–162.
Morais, M.F. 2017. Criatividade: desafios ao conceito. In: *Congresso Internacional de Criatividade e Inovação, 1., 2011, Manaus. Anais... Manaus*: UFMA, 2011. p. 8–28. Disponível em: http://www.criabrasilis.org.br/arquivos/pdfs/122_ anais_trabalhos_completos.pdf>. Acesso em: 28 mar.
Ostrower, F. 2007. *Criatividade e processos de criação*. Petrópolis: Vozes.
Tackara, J. 2008. *Plano B: O design e as alternativas viáveis em um mundo complexo*. São Paulo: Saraiva.
Vasconcellos, E. 2007. La empresa innovadora. In: *BID-SECAD-CINDA. Gestion tecnologica en la empresa*. Santiago: [s.ed.].
Yin, R.K. 2005. *Estudo de caso: planejamento e métodos*. 3. ed. Porto Alegre: Bookman.
Zanluchi, F.B. 2005. *O brincar e o criar: as relações entre atividade lúdica, desenvolvimento da criatividade e educação*. Londrina: O Autor.

Miramar embroiderers, design and territory

F.P. Vallejos
Pontificia Universidad Católica de Chile, Santiago, Chile

ABSTRACT: Chile is a long and narrow territory, a condition that grants it an infinite wealth of geographic diversity, climate, and landscape. In these environments, diverse human groups have given place to a great variety of cultural expressions, among them a wide range of textile handicraft, some of which go back to pre-Columbus times to textile contemporary artistic expressions. The work that is communicated here is based on the activity done by an artisan's group from the south of Chile, "Miramar Embroiderers", women that live in Niebla, in the coastal zone near the city of Valdivia. An analysis of their work is done and the possibility of contextualizing handicraft embroidery in the contemporary urban clothing field is explored, therefore the creation of a collaborative work bond between Artisans and Designer, suggesting so a new market of insertion for these textile handicrafts, starting from a work methodology based from local to global.

1 INTRODUCTION

Chile's peculiar geographic features, consider a high plateau territory, a long mountain chain on one side and a wide sea on the other one, deserts, fertile valleys, great extensions of woods and steppe that define a climate diversity and habitability spaces that have defined different knowledge and skills in their inhabitants with the resulting diversity of their cultural expressions. However, these vernacular demonstrations have been influenced continuously among history by foreign cultural agents and patterns, taking as a reference of development and inspiration what is done in countries in Europe, North America and, lately, in Asia.

As a consequence, a national identity has been developed, of which there is not much awareness. Historically, the exotic features have been overvalued and a vision of production and development with a focus on the foreign has existed, leaving aside the importance of local manifestations.

If we look at the territory as a diverse zone rich in culture, as well as in terms of tradition and patrimony, the self-defining character from each of these regions emerges. In every locality there's a collective construction that goes with a community or a determined culture, that has, in many occasions, direct relation with the artisanal practices, one of those areas is crafts, expressions that possess a manual and occupational character that exposes necessities, motivations and experiences made from tradition and the ancient ancestors of those who made them (Chile Artesanal Patrimonio hecho a mano, 2008).

The *Consejo Nacional de la Cultura de Chile* (CNCA, State organization in charge of implementing public policies for cultural development), under the 2010–2015 Encouragement Policy says that nowadays there's a craft's development, who's character resides in the transmission of knowledge and learning from ancient generations to newer ones, character that experiences modifications to be able to settle down on the demands of today's world, but that it has little recognition due to the depreciation of oral transmission, generally in the families and non-formal education, as a consequence of community life, the influence of mass media and future references. (p. 13–23).

Today, there's a worldwide boom around these activities, which goes oriented towards industrially manufactured products and these having an "artisanal character". It proposes interrogations in terms of the loss of symbolic wealth and its homogenizing power, that has little to do with location, the land, it's health and the health of the people, in contrast to artisanal inherent expressions from every territory.

During the last years, a change has been seen in this kind of thinking and from some areas it has begun to reappraise what is bred and created inside national territory in different areas, sticking to trends that promote national production and revalue the artisanal process or the handmade property. This tendency is also coherent with the increasing sustainability commitment that encourages projects that use local raw material and that take carbon footprint into consideration.

In this investigation the aim was to reflect about the identitarian wealth of designing from a territory, therefore proposing a search for inspiration from the inside, making wealth and value important, and on the other hand being able to analyse the relationship between Artisan and Designer as a proper way to approach new solutions to local handicraft depreciation.

According to Local Handicraft-Global Design concepts, this work is looking to position a textile national handicraft in a contemporary clothing context, proposing a new way of showcasing and valuing occupations that are on the way of extinction due to the lack of interest and diffusion. In this sense, it is possible to offer a new insertion market for the artisans and at the same time to activate the craft's work and preserve traditions. Also, providing an innovative and contemporary offer of clothing with a local identity that is different from the foreigner offer, it helps to make the national cultural patrimony known and to enhance its handicraft quality and current expression.

A theoretical framework is exposed in parallel with the collaborative work development made during 2015 and 2016 between the group of Embroiderers and the Design discipline, to finally give place to the specific context in which the project is settled in, the national Chilean territory characterization as a distinction label from other Latin American expressions and on the other hand, clothing and textile as a support, communicator and identity carrier, and as an appropriate way to make this group's occupation relevant.

2 MIRAMAR EMBROIDERERS: CRAFTS AS A WAY OF TELLING THE STORY OF A TERRITORY

The concept and the world that surround the artisan acknowledges that the motives, techniques and used materials speak about an identity and an experience to which it relates directly, the geography is shown then as a key factor inside the handicraft expression and the products that come from it.

Figures coming from the CNCA make a distinction in diverse areas covered by handicraft like textile, woodwork, pottery and ceramic, basketwork, jewellery craftwork, stonework and leather work. 48% of artisan's work in Chile goes to Textile, where the majority of the labour is done by women. (2015)

As a line of work, in the textile area we find Embroidery, a technique that has been present in almost every ancient civilizations, its presence is proven in Asia, Europe and America, in cultures like Incas, Mayans and Aztecs. In the Andean American context the main embroidery is found in pre-Columbine cultures like Paracas, Nasca, Tiwanaku, Wari and Inka.

The history of embroidery in Chile contains two particularly important aspects. On one side the early manifestation of textile embroidery techniques; finishes in the pieces of pre-Columbine Andean people have been registered (mostly of Tiwanaku influence), in applications as design embroidery, knitting finishes and stitches embroidered. (Hoces de la Guardia y Brugnoli, 2006)

Another point is the arrival of European settlers and with them new techniques and style (Rococó), rearranging embroidery techniques that had been already installed. The transmission of these techniques landed in the hands of religious women who lived in convents and the feminine elite that had access to the learning of the trade. Later, the transfer of knowledge of this manual expression its projected through formal education in schools directed exclusively to women. This way embroidery of hybrid characteristics with Andean and European influences takes place. Inside this category the Burlap Embroiderers are situated, who belong to an urban expression of this handicraft and it's where this project is settled in.

Formerly, in Chile, women who lived in rural areas, facing the lack of materials they reused the flour and potato sacks to make bags, sheets, pillows, etc. Then they found a new sense in the reuse of this material, giving place to the burlap, truly embroidered paintings that are set as a mean of expression of their feelings and emotions. One of the first exponents in doing this kind of pieces was the renowned poet and Chilean artist Violeta Parra, who exposed her work in Paris. Later, in 1966, the first burlap artist's group was formed, *Las Bordadoras de Isla Negra* (Isla Negra's Embroiderers), whom later influenced with their knowledge, other embroidery groups of the country. A particular case is the burlap made by wives and family of political prisoners, during the military regime in Chile, these were supports that accomplished a manifestation and social denunciation character and being a means to express the complex moments and the search undergone by the families of the detained and disappeared. Some of these groups are still active like the ones in Peñalolén, South Santiago, Melipilla among others. Currently, the most important groups at a national level are: Macul's, Lihueimo's, Huilquilemu's, Ninhue's, Copiulemu's and Niebla's embroiderers.

The focus of this experience is in the Collective *"Bordadoras de Miramar"* (Miramar Embroiderers) that is settled in this mean of contemporary expression. The collective consists of a group of women that have been together for 18 years, brought together at the beginning by Hilda Gallegos-Bertuline in the locality of Miramar, Niebla (Fig. 1). These women are from the Los Ríos Region, mostly from Valdivia and Niebla, in the southern zone of Chile that is known for its dense woods, the vegetation of the "Valdivian jungle" and wetlands caused by the *Calle Calle* River that flows out on the Pacific Ocean. They gather to embroider every week in the neighbourhood venue, they are all married to fishermen of the near local cove "*El Piojo*", consequently their life has always been linked to artisan fishermen's world, as well as the motives of their embroideries (Fig. 2).

Figure 1. The Collective working in neighbourhood venue. [Personal Record].

Figure 2. Left: Map of Chile indicating Niebla's location. Right: Niebla landscapes. [Personal Record].

Twelve women constitute the group, where there's a role designation inside the group that establishes a collective organization in terms of resource allocation and work periods. There are three permanent and fundamental aspects inside their work that consciously characterizes them: patrimony, cultural identity and family tradition remembrance.

2.1 *A look inside: Cultural and geographic identity of Niebla as an added value*

The technique used by the collective is defined by them as Lanigraphy, antique technique, of local tradition, in which the drawing is overpowered, it is worked with a needle and wool by means of embroidery over the burlap. They initiate embroidering in osnaburg, then sacred art in tapestry, and later in linen-jute, a material they use to this day (Fig. 3).

At the beginning of the investigation in 2015, their embroidery expression is analysed, with a sample of 45 embroideries, doing an iconographic taxonomy of their work and a study about the use of colour, shapes, representation and motives. One of the main characteristics of their embroidery is the capacity of creating from their own experiences, it's perceived as an expression linked to naïve art. The embroiderers manage to transpose their experiences and thoughts spontaneously to the fabric, the drawing or the lack of technical knowledge of it, determines a work spontaneity that represents its great final quality. In this regard, Sennet says that: *"The artisan work embodies the great paradox that an activity of great refinement and complexity arises from mental acts as simple as the detailed description of the facts and their subsequent inquiry"* (2009, p. 329).

Likewise, the theme of patrimony and cultural identity is always present in their works, portraying the territory: landscapes, lived experiences, representation of their families and fishermen husbands, thus, artisan fishermen's world, doing a critic about environment's abuse, because they observe how deforestation and other factors of globalization are affecting their land (Fig. 4).

In terms of technique, they use two types of stitches, backstitch or also called industrial and chain stitch. Alternating between one or another allows that, with the help of color usage, different textures are formed over the fabric, creating patterns and diverse textures with the minimum resource of two stitches (Fig. 5).

The embroiderers do a quality and great trade work that only 4 years ago they decide to commercialize, showing their embroidery's in the area around Niebla or in Valdivia. Due to the limited accessibility to commerce spheres there are only a few people that can actually get

Figure 3. Two recent works from the Collective Miramar Embroiderers. [Personal Record].

Figure 4. Comparison of their works and the landscapes. They reflect geography and experiences that surround them in their work. [Personal Record].

Figure 5. Constrained summary of iconographic taxonomy and textures done over the collective's work. [Personal Record].

to see or buy their products. They are located far away from marketplaces that could buy this kind of handicraft. This evaluation added to the acknowledgment of their proposal triggers the search for a strategy to value their work.

2.2 *Collaborative work between Artisan and Designer*

After studying their technique, it is understood that doing an assembled work with artisans, from design, it must be paid attention to what kind of intervention and action wants to be made and which are the limits in it. To fully understand what kind of intervention it's appealed in this investigation the relation between H+D (Handicraft and Design) is exposed and then the levels of intervention in the handicraft.

In 2005 the "Taller de Artesanos y Diseñadores" (Artisans and Designers workshop) is done in Santiago, Chile with the final objective of clarifying the limits in terms of application field, the forms and the intervention's process of design in handicraft. This is, thus, a dialogue and reflection instance about the state of handicraft at the moment, and how design is a strategic tool that can cooperate with the development of handicraft, proposing a collaborative labor between those two, adding that:

> *Any intervention in the handicraft's work from design has to have the artisan as a starting and ending point. The designer must understand that its endeavor on A+D relationship will have a direct impact on the culture that the artisan is inserted in.* (Alfaro y Rodríguez, 2009, p. 31).

Another important reference to this project was Eduardo Barroso's Pyramid model (2009), in which he exposes that the closer the products are to the top of the pyramid, bigger is their cultural value and the use of industrialized means is lower. It is essential to understand the handcrafted product that is going to be intervened in a deep non-superficial way, in order to understand its richness and to extrapolate these to other areas and products.

In this area the level of intervention that is proposed on this project is contemporary or urban handicraft produced by individuals with a wider cultural and technologic base, also known as, "Creation Handicrafts". Its commercial value is, in great measure, determined by the balance between the expressive value (esthetic and cultural references) and the usage value. It is concluded that the level of intervention of Design in Handicraft will be equal to having the possibility to make a modification to the materiality used for the project. It's proposed to experiment with different textile materials, yarns and wools.

2.3 Garments as a communicator

Based on the study done about the collective's work, it is proposed to cover the collaboration between Artisan-Designer from the garment's perspective, the body as a carrying structure of clothing, and clothing as a tool to show the handicraft.

Although all design disciplines revolve around the body, in clothing the body is the base structure of the object that is projected because clothes do not carry themselves, but they take shape according to the body-user relationship, therefore both, body and garment establish a dialectical relationship in which both of them modify one another. (Saltzman, 2004)

As said by Andrea Saltzmann in her book, *El Cuerpo Diseñado*, the body must be conceived as an individual and collective perception place, as a user that perceives the world through and from the garment, an identity carrier, the body is a member of a culture and a context. Garment then its configured as a powerful expression territory that gives identity to the design. The garment represents a signs system that becomes in sense, it is a visual and spatial code that is configured and determined by context and situation (2004).

Likewise, according to what Kate Fletcher and Lynda Grose propose, designing from a local perspective, in a dialogue with local artisans and that way getting fashion nearer to what is local, under a low scale work, is a way of getting to sustainability. Fashion has found inspiration on the conjoint work with handicraft, countries and communities, emphasizing at the same time the work between the Designer and the Artisan, promoting that way taking local plans to the fashion area to promote aesthetic and cultural diversity. The designs should then have a special sensitivity to the local agent's knowledge, their myths, symbolism, materials, colors and processes, in order to be projected. (2012, p. 108–110).

2.4 The Design project as a communicational tool for artisans

The study and project that origins in 2015 and ends in 2016, concludes with the creation of a clothing collection called *Niebla*, inside a collaborative working system between Artisans and Designer, where from the beginning, specific roles were established for Artisans and Designer. The process consists of reunions every two weeks in the neighborhood council's venue and the weeks of work are planed altogether. This way the embroiderers express their apprehensions in terms of the requirement's difficulties, the risks and the possibilities of completing the task. The process initiates with the Designer's thorough analysis of their work, to eventually go to a stage of inspiration, choosing what studied icons will be embroidered and therefore choosing a guiding thread for the collection's design.

Following this process there is an experimentation stage with new materials, in which, the artisans embroider with different fabrics and formats, the artisans give feedback and a selection of the embroidered fabrics is made, where the result of the embroidery in different fabrics can be evaluated. Finally, there is a creation of garment and embroidery proposals, to

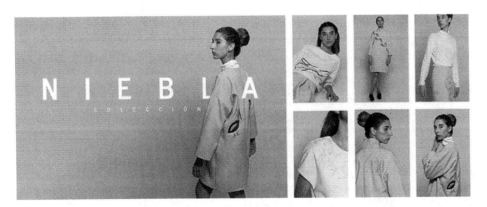

Figure 6. Final collection by Niebla. [Personal Record].

finally carry out the embroidery's production process in the collective workshop and then on the sewing workshop. The final product, on its last stage is complemented with the creation of the brand's identity, a logo, photography and a business plan to position the product on a real design and handicraft context in the Chilean market (Fig. 6).

3 CONCLUSIONS

Concluding the project and the gatherings at the workshop with the Collective, an analysis of the last 12 months is made, evaluating the challenges, difficulties and success of the task.

It is concluded that during the development of the project a mutual learning takes place, the Designer must understand the artisan's technique thoroughly, therefore the people and the territory's identity. On the other hand, the artisans learn aspects about the confection of clothing, achieving to expand their vision on the possibilities of positioning their handcrafted product on today's market. They amplify their understanding towards the work's diversification opportunities, observing their work from another perspective.

At a methodological level, one of the primary challenges working collaboratively must be getting to an understanding of the role that each one of the parts involved must take during the project. There are dialogues and communication from every involved discipline, therefore it is relevant to understand the *level of intervention* that is going to be done on the handcrafted product, the communication must be crystal clear from the very first moment, to avoid misunderstandings that could obstruct the development of the final project.

After looking at the project's results and the audience's perception of the finished final product, it manages to establish the importance and added value, on one side, of a product thought and bred from a locality that has its own identity, and on the other hand the Design's role as a bridge tool between a handcrafted quality technique and an unknown audience, creating a handcrafted product that contains a new language and that, through garment it can show other possibilities of insertion. The collaborative modality of work between Design and Handcraft constitutes a way that can add positively on the creation of products rich in identity, proposing the design from within the territory, with an inside look and not necessarily with a vision that sets on the exterior.

The main challenge in the future is to systematize work between Designer and Artisans to plan other collaborative projects, and besides spreading the trade, manage to insert the product fully on a contemporary market where national cultural patrimony is appreciated.

REFERENCES

Alfaro, E. & Rodríguez, C.; 2009. *Dossier no2: Taller A+D Encuentro en Santiago de Chile.* [ebook] Santiago, Available from: http://www.unesco.org/new/leadmin/MULTIMEDIA/FIELD/Montevideo/pdf/Dossier_UNESCO_completo_feb11_01.pdf [Accessed 3rd february 2016].

Barroso, E.; 1999. *Diseño y Artesanía: límites de intervención.* Brasil: s.n.

Consejo nacional de la Cultura y las Artes; 2008. *Chile Artesanal Patrimonio hecho a mano.* [ebook] Santiago. Available from: http://www.memoriachilena.cl/archivos2/pdfs/MC0053875.pdf [Accessed 5th May 2016].

Consejo nacional de la Cultura y las Artes; 2015. *Política de fomento de las Artesanías.* [pdf] Santiago. Available from: http:// www.cultura.gob.cl/wp-content/uploads/2011/09/politica_artesania.pdf [Accessed 30th march 2016].

Fletcher, K. & Grose, L.; 2012. *Gestionar la sostenibilidad en la moda, diseñar para cambiar.* España: Laurence King Publishing, Londres.

Hoces de la Guardia Ch, M. & Brugnoli, P., 2006. *Manual de técnicas textiles andinas.* Santiago de Chile: Museo Chileno de Arte Precolombino.

Saltzman, A.; 2004. *El cuerpo diseñado.* Buenos Aires: Paidós.

Sennett, R.; 2009. *El artesano.* Barcelona: Anagrama.

Strategic approach to implement sustainability in the joineries of the city of Uberlândia, Brazil

F. Moreira da Silva & J. Cardoso Braga
CIAUD, Lisbon School of Architecture, Universidade de Lisboa, Portugal

ABSTRACT: Recent studies focusing on sustainable design have presented alternatives in what concerns sustainable strategic design applied on local contexts. In Brazil, the wooden furnishings industry is comprised of approximately 16,000 companies of various sizes. However it is estimated that most furniture companies (84%), consist of MSEs – Micro and Small Enterprises which are cabinetmakers performing customized work. Micro businesses companies are those with up to 19 employees, whereas small businesses have 20–99 employees (SEBRAE 2013). Currently, these companies use MDF – Medium-Density Fiberboard as their main raw material. Brazil is the world's seventh largest producer of reconstituted wood panels in the world. This paper presents partial results of a research project, from the diagnosis phase performed in joineries, small furniture companies in the city of Uberlândia, Minas Gerais State, Brazil. The production process of these joineries is not sustainable and results in a large volume of discarded waste on the outskirts of the city. The progress of these furniture MSEs depends fundamentally on minimising the generated waste of materials. This can only be achieved by a sustainable strategic approach, with the implementation of new practices and cultural/mentality changes.

1 INTRODUCTION

According to Movergs (2014), in Brazil, the wooden furnishings industry is comprised of approximately 16,000 companies of various sizes. It is estimated that most furniture companies (84%), consist of MSEs – Micro and Small Enterprises (FIEMG 2002; Garcia 2005) many of them are joineries performing custom work (Galinari et al. 2013). Currently, these companies use MDF – Medium-Density Fibreboard as their main raw material. Brazil is the world's seventh largest producer of reconstituted wood panels in the world; in 2015, the production was 7.98 million cubic meters. A large part of this production, 95%, goes to the domestic market (IBÁ 2015). An expressive number of custom furniture Micro and Small Enterprise – MSEs, around eight hundred, is located in the city of Uberlândia (669,672 inhab.), in Minas Gerais State, Brazil (IBGE 2015). The production process of the joinery is not sustainable and results in a large volume of discarded waste on the outskirts of the city. In July 2015, there were ninety-nine points of irregular disposal of waste of various types (verbal information). This practice occurs, mainly, due to the low social and environmental awareness of cabinetmakers, and due to the lack of proper management of industrial waste by the municipality. To obtain more efficient results, we have to face the problem of the waste generated by these small familiar joineries focusing on people, integrating them in the system and not only by producing laws and guidelines. We present the partial results of CACO research project, from the diagnosis phase developed in 15 customized furniture companies in the city of Uberlândia, Minas Gerais State, Brazil. It is believed that the insertion of a set of ecodesign strategies in MSEs can generate greater competitiveness for companies generating new business opportunities, creating more favourable conditions of competition among companies of the sector in the region, improving their income while simultaneously promoting commitment and the sharing of responsibility (social, environmental, cultural) by all stakeholders. In this case, it is noted that the progress of furniture MSEs depends fundamentally on minimising the waste generated and the wastage of materials. The project we have been developing with the joineries and the local furniture unions shows that with some specific changes and the use of a language that the cabinetmakers understand, we can achieve our goals for a more sustainable regional development and social integration.

1.1 *Furniture waste*

If on one hand the furniture sector generates a large number of jobs, on the other, it is also considered a major consumer of raw materials and waste generator. The industry makes integrated use of materials that, due to their distinct nature and lack of an adequate management plan, end up hampering reuse, recycling and other forms of appropriate final disposal (Kozak et al. 2008, p. 205). The most commonly used materials are wood derivatives (solid wood, wood veneers, plywood or particle board, reconstituted wood panels such as MDF and veneers); metals used in handles, hinges and slides; chemicals (paints and varnishes, glues and resins); plastics (edge ribbons, laminates, handles, etc.) and textile fabrics and leathers (natural and synthetic) (Nahuz, Figueroa & Lelo 2002). In addition to the diversity of materials, the production of customized furniture requires a number of productive processes such as machining (cutting, drilling, milling); sanding; assembly (fixing and gluing); transport; cleaning. These processes generate a large volume of waste or various types (sawdust, dust, panel patches, edge ribbons, packaging debris, bolts, glues and tows). The estimated annual volume of waste from the furniture sector of Uberlândia and region corresponds to approximately 22,000 m^3, of which a significant part is composed of residues of wood (dust, sawdust and MDF flaps). The presence of urea-formaldehyde or phenol-formaldehyde used in the composition of reconstituted wood panels, requires special attention due to the fact that this resin is part of the International Agency for Research on Cancer (IARC) list of carcinogenic products. For all this, the joineries of the Triângulo Mineiro and Alto Paranaíba have an expressive environmental liability, mainly due to the irregular disposal of residues on periphery lands of the cities of the region. This practice stems from factors such as: municipal policy, not sensitive to environmental issues; the lack of municipal management aimed at the adequate disposal of this type of waste; the resignation of the most needy people who live in the neighborhood of these ditches; the reduced or no interest in assuming the share of environmental responsibility on the part of the entrepreneurs of MSEs in the region and the weak social and environmental awareness of designers about the environmental impacts of their projects.

According to the diagnosis made by SENAI (2006), 90.8% of the furniture manufacturers in the region rest industrial waste, 3.7% sell them and only 5.5% recycle. In July 2015 the municipality had ninety-nine critical points of irregular dumping of indistinct waste, distributed mainly in public areas of the city. It is understood that these facts do not exhaust a very complex conjuncture, difficult to transform. Targeting the generated waste in an environmentally sound way is not enough to solve the problems dealt with: "reducing the source of resource consumption is fundamental, aiming, fundamentally, to reduce waste and, consequently, reduce the environmental impact" (Pereira, Carvalho & Pinto 2010). However, in the specific case of the Triângulo Mineiro region, few actions have been taken to combat the advance of environmental degradation. In addition, there is a lack of tools to train, educate, guide and foster a new local culture that favors responsible positions on the part of professionals and micro and small industries, regardless of the sectors of activity. Special control and concern should also be devoted to the health of joineries' workers due to the risk of inhalation of small particles of reconstituted wood containing formaldehyde. In addition, the accumulation of residual dust generates fire hazards, if neglected, due to its flammable characteristic. Joineries also suffer financial losses due to the lack of adequate management of their waste and, mainly, the waste of raw material caused by the lack of optimization of the wood panel cuts, inadequate separation and storage of the parts that can be reused. Environmental, economic and social issues must be considered when analyzing the consequences of the practice of irregular waste disposal by the woodworkers on the outskirts of cities aggravated by the common practice in the region of firing with the intention of carrying out land cleaning. The problem of the conjugation of these practices consists in the emission of particulates and GHGs in the atmosphere that increases the risks of uncontrolled fire due to the high flammability of the waste in the form of sawdust. This factor is very significant for the worsening of the populations' living conditions that subsist on the poverty line or below that threshold.

2 LITERATURE REVIEW

2.1 *Strategic thinking*

The term strategy has been used differently over the years, and in different fields (administration, marketing, military, design) but there is no universally accepted definition. In recent works, Mintzberg, Ahlstrand and Lampel (2008) and Mintzberg et al. (2002) investigated the origins of the strategy and the views of different currents of thinking from five definitions: strategy as a plan, as a pretext, a standard, a position or a perspective. The first and most traditionally used definition is strategy as a plan, understood as an intention, a guideline, or a set of guidelines designed to deal with a specific situation. Strategies are formulated consciously, before the actions to which they will be applied. Therefore, they are also considered as future plans of the organizations, ie, intended strategies (Mintzberg et al. 2002). The definition presented by Glueck (1980), from the administration point of view, reinforces this vision. For the author, strategy is a unified, broad and integrated plan that can determine, through a normative model, which actions and/or human and material resources are needed to ensure that the company's goals are achieved. This vision is also strengthened in the field of design. For Best (2010), the strategy is a plan that describes how the company intends to fulfill its mission and vision vis-à-vis different parts of the organization. However, Mintzberg, Ahlstrand and Lampel (2008) and Mintzberg et al. (2002) criticize this point of view because they consider that the desired strategies are not always realized. In this sense, they believe that, in practice, strategies are composed of a combination of "deliberate strategies" (characterized by fully implemented guidelines), "unrealized strategies" (characterized by unfulfilled directives) and "emerging strategies" (irrespective of the guidelines, and it was not provided for in the guidelines). Most of the implemented strategies involve both deliberate and emerging issues. Zurlo (2012) shares this position by stating that in practice there is often a gap between a strategic decision and its achievement. The progressive formulation of the strategy is more useful for the incorporation of design in the company in the long term, because the emerging strategies allow the insertion of design ideas in a collective process of consensus building among stakeholders. For Mozota (2003), strategic design is a factor of competitiveness and differentiation for companies, it means contributing to the organizational culture of companies, seeking opportunities for innovation and identity, offering a unique combination of value. According to Zurlo (2012), strategies are interpreted in different ways and can be associated with a plan or set of followed actions and precise results; as a model, where entrepreneurs seek a reference in other organizations that represent the best in certain areas; and as positioning, where companies must find a place in the market and differentiate themselves from competitors, in line with Porter's (2008) competitiveness vision. However, when the strategy is associated with a plan, it reiterates the necessity of adapting the guidelines, since the conception of strategies does not always coincide with reality, and in this case, it is necessary to change direction. Strategic design is seen as an interdisciplinary field that can interact on various business activities with the purpose of generating value, allowing companies to enter a new market or increasing their participation in it and, for this reason, plays an important role in the transition process towards sustainability (Mozota 2003). According to Best (2010), design acts in an integrated way with other disciplines and professions and provides an integrative and holistic approach on the cultural, environmental, political and social impact of the organizations. In addition, strategic design can generate eco-projects that are technically and economically viable, reorienting production and consumption systems to the formation of a more sustainable mindset. (Manzini & Vezzoli 2008).

Based on studies of European design-oriented MSEs, Mozota (2006) presents a Design value model as differentiator, integrator and transformer, called the four powers of design: 1. Design as differentiator – design as a source of competitive advantage in market by creating value for the brand, higher price, loyalty or customer orientation; 2. Design as an integrator – design as a resource to improve product/process development, developed to favor modular design, to develop user-driven innovation models and project management; 3. Design as transformer – design as a resource to create new business opportunities, to improve the company's ability to face changes, or as expertise to better interpret business and the market; 4. Design as good business – design as a source of sales increasing, better profit margins, greater value for the brand, greater market share, better investment return; design as a resource for society (inclusive design, sustainable design).

2.2 Challenges and opportunities in the insertion of the strategic design in the furniture MSEs

In Brazil there is a very complex and difficult situation to transform and overcome for the implementation of strategic design in the business sector. This is due to facts such as lack of interest and commitment from relevant partners (local public leaders, associations, NGOs). In addition, in the specific case of the furniture sector of the city of Uberlândia and region, it was possible to identify some of the main obstacles presented by the MSEs themselves: 1. Lack of interest of the stakeholders to dedicate a coordinated effort to solve common problems together; 2. Lack of motivation induced by poor understanding of sustainability aspects or lack of recognition of the importance of design for their business; 3. Difficulty in continually taking on the proposed practices, modifying internal policies and assuming co-responsibility with others; 4. Low confidence, transparency and belief in mutual efforts to achieve group goals; 5. Cultural and behavioral resistance to organize and carry out changes in productive processes; 6. Lack of understanding about the value or need to invest time and financial resources in the design process. However, it is important to consider that these companies, mainly composed of MSEs of family origin, with low degree of specialization and production of low technological level, most of them face many difficulties to remain in the market. Therefore, factors such as the economic crisis faced by Brazil, especially from 2015 onwards, with a reduction in credit lines, difficulties in obtaining financing and an exorbitant increase in interest rates, a rise in prices and, above all, an increase in energy, fuel and water, have hindered the very survival of these companies. It should also be considered that there is still a dichotomy between the perspectives of design application in the business sector and its implementation. There is also a general lack of conviction about the application of design as a tool for business management and as a factor of innovation. The fact is that the importance of design for organizations is still little recognized by corporate managers (Mozota 2006).

One of the biggest problems hindering the insertion of the design in the companies "resides in the lack of knowledge on the part of the entrepreneurs, of the design meaning and what can be changed in the company, besides some prejudices, as, for example, that the design is expensive" (Santos & Menezes 2009, p. 97). In addition, it must be considered that each company has its own needs and therefore, the process of design implementation is particular to each organization (Freitas & Merino 2011) and, therefore, it becomes more difficult.

2.3 Strategies for the development of sustainable products

The insertion of sustainable design strategies in local contexts and non-industrial production scales is very relevant for companies and indispensable for reducing the generation of solid waste, material waste, pollution and environmental damage caused by production of customized furniture. According to Fiksel (2009), sustainable design practices not only represent ethical behavior and good citizenship but, above all, a business value strategy. This principle of generating value and economic benefits for companies is known as eco-efficiency—doing more with less. It is part of this perspective to reduce the consumption of resources (energy, materials); to reduce environmental impact (reduce emissions, waste and use of renewable resources); and to create value (providing better products, additional services, offering benefits to customers). Several corporations around the world already recognize sustainability as a way of developing, growing and surviving businesses and their ability to stimulate new business opportunities, improve competitiveness, increase customer confidence, and motivate employees, local communities, partners, etc. (Fiksel 2009). Organizational changes for the implementation of eco-efficient design strategies also favor innovation and can offer a greater capacity to attract customers, as it improves the image of the organization and allows them greater profitability through the reduction of environmental impacts, among other issues. All these benefits obtained by adopting the principles of design and sustainability need to be integrated at a strategic level so that they can act more effectively (Bhamra 2007). Eco-design strategies aimed at a systemic approach involving all stages of product life should be considered during product design. In this sense, whenever possible, designers should apply a set of eco-design strategies involving the entire product life cycle, i.e., considering the stages of pre-production, production, distribution, use and disposal. According to strategies presented by several authors (Barbero & Cozzo 2009; Behrendt et al. 1997; Bhamra 2007; Brezet & Hemel 1998; Fiksel 2009; Gilsbert & Garcia 2004; Manzini & Vezzoli 2008), it was possible to identify seven key strategies and their tactics for the development of sustainable products: 1 – the choice of low impact materials; 2 – reduction in the consumption of materials (raw materials and energy); 3 – optimization of production; 4 – optimization of the distribution; 5 – optimization during use; 6 – extension of product life; 7 – product end-of-life optimization.

3 CACO PROJECT

The CACO research project involves the development of a sustainable strategic design model aiming to minimise the waste produced by furniture manufacturers in the city of Uberlândia. The word "CACO" (shard) was proposed due to its meaning: a piece, small fraction, part or remainder of something (Michaelis 2009). The use of this term is meant to represent, symbolically, the need to connect the different areas necessary for the development of the proposed project and integrate "fragmented" knowledge in favour of more sustainable actions for the furniture sector. Strategic design proposals should generate direct and indirect benefits for companies. With the reduction of waste, it is expected that there will be an increase in the firm's profitability. Therefore, environmental and economic progress, if achieved, can also generate social and cultural benefits through the maintenance of jobs, the preservation of this millenarian craft (know-how), and the permanence of some cabinetmakers in the market. The model will be comprised of a set of ecodesign strategies that will be implemented systemically in all phases of the product's life cycle (pre-production, distribution, use and disposal). The strategies are being defined from the conclusion of the diagnosis phase. The model intends to supply, firstly, the needs and collective problems—already diagnosed as being common issues among the companies in the sector. However, according to Freitas and Merino (2011), it must also be considered that each company has individual needs and, therefore, the implementation process must meet the geographic, economic, social and cultural characteristics of each one of them.

3.1 *CACO diagnosis phase*

The diagnosis was made in fifteen MSEs (with a minimum number of 06 and a maximum of 20 employees) who produce customised furniture and use MDF as the main raw material. The data was collected through: (i) semi-structured interviews; (ii) direct observation with mechanical support; (iii) documentary research; and (iv) data collection (toolkit) on the volume of waste generated during the production process. According to Hubert Rampersad (2007), waste is usually hidden by the workers because it is considered as a normal symptom in the workplace. Therefore, the diagnosis was made to detect the main causes of waste generation in the joineries by analysing all phases of the customised furniture life cycle. In order to evaluate the generation of residues during the production of furniture, a toolkit was created with the objective of detecting the amount of used material (material input). In this sense, the employees were guided and monitored during the use of the toolkit, which aimed to register: (i) the form used for the planning of the cut; (ii) and the consumption of material (new and reused MDF sheets). In addition, during the furniture production process, the generated waste was separated, identified, weighed and measured in order to identify the percentage of material loss during the production process waste. The joineries' owners declared that in general they use an average of 110 MDF sheets monthly (white and coloured), which are purchased from resellers. According to the carpenters, the MDF is responsible for approximately 42% of the furniture manufacturing costs. This data demonstrates the importance of MDF in economic spending and reinforce the need for optimisation and better use of MDF sheets even for the financial health of the companies. The diagnosis resulted in the identification of several common causes for MDF waste.

1. Pre-production: All the companies work with approximately five different MDF brands. The wide colours and brands variety has increased losses, reduced the possibilities of reuse, and contributed to the generation of a large volume of waste.
2. Production: lack of concentration of employees due to the use of the cell phone while working; lack of employees' commitment; absence of verification protocols (of material, tools and procedures) for assembling the product. It was also verified that all the owners of the studied companies accumulate a significant amount of activities (elaboration of the budget, negotiation with clients, surveying of measures, the realisation of production order, production of furniture and, when necessary, administrative activities). The accumulation of functions also contributes in part to reworking caused by measurement error and communication problems between management and production. 80% of the furniture companies have a production process per project, being each joiner responsible for the complete execution of the furniture. The practice of production by the project is causing a lot of waste due to the lack of planning, especially during the cutting stage of the sheets. In 100% of cases, MSEs cut the material without performing at least surface calculations or analyse the best way to make good use of the sheet. Only one of the companies uses software for sheet optimisation. Although employees are systematically advised to reuse MDF scraps to produce smaller items (drawers, shoe racks, slats, etc.), MDF leftovers accumulate very quickly. Due to overlapping MDF scraps, disorganisation, dust accumulation, and lack of inventory control, the workers prefer to use new sheets, even to make smaller objects. In all studied companies there are no spaces or shelves organised to receive leftovers. The most common practice consists of stacking pieces that are pressed against the walls of carpentry (taller pieces are placed on the bottom and lower ones on the front). However, the arrangement of pieces without any classification by size, thickness, or colour makes it difficult and doesn't encourage reuse, since it prevents the quick identification of the desired peace. Added to this the leftovers accumulate a large amount of dust.
3. Distribution: The transportation of goods, in 80% of cases, is carried out by own truck. The finished products are packed with blankets, cardboard and/or bubble wrap. With the exception of the blankets, the other packaging materials are usually used once or twice and then are sent to the municipal landfill. These packages do not always protect furniture sufficiently during transport, causing havoc and rework. In addition, recyclable packaging is usually mixed with other non-recyclable ones (such as MDF), making it impossible to dispose correctly the final product.
4. Use: The companies do not offer guidance to the users on the correct ways of maintenance and cleaning of the furniture. This is done verbally, and only when asked by users. In addition, the studied companies do not offer maintenance or repair services for furniture. This type of service is only provided, in special cases, to established clients. It is concluded that the lack of information on the correct maintenance of furniture and the lack of repair services can lead to a reduced shelf life.
5. Disposal and/or recycling: Among the fifteen studied companies, four joineries carry out the final destination of the waste in the municipal landfill. The other companies dispose their waste irregularly. These practices cause public health risks arising from the toxicity of urea-formaldehyde or phenol-formaldehyde additives used in the production of reconstituted wooden boards, these substances are on the list of carcinogens issued by the International Agency for Research on Cancer (IARC). The companies don't carry out the separation of residues (MDF, plastic, cardboard, screws, etc.). There is also no suitable place to receive recyclable materials in the municipal landfill for recycling. Therefore, in general, companies have been using this fact to justify the lack of separation of furniture residues.

3.2 *Preliminary results*

According to PNRS – National Policy on Solid Waste, management should be prioritised by the following order of priority: non-generation, reduction, reuse, recycling, solid waste treatment and the environmentally safe final disposal of waste (BRAZIL, 2012). So, to encourage the non-generation, reduction and reuse of waste, some preliminary strategies have already been proposed: To develop, in partnership with designers from the region, annual catalogues of colour trends and MDF standards, with the objective of significantly reduce the number of colours recommended by the woodworking shops to users and, consequently, increasing the possibilities of reuse of raw materials. Prioritising the local economy through raw materials; to improve the layout of the company and the organisation of workbenches in order to reduce production time, increase worker concentration, and reduce material losses caused by errors; to create a checklist for measures surveying in order to minimise errors that occur during the measurement conference; to designate a single employee to perform the sheet cutting, facilitating the flow optimisation and control of new and/or reusable raw materials; to provide training on cutting planning to optimise the use of MDF panels, or even using software to reduce losses and waste; to encourage the use of software which enables wood panel planning and optimisation of cut-outs, reducing waste and the cost of purchase of equipment and material; to carry out the planning of cutting of the furniture from a modular approach (when possible) aiming a greater use of the sheet; to create a suitable shelf for facilitating reuses with stock organisation and classification of the MDF pieces according to the dimensions and colour of the pieces of panels; to replace disposable packaging, used during transport, with reusable packaging; to create a checklist of delivery and assembly of the furniture in order to avoid errors and rework, reducing unnecessary travel, fuel consumption, CO_2 emission and delays in delivery; to create an instruction manual with the purpose of informing users about the correct maintenance of the furniture (optimising life span); to indicate the collection system (environmentally correct final disposal), reuse and/or recycling (donation and/or renovation) of the furniture after use; to guide companies on disposal methods for environmentally sound waste; to provide repair services; to promote the recycling of materials such as plastic, cardboard, aluminum and others; to create a communication platform for buying, selling and exchanging material between manufacturers. The platform will be developed and made available to MSEs in order to sell debris material with potential for reuse, for example, medium or large indentations reconstituted wood plates. This communication channel intends to confirm the reduction of waste, encouraging the separation and proper storage of materials, while demonstrating the value of products that are commonly disposed of by such factors as reduced demand for very specific colour panels, and improper storage.

4 CONCLUSION

The diagnosis made by this investigation verified that MSEs of customized furniture in the city of Uberlândia have discarded annually a large volume of residues composed mainly of MDF. The average amount of waste per company is 14 tonnes per year, so if we consider the eight hundred companies in the city, the volume of discarded waste is approximately 11,000 tonnes per year. It has also been found that this waste of materials has generated, not only serious environmental problems but has also hampered the development and progress of these joineries. The diagnosis pointed out that the highest cost of the companies (around 42%) is related to the acquisition of raw material (MDF). The investigation found out that the serious political and economic crisis faced by Brazil since 2015, affected the companies in the sector and caused a significant slowdown in growth. In the Uberlândia joineries, the crisis has considerably reduced the production of furniture by order, has caused layoffs and is putting at risk the very subsistence of these micro and small companies. After the diagnosis, it is intended to carry out a set of measures of sustainable strategic design in the furniture MSEs of the region, encouraging the change of culture with the objective of reducing the consumption of resources and to minimise the waste. The entrepreneurs are motivated to implement the strategies proposed by CACO Project in the joineries because we could prove them that they are losing a lot of money (at least 1/3 of the profit) without the implementation of basic measures. The insertion of a set of ecodesign strategies in the MSEs can promote greater competitiveness for companies generating new business opportunities, creating more favourable conditions of competition among companies of the sector, improving their income while simultaneously promoting commitment and the sharing of responsibility (social, environmental, cultural) by all stakeholders.

ACKNOWLEDGMENTS

This work was carried out with the support of the Research Centre for Architecture, Urbanism and Design—CIAUD, Lisbon School of Architecture, Universidade de Lisboa and FCT.

REFERENCES

Barbero, S., Cozzo, B. 2009. Ecodesign. Barcelona: HF ULLMANN.
Behrendt, S. et al. 1997. *Life Cycle Design: A manual for small and medium sized companies.* New York: Springer.
Best, K. 2010. *The Fundamentals of Design Management. Lausanne*; La Vergne, TN: AVA Publishing.
Bhamra, T. 2007. *Design for sustainability: a practical approach.* Aldershot, England : Burlington.
Brasil 2012. Lei 12.305 de 02 de agosto de 2010. Política Nacional de Resíduos Sólidos. Brasília, p. 73.
Brezet, H. & Hemel, C.V. 1998. Ecodesign: A Promising Approach to Sustainable Production and Consumption. Paris: UNEP.
Bruce, A. & Llangdon, K. 2000. *Essential Managers: Strategic Thinking.* 1st edition ed. New York: DK.
FIEMG 2002. *Diagnóstico do pólo moveleiro de Ubá e região Belo Horizonte*: IEL/MG.
Fiksel, J.R. 2009. *Design for environment: guide to sustainable product development.* NY: McGraw-Hill.
Freitas, R.F. & Merino, E.A.D. 2011. *A gestão de design como estratégia organizacional.* RJ: Rio Books.
Galinari, R. et al. 2013. *A competitividade da indústria de móveis do Brasil: situação atual e perspectivas.* BNDES Setorial, n. 37, mar. 2013, pp. 227–272.
Garcia, A. 2005. *Estudo de mercado e pesquisa de benchmarking para o setor moveleiro* DF. Brasília- DF: SEBRAE.
Gilsbert, P.F. & Garcia, R.L. 2004. Las estratégias de disenõ respetuoso com el medio ambiente. In: RIZO, T. G. N. E. S. C. (Ed.). *Ecodiseno: Ingenieria del ciclo de vida para el desarrollo de productos sostenibles.* Valencia: Ed. Univ. Politéc. Valencia.
Glueck, W.F. 1980. *Business Policy and Strategic Management.* New York: Mcgraw-Hill.
Ibá (ed.). 2015. Cenários Ibá: Estatísticas da indústria Brasileira de Árvores Ibá.
IBGE. Cidades. In: http://cidades.ibge.gov.br/xtras/perfil.php?lang=&codmun=317020. Acesso em: 22 september. 2015.
Kozak, P. et al. 2008. Identificação, quantificação e classificação dos resíduos sólidos de uma Fábrica de Móveis. Ver. *Acad., Ciênc. Agrár. Ambient.*, Curitiba, v. 6, n. 2, pp. 203–212.
Manzini, E. & Vezzoli, C. 2008. *Desenvolvimento de produtos sustentáveis: os requisitos ambientais dos produtos industriais.* São Paulo: EDUSP.
Michaelis 2009. *Moderno dicionário da língua portuguesa.* São Paulo: Companhia Melhoramentos.
Mintzberg, H. et al. 2002. *The strategy process: concepts, context, cases.* NJ: Prentice Hall.
Mintzberg, H., Ahlstrand, B. & Lampel, J. 2008. *Strategy Safari: The complete guide through the wilds of strategic management.* 2 edition ed. Harlow, UK: Pearson Education Canada.
Movergs 2014. *Relatório setorial 2014 polo moveleiro do Rio Grande do Sul* IMEI.
Mozota, B. 2003. *Design Management: Using Design to Build Brand Value and Corporate Innovation.* 1 edition ed. New York, NY: Allworth Press.
Mozota, B. 2006. The four powers of design: A value model in design management. *Design Management Review*, v. 17, n. 2, pp. 44–53.
Nahuz, M.A.R., Figueroa, F.M.Z. & Lelo, P.K.Y. 2002. *Perspectiva tecnológica da cadeia produtiva: madeira e móveis.* São Paulo: IPT – Divisão de produtos florestais.
Pereira, A.F., Carvalho, L.S. & Pinto, A.C. 2010. Resíduo de madeira: limites e possibilidades de seu uso como matéria-prima alternativa. In: *Anais do 9º Congresso Brasileiro de Pesquisa e Desenvolvimento em Design*, São Paulo.
Rampersad H. 2007. *Balanced Scorecard pessoal – Caminho para a Felicidade Individual.*, Brasília: Qualitymark.
Santos, C. & Menezes, M. 2009. Design para micro e pequena empresa: o desenho como abordagem do projeto. In: Menezes, M. & Paschoarelli, L. (Eds.). *Design e planejamento: aspectos tecnológicos.* São Paulo: Cultura Acadêmica.
SEBRAE (ed.) 2013. *Anuário do Trabalho na micro e pequena empresa.* Brasília- DF: DIEESE.
SENAI 2006. *Diagnóstico empresarial das indústrias moveleiras de Uberlândia e região.* Uberlândia: FIEMG.
Zurlo, F. 2012. *Le strategie del design. Disegnare il valore oltre il prodotto. Milano*: Libraccio Editore.

A proposed procedure to develop clothing for pregnant women

D.A.N. Mentone, R.H. Osava, S.H.A. Gomes & R.A. Sanches
University of São Paulo, São Paulo, Brazil

A.Y.S. Duarte
Nossa Senhora do Patrocinio University Center, Salto, Brazil

ABSTRACT: This work describes the development of a maternity garment prototype that meets comfort and well-being needs of this special group of users. The "design thinking" method was adopted, which started from understanding user needs, based on interviews conducted with thirty pregnant women. With the collected material, the need of clothes that favor comfort in the breast region was identified as a common issue. A prototype of a blouse with synthetic knitted fabric containing elastane was developed. The product was proposed as an outdoor garment for pregnant women, that could also dispense bra use. The prototype was then created and produced. A smaller group of pregnant women undertook a usability test in order to assess functionality of the garment. The prototype was well evaluated, above all, in the matter of breast support. Seamless technology pleased the pregnant women in this study due to dispensing seams that often cause discomfort and allergies.

1 THE FASHION INDUSTRY IN THE UNIVERSE OF PREGNANT WOMEN

Innovations and advancements of the textile industry enabled improvements on fabrics, making clothes more comfortable and functional for users (Sanches, 2006). The development of new materials makes the production of comfortable clothing possible, resulting in changes in the well-being of individuals. Seamless technology emerges convening versatility. It is a technology without lateral seams, which enables the production of a seamless knitted tube. The circular knitting machine for seamless garments is an evolution of the machines employed in the making of socks, by using different diameters suitable for other body measurements (Magnus et al, 2010). Technology makes the production of different structures in the same product possible by using yarns of different raw materials, elastane among them.

The term clothing, culturally, encompasses mainly the idea of protecting the body. It is necessary to emphasize that it has the equally important function of providing users with psychological comfort and a feeling of good appearance and, likewise, to act as an intermediary between human body psychological comfort and the environment in which each person lives. With this understanding, garments play a significant role in determining the subjective perceptions of the state of comfort of individuals (Choudhury et al., 2011).

Similarly, based on the importance of the fashion industry and its specialties, clothing can be considered an important element in the universe of pregnant women. Due to changes occurring in a woman's body throughout her pregnancy, this theme requires attention focused on the development of a garment capable of providing greater comfort to these women, since materials, patterns and models of conventional fashion do not always meet the needs of these women whose bodies are in transition.

The present article is an excerpt from the dissertation "Proposed procedures for the development of knitwear products for pregnant women using seamless technology", which had the main objective of developing a maternity garment, made from synthetic fibers of polyamide and polyester (conventional and microfibers), using seamless technology, aiming to provide greater comfort to pregnant women while wearing the proposed garment. This excerpt presents the key stages performed in order to select the knitted structures used for manufacturing the prototypes, along with the technology used to produce the articles and the usability test applied to the prototypes.

2 PREGNANCY AND ITS CONTEMPORARY PORTRAYAL

In studying the relationship of clothing and pregnancy, Marinho and Rocha (2015) say that pregnancy is a period that brings physiological, psychological and social alterations; generates new meanings and requires adaptations in the daily life of the expectant woman. These alterations provoke a restructuring that reaches several dimensions such as the redefinition of her role as a woman and a change on her identity. In general, pregnancy is also marked by new habits, a new image of the woman's own body, and the emergence of challenges, such as the choice of clothes to wear when she has her body measurements altered (Ogle et al., 2013). The body gradually undergoes changes in certain proportions, weight and shape. In the belly region, measurements increase up to 80% in the lateral dimension during the last month of pregnancy, for example, changing from 16.5 cm to 29.7 cm (Itiro, 2005).

In the context of contemporary culture, during pregnancy, women relate the image of their own body with the use of pregnancy garments for different functions and consider clothing as an extended dimension of the body itself (Sohn and Bye, 2014). The same authors add that this theme requires attention when developing a product suited to the specificities of pregnant women, since fabrics, patterns and models do not always meet the needs of the target audience.

Studies have found that maternity garments, when found and acquired, and when used, play an important role in the well-being and self-esteem of pregnant women (Ogle et al., 2013 and Sohn and Bye, 2014).

2.1 *Synthetic fibers and seamless technology*

The use of synthetic fibers is considered important in the textile field, both because it is an alternative to natural fibers, but also because it offers benefits in terms of versatility, practicality, durability and contemporary finishing (Kandolf, 2007). With regard to the different types of fiber, each of them has its characteristics and properties.

Seamless technology is an enhancement of circular knitting machines that produced stockings, adjusted to the measurements of the body, obtaining a garment, without going through the cutting process and dispensing the use of lateral seams. This evolution of fibers and seamless technology in the textile industry enabled human beings to dress in body-adjusted and more comfortable clothes, minimizing the inconvenience of side seams (Magnus et al. 2010 and Leite, 2011). Seamless machines have the capacity to produce various knitted structures and to use different raw materials in predetermined areas; they can also produce pieces of clothing that meet both aesthetic and functional requirements, with the possibility of adding elastane threads in specific areas for compression and support.

In tables of degrees of compression used by manufacturers of medical socks and commonly practiced among seamless manufacturers, soft compressions with pressure up to 20 mmHg are indicated, as well as moderate compressions ranging up to 30 mmHg. Compressions varying up to 40 mmHg are considered strong pressures and likewise extra strong compressions for pressures reaching 60 mmHg. For the specific case of maternity wear, seamless articles may accompany the entire gestation period, guaranteeing comfort and mobility combined with support (Leite, 2011). There are other attributes such as appealing design in the garments and providing further support in different regions depending on the seamless technology applied. Products for pregnant women containing this technology have different levels of compression indicated for each user need. Functional clothing allows even weight distribution without leaving marks. Additionally, such products adjust to the body without compromising any movements.

2.2 *Design associated with functionality, ergonomics and comfort*

Regarding the use of garments, comfort is defined by four aspects: thermal, psychological, physical or sensorial and ergonomic (Soutinho, 2006). The comfort provided by a garment that fits the body is related to the existence of some important characteristics: performance of different fibers and yarns, the means of creating the design, sewing technologies and finishes, and product quality (Sanches, 2006).

Ergonomics is a discipline that applies principles, data and methods to design products in order to bring comfort and well-being to individuals, considering limitations and characteristics of users (Bezerra and Martins, 2006). In this understanding, Gonçalves and Lopes (2006) add that besides considering well-being, ergonomic studies still contribute to a better development of fashion products.

Using ergonomic factors in the product definition stage requires the design professional to know the consumer prior to product design. Therefore, it is necessary to investigate human anatomy, anthropometry, preferences of textile materials and consumer behavior (Silveira, 2008). Ergonomics is at the core of usability. The challenge is to develop an easy-to-wear garment for different situations. This challenge is overcome through technical and methodological competences that, in the end, aggregate knowledge to work on the development of a specific product for users (Cybis et al., 2010).

Usability of fashion products is a property that causes interaction between wearer and garment; a combination of factors that affect the individual's experience with clothing, such as efficiency, ease of wear, and comfort while using the product. Hence why Jordan (1998) cautions: lack of usability may compromise the use of a product. Along with this warning, the connection with design is highlighted by stating that usability played an important role in fashion design. It is indispensable to understand who will wear the clothes and their characteristics. Design for usability means developing an article for those who will utilize the product in question. Jordan (1998) further emphasizes that the goal of design is to develop a product that is usable by the largest number of people in the group in question. Additionally, the most important aspect used to assess usability is satisfaction, which indicates how acceptable a product is, not only in order to achieve a certain goal, but also taking into consideration the level of comfort that the individual feels when using a particular product.

3 METHOD

An adapted Design Thinking process (D. School, 2011) was used throughout the qualitative data research process and in the development of seamless clothing. Table 1 shows the main stages of the study.

In the first stage, a bibliographic review on changes that occur in the bodies of pregnant women was carried out. In the second stage, in order to better understand women's reality regarding clothing and fashion during the gestation period, semi-structured interviews were conducted with the purpose of revealing and analyzing this group and understanding the discomforts generated by pregnancy changes. Thirty interviews were completed and the relevant aspects served as reference to verify parameters of comfort and product functionality. The third stage was based on the first two stages of the adapted Design Thinking method, and knitted structures, raw material as well as production technology were selected. In the 4th stage a blouse was developed, after detailed discussion concerning the prototype, predicting the production of a garment with the characteristics arisen in the interview results. The sample was manufactured with different knitted structures and produced in the seamless equipment (Mentone, 2018).

The production of the seamless garment was carried out in an SMT-TOP2 SANTONI electronic circular knitting machine with medium diameter, made available by an educational institution. Afterwards, the pieces were cut and sewn and then the textile finishing (dyeing) was carried out.

Table 1. Study stages based on design thinking (adapted from D. School, 2011).

	METHODOLOGY	STAGES OF RESEARCH
1st Stage	EMPATHY Develop understanding about the subject	The group chosen for this study was composed of pregnant women. It was necessary to understand modifications caused by pregnancy.
2nd Stage	DEFINITION Articulate problem that needs to be solved	Thirty women were interviewed in order to comprehend how they dealt daily with issues of clothing, fashion and comfort during pregnancy and to understand the discomforts generated when dressing.
3rd Stage	IDEATE Brainstorm of potential solutions	Based on the qualitative data analysis, the following were selected: raw materials, fabric structures for the production of prototypes and the technology to be used for producing the articles.
4th Stage	PROTOTYPE Design one or several prototypes for testing	A drawing was developed, based on the reports of pregnant women; four prototypes were produced. All the same model but with different raw materials selected. All with the same knitted structures manufactured in the seamless equipment.
5th Stage	TEST Articulate a continuous and small cycle of innovation in design	After the production of prototypes, five pregnant women took a usability test and a semi-structured questionnaire was conducted with the purpose of analyzing reports regarding comfort and functionality characteristics of the materials tested in the physical experiments.

4 RESULTS AND DISCUSSION GETTING STARTED

Thirty women were allocated to answer a semi-structured survey containing variables which would indicate discomfort related to the use of clothing during pregnancy, as well as select textile material and product development. The women in this study attended a private physician's office in the city of Gramado, State of Rio Grande do Sul, Brazil. With the exception of two women, all others worked in stipendiary activities in the service sector. 10 women were interviewed for each gestational trimester. It is understood that the first gestational trimester encompasses the period within the 1st day from the beginning of gestation to the 13th week of gestational age; the second quarter is from the 13th to the 26th week; and third trimester, from the 26th week to the baby's birth.

As a result of physical changes during pregnancy, alterations of lifestyle and daily activities, the research addressed issues of habit and behavior regarding the use of clothes. It was verified that, in reporting discomforts generated by the use of clothes, 100% of the pregnant women presented discomfort when dressed, uncomfortable due to modifications in their bodies; also, 100% of the women reported that they had acquired some new garment during pregnancy and that the acquisition was associated with comfort and well-being.

When asked about clothes that caused the greatest discomfort during pregnancy, two garments were mostly mentioned: jeans and bra. Regarding gestational age, jeans were uniformly indicated as causing discomfort in the three trimesters. Regarding bra, although the discomfort was reported throughout pregnancy, it is in the third trimester that the complaints are more often referred, with 59% of mentions. Figure 1 shows the proportion in which these garments were rejected over the three trimesters, evidencing that the bra is mostly rejected during advanced gestation, while jeans were more often discarded in the initial pregnancy.

When asked about certain characteristics regarding clothing comfort, discomforts related to pattern cutting were reported, especially concerning seams, as well as garment size and information about textile fibers contained in the clothes. Regarding the materials used in the production of their clothes, almost all pregnant women reported not knowing what these materials were, but indicated preferences related to comfort and well-being. It was possible to observe rejection towards materials that are related to discomfort, with an expressive association of knitwear and elastane with the concept of comfort and well-being. Figure 1 shows reported discomforts associated with the use of garments during pregnancy.

The possibility of wearing body-adjusted clothing and the comfort of wearing garments that do not tighten parts of the body were associated with the presence of elastane. Knitwear and non-elastic garments appear to be very relevant features when women were asked about material preferences. Most women reported choosing a garment by touching the fabric and only 3% stated choosing a garment by its composition label.

Only 20% of the women in the sample gave importance to the use of specialized maternity wear. There were no significant data on the presence of psychological changes due to pregnancy, however, 27% of the surveyed stated that comfortable and elegant clothes contributed to their self-esteem, with a higher prevalence of this type of mention in the final stages of gestation.

Considering the preponderance of reports related to comfort in the upper chest, an innovative product was proposed: a blouse for pregnant women for external use with the purpose of dispensing the need of bra, respecting and using knitwear and seamless technology for the creation of a functional product; a basic product that, after tested, could become aesthetically attractive and contemporary. Figure 2 shows the garment production process until the usability test.

In the process of detailing the design for product construction, garment samples from Brazilian brands were used as reference for generating information regarding the model. The selected samples were underwear garments (tops and panties) and seamless leggings from the conventional fashion segment, and also a specific dress for pregnant women (Figure 2A).

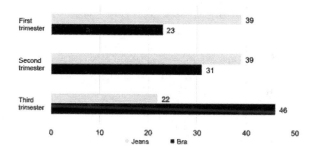

Figure 1. Proportion (%) of reported discomforts associated with the use of jeans and bra regarding gestational age (Mentone, 2018).

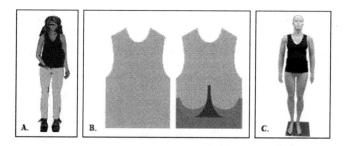

Figure 2. Steps from the production process to the final product (Mentone, 2018).
A. Design of a basic blouse for pregnant women that served as reference for the development of the research.
B. Knitted structures used in the production of prototypes (each color is equivalent to a knitted structure in the manufacturing process of prototypes).
C. Final prototype produced and employed in the usability test.

These pieces of clothing were used to define fabric weight and density of knitted fabrics (number of rows and columns) normally used when manufacturing this type of clothing. It was necessary to select a clothing measurement standard used in women's fashion in order to build the prototypes. Once the standard was defined, a pregnancy weight range table was employed, predicting both a dimensional increase in breast, belly and hip regions, as well as the different knit structures necessary to fulfill the functionality of breast support.

Tensions applied in different parts of the garment were based on results from interviews with the women. In Figure 2B, it is observed that the lighter color represents a smooth structure, allowing a loose blouse on the back, breast and belly, providing space to accompany the dimensional changes during pregnancy. The grey colored band represents a 3×1 rib knitted structure, adjusting to the body and allowing greater appreciation of the body contour. The black color represents a structure that was built between the breasts: creased, with the same 3×1 rib structure—also used in the band—between the padding cups of the blouse, which help accommodate and separate the breasts and enhance body contours (Figure 2B). A similar structure was used in the section that gives support to the padding cups and in the anatomical light gray part above the band, with a $2\times$ rib configuration. Structure and compression were also analyzed in previously selected seamless garments manufactured by companies already operating in this market. In order to apply the desired tension in different parts of the garment, elastane was tighter in the bust and back regions so as to obtain a moderate and lighter support function on the belly region. A band with higher tension was applied under the chest so that the function of breast support could be achieved.

A usability test was performed to evaluate comfort and well-being results and articulate a continuous cycle of innovation in design. The usability test was conducted with five women in different gestational trimesters. Information regarding the current body characteristics of each woman was collected, having a significant relevance to the experiment. This data will help determine clothing dimensions of pregnant women and proportions between a general fit and measurements in the breasts and belly regions.

To understand purchase behavior of pregnant women, we investigated the way in which they would choose blouses in a store and how the process of choice would proceed. The women were asked to express their feelings of comfort and well-being during the experiment in written form. Design and body appreciation and factors such as practicality, aesthetics, fashion and durability were also discussed.

One of the objectives of the experiment was to attain seamless product functionality. The proposal of the blouse is to provide well-being to a pregnant woman at the end of her day, through breast support functions, avoiding the use of elastics and bras. Because it is a blouse without side seams, tightness and skin markings are avoided. The product should remain dry and pleasant, considering that pregnant women sweat more due to an increase in general circulation and greater need of heat dissipation. Participants had a positive reaction towards the developed prototype and contributed with additional comments that enriched the experiment. They also added some considerations concerning the prototypes, such as measurement corrections for size P. It was suggested to maintain the various types of knitted structures and enhance compression in both the bust region and breast support band as well as increase the size of bust region. For greater psychological comfort, it was also evident that—after corrections are made in the prototype—new proposals on textures and models will better serve the possibilities of use due to lifestyle, climate and aesthetic requirements. With this model approved, varied necklines, sleeves, Jacquard, fish net stitches and assorted colors can be added.

5 FINAL CONSIDERATIONS

According to the usability test, it is concluded that when choosing clothes for personal use, pregnant women consider color and fabric softness to be important.

The prototype produced was approved regarding material, total length, fabric hand feel, color and, especially, seamless technology. It can be concluded from the usability test reports that seamless technology contributes to the comfort and well-being of pregnant women by avoiding the use of seams and elastics, particularly in the third trimester when skin becomes more sensitive.

Finally, it should be noted that the validity of this study is restricted to the investigated group, but it undeniably serves as a starting point, since data presently collected can be of great value and thus serve as a reference to designers interested in elaborating new product projects focused on the development of maternity wear.

ACKNOWLEDGEMENTS

The authors would like to thank the School of Arts, Sciences and Humanities of the University of São Paulo, the Research Support Program sponsored by SENAI-SP, the University Center of FEI for the use of their quality control laboratories, RHODIA and UNIFI for material supply. They would also like to thank all the pregnant women who, through their involvement, enriched this experiment.

REFERENCES

Bezerra, G.M.F. & Martins, S.B. 2006. Equação da ergonomia no design de vestuário: espaço do corpo, modelagem e materiais. In: *Colóquio da Moda, 2. Anais...* [online]. Available at: <http://www.coloquiomoda.com.br/anais/anais/ edicoes/2-Coloquio-de-Moda_2006/artigos/107.pdf> [Accessed 20 May 2017].

Choudhury, A.K.R., Majumdar, P.K. & Datta, C. 2011. *Government college of engineering e textile technology*. Factors affecting comfort: human physiology e the role of clothing, Capítulo 1, pp. 3–60.

Cybis, W., Beiol, A.H. & Faust, R. 2010. *Ergonomia e usabilidade*: conhecimentos, métodos e aplicações, 2nd ed., São Paulo: Novatec.

D. School. 2011. *Bootcamp Bootleg*. Escola de Design Thinking da Universidade Stanford. [online]. Available at: <http://dschool.stanford.edu/wpcontent/uploads/2011/03/BootcampBootleg2010v2S-LIM.pdf> [Accessed 10 November 2015].

Gonçalves, E. & Lopes, L.D. 2006. *Ergonomia no vestuário*: conceito de conforto como valor agregado ao produto de moda. Moda palavra, Estação das Letras, Vol. 4, Florianópolis, UDESC/CEART.

Itiro, I. 2005. *Ergonomia*: projeto e produção. 2nd ed., São Paulo: Edgard Blucher.

Jordan, P.W. 1998. *An introduction to usability*. London: Taylor & Francis.

Kandolf, S.J. 2007. *Textiles*. New Jersey: Person Prentice Hall.

Leite, D. 2011. *Influência da estrutura de malha e do elastómero na compressão das malhas seamless*. Dissertação (Mestrado em Engenharia Têxtil), Universidade da Beira Interior, Covilhã (Portugal).

Longhurst, R. 2005. (Ad)dressing pregnant bodies in New Zealand: clothing, fashion, subjectivities e spatialities. *Gender, Place & Culture*, Vol. 12 No. 4, pp. 433–446.

Magnus, E.B., Broega, A.C. & Catarino, A.P. 2010. Tecnologia seamless: perspetivas futuras. In: Congresso Brasileiro de Pesquisa e Desenvolvimento em Design, 9., São Paulo, *Anais...* São Paulo, Anhembi Morumbi.

Maldonado, M.T. 2002. *Psicologia da gravidez*: parto e puerpério. São Paulo: Saraiva.

Marinho, N.N. de & Rocha, M.A.V., 2015. Desenvolvimento de produto para consumidoras grávidas: reflexões por meio da roupa predileta. *Moda Palavra e-Periódico*, Vol. 8 No. 17, pp. 250–267.

Mentone, D.A.N. 2018. *Proposta Proposta de procedimentos para o desenvolvimento de produtos de malha para gestantes utilizando a tecnologia seamless*. 185 f., Dissertação de Mestrado – Escola de Artes, Ciências e Humanidades, Universidade de São Paulo, São Paulo.

Ogle, J.P., Tyner, K.E. & Schofield-Tomschin, S. 2013. The role of maternity dress consumption in shaping the self e identity during the liminal transition of pregnancy. *Journal of Consumer Culture*, Vol. 13 No. 2, pp. 119–139.

Sanches, R. 2006. *Procedimento para o desenvolvimento de tecido de malha a partir de planejamento de experimento*. 221 f., Tese (Doutorado em Engenharia Mecânica) – Faculdade de Engenharia Mecânica, Universidade Estadual de Campinas, Campinas.

Sanches, R.A. 2011. *Estudo comparativo das características das malhas produzidas com fibras sustentáveis para fabricação de vestuário*. Tese (Livre-Docência) – Escola de Artes, Ciências e Humanidades, Universidade de São Paulo, São Paulo.

Santoni. 2017. *Monofrontura SM8-TOP2*. [online]. Available at: http://www.santoni.com/es-macchine-sheet.asp?idm=283> [Accessed 17 February 2017].

Silveira, I. 2008. Usabilidade do vestuário: fatores técnicos/funcionais. *ModaPalavra e-periódico*, Vol. 1, pp. 21–39, janeiro/julho.

Sohn, M. & Bye, E. 2014. Pregnancy e Body Image Analysis of Clothing Functions of Maternity Wear. *Clothing e Textiles Research Journal*, Novembro 10.

Soutinho, H.F.C. 2006. *Design funcional de vestuário interior*. 223 f. Dissertação (Mestrado em Design e Marketing) – Escola de Engenharia, Universidade do Minho, Braga (Portugal).

Functional fashion focused on the needs of people with disabilities

L.N. Souza, S.H.A. Gomes, C.R.G. Vicentini & R.A. Sanches
School of Arts, Sciences and Humanities, University of São Paulo, São Paulo, Brazil

A.Y.S. Duarte
Nossa Senhora do Patrocínio University Center, São Paulo, Brazil

ABSTRACT: According to the Brazilian Institute of Geography and Statistics (IBGE, 2012), in Brazil alone there are 45,600,000 people with at least one type of disability, corresponding to 23.9% of the Brazilian population. Each impairment has its needs that must be considered in the product development process. The objective of this work is to present a prototype of pants for women, adapted from a model found in the market, aiming to improve the daily life of people with disabilities and bring long-term benefits. The prototype was developed based on an analysis of a survey applied to patients and professionals from the Lucy Montoro Rehabilitation Network. The finished product was made following ergonomics concepts, to facilitate dressing and undressing for wheelchair users, and to guarantee greater comfort when wearing the garment. The results obtained in the usability test showed a reduction in the average time to dress and undress the pants.

1 INTRODUCTION

Currently, there is an increase in the number of consumers who are better informed, fashionable and demanding about safety, well-being, aesthetics, comfort and functionality of textile articles, perceived before and during product use. These requirements can be achieved by combining properties of raw materials, fabric structure (machine settings) and finishing operations and modifications made throughout the production of garments (Sanches, 2011).

The behavior of textile articles and their applications can be better understood through properties of textile fibers. Final products must meet consumer needs and contain functionality. According to Kadolph and Langford (2006), clothing should offer: attractiveness (such as hand feel, hang, appearance), protection (against heat, water, cold), ease of maintenance, comfort and durability.

This article is an excerpt from the dissertation "Proposal of methodology for adapting of garments for disabled people (wheelchair users)", whose main objective was to propose procedures to facilitate the development of clothing adapted for people with physical disabilities (wheelchair users), with similar characteristics to common garments, using woven (denim) and knitted fabrics as raw material, aiming to increase self-esteem, safety and quality of life of these individuals. In this excerpt, a prototype of pants for women is presented. It has characteristics similar to ordinary clothing, yet aiming to reduce the average dressing and undressing time and increase safety and comfort of the wheelchair user.

1.1 *Fashion communication across various body types*

The interrelation between body and fashion is directly linked to social and cultural practices of a specific time. It is possible to view fashion as the way we connect with others through the dressed body. The body itself carries meanings and languages that are externalized to the world from the moment of birth. In this sense, the human body can be considered an object of significations.

Castilho and Martins (2008) say that body languages are potentialized through interferences, as well as by overlapping materials, which allow the construction of new shapes and volumes which atribute nem values and alter their meaning. Human beings naturally want to adorn the body, establishing relations between the individual and the collective. A frequent desire to rebuild oneself encourages the use of mediums such as paintings, tattoos, plastic surgeries, mutilations, makeup, cosmetics, among other processes.

Campelo (1997) states that an individual's cultural identity is inscribed in its body, clearly visible through information emanating from that body. Avelar (2011) establishes an oscillating relation between what can be understood as a person with or without a physical disability and the perspective of disabled bodies before society. It is cited, as an example, that a woman who undergoes a surgical procedure to enlarge her breasts, acknowledges her body as obsolete and requires some type of intervention. Clothing appears related to this action, since it will exercise her social body. In these conditions, humans manipulate their bodies to suit what they expect of themselves, clothing being a constituent element of their individuality.

Fashion is a specific form of social change, independent of any particular object; it is first and foremost a social mechanism characterized by a particularly brief timespam and by more or less fanciful shifts that enable it to affect quite diverse spheres of collective life. (Lipovetsky, 1987, p. 16).

The author states that fashion is part of a historical and social context of a given time and that clothing is only the consequence of a certain event. The authors of this article do not fully agree with the idea of "consequence", but rather perceive the idea of a constituent part of subjectivity, since discussions that are not abided only in hierarchical formulations are open. Thus, everything that belongs to the process of individual construction is pervaded by products that together generate values and symbols.

1.2 *Disability and fashion*

Some historical facts contributed to the idea of a more inclusive fashion. Brown (2010, p. 147) relates that:

(...) emergencies in times of war led to a huge and unprecedented public investment in the post-war era, the initiative passed into the private sector (...) R&D laboratories flourished in all sectors—including agriculture, automobiles, textiles and telecommunications (...) the belief that science would solve all problems and technology would translate them into goods to satisfy all needs.

In the clothing sector there is still a need for development and study. The demand for products that meet this public are unquestionable. Fashion is based on the differentiation of individuals and in a certain way on a critical look upon the appearance of others.

[...] fashion is not only one of the most important social—and economic—phenomena of our time, as well as one of the safest standards for measuring psychological, psychoanalytic, socioeconomic motivations of humanity. It is also one of the depositaries of a certain style in the sense that, at a certain epoch, fashion guides design when applied to clothing, decoration, fabrics, furniture. It is, therefore, one of the most sensitive indicators of the particular taste of that time. (Dorfles, 1990, p. 14).

These changes can be perceived through garments and body modifications that contextualize a current historical movement. Fashion can define the social involvement of users. Peclat (2002) reinforces: "clothing speaks substantially about a person, usually having the ability to indicate social and economic status, the class or group to which a person belongs or would like to belong". Grave (2010) emphasizes the ease with which fashion incorporates different voluntary presentations, but an indifference when physical transformations happen involuntarily.

For people with disabilities, social inclusion is desirable because of the limitations that the body expresses. For Kawamura (2005) fashion has four dimensions: the imitation of a certain standard or trend; economic class distinction; social habit, while belonging to a social group; and collective selection, which determines who may or may not belong to a certain group.

1.3 Stigmatization and inclusive design

Stigmatization in inclusive design has two aspects: one that is due to the product's own functionality, which often excludes consumers by not meeting their physical limitations; and also due to visual aspects—while there are adaptations favorable to people with disabilities, if a product does not have an attractive visual aspect, it will not meet emotional needs of consumers (Souza, 2016).

Jordan (1999) sets the importance of studying the physical needs of a population as a starting point in the conception of design. Social pleasure must be linked to products in a way that purchases made by people with disabilities do not focus only on physical benefits, but that they have the power of choice, thus forming their own identity. Care must therefore be taken when developing products that are adaptable to the needs of people with disabilities, as there is a thin and precarious line that can hold off consumers. Mainstream products should be made without definitions that diminish individuals. A more inclusive design can be developed by employing principles of universal design (such as functionality and usability) linked with principles of semiotics, which correspond to the emotional relationship between people and products.

Monge (2006) exemplifies this stigmatization through sound equipments: people with and without hearing difficulties make use of these devices for different purposes. However, products marketed to people without disabilities carry beyond functionality, fashion information giving possibilities and choices on materials and complementary functions, while those for people with disabilities have singularly the "hearing aid" function, hence the power of choice is limited to the product's functionality.

A reflection on inclusive design therefore challenges designers to develop products that can be discreetly adapted to the needs of people with disabilities, giving power of choice to their users on how they want to be viewed before society. If desired, their aspirations must be recognized through their goods, guaranteeing autonomy in the construction of their own identity. Souza (2016) points out that adaptations towards purchase environments should also be prioritized, such as: placement of ramps in physical stores, garment layout in stores, specialized services, availability of color reading applications and other conveniences during the act of purchase.

1.4 Comfort

Comfort is one of the most important attributes in life. Human beings are always, consciously or unconsciously, trying to maintain or improve their state of comfort, whether physical or psychological, in the course of their actions.

According to Soutinho (2006) comfort is defined by four aspects regarding the use of clothing:

a. Thermophysiological (thermal) comfort historically has the most important reason for the existence of clothing, which should protect people from cold and heat, and simultaneously must allow moisture transfer through its layers.
b. Psychological comfort is achieved through a combination of factors such as: ease of maintenance, durability, aesthetics, fashion, social and cultural environment, meaning it is predominantly related to fashion trends followed by society.
c. Physical (sensory) comfort is caused by mechanical and thermal contact between fabric and skin, that is, it is related to the sensations caused by the contact of fabric with skin. Contact can be static or dynamic. Static contact pressure depends on construction, size and weight of the garment and the flexibility and compressibility of fabrics. In dynamic contact, mechanical effects are more complex during movement. The importance of the fabric cutting employed in clothing manufacture is increased, as well as elastic properties

in the garment. The weight of the final product and friction between threads result in mechanical energy loss, decreasing the sensation of physical comfort (Neves et al., 2008). To provide this kind of comfort, clothing needs characteristics such as softness and flexibility during use and must not scrape, irritate or constrict the body.

d. Ergonomic comfort is linked to garment shape. Factors that have the most influence on this type of comfort are: cut, seams, pattern shape and anthropometric tables. Factors associated with the ability to perform body movements related to the type and structure of materials used are also important.

The comfort features that a consumer looks for in garments can be viewed as their functional and aesthetic specifications. Aesthetic specifications are those that will give the consumer psychological comfort, while functional specifications are those that will guarantee physiological and sensorial comfort (Broega & Silva, 2008).

2 MATERIALS AND METHODS

For this study, 4 types of woven fabric (denim) were selected from the market.

2.1 *Material*

Figure 1 shows the front and back sides of the fabric chosen to make a pair of female flare pants. The selected material has the following characteristics: Woven fabric (denim) - composition: 56% cotton 42% polyamide 2% elastane; linear density 344 g/m².

Fabric choice was made according to results obtained in physical tests and in the statistical analysis in the methodology proposed by Souza (2016). The following results were obtained from the chosen sample: high abrasion resistance, good percentage of elongation in the transverse direction, low dimensional change percentage in the longitudinal and transverse directions and high speed in moisture transport. In relation to the other fabrics studied, the selected fabric gives the final article high durability, guaranteeing conditions of product use for a period of time, and easy maintenance, which can be understood as the guarantee of conservation of the textile article during use and after washing and greater moisture transfer rate, meaning a greater sense of comfort when wearing the garment.

2.2 *Method*

The prototype was developed from a sequence of procedures adapted by the authors from Design Thinking methodologies and from the methodology proposed by Alvarenga (2006). The main steps used in the development of adapted clothing were the following:

a. Choice of disability type to be studied
The authors chose to study wheelchair users for this work.

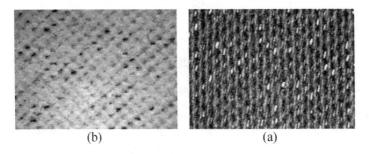

Figure 1. Back side (b) and front side (a).

b. Survey

For this research, patients and healthcare professionals were interviewed at the Lucy Montoro rehabilitation network. Several problems were identified in "common" garments worn by the wheelchair users, as well as the need to develop adapted clothing.

At this stage, an authorization was requested to the Ethics Committee for Human Research—School of Arts, Sciences and Humanities (EACH) of the University of São Paulo (USP). According to the Committee, all research carried out on human beings should comply with the current legislation for the category, especially Resolution No. 196 of October 10, 1996, of the National Health Council. Thus, for the application of survey or interview guide, a Term of Consent was delivered with the questionnaires guaranteeing the rights of the interviewees.

The interview guide ensures that the information is developed efficiently. The questionnaire establishes a pattern of information that enables the follow-up of the research. However, freedom given to the interviewees contributes in an effective way to obtain meaningful information to the project.

Ten questionnaires were applied: 5 to wheelchairs users (patients) and 5 to healthcare professionals (a physician, a physiotherapist, a nurse, a psychologist and an occupational therapist).

c. Development of possible clothing adaptations

From the analysis of results from the survey, some demands were identified, such as: pants with pattern cutting appropriate to the physical type of wheelchair users; fabrics with high elasticity index; jackets/blazers/suits without lining and with easy closure system for those who cannot execute a pincer grasp or with limited hand control. From these demands some possibilities of clothing adaptations were selected.

d. Prototype production

In this stage, after choosing the most appropriate fabric and initial pattern cutting tests, a final garment specification sheet was made in order to guide the pattern cutting development of the adapted clothing. The garment specification sheet contains the following information: garment description; fabric used; season (autumn/winter or spring/summer); technical drawing with specifications on trims, laundering, sewing details; thread colors used; operational sequence; among other items.

e. Usability test

The prototype was made with measurements from one of the interviewees. This person has a motor physical disability (wheelchair user), needs to get dressed in bed and has difficulty when zipping pants, since zippers are usually very small. Her main desire is to own pants that are easy to wear and undress.

3 RESULTS

A flare pair of pants for women was developed with the purpose of establishing a relation between garment functionality and aesthetics according to the needs of the studied group. Figure 2 shows the garment specification sheet, while Figure 3 shows the technical drawing for the pants.

The prototype pattern cutting has several modifications regarding traditional jeans. Initially, alterations were made in the upper part of the pants where the opening lies. Unlike pants commonly found in the market, this prototype has side openings, which were dislocated to where the front pockets (slant pockets) would be cut, so that the total opening is wider, facilitating dressing and undressing. Zippers with ring shaped pull tabs also provide a clothing adaptation for people with disabilities who have reduced mobility or atrophy in the hands.

The user can open and close zippers with a lever movement using one finger. Its buttoning comprises 2 magnetic side buttons and the back waistband has an internal elastic to better fit the body.

The pants do not have back pockets to facilitate moisture transfer. Besides lacking functionality, fabric accumulation may increase chances of developing pressure ulcers, harming the wheelchair user's health.

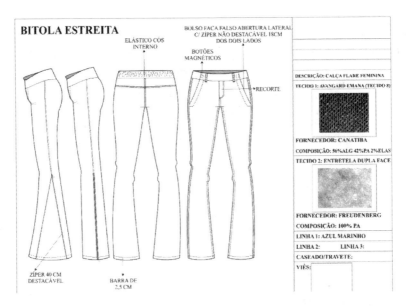

Figure 2. Garment specifications sheet (Souza, 2016).

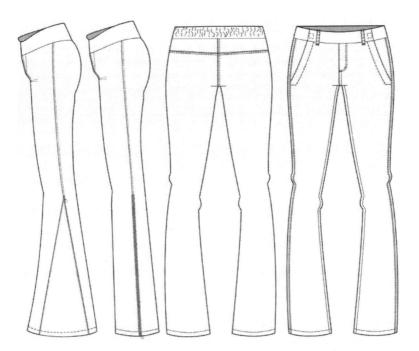

Figure 3. Technical drawing of female flare pants (Souza, 2016).

The sewing process operational sequence has been optimized, so the prototype has the minimum possible number of stitched seams (lockstitch sewing machine) and the maximum possible of threaded seams (overlock machine), reducing fabric friction with the skin.

The pants have a double pattern cut, and can be used with open zippers (flare pattern) or closed zippers (straight pattern). The side zippers also have ring shaped pull tabs and facilitate the passage of feet, a difficulty that causes many people with disabilities to stop wearing jeans.

Figure 4 shows the final product (prototype).

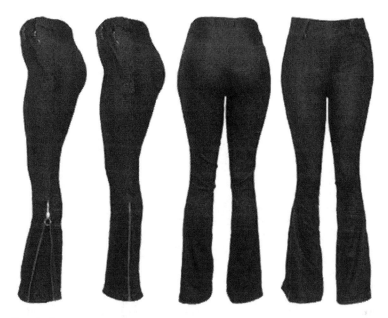

Figure 4. Female flare pants (Souza, 2016).

4 CONCLUSION

The clothing adaptation was developed according to the analysis of interviews conducted in the Lucy Montoro Rehabilitation Network and to previous observations and experiences of the authors.

The comfort features that consumers seek in garments can be viewed as their functional and aesthetic specifications. Aesthetic specifications are those that will give the consumer psychological comfort, while functional specifications are those that will guarantee physiological and sensorial comfort.

The prototype presented in this article was adapted from a pattern found in the market, in order to meet the needs of the group of wheelchair users interviewed, guaranteeing an increase in self-esteem, safety and quality of life to these individuals.

The adapted pants went through usability tests. The average time a wheelchair user takes to dress a pair of jeans is 25 minutes. With the adjustments made, the average time to dress the pants was 5 minutes.

The validity of this study is restricted to its sample, not allowing a generalization, since the sample was not probabilistic. Still, the data presented are of great value to guide designers in developing adapted apparel.

REFERENCES

Alvarenga, F.B. 2006. *Uma Abordagem Metodológica para o Projeto de Produtos Inclusivos,* Campinas: Faculdade de Engenharia Mecânica, Universidade Estadual de Campinas. 218 p. Tese de Doutorado.

Avelar, S. 2011. *Moda, globalização e novas tecnologias.* 2 Ed. São Paulo: Estação das Letras e Cores Editora. 166 p.

Broega, A.C. & Silva, M.E.C. 2008. *O conforto como ferramenta do design têxtil.* 3º Encuentro Latinoamericano de Diseño. Buenos Aires (Argentina).

Brown, T. 2010. *Design Thinking – uma metodologia poderosa para decretar o fim das velhas idéias: 1ª Edição.* Rio de Janeiro: Editora Elsevier. 147 p.

Campelo, C.R. 1997. *Cal(e)idoscorpos.* São Paulo: Annablume.

Castilho, K. & Martins, M.M. 2008. *Discursos da moda: semiótica, design, corpo*. 2. Ed. São Paulo: Editora Anhembi Morumbi. 35 p.

Dorfles, G. 1990. *Modas & modos*. Trad. Antonio J. Pinto Ribeiro. 2.ed. Lisboa: Edições 70. 14 p.

Grave, M. de F. 2010. *A moda – Vestuário e a ergonomia do hemiplégico*. São Paulo: Escrituras. 31 p.

IBGE – Instituto Brasileiro de Geografia e Estatística, 2012. Disponível em: <http://www.ibge.gov.br>. Acesso em 24 Jan 2014.

Jordan, P.W. 1999. *" Inclusive design – Design for All", e "Pleasure with Products – Human Factors for Body, Mind and Soul"*. Human Factors in Product Design – Current practice and Future trends. London: Taylor & Francis. pp. 171–181 e 206–217.

Kadolph, S.J. & Langford, A.L. 2006. *Textiles*. Ed. Prentice Hall. New Jersey.

Kawamura, Y. 2005. *Fashionology*. Oxford: Berg.

Lipovetsky, G. 1987. *O império do efêmero: A moda nas sociedades modernas*. São Paulo: Schwarcz. 16 p.

Monge, N.2006. *Design de produtos inclusivos, satisfatórios: a abordagem holística ao design inclusive*. Edições Universitárias Lusófonas, 2003. Caleidoscópio: Revista de Comunicação e Cultura. n° 07.

Neves, M.M. et al. 2008. *Projecto de vestuário termicamente confortável*. 5° Congresso Luso-Moçambicano de Engenharia e 2° Congresso de Engenharia de Moçambique. Maputo.

Sanches, R.A. 2011. *Estudo comparativo das características das malhas produzidas com fibras sustentáveis para fabricação de vestuário*. São Paulo, Escola de Artes, Ciências e Humanidades, Universidade de São Paulo. Tese de Livre-docência.

Soutinho, H.F.C. 2006. *Design funcional de vestuário interior*. Braga (Portugal): Escola de Engenharia, Universidade do Minho. Dissertação de mestrado.

Souza, L.N. de. 2016. *Proposta de metodologia para adaptação de vestuário para pessoas com deficiência física (cadeirante)*. São Paulo, Escola de Artes, Ciências e Humanidades, Universidade de São Paulo.

Designing for emotions: Evaluation of the drooler, a toy for preschoolers

B. Providência
Lab2PT / School of Architecture, University of Minho, Guimarães, Portugal

R. Brandão
Master in Product Design and Services, University of Minho, Braga, Portugal

P.B. Albuquerque
School of Psychology, University of Minho, Braga, Portugal

ABSTRACT: This article addresses the process of emotional evaluation of a product, its relevance and suitability to a group of children from 3 to 5 years of age. It focuses on the role of emotions and feelings in human growth and emphasizes the notions of emotional design and positive design, which play a very important role in the stimulation of user experience and in the production of emotions and feelings that encourage civilizational progress. This is an exploratory work that aims to understand the emotions and feelings with a handmade toy.

1 INTRODUCTION

In the last decades, emotional design has become a research field that focuses on user interface, namely on sensorial and cognitive experiences as stimuli to the perception of pleasure and affection. In this sense, experts from diverse areas, such as Damásio, Jordan, Norman, or Desmet have worked on this problem from different points of view and offering different theoretical and methodological contributions, however with a common denominator: the analysis and evaluation of emotional perception as an approach that enhances life quality in user interface.

This case-study's two analytical lines were, on the one hand, the design for emotions, for affections, and, on the other, the emotional evaluation. The goal of this theoretical framework was to understand the response to emotional stimuli via an object.

2 DESIGNING FOR EMOTIONS

2.1 *Positive design and the affections*

Affection was defined as the set of emotions, feelings, and mechanisms responsible for the production of experiences that become feelings.

> Feelings accompany the unfolding of life in our organisms, whatever one perceives, learns, remembers, imagines, reasons, judges, decides, plans, or mentally creates. Regarding feelings as occasional visitors to the mind or as caused only by the typical emotions does not do justice to the ubiquity and functional importance of the phenomenon. (Damásio, 2018, pp. 119).

Although the two terms are commonly conflated, it is important to distinguish between Emotion and Feeling. According to neuroscientists such as António Damásio, emotions are something physical that happens in our body whereas feelings are the mental reaction to such event. Though triggered by the mind, an emotion is something visible. It happens in our body through endocrine reactions and is, therefore, difficult to hide from others: internal physiological changes, facial expressions, changing skin tones, body movements.

As a result, an emotion is a change of varying intensity. It may be more or less positive or negative and it arises after a certain confrontation with an internal or external agent. On the other hand, because the feeling is the outcome of such change, it cannot be seen and can, for this reason, be more easily manipulated. A person may have a certain feeling and be able to hide it behaving in the opposite way, but it will be more difficult to mask an emotion.

> *Feelings can annoy us or delight us, but that is not what they are for if we are allowed to think teleologically for a moment. Feelings are for life regulation, providers of information concerning basic homeostasis or the social conditions of our lives. Feelings tell us about risks, dangers, and ongoing crises that need to be averted. On the nice side of the coin, they can inform us about opportunities. They can guide us toward behaviors that will improve our overall homeostasis and, in the process, make us better human beings, more responsible for our own future and the future of others.* (Damásio, 2018, pp. 163).

Design for Emotions aims to manufacture, or create products whose visual, tactile, acoustic, palatable or olfactory stimuli arouse a certain emotion, and consequently a feeling in the user, whether negative or positive. User experience with a product is not only determined by its functionality but also by its aesthetic, affective, personal, social, and cultural significance. This relationship is also shaped by the context, type of user, as well as by his/her mood at the time of interaction. Consequently, a user interacting with an object will have an emotional experience, a cognitive experience, and an aesthetic experience (Norman, 2004).

The interpretation of the relationship between the user and the object has been analyzed by several authors. While with different denominations they all pay attention to the Aesthetic, Functional and Memories stages. Paraphrasing Donald Norman (2004), the cognitive process takes place at three levels: Visceral, Behavioral, and Reflective.

The visceral response to an object occurs on a preconscious, pre-thinking level. It is the immediate response to the stimulus and focuses primarily on its visual appearance and touch. However, there are substantial differences in the ways in which different individuals respond to that stimulus. In fact, although the visceral system has developed to protect the body from danger, many experiences intentionally sought by human beings include feelings of horror and danger. Examples include viewing horror films, or practicing extreme sports, where the production of adrenaline results in pleasure.

At the next level, we find the behavioral response that focuses on the lived experience of the user when interacting with the product. That experience includes several levels: functional, linked to the activities it performs; effectiveness in performing the function; and usability, linked to the ease with which the user interacts with the product and makes it work.

Finally, there is the reflective level, which satisfies the user either because s/he identifies with the object since it brings back good memories, or because s/he feels that interacting with the object fosters positive perspectives about the future.

An object can trigger these three response levels in different proportions, not all of which have the same preponderance for all individuals. Different (social, cultural, individual personality) circumstances and contexts can value some aspects more than others and attribute certain characteristics to the object that place it on a certain interpretation level. It makes sense, therefore, to analyze and measure the relationship between subject and object by resorting to an emotional evaluation to understand how the first interacts with the latter and how are emotions prompted by such interaction.

2.2 Positive design

> *Emotions are an expression of our values; to design for emotion is to design for values.* (Desmet, 2015, p. 37).

Considering that Positive Design is the Emotional Design for Happiness, or the Design of Happy Feelings, and that happiness is paramount to the life and evolution of Humankind, the question is: What makes someone happy?

According to Lyubomirsky, happiness has three fundamental principles: 50% correspond to genetics, 10% to life circumstances, and 40% to the intentional activities of daily life (Lyubomirsky et al, 2004). It is mainly on these 40% that Positive Design works.

We live in a post-materialistic age, which means that we realize that it is not only objects that make us happy. Human beings take pleasure in the acquisition and manipulation of objects, but also, and with a longer term impact, in the creation and enjoyment of non-material experiences. Ephemeral pleasure, the avoidance of momentary pain, and hedonism are not the only things that matter. Eudaimonia, which corresponds to the pleasure we derive from our personal development and the general sense of life is also important.

Thus, Positive Design will be the one that results in a positive emotion, a better day for the user, the prospect of a brighter future and, ultimately, in the growth of Humankind.

The designer can work and accentuate the stimuli of a product or service to intensify the user's emotion when interacting with it. According to Pieter Desmet (2015), Positive Design must obey three basic principles: drawing for pleasure, for virtue, and for personal meaning. In other words, it must make us feel good, help us to be morally more correct, and have some meaning, whether relative to the past or a future life purpose. It must also have a long-term impact and balance pleasure and meaning, short- and long-term goals, individual and social concerns. It should not be a quick and transient solution, but rather contain adaptive support possibilities. On the other hand, it should offer a customized solution to a problem, not a generalist response that suits everyone since this would prevent us from including context concerns in the relationship. Finally, it should actively involve the user in problem-solving. It is fundamental to realize that the whole object exists in a social, cultural, geographical, and temporal context, thus assuming a positive design in the eyes of user A at a time X, but being potentially obsolete, useless, and meaningless for user B at a time Y. Therefore, it is up to the designer to analyze such contexts and decide how to respond to the challenge in a more effective and customized way. As such, the emotional evaluation of the subject/object relationship is also relevant since it is the only way of knowing how the interaction unfolded and the emotions it arose.

> *... it is not possible to talk about thinking, intelligence, and creativity in any meaningful way without factoring in feelings. Feelings play a role in our decisions and permeate our existence.* (Damásio, 2018, p. 163).

2.3 The emotional evaluation of the object

In 1897, Wundt suggested that emotions could be classified in terms of "pleasurable or unpleasurable," "arousing or subduing," and "strain or relaxation". In fact, most classification models order emotions according to three factors: valence (positive or negative), activation (calming or exciting), and dominance (dominant or dominated). Products can also be classified according to these parameters depending on the emotions that they awaken.

While the classification of emotions is somewhat more linear, the quadrant where we would place the products associated with them is not universal since the same product can arouse different emotions in different people or in different contexts. Designers may want a product to stimulate a particular emotion among a particular audience, but only when faced with user/object interaction can they draw firmer conclusions about its effectiveness. Understanding this dynamic requires an emotional evaluation. This can be done in different ways, with different means, to achieve different goals.

Among other examples, it is possible to carry out an emotional analysis of the object itself (shape, texture, color), of object/user interaction (functionality, effectiveness, usability), of the user's state of mind before or after interacting with the object, or even of the user's emotional state when s/he is deprived of the object after interaction.

Evaluations can be based on verbal, gestural, facial, vocal, motor, postural, or physiological expressions. We can analyze and measure these reactions by means of self-report psychological evaluation scales—questionnaires with nominal, ordinal, interval, or proportional measurement levels, for example; through an analysis of peripheral measures (biofeedback, facial temperature, muscular responses, heart rate, etc.); or by resorting to measures of central cerebral functioning (functional magnetic resonance imaging or evoked potentials recording).

2.4 *Methodology and case-study*

For the emotional evaluation of this product, a session was held at the OSMOPE School in Porto, in a mixed kindergarten room with children from 3 to 5 years of age. An evaluation questionnaire based on psychological self-report scales was used. The study was based on preexisting methods of emotional evaluation, such as POMS or PANAS-C. The content of the questions was adapted.

A toy was built to test emotions. The toy was called the drooler. It was handcrafted to enhance the emotional interaction of pre-school children. With a simple and round volume, its dimensions were designed to fit a child's hand. It was made of white birch wood, which has a light color and a smooth texture and touch, features that suit its intended user. This material provides weight, strength, and durability to the object and offers an emotional connection to nature. The natural wool cap is hand-knitted, awakening memories and strengthening the relationship between the object and the family, namely the association to home-knitted caps. As Walter Aarron points out in his 2011 "Designing for Emotion", the handmade component gives heart and character to products:

> *As mass production expanded in the mid-nineteenth century, the Arts and Crafts movement sought to preserve the craftsman's role in domestic goods production, and with it the human touch. The founders of the Arts and Crafts movement revered the things they designed, built, and used every day. They recognized that a craftsman leaves a bit of themselves in their work, a true gift that can be enjoyed for many years.* (Walter, 2011, p. 2).

It was assumed that the child/object interaction is enhanced by two changing elements: the eyes and a chromotropic ink symbol on the chest that reacts to the warmth of the child's hand touch. When activated by touching the drooler, the eyes close—and the heart/thunderbolt/ball appears by chromatic contrast – Figures 1 and 2.

In order to understand the child's emotional responses to the drooler, the following interactions were tested: Closing the eye and chest symbol with three variables i) heart, ii) ball, iii) thunderbolt.

The POMS (Profile of Mood States) is a scale developed by McNair, Droppleman, and Lorr in 1971. This method measures six different mood dimensions over a specific time period. It resorts to a 5-point Likert scale included in a questionnaire of 65 questions in its full version, or 37 questions in its reduced version, created by S. Shacham in 1983.

Figure 1. Initial stage.   Figure 2. Drooler after activation of the chromotropic pigment.

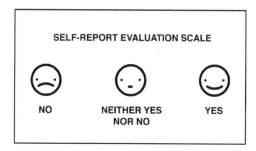

Figure 3. Scale adapted to the POMS and PANAS-C methods.

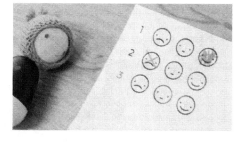

Figure 4. Children's answer sheet.

The PANAS-C (Positive and Negative Affect Schedule-Children) is a derivation of the PANAS method for infants. This was created in 1988 by researchers from the University of Minnesota and the Southern Methodist University of Texas and measures the negative and positive affect of participants in a given situation. It also uses a self-report questionnaire with a 5-point Likert scale. The PANAS-C was developed as an attempt to differentiate the affective expressions of anxiety and depression in children. It is a simplified, shorter version of the original.

A very simple questionnaire adapted to the children's age was used. It was based on a Likert scale reduced to 3 points, so that children could answer "no", "plus or minus", and "yes" (Figures 3 and 4). There was also room for children to freely express their opinions.

3 RESULTS OF THE EMOTIONAL EVALUATION – OSMOPE SESSION

On February 1, 2018 we visited the school and organized one session with the children (Figure 5).

Twenty-five children aged 3 to 5/6 years were evaluated. Not all children were present from the beginning to the end of the session, which is why the total sums vary.

The evaluation addressed the visceral, reflexive, and behavioral levels of the cognitive process.

3.1 Stage 1 – Visceral level of the cognitive process

– Evaluation of the child's interaction with the toy in its initial stage.
– Purpose: To evaluate the child's reaction to the toy before interaction.
– With the whole group together at first, and before interacting with the toy, questions were elaborated to which the children responded by painting on the answer sheet the corresponding icon (Figure 4).

The answers can be seen in Figure 6:
Some interpretations of Figure 6:

– Most children (63%) said they liked the color black. This study could be further elaborated to analyze the children's responses to different color toys in order to effectively perceive their chromatic preferences.
– Although 18 children (95%) said that they liked the toy, only 14 (74%) replied that they really wanted to play with it. However, it is noteworthy that the great majority liked and wanted to play with the toy even before knowing its ability to change due to the effect of the chromotropic ink.
– Most children thought the toy was neither sad, nor happy. In a total of 19 children, 11 (58%) answered "more or less" compared to 8 "yeses" and 0 "nos". Is this answer due to the lack of expression on the toy's face?

Figure 5. OSMOPE session.

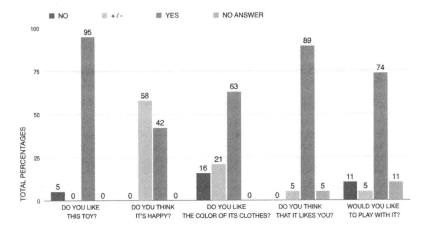

Figure 6. Visceral evaluation of the cognitive process.

3.2 Stage 2 – Behavioral level of the cognitive process

– Evaluation of the child/object interaction with activation of the eyes by touching the toy.
– Purpose: To evaluate the child's interaction with the toy in its eye-closing component as an effect of the chromotropic ink.
– The children are divided into 3 age groups. The same questions are asked to each group.
– With the heat on the face of the toy, the design of the eyes changed and it looked asleep. There was a reaction of astonishment to what had happened. The older children tried to find out how this happened. When the toy's eyes closed, the children amazed and, at the same time, enthusiastic were immediately silent so that the toy would not wake up. A child volunteered to very carefully and quietly put the toy on a table so that it could sleep. This reaction can also be analyzed and read as positive in the relationship that children have with the experience.

Results are represented in Figure 7.

It is possible to conclude that the children clearly found that the toy had been content to have been put asleep, that it had not been sad because of it, and that they enjoyed participating in this transformation process.

3.3 Stage 3 – Reflective level of the cognitive process

– Evaluation of the toy with activation of the symbol on the body.

- Purpose: To evaluate the child's interaction with the toy when the symbol on its body appears. To understand the relationship between the symbol and the child's memories and cultural and social associations. To analyze the child's emotional relationship with the object and compare it to the initial assessment.
- The children are divided into 3 age groups. Three-year-olds are given the toy with the chest circle; four-year-olds are given the toy with the heart; and five-year-olds the one with the thunderbolt. The same questions are asked to each group.

In addition to the overall results in Figure 3, it is also important to analyze the detailed results according to age group shown in Table 1.

Since the children were already tired, this part of the questionnaire was collectively answered. In other words, the questions were asked to the group of children who put their arms in the air. The number of answers was then recorded by the adults.

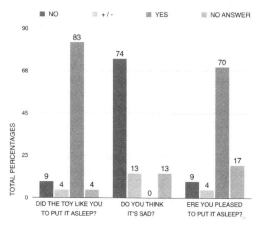

Figure 7. Behavioral evaluation of the cognitive process.

Table 1. Age group show.

PART 3		3-years-old (9 children) TOY WITH BALL				4-years-old (10 children) TOY WITH HEART				5-years-old (5 children) TOY WITH THUNDERBOLT			
		NÃO	+/-	SIM	N/R	NÃO	+/-	SIM	N/R	NÃO	+/-	SIM	N/R
1	DID YOU LIKE WHAT HAPPENED?	1	2	5	1		1	9				5	
2	DO YOU THINK THE TOY'S CHEERFUL?	2		7				10				5	
3	DO YOU THINK THE TOY RUNS FAST?	2		7		1		9		5			
4	DO YOU THINK THE TOY'S IN LOVE?	2		7		1		9		1		4	
5	DO YOU LIKE THIS TOY?	1		8		6	1	3				5	
6	DO YOU THINK THIS TOY LIKES YOU?	2		7		3		7				5	
7	DO YOU FEEL LIKE PLAYING WITH THIS TOY NOW?"			9				10				5	

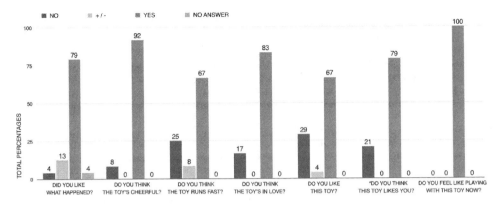

Figure 8. Reflective evaluation of the cognitive process.

I wanted to find out if there were different perceptions of the toys according to their different icons. However, I believe that, at this stage, the answers were largely shaped by the children's emotional state. The activity had lasted a little longer than desirable and some children were obviously weary. Another factor limited the results: the three and five-year-olds were sat on tables whereas the more numerous four-year olds were sitting on the floor, which is a more distracting position. There was a growing lack of willingness of the group to cooperate during this last stage.

If in the first stage we only have 1 child saying that s/he does not like the toy, now we have 7 who say "no", 1 who says "more or less" and 16 who say "yes". I believe that these answers do not have statistical significance as regards the object since all the children eventually say they want to play with the toy, but it may be significant as regards the questionnaire itself.

On the other hand, concerning the symbols, it is curious to note that there is no relationship between the thunderbolt and "fast running" – "It has no legs," said one of the children, "so it cannot run!" –, or a direct link between the heart and "being in love" – regardless of the symbol, most children say that the toy is in love.

It is also interesting to notice the different answers to the question "Do you think the toy is happy?" between the first and the third stages. If first we had 11 "more or less" answers and 8 "yeses", results now turn to 2 "nos" and 22 "yeses". Should we conclude that this change was due to interaction with the toy?

Finally, the children were asked if they wanted to keep one of the toys in the room, which is another way of asking them if they liked the object, and whether they wanted to name it. The answer was unanimous, so the choice of name followed. Each child made a suggestion, which was then voted on by all children. They were also asked which toy they wanted. From a total of 24 children, 8 liked the toy with the thunderbolt, 13 preferred the toy with the heart, and 3 the toy with the ball.

4 CONCLUSION

It was concluded from the emotional evaluation that the children liked the toy and they wanted to interact and play with it. The experiences undoubtedly accentuated the affective relationship between users and the object. The enthusiasm shown at the end, when they realized that the toy would stay in the room, along with the moments during the completion of the answer sheets when they were picking up and interacting with it, especially when they were not asked, are perhaps be the most solid proofs that the evaluation was very positive.

As for the evaluation method developed for children, I believe it was too long for the younger ones (3 and some 4-year-olds). A mixed group has very specific requirements since a difference of 1 or 2 years at these ages requires very different strategies. This should be taken into account in future interventions.

On the other hand, the icons chosen for "no", "yes", "more or less", may have been sometimes confusing. For example, in the question "do you think the toy's sad?" the response meter was confused with the answer itself, leading to false conclusions. The children who wanted to say that the toy was not sad and was therefore happy painted a happy face when, in fact, they should paint a sad face because the answer was actually negative.

In spite of these less positive aspects, the children were very receptive, cooperative, and participative. Let us now wait for some time before asking if the children spontaneously play with the toy. This will be the final product approval test.

There will be a follow-up of this project. This initial study and the session with the children will be a basis for future work. The less achieved aspects will be reformulated and re-oriented in the following phases.

AKNOWLEDGEMENTS

To OSMOPE—the entire team and Educator Alzira Mendes, in particular—for the enthusiastic availability and collaboration in the evaluation session.

The authors would like to thank all those collaborating in CIMODE and all the participants who will contribute the event.

This work has the financial support of Project Lab2PT—Landscape, Heritage and Territory Laboratory—AUR/04509, funded by the FCT/MCTES through national funds (PIDDAC) and co-funded by the European Regional Development Fund (ERDF) POCI-01-0145-FEDER-007528 under the new partnership agreement PT2020 through COMPETE 2020 – Operational Program Competitiveness and Internationalization (POCI).

REFERENCES

Damásio, A. 2018. *The strange order of things: life, feeling, and the making of cultures.* Pantheon Books; New York.
Desmet, P.; Jimenez, S.; Pohlmeyer, A. 2015, *Positive Design, Reference Guide*, The University of Technology.
Desmet, P, 2002. *Designing Emotions*, Delft, the Netherlands: Delft University of Technology.
Desmet, P.; JimePohlmeyer, A. 2015 *Positive Design: An Introduction to Design for Subjective Well-Being*, t University of Technology.
Hughes, A.A and Kendall, P.C. 2009. *Psychometric properties of the Positive and Negative Affect Scale for Children (PANAS-C)* in children with anxiety disorders. Child Psychiatry Hum Dev., 40:343–352.
Lyubomirsky, S., Sheldon, K., Schkade, D. 2005. *Pursuing happiness: The architecture of sustainable change.* Review of General Psychology, 9, 111–131.
McDonagh, D., Hekkert, P., van Erp, J. and Gyi, D. 2004. *Design and Emotion.* London: Taylor & Francis.
McNair D.M. Heuchert J. Shilony E. 2003. *Profile of Mood States (POMS): Technical Update.* Tonawanda, NY: Multi-Health Systems, Inc.
Norman, D. 2004. *Emotional design: why we love (or hate) everyday things*, Basic Books, New York, NY.
Sennett, R, 2008. *The craftsman*, Yale University Press, New Haven & London, 2008.
Walter, A. 2011. *Designing for Emotion*, A Book Apart.
WUNDT, W.M., 1897. *Outlines of Psychology*, in: Classics in the history of psychology, York University, Toronto, 2010.

Creativity and industrial fashion design—reviews, analyses and connexions

J.A.B. Barata & R. Miguel
University of Beira Interior, FibEnTech Research Unit, Covilhã, Portugal

S. Azevedo
University of Beira Interior, CEFAGE Research Unit, Covilhã, Portugal

ABSTRACT: Creativity plays an important role in nowadays live. When it comes to the industrial context, this creativity is meant to be successfully implemented, innovation. It is fundamental to be prepared for the actual instable economies and creativity (and than innovation) is considered essential pillars. Psychology is the main area studying creativity but once it extends to all human activities, there is a need to observe creativity in other domains.

The main goals of this study are to (1) provide a literature review on creative research, (2) explain the major differences between creativity and innovation, (3) understand design as a vehicle driving creativity to innovation and (4) present an insight on the actual panorama of the fashion industry as well as what is the fashion designer's job in the current framework. In the end of this article is presented a connection between an approach from psychology on creativity (Csikszentmihalyi 2015) and a study about the creativity on the fashion industry context (Ruppert—Stroescu & Hawley 2014).

1 INTRODUCTION

Even with some decades of research, the literature review on creativity has not shown a consensual definition on what is creativity. 'New' and 'useful' are, among other features, the ones gathering some accordance among scholars. From the revision, some approaches, models and paradigm-change visions are presented in the first section. More recent works point out that creativity flourishes cording to favourable physical and non-physical contexts.

Innovation differs from creativity once it is the implementation of the creative outputs. Design, as a project, becomes a bridge connecting creative to its successful implementation. In this document 'design' is seen as 'planning' the activities in order to answer the consumer's needs and demands. This design follows trends and market research.

Fashion products (e.g. apparel) need to have a function, a shape and needs to establish symbolic connections. Industrial fashion design is a system and so it will be analysed using Csikszentmihalyi's systems approach.

2 CREATIVITY: PARADIGMS, VISIONS, MODELS AND APPROACHES

Creativity drives from Latin '*creare*' and from Greek '*krainen*', meaning to create (Bahia 2008; Pereira et al. 1999). And this etymological definition seems to be the only one commonly accepted by scholars. Beyond this, the literature on the subject is filled with approaches and points of view, never to achieve a consensual definition of what is creativity. In this scenario, research touched "different parts of the same beast and derive distorted pictures of the whole" (Wehner et al. 1991, p. 270). In fact, this literature review presents visions and stories (Gurgel 2006; Albert & Runco 2003; Sternberg & Lubart 2003) with models and methodologies from different approaches. Creative outputs may be seen in an anthropologic perspective since it seams to be connected with problem solving, heuristic, activities (Bahia 2008). This characteristic has to do with the species' survival. When taking a look at Flusser's analises (2010), it is possible to perceive creativity as a design, a mechanism to overlap the human's limitations (e.g. the spoon). Glăveanu (2010) presents these views as paradigm changes during the track of studies: the He, I and We paradigms. Here follows some significant Western approaches and findings from cognitive and social psychology on this subject.

The mystic approach or the He-paradigm—He because the creative act was originally connected with male entities (Albert & Runco, 2003) – associates creativity to supernatural creatures guarding the spirit (Glăveanu 2010, p.80) or demons living inside the creator (Sternberg & Lubart 2003, p.5), channelling clairvoyance to some privileged men. This *ex nihilo* process had no scientific approach (Oliveira 2012; Wechsler 1998; Gurgel 2006; Valente 2015; Glăveanu 2010; Albert & Runco 2003; Von Stamm 2008).

During the Renascence, the Human figure was the centre of the Universe and so creativity starts to be seen as humanistic characteristic, opening the path for scientific approaches.

The I—paradigm revels the beginning of the real democratization of creativity (Glăveanu 2010). The literature review shown that the first attempts to understand the origins of creativity are linked to Freud (1910). Here the study seeks to understand why and how creativity happens.

The idea behind this approach is that the creative outcome is born from a conflict between the consciousness and the subconscious.

The Psychodomainance approach introduced the concepts of (i) regressive adaptation and (ii) elaboration to the study of creativity. In the (i) regressive adaptation, ideas appear in abstract, or amorphous formats; this is the first process in the creative development, it occurs both in the attempt to solve problems, during sleep, using drugs and in a state of neurosis. (ii) Elaboration is the second stage of this process, and here, lucidly, cultural and social moulded, the creator shapes what was abstract in the first stage (Albert & Runco, 2003; Glăveanu, 2010; Gurgel, 2006; Oliveira, 2012).

With the foundations provided by the earlier approach, many have attempted to study the cases of eminently creative individuals, as Freud (1910) did with Leonardo da Vinci.

Guilford (1950) indicated in the inaugural speech at the American Psychological Association, that this study of eminently creative individuals could be extended to other samples using paper and pen tests. Here the focus of the study is related to the divergent thinking (Sternberg & Lubart, 2003; Guilford, 1956, 1959; Sternberg & Grigorenko, 2001).

The Psychometric Approach becomes relevant to the study of creativity because here the first instruments to evaluate individuals whose creativity was eminent were born. From a set of variables, these tests operate at the level of biographical inventories, personality inventories (predisposition for creativity) and behavioural tests (Amabile, 1996; Finke, Ward, & Smith, 1996; Galton, 1869; Gardner, 1988; Lima, 2006; Runco, 2007).

If until now the attempts were to understand the origin of creativity, the cognitive approach sought to understand the processes by which it happens.

Finke, Ward, & Smith (1996) proposed the Geneplore model, *gene-* from generate and *-plore* from explore (Runco 2007, p.31), which indicated that the creative process occurred in two phases: the (i) generation and the (ii) exploration of ideas.

In the (i) generative phase the first ideas arise through mental constructions, these do not yet appear as 'the one' but are considered the promising 'gene' for the final discovery. In the (ii) exploratory phase, the individual uses constructs created in the first step to work them through mental processes (e.g. associations, synthesis, analogic transfer, categorical reduction).

Also in light of cognitive studies, Weisberg (1993, 2003) indicates that creative outputs are the result of a cognitive processes (analogical transference) associated with the knowledge of those who seek creative solutions. Weisberg suggests that most of the results (insights) depend on the changes on how the problem is initially interpreted or on a different approach to the same problem's representation (Runco, 2007, p.12). In this way, the process is observed as a search for re-connection between elements of the individual's universe and his/hers cognitive ability to solve the problems (Runco, 2007, Sternberg & Lubart, 2003, Weisberg, 1993, 2003).

At the same time, academic work began to emerge investigating personal/individual variables. Barron and Harrington (1981) presented some personality traits that appear to be common on eminently creative individuals.

There is a vast list of characteristics presented in the literature, however, here are some of the features in common in the review of characteristics about creative individuals: creative people tend to be/have tolerance for uncertainty and ambiguity, self-confidence, unconventional, originality, intrinsic motivation, above-average intelligence, determination for success, humour, unconventionality (Sternberg & Lubart 2003, Runco 2007, Glăveanu 2010, Amabile 1996, Sternberg 2012; Amabile 1983; Terra 2000).

Wehner et al. (1991) indicated that studies on creativity presented a 'parochial isolation' in the studies. Implying that several disciplines have dealt with the subject with an isolated perspective. This fact may be revised, for example, in the divergence of classifications. The economics theses referred to creativity as 'innovation' and the area of psychology referred to the same object of study with 'creativity' (Sternberg & Lubart, 2003).

These approaches include a point of view involving social psychology in a way that creativity can and is enabled by a sett of contexts. The literature review shows an effort regarding the multiple layered approaches converging to creativity. Here is the start of the We-paradigm and the pinnacle of the creativity's democratization. Next are presented two models of this approach. Not to say they are better than others, this review is short for the article's purpose.

Amabile's (1983; 1996) componential model shows creativity as a result of (i) domain-relevant knowledge and abilities, (ii) creativity-relevant skills and (iii) intrinsic motivation (rather than extrinsic motivation alone, which derives from recognition and/or rewards) have a foundational role in creative outputs (Sternberg & Lubart 2003; Amabile 1996; Amabile 1983; Hennessey & Amabile 2010). The studies conducted by Amabile and Henessey have demonstrated that task intrinsic motivation is extremely important to enhance the creativity once one "rarely do creative work in an area unless they really love what they are doing and focus on the work" (Sternberg 2012, p.6). In this way, motivation is observed not only as essential for a creative response, but also as an element that can boost the final result.

Table 1. Review's summary—self devised.

Paradigm change Glăveanu (2010)		Approach	Reviewed authors	Contribution
HE-paradigm		Mystic		
I-paradigm	Democratization of creativity	Psychodominance approach	Freud (1910)	Why and how? (i) regressive adaptation (ii) elaboration
		Psychometric approach	Gulford (1956, 1959)	First instruments to evaluate individual creativity
		Cognitive Approach	Finke, Ward, & Smith (1996)	Creative process (i) generation phase (ii) exploration phase
			Weisberg (1993, 2003)	Re-connection between elements of the individual's universe
WE-paradigm		Personality traits	Barron and Harrington (1981)	Characteristics on creative individuals
		Componential model	Ambile (1983; 1996)	(i) domain-relevant knowledge and abilities (ii) creativity-relevant skills (iii) motivation
		Domain—Person—Field systems approach	Csikszentmihalyi (1988; 2015; 2014)	Relations between culture, people and society.

Csikszentmihalyi (2015; 2014) proposed the well-known Domain—Person—Field interaction (DPFi) systems approach (firstly publish in 1988 under the title of "Society, culture, and person: a systems view of creativity" by Cambridge University Press, also presented in the 2015 collection of Csikszentmihalyi's works). In this systemic model, the creator drains information from the domain or domains within the culture to propose novelties to society. In this panorama, the critics or gatekeepers will approve or reject the proposal. If the outcome is well receive, it will intergrade the culture and domain's knowledge (Glăveanu 2010; Sternberg & Lubart 2003; Sternberg 2012). A more detailed description of the DPFi will be presented in the section 6.

The Table 1 presents a brief explanation of the approaches mentioned above as well as their contributions.

3 INNOVATION: WHAT IS IT AND WHERE IT DIFFERS FROM CREATIVITY

Innovation derives from the Latin word *'innocatione'*, meaning renewal. This action results from a systemic approach involving creativity—fuelling the innovation pipeline (Mclean 2005) - and a management's conscious decision-making process. "This is reflected in the now widely accepted definition of innovation equalling creativity plus (successful) implementation." (Von Stamm 2008, p.1). Rarely creative outputs alone generate profits and so it is fundamental to have an organized and supportive management do implement these same new and useful ideas (Hollanders & Cruysen 2009; Ferreira 2012).

The Oslo Manual (OECD 2005; OECD 1997) indicates four types of innovation: product, process, organizational or marketing (OECD 1997; Europeia 1995; Von Stamm 2008; OECD 2005). When it comes to the innovation outputs they can be (i) radical or (ii) incremental, meaning a (i) new product or service in the market, absolutely new to the consumer's universe or (ii) a new version of an already existing product or service to be perceived as a better version of the artefact or experience (Monteiro-Barata 2005; Von Stamm 2008). This implementation involves the whole process from ideas' selection, product's development and its final distribution.

4 DESIGN: THE PROCESS FROM IDEA'S GENERATION UNTIL IMPLEMENTATION

Design may be studied as a name/verb (e.g. Flusser 2010), as an activity or as a methodological planed activity. In the last context, design became a bridge connecting the creative production until its successful implementation (Von Stamm 2008; Hollanders & Cruysen 2009; Ferreira 2012; Lima 2006; Tironi 2014).

In the 1950s, design began to be seen in its quality of planning and so it can be divided into stages of its methodology (Bayazit 2004).

General design methodologies involve briefing, problem's division into logical phases throughout the procedure (e.g. Archer 1979; Munari 2014), overview of processes and products, creation of conceptual images and analyses of the acceptable outcomes (Zeisel 1984). In the fashion and apparel industry, to these methodologies are added other elements that are not part of the process but influence it as they determine the procedures and the final result—product (Lima 2006; Neves & Branco 2000; Agis et al. 2010). In the model proposed by Gaskill (1992) the starting point is the trend's analysis: influenced and influencing major factors such as the definition of target markets. The exchanges of information therefore seem to be one of the most important factors in today's industrial economy. "Information has become the key" (Sinha 2002, p.1).

5 FASHION PRODUCT AND FASHION DESIGNER'S JOB

The fashionable product should be/have (i) new, (ii) useful (i.e. the same that is common in the definition of the creative product), symbolic value, and should pull the consumer's

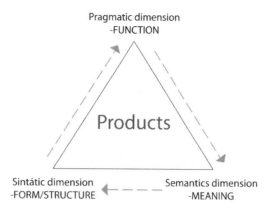

Figure 1. Function, structure and semantics of fashion design products—Self-devised, based on (Neves & Branco 2000, p.36).

desire (Ruppert- Stroescu & Hawley 2014; Lima 2006; Neves & Branco 2000). In this sense, the fashion project's outcome must comprise three dimensions, namely, pragmatics (function), syntactic (structure—form) and semantics dimension (meaning—sign – *a liquid stat pró aliquo*). These are presented in the Figure 1 below.

The fashion industry represents a complex context of industries; here each partner produces fragments of the whole. The different fields of action are in the production and marketing of goods targeting different markets. It is known that the procedures always start with trend's analyses followed by textiles' and artefact's production to be distributed in different channels (Jones 2006; Neves & Branco 2000; Gaskill 1992).

In today's fashion industry landscape, companies have to face fierce competition (Cardoso & Monteiro 2017; ATP—Associação Têxtil e Vestuário de Portugal 2015; ATP—Associação Têxtil e Vestuário de Portugal 2014; Sinha 2002; Ruppert—Stroescu & Hawley 2014). The democratization of access to trend information has led to an increase of the consumer's consciousness and demanding and the competition increases with, for example, the quick response to these demands.

Some studies point to a set of characteristics and roles for nowadays fashion designers. Beyond someone who chooses colours and materials, who draws illustrations, precise designs and technical drawings and supervises the production (Lima 2006; Gaskill 1992), the fashion designer needs to be a trend analyser, a market researcher of visual and qualitative data, must understand market statistics and must be an interpreter of meanings (Sinha 2002, p.4).

6 CREATIVITY AND INNOVATION: ANALYSING THE CONNEXIONS BETWEEN CSIKSZENTMIHALYI'S APPROACH AND A STUDY ON FASHION DESIGN INDUSTRY

This final section presents Csikszentmihalyi's DPFi systems approach (Figure 2) and the empirical findings from Ruppert—Stroescu & Hawley's (2014) study on the Fashion Industry.

The DPFi approach offers a schematic representation on how cultures', creator's and society's systems are linked in the creative process. Systems are formed by inputs and outputs.

Culture is filled with different domains and these are made of different types of knowledge regarding the domain's specific area (e.g. history and research). People receive information from the domain in order to understand it, learn from it, and also as an inspirational starting point. From here, people may produce a creative output for society. Finally, it is society's job to evaluate these productions. Acceptance might be a difficult part of the creative process, "99% of all new ideas are garbage, regardless of the domain or the status of the thinker."(Csikszentmihalyi 2015, p.52). If it is well received by the gatekeepers of the field (e.g. art critics), the creative outcome will integrate the original domain's knowledge.

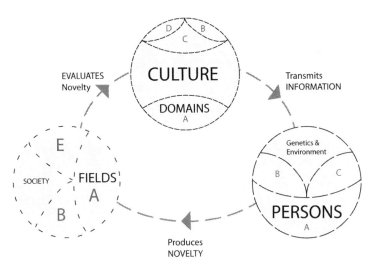

Figure 2. Csikszentmihalyi's DPFi systems approach—Self-devised, based on (Csikszentmihalyi 2015, p. 211).

Table 2. Connections between Csikszentmihalyi (2015) and Ruppert-Stroescu & Hawley (2014) – Self-devised.

Csikszentmihalyi (2015)		Ruppert-Stroescu & Hawley (2014)	
"Creativity is a phenomenon that results from interaction between these three systems, (…)" (Csikszentmihalyi 2015, p. 47 and 48)		*"Today's fashion industry endures because of change as it simultaneously reects and affects the culture in which it exists."* (Ruppert- Stroescu & Hawkley 2014, p. 12)	
		Leadership creativity	Adaptive creativity
DOMAIN	Culture has Domains; People receive information from Domain (or domains)	Fashion designers search abstract references or crossed references from different domains.	Fashion designers use secondary sources or knowledge from the specific domain of fashion; Uses outputs from Leadership creativity that were already accepted by society.
PEOPLE	People produces novelty to present to the Field	Fashion designer produces new (or improved) products for the field; Previous research is made to reduce de disapproval from the field (target audience).	
FIELD	Field accepts: novelty goes domain knowledge; Field rejects: novelty goes to domain's archive.	Field accepts: novelty becomes trend; Field rejects: novelty becomes alternative trend.	

The study developed by Ruppert- Stroescu & Hawley (2014) proposed two typologies of creativity in the industrial fashion design development, the (i) Leadership Creativity and the (ii) Adaptive Creativity as two extreme ends of the continuum. The main differences were appointed in terms of the (a) research and development (R&D), (b) selling price, (c) nature of the product, (d) consumer taste level, (e) technique, (f) number of designs created and reproduced in a season, (g) consumer perception and life cycle of the product and (h) source of design inspiration.

Here are some major differences: Regarding (a) R&D, the (i) leadership creativity has demonstrated high levels of interest in the production of entire new and differentiated products, while (ii) adaptive creativity tends to use these new outputs as a staring point, transforming the existing knowledge/information. With this, the (c) nature of the product from leadership creativity is radical innovation and adaptive creativity develops incremental innovation. The staring point for the creative process, (h) source of design inspiration, for the (i) leadership creativity are abstract representations and (ii) adaptive creativity uses secondary sources of inspiration; it means that adaptive creativity might start from the outputs provided from leadership creativity. The Table 2 shows the connections between these empirical finding and the DPFi approach.

7 CONCLUSIONS

Although a common definition on creativity was not found, 'new' and 'useful' characteristics are the ones in common in the majority of research on the subject.

In this study, design was used as a mean driving creative ideas to their implementation. Design therefore is linked to creativity and innovation. Like general creative products, fashion outcomes need to be 'new', 'useful'. In the fashion context there is a need for the symbolic value and desire features.

Innovation can be radical or incremental and fashion products follow the same path. Leadership creativity usually introduces entirely new products and adaptive creativity tends to follow the repetition and redesign of already existing products.

In the final section is proven that the creative process in fashion design (more precise in the industry) can be study by using knowledge from the Psychology domain. In this context, Domain is an area with knowledge associated to it. Fashion designers get information about one or more domains to produce novelty. The accepted novelty will convert to domain's knowledge; creativity/innovation are to become future information on the domain. To note that leadership creativity is not perceived as better than adaptive creativity because both are related with the Domain's knowledge as shown on the section 6.

As a final remark, the authors indicate the need for more research on creativity directed to the fashion design practice.

REFERENCES

Agis, D. et al., 2010. *Vestindo o Futuro Microtendências para as indústrias têxtil, vestuário e moda até 2020.*
Albert, R.S. & Runco, M.A., 2003. A History of Research on Creativity. In *Handbook of Creativity*. pp. 16–31.
Amabile, T., 1996. *Creativity in Context*, Oxford: Westview Press.
Amabile, T., 1983. *The Social Psychology of Creativity* 1st ed. R.F. Kidd, ed., Springer Series in Social Psychology.
Archer, B., 1979. Design as a Discipline. *Design Studies*, 1(1), pp. 17–20.
ATP – Associação Têxtil e Vestuário de Portugal, 2015. *ATP 50 Anos*, Vila Nova de Famalicão.
ATP – Associação Têxtil e Vestuário de Portugal, 2014. *Têxtil 2020: Projetar o Desenvolvimento da Fileira Têxtil e Vestuário até 2020*, Vila Nova de Famalicão.
Bahia, S., 2008. Promoção de ethos Criativos. In M. de F. Morais & S. Bahia, eds. *Criatividade e educação: conceitos, necessidades e intervenção.* Braga: Equilibrios, pp. 229–250.
Bayazit, N., 2004. Investigating Design: A Review of Forty Years of Design Research Nigan Bayazit. *Design Issues*, 20(1), pp. 16–29.
Cardoso, F. & Monteiro, I.P., 2017. *Liderança de Equipas na resolução de problemas complexos* 2ª edição., Lisboa: Edições Sílabo.
Csikszentmihalyi, M., 2015. *The Systems Model of Creativity – The Collected Works of Mihaly Csikszentmihalyi*, Springer Netherlands.
Csikszentmihalyi, M., 2014. The Systems Model of Creativity and Its Applications. In D.K. Simonton, ed. *The Wiley Handbook of Genius*. John Wiley & Sons, Ltd.
Europeia, C., 1995. *Livro Verde sobre a Inovação.*
Ferreira, M.C., 2012. *Design como indicador de inovação: estudo sobre as atividades de design na economia portuguesa*. ISCTE-Business School/Instituto Universitário de Lisboa.
Flusser, V., 2010. *Uma filosofia do design: A forma das coisas* Relógio D.

Freud, S., 1910. *Five Lectures on Psycho-Analysis: Leonardo da Vinci and Other Works*, London: The Hogarth Press.
Gaskill, L.R., 1992. Model of Retail Product Development: A Case Study Analysis. *Clothing and Textiles Research Journal*, 10(4), pp.17–24.
Glăveanu, V., 2010. New Ideas in Psychology Paradigms in the study of creativity: Introducing the perspective of cultural psychology veanu. *New Ideas in Psychology*, 28(1), pp. 79–93. Available at: http://dx.doi.org/10.1016/j.newideapsych.2009.07.007.
Guilford, J.P., 1956. The Structure of Intelect. *Psychological Bulletin*, 53(4), pp. 267–293.
Guilford, J.P., 1959. Three faces of intellect 1. *American Psychologist*, 14(8), pp. 469–479.
Gurgel, M.F., 2006. *CRIATIVIDADE & INOVAÇÃO: Uma Proposta de Gestão da Criatividade para o Desenvolvimento da Inovação*.
Hennessey, B.A. & Amabile, T.M., 2010. Creativity. *Annual Review of Psychology*, 61(1), pp. 569–598. Available at: http://www.annualreviews.org/doi/abs/10.1146/annurev.psych.093008.100416.
Hollanders, H. & Cruysen, A. Van, 2009. *Design, Creativity and Innovation: A Scoreboard Approach*,
Lima, E., 2006. *Um Contributo para Potenciar a Criatividade no Design de Vestuário*. Universidade Técnica de Lisboa.
Mclean, L.D., 2005. Organizational Culture's Influence on Creativity and Innovation: A Review of the Literature and Implications for Human Resource Development., 7(2), pp. 226–246.
Monteiro-Barata, J., 2005. Innovation in the Portuguese Manufacturing Industry: Analysis of a Longitudinal Company Panel. *International Advances in Economic Research*, (11), pp. 301–314.
Munari, B., 2014. *Das Coisas Nascem Coisas*, Edições 70.
Neves, M. & Branco, J., 2000. *A Previsão de Tendências para a Indústria Têxtil e do Vestuário.*, TecMinho.
OECD, 1997. *Manual de Oslo: Diretrizes para a Coleta e Interpretação de dados sobre Inovação Tecnológica*, Available at: http://www.oecd-ilibrary.org/science-and-technology/manual-de-oslo_9789264065659-es.
OECD, 2005. *Oslo Manual – GUIDELINES FOR COLLECTING AND INTERPRETING INNOVATION DATA*, Available at: http://scholar.google.com/scholar?hl=en&btnG=Search&q=intitle:Oslo+Manual#0.
Oliveira, R.S.R. de, 2012. *Um Programa de Treino da Criatividade Estudo exploratório com alunos do 1° Ciclo*. Universidade da Madeira.
Pereira, B., Mussi, C. & Knabben, A., 1999. Se sua empresa tiver um diferencial competitivo, então comece a recriá-lo: a influência da criatrividade paea o sucesso estratégico organizacional. *23° Encontro Anual de Associação Nacional dos Programas de Pós-Graduação em Administração*, 1, pp. 1–10.
Ruppert- Stroescu, M. & Hawley, J.M., 2014. A Typology of Creativity in Fashion Design and Development. *Fashion Practice*, 6(1), pp.9–36.
Sinha, P., 2002. Creativity in fashion. *Journal of Textile and Apparel, Technology and Management*, 2(IV), pp. 1–16.
Von Stamm, B., 2008. *Managing innovation, design and creativity* 2°., Chichester: John Wiley & Sons Ltd.
Sternberg, R.J., 2012. The Assessment of Creativity: An Investment-Based Approach., 24(1), pp. 3–12.
Sternberg, R.J. & Grigorenko, E.L., 2001. Guilford's Structure of Intellect Model and Model of Creativity: Contributions and Limitations., 13, pp. 309–316.
Sternberg, R.J. & Lubart, T.I., 2003. The Concept of Creativity: Prospects and Paradigms. In *Handbook of Creativity*. Cambridge: University press, pp. 3–15.
Terra, J.C., 2000. Gestão da Criatividade. *Revista de Administração*, 35(3), pp. 36–47.
Tironi, M.R., 2014. "Conexões" Design Estratégico E Economia Criativa: Inovação Além Do Design De Moda. *10° Colóquio de Moda*, (7ª edição), pp. 1–10.
Valente, L.F., 2015. *O livro do artista; um caminho para a criatividade*. Universidade de Lisboa.
Wechsler, S.M., 1998. Avaliação multidimensional da criatividade: uma realidade necessária. *Psicologia Escolar e Educacional (Impresso)*, 2(2), pp.89–99. Available at: http://dx.doi.org/10.1590/S1413-85571998000200003.
Wehner, L., Csikszentmihalyi, M. & Magyari-Beck, I., 1991. Current Approaches Used in Studying Creativity: An Exploratory Investigation. *Creativity Research Journal*, 4(3), pp. 261–271.
Zeisel, J., 1984. *Inquiry by Design: Tools for Environment-Behavior Research*, New York: Cambridge: University press.

Procedures and guidelines for the instruction and execution of pattern making: An analysis

P.A.A. Spaine, D.M. Brito, L.M. Pereira, N. Pinheiro & R.R. Andrade
Federal University of Parana—UTFPR, Paraná, Brazil

ABSTRACT: The process of teaching and execution of pattern making begins with the use of procedures and guidelines on the development of a product's pattern that will result in a piece of clothing. Such thought is led based on the formulation of a methodological sequence of construction. Therefore, the current study presents an analysis on some authors that have discussed such matters and their importance to the construction of pattern making, either flat or three-dimensional.

1 PROCEDURES AND GUIDELINES FOR THE INSTRUCTION AND EXECUTION OF PATTERN MAKING

Pattern making, in garment's design, consists in the process of adjusting a flat fabric, which is two-dimensional, into a product that should embrace a three-dimensional body, and this process is only possible through the use garment's pattern making techniques.

The methods used to elaborate garment's pattern making could be defined as techniques that are able to recreate the shapes of human body, based on the wearer's body measurements, that can be developed either two-dimensionally or three-dimensionally, in paper or fabric.

In this aspect, Fischer notes that:

Construction is the basis of garment and fashion design—it is crucial that fashion designers know and understand the techniques of creating three-dimensional clothes from a two-dimensional pattern in order to create shape and aesthetic fit to the body in movement (Fischer, 2010, p. 07).

In addition, the author highlights that the construction of a garment involves not only technical matters but also creative ones, and they are both fundamental to the elaboration of a garment that is suitable for the customer.

For that matter, Osório (2007, p. 17) states that the art of pattern making consists in the construction of anatomic geometrical blocks, in which the purpose is to reproduce the shapes of the studied body, through the use of a measurement table and of lines, curves and dots.

The process of pattern making carries technical aspects and should present an accurate engineering, respecting, simultaneously, the boundaries corresponding to ergonomics and anthropometric aspects and its variations, in the execution of a pattern (Fraga, 2012, p. 26).

In this sense, pattern making may be considered the technique applied on the construction of clothes, that could be presented in two ways: Drafting 2D or two-dimensional; and Draping 3D or three-dimensional.

Two-dimensional pattern making comprehends the study of height and length of the body's measurements, although it does not enable a preview of the depth of the product being created, since it is made on a flat surface. This method can be handmade over a level surface, and/or through computerized systems.

Three-dimensional pattern making enables the preview of the height, width and length of the product, since it is developed, directly, in mannequins or even on the wearer's body. It is a tool that enables to visualize the product throughout the pattern making development process, which simplifies its creation.

1.1 Procedures to pattern making development

Lacchi, Biégas and Vieira (2013, p. 15) describe the process of pattern making development, step by step, during the execution of a new pattern, and proposes an "integrated way to approach these aspects, correlating the elements involved in the pattern making process, since all the integrated aspects cannot be found all at once on literature, only individually", which are:

a. Interpreting the design, on which the purpose is to select the best method to develop the pattern making;
b. Identifying the requirements of the product's pattern making, that includes the measurement table, the fabric's traits and the sewing and finishing process of the item being manufactured;
c. Developing the pattern making, when the product is reproduced to a flat surface (pattern);
d. Characterizing the set of patterns, when the patterns are identified, in order to enable the product's industrialization;
e. Correction of the patterns, it is characterized as a necessary measure, when the pattern making was not properly applied or when there was an interference from the fabric and/or process applied;
f. Grading, step that provides bigger and/or smaller sizes, from a basic pattern, using the measurement table as reference.

The step by step, described above, can be complemented with the chronological sequence of pattern development, formulated by Araújo (1996, p. 97), which, as illustrated in Figure 1,

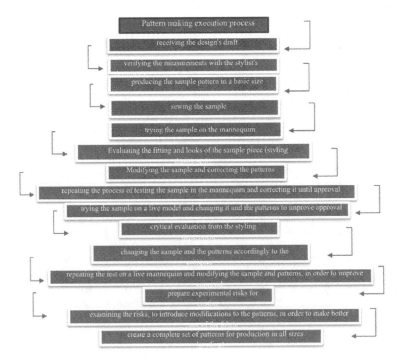

Figure 1. Diagram of the process of pattern making.
Source: Araújo, 1996, p. 97, adapted by Patrícia Ap. de Almeida.

describes, precisely, the process of execution of pattern making, from the stage of receiving the design to the pattern grading.

Souza (2006) in another analysis also addresses the procedures to elaborate a pattern making, as well as a way to execute the patterns through de presentation of the steps that comprehend this sequence of garment's construction, as shown in Figure 2.

Spaine (2010, p. 40) analyzes the processes of construction of flat pattern making and identifies the basic steps that should be considered in this process and for that, the author presents a diagram of the activities performed in the construction of the patterns, as demonstrated in Figure 3:

Osório (200, p. 36) presents a set of technical operations and procedures (Figure 3) that every pattern maker, pattern engineer, should know in order to develop pattern making, as a way to facilitate the construction of a garment's pattern. According to Osório:

> *The objective of the pattern engineer goes beyond tracing pattern pieces. It is necessary to interpret the fashion illustration in order to transform it into parts of the pattern, capturing not only the essence of aesthetic visual and the functionality of the garment as a whole, but also emphasizing production factors and fabrics' characteristics to achieve a productive fashion design.* (Osório, 2007, p. 35).

Such operations and procedures (Figure 4) are extremely relevant, because, aside from mastering essential matters about pattern making, about the usage of measurement tables and about executing patterns, among others, it is necessary that every pattern engineer knows, thoroughly, every aspect that involves the elaboration of a product's pattern making.

Therefore, the execution of a garment's pattern making begins with the expertise of stages that should be considered to achieve pattern making. The understanding of all the procedures that involve the accomplishment of the pattern making facilitates the process of pattern's teaching and construction.

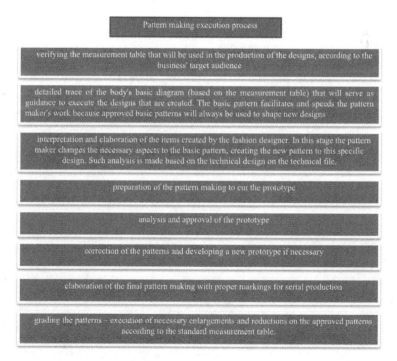

Figure 2. Diagram for pattern making development.
Source: Souza, 2006, p. 24, adapted by Patrícia Ap. de Almeida.

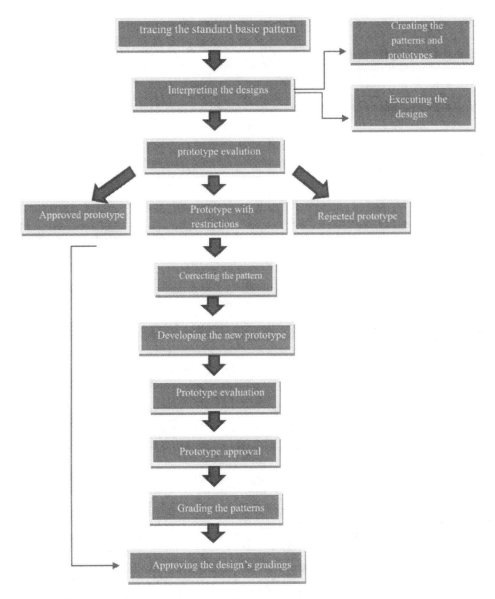

Figure 3. Diagram of the steps for pattern making creation.
Source: Created by Patrícia Ap. de Almeida.

1.2 *Methodological guidelines for the instruction of pattern making*

Methodological guidelines can be defined as paths that lead to an attempt to organize a trajectory, by which it may be possible to achieve a certain goal. In the process of learning and executing pattern making, using guidelines may assist in the comprehension of this stage of garment's construction.

Spaine (2010, p. 96) presents the Methodological guidelines for the process of instruction of industrial flat pattern making (Table 1), which purpose is to support the information that should be taught before the process of instruction of the pattern's trace. These guidelines can be used and applied on the stage of teaching of pattern making as well as during the execution of patterns by students and pattern makers.

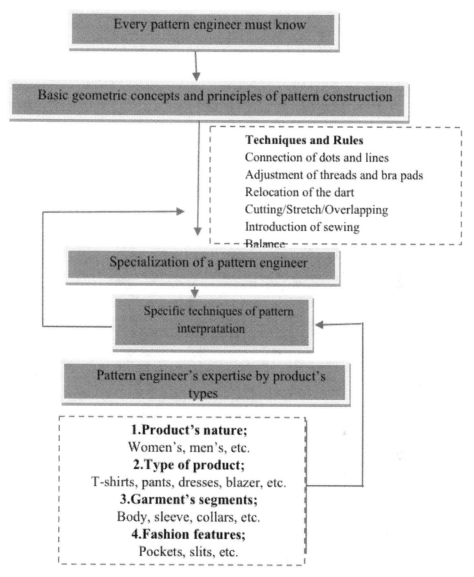

Figure 4. Diagram of technical procedures for the pattern engineer.
Source: Osório, 2007, p. 36, adapted by Patrícia Ap. de Almeida.

The guidelines suggest that, in the process of teaching flat pattern making, the aspects may be addressed by the instructor of the Basic Pattern making subject or the responsible for the Ergonomics course, if it is a part of the school program. What is relevant, in this matter, is to address these contents when teaching and executing a pattern making.

The guidelines exposed by the author present a targeted study with a methodological sequence of key matters to the process of learning and elaborating a garment's flat pattern making, and its purpose is to support the training of the pattern maker or the student, in the execution of patterns.

Beduschi (2013, p. 152) made an analysis on garment's pattern making and, from that, suggested "an instruction that involves unified thought regarding this area, a complex thought,

Table 1. Methodological guidelines for the process of teaching industrial flat pattern making.

FACTORS	ASPECTS TO BE CONSIDERED
ERGONOMICS	Confort: physical, physiological, psychological, tactile, thermal, visual
	Usability: effectiveness, eficiency and satisfaction
	Safety
	Wearabality and functionality
	Needs: physical and psychic
	Freedom of motion
	Choice of materials
	Shape and fit
ANTHROPOMETRICS	Compatibility between the wearer's body measurements and the product
	Types of body structures: sizes
	Body proportion
	Body volume
	Body shape
WEARER'S BODY	Movements
	Joints
	Body expressions
	Body languages and senses
	Sentitive, cultural and social relations
	Anatomy and physiology
	Basic body structures
	Plastic constitution
	Human proportion
	Formal adequacy
	Relation between body/garment
	Body and clothing space
GEOMETRICS	Mathematics notions
	Angular notions
	Geometrical notions

Source: Spaine, 2010, p. 96.

in order to be possible to link flat and three-dimensional pattern making and complementary theoretical content, such as ergonomics and geometrics." In this regard, the author suggests the Methodological guidelines for the teaching of pattern making, that are as follows:

- Pattern making instruction integrated with its aspects: the guideline proposes that the teaching of three-dimensional and flat pattern making are complementary, to facilitate the learning process for the students and reduce difficulties in the techniques elaboration;
- Teaching based on some practical aspects added to the theoretical: indicates the importance of the insertion of theoretical contents concerning pattern making teaching materials, to facilitate the understanding of the execution of a pattern;
- Outlined step by step content: reassures the importance of describing the contents in stages, with items indicated by letters and/or numbers, to demonstrate how the pattern making is made.
- The use of images during the explanation of a trace and/or of an elaboration of three-dimensional pattern making: highlights the importance of images, such as sketches, pictures, diagrams or drawings, to complement the textual explanation during the execution of draping.
- The inclusion of complementary information to the theoretical contents, to assist in the moment of execution of each diagram: points out the relevance of teaching theoretical contents at the beginning of the learning process of pattern making, as well as, some specific

relevant contents, either related to the materials or complementary to the teacher's explanation, when elaborating each type of diagram.

As addressed by Spaine (2010), Beduschi (2013) also highlights the importance of studying matters related to anthropometry, ergonomics and geometry, as fundamental support to the process of teaching and to the conception of pattern making, in order to facilitate the performance of the pattern makers and the student's learning process.

Spaine (2010) and Beduschi (2013) present guidelines that suggest certain aspects that should be considered and encompassed through the stages of teaching and executing a pattern making, however it still leaves open the possibility of identifying and gathering new ways of interactions in the execution of the pattern.

2 FINAL CONSIDERATIONS

This study presented procedures and guidelines that assist in the understanding and elaboration of a pattern. In this sense, it addressed the relevance that the construction of pattern making comes from the knowledge of the stages that should be considered for the execution of a pattern, as well as the understanding that all procedures involving the execution of pattern making facilitates the process of teaching and constructing patterns.

Therefore, it approaches this theme as a source of future discussions and researches in the area of garment's pattern making, and the possibility of developing proposals related to: elaboration of a product's pattern; constructing guidelines and teaching procedures; and the execution of pattern making.

REFERENCES

Araújo, M. 1996. *Tecnologia do vestuário*. Lisboa: Fundação Calouste Gulbenkian.
Beduschi, D.P. 2013. *Diretrizes para o ensino de modelagem do vestuário*. São Paulo, 2013. Dissertação (Mestrado em Têxtil e Moda) – Universidade de São Paulo.
Fischer, A. 2010. *Fundamentos de design de moda: construção de vestuário*. Tradução: SCHERER, Camila Bisol Brum. Porto Alegre-RS: Bookman.
Fraga, D.G.F. 2012. *O pulo do gato: modelagem industrial feminina, método de planificação do corpo desenvolvimento de bases*. Minas Gerias- MG: Casaoito.
Lacchi, T.; Biégas, S. & Vieira, A.M. 2013. Tecnologia da modelagem na indústria de confecção: abordagem de ensino e prática. *Revista Tecnológica*, Maringá, edição especial SIMEPRO, 2013, p. 13–23.
Osório, L. 2007. *Modelagem: organização e técnicas de interpretação*. Caxias do Sul-RS: Edusc.
Souza, P.M. 2006. *A modelagem tridimensional como implemento do processo de desenvolvimento do produto de moda*. Bauru, 2006. Dissertação (Mestrado em Desenho Industrial) – Universidade Estadual Paulista.
Spaine, P.A.A. 2010. *Modelagem plana industrial do vestuário: diretrizes para a indústria do vestuário e o ensino-aprendizado*. Bauru-SP, 2010, 109 p. Dissertação (Mestrado em Design) – Universidade Estadual Paulista.
Spaine, P.A.A. & Vicentin, J. 2016 *Desenvolvimento e adaptação das Figuras*.
Spaine, P.A.A. 2016. *Diretrizes para o ensino e construção da modelagem: um processo híbrido*. Bauru, 2016, 188 p. Tese (Doutorado em Design) – UNESP- Universidade Estadual Paulista "Júlio de Mesquita Filho", FAAC – Faculdade de Arquitetura, Artes e Comunicação.

Anthropometry and clothing for overweight and obese children: A literature review

R. Campos
Federal Institute of Rio Grande do Sul, Erechim, Brazil

M.A. Carvalho
University of Minho, Guimarães, Portugal

ABSTRACT: This paper presents a literature review focused on the clothing for overweight and obese children and the considerations of how anthropometric studies, which are the basis for the development of measurement tables, can contribute to improve the garment fit and comfort. Furthermore, the impact of 3D body scanning technologies in fulfilling this goal is considered.

1 INTRODUCTION

Companies specialized in children's clothing develop their collections from measurement tables based on the average population. This can be observed in the case of Portuguese companies, as well as in companies from other countries. Commonly, they do not consider the specific characteristics and needs of overweight and obese children (Campos et al., 2017). There are few studies aimed at improving clothing for these children (Nafiu & Burke, 2013) and the market offer for this population niche is still scarce.

In order to avoid negative experiences and psychological pressures on such children, this problem must be solved (Nafiu & Burke, 2013) or at least minimized. For many authors, the improvement of quality and comfort of fashion products is directly related to ergonomic and anthropometric studies focused on specific target populations (Iida, 2005; Heinrich, Carvalho & Barroso, 2008; Silveira, 2008; Boguslawska-Baczek, 2013; Grave, 2010; Martins, 2008).

2 PROBLEMATIZATION: CHILDHOOD OBESITY AND CLOTHING

2.1 *Childhood overweight and obesity*

Recent studies point out an increase in the body size and childhood obesity (WHO, 2009) due mainly to eating habits and lifestyle changes. Stone (2007) states that children are becoming heavier, and the Boguslawska-Baczek (2013) study shows an average weight increase of 3 kg and 10 to 20 cm in height. Sybilska and Mielicka (2011) also report a tendency of fat accumulation on the children waist and hip circumferences.

2.2 *Clothing and well-being*

Clothing can affect the physical and emotional well-being of people, mainly due to the discomfort caused by their fit to the body (Heinrich, Carvalho & Barroso, 2008; Silveira, 2008; Nafiu & Burke, 2013). Some authors report embarrassments and psychological problems that overweight and obese children feel related to their own bodies and clothing.

The need to use larger sizes may cause fear of criticism by peers (Brixval et al., 2012). Another study reveals how a poor quality regarding the fabrics, design, and fit of sportswear, can lead to negative experiences or even discourage the practice of sports (Best & Harmon, 2015). This means that clothing has an impact on children's sense of well-being.

In this sense, overweight and obese children may end up feeling helpless in a society that imposes specific beauty standards and ideals. Often, these children are stigmatized because their clothing does not fit their bodies or because it is different from the ones used by other children.

Children's clothing available on the market are inappropriate for overweight and obese children, and usually hamper movements, as they do not consider the ease values necessary for comfort. Shin and Istook (2008) report that these children commonly wear larger sizes than those designed for their age. On the other hand, adult clothing does not present appropriate measurements, proportions, and ease values, being in most cases too large and not appropriate for children. In addition, such clothes commonly need to be modified, implying time and money (Nafiu & Burke, 2013). Tailored clothing is also costly and limited in terms of design offer.

3 CLOTHING FIT AND COMFORT

Currently, the fashion market still faces numerous challenges: the difference among brands pattern sizes, diverse standard sizing systems in different countries, and so on. These challenges are also due to the fact that the design, manufacturing, distribution, and consumption are commonly carried out in different countries. Boguslawska-Baczek (2013) also points out the lack of standards or use of outdated measurement tables. The clothing fit and comfort to different body shapes is also one of these challenges.

In many cases, people end up adapting themselves to the clothes offered on the market, highly dependent on the ideals of beauty and fashion (Reddy-Best & Harmon, 2015; Romeu & Lee, 2015).

3.1 *Ergonomics in clothing*

Ergonomics is defined by the International Ergonomics Association (IEA, 2016) as the scientific discipline concerned with the understanding of interactions among humans and other elements of a system, and the application of theory, principles, data, and methods to design in order to optimize human well-being and overall system performance. Practitioners of ergonomics, or ergonomists, analyze work globally, including physical, cognitive, social, organizational, environmental, and other aspects. Among its areas of expertise, it is the Physical Ergonomics that studies the human anatomical, anthropometric, physiological, and biomechanical characteristics (Iida 2005). Nagamachi (1995) states that ergonomic solutions were initially intended for building projects, equipment, and machines, and were subsequently used for the design of fashion products.

According to Iida (2005), clothing companies can achieve success if their products present technical quality (product functionality and efficiency), ergonomic quality (design that facilitates movements) and aesthetic quality (factors that make the product desirable). For other authors, the success of clothing companies is directly related to the design, materials, fit and seams, which must provide comfort (Silveira, 2008), and be in accordance with the psychological, physiological and physical factors of the consumers (Broega, 2008; Braga, 2008).

3.2 *Measurement tables*

Measurement tables are the basis of clothing design, giving support to the development of pattern blocks. Such tables contain numbers or acronyms and vary from countries or from the specific market segment to which the apparel is intended (Silveira, 2008; Capelassi, 2010; Boueri, 2008; Heinrich, Carvalho & Barroso, 2008). They aim to assist ready-to-wear clothing companies, as they reduce production costs and wastes.

However, some companies do not use such tables, as they are not mandatory in some countries (Capelassi, 2010; Boueri, 2008; Boguslawska-Baczek, 2013). In many cases, the measurement tables used by companies are based on technical standards that consider the average population of a given country. These technical standards provide body measurements from specific real-body groupings and are developed considering average measurements (Nafiu & Burke, 2013). However, the use of average measurements has higher impacts on individuals that are more distant from the average values. In the end, such people are the most affected by the limited market offer. This is also observed in the case of children's clothing.

In addition to the variation of body types and nutritional conditions between countries, there is also a difference in the way the measurements tables for children's clothing are developed. Some companies use height measurements as a reference, while others prefer to use age to define the sizes. According to Boguslawska-Baczek (2013), this is not appropriate because children of the same age may have different heights. For this author, in order to facilitate commercial transactions, it is necessary to find the most adequate way to develop such tables and find solutions aiming their standardization (Boguslawska-Baczek, 2013).

3.3 Clothing design

According to Heinrich, Carvalho, and Barroso (2008), the added value of clothing is directly related to modeling. There are different garment modeling methods: traditional 2D flat modeling, computer modeling in CAD/CAM, and 3D modeling. All these methods are based on volumes and recesses of the body (Radicetti, 1999 apud Boueri, 2008). Flat and computer modeling methods are the most used in the industry. Generally performed by a modeler, the modeling process starts from the measurement tables (Boueri, 2008), which support the development of pattern blocks.

Betti (2014) states that modeling clothes for overweight and obese people is complex since they present different specificities and body types from those with an ideal weight. It is from appropriate measurement tables that modeling can meet the needs of overweight and obese children, considering their body shapes and anthropometric measurements, and providing appropriate ease values that can lead to comfort, safety, quality, and fit (Silveira, 2008; Martins, 2008).

4 ANTHROPOMETRY AND CLOTHING INDUSTRY

Anthropometry is the science responsible for the study of body dimensions and variations: sizes, proportions, volumes, shapes, movements, and joints (Iida, 2005). Anthropometric studies attempt to evaluate the body dimensions of the populations (Iida, 2005) and are performed through scientific norms and procedures (Silveira, 2008).

From this, anthropometric studies are of paramount relevance for the clothing industry, as they assist in solving problems related to clothing fit and comfort. Such studies provide relevant data for developing measurement tables and final products in accordance with the users' bodies (Iida, 2005; Silveira, 2008) and their limitations and age (Iida, 2005; Boueri, 2008).

4.1 Anthropometric studies

In the clothing industry, anthropometric studies are carried out through static measurements, in the case of casual clothing, and dynamic and functional, in the case of sportswear (Boueri, 2008; Silveira, 2008). Body measurements can be single evaluated or related with weight, height, skinfolds and different body circumferences (Nacif & Viebig, 2007 apud Capelassi, 2010). According to Silveira (2008), the anatomical position of the body to perform an anthropometric study should be orthostatic, with the head adjusted to the Frankfurt Horizontal Plane (face forward and horizontal line of vision), arms hanging freely by the sides of the trunk with palms front facing, legs slightly apart, and feet forward (Silveira, 2008).

Anthropometric studies can be performed mechanically (manual), using anthropometers, scales, body calipers, measuring tapes, etc., or aided by 3D body scanning technologies. The most appropriate method depends on the required data, and both have advantages and disadvantages and require training.

4.2 3D body scanning technologies and anthropometric studies

A "manual anthropometric" data collection, besides being time-consuming, does not guarantee a precise database for developing the measurements standards required by industry (Silveira, 2008). According to some authors, the development of 3D body scanning systems has facilitated such studies, as they allow a quick data collection and a greater measurement precision (Chun, 2007; Silveira, 2008; Bragança et al., 2014). Furthermore, it is possible to check results, and physical contact between the researchers and individuals is not required. They are able to scan the body using one or more optical lightening sources (Daanen & Water, 1998 apud Bragança et al., 2014), collecting body measurements and showing a 3D body images in a computer screen (Silveira, 2008). Currently, there is a considerable selection of 3D body scanners in the market, presenting different characteristics and advantages (Daanen & Haar, 2013 apud Bragança et al., 2014).

Among several anthropometric studies carried out using 3D body scanning systems for clothing design, it is possible to highlight some pioneers. The CAESAR (Civilian American and European Surface Anthropometry Resource) study, conducted between 1998 and 2000, in a partnership among garment industries, and the United States government and Air Force, aimed to develop ergonomic military uniforms (Bragança et al., 2014). Another relevant study was conducted by the German company Human Solutions, which used 3D body scanners to collected anthropometric data in nine European countries to validate the EN 13402/2006standard for clothing sizes in Germany (Heinrich, Carvalho & Barroso, 2008). Other initiatives of body scanning technologies applied to the clothing are the following: SizeUSA, SizeUK, French National Size Survey and Size Korea (Chun, 2007). In Brazil, the SizeBR study (Bastos et al., 2013) also focused on the standardization of measurement tables.

5 FINAL CONSIDERATIONS

Currently, the increase in children's weight as a result of changing eating habits and lifestyle is a reality described by international organizations such as the World Health Organization (WHO, 2009) and the International Obesity Task Force (IOTF, 2016). The acceptance and inclusion of overweight and obese children in society have several aspects. One of these is related to an appropriate clothing offer.

These children may end up wearing larger clothes than those indicated for their age or even adult sizes. This has a direct impact on the well-being of these children and can lead to psychological and social consequences.

The solution passes through a market offer of appropriate clothing, suitable to the body shapes and dimensions of such children. This demands the development of measurement tables to better assist the clothing modeling. Thus, anthropometric studies should be conducted with the purpose of understanding the specific anthropometric characteristics of these children.

Within this framework, 3D body scanning technologies can assist in fulfilling such goals. Such technologies allow a faster data collection, provide more accurate data and require affordable investment costs.

Through anthropometric studies, appropriate measurement tables can be developed and used by children's clothing companies, assisting them to design products in accordance with the needs of overweight and obese children. In the end, this may increase their self-esteem and well-being.

The literature review presented in this paper highlights the relevance of this issue and emphasizes the importance of aligning the efforts and knowledge of the disciplines of Ergonomics, Anthropometry and Fashion Design, and the application of new technologies for the development of appropriate clothing for such children.

ACKNOWLEDGMENTS

This work is financed by the Federal Institute of Rio Grande do Sul, Brazil and FEDER funds through the Competitive Factors Operational Program (COMPETE) POCI-01-0145-

FEDER-007136 and by national funds through FCT-Portuguese Foundation for Science and Technology, under the project UID/CTM/000264.

REFERENCES

Bastos, S.F. et al. 2013. *SizeBR – O Estudo Antropométrico Brasileiro*. SENAI CETIQT, Rio de Janeiro, Brasil.
Betti, M.U. 2014. *Beleza sem medidas: corpo, género e consumo no mercado de moda plus size*. Dissertação de mestrado em Antropologia Social. Universidade de São Paulo.
Boguslawska-Baczek, M. 2013. Analysis of the contemporary problem of garment size. Conference: *The 7th Textile Science 2013 Conference*, At Liberec, Czech Republic.
Boueri, J.J. 2008. Sob Medida: antropometria, projeto e modelagem. Pires, D.B. (Org.). *Design de Moda: olhares diversos*. São Paulo: Estação das Letras e Cores, p. 347–369.
Braga, I.M.S. 2008. *Optimização do design do vestuário cirúrgico através do estudo do conforto termofisiológico*. Universidade do Minho. Tese de mestrado.
Bragança, S. et al. 2014. Validation Study of a Kinect Based Body Imaging (KBI) Device System Based on ISO 20685:2010. *5th International Conference on 3D Body Scanning Technologies*, Lugano, Switzerland, 21–22 October.
Brixval, C.S. et al. 2012. Overweight, body image and bullying—an epidemiological study of 11- to 15-years old. *The European Journal of Public Health*, 22(1), 126–130.
Broega, A.C. 2008. *Contribuição para a Definição de Padrões de Conforto de Tecidos Finos de Lã*. Tese de Doutoramento. Universidade do Minho.
Campos R. et al. 2017. Anthropometric data collection of Portuguese children with overweight and obesity. Los Angeles: *Proceedings of the 8thInternational Conference on Applied Human Factors and Ergonomics*.
Capelassi, C.H. 2010. *Metodologia projetual para produtos de moda e a sua interface com as tabelas de medidas do vestuário*. Dissertação de Mestrado em Desenho Industrial. Universidade Estadual Paulista, Bauru.
Chun, J. 2007. Communication of sizing and fit. ASHDOWN, S.P. (Org.). *Sizing in clothing*. The Textile Institute, p. 220–245.
Grave, M. de F. 2010. *A moda-vestuário e a ergonomia da hemiplégico*. São Paulo: Escrituras.
Heinrich, D., Carvalho, M. & Barroso, M. 2008. Ergonomia e Antropometria aplicadas ao vestuário–discussão analítica acerca dos impactos sobre o conforto e a qualidade dos produtos. *Diseño en Palermo*, Universidad de Palermo, Julho.
IIDA, I. 2005. *Ergonomia: projeto e produção*. São Paulo: E. Blucher.
International Ergonomics Association (IEA), 2016. Available in: http://www.iea.cc/.
International Obesity TaskForce (IOTF), 2016. Available in: http://www.worldobesity.org.
Martins, S.B. 2008. Ergonomia e moda: repensando a segunda pele. Pires, D.B. (Org.). *Design de Moda: olhares diversos*. São Paulo: Estação das Letras e Cores.
Nafiu, O. & Burke, C. 2013. *Large Clothing Size in Children Is Associated with High Body Mass Index and Clustering of Medical Comorbidities*.
Nagamachi, M. 1995. Kansei engineering: a new ergonomic consumer-oriented technology for product development. *International Journal of industrial ergonomics*, 15, pp.3–11.
Reddy-Best, K.L. & Harmon, J. 2015. *Overweight boy's and girl's experiences with and perception of athletic clothing and its relationship to physical activity participation*.
Romeo, L.D., & Lee, Y.A. 2015. Exploring apparel purchase issues with plus-size female teens. *Journal of Fashion Marketing and Management*: An International Journal, 19(2), 120–135. doi 10.1108/JFMM-02-2014-0012
Shin, S.J.H. & Istook. C.L. 2008. *Sizing Systems for Children's Wear in the United States*. Texas Tech University, Lubbock TX, USA.
Silveira, I. 2008. Usabilidade do vestuário: factores técnicos-funcionais. *Moda Palavra*- periódico. Ano 1, n. jan-jul-p.21–39.
Stone E. 2007. *Infashion Fun Fame Fortune*. Fairchild publications, Inc. NY. Ulrich, P., Connell.
Sybilska, W. & Elżbieta Mielicka, E. 2011. Applying a 3D Body Scanner to Qualify the Postures and Direction of Changes in Human Body by Children Example. Lodz, Polônia *2nd International Conference on 3D Body Scanning Technologies*, Lugano, Switzerland, 25–26 October.
World Health Organization. 2009. *Childhood overweight and obesity on the rise*. Geneva: WHO.

The materialization of fashion products: An experience of the textile behavior

P.M. Souza, M.M. Otani & P.P. Silva
State University of Londrina, Londrina, Paraná, Brazil

I.C. Italiano
University of São Paulo, São Paulo, Brazil

ABSTRACT: This paper addresses the relevance of the materials in fashion design projects and the importance of the experimentation and the physical contact with textiles to increase the sensorial perception on their behavior. This investigation included the measuring of the samples' drape index and a comparative analysis of products—the same prototype developed with different fabrics, subsequently transferred to plan view (two-dimensional plan). Tridimensional modeling techniques enabled the development of prototypes and the evaluation of both the process and the results. Thus, it is possible to conclude that different materials with singular characteristics can—in certain situations—be applied to design the same product.

1 INTRODUCTION

The project process in design consists of several non-linear procedures and phases—it can be described as a holistic system—in which all activities are interconnected and influence one another. The decisions made throughout the project help meeting the product requirements, until the process is completed—that is, until the product is materialized. In this sense, each step is important within the design project, which means one step cannot be isolated from the others.

Therefore, the material selection phase is extremely relevant: it may interfere in the other phases by causing considerable impacts in the subsequent operational activities. Besides, the material selection is responsible for achieving high-quality structural results in the project. According to Calegari & Oliveira (2014), the textiles materialize the ideas and drawings developed by designers, thus representing a major step within the design process. In addition, the authors suggest that the materials contain signs—which can convey countless meanings to those who wear them –, since they have a great influence on the structural arrangements of the clothes. This is because the formal structure of a product is strictly related to the material used and its intrinsic characteristics.

The decisions related to material choice—which sometimes take place just before the production phase—should ideally occur in the beginning of the design project, during the conceptualization phase.

The visual judgment is one of the most important abilities required from fashion designers, as it may help them understand and predict the textile behavior and its possible applications. A consistent and thoughtful analysis of the available materials assists the identification of their visual, tactile and structural characteristics, so that designers can decide on the product attributes and determine its functional aspects. Aldrich (2010) argues that students and professionals who deal with the development of clothing products need a practical verification method that is capable of promptly providing answers to facilitate the material selection.

Deciding between one textile or another can cause significant differences in the modeling process and, therefore, affect not only the fitting and cutting of the modeling parts in the fabric, but also have an impact on the product consumption and on materials waste.

Along the process, designers can reach different solutions for the project (Lawson, 2011). Even if all the solutions indicate the use of the same material, it is important to bear in mind that the same product can be materialized through different techniques—it only requires the establishment of a proper relationship between the fabric and the body shape.

According to Aldrich (2010), the best technique to learn how the fabric behaves on the body is to work directly on a mannequin: this experience may help designers overcome the difficulties in transferring the three-dimensional mental schemes into two-dimensional patterns. Souza (2013) is equally supportive of the experimentation as a means to improve the understanding on the textile behavior and, therefore, highlights the benefits of applying tridimensional modeling techniques in the product development. The author argues that the real-world experimentation allows the perception of proportion and scale through the exploration of spatial relationships and the possible silhouette arrangements depending on the available materials. The comprehension on the body proportions increases as the designer experiences the process, thus improving their ability of analysis regarding the conformity between the body structure and the garment. This leads to a correct measuring of the spaces between body and product, also considering the specific attributes of the material and the structural/formal arrangement required.

In this context, this study presents an investigation on the behavior of textile materials, through a comparative analysis of fashion products—the same product developed with different fabrics and the application of tridimensional modeling techniques, which enabled the prototyping of solutions and the evaluation process. We present a variety of silhouettes and structural/formal arrangements developed, as well as the two-dimensional, plan view of the prototypes and their relationship with the textile characteristics. Additionally, we measured the materials drapability, according to the method proposed by Winifred Aldrich (2010). First, this paper proceeds to the measurement of the fabric's samples and, afterwards, to the analysis of the final prototypes—that is, the materialization of clothing pieces that addresses the aspects included in this study.

2 THE IMPORTANCE OF THE TEXTILE MATERIAL

According to Sanches (2017), the shape of the fashion apparel results from the integration of structural, productive, and interactive aspects. Besides that, the author highlights three fundamental types of knowledge that provide students with support to develop formal/visual solutions: material resources—especially the textile materials; structural resources—related to the modeling and the production technology, responsible for transferring plan view ideas to tridimensional compositions; and the syntactic principles—those that facilitate the perception over aesthetical-symbolic information. The author argues that the material resources possess intrinsic properties that interconnect to the product's aesthetics, and directly influences the perception of the artifact.

In this sense, the fashion designer must consider not only the technical aspects of the materials, but also reflect on the attributes that affect how they are perceived. According to Del Curto, Fiorani & Passaro (2010), the vision and the touch are the most important senses in the design interaction—even though the authors reinforce that all the senses must be considered in order to develop a completely meaningful project. The vision is the first sense to perceive the objects, thus revealing its characteristics immediately. But, according to the authors, this first impression is improved when the touch confirms what we see. For instance, in order to perceive the softness and lightness of a fabric, the touch must confirm these properties. Both vision and touch interact continuously. If they work apart, it is not possible to accurately perceive the material properties, the tridimensional composition, or even the spatial arrangement.

According to Ashby & Johnson (2011), the information gathered with the observation and the perception is extremely important for the creative thinking. The designer—provided with the knowledge on the materials since the beginning of the project process—will be able to process the information through the visual reasoning and visualize the ideas as a physical product.

Aldrich (2010) argues that the designer must recognize the textile characteristics that are more likely to influence the development process of a product and, therefore, impact on the correct integration between shape, material, and modeling aspects. According to the author, weight, thickness, distortion, drapability, and elasticity are the properties with most impact on the visual appearance of the products. She suggests that those properties can help designers increase the intuitive integration of shape and fabric—that is, by applying their perception abilities, they will be able to intuitively (and quickly) de-codify and compare the textiles.

In this context, the study on the properties and formal arrangements of the materials must be (and they have been, in fact) included within the project process, in order to improve the designer performance.

3 THE EXPERIMENTATION PROCESS

This study has an exploratory and descriptive nature. It also applies an action research approach, since the researchers—a Scientific Initiation student, her supervisor professor, and other professors who collaborate with the research project—were in touch with (and participated in) the activities developed within this investigation. The bibliographic survey, the measuring procedures, the experimentation, and the development of prototypes took place in two contexts: first, through the assignments developed in the course Advanced Laboratory of the Shape, which is a course within the third grade of the Fashion Design program at the State University of Londrina (lectured by the first author of this work); and second, during mini-projects developed within Scientific Initiation projects (research projects developed by undergraduate students under the supervision of the author, aiming at improving their scientific thinking skills).

During the research, experimentation activities were performed and documented to investigate the materials' behavior through practical applications in the development of multiple fashion products. The main objective was to find a standard in the behavior of certain materials, in order to predict their performance in real products and improve the accuracy of the solution proposed by the designer. In the course, a common practice among professors is to perform these assignments—the conception and/or interpretation and the materialization processes—through a tridimensional modeling approach, using a 1:2 scale mannequin to support the development.

Prior to starting the assignment, parameters of formal similarity were established to guide the project process. First, students were asked to choose a reference model (a real piece of clothe of their choice), and then select six different materials to develop the prototypes directly in the mannequin. After the experimentation and the definition of the most appropriate composition—the shape and the textile arrangement to reach the required formal similarity –, students should transfer their prototype to plan view, two-dimensional modeling.

Students were divided in four groups. Each group should develop the prototypes based on the reference model chosen previously. Prior to starting the process, they were encouraged to discuss about the reference model selection, the best approach to begin the development of the pieces, and the resources needed to perform the assignment. This contributed to improve the team communication, encouraged students to think of multiple solutions for the same context, and provided additional knowledge over the situation. Despite the fact that they discussed in groups, students were asked to develop their prototypes individually. Thus, each one of the prototypes conveys the particular, unique perspective of each student. This paper presents the work developed by six students (belonging to the same group).

The materials selected should be intentionally diverse, so that the differences in their behavior could be clearly noticed by the students. The group selected the following materials: Viscolycra (97% viscose, 3% elastane), mousseline (100% polyester), Tricoline (67% cotton, 30% polyester, 3% elastane), satin (100% polyester), raw cotton (100% cotton) e twill (100% cotton).

Parallel to the development of the prototypes by the students, those materials were subjected to measuring procedures in the Scientific Initiation mini-projects—according to the properties listed by Aldrich (2010). The author argues that the characteristics most likely to change the behavior of the textiles—and, thus, influence the final shape of the products—are: weight, thickness, drapability, distortion, and elasticity. Among them, the drape property was selected for analysis, since this aspect is essentially related to the visual perception, which is a primary concern of this study and impacts on all the other properties.

The following steps guided the preparation of the drape index measuring tool: starting from a single dot, a central line was drawn towards the ground; two lines were traced in 45 degrees, forming a semicircle, which was then divided in five parts to the right and five parts to the left of the central line. Prior to starting the measuring, the paper was fixed with tape in a flat surface. The fabric sample of 20 cm × 20 cm—cut along the warp and the weft directions—was attached with a pin by one of the tips in the central point of the paper, according to Figure 1.

The five divisions represent the variation between the highest and the lowest drape index, measured visually, considering the area occupied by the sample during the measuring. Aldrich (2010) explains them: (1) high-drape; (2) medium-high drape; (3) medium drape; (4) medium-low drape; and (5) low-drape.

By the end of the activity, six non-identical prototypes had been developed—even though they were based on the same reference model—resulting in six different modeling schemes, depending on the material used and how it was manipulated. The different prototypes and modeling schemes also reflected the technical and cultural repertoire of the student who developed them.

Based on the comparison between the prototypes, the researchers then analyzed the results, considering: the adaptability of the materials to the reference model; the formal similarity of the pieces in relation to the reference piece; and the drape index of each textile—which influences the formal structure of the silhouettes. Except for the index measuring, all the other aspects refer to a subjective analysis of the results—that is, the analysis of the fit was performed through a sensorial, perceptive approach.

Lerma, Giorgi & Allione (2011) explain that the sensorial evaluation is a scientific method to measure, analyze and interpret the sensations perceived by our senses. The perception is decoded according to two major types of evaluation: the analytic and the hedonistic. The process of "measuring the perception" applies different procedures depending on the field of

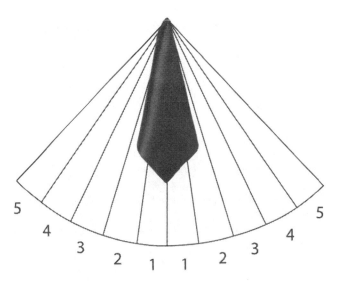

Figure 1. Caption of drape index measuring tool.
Source: The author (2017), Adapted from Aldrich (2010).

knowledge and the object of study. The discriminatory method adopted in this study is an analytic approach that allows a qualitative analysis of the samples without, however, having to indicate the countless aspects in which they are equal or different. Therefore, the analysis of the results in this study is based on comparisons.

4 DISCUSSIONS AND RESULTS

Table 1 presents the results collected after the assignment conclusion. The table is structured to allow the visualization of the original reference model and to establish comparative relationships between the drape indexes of each material, the prototypes developed, and finally, the two-dimensional modeling of each prototype. The reference model selected by the sample group to proceed with the development—a twisted blouse with double straps—properly allowed the analysis of the characteristics and the textile behavior.

Also, the model was adaptable to multiple approaches in relation to the constructive thinking on the prototype's materialization, that is, it provided students with opportunities to explore various structural/formal arrangements, which can be confirmed by the variety of two-dimensional modeling schemes developed. Of course, this can be related to the material's properties, since each textile requires specific types of manipulation and different fabric sizes to create volume, looseness and certain shapes. We cannot despise, however, the particular approach each student applied along the design process, which is related to their technical and cultural repertoire. Besides that, experimenting with tridimensional modeling provides multiple paths to follow, and many times the ideas presented in the beginning of the process undergo some adjustments, thus resulting in novel solutions that may be completely different from the original sketches.

Regarding the materials adaptability, the most consistent fabrics used for development were viscolycra and mousseline, because they presented greater similarity in terms of fit and properly adapted to the shape of the reference model, thus forming the "twist" in the front part of the blouse. Apparently, viscolycra adapted even better than the mousseline, although this cannot be confirmed. This impression may be due to the fact that the length of the mousseline prototype is a little shorter than in the original reference model. Also, among all the samples analyzed, these materials are the ones with the highest drape index, thus confirming what the previous measuring procedures suggested.

This also applies to the raw cotton and the twill: they present lower drape index, which reflects in the products developed with these fabrics—the prototypes presented more geometric lines and shapes, resulting in a harder, non-flexible, less fluid "twist" effect. Also, the lines in the silhouette are not as organic as in the reference model. This impacts on the creation of volume (almost none) and causes the product to be more static, as if the folds were too stiff. In the original model the vertical lines present a more natural fit, protuberant waves of fabric and more volume.

Tricoline can be considered an intermediate choice among the materials: it does not present fluid waves as the original model, but it is not as hard as the raw cotton and the twill either. Tricoline presents a medium drape index, it is light and flexible to a certain extent. However, it is more robust when compared to the first two—it does not form many lines when twisted and lacks fluidity.

Finally, the satin presented a behavior that is far distant from the characteristics of the reference model, especially in relation to the direction of lines and waves. Even though the satin drape index is very similar to tricoline's, the prototypes developed with these fabrics do not resemble one another—mainly because of the differences in the fit. In this specific case, we can assume a technical inability of the student during the development of the prototype. The satin is light and usually creates a fluffy-looking volume, which, in theory, would have helped the development of a prototype that was more similar to the reference model.

Regarding the relationships between the drape indexes measured in the samples and the set of two-dimensional modeling schemes, the prototypes made in raw cotton and twill—which presented lower drape index—required three modeling parts to assemble the piece of clothe,

Table 1. Caption of comparison between the drape measures, the fashion products and the patterns developed.

Reference model	Material	Drapability	Prototype	Two-dimensional modelling
	Viscolycra			
	Mousseline			
	Satin			
	Tricoline			
	Raw cotton			
	Twill			

Source: The author, 2017.

whilst the others required only two. Although the modeling parts were very different in shape, both required two pieces of fabric (stitched) and the application of straps so that the front part could be assembled.

In the clothing pieces using medium drape index fabrics, the modeling parts were more geometric (almost square-shaped), and the position of the arm opening in opposites sides and directions was visible, as a consequence of the twist. However, further manipulation—such as folds and internal sewing—was needed in order to align the prototype with the reference model. In turn, the modeling parts derived from higher drape index prototypes presented unusual shapes—completely different from the conventional ones used in the industry, which complicates the identification of the parts in the modeling scheme.

The differences perceived with the comparison of the modeling parts also raised questions about the influence of the material selection in the following steps of the process: some steps may certainly require a more effective planning of fitting and cutting; others may demand a higher fabric consumption, that perhaps does not justify the production of that specific piece; some steps may facilitate the production process, while others may add complexity to it.

5 FINAL CONSIDERATIONS

The importance of the material selection is undeniable. A proper material may assist the development of the ideas proposed by the designer. On the other hand, a wrong decision can negatively impact on the prototyping stage, which also reflects in the subsequent phases of the development process. In the context of a design project, the selection of textile materials must be seen as a strategic action in the beginning of the process, in order to prevent errors and decrease risk of failure in the following stages. It also influences operational activities, since it extends to the subsequent phases of modeling, fitting, cutting, and assembling the piece.

The tridimensional modeling approach in the development process—which considerably improves the possibility of developing modeling parts that differs from the conventional shapes—requires careful planning, so that the prototyping and production stages can be optimized.

In addition to that, the aforementioned technique improves the development of the tridimensional perception, through the promotion of the sculptural activity on real-world resources and the experimentation/manipulation of textile materials. The materialization process is the result of a gradual understanding of the tridimensional elements, which reflects in the product structural/formal composition. Through the experimentation with the available materials and their relationship with reference models and resources, designers understand the importance of the textiles in relation to the body structure and the clothes' fit. Besides, the visualization of a product in three dimensions allows the immediate evaluation of wearability and fit aspects.

In relation to the adaptability of the materials and the reference models, it is possible to conclude that materials with different characteristics can, in certain situations, be applied to develop the same products, achieving a high level of formal similarity.

Also, the drapability of some materials—checked in small samples—kept unchanged when the fabric was transformed into a product: this is clearly perceived in the fit of some of the prototypes. Therefore, the sample behavior—depending on the specific requirements of the piece to be developed—can be used as a prevision of how the fabric will interact with the body.

Besides that, the particular perception of each individual—students or professional designers—and their repertoire also influences the final result of the project, that is, the project depends on the types of knowledge possessed by designers and/or students.

Finally, the experimentation and the frequent contact with textile materials along the project process are proven to be relevant as a means to improve the tactile and the practical knowledge. Also, this approach increases the intuitive thinking about the material behavior in the products development—even before they are developed.

REFERENCES

Aldrich, W., 2010. *Tejido, forma y patronaje plano.* São Paulo: Gustavo Gili.

Alves, A.S., 2010. *Design do vestuário: protótipo funcional para o encaixe de moldes no tecido.* Dissertação (mestrado) – Universidade Federal do Rio Grande do Sul, Escola de Engenharia, Faculdade de Arquitetura, Pós-graduação em Design, Porto Alegre.

Alves, A.S., 2016. *Algoritmos para o encaixe de moldes com formato irregular em tecidos listrados.* Tese (doutorado) – Universidade Federal do Rio Grande do Sul, Escola de Engenharia, Faculdade de Arquitetura, Pós-graduação em Design, Porto Alegre.

Ashby, M. & Johnson, K.; 2011. Materiais e design: a arte e ciência da seleção de materiais no projeto de produto. Rio de Janeiro: Campus.

Barauna, D. et al., 2015. *Seleção de materiais no design: informações necessárias ao designer na tomada de decisão para a conceituação do produto.*

Bellavitis, A. Dell' Acqua., 2013. I diversi intrecci del design contemporâneo. In Conti, G. M. *Design della maglieria: strumenti e metodologie progettuali.* p.11–21. Milano: Lupetti.

Brehm, L.S., 2011. *Contribuição para classificação e descrição do caimento dos tecidos de seda 100% empregados no vestuário.* 2011. Dissertação (mestrado) – Universidade Federal do Rio Grande do Sul, Escola de Engenharia, Faculdade de Arquitetura, Pós-graduação em Design, Porto Alegre.

Calegari, E.P. & Oliveira, B.F., 2014. *Aspectos que influenciam a seleção de materiais no processo de design.* Arcos Design. Rio de Janeiro, V.8 N. 1, junho 2014, pp. 1–19.

Del Curto, B. et al., 2010. *La pelle del design*: progettare la sensorialità. Milano: Lupetti.

Dominoni, A. & Tempesti, A., 2012. *Forma e materia*: design e innovazione per il tessile italiano. Milano: Maggioli.

Lawson, B. 2011. Como arquitetos e designers pensa. São Paulo: Oficina de Textos.

Lerma, B. et al., 2011. *Design e materiali*: sensorialità_sostenibilità_progetto. Milano: Francoangeli.

Sanches, M.C.F., 2017. Moda e projeto: estratégias metodológicas em design. São Paulo: Estação das Letras e Cores.

Souza, P.M., 2013. *Estratégias de construção para estruturas têxteis vestíveis.* 2013. Tese (Doutorado em Design) – Faculdade de Arquitetura, Artes e Comunicação, Universidade Estadual Paulista, Bauru.

Design and redesign

G. Montagna & M.J. Delgado
CIAUD, Lisbon School of Architecture, University of Lisbon, Lisbon, Portugal

ABSTRACT: In this article some lines are drawn for the understanding of the limits that are imposed to the reverse design and the redesign in the process of development of design products. Reflecting on the relations between the academy and industries, along with some case studies of companies and brands that stand out in the world of fashion, we try to relate the central values of the discourse of design, such as aesthetic, functional, social, cultural and ethical, emphasizing creativity of the product and the value of its authorship. In the era of globalization, it is therefore essential that designers, aware of the strategies defined by the various brands, incorporate in their projects the theoretical foundations of design, oriented to respond to sociocultural differences, ergonomic needs of users, technological innovations and market requirements in a respect consistent with the ethical principles that guide the projectual practice of design.

1 INTRODUCTION

Design is still considered a new-found discipline, which seeks not only a sustained design thinking associated with a design practice, but also and especially the adoption of a design culture, which is still far from what many designers and industries idealize and need. If, on the one hand, design is an element that aggregates the different disciplinary areas that participate in the construction of its object of study, with different levels of performance and representation, on the other hand, to design, and in particular to fashion design, are often associated with significant representations of aesthetic value, which only partly defines and characterizes it. In turn, this aesthetic representation is too often used by the fashion industry to present products that may be appealing but neglect all other dimensions, especially the social, cultural, and ethical dimensions that are central to design discourse.

From the literature review and resorting to case studies we intend to confront the methodology of the project design of the design used in the academy, which promotes among other values the authenticity of the product and the creativity to achieve an unprecedented result, with the reality of many internationally recognized large distribution design and fashion design companies that use existing object reproduction methodologies, applying small aesthetic variations but which are disruptive to the theoretical foundations of design and its approximation to the needs of users.

2 LITERATURE REVIEW

2.1 *The design prospective*

The relationship between the academy and the business community is a complex relationship that over time has been achieving some approximation, but still requires greater integration and dynamism. Although it is sometimes viewed as antagonistic or even positioned in opposing assumption, this relationship arises from the increasing need to optimize the production process in response to consumer needs, which anticipates the collaboration and complementarity of premises that must be assumed as "two faces of the same coin". If, on the one hand, academia studies, theorizes, designs and develops studies and products based on a holistic view of design, as a discipline that tries to interpret and respond to the needs of its users, on the other hand, the industry boosts its productive capacity, managing all its resources in a sustainable way, with the objective of responding quickly and effectively to its consumers.

In fact, this relationship between academia and industry, which is apparently distinct and sometimes conflicting, is imperative. Although they are based on different assumptions, they contribute to achieving the same objective, in a complementary process that stimulates, almost as a continuous challenge, the development of constant improvements in each of these areas. In this sense, the authors Tanure and Kistmann (2010, apud A2D Org) propose, for the implementation in the business market, the tactical design manager that "acts mainly as a facilitator between the design sector and the other sectors of the company, promoting the integration and exchange of knowledge and information in product development". Fashion design is a discipline that is based on the production of fashion artifacts aimed at satisfying the physical and sensorial needs of the users, led by the social and cultural dimensions that provide a greater and better identification of the users with the products. Ana Paula Miranda (2014) refers that cloths are imperative items in the social construction of identity, and a silent language where individuals communicate through the use of these visual and nonverbal symbols and signs. Some aspects of design, where the user recognizes himself and considers being able to assume them as elements that represent his personality, as a specific aesthetic, for example, are essential aspects in design, which often overlap with other equally important but less visible and socially less important.

In this sense, fashion design seeks to offer users the product they need, so that they feel their needs satisfied, both physically and physiologically, socially or even emotionally. In terms of human physical performance, whether in sport and/or occupational activity, comfort and usability will be influenced by physical, physiological, psychological and technical factors (Laing and Sleivert, 2002; Martins, 2012). These products seek to respond to different levels of human needs that interact directly with the individual, the ecosystem where they are inserted and, finally, the factors that these two levels can produce in their interaction (Montagna, 2015).

International designers and big brands are very attentive to these needs of the users and constantly promote the exploration of new configurations, details, textures, association of colors, etc., in order to try to interpret the present society and to be able to satisfy the users' needs. (Figure 1).

If today's society is increasingly identified with a voracious need for knowledge, experience and change, the area of design and fashion design follow these challenges, producing more and offering different experiences through their products and communication of an image that pushes most of the daily felt realities.

Figure 1. L to R: Vaquera Spring/Summer 2018, Charles Jeffrey Loverboy Spring/Summer 2018, Gosha Rubchinskiy Spring/Summer 2018.
Source: businessoffashion.com apud InDigital [Retrieved 9 March 2018].

In an age of globalization where all are mixed and where everything can be acquired every where, it is fundamental that the new designers and the academy should aim at high levels of education, in order to understand and interpret different cultures and sub-cultures, new designers have practices in the new technologies and are able to evaluate and manage the different resources that are available to them in this total opening of the global world. As a result, higher education curricula are becoming increasingly influenced by ethical issues, philosophy, innovative technology, and an increasing sensitivity to environmental issues and different cultures (Marshall, 2009).

Fashion proposes and pursues the class struggle of its users. Each individual wish to be able to possess objects belonging to a higher social class, in order to be able to exhibit them, in the assumption of acquiring a greater status and social projection.

There is no state-hierarchical society, stated Lipovetsky and Roux (2012), without escalation of sumptuous signs of social inequality [...]. Democratizing access to certain products has become in the last decades a strategy of the great brands of haute couture. In order to support the presentation of a collective imagery created by the haute couture collections for a small group of people with the economic power to acquire them, the great couture brands produce accessories and perfumes with more accessible prices, thus offering all other users the possibility of acquiring a little of this special world.

Another business model that has invaded the clothing market is that adopted by mass market brands that in recent years have come to democratize the market, offering products of medium or low quality but at competitive prices. These are brands present business models focused on quantity rather than quality, which elect relocation as the great financial surplus of their productions and which resort to programmed obsolescence as a guarantor of a short and controlled product life cycle. Simultaneously, at the offer level they promote the creation of aesthetically up-to-date products, based on proposals previously launched to the market by the big brands, betting on the re-styling of products from the big brands to the big mass market (Figure 2).

There is a great aesthetic and thematic appropriation of the proposals formulated by the big brands that, through market research and attention of the needs of its users, present aesthetically and socially adequate current products, as yet economically accessible, but without the investment in research and developments of social aspects necessary for this product to be considered a true design product.

Figure 2. Gucci AW15 vs Mango. http://www.dazeddigital.com/fashion/article/28574/1/how-the-high-street-gets-away-with-fashion-robbery [Retrieved 9 March 2018].

3 WORKING MODELS

If the major fashion design brands implemented in international markets promote research and development work for the launching of their collections, "whose main rule is innovation and whose task will be, consequently, to introduce unceasing revolutions of forms, materials and specific reasons" (Monneyron, 2007, p.27), the work carried out by large-scale brands does not follow these lines of action.

These companies, aiming at large-scale distribution, knowing the desire of many consumers to acquire iconic garments and models, and based on the technological evolution, the variety of materials and the possibilities of creation, are inspired by these models, and produce similar very attractive garments at more affordable prices. It is not difficult to recognize models marketed by high-end brands and whose resemblance to big brand models is easily recognizable by many. These are garment clothes that are reproduced but differentiated in some detail using more economical materials whose general quality of the product is usually lower than the original high street brand garment.

Redesign and reverse design in fashion have existed since the product became an industrial product and, due to its seasonal rhythm, its fast obsolescence and in the current context of fast fashion, they needed to develop more competitive business models.

Companies use methods that lie between redesign and reverse design to easily obtain work bases and offer products that not being new, they could be aesthetically different, up-to-date and appealing. With this new vision of the fashion market, mass market fashion brands develop products for the "democratization" of the product of the great international brands, which intends to make the exclusive object accessible to all, went from the exclusive to the mass market commercialization. As Lipovetsky and Roux (2012) say, "the luxury or semi-luxury sector itself has engaged in a real change by proposing its own fashion objects in lower and differentiated market lanes in order to give their ideas to as many customers as possible. In this sense, we will be here to see the reduction of the psychological distance between the two segments, 'luxury' and "mass market". Lipovetsky and Roux (2012).

Many design objects, which run their products through large generic shops and markets, feature "copies" of large-brand design products at much more competitive prices and quality. In fashion design, companies such as Massimo Dutti, Zara, Next, H & M, among others, adopt redesign/reverse design techniques to produce more seasonal product and do not explore ideas and concepts that should originate in users and environments where they live and operate. The Spanish retailer Zara has approximately two hundred designers who develop forty thousand styles each year, of which twelve thousand are produced (Faerm, 2012 apud Seigel, 2011). The production cycle for the Swedish retailer H & M, design-to-retail, is just

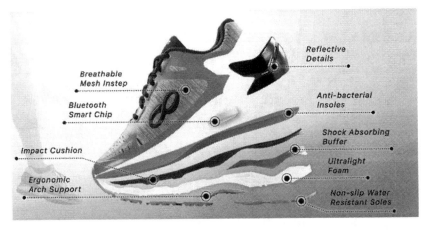

Figure 3. Runtopia-reach-exploded. http://www.gadfit.com/runtopia-reach-crowdfund/runtopia-reach-exploded/ [Retrieved 9 March 2018].

three weeks and involves a highly choreographed network of chain management around the world (Faerm, 2012 apud Seigel, 2011).

The reverse design methodology, which allows the inverse observation of design processes, or redesign methodology, that provides a functional and aesthetic review of an existing product, are methodologies of observation of a product that allow one to understand how and in what form objects were constructed and produced. The opening and/or disassembly of a design object allows quick access to the elements that allow to obtain the basic functionalities of the object itself, but also to perceive the conditioning and the options taken for its production, allowing the perception of eventual specific issues of the study object (Figure 3).

For the adoption of this type of methodologies in some of the commercial brands present in the large distribution market, we can find at least two important factors: on the one hand the security of being able to produce objects that are of great demand and that already are appetizing by mass market users' reducing the risk of rejection by the market; on the other hand, to reduce or cancel out the huge investment in R & D that is responsible for greatly increasing corporate spending, thus maximizing profit margin.

4 DISCUSSION

Design and companies face the development of the product from two points of view that are based on the objectives of their own work. If for the designer the user and his formal and functional needs are the basis for the development of the project, being less concerned with the issues related to the economic sustainability aspect, companies assume this aspect as central in their productive chain and always point out their needs for fast trouble-free production and profits maximization. The future of our design practices is to "produce ideas and design solutions that demand high level of education, skills, and creativity" (Skjold, 2008).

The reverse design and the re-design processes that large-scale companies adopt to develop many of their products offer objects whose design dimensions are largely neglected, focusing almost exclusively on issues that allow the reproducibility of the object itself. Here we see a distortion of the design process which is promoted by the companies themselves, which produces the design objects and places themselves on the market as a guarantee of current design proposals aimed at its users. In reality these companies do not produce design objects based on the users' needs but rather on the basis of the reproduction of the object itself. As referred by Waddell (2004) "As with most industries, fashion needs innovation to survive and prosper; innovation relies on the inventive and original ideas of designers to promp change, and change is what keeps the market healthy and interested. (…) without the impetus of new ideas, industries like fashion soon shrivel and die."

In academic terms this type of methodologies can be an important complement for learning and a fundamental approach to the materials and construction techniques of different types of product. Taking into account the need to bring the design project closer to technological production, material innovation and sustainability, these methodologies could be useful for a better understanding of the needs of industrial production and logistics.

The industrial design profession, nowadays officially contains design of services and experiences besides industrial products. The expansion of the edges fringes of design practice increased the complexity of design, locating it theoretically on an inter-epistemological basis, that is, at the intersection of intellectual areas (Tabak and Farbriarz 2012). This type of adaptation means that designers will have expanded social responsibilities in the development of their design projects. As said by Norman (2014), classical industrial design is a form of applied art, requiring deep knowledge of forms and materials and skills in sketching, drawing, and rendering.

Reverse design and re-design methodologies that may be useful instruments for the perception and improvement of some functional aspects of design products, may be perceived in some cases as a state of the art of some technical principles, and are used in many cases to promote copies of products that present themselves with some necessary changes to avoid legal implications.

5 CONCLUSIONS

Taking into account the issues related to reverse design and redesign practices; we emphasize the need to strengthen the design culture between the academy and industry, from a greater and better integration of designers with the companies. The presence of designers in the industry with a direct link with the technological production and materials innovation systems contributes to the success and affirmation of design products, which included the participation of users in the different phases of the product creation process, using reverse design and re-design methodologies, ensuring a better ergonomic adaptation and usability of the final product.

From the distinction between the design processes and the production processes and their individual and common assumptions, it will be possible to establish relations between academia and companies, creating relationships that can promote the education of integrated working groups for the development of ideas and more targeted to final consumers. Designers are professionals who are able to include in the design of their products aesthetics, functionality, ergonomics and social, cultural and ethical values that underlie end-user options, dimensions often neglected in many reverse design and redesign processes that focus exclusively in the product, in its technical, technological and commercial dimension.

ACKNOWLEDGMENTS

The authors of this paper wish to thank the Centre for Research in Architecture, Urbanism and Design (CIAUD) of the Lisbon School of Architecture of the University of Lisbon and FCT for founding this project.

REFERENCES

Faerm, S. 2012. Towards a Future Pedagogy: The Evolution of Fashion Design Education. *International Journal of Humanities & Social Science*; Dec 2012, Vol. 2 Issue 23, p210.
Laing, R.M., & Sleivert, G.G. 2002. *Clothing, Textiles and Human Performance* (Vol. 32). Manchester: The Textile Institute.
Lipovetsky, G. & Roux, E. 2012. *O luxo eterno: da idade do sagrado ao tempo das marcas*, Edições 70 lda., Lisboa.
Marshall, T. 2009. *Designing design education*. Form, 224. [Online] Available: http://www.icograda.org/education/education/articles1397.htm.
Martins, S.B. Ergonomics and Fashion: the OIKOS methodology for usability and comfort evaluation in clothing and fashion. *Work: A Journal of Prevention, Assessment and Rehabilitation*, 41, 6059–6067.
Miranda, Ana Paula de. *Consumo de moda: a relação pessoa-objeto*. São Paulo: Estação das Letras e Cores, 2014.
Monneyron, F. 2007 *A moda e seus desafios: 50 questões fundamentais*. São Paulo: Editora Senac São Paulo.
Montagna, G., 2015. Multi-Dimensional Consumers: Fashion and Human Factors. Procedia *Manufacturing*, 3, pp. 6550–6556.
Norman, D. 2014. *State of Design: How Design Education Must Change*. Retrieved from: https://www.linkedin.com/pulse/20140325102438-12181762-state-of-design-how-design-education-must-change?trk=mp-reader-card.
Skjold, E. 2008. *Fashion research at design schools*. Kolding, Denmark: Kolding School of Design.
Tabak, Tatiana; Farbiarz, Jackeline Lima; "(not)Solving (non)problems: Design contributions to Education in a complex world", p. 56–59. In: Farias, Priscila Lena; Calvera, Anna; Braga, Marcos da Costa & Schincariol, Zuleica (Eds.). *Design frontiers: territories, concepts, technologies* [=ICDHS 2012 – 8th Conference of the International Committee for Design History & Design Studies]. São Paulo: Blucher, 2012. São Paulo: Blucher, 2014.
Tanure, R.L.Z. and Kistmann, V.B., 2010. A Apropriação da Gestão do Design: Um Estudo de Caso no Setor do Vestuário. *Design & Tecnologia*, 1(02), pp. 22–28.
Waddell, G., 2004, *How Fashion Works, couture, ready-to-wear & mass production*, 1ªed, Blackwell Publishing.

Marketing and consumption

Psychographic segmentation of female fashion consumers in Portugal

B. Moreira
Master in Industrial and Firm Economics, School of Economics and Management, University of Minho, Braga, Portugal

A. Azevedo
School of Economics and Management, University of Minho, Braga, Portugal

ABSTRACT: This study aims to identify the criteria of psychographic segmentation of the female consumer in the fashion sector in Portugal. It also aimed to contribute to a better understanding of the female fashion consumer and thereby help industry managers adopt the best strategies in the textile and clothing industry. For the accomplishment of this study, an online questionnaire obtained 220 answers. A cluster analysis identified four distinct segments: A – "fashion leaders"; B – "low fashion consciousness pragmatics"; C – "Sophisticated shoppers" and D – "Classic style lovers".

1 INTRODUCTION

The fashion industry has to look for market differentiation, have to be always aware of people's consumption behavior and invest in research to delineate efficient market strategies to win more and more consumers and thus ensure the success of sales. Segmentation is a process of classification of potential consumer into target groups with similar needs and characteristics that are likely to have similar buying behavior.

Market segmentation turns out to be the ideal strategy, as it promotes in-depth knowledge of customers and the identification of opportunities, thereby satisfying better the customer needs. Therefore the main research question of this study is: what are the variables that can serve as criteria of segmentation of the female fashion consumers in Portugal? Figure 1 shows the theoretical model of the study.

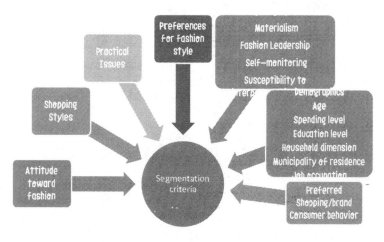

Figure 1. Theoretical framework.

2 LITERATURE REVIEW

2.1 *Factors influencing consumer behavior in fashion*

From very early on, Hurlock (1929) studied the concept of "need or motivation to dress". For example, Cassill & Drake (1987) recognized three factors of assessment criteria used for the purchase of social occasion apparel among female consumers: suitability, economy and orientation to others approval.

The consumer is influenced by cultural factors, social factors, individual factors and psychological factors. According to Solomon (1998), culture is the agglomeration of meanings, rituals, norms and traditions shared among the members of an organization or a society. On the other hand, social classes consist of relatively permanent and homogeneous divisions in a society. The consumer behavior is also influenced by social factors such as reference groups, family and social positions.

Consumer decisions are also influenced by personal characteristics, which include age, life cycle, occupation, but also by psychographic factors such as learning, attitude, motivation and involvement, self-esteem, personality, and lifestyle (Azevedo, 2006). Lifestyle refers to how people live, how they use their time, and how they spend money (Mowen & Minor, 2003), or alternatively a stated standard of living in terms of interests, activities and opinions. Values are consumer beliefs about life and plausible behavior, they reveal goals that motivate people, and appropriate ways to achieve those goals (Engel, Blackwell & Miniard, 2000). Moreover, the personality consists of a set of different psychological characteristics of a person that lead to consistent and lasting responses in their environment.

Motivation has been designated as an active-state need that provides targeted behavior. The need may be hedonic, in that it consists of the need for experience, involving responses, pleasures or emotional fantasies. However, it can also be utilitarian, which corresponds to the desire to obtain some practical or functional benefit, where it takes into account the functional attributes or benefits, objectives of the product.

Another important factor that shapes the type of decision-making behavior, according to Engel et al. (2000), is the degree of personal involvement. This consists of the reflection of strong motivation in the form of perceived personal importance of a product or service in a special context. Involvement becomes active and meaningful when intrinsic personal characteristics such as values are confronted with appropriate marketing incentives within a given situation.

2.2 *Market segmentation*

The Experian study, conducted in 2002 in the United Kingdom, was the great source of inspiration for this work. This study identified about 20 different types of female consumers (see Figure 2).

The concept of market segmentation arises from the publication of an article by Smith (1956) quoted by Wedel & Kamakura (2000). According to Bagozzi et al. (1998), segmentation allows the recognition of different groups that demonstrate relatively homogenous needs. Paço (2007) was able to demonstrate the importance of five main motives and seven

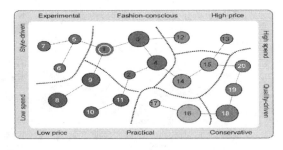

Figure 2. Psychographic segmentation proposed by the Experian's study.
Source: http://www.experian.co.uk/assets/business-strategies/brochures/fashion_segments_handbook_females_small[1].pdf

factors related to the dimension of attitudes towards fashion. This study obtained four different clusters, the first and the highest percentage designated "the sociable", the second the "exhibitionists", the third "the conventional" and finally "the sensitive." Another study by Hota (2002) was based on the segmentation of Spanish consumers, based on their personal values connecting to the typologies of commercial fashion styles.

It may also be considered that there are four types of criteria for market segmentation: 1) the *geographical* segmentation suggests the division of the market into distinct geographic units; 2) the *demographic* segmentation divides the market into groups, based on demographic variables such as age, sex, family size, family life cycle, income, among others; 3) the *psychographic segmentation*, divides buyers into different groups, based on their lifestyle and/or personality; 4) and finally, the *behavioral* segmentation, in which the buyers are divided into groups based on their knowledge, attitude, use or response in relation to a product, also trying to understand the consumer behavior, from the users and non-users perspectives.

2.3 *Psychographic segmentation: Fashion leadership, self-concept, materialism, buying styles and self-monitoring*

The fashion leader is a consumer who has a greater interest in fashion than the rest, who buys new trends earlier thus influencing other consumers. Several authors, such as Kaiser (1990), studied this concept. The type of clothing that people wear is closely intertwined with their self-concept, as it seeks to reflect how they see themselves through the clothing they wear, as Morganosky & Vreeman (1986) refer.

Self-concept consists of the perceptions and attitudes that people have of themselves as objects. According to Onkvisit & Shaw (1987), people know their self-concept through their interactions with the external environment, especially with other people, and these thoughts and feelings about themselves are stable and constant. Self-concept has a complex structure with several dimensions, embracing the idea of a current self, an ideal self, and a social self, but is usually thought of in two ways, an "ideal" self-concept consisting of how the individual would like to be, and a "real" self-concept that consists of how the individual really is (Sirgy, 1982).

On the other hand, according to Triandis (1979), the attitudes reveal the beliefs and feelings about the products, the purchases and the consequences in the purchase of certain goods. The purchase of clothing is stimulated by factors that go beyond the basic purpose of protective clothing of certain elements (McCracken, 1986). For Morganosky & Vreeman, (1986), fashion clothing provides a link to express the individual's concept or self-esteem and the values pursued by consumers.

Belk (1985) defined *materialism* as a posture that encompasses traits related to the personality of possessiveness, lack of generosity and envy and contemplates the importance of the consumer in relation to the possession of goods. Next, Richins & Dawson (1992) explained materialism as a personal value of attaching importance to the possession of materials and divided it into three dimensions: centrality, success, and happiness.

Furthermore, Sproles & Kendall (1986) developed a scale under the name of *Consumer Styles Inventory* (CSI), which identifies eight major shopping styles: quality, brand, fashion, recreation, price, impulsiveness, confusion by overchoice, and loyalty. Several other studies have validated CSI methodology in other cultural contexts (Walsh et al., 2001).

Self-monitoring is a psychographic construct related to the control of human behavior, such as facial expressions and gestures, according to observed behaviors (making "good looks" in front of others). Individuals with high self-monitoring are concerned with their presentation and regulate this presentation according to public presentations, while low self-monitoring does not care so much about their expressive behavior, thus reflecting their own attitudes, emotions and (Snyder, 1974).

The *social values* are related to the high need of affiliation and identification to a respective group, conformity and to a greater emphasis on the unilateral aspects and the exhibition of the clothing. According to Kahle (1983) and Rokeach (1973), the values correspond to desirable and enduring end states that persuade both attitudes and behavior. As desired ends, the values outweigh specific situations. Valuation methods used include the Rokeach Values survey, the

Values and Lifestyles system, and the List Of Values (LOV). The *susceptibility to interpersonal influence* refers to the influence of third parties on the behavior of the individual, varies from person to person and is considered the propensity to identify with a group, through the purchase of products or brands for compliance, or the tendency to learn about products and services by observing others or seeking information from others (Bearden, Netemeyer & Teel, 1989). Finally, the *desire for exclusive products* is a propensity for people meet their uniqueness needs through the acquisition and consumption of unique products (Lynn & Harris, 1997).

3 METHODOLOGY

In order to answer the research question (what variables can determine the targets based on psychographic criteria?), an empirical study was conducted through an online questionnaire with the following questions:

A. Shopping styles, adapted from CSI (Consumer Styles Inventory) by Sproles & Kendall (1986), is a group consisting of ten questions, which aim to understand essentially the influence of quality, brand and price.
B. Practical Aspects, adapted from the Experian study. It consists of six questions, which essentially intended to understand the importance of durability, comfort and purchase by necessity.
C. Attitudes towards fashion, adapted from the Experian study. Group consisting of 14 questions assessing if the respondents make frequent purchases, if they consider themselves independent of the fashion, if they care about the details, with a sophisticated look, a new look and with the expense.
D. Style Preferences, adapted from the Experian study. Group composed of seven questions, aiming to understand whether the respondents prefer a classic or casual appearance, if they care about the style and its prominence.
E. Materialistic attitudes, adapted from Moschis & Churchill (1978), consisting of five questions analyzing the importance of money both personally and professionally, and the importance of products and brands.
F. Fashion Leadership adapted from the study of Goldsmith, Freiden & Kinlsheimer (1993). This group comprised six questions intending to understand the importance of being a leader of fashion, fashion trends and individuality.
G. Self-monitoring adapted from Snyder's study (1974) comprising 15 questions for assess the personality/behavior of individuals.

The second part of the questionnaire included the sociodemographic characterization of consumers (gender, education level, age, household, profession, parish and municipality of residence, average monthly expenditure) and Consumer Preferences (brands, stores). Its spread was through the Facebook from February until April of 2016.

A convenience sample of 220 female respondents, with an age's average of 24.23 years (SD = 6.67) with an average household of 3 persons per family. The monthly expenditure on clothing is on average 53.24 euros (SD = 52.04). The majority of respondents live in the north of Portugal in cities such Braga (N = 63) and Guimarães (N = 51).

4 DISCUSSION OF RESULTS

4.1 *Psychographic segmentation: Cluster analysis*

According to Hartigan (1975), the cluster analysis makes it possible to join groups of observations in order to maximize the variance between groups and to minimize it within them, that is, entities belonging to a group are the most identical between them and differentiated from the rest. In order to determine which variables present significant differences among the four clusters identified by the analysis of k-means clusters, we performed a variance analysis (ANOVA), the results of which are shown in Table 1.

Table 1. Cluster analysis and segmentation criteria.

	Cluster A Sophisticated N = 38		Cluster B Classic style lovers N = 60		Cluster C Low fashion consciousness N = 103		Cluster D Fashion leaders N = 14		Total N = 215		F (p)
	M	DP	M	DP	M	DP	M	DP	M	DP	
Attitudes towards fashion [1. I always buy at least one piece of clothing that is fashionable in every season.]	4,42	1,004	3,72	1,209	3,21	1,419	4,64	,497	3,66	1,347	11,918***
Attitudes towards fashion [2. It is important that my clothes and shoes are always up to date with current fashion trends.]	3,50	1,225	2,90	1,069	2,47	1,101	3,93	,730	2,87	1,186	13,217***
Attitudes towards fashion [3. Fashion details are very important to me.]	3,63	1,172	2,93	1,118	2,48	1,170	3,50	1,019	2,87	1,226	11,102***
Attitudes towards fashion [4. I'm not interested in following fashion trends.]	2,55	1,245	2,93	1,039	3,31	1,094	2,79	1,477	3,04	1,164	4,703**
Attitudes towards fashion [6. I'm more interested in fashion than most people I know.]	2,68	1,233	2,32	,965	1,79	1,035	3,07	1,207	2,18	1,138	11,122***
Attitudes towards fashion [9. I really like shopping for clothing and footwear.]	4,53	,893	4,33	,655	3,75	1,218	4,50	,855	4,10	1,061	8,217***
Attitudes towards fashion [11. I spend a lot of money on clothes, shoes and accessories.]	3,21	1,212	2,42	,926	2,02	1,075	3,79	,802	2,46	1,179	20,111***
Attitudes towards fashion [14. I buy clothes that make me look sophisticated.]	4,05	,769	3,43	,909	3,14	1,048	4,07	,730	3,44	1,012	10,892***
Influences of purchase [2. The brand name of the clothes and shoes I buy is very important to me.]	2,53	1,109	2,27	,954	2,03	1,024	2,50	,941	2,21	1,028	2,760*
Influences of purchase [5. The price is not important if I really like something.]	3,05	1,207	2,90	1,100	2,32	1,087	3,50	1,160	2,69	1,172	8,256***
Purchase influences [8. I have tendencies to buy on impulse instead of planning]	3,03	1,174	2,60	1,045	2,30	1,119	3,50	1,225	2,59	1,164	7,350***

(*Continued*)

Table 1. (Continued).

	Cluster A Sophisticated		Cluster B Classic style lovers		Cluster C Low fashion consciousness		Cluster D Fashion leaders		Total		F (p)
	N = 38		N = 60		N = 103		N = 14		N = 215		
	M	DP	M	DP	M	DP	M	DP	M	DP	
Influences of purchase [10. Shopping puts me in a good mood.]	4,13	1,018	3,90	,933	3,51	1,236	4,64	,497	3,80	1,127	6,536***
Practical Aspect [1. Durability is more important than clothes that are fashionable.]	3,84	1,001	3,45	1,048	3,89	,938	2,86	1,167	3,69	1,032	6,144**
Practical Aspect [2. I only buy new shoes mainly to replace something I have worn.]	2,82	1,333	3,03	1,178	3,54	1,274	2,64	1,216	3,21	1,290	5,024**
Practical Aspect [5. I only buy clothes if I really need them.]	2,58	1,266	2,87	1,065	3,29	1,210	2,29	,994	2,98	1,207	5,760**
Style Preferences [3. I like classic appearance, but only when it is updated and reflects the new trends.]	3,61	1,001	3,67	,914	3,09	1,095	3,64	,929	3,38	1,051	5,325**
Style Preferences [6. My clothes help me to express my personality.]	4,29	,732	3,97	,663	3,78	1,066	4,14	,535	3,94	,900	3,405*
Style Preferences [7. I like to try new looks and styles.]	4,00	,930	3,52	1,049	3,25	1,178	4,00	,784	3,51	1,114	5,514**
Materialist attitudes [1. I really believe that money can buy happiness.]	2,50	1,289	2,62	1,223	2,20	1,216	3,14	1,512	2,43	1,269	3,123*
Materialist attitudes [2. My dream in life is to be able to possess expensive things.]	1,89	1,110	1,88	,940	1,80	1,132	2,71	1,437	1,90	1,114	2,868*
Materialist Attitudes [4. Money is the most important thing to consider when choosing a job.]	2,50	1,059	2,37	1,221	2,10	1,062	3,14	1,231	2,31	1,144	4,245**
Fashion leadership	16,29	5,162	14,10	4,391	11,90	4,324	19,29	5,650	13,77	5,059	16,341***
Self-monitoring [5. I think I could do a show to impress or entertain people.]	3,84	1,386	3,80	1,219	4,13	1,177	3,00	1,414	3,91	1,267	3,734*
Monthly expenditure on clothing,	89,87	12,33	48,33	5,42	19,33	8,70	167,86	24,862	49,56	41,860	1070,80****
Age	25,58	7,168	23,75	5,239	24,03	7,633	24,00	3,234	24,22	6,728	,644 Ns
Household size	3,21	1,212	3,42	1,078	3,49	1,047	3,79	,975	3,44	1,083	1,116 ns

Legend: ***p < 0,001; **p < 0,01; *p < 0,05.

This analysis identified four different groups. Starting with the group D (N = 14), it is the group with the highest number of variables with the highest mean. You can see that this is the group where people are most interested in fashion, who value more clothes, shoes and accessories and always try to be fashionable. They tend to buy on impulse, and they do not care about money if they really like something. Being that they consider that money can buy happiness and that is the most important thing to consider when choosing a job, also considering that the dream in life is to be able to own expensive things. With grouping of these variables, we can define group 4 as the "*fashion leaders*", since it is the group that is most interested in fashion.

With the analysis of group C (N = 103) we find that it is the group opposite to group 4. In this group, people are not interested in following fashion trends, considering that durability is more important than clothes be fashionable. They only buy new clothes or shoes when they really need or replace something that has worn away. This group was defined as the "*Low fashion consciousness pragmatics*".

Group A (N = 38) is the closest to group D. It has been found that in this group people give importance to fashion details, that they really like to shop and that they consider their clothes to help express their personality. This group was defined as the "*sophisticated shoppers*", that is, as people who like to be fashionable and looking for a bold look.

And finally, the group B (N = 60) that presents only one variable with the highest average. In this group, people like to look classic, but only when it is updated and reflects the new trends. However, it also has high averages (although not the highest) in relation to a lack of interest in following fashion trends, and only buying shoes and clothing when you really need it or if something has worn out. This was the most difficult group to define, because of its lack of highlighted variables, and the discordance of the highlighted variable with the remaining variables. However, it was called group B as the "*classic style lovers*", would be the most appropriate, since the classics are always fashionable, even though it is not keeping up with fashion trends. In order to confirm the differences between the clusters, authors conducted a discriminant analysis (Wilks Lambda = 0,038; chi-square = 651,662; df = 78; p < 0,001) which classified 99,5% of original cases (see Figure 3).

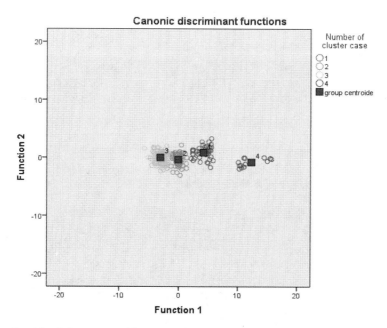

Figure 3. Graphic of cluster centroids.

5 CONCLUSIONS

Fashion leaders spend more money, are more interested in shopping, brands and style, their impulse tends to be bigger. They believe that money can buy happiness and that is the most important thing to consider when choosing a job and they care less about durability.

Women, who spend more money, also make more purchases for fun, their drive and interest in brands tends to be higher, consider that the clothes they wear and tend to look for a more sophisticated look reinforce self-esteem. Those who value brands more make more purchases for fun, their impulse tends to be bigger. They care more about quality and tend to spend more money.

This study is important for marketing managers in the fashion industry, to develop a psychological profile of consumers. It also contributes to the entrepreneurs in the construction of their own brand, where it allows the knowledge of the consumers and their preferences, demonstrating what types of clothing styles they should adopt depending on the target audience. One of the recommendations that is suggested is the bet on fashion leaders.

There are many people fascinated by fashion, being the most willing to spend money and buy on impulse. Thus, brands must be up-to-date on new trends. It is also recommended that brands should always bet on the psychographic segmentation, considering a fundamental means for the study of the consumer.

This study allowed collect better information of the female consumer through the identification of the positioning of products for the market based on assessed needs and preferences. It allows fashion brand managers to develop a marketing plan that meets the specific needs of the market, but also allows the knowledge of the products competing in your specific market. In order to better identify the four segments found in this study, we chose to analyze the marketing mix tailored to each segment. For the "Fashion Leaders", brand managers must bet on recreation/creation and anticipate new trends. For the "Sophisticated shoppers", brand managers must bet on the detail, on the differentiation with refinement. While for the "Classic style lovers" one must take into account the price, since they attach too much importance to the savings. Finally, for the "Low Fashion Consciousness pragmatics" in addition to the attention to the price, managers must bet on comfort and durability.

We also compared the segments of this study with the Experian study (see Figure 2), denoting that: group A "sophisticated shoppers" is similar to type 12, where women have a high spending capacity and dress with the best brands; group B "Classic style lovers" is similar to type 18, which refers to women who only like to shop by catalog; the group C "Low fashion consciousness pragmatics" is similar to type 16, which includes women who only make purchases to replace worn clothes; and group D "Fashion Leaders" is similar to type 5, which corresponds to women who care about fashion. This study presents some limitations, namely the use of a convenience sample. In this way, it is recommended that future studies be conducted with samples that are more representative of the Portuguese fashion consumer.

REFERENCES

Azevedo, A. 2006. Influência das variáveis psicográficas no processamento da publicidade. In P.Cardoso, S.Gaio and J. Seoane (orgs), *Jovens, Marcas e Estilos de Vida:* 187–200. Porto: Edições Universidade Fernando Pessoa.

Bagozzi, R.P., Rosa, J.A., Celly, K.S. & Coronel, F. 1998. *Marketing Management*. Prentice-Hall International, Inc.

Bearden, W., Netemeyer, R. & Teel, J. 1989. Measurement of Consumer Susceptility to Interpersonal influence. *Journal of Consumer Research*, 15: 473–81.

Belk, R. 1985. Materialism: the trait aspects of living in the material world. *Journal of Consumer Research*, 12: 256–281.

Cassill, N. & Drake, M., 1987. Apparel selection criteria related to female consumers' lifestyle. *Clothing and Textiles Research Journal*: 20–28.

Engel, J., Blackwell, R. & Miniard, P., 2000. *Comportamento do Consumidor*. Ed. Rio de Janeiro: LTC.

Hartigan, J. 1975. *Clustering Algorithms*. John Wiley and Sons, Inc.

Hota, F.J.-S. 2002. Using values and shopping styles to identify fashion apparel segments. *International Journal of Retail and Distribution Management,* 40(3): 180–199.

Hurlock, E. 1929. *Motivation in fashion.* New York: Archives of Psychology.

Kahle, L. 1983. *Social values and social change.* New York: Praeger.

Kaiser, S.B. 1990. *The social psychology of clothing (2nd).* New York: Macmillan.

Lynn, M. & Harris, J. 1997. The desire for unique consumer products: a new individual differences scale. *Psychology and Marketing.*

McCraken, G. 1986. Culture and consumption: A theoretical account of the structure and movement of the cultural meaning of consumer goods. *Journal of Consumer Research,* 71–84.

Morganosky, M. & Vreeman, A.L. 1986. The involvement concept: Relationships to the apparel consumer. *Unpublished manuscript, University of Illionis.*

Moschis, G. & Churchill, G. 1978. Consumer socialization: a theoretical and empirical analysis. *Journal of Marketing Research,* 15: 599–609.

Mowen, J. & Minor, M. 2003. *Comportamento do Consumidor.* São Paulo: Prentice Hall.

Onkvisit, S. & Shaw, J. 1987. Self-concept and image congruence: Some research and managerial implications. *Journal of Consumer Marketing,* 13–23.

Paço, A.M., 2007. O "consumidor" de moda - um estudo de segmentação de mercado. retrieved from: http://convergencias.esart.ipcb.pt/artigo.php?id=3.

Richins, M. & Dawson, S. 1992. Consumer values orientation for materialism and its measurement. *Journal of Consumer Research,* 19: 303–317.

Rokeach, M.S. 1973. *The nature of human values.* New York: Free Press.

Sirgy, M.J. 1982. Self-concept in consumer behavior: A critical review. *Journal of Consumer Research,* 287–300.

Smith, W.R. 1956. Product Differentiation and Market Segmentation as Alternative Marketing Strategies. *Journal of Marketing, July:* 3–8.

Snyder, M. 1974. Self-monitoring of expressive behavior. *Journal of Personality and Social Psychology,* 30: 526–537.

Solomon, M. 1998. *Consumer Behavior: buying, having, and being.* New Jersey, USA: Prentice-Hall: 4th ed.

Sproles, G. & Kendall, E.L. 1986. A Methodology for Profiling Consumer's Decision-Making Styles. *The Journal of consumer affairs,* 20(2): 267–270.

Triandis, H. 1979. Values, attitudes, and interpersonal behavior. *In Nebraska Symposium on Motivation:* 195–259. *Lincoln: University of Nebraska Press,*

Walsh, G., Mitchell, V. & Hennig-Thurau, T. 2001. German consumer decision-making styles. *Journal of Consumer Affairs,* 35(1): 73–96.

Wedel, M. & Kamakura, W. 2000. *Market Segmentation. Conceptual and Methodological Foundations.* 2nd Edition, ISQM - International Series in Quantitative Marketing, Kluwer Academic Publishers.

Taxhion. Model taxonomy clothing and accessories e-commerce

B. Alcoceba, M. García-Ergüín, R. García, D. de las Heras & M. Adsuara
ESNE, Escuela Universitaria de Diseño, Innovación y Tecnología, Madrid, Spain

ABSTRACT: Taxhion is a research project part of Tagsononomy– "Catch what you watch; a game-changer technology that will change how people watch TV and revolutioniseinmovie advertising and product placement"–, which received a H2020 grant agreement from the European Union in September, 2017. Nowadays, Tagsononomy – also known as Dive-, attains leadership in the integration of e-commerce in audiovisual, television and cinematographic products. In this way, its role is to identify garments in the above-mentioned audiovisual products and relate them to similar ones in the online market, to be offered to the viewer. Therefore, Taxhion is developed as a system to offer a method of clothing classification which can be adapted to the necessities of Tagsononomy. The proposal outlines 62 clothing typologies, a list of 7 attributes and 17 elements of definition. This study shows the nature and results of its implementation.

1 INTRODUCTION

Nowadays e-commerce is a business in full expansion. The term virtual store was adopted from catalog sales in the United States at the beginning of the 20th century. This is no longer utopian. It has progressively become the most popular shopping system used by fashion consumers worldwide. As a consequence, modern-day street shopping has been replaced by e-commerce. In this context, Taxhion arises as a taxonomic research project for Tagsononomy, which provides the viewer with additional interactive information in real time, thus applying e-commerce to the viewing of movies or series.

The sponsor company, Tagsononomy S.L. –specialist in generating content of Artificial Intelligence, Analytics, Big Data and Humanities that operates from the semiautomatic identification of the image- received a Horizon 2020 grant agreement (SME Instrument, phase 2), the biggest EU Research and Innovation programme, in September 2017. Taxhion, as a tool able to exploit its resources, has helped in this development.

The main objective of Taxhion is to create a document for clothing analyisis, recommendation and evaluation, which would establish a structure of typological classification of the different garments and accessories. Besides, it unambiguously determines the namings and definitions, so it can be processed in a format applicable to e-commerce using video images. The methodological procedure used to the creation of the tool and the results obtained are presented below.

2 METHODOLOGY

In order to develop the taxonomic tool, a mixed methodology, which combines both the documentary and the experimental analyses, is used. The method adopted can be divided into three sections:

- Case study analysis
- Taxonomic proposal generation
  - Taxonomy
  - Catalog of parameters
- Data sheet

2.1 Case study analysis

The search for taxonomic information is conducted by means of the state of the question analysis and the frame of reference, thus adopting different approaches:

2.1.1 State of the question

Phd Dissertations (Alcoceba, 2015) (Hobbs, 2015) (Rissanen, 2013) and international articles (Bossard, 2013) (Yamaguchi, 2012) (Zi Jian Xu, 2006) are employed in the scientific research process. The classifications and clothing parameters analysed in each case are not coincident; therefore, an own proposal for structure and attributes is generated. Nevertheless, it can be affirmed that the common structure is: Gender/Typology/Textile structure/Material/Color.

2.1.2 Reference framework

Personal interviews, documentary analyses and web research lay the foundations for the non-scientific research.

– Bibliographic

Books published by specialised authors and editorials have been considered (Calasibetta, 2003) (Clarke, 2011) (De Margerie, 2011) (Estany, 1987) (Fashionary International Ltd, 2016) (Feyerabend, 2012, 2016) (Fogg, 2014) (Ministerio de Cultura. Ed, 1993) (Newman, 2010) (Rivière, 1996) (Saltzman, 2004). The most repeated structure is Gender/Typology/Detail.

– Institutional

Specialised institutions from the clothing industry have provided professional advice (Ballet Nacional de España) (Europeana) (Le Musée des Arts Decoratifs) (Museo del Traje de Madrid) (PallaisGalliera). The most repeated concepts are the Historic moment/Author/typology/Material/Body placement/Use. The use of concepts such as "search tool" and "factsheet" are also taken into account.

– Normative

Institutions that provide normative data as regards manufacturing (ISO)(UNE), property (WIPO) or standardization (CEN) have also contributed to the present research. The most coinciding information is Item/General Typology/Specific Typology.

– Commercial

Together with the scientific view, e-commerce cataloguing (Asos) (H&M) (Macy's) is closely related to the activity developed in Taxhion. The online version of most companies has a common structure: Gender/Typology/Filter. In order to be as flexible as possible when customizing the search, the order of the parameters is variable. As regards the filters, several attributes are identified, Color/Size/Brand being the most common ones.

2.2 Taxonomic proposal

The data obtained in section 2.1 are analyzed in order to outline a definite taxonomic proposal. The tree diagram is regarded as a key tool during this process, since it offers the possibility of displaying the basic concepts selected and verifying the suitability in an efficient way.

To complement this, a catalog of parameters is designed of each typological classification. However, attributes such as color and pattern are omitted, since these are obtained automatically by means of image recognition software.

With the objective of encoding the search, both the taxonomy and the parameters catalog are complemented with data tables compiled in a binary code. In this way, the objective of the present study is facilitated: the identification of items in the broadcasting and the tracing of similar e-commerce products.

2.2.1 Taxonomy

The taxonomy serves to establish the basic classification of the item; it is based on a tree diagram and expresses a horizontal hierarchy, from the general to the particular (Figure 1):

Figure 1. Main diagram of the taxonomy.

- FASHION. Field of action; it encompasses those items associated with dressing and decorating the body.
- GENDER (4). Cultural meaning assigned to a garment depending on the user's gender. A taxonomy type that can be associated with all the genres contemplated is developed.
  - Woman. Adult female person.
  - Man. Adult male person.
  - Girl. Young female person.
  - Boy. Young male person.
- GROUP (4). Items grouped according to the use and relationship with the body:
  - Garment. Article that covers, totally or partially, the body. Its length is measured from top to bottom, except in hosiery, which reverses the order (as in shoes).
  - Footwear. Article that covers the foot and has a sole. Its height must be read from bottom to top.
  - Bag. Object that serves to transport objects and is part of the attire.
  - Accessory. Object used to decorate the body or complement the outfit, but that is not necessary to dress the body.
- BODY PLACEMENT X (3). Group according to its horizontal placement on the body, following an order from outward to inward depending on the layers of clothing:
  - Outerwear. Garment that is placed on other garments, covering all of them. It is considered an overcoat.
  - Wear. A garment worn between underwear and outerwear, making it the most common clothing.
  - Underwear. Intimate garment that has direct contact with the body. This concept is also associated with homewear.
- BODY PLACEMENT Y (3). Group according to its vertical placement on the body, following an order from top to bottom:
  - Complete. A garment that covers the body entirely so that it is not necessary to use another complementary garment. The shoulders are usually taken as the upper reference point, and the crotch as the minimum lower reference point.
  - Upper. Garment that is placed on the upper part of the body, taking the waist as the lower reference point.
  - Lower. Garment that is placed on the lower part of the body, taking the waist as the lower reference point.
- TYPOLOGY. Typological grouping determined by the morphological nature of the items.
  - General. Primary category
  - Specific. Secondary category

The sixty-two typologies found are ordered according to the criteria mentioned in Figure 2. Subsequently, a specific definition for each of the above groups, obtained from the data tables mentioned in section 2.2, is provided. An example of a data table is included below:

2.2.2 *Catalog of parameters*

The catalog serves to establish the specific classification of each group of items. It is composed of attributes and constructive elements. A summary of the suggested catalog can be seen in Figure 4. However, this information is thoroughly developed in the research project, as Figure 5 shows.

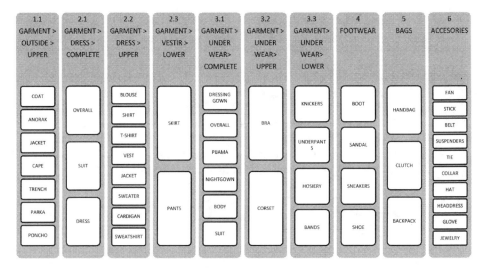

Figure 2. General typologies.

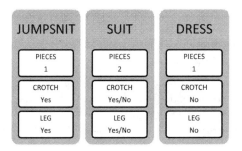

Figure 3. Example of one of the ten data tables: GARMENT > WEAR > COMPLETE.

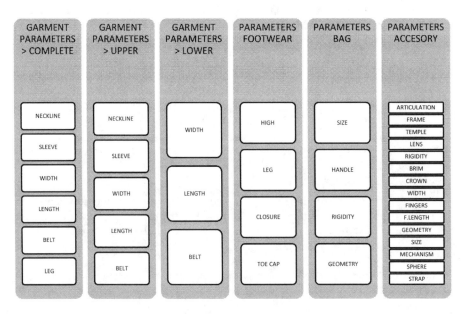

Figure 4. Catalog of parameters summary.

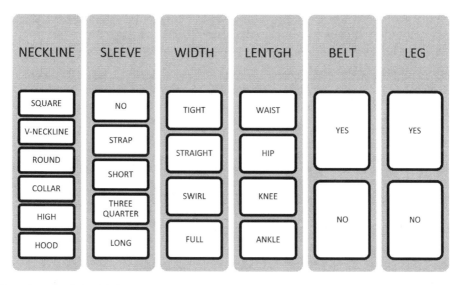

Figure 5. Detailed table for PARAMETERS GARMENT > COMPLETE.

Figure 6. Data sheet type.

2.3 *Data sheet*

In order to check the suitability of the taxonomy and obtain the results of its functioning, empirical data sheets are elaborated. Television references are obtained, and classified according to Taxhion -the taxonomic proposal -, and they are further compared with e-commerce results found with the same strategy employed.

As Figure 6 shows, the result is not precise, but approximate, thus widening the clothing options for a potential consumer.

3 CONCLUSION AND DISCUSSION

A great variety of clothing typologies can be observe din the fashion industry, their classification being of considerable complexity, not only because of the quantity, but because of the subtlety which differentiates the garments. Each garment is composed of different elements, materials and formats. In addition, they perform different roles. Therefore, an exhaustive categorization would imply the production of a document with hundreds of denominations and descriptions which would increase as fashion evolves. With the objective of simplifying such task, and focusing on the practical implementations of Tagsononomy, a unique taxonomy has been elaborated, which starts from the concept of Fashion, and spreads into a classification of 62 typologies.

The typological division in groups is:

– Garments: 33 typologies
– Footwear: 4 typologies
– Bags: 3 typologies
– Accesories: 22 typologies

The criterion employed to define them has not been uniform, since the more morphological variety, the greater the typological extent. As a consequence of the greater variety of formal variables, a more exhaustive categorization in upper garments than in lower ones has been considered.

The taxonomy described is complemented by a highly intuitive catalog of parameters (attributes and elements), which allows—once the item has been classified- the description of the garment even by a person without prior knowledge of fashion.

The general attributes which have been found are: size, length, width and geometry. Other specific attributes for particular cases such as articulation, flexibility and rigidity have been included (others, such as color and pattern, have been omitted from the current study, as they are deduced from an automated system by means of algorithms, and are not part of the research of the current study).

Finally, and more specifically, elements of definition have been included: neckline, sleeve, belt, trouser leg, boot leg, closure, toe cap, handle, frame, temple, lens, brim, crown, fingers, mechanism, sphere and strap.

REFERENCES

Bibliography

Alcoceba, B. 2015. *Piel artificial. Metamorfosis arquitectónica del cuerpo a través de la superficie*. PhD Thesis, Universidad Politécnica de Madrid.
Bossard, L., Dantone, M., Leistner, C., Wengert, C., Quack, T. and VanGool, L. 2013. Apparel Classification with Style. In: Lee K.M., Matsushita Y., Rehg J.M., Hu Z. (eds) *Computer Vision – ACCV 2012. Lecture Notes in Computer Science*. Berlin, Heidelberg: SPRINGER, vol 7727.
Calasibetta, C. 2003. *The fairchild dictionary of fashion*. London: Laurence King.
Clarke, S. 2011.*Textile design*. London: Laurence King.
DeMargerie, G. 2011. *Dictionnaire du look: une nouvelle science du jeun*. Paris: Robert Laffont.
Estany, M. 1987. *Diccionario textil y del vestir*. Barcelona: Manuel Estany.
Fashionary International Ltd. 2016. *Bag Design*. Hong Kong.
Fashionary International Ltd. 2016. *Fashionpedia*. Hong Kong.
Fashionary International Ltd. 2016. *Shoe Design*. Hong Kong.
Feyerabend, F.V. 2012. *Accesorios de moda*. Barcelona: Gustavo Gili.
Feyerabend, F.V. 2016. *Ilustración de moda*. Barcelona: Gustavo Gili.
Fogg, M. 2014. *Moda: toda la historia*. Barcelona: Blume.
Hobbs, K. 2015. *Defining garments through details*. MA Thesis, University of Borås.
Ministerio de Cultura. Ed. 1993. *Anales del Museo del Pueblo Español V*. Madrid.
Newman, A. 2010. *Moda A-Z: diccionario ilustrado*. Barcelona: Blume.
Rissanen, T. 2013. *Zero-waste fashion design: a study at the intersection of cloth, fashion design and pattern cutting*. PhD Thesis, University of Technology of Sydney.
Rivière, M. 1996. *Diccionario de la moda: los estilos del siglo XX*. Barcelona: Grijalbo.
Saltzman, A. 2004. *El cuerpo diseñado: Sobre la forma en el proyecto de la vestimenta*. Buenos Aires: Paidós.
Yamaguchi, K., Kiapour, H., Ortiz, L., and Berg, T.L. 2012. *Parsing Clothing in Fashion Photographs*. CVPR

Zi Jian Xu, Song Chun Zhu, Hong Chen, ZiQiang Liu. 2006. Composite Templates for Cloth Modeling and Sketching, In: *2013 IEEE Conference on Computer Vision and Pattern Recognition, vol. 01, no., pp. 943–950, 2006, doi:10.1109/CVPR.2006.81.*

Web pages

Asos. [Online]. Available from: http://www.asos.com/es/hombre/?r = 1. [Lastaccessed 25/07/17].
European Committee for Standardization (CEN). [Online]. Available from: https://standards.cen.eu/dyn/www/f?p=204:29:0::::FSP_ORG_ID,FSP_LANG_ ID:6229,25&cs=1EF3C196F97138C6F7027E35E87F3C8DB#1. [Last accessed 20/07/17].
Europeana Collections. [Online]. European Union. Available from: http://www.europeana.eu/portal/es/collections/fashion. [Accessed 11/07/17].
H&M. [Online]. Available from: http://www2.hm.com/es_es/index.html. [Last accessed 26/07/17].
International Organization for standardization (ISO). [Online]. Available from: https://www.iso.org/standards-catalogue/browse-by-ics.html. [Last accessed 20/07/17].
Le Musée des Arts Decoratifs. [Online]. Available from: http://www.lesartsdecoratifs.fr/. [Last accessed 15/07/17].
Macy´s [Online]. Available from: https://www.macys.com/. [Last accessed 26/07/17].
PalaisGalliera. [Online]. Available from: http://palaisgalliera.paris.fr/. [Lastaccessed 15/07/17].
Una Norma Española (UNE). [Online]. Availablefrom: http://www.aenor.es/aenor/normas/ctn/fichactn.asp?codigonorm = CTN%2040&pagina = 1#.WVzz0YTyiUk. [Lastaccessed 21/07/17].
World Intellectual Property Organization (WIPO). [Online]. Available from: http://web2.wipo.int/classifications/nice/nicepub/en/fr/edition-20130101/taxonomy/class-25/. [Lastaccessed 21/07/17].

Softwares

Domus [Software]. Madrid: Museo del traje de Madrid.
Trajescena [Software]. Madrid: Ballet Nacional de España.

Cultural values of minimalist fashion

A.P. de Miranda, I. Domingues & N. Souza
Federal University of Pernambuco, Caruaru, Pernambuco, Brazil

ABSTRACT: This study examines the culture of consumption of minimalist fashion bloggers, who are not consumers, but also act as multipliers, aiming at identifying cultural values that may explain this minimalist movement. The main objective was to detect the ways by which the minimalist identities of such digital multipliers are built, assuming they play the role of fashion behavior disseminators. In order to achieve so, a qualitative research was conducted using the technique of structured in-depth interviews with five minimalist bloggers. The research findings were examined applying discourse analysis, concerning both the interviews' narratives and the images captured in their blogs, the latter following principles proposed for denotative analysis of images. The results on cultural values points to an understanding that values such self-determination and philanthropy guide this behavior.

1 INTRODUCTION

Clothing is more than simply choosing something to wear: outfits are non-verbal communication platforms that allow the observers to deduce aspects regarding their users, such as social class, personality traits, and values, among others.

The present article analyzes minimalism fashion consumption, with the object of study being bloggers self-proclaimed minimalists of diverse Brazilian regions. The study investigates minimalist fashion behavior as a consumption culture and looks for answers to the following research guiding question:
– What are the cultural values of minimalist fashion consumption?

1.1 *Culture, fashion and consumption*

According to McCracken (2003), culture comprises a set of ideas and activities through which individuals construct and constitute the world: a system of meanings, attitudes, and shared values. These meanings are characterized by two concepts: cultural categories and cultural principles. Cultural categories are the result of world segmentation, enabling the classification of distinct elements, such as child/adult, young/old, yesterday/today/tomorrow, among others (McCracken, 2003, p.101). On the other hand, cultural principles are ideas or values through which segmentation is performed: "delicacy of women" and "strength of men", for example (McCracken, 2003, p.105). In this sense, we may understand that categories are the fact of being/belonging, though what leads to being/belonging are the principles. For the purpose of this study, it's important to understand the role of the fashion system as an instrument for transferring meanings from the culturally constituted world to consumption goods, i.e. the assignment of principles to the categories. According to McCracken (2003), the fashion system creates new cultural meanings carried out by "opinion leaders," those who help shape and refine existing cultural meanings, encouraging a revision of cultural categories and principles.

Social and cultural factors influence and shape the consumer behavior and they are related to the environment where people live in, to the relationships they establish and to the habits they acquire. Some preferences reflect their family interactions, others their professional life, their relationship with friends, religious beliefs, and so on (McCracken, 2003, p.110). This structure of thinking called "movement of meaning" has been updated with the concept of "sources of meaning for goods and services," where fast fashion, fashion and trend are cultural domains and each domain holds its own characteristics, structures and logics (McCracken, 2005, p.177). The consumer culture determines the overall priorities individuals associate with distinct activities and products, what corroborates the presuppositions previously reviewed and points to a path for the comprehension of cultural values as behavior guidelines (Solomon, 2002).

Regarding cultural values, Tamayo and Schwartz (1993, p.330) note that they represent the force that impels individuals to search, regularly and throughout their lives, for the achievement of certain goals. According to the authors, values reflect distinct motivational types which are, in Brazilian culture, split into nine types and their related goals: 1) Hedonism: searches for pleasure and sensual gratification; 2) Self-Realization: looks for personal success obtained through the demonstration of competence leading to social acknowledgement; 3) Social Power: looks for social status, prestige, and control over people and resources; 4) Self-Determination: aspires independent thinking, action, and options; 5) Conformity: aims at controlling impulses and personal behavior to conform to social expectations; 6) Benevolence: shows preoccupation with the well-being of close people; 7) Safety: displays concern with personal integrity, as well as the one of identifying groups; 8) Tradition: is motivated by respect and acceptance of society ideals and traditions; 9) Philanthropy: looks for comprehension and acceptance by others, as well as overall well-being.

1.2 *Minimalism*

In order to understand minimalist consumption it's necessary to understand that minimalism was a 20th century artistic strand that proposed the expression through essential elements, with greater visibility within fine arts, though it also influenced literature and music. The movement was marked by the renowned expression "less is more," developed by the German architect Mies van der Rohe. Minimalism looks for the identification of the essence in order to eliminate any excess. It serves as a tool to get rid of the superfluous and become focused on what is relevant. In other words, it means to have what is sufficient; nothing more and nothing less. This concept may be applied to everything that surrounds us, not only objects, but also commitments, relationships, work, and lifestyle (Mota, 2013).

The intense production of services and goods in industrial scale throughout the 20th century led to an increasing reflection on the impacts of consumption and disposal of non-recyclable products, creating a reaction that gained strength as a cultural movement. Latouche (2009) contributes to the understanding of such issue with the concept of gradual degrowth. According to the author, degrowth is a concrete utopia in productivist societies and it concerns the avoidance of ecological and humane catastrophes. Minimalism as a cultural movement reflects current consumers' perspective on the search for degrowth in their consumption patterns.

Buckingham (2012) suggests that initially it's necessary to get liberated from societal external impositions, as well as from internal dissatisfactions as desires, emotions, and fears. It then becomes possible to adopt a minimalist consumption attitude. In the first decades of the 21st century it could be observed the development of a reflection movement regarding quotidian consumption decisions, taking into consideration the need for degrowth pointed out by Latouche (2009). The internet is the locus for the meeting and proliferation of these ideas where minimalist consumption takes place. The core purpose regards a voluntary simplicity, starting with a reduction in the consumption, and reaching all aspects of an individual's life: work, food, interpersonal relationships, and health. The target is to live with less, possessing only what is necessary and leaving aside the superfluous. In this sense, minimalism may be understood as a consumer behavior, once these consumption patterns are a reflex of a consumption culture. In a broader perspective, it's also a way these individuals have found to pursue a life that makes more sense, one that brings them more happiness (Negretto, 2013).

Bringing those ideas into the field of fashion consumption, it's possible to understand the principle of planned obsolescence, inherent to the fashion system, and challenged by the development of minimalist consumption. As Barthes suggests, fashion is based on the idea that consumption is greater than deterioration, i.e. people will always dispose goods before they are entirely worn out due to the desire of acquiring new fashion items (Barthes, 1979). The movement of minimalist fashion denies the category cultural principle of "futility" deriving from its fleeting nature. For minimalist consumers, consumption becomes a source of emotional comfort at the shopping moment, as well as during the consumption itself, which for them is not ephemeral, but long lasting.

2 METHODOLOGY

The objective of this study is to unveil the cultural values behind minimalist fashion consumption behavior. In order to achieve so, a qualitative methodology was adopted and in-depth structured interviews were conducted using a guideline. The research findings were examined using discourse analysis, as proposed by Bauer and Gaskell (2008). The analysis of fashion images followed the protocol developed by Maciel and Miranda (2009). The authors propose the following pre-established criteria for the analysis of images: 1) Shape: common aspects regarding the outfits – cut, length, and volume of the pieces under analysis; 2) Color: common color prevalence and distribution along the outfit; 3) Materials: common elements used for manufacturing the pieces and accessories, such as fabrics and trimmings; 4) Combination: repetitions concerning the manner to mix and combine pieces and accessories, even if they are updated or replaced by other symbols when composing the outfit; 5) Gesture: common ways of using the pieces and behaving in the moment the images are captured.

The study covered five Brazilian bloggers active on the internet in 2017. They are: 1) Alana Ruas. With a degree in Fashion Design, she has the blog http://alanaruas.com/blog/ focused on women's fashion, "look of the day" and professional experiences; 2) Bruna Aureliano. Design student, she keeps the blog http://www.bonecadeplatina.com/ targeting women's fashion, lifestyle, music, books, home design, among others. 3) Fabiano Gomes. With degrees in Library Science and Information Science, he is a Fashion Design student and keeps the blog http://www.ocarafashion.com/, dedicated to men's fashion and beauty tips; 4) Juliana Barros. Design student, developer of the blog http://ourbag.com.br/ which covers music, design and travel; 5) Leonardo Leal, also known as "Coloral." With a degree in Marketing, he keeps the blog http://www.machomoda.com.br/ which covers men's fashion and beauty.

3 RESEARCH FINDINGS

Although slightly diverging opinions were collected, several convergent elements were detected and became the focus of analysis. Taking into account what Mota (2013) states in regards to minimalist consumption as a philosophy of life that looks for the "identification of the essential and disposal of the remainder," avoiding the excess and the superfluous, we may assert that there are two cultural values which serve as a guideline for this behavior nowadays:

3.1 *Self-determination*

As suggested by Tamayo and Schwartz (1993) it's reasonable to assume that the cultural value of self-determination – which aims at independent thinking, actions, and options – is one of the values guiding minimalist consumption in contemporary life. The following statements illustrate the point: Blog *O Cara Fashion* – "instead of buying, I invest in one piece that I will really use and that will last a long time;" Blog *Macho Moda* – "I like to shop and I try to balance my needs and wishes;" Blog *OurBag* – "to consume only what is necessary;" Blogger Alana Ruas – "it's the search for things that are really basic and necessary in your life;" Blog *Boneca de Platina* – "I believe that being surrounded by objects with symbolic meanings is important, it's part of building your identity."

The accounts make evident that despite the temptations inherent to fashion, oriented by fast consumption with disregard to the natural wear and tear resulting from prolonged use of products, bloggers that self-identify as minimalist consumers argue that it's possible to take pleasure consuming less and valuing the extensive use of their goods, including fashion ones. Symbolic function of minimalist consumption is also noticeable, given that the option for this kind of conscious consumption contributes to an identity built with symbolic values meaningful in the present days. By being self-determined and rationalizing their purchase decisions, these digital influencers serve as examples for their audience, guiding consumption behaviors that propose a balance between "desire and needs."

3.2 Philanthropy

Still according to Tamayo e Schwartz (1993), we may identify that the second cultural value guiding today's minimalist fashion consumption is philanthropy. The goal related to this value is the understanding and acceptance of others, as well as a concern for overall well-being. Since these individuals are digital influences, with thousands of followers in the internet, bloggers care for the collective, denoting a degree of altruism, in contrast to the individualism disseminated as a core cultural value in the hyper-consumption society (Lipovetsky, 2007).

The interviewees' philanthropic profile may be illustrated by the following statements: Blog *Macho Moda* – "I believe that the consciousness one gains when thinking about consumption in this way is very cool and important, because we may apply it in other quotidian things;" Blog *Boneca de Platina* – "This means to consume things that really add something;" Blog *Our Bag* – "I think that the main value is consciousness. The moment someone stops for a while to analyze a certain product, how it was produced, if it will be useful, its value, this is already a major accomplishment of minimalism;" Blogger Alana Ruas – "I tend to value handmade products, artisanal ones, offered by small business." The testimonies make apparent that minimalist consumption pertains to a broader, and more conscious, consumption movement, also articulated with solidary trade and political consumerism, raising consumption to a level of political decision making that goes far beyond the pleasure of shopping, possessing, using, and showing off the acquired objects (Domingues, 2013).

4 IMAGE ANALYSIS

According to Miranda and Rocha (2016), clothing signs make it possible for the observer to deduce relevant aspects regarding social class and cultural values, among other aspects that build an identity. The authors suggest that when analyzing groups and subcultures by their clothing style it's important to keep in mind underlying aesthetic values of the subjects under observation. Those are values shaped by historical, economic, social, and cultural facts, and they have a clear (or veiled) reflex on the way a certain group or subculture dresses.

The analysis of the images revealed how this discourse is materialized in the minimalist look, leading us to identify two types of minimalist styles within the collected data: the minimalist modern look and the minimalist classic look.

4.1 Category 1 – minimalist modern look

The category of minimalist modern look was frequently observed within the examined images, thus becoming relevant to the study. The illustration below exemplifies this category and shows how the image analysis was conducted following the presuppositions established by Maciel and Miranda (2009):

As depicted in the following table, it's relevant to notice that the Minimalist Modern Look category aims at transmitting a modern, relaxed and young dressing style. This translates into the selection of fabrics that last longer and have neutral and solid colors, without prints or exaggeration. The chosen materials reveal the conception of a "minimalist modern self" through distinct cuts, conveying a contemporary, classic (not flashy) and elegant style.

Aside from discretion at color and texture selection, the gestures also emphasize consumer's minimalism, as illustrated in Figure 1, where the young man displays a discrete smile, not an exaggerated or effusive one.

Figure 1. Minimalist modern look.

Table 1. Image analysis of minimalist modern look.

	Denotation	Connotation
Shape	Slim pants, long and asymmetrical shirt, and sneakers.	Relaxed and young. The long shirt implies innovation and transgression.
Color	Washed jeans, dark blue shirt with light blue striped bottom, white sneakers.	Modern, relaxed and young.
Material	Jeans and/or mesh.	Practical and young.
Combination	Straight shirt and asymmetrical cut, slightly tight at the bottom, neutral colors, durable fabrics, wristwatch and bracelets.	Innovation and modernity. The long shirt points to gender-neutral fashion.
Gesture	Bent arms, one hand holding the other, one leg crossed with the tip of the foot supporting the body, leg, head facing ahead, and a half-smile.	Informality.

4.2 Category 2 – minimalist classic look

Minimalist classic look was also frequently observed among the interviewees. As we may observe in Picture 2, these consumers also favor a relaxed, classic (not flashy) and a clean aesthetic look without exaggeration or pretension. Their dressing style dialogues with classic fashion combinations, such as black and white.

Minimalist classic look is associated with consumers that favor pieces, accessories and timeless gestures, vastly established within the fashion world, such as the combination of black wide-legged short pantaloons and white shirt, along with sandals and retro style sunglasses, as seen in Figure 2.

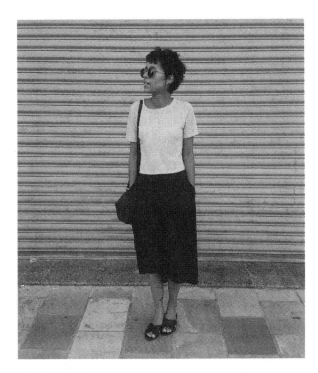

Figure 2. Minimalist classic look.

Table 2. Image analysis of minimalist classic look.

	Denotation	Connotation
Shape	Wide-legged short pantaloons, shirt, and flat sandals.	Simple, young. The shapes point to comfort and naivety.
Color	Black pantaloons, sandals and bag; white shirt.	Relaxed, classic (not flashy) and young.
Material	Twill and mesh.	Practical and young.
Combination	Slightly tight shirt, wide-legged short pantaloons, neutral colors, durable fabrics, sandals, retro sunglasses, and shoulder bag resting on the side.	The sandals and the sunglasses in combination with the short pantaloons and the shirt display a sense of style.
Gesture	Both hands in the pockets, head inclined to the right, feet firmly set on the ground.	Introspection and observation.

It's worth noticing that aside from using classic fashion elements, the minimalist classic consumers also tend to adopt firm and assertive gestures. The clothing pieces, the accessories and the combination are classic precisely because they have been widely accepted for a long time within mainstream fashion. Consumers that adopt the style rest assure of their choices as they are already accepted and approved.

5 FINAL CONSIDERATIONS

The values of self-determination and philanthropy are the cultural values guiding minimalist consumption, which is part of broader context of increasing consciousness regarding consumption these days. In this context, consumption becomes a political gesture and an exercise of citizenship, as there is a concern about how the clothing has been produced, which materials are used and how they are obtained, and what impact they may have on the society and the environment. These consumers often make the choice for pieces that last longer and that are in accordance to current legal and ethical aspects.

This study, aside from examining discourses, aimed at identifying how minimalist fashion bloggers identity is materialized, taking into account Mota's (2013) suggestion that the blogosphere is today's main vehicle for the expansion of minimalism. With an exponential number of fashion blogs widely accessible over the internet, these blogs are accessed by several types of consumer, with distinct age, social class and behavior profile. As a consequence, it's reasonable to assume that by examining the bloggers' looks we may identify what is disseminated and consumed as a minimalist identity. The analysis indicates that the minimalist look is generally composed by pieces without print or that use timeless patterns, such as stripes and polka dots. They could have a simple or more elaborated cut, though always conveying discretion and lightness in their composition. The same applies to other clothing elements and even the gestures of these individuals, therefore pointing to a consistency between the discourse and the practice of minimalist fashion.

Miranda (2008) highlights that "real objects" go beyond their utilitarian meanings, carrying symbolic meanings as well. In this sense, in regards to minimalist consumption the symbolic function becomes a decisive aspect in the choices made by these consumers. Minimalist consumption seems to oppose pretentious consumption as proposed by Veblen (1899). Nevertheless, a closer look, taking into account contemporary western cultural values and the spirit of the age, it seems reasonable to consider that the minimalist consumption itself is perhaps something to show off nowadays. In the 21st century consumerism society, when everything is widely accessible and economy is global, possessions don't bring status to certain groups as it did in the past decades. In the post-industrial world where the production of non-material goods intensifies, minimalist consumption seems to be a sign that carries great value to exhibit.

REFERENCES

Barthes, R., 1979. *Sistema da moda*. São Paulo: Editora da Universidade de São Paulo.
Bauer, M. W.; Gaskell, G., 2008. *Pesquisa qualitativa com texto, imagem e som: um manual prático*. Rio de Janeiro: Ed. Vozes.
Buckingham, W., 2011. *O livro da filosofia*. São Paulo: Ed. Globo.
Domingues, I., 2013. *Terrorismo de marca. Publicidade, discurso e consumerismo político na rede*. Rio de Janeiro: Ed. Confraria do Vento.
Lipovetsky, G., 2017. *A felicidade paradoxal. Ensaio sobre a sociedade do hiperconsumo*. São Paulo: Companhia das Letras.
Maciel, E. J. C.; Miranda, A. P. de., 2009. DNA da imagem de moda. In: *V Colóquio Nacional De Moda, Recife. Anais do V Colóquio Nacional de Moda*, 2009.
Mccracken, G., 2003. *Cultura & Consumo—Novas Abordagens*, Rio de Janeiro: Ed. Mauad.
Mccracken, G., 2005. *Culture and Consumption II. Markets, Meaning, and Brand Management*. Indiana University Press.
Miranda, A. P. de., 2008. *Consumo de moda: a relação pessoa-objeto*. São Paulo: Estação das Letras e Cores.
Miranda; A. P. de, Rocha; A. R., 2016. Consumption and Constraint: Muslim Women, Fashion And Identity, *Cimode Proceedings*, Buenos Aires.
Mota, D., 2013 disponível em http://www.dn.pt/revistas/nm/interior/gente-feliz-com-menos-3275609.html acessed 10 june de 2016.
Solomon, M. R., 2002 *O comportamento do consumidor: comprando, possuindo e sendo* / Michael R. Solomon; trad. Lene Belon Ribeiro. – 5.ed. – Porto Alegre: Bookman.
Tamayo, A. & Schwartz, S. H., 1993. Estrutura motivacional dos valores humanos. *Psic.: Teor. e Pesq.*, Brasília, Vol 9, N° 2, pp. 329–348.
Veblen, T., 1899. *The theory of the leisure class*. New York: MacMillan.

Emerging platforms for fashion design entrepreneurs: An assessment

C.E. Fernandes, L. Ribeiro & M.S. Silva
University of Beira Interior, Covilhã, Portugal
FibEnTech Research Unit, Covilhã, Portugal

M. Pereira
IADE – Universidade Europeia, Lisbon, Portugal
UNIDCOM Research Unit, Lisbon, Portugal

M.J. Madeira
University of Beira Interior, Covilhã, Portugal
NECE Research Unit, Covilhã, Portugal

ABSTRACT: In Portugal, fashion designers are thriving to succeed in an ever-changing industry, evolving into a new fashion design entrepreneur generation, filled with business opportunities in the post-crisis market. However, the lack of information, networking and business-creation knowledge can be a hard wall to come across, as fashion designers are not prepared in that sense by higher-education courses. The purpose of this study is to determine which solutions are being created and who are the people behind them. Using a qualitative methodology, interviews were held, directed to creators of solutions and platforms created in Portugal and in the UK, as well as experts of the fashion industry. Further analyzed through the qualitative analysis software Nvivo, the results obtained show the main difficulties felt by fashion design entrepreneurs, as well as recommendations for the creation of more platforms of the kind, in order to create sustainable and ethical fashion on smaller scales.

1 INTRODUCTION

The Portuguese government has made it a priority to face youth employment, following the directives of the European Commission. The European organization has approved in the last few years, a set of financial incentives aiming to correct this problem (Martins, 2015). Education institutions are also seeking to stimulate creation and innovation, as well as raising entrepreneurial thinking among their students, as it has been considered a goal to reach in the next years, to create stronger relations between enterprises and higher education institutions (Cruchinho, 2009, p. 246).

Entrepreneurship scholarships and other incentives have been designed around the country for students to concretize their projects in many fields (Souza, 2010, p. 82). The creation of such incentives by Universities is directly linked to the vocation of design by teaching students to use design thinking and methodologies to be able to practice what will be their profession, but also to create new forms of investigation and new resources that could become tools (Nielsen & Stovang, 2013; Evans, 2011). However, the issue with the creation of new businesses in the fashion industry remains, as being a fashion design entrepreneur can be very overwhelming, considering this fast-paced industry and its implications in terms of competitiveness.

2 LITTERATURE REVIEW

In a social and economic paradigm where the traditional lifetime job has become utopic, the creation of incentives and ventures that aim to augment youth entrepreneurship has turned out to be vital. The nature of self-employment among young people can become a solution against the problem of youth unemployment (Schoof, 2006), as unemployment itself is an opportunity to become an entrepreneur (Davidsson, 1992). Developed countries like the United States of America are the living proof that entrepreneurship is an alluring solution for growth and evolution (Ozaralli & Rivenburgh, 2016). Many are the influences and reasons that can lead young people to become entrepreneurs, as experiences and upbringing have been proven to be a massive part of entrepreneurial behaviors (Newton & Shreeve, 2002). Any fashion design student with entrepreneurial intentions should be prepared for the reality of being a fashion designer-entrepreneur, as the amount of work required in such profession goes beyond the creation of garments since the time needed to take care of the business is crucial (Sousa, 2015). However, fashion design entrepreneurs have a hard time to mix design thinking and entrepreneurial thinking, Kurz states the following (2010, p. 52): "They need to position themselves as entrepreneurs in order to present themselves to investors and banks and to facilitate the access to external finance. Their performance in front of investors or bank employees is crucial for accessing financial support. Therefore, a relevant and well thought out business plan is necessary. The business plan needs to be an integral part in the education of a fashion designer, who wants to start a company. This should be taught at universities, at incubators and in post-academic programs. Bank loans should be easier to achieve for creative people and adjusted to an adequate scale of finance." It is then, imperative for fashion design entrepreneurs to be capable of making the product but also comprehend the market to be capable of selling it, for that matter, understanding the consumer is crucial. Furthermore, young fashion design entrepreneurs must grasp the importance to create value, as it will reflect the real image of the brand and its identity, a quality that is crucial for the creation of the business plan (Testa & Frascheri, 2015).

Platforms and other offline solutions can help to inform and provide tools for aspiring fashion design entrepreneurs, incubators have evolved a lot since their first appearance in Portugal, in 1987 (Shev, 2015). However, as different types and sizes are available, this study will only focus on a sample available of the creative industries incubators in Portugal. According to Junior & Andrade (2017), incubators have become such a phenomenon in Portugal and abroad, that it is crucial to observe some of the existing models of incubators dedicated to creative industries in Portugal, as they aim to support creativity and innovation in the community in which they are inserted. Moreover, the research on online platforms aiming to guide fashion design entrepreneurs is also approached in this analysis.

3 METHODOLOGY

After an extended literature review on fashion design entrepreneurship and the different platforms available in Portugal and abroad, the use of qualitative data is explored for this study. Interviews are the main tool for this approach, directed to creators of solutions and platforms aiming to help fashion design entrepreneurs (Coutinho, 2015). On the other hand, another group of professionals was also interviewed, to explore their take on fashion design entrepreneurship and the solutions created to help their new ventures.

Interviews were held via video chatting devices and recorded to be further analyzed with QSR Nvivo (Qualitative analysis software). Through the use of a codification approach, the software enables the possibility to define patterns in the interviews.

Open questions were made on these professionals' visions of the industry, their past experiences and expectations for the future, as well as their goals and motivations in the creation of solutions for fashion design entrepreneurs.

4 EXPLORATION OF THE SOLUTIONS

Starting with a precise research on existing platforms, which aim to help fashion design entrepreneurs in the creation of a new venture in the industry, the research contemplated all types of platforms, offline and online. It is important to consider that some platforms are inserted in fashion design higher-education programs and were not considered in this particular study, as it is the case for the Fashion Innovation Agency, promoted by the London College of Fashion, or scholarships provided by these institutions to reward innovative-driven projects, such as the "Future Fashion Designer Scholarship". This incentive was created in 2014 by Bruno Pieters, Belgium-based designer and founder of "Honest By[1], a fashion brand where transparency is key. "Future Fashion Designer Scholarship" follows the same example, as the website explains that FFDS main goal is "to offer financial support to exceptional students who want to develop their collection in a sustainable, vegan and transparent way" (FFDS, 2017)[2].

Nevertheless, this study is based on the creation of solutions for fashion design entrepreneurs who are already working in the industry or have recently graduated. One of the main issues felt by these fashion design entrepreneurs is the difficulty to do everything, as their basic formation does not comprehend skills such as business creation, administration, and other abilities that could come in hand in the creation of a new venture. Another preoccupation resides in the fact that existing solutions for young entrepreneurs are sometimes not adapted for creative industries and are more focused in backing projects related to health or engineering. Moreover, the application forms to such entrepreneurship incentives are sometimes confusing for fashion designers, as they are not familiar with business plans and other information that are mandatory to participate.

Starting with these ideas, the search for solutions created for fashion design entrepreneurs begins with two well-known platforms in Portugal, worth mentioning; *Portugal Fashion* and *Moda Lisboa*. However, as this study seeks to assess recently created solutions, these were not considered. Now observing the four solutions chosen in this study, it is important to mention that two of these entities are located in Portugal, as for the others, both are online platform and their headquarters are located in the United Kingdom.

4.1 *AwayToMars*

First introduced in March of 2015 during Moda Lisboa's 44th edition, AwayToMars is a new concept of fashion, created by Alfredo Oróbio and Carlo Valentini, this platform allows anyone to pitch an idea or inspiration on the website, idea that can be picked up by the AwayToMars community in order to be produced and commercialized. AwayToMars was conceived to disrupt the current model of fashion, create balance and fairness among creation and production, as the creators believe that anyone can have a good idea but not everybody has the tools to turn them into reality (Cardoso, 2015). Oróbio has also participated in the interview process; his answers were used in the results of this study.

4.2 *Mastered*

"Mastered" is an online accelerator program for creators of the fashion industry, as many different angles are contemplated, such as fashion stylist, accessories, womenswear designer, menswear designer, etc. Launched at the beginning of 2014 by Perri Lewis, Adil Abrar and Cheryl Adamson (Bobila, 2015), this paid online accelerator allows support and guidance for creative careers, with the help of coaches and industry experts who give feedback to the trainee's works. The 10-month program is made for the masses but customized for each trainee, as it promotes self-empowerment and learning. Mastered's logic resides in the ever-changing industry of fashion, preparing these professionals for an improvement in themselves and their careers.

4.3 *Mouraria Creative Hub*

Situated in Lisbon, the "Mouraria Creative Hub" is the first creative industry specialized hub of the capital. Created by the city and for the community in 2015. Projects involving

---

1. More information about "Honest By" on the website: http://www.honestby.com/
2. Quote retrieved from the FFDS website, available at: http://www.futurefashiondesignerscholarship.com/

design, media, fashion, jewelry among others are being incubated at Mouraria. The hub can welcome ventures for 24 months. In 2015, the incubator's jury composed of collaborators of the Lisbon city hall and Mouraria officially accepted nine projects out of the original 39 candidates. Of these nine projects, five of them involved fashion design.

4.4 Minty Square

In the text, place the authors' last names (without initials) and the date of publication in "MintySquare" is an interesting case of Portuguese brand promotion. Created by Ana Cravo and João Figueiredo in 2014, the online platform has grown into a reference for emerging fashion designers. As explained in their interview, Cravo and Figueiredo aim to promote "Made in Portugal" brands and use this strategy to grow internationally. However, as their goal was to be very selective on the brands they promoted, it can be observed on the website that this continues to be the case, as Cravo & Figueiredo had previously stated their will to only promote "emerging talents and renowned fashion designers", they also announced their restrictive list of acceptances, as they admitted that they would only promote young fashion designers "if they had already shown their work at least on one of the main Portuguese catwalks".

After this brief approach on each solution, a table was made to complement these elements, and present the people interviewed from each entity. The table is separated by category, since two experts of the Portuguese fashion industry were also interviewed, as previously explained (see Table 1).

In Table 1, it can be observed that in the first category, called "solvers", the four solutions approached previously are presented, as each is represented by someone, interviewed for this study. On the other hand, the second category was used to complement the results obtained from the first category, as both Paulo Vaz and Manuel Serrão can be considered "experts" of the Portuguese fashion industry, due to their long and successful experiences in the field. Results obtained in the interviews were processed in the qualitative analysis software Nvivo, in order to find repetitions, coherence, and cross-references in the various speeches analysed.

4.5 Results and discussion

For MINTY Square's Ana Cravo: "The lack of support for young fashion designers inspired our platform, particularly the lack of existence of a physical space for the to expose their work. Our online platform grants visibility for their creations to the public. Being attentive to the issues for emerging designers, this opportunity came up to value the quality of designs made in Portugal".

As for AwayToMars' Alfredo Oróbio, the main difficulty to create his project was the business plan: "being able to show a viable business model, show that it's possible to create fashion goods with any technical knowledge, it is a real barrier in the industry", a view verified by authors like Hartley & Montgomery, in the sense that: "Fashion Design is an evolutionary

Table 1. Interviewed people per category.

Categories of the interviewed people:	Name	Role	Entity Represented	Country
Fashion design "Solvers"	Alfredo Oróbio	Co-founder	AwayToMars	UK
	Samantha Southern	Expert producer	Mastered Accelerator	UK
	Bernardo Gaeiras	Director	Mouraria Creative Hub	Portugal
	Ana Figueiredo	Co-founder	MintySquare	Portugal
Fashion design "Experts"	Paulo Vaz	Director	Associação Têxtil e Vestuário de Portugal (ATP)	Portugal
	Manuel Serrão	Director	Associação Seletiva de Moda (ASM)	Portugal

process, involving amplification or scaling up of the successful designs to the next stage of the process" (2009, p. 63), as also explained by Ostwalder et al. in their "Value Proposition Design" (2012). This view was also shared by Mouraria Creative Hub and FabLab Lisbon manager Bernardo Gaeiras, who explained that "the business plan, how innovative the idea is, market potential and profiles of the team members" are fundamental points to consider when it comes to choosing a new project for the hub, as this opinion is also shared by authors such as Eliza Kurz (2010), Testa & Frascheri (2015), however, authors Hodges et al. consider it important but also highlight the importance of traits "such as empathy and a global outlook are as important to entrepreneurial competency within the global apparel industry as developing a business plan, and therefore should be addressed in the classroom" (2016, p. 80).

Gaieiras also stated that the "sustainable preoccupation" of the project is essential, considering socially and environmentally ethical practices as well as the co-existence with the environment and community in which the project will insert itself, an opinion already referred by participants of the previous group. Mastered's expert producer, Samantha Southern, also identified that higher education programs "(...) were falling short when it came to equipping design students with the relevant entrepreneurial skills to successfully set up their brand or business, particularly in such a fast-paced industry as the fashion industry", also highlighted by Tedeschi et al. (2015), in their study on the importance to implement teaching methods to encourage entrepreneurial thinking.

When asked about the creation of these new platforms, experts Manuel Serrão and Paulo Vaz had interesting takes on the subject. ASM's CEO, Manuel Serrão, referring that "a project that presents sustainability and innovation" are elements of distinction among other fashion design propositions, also highlighted by author Jenifer Craik as an essential issue among consumption and production habits (Craik, 2009), also related to the new generation of consumers exposed and their perceptions of sustainability by Hill & Lee (2012). The entrepreneur also highlighted the recently created "Fashion and Technology Cluster" as a potential solution for fashion design entrepreneurs.

For ATP's Paulo Vaz, the creation of a platform able to help fashion design entrepreneurs is: "very important, if it does not exist yet, it should be created in the close future, it is definitely something that should be created". As the visions of these professionals clearly reflect the responses obtained through the interview process with the solvers group, the data obtained also mirrors the opinions of several authors referred in this study.

5 CONCLUSIONS

Interviews made it possible to comprehend more about fashion design entrepreneurship, the need to create more small businesses in the industry, as well as the definite need for sustainable and ethical solutions within the industry. As the goal was to observe existing solutions created to help fashion design entrepreneurs in need of a push, this study has also presented the people behind these different platforms, as well as the opinions and views of experts of the Portuguese fashion industry.

The main difficulty felt by fashion design entrepreneurs is to produce conscious items on a reduced scale, as manufacturing business have a hard time producing that way. Trust issues and lack of communication were also referred as a conflicting topic between young fashion design entrepreneurs and manufacturers.

Results of this study highlight the need for further exploration of this multidisciplinary field, as many variables must be considered, such as the actual fashion design education process, and very few studies focus on fashion design entrepreneurship.

Moreover, the creation of other solutions to help, guide and inform fashion design entrepreneurs are appreciated, as fashion design courses are just now starting to change curricula to follow this new trend, however, this change is still very shy and late in comparison with the paradigm of fashion design entrepreneurship.

It is crucial to note that even if not explored fully in this study, the creation of incentives for fashion design students inside the institutions is a priority, as it provides a real motivation for students to come forward with innovative and sustainable new projects for the fashion industry and grants them a first step in the industry as fashion design entrepreneurs. This research observed the actual need for the creation of these specific platforms, either online or offline, as they are more than ever, needed by fashion design entrepreneurs starting a project.

Moreover, this study has established the need for incentives and platforms specially directed to fashion design ventures, as general associations for entrepreneurs cannot always help the specificities experienced in the fashion industry (Wenting, 2008). On the other side, already-existing physical platforms like *Moda Lisboa* or *Portugal Fashion* can only help emerging designers to project their collections for a limited time and cannot help fashion design entrepreneurs on specific matters like IP, funding, administrative processes, commercialization, marketing, etc., a reality highly evidenced by the qualitative data collected during the interviews.

REFERENCES

Bobila, M., 2015. *Mastered offers online courses taught by Tim Blanks*, Nick Knight, Katie Hillier and more. Fashionista.
Cardoso, J.A., 2015. *Um Site Que Nos Ajuda a Ser Designers De Moda*. Público.
Coutinho, C. P. 2015. *Metodologia de investigação em ciências sociais humanas: teoria e prática*. Coimbra: (ed.) Almedina.
Coutinho, C. P., & Chaves, J. H. (2002). *O estudo de caso na investigação em Tecnologia Educativa em Portugal*. Revista Portuguesa de Educação, 15(1), 221–243.
Cruchinho, A.I.B.G., 2009. Alexandra Isabel Cruchinho Barreiros Gomes Design - *A construção contínua de competências*. Alexandra Isabel Cruchinho Barreiros Gomes contínua de competências. Universidade do Minho.
Davidsson, P., 1992. Researching Entrepreneurship Conceptualization and Design 2nd ed. J. Zoltan & D. B. Audretsch, (eds.), Springer.
Evans, M., 2011. *Case Studies in the Evaluation and Evolution of Tools to Support Design Education*. In Researching Design Education 1st International Symposium for Design Education Researchers CUMULUS Association//DRS SIG on Design Pedagogy. Paris, France, pp. 168–186.
Hartley, J. & Montgomery, L., 2009. *Fashion as consumer entrepreneurship: Emergent risk culture, social network markets, and the launch of Vogue in China*. Chinese Journal of Communication, 2(1), pp.61–76.
Junior, T.B. & Andrade, M.M.G. de, 2017. *A creative incubator model to promote innovations' solutions in e-Gov*. In 17.a Conferência da Associação Portuguesa de Sistemas de Informação. Guimarães, pp. 207–217.
Kurz, E., 2010. *Analysis on fashion design entrepreneurship: Challenges and supporting models*. University of Boras.
Martins, R., 2015. *Portugal vai receber 48 Milhões de Euros para Combater o Desemprego Jovem*. Público.
Newton, J. & Shreeve, D.G., 2002. *An Investigation Into The Relationship Between The Characteristics And Life Experiences Of Entrepreneurs*. Journal of Research in Marketing & Entrepreneurship, 4(1), pp.16–36.
Nielsen, S.L. & Stovang, P., 2013. *DesUni: university entrepreneurship education through design thinking*. Education + Training, 57(8), pp.977–991.
Ostwalder, A. et al., 2014. *Value Proposition design*, Hoboken, New Jersey: Wiley & Sons, John.
Ozaralli, N. & Rivenburgh, N.K., 2016. *Entrepreneurial intention: antecedents to entrepreneurial behavior in the USA and Turkey*. Journal of Global Entrepreneurship Research, 6(3).
Schoof, U., 2006. *Stimulating Youth Entrepreneurship: Barriers and incentives to enterprise start-ups by young people*.
Shev, I., 2015. *Incubadoras de Empresas em Portugal*. PME.pt.
Sousa, G., 2015. *Empreendedorismo e(m) Design de Moda: uma visão estratégica para o Ensino Superior*.
Testa, S. & Frascheri, S., 2015. *Learning by failing: What we can learn from un-successful entrepreneurship education*. International Journal of Management Education, 13(1), pp.11–22.
Wenting, R., 2008. Spinoff dynamics and the spatial formation of the fashion design industry, 1858–2005. Journal of Economic Geography, 8(5), pp.593–614.

The use of online advertisement as a TXM branding methodology's tool: FARM's case study

L.W. Ribeiro
Blumenau's Regional University, Blumenau, Santa Catarina, Brazil

L.S. Gomez
Santa Catarina's Federal University, Florianópolis, Santa Catarina, Brazil

ABSTRACT: The evolution of branding processes acts to closen up consumers and brands even more. Online advertising stands out as a tool capable of creating and managing a digital atmosphere, connecting the "DNA" of a brand to its consumers. For this tool to fulfill its function, consumers and stakeholders need to be involved in this process. Advertising can bring fashion brands and consumers together, creating a digital atmosphere. This work aims to present the importance of online advertising as a TXM Branding methodology's tool. In order for this objective to be met, the methodology developed by the Organizational Genesis Orientation Laboratory—LOGO—of Santa Catarina's Federal University—UFSC—will be presented. Afterwards, the performance of online advertising in brand management will be contextualized, acting mainly in the methodology's last stage. The FARM brand uses this digital atmosphere tool to put consumers in touch with their essence.

1 INTRODUCTION

Presented to the public for the first time in 1997, in the city of Rio de Janeiro, FARM was born with an authentic "DNA" that charmed its consumers (Tamamar, 2015). With its well-structured branding process, the brand differentiates itself from the competition for its intangible attributes and for the valorization of Rio de Janeiro city, for Rogar (2015) what made FARM "leverage" was its branding process. To be a well-designed brand and to have the desirable visualization, it needs to be structured, not just corresponding to a visual signature or symbol, but rather containing an essence intrinsically linked to it (Ellwood, 2004). This process can be accomplished through several methodologies, among them the *TXM Branding*, developed by the Organizational Genesis Guidance Laboratory—LOGO—of Santa Catarina's Federal University—UFSC—in Brazil. In addition to building a brand and aligning its essence with the needs of the market, it becomes important to put this essence in contact with the consumer. Advertising can be used as a tool to put the consumer in touch with the essence and purpose of the brand, for Jones (2005, p. 12) "*É a publicidade que transforma um produto com benefícios apenas funcionais em uma marca que ofereça ao consumidor recompensas psicológicas além daquelas funcionais*"[1]. By communicating this purpose to consumers in the digital environment, brands don't need to only communicate about their products, but also enhance their essence, in order to connect consumers and admirers of it's cause. This article aims to present the importance of online advertising as a *TXM Branding* methodology's tool, this tool can create a digital atmosphere where the consumer is able to recognize the essence of a fashion brand. In this sense, *TXM Branding*'s methodology, through its last step *Manage*, can be used in online advertising to create a digital atmosphere.

---

1. Author's free translation to: "It is the advertising that transforms a product with only functional benefits into a brand that offers the consumer psychological rewards beyond those functional."

2 BRANDING

Derived from the Nordic Brander which means to burn/mark, Sampaio (2013, p. 191) affirms that this term itself *"foi cunhada para designar a ação de marcar o gado nas fazendas e pradarias, quando os fazendeiros marcavam com suas iniciais os animais que haviam nascido em suas propriedades"*[2]. For Gomez (2012, p. 13) *"branding, ou gestão de marcas, é mais do que apenas se certificar de que os clientes reconheçam o logotipo ou o nome de um produto; significa criar uma associação emocional entre o cliente e o produto, serviço ou empresa"*[3]. The metaphor of "DNA", as Gomez (2012, p. 02) continues to explore, comes to exemplify the gathering of some *"conceitos fundamentais que uma marca precisa para transmitir aos consumidores o que ela é com autenticidade e posicioná-la adequadamente na mente do consumidor obtendo sua fidelidade"*[4]. The quest for the essence in an organization is the initial step towards building a brand and its entire brand management process. For Elwood (2004, p. 174) *"o DNA da marca é a essência do que a marca significa e deve estar presente em todas as formas de expressão da marca"*[5] and for Kotler (2010, p. 33) *"Para estabelecer conexão com os humanos, as marcas precisam desenvolver um "DNA" autêntico, o núcleo de sua verdadeira diferenciação"*[6]. This connection in the consumer's mind can be made through the brand's purpose, uniting its "DNA" with its line of communication. Batey (2008, p. 189) reports that *"o significado da marca é estabelecido a partir de como ela é percebida pelo público em nível consciente e de como ela age dentro dele no nível semi ou subconsciente"*[7], this process of establishing relationships between the brand and the consumer can be done through advertising, for Jones (2005, p. 32). *"Marca é um produto que oferece benefícios funcionais além dos valores agregados que os consumidores valorizam o bastante para adquiri-lo"*[8]. It is now known that consumers stopped buying a product and started to seek a concept, a lifestyle allied to the brand. For Lipovetsky and Roux (2005), the brand came to prevail over the product and for Lipovetsky (2007, p. 40) *"o que se vende já não é um produto, mas uma visão, um "conceito", um estilo de vida associado à marca"*[9]. The brand offers something beyond functional purpose. Campanholo and Bevilacquar (2011, p. 09) add that *"A marca pode possuir mais valor do que o próprio produto em muitas situações"*[10]. For Gomez and Santos (2010, p. 51) more than consumers, *"passam a existir "admiradores" das marcas, pessoas que estão apaixonadas e desejam demonstrar que usam aquela marca e se expressam por meio daquele "logo"*[11]. Kotler (2017) notes that in the digital universe, consumers are acting as lawyers for the brand. This brand power makes consumers more loyal and they maximize the return on their investment, this return can be in the form of pleasure, appearance or social currency (Gerzema, 2009). Thus, it is understood

---

2. Author's free translation to: "was coined to designate the action of marking cattle on farms and prairies, when farmers marked with their initials the animals that had been born on their estates".
3. Author's free translation to: "Branding, or brand management, is more than just making sure that customers recognize the logo or a product's name; it means creating an emotional association between the customer and the product, service or company".
4. Author's free translation to: "fundamental concepts that a brand needs to convey to consumers what it is with authenticity and position it properly in the consumer's mind getting their loyalty".
5. Author's free translation to: "brand's DNA is the essence of what the brand means and must be present in all forms of brand expression".
6. Author's free translation to: "To connect with humans, brands need to develop an authentic 'DNA', the core of their true differentiation".
7. Author's free translation to: "The brand's meaning is established from how it is perceived by the public on a conscious level and how it acts within it at a semi or subconscious level".
8. Author's free translation to: "Brand is a product that offers functional benefits beyond the added value that consumers value enough to acquire".
9. Author's free translation to: "what is sold is no longer a product but a vision, a 'concept', an associated lifestyle to the brand".
10. Author's free translation to: "The brand may have more value than the product itself in many situations".
11. Author's free translation to: "there are 'admirers' of brands, people who are in love and want to demonstrate that they use that brand and express themselves through that 'logo'".

that a well-constructed and managed brand can achieve a much larger concept than representing a product, since that essence reaches the consumer. For a brand to reach its potential for awareness and fixation in the consumer's mind it is important that this identity is well developed and managed. For this process we can highlight the *TXM* Branding methodology that is developed by the LOGO of UFSC.

3 TXM BRANDING

With the increasing demand to connect brands and consumers, it is becoming increasingly important to apply a brand management project, to Aaker (2015, p. 11) "*um dos objetivos da construção de marca é ampliar o tamanho e intensidade de cada segmento de fidelidade, tornando a base do relacionamento com o cliente mais consistente no longo prazo*"[12]. A brand must have its concept of "DNA" well understood and contextualized by all involved. For Ribeiro and Gomez (2016, p. 56) "*sabe-se que uma marca deve ser gerida e acompanhada desde o seu início para que o seu real "DNA" entre em contato com o consumidor*"[13]. For this essence to be extracted from the company and put to test with it's public some studies must be done. The methodology of *TXM Branding* has been constructed based on applied academic and market studies. This brand building process seeks to identify the "DNA" of the company, generate real experiences and put those experiences in contact with consumers. For Braun et al. (2014, p. 5), "*esta metodologia tem por objetivo a criação e o desenvolvimento de identidades de marcas de forma co-criativa, ou seja, integrando além dos designers nesta construção, os representantes da empresa que dão suporte a referida marca*"[14]. For Stodieck and Gomez (2012) the brand's "DNA" is what will differentiate one brand from the other, just as in living beings the brand's "DNA" is unique, and serves as both an element of differentiation and as a tool for brand loyalty towards its consumer, for Jones (2005, p. 18) "*fabricantes fazem todo o possível para diferenciar suas marcas*"[15]. Looking for this differentiation the methodology was divided into 3 stages. The first step aims to meet the search for "DNA" and positioning the brand, the second step is to develop experiences and finally, the third step, puts these experiences in contact with the consumer.

The first step in the methodology, as presented on the Figure 1, is called *Think*. This step is responsible for thinking the brand and pursuing its real essence, its purpose and positioning. The *Think* stage is the initial phase of the methodology, for Braun et al. (2014, p. 42) "*nesta primeira etapa busca-se principalmente estabelecer as bases conceituais que direcionarão a construção da identidade sensorial (sua parte tangível) e as estratégias de atuação da marca no mercado*"[16]. The second stage is called *Experience* and is responsible for creating sensory experiences for the brand. These sensory experiences must be aligned with the positioning and "DNA" of the brand. Lindstrom (2007, p. 05) reports that, in the new *branding*, brands must create something that approaches a religious worship or faith for the brand. Lindstrom (2007, p. 03) also states that "*uma marca tem de se transformar em uma experiência sensorial que vai muito além do que vemos*"[17]. The *Experience* step of the *TXM Branding* methodology

---

12. Author's free translation to: "One of the goals of brand building is to increase the size and intensity of each loyalty segment, making the customer relationship base more consistent in the long run".
13. Author's free translation to: "it is known that a brand must be managed and monitored from its inception so that its real 'DNA' contacts the consumer".
14. Author's free translation to: "this methodology aims at the creation and development of brand identities in a co-creative way, that is, integrating in addition to the designers in this construction, representatives of the company that support the brand".
15. Author's free translation to: "manufacturers do everything possible to differentiate their brands".
16. Author's free translation to: "in this first stage, the main objective is to establish the conceptual bases that will guide the construction of the sensorial identity (its tangible part) and the strategies of the brand's performance in the market."
17. Author's free translation to: "a brand has to transform itself into a sensorial experience that goes far beyond what we see".

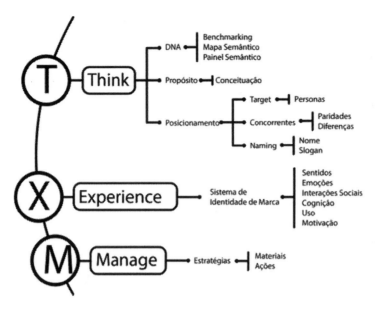

Figure 1. TXM branding methodology.

boils down to creating all the experiences that will be put in touch with consumers through the last step of the methodology, *Manage*.

4 MANAGE

The third stage of the TXM Branding methodology, the focus of this study, is responsible for managing the communication channels where the experiences will be put in contact with the consumers. In its literal translation we can define the word *Manage* as the intention to make a management, in this case, a brand. The branding process goes beyond just the "create" a brand barrier. *Branding* is brand management, it is a continuous process of maintaining the values identified and renewing them (Ribeiro and Gomez, 2016). The *Manage* stage appears to make a link between company, brand and consumer. You can make use of publicity plans and tools like advertising and commercials, visual merchandising, online advertising, among others. For Morais (2008, p. 108) "*As percepções que os consumidores têm de uma marca são uma síntese do conjunto de mensagens que recebem*"[18]. The goal is to put and manage the "DNA" of a brand in direct contact with its public in several channels always following a strategy. It becomes important in the last step of this methodology, the use of online advertising as a way to insert the consumer in contact with the experiences created from the "DNA". For Araújo and Rosas (2012, p. 2) "*As empresas passaram a incluir nas ações de marketing o desafio de posicionar suas marcas na mente dos consumidores, buscando fidelidade dos clientes e agregando valor ao seu produto.*"[19] Experiences can begin in web-sites, promotional materials, brand interaction chats, and even in points of sale, for Gomez (2012, p. 6). "*Do mesmo modo, recentemente, a gestão de marcas, ou branding, passou a atentar aos espaços onde os consumidores "experimentem" as marcas*"[20], physical or virtual spaces. The way these experiences

---

18. Author's free translation to: "The perceptions that the consumers have of a brand are a synthesis of the set of messages that they receive".
19. Author's free translation to: "Companies started to include in the marketing actions the challenge of positioning their brands in the minds of consumers, seeking customer loyalty and adding value to their product".
20. Author's free translation to: "Similarly, now a days brand management, or branding, began to pay attention to spaces where consumers 'try' brands".

are put to the consumer can encompass all fields of communication, acting both in online and offline environments, simultaneously and following the "DNA" of the brand. According to Lindstrom (2007, p. 14), *"estudos mostram que as marcas que incorporam som em suas páginas virtuais têm 76% a mais de chance de ter maior tráfego de internet'e que marcas com músicas que "se encaixem" em sua identidade de marca têm 96% mais chance de memorização imediata"*[21]. The Manage stage can use online advertising as a way of pinpointing the brand's purpose in the consumer's mind. According to Jones (2005, p. 81)" *A tarefa crucial para quem está envolvido no planejamento e na avaliação da publicidade é determinar precisamente como isto acontece – sem este conhecimento estamos condenados a nos confundirmos e a fazer um trabalho de adivinhação"*[22]. Following the same line of thinking Lipovetsky and Roux (2005, p. 2) say *"a publicidade amplia a aspiração ao bem-estar"*[23]. These experiences and this "DNA" must be built over time and always revalidated. For Carrilho (2012, p. 01) *"o novo consumidor quer ir além da compra, quer um produto que seja emocional. Um aroma que tenha significado, uma roupa que reflita a vida"*[24]. These new consumers seek more than the purchase, they want functional, emotional and spiritual satisfaction in the purchase (Kotler, 2010). With increased connectivity the vast majority of brands are connected and offering varied digital experiences to their consumers, to differentiate one should create a unique environment that reflects what the brand really does (Carvalhal 2016). By creating this digital atmosphere, regardless of the channel of performance, brands are able to distance themselves from the competition and enhance their essence in different channels.

5 DIGITAL ATMOSPHERE

Traditional advertising no longer serves to connect consumers, for Lindstrom (2007, p. 03) *"se os produtos e as propagandas quiserem sobreviver por mais de um século, terão de mudar inteiramente de direção"*[25]. This change directly refers directly to the association of a marketing plan that uses tools such as online advertising to manage aspects related to the brand's "DNA" and to put the experiences created in contact with the consumer. With the advancement of the digital age, consumers have become not only viewers, but the most active participant in the process of managing a brand. Advertising as it was done up until then can no longer engage consumers, this way, Lindstrom (2007, p. 03) says *"um cartaz pregado em um outdoor na Times Square não dará resultado"*[26]. According to Kotler (2017, p. 66), *"comunicar sistematicamente e de modo repetitivo a identidade e o posicionamento tradicional da marca [...] pode não ser mais suficiente para o sucesso no marketing tradicional"*[27], it is necessary to involve the consumer, and place him inside the brand's essence through a different experience, Kotler (2017, p. 66) also states that *"o que deve permanecer constante, porém, são a personalidade e os códigos da marca"*[28]. The biggest challenge for brands is to create a brand atmosphere in

---

21. Author's free translation to: "studies show that brands that incorporate sound into their virtual pages are 76% more likely to have higher internet traffic—and that brands with songs that 'fit' into their identity brand have 96% more chance of immediate memorization".
22. Author's free translation to: "The crucial task for anyone involved in advertising planning and evaluation is to determine precisely how this happens—without this knowledge we are condemned to confuse ourselves and to do a work of divination".
23. Author's free translation to: "advertising broadens the aspiration of the well-being".
24. Author's free translation to: "the new consumer wants to go beyond buying, it wants a product that is emotional. An aroma that has meaning, a clothing that reflects life".
25. Author's free translation to: "if products and advertisements are to survive for more than a century, they will have to completely change direction".
26. Author's free translation to: "a poster nailed to a billboard in Times Square will not work".
27. Author's free translation to: "systematically and repetitively communicating the traditional identity and positioning of the brand [...] may no longer be sufficient for success in traditional marketing".
28. Author's free translation to: "what should remain constant, however, are the personality and codes of the brand".

Figure 2. Reserva website.

the digital world, where all their essence and authenticity are connected to their consumers in multi-channel experiences. Connecting websites, blogs, social media, online videos to create a digital atmosphere that refers to the concept of a brand's "DNA" is already considered a necessity for survival (Aaker, 2015). As in the physical environment, brands create digital atmospheres that reveal their essence, impregnating, as attested for Gomez (2012, p. 2) "*alma, personalidade e apelo sensorial, transmitindo os próprios valores das marca*"[29]. They can use social media to leverage their brand power and engagement by creating revolutionary ads that can extend their lifespan in the digital world (Aaker 2015). Adopt sites that, in addition to transmitting their essence, act as a central "knot" of the entire network of programs, thus strengthening all other platforms (Aaker 2015). The most important role of this atmosphere is not intrinsically linked to the product, but rather to the purpose and timing of the "DNA" of the brand. By connecting more and more consumers, discourses leave the scene and enter the actions that evidence the speeches.

The men's fashion brand Reserva, enhances its purpose through its website and its social networks. Figure 2 shows three screens taken from the brand site where all are related to the culture and purpose of the company, either in customer service, actions that involve sustainability or the affection with its customers and suppliers, legacy that the brand seeks to pulverize. By relating and dedicating three different sections of your site to promote your purpose with real photos and reports, the brand highlights that creating a digital brand atmosphere can connect consumers and integrate them with the adepts of their purposes' causes. In this sense, it is observed that without the creation of a digital atmosphere the dissemination of this purpose will be limited to isolated *insights* or traditional communication.

6 CASE STUDY: FARM

FARM is focused on the young female segment, focused on exploring colors in its product construction processes and has become one of the leading fashion brands in Rio since the end of the 1990s (Rogar, 2015). With a brand essence totally focused on the lifestyle and culture of the Rio de Janeiro girls FARM seeks to take in its brand DNA, in addition to the aspects of Rio de Janeiro, all the Brazilian nature and colors that this country offers.

It is observed that in the brand's website this essence is valued and is shown from the beginning to make contact with its consumers. In Figure 3 it is noted that the brand dedicated a tab of its website, entitled "*Adoro!*", for the propagation of its purpose. In order to reveal its real essence and source of inspiration, the brand has putted in contacted all its "DNA" with its consumers, seeking to divulge scenarios, local artists, films, cinemas, travel and other subjects. Using the Brazilian scenario as a source of inspiration for its photographic campaigns and marketing material, FARM seeks more than to promote its products on digital platforms. It seeks to promote its purpose and make its essence clear, valuing much more the settings and culture of Rio de Janeiro than its own products. When looking at the background image it is noticeable that the landscape occupies a much larger space than the product itself, creating a digital atmosphere much more connected to its "DNA". This disclosure also aims to create an entertainment portal, shortening the virtual displacement of consumers, according to Kotler (2017, p. 77) "*com o aumento da mobilidade e da conectividade, os consumidores já dispõe de*

---

29. Author's free translation to: "soul, personality and sensorial appeal, conveying their own brand values".

Figure 3. FARM essence—ADORO! Webpage.

Figure 4. Compiled from FARM's Instagram.

tempo limitado para examinar e avaliar as marcas"[30]. The brand uses this tool to enchant and hold more consumers' attention in their online environments.

The creation of a digital atmosphere in the construction of the FARM brand seeks to evolve into all its channels of dissemination. Figure 4 shows two images captured from the official brand's *Instagram*, where none of them seeks as main focus the dissemination of a product, but rather to raise the purpose of the brand that is to divulge and maintain its essence in the city of Rio de Janeiro. The images shown here refer to landscapes of the city and had no production done. These images seek to reveal and make ever clearer the essence of the brand. By publicizing these images the brand does not seek to offer any product, the goal when using this type of photography in official channels is to connect consumers with their atmosphere. It is noted that both on the brand's website as in other official channels of publicity from FARM the intention is always to highlight the atmosphere and the purpose in which the brand is inserted. In this sense, it can be observed that the brand seeks to give much more value to its "DNA" and to reinforce this concept with its consumers than just to show products. This point of differentiation is remarkable throughout its line of communication, differentiating FARM from other brands.

7 CONCLUSION

It can be understood from the data collected in this study that branding actions should always be geared towards the real essence of the brand. By placing the "DNA" of a brand in contact with its consumers, they begin to stop buying a product and become advocates of a cause through a brand. For Kotler (2017, p. 9) "*O papel mais importante do marketing digital é promover a ação e defesa da marca*[31]. Thus, by connecting through branding actions in the

---

30. Author's free translation to: "with increased mobility and connectivity, consumers already have limited time to examine and evaluate brands".
31. Author's free translation to: "The most important role of the digital marketing is to promote the action and defense of the brand".

digital area, brands become more active and adept at their essence. In order for this essence to be available to the target audience, the *TXM Branding* methodology can act with the function of creating and managing brands. The last stage of the methodology, the *Manage* step, can be used for online advertising in a way that consumers can connect with the brand's essence. By creating digital atmospheres the target audience can interact with this essence and so become an admirer, or advocate, of the brand. Digital atmospheres must be created in order to bring consumers closer to the real essence of a brand. In this sense, the environment created must show the whole essence of a brand, its real "DNA" and not only products that exploit this bias. When looking at the cases presented here, it is understood that the creation of a digital atmosphere can be the link between brands and consumers, highlighting the points of contact. FARM uses as its main legacy and brand essence the divulgation of the "girl from Rio de Janeiro" lifestyle and all its *"brasilidade"*. In its digital environments the brand creates a unique atmosphere, where all the essence is put in contact with its consumers. Consumer experience in a digital atmosphere can connect you even more with brand DNA by generating an interest not only in the product, but also in the content that surrounds the product. Based on this atmosphere created consumers can get more involved with the experience that the brand wants to hand and so, staying more time in this space.

REFERENCES

[Figure 1. TXM Branding Methodology.] n.d. [image online] Available at: <http://logo.ufsc.br> [Accessed: 14 January 2018].

[Figure 2. Reserva website.] n.d. [image online] Available at: <http://www.usereserva.com.br> [Accessed: 14 January 2018].

[Figure 3. FARM essence—ADORO! Webpage.] n.d. [image online] Available at: <http://www.farmrio.com.br> [Accessed: 14 January 2018].

[Figure 4. Compiled from FARM's Instagram.] n.d. [image online] Available at: <http://www.instagram.com/adorofarm> [Accessed: 14 January 2018].

Aaker, D., 2015. *On branding: 20 princípios que decidem o sucesso das marcas*. Porto Alegre: Bookman.

Araújo, A. and Rosas, J., 2012. A estratégia do marketing da sandália havaianas. *Congresso Norte Nordeste de Pesquisa e Inovação*, 10(08), pp. 30–42.

Assunção, H., Gomez, L.S.R., Reis, P.F. and Stodieck, W.F. 2011. DNA de marca: Um estudo prático do processo de identificação. *Revista LOGO*, 02(01).

Batey, M., 2010. *O significado da marca: como as marcas ganham vida na mente dos consumidores*. Rio de Janeiro: Best Business.

Braun, J., Lopes, D., Werner, L., Perassi, R. and Gomez, L., 2014. *O positivismo percebido nas etapas de desenvolvimento do DNA da marca*, Revista Arcos Design, [e-journal] 8(1) pp. 36–48. https://doi.org/10.12957/arcosdesign.2014.13020.

Campanholo, T. and Bevilacqua, S., 2011. *Luxo: is not for all*. Porto Alegre: Revista da Católica.

Carrilho, A., Pinto, T. and Gomez, L., 2012. Design de loja como experiência de marca. *Ii Conferência Internacional De Integração Do Design, Engenharia E Gestão Para A Inovação*. 05(02), pp. 01–07.

Carvalhal, A., 2015. *A Moda Imita A Vida: Como Construir Uma Marca De Moda*. Rio De Janeiro: Senac.

Carvalhal, A., 2016. *Moda Com Propósito: Manifesto Pela Grande Virada*. Rio De Janeiro: Companhia Das Letras.

Ellwood, I., 2004. *O Livro Essencial Das Marcas*. São Paulo: Clio.

Feijó, V. Branding Digital: O Desafio Das Marcas Na Atualidade. *Intercom*. 13(13), pp. 01–14.

Fonseca, M., 2016. Identidade e a comunicação da marca @adorofarm. *Intercom*. 29(29), pp. 01–13.

Gerzema, J., 2009. *A Bolha Das Marcas*. Rio De Janeiro: Campus.

Gomez, L.S. (in press) *Espaços para viver as marcas de moda*, Florianópolis: Revista Expressão Gráfica.

Jones, J.P., 2005. *A publicidade na construção de grandes marcas*. São Paulo: Nobel.

Kotler, P. and Keller, K., 2013. *Marketing essencial: conceito estratégias e casos*. São Paulo: Pearson Education Do Brasil.

Kotler, P., 2010. *Marketing 3.0: as forças que estão definindo o novo marketing centrado no ser humano*. Rio De Janeiro: Elsevier.

Kotler, P., 2017. *Marketing 4.0: do tradicional ao digitais*. Rio De Janeiro: Sextante.

Lindstrom, M., 2007. *Brand sense: a marca multissensorial.* Porto Alegre: Bookman.
Lipovetsky, G. and Roux, E., 2005. *O luxo eterno: da idade do sagrado ao tempo das marcas.* São Paulo: Companhia Das Letras.
Lipovetsky, G., 2007. *A felicidade paradoxal: ensaio dobre a sociedade de hiperconsumo.* Lisboa: Edições 70.
Martins, J., 2006. *Branding: um manual para você criar, avaliar e gerenciar marcas.* São Paulo: Global Brands.
Moraes, D., 2008. *Moda, design e complexidade.* Barueri: Estação das letras e cores.
Ribeiro, L. and Gomez, L., 2016. A percepção de uma marca de moda pelo consumidor. Florianópolis: Intercom.
Rogar, S., 2015. Borogodó carioca: o fenômeno da farm. Vogue [online] Available at: <Https://Vogue.Globo.Com/Moda/Moda-News/Noticia/2015/11/Borogodo-Carioca-O-Fenomeno-Da-Farm.Html> [Accessed 21 November 2015]
Sampaio, R., 2013. *Propaganda de a a z: como usar a propaganda para construir marcas e empresas de sucesso.* Rio De Janeiro: Elsevier.
Santos, C. and Gomez, L., 2007. Logomania: as marcas de luxo e o logo. *Modapalavra* 05(05) pp. 46–57.
Stodieck, W.F. and Gomez, L.S., 2012. A Utilização do Brand DNA Process na Criação de uma Marca de Empresa de Cervejas Diferenciadas. In: *10º* CBPDD *(Congresso Brasileiro de Pesquisa e Desenvolvimento em Design)*. São Luis, Brazil, 08–10 October 2012. São Luís: CBPDD.
Tamamar, G., 2015. Antes De Criar A Farm, Sócios Perderam Dinheiro Com Franquia. Estadão [online] Available at: <http://Pme.Estadao.Com.Br/Noticias/Noticias, Antes-De-Criar-A-Farm--Socios-Perderam-Dinheiro-Com-Franquia,6086,0. Htm>. [Accessed 29 October 2015].
Tybout, A. and Calkins, T., 2006. *Branding.* São Paulo: Editora Atlas.
Upshaw, L., 1997. Transferable Trhths Of Brand Identity. *Design Management Journal.* 09(14) pp. 01–40.

The influence of social network sites on digital branding

M. Barreto
IADE – Universidade Europeia, Lisboa, Portugal
Department of Science and Textile Technology, University of Beira Interior, Covilhã, Portugal

L. Ribeiro
IADE – Universidade Europeia, Lisboa, Portugal
FibenTech Research Unit, Department of Science and Textile Technology, University of Beira Interior, Covilhã, Portugal

ABSTRACT: Currently, the way human being uses their social time is a reflection and undeniable associated with technological evolution and changes. Meaning that more time is spent around a specific social phenomenon that are digital social networks, or in other words Social Network Sites (SNSs). What began as a form of digital contact between people is now a means of disseminating information in the most diverse contexts. On the way to develop online strategies for creating lasting relationships with customers, SNSs represent one of the main digital powers in the interaction and building of users' affection with brands. The present study addresses the challenges faced by brands in the development of a digital branding strategy, the relevance of using SNSs as a tool for collecting information as well as to communicate, and an exploratory study in the Portuguese market is discussed.

1 DIGITAL BRANDING

This new world reinvented the traditional relations and forms of communication (Las Casas, 2010). Constant technological changes have allowed great evolutions in the way a brand can communicate and interact with the consumer, namely in the digital environment. A brand is now defined as an autonomous communication entity, where consumers are no longer attracted only by the intrinsic attributes of a product, but also by their broader discourses, more information, and more seductive stimuli (Keller, 2013; Plascak, 2009). Currently, the consumer, in addition to assuming the role of buyer, wants to be part of a company, interact with, and be heard (Bretzke, 2000). All its interests are placed in the experiences and benefits that brands have to offer, as providing a sensory journey to a place that goes beyond conventional routines, hence the importance of digital branding in the way a brand is interpreted and built by the consumer.

Digital communication is increasingly important for companies in a market where the key words are: interactivity, integration, globalization, approximation, convergence and democratization of information. The present context creates a new concept of social ties, where the multiple interactions contribute to the structuring of a new public space, a digital environment where the social becomes ate the same time more fragile and dynamic (Wolton, 2010). Digital communications have become a social norm, especially for the younger audience, and the absence of a product or service in the digital environment can negatively qualify a brand. The presence of a digital brand represents, today, one of the most striking ways of a brand to communicate and disseminate information and its personality (Dutta, 2010). These aspects have led companies to bear diverse challenges, risks, and opportunities, without being able to hide any previously unknown aspects (Hernandez, 2002).

Meanwhile, Internet is the first mass media that enables interactive feedback in the process of communication between a consumer and a brand, a relationship based on a set of experiences, exchanges, and perceptions, all simultaneously in the same field (Perrota & Toledo, 2006). Digital branding assumes the management function of an interactive environment, with consumers who, in addition to potential buyers, are also co-creators (Ciaco, 2008).

Is also important to state that a company that succeeds in its digital branding strategies does not mean that it has made big investments in the digital places, most of all it has understood the environment and surrounding beings and realized the influence of the real world in the perception of the virtual world. The strength of a brand lies in what is retained in consumer's mind, and Internet appears as one of easiest and fastest way absorb brand information (Machado & Keller, 2006). The role of branding is not only reflected in the construction of a brand, but in its management in interactive environments (Ciaco, 2008).

2 SOCIAL NETWORK SITES (SNSS): ADVANTAGES AND DISADVANTAGES FOR BRANDS

The relationship consumers have with existing brands is more personal and long-term way, requiring an authentic, committed, and real-time response, as for that, SNSs have generated in the brands a great interest, by the possibility to promote their products and gain the confidence of consumers, and encouraging them to buy online but also offline.

SNSs sites can be defined as "web-based services that allow individuals to (1) construct a public or semi-public profile within a bounded system, (2) articulate a list of other users with whom they share a connection, and (3) view and traverse their list of connections and those made by others within the system." (Boyd & Ellison, 2007, p. 211). In this virtual place, consumers express their personal preferences and share recommendations and opinions, being a major influence in the purchasing decision process of other consumers (Kim & Srivastava, 2007). SNSs turns out to be the junction of "real" and "virtual", and can function as a network of professional, community or political relationships, without ever ceasing to be a way for consumers to exhibit and satisfy their social and consumption needs (Silveira & Soares, 2011). These online communities have enabled the share of passion for a brand, by proud customers with other potential consumers generating more useful, credible, and far-reaching content compared to other media. Through effective management, companies can gain a competitive advantage in sales, customer support, research, development, and recruitment, without restraining the need for socialization and differentiation (Ray et al., 2011). SNSs are harnessed to achieve notoriety and to gain information on consumers and competitors, and in the end as a method of managing brand reputation. It is also important to emphasize the fact that members of these communities, as informed and attentive consumers, have a dual role, which will make them less receptive and tolerant to advertising (Palmer & Koenig—Lewis, 2009).

On this manner, it is possible to measure the audience's level of interest in content, where sharing or "liking" a publication is an indication or not of its interest. According to Afonso & Borges (2013), there are five objectives that justify the presence of a company in SNSs: listening, talking, revitalizing, giving support and engaged. SNSs allow a real-time interaction with the consumer and, through the monitoring and analysis of criticism, opinions, and behaviours, it is possible to obtain information that can optimize the process of co-production. These are representations established within the Internet about relationships and interactions between individuals in a group, which the consumer increasingly uses in the pre-purchase phase of a product, for information and advice, and in the after-use phase, for share experience of pleasure or dislike.

However, a SNSs should not replace existing communication channels within a brand, it only arises as a new approach to adding value. It is a phenomenon that should be planned to use according to a company's communication strategy, and not only because it is adhered to by most brands (Afonso & Borges, 2013). According to Kapferer and Bastien (2012), to create an impression of perpetuity in a brand, it must be associated with a defined identity, giving the brand feeling, uniqueness, timelessness and authenticity. Thus, the challenge in the digital environment is to work on the personality of a brand to meet those characteristics.

There is a balance of "forces" between consumers and companies, where market knowledge, which was previously only available to companies, has also become available to consumers (Pires, Stanton, & Rita, 2006), leaving companies more exposed and vulnerable. Companies have been reneged on as mere observers, where changing a publication or comment from a consumer is unthinkable (Kaplan & Haenlein, 2010). SNSs have the ability to listen to the market in real time, are useful in improving customer service and are a good way of interaction with consumers, nonetheless manage a brand in a SNSs requires time and availability, specially managing brand reputation, due to possible negative comments from users.

SNSs also allow the segmentation of consumers, and it is possible to direct messages only to the most relevant audience (Dionísio, Rodrigues, Faria, Canhoto, & Nunes, 2009). A network user seeks informality and entertainment, which are the conditions for personal satisfaction and for the continuation of knowledge sharing. Thus, SNSs appear as a medium that has advantages and disadvantages. A company needs to keep in mind its strategy inside this new reality, because as Evans (2012) points out, today's customers and fans are potential competitors tomorrow. Therefore, the presence of a brand in a SNSs represents a good communication channel, multilateral and participative with the public, but it takes time, learning new concepts and a better understanding of their functioning, as well as the establishment of concrete objectives. If companies ignore the potential of the tools that are available, they end up losing the next generation of consumers (Weber, 2011).

2.1 *Facebook and its impact on brand communication*

Facebook was created in 2004 by Mark Zuckerberg and is currently one of the most successful SNSs with around 1.94 billion users worldwide (Statista, 2017). It is a network used by several brands, namely in the fashion industry, as a way of communicating directly with the consumer and obtaining their feedback. Hei-Man (2008) qualifies Facebook as a social platform that combines several tools sought by users on the Internet. It is an innovative and attractive space for socialization, where the great objective is to listen, inform, observe behaviours, entertain, and participate as a member of a community. It represents a great opportunity for a brand to engage the consumer in its process of value creation and generation of ideas (Palmer & Koenig-Lewis, 2009). One of its great advantages is the fact that it is made up of several communities, which allows companies an easy access to their target consumers. For Carter (2013), in general consumers are not on Facebook to shop, but to keep up with news and "exciting things", depicting a typical user behaviour in a scheme: "Login—Like or comment—View photos—view the news feed—Return to the news feed". Facebook, featuring brand-ready pages, makes it easy for brands to interact with the audience. With the incorporation of a bigger space for the logo, with the availability of data about how users get involved in publications and with the possibility of statistical analysis of data, it is possible for the brand to draw conclusions from consumer opinions and behaviours. And when a company creates a presence on Facebook intends to get financial return on the investment made, so it uses the creation of ads as a tool for simple and economical marketing and advertising. The use of this network should be planned in such a way as to promote the natural involvement of the consumer, both through the posted messages and the tools provided. Facebook is a collaborative platform that allows people, brands and products/services to contact each other, access news and share content.

3 METHODOLOGY

As stated previously, SNSs as well as mass media, are a source of information about consumer behaviour. In Portugal, Facebook is the most used network, according to a study by Marketest, about 96% of Portuguese's use it (Marketest, 2017). However, the impact that the presence of a brand on Facebook has on the Portuguese consumer, is still somewhat unknown, which makes it difficult for a brand to decide whether has or not advantages to be there. Given this problematic, which has not been accounted in many studies, it was found relevant to carry out a questionnaire and analysis of data on the subject, namely the dimension and expression of Facebook in Portugal, as well as the relationship between the Portuguese consumer and a brand in the respective network. Thus, a questionnaire was created to collect data about Portuguese consumers inserted in the SNSs Facebook. The questionnaire, addressed to Facebook users, contained eighteen questions, developed as a result of a prior analysis of the tools and content offered by the network concerned. Examples of these questions are: *"What are the reasons why you joined the Facebook SNSs"; "What kind of publications attract you in a SNSs?"; "Can brand publications on Facebook change your relationship with it?".*

The questionnaire was created with Google forms, available on Facebook from December 10, 2017 to January 10, 2018 and disseminated online on the network itself, as well as by individual messages for network users.

4 RESULTS

The questionnaire released on the SNSs Facebook was answered by 108 Portuguese users. Considering descriptive analyses of socio-demographic data, gender and age information was gathered. On gender, 55% respondents were females and 45% males. Respondents aged between 16 and 52 years, yet 70% are young consumers between the ages of 20 and 29.

The first questions approach how much time consumers spend on Facebook, and the reasons that led them to join the network, to later analyse their behaviour in the presence of a brand, and how it is appreciated and evaluated by the national user. The data gathered show that most users access the SNSs more than three times a day (66.7%), and that only a small percentage access Facebook once a day (5.6%). In all these accesses, it was also concluded that more than half of the individuals did not use the network for more than one hour (60.2%), and that about one quarter spent more than two hours on Facebook (21.3%).

The reasons that lead an individual to choose to Facebook can be varied, but the main reasons demonstrated by our sample of Portuguese consumers are interaction with people (77.8%), curiosity (42.6%) and knowledge updating (31.5%). Few use Facebook as a form of power (0.9%), altruism (0.9%) and increased self-esteem (1.9%).

After knowing the amount of time that an individual spends on Facebook, remains to know how he spends it. Thus, it was concluded that the major interest of the consumer in the SNSs are in news (79.4%), photographs (77.6%), in videos published (50.5%), events (40.2%) and other formats of entertainment (41.1%). Advertising (6.5%) appears in the bottom of the less chosen contents visualized in Facebook. The online relationship between Portuguese consumers and brand online is low, since only 25.9% of users follow between 10 to 20 brands on Facebook. Most respondents follow less than 10 brands (40.7%). Also, follow-up does not fit into a routine activity, since about half of the individuals (53.7%) read and observe the publications of a brand, only sporadically. Despite this, there are still users who follow the brands, most of the time (27.8%) and whenever they publish new publications (7.4%).

Most respondents considered the presence of a brand in Facebook relevant (87.9%), and the reasons they describe are varied. The number of individuals who affirmed the opposite is minimal (13.1%), some implausible justifications were presented, and of 13 negative answers, only 4 presented a reasoning. For a brand it is important to know what the consumer most appreciates, in this case content update (38.1%), which is put as the best strategy to be applied in the activity of a brand in Facebook, followed by content diversity (23,8%), quick answer (19%) and permanent availability (19%).

On a brand's Facebook page there is always something that catches up consumer attention, the content of the posts (76.9%) and comments from other users about the brand (44.4%). The least relevant feature is the cover photo (9.3%) and the number of likes (10.2%) and followers (16.7%). The usage of Facebook is mainly a mean to get information about the brand, and about their products or services, but not in a regular basis (17,8%). It can also be observed that some users who do not consider the presence of a brand in Facebook to be relevant, also do not seek information about it on the network (7.5%). For most respondents the relationship between the user and a brand can sometimes be called into question by the activity of a brand on Facebook (75.2%), 11% answer that the publications of a brand do not alter its relationship with the brand, while 13.8% fully agree with the influence brand content has on its relationship with it.

82.6% of consumers prefer to buy products in a physical store and not through the online medium, yet 17.4% of the individuals choose the online purchase service. Although Facebook presents a large sample of products, 32.4% of users never bought an article after it was published on the net, only 65.7% have bought occasionally the product after it has spread. Still, some people always buy a product after discovering it on the SNSs (1.9%). When deciding to buy a product, few people turn to Facebook as a means of help making decisions (1.9%), 12% of users research most of the time and 40.7% rarely do it. Most users do not use the network as a security source for their purchase (45.4%). Although, the opinions and comments published on SNSs are one of the major influences on the way the user sees the brand, for some (31.5%) the interest on those opinions is continuous and the great majority (63%) identified that it happens occasionally. The percentage of individuals who are not interested in the opinions posted on Facebook is small (5.6%). Finally, on the analysis of the number of users who share their shopping experience with others on Facebook, was found that, in general, users never share their experience on the network, 59.3%, 38.9% only share a few times, and only 1.9% always share with others. Through the analysis and evaluation of all the results and data provided by the study, some conclusions can be draw.

4.1 Results discussion

First, the number of Portuguese Facebook users corresponds to 5.8 million (Ferreira, 2013), placing Facebook in one of the most comprehensive media in the national market, reaching more than half the number of Portuguese citizens residing in Portugal, or 61.8% of the 9 387 932 citizens (Pordata, 2018). However, given the number of users of Facebook, the number of respondents obtained for this study, is a very small sample compared to the total universe, and therefore not representative. Through the analysis of the behavioural criteria of the user, it was concluded that the majority accesses the SNSs more than 3 times a day and for this, does not spend much more than 1 hour. Thus, it is important for a brand to invest daily in publications, to improve brand recall, to attract and interact with consumers, arousing possible interests in products or services, campaigns, events, or promotions. This strategy also represents a means to make the consumer a frequent follower of the brand. A brand should then bet on launching news, photos, videos, events, and other forms of entertainment, without ever showing the idea of advertising. All publications should be interactive and interesting for the consumer, with the aim of transmitting the identity and personality of the brand in a natural way, and above all, make the user first to fall in love with the brand and its surroundings, products or services. The secret is in the offer to the consumer of everything that made him turn to Facebook, is in the interaction with people, in updating knowledge, in accessing curiosities and in valuing all the feedbacks obtained. The true challenge of a brand in Facebook is in the implementation of the consumer's will and desire to follow, since in Portugal, most users follow few brands in the SNSs, which leads to a lesser monitoring of brands by consumers. At this level, in the competitive SNSs, Instagram, brands have higher rates of followers, justified by the presence of a younger audience, which reveals a greater demand for brands, trends and novelties. Nonetheless, exist user-friendly brands on Facebook, a work that results from the use of creative, engaging, and trustful communication and interaction strategies.

Despite the lack of follow-up of the brands, consumers consider important and advantageous for a brand to be present on Facebook. A brand in the SNSs also achieves greater interactivity, visibility, and exposure, as well as greater dissemination of its content without the need for consumer demand. Facebook also allows the resolution of direct issues between the brand and the customer, which contributes to the process of consumer loyalty with the brand. For a brand on Facebook, is important to know how answer to user expectations and preferences, to foster expansion and effective communication, in this case with content update and diversity. The constant launch of different publications keeps the consumer aware of the brand, arousing their interest and curiosity. In addition, it is prudent for the brand to always pay particular attention to the greater interests of the user in opening its page, it is necessary to be careful with the content of the publications and comments of other users about the brand, which the brand should always respond with a sympathetic attitude and superior to the negative criticisms, without ever devaluing the opinion of the consumer. These are all situations that can sometimes put into question the brand—consumer relationship.

Facebook, in general, is not the preferred medium in the process of buying a product, since more than half of the users prefer the purchase in a physical store, for the convenience, possibility of experimenting and direct contact with the product, which enables a better perception and evaluation of the same, by the all experience of purchase, security, confidence and reliability, as well as the immediate access to the product. However, purchases through the SNSs Facebook has been gaining more and more reach, not only due to the inexistence of a physical store in some brands, but also because of the practicality, possibility to save time and money spent on the trip, the comfort of the purchase from anywhere without confusion, efficiency and speed, and also for greater ease in price comparison. Indeed, Facebook works nowadays not only as a medium of communication, but also as channel to buy and sell online. Even if in cases where a brand that does not have an online shopping service, can always advertise their products and get in touch with possible future customers, since a large number of users have already bought some products after their propagation in the network. Thus, Facebook becomes a platform for brands, where any user can consult information related to a product before the act of purchase.

Although, one of the major disadvantages of this type of online presence is the lack of security and reliability that the user demonstrates to the content posted on Facebook, being essential the transparency and honesty demonstrated by a brand. Brands should be prepared to respond to any less pleasant comments, as consumer confidence is increasingly being deposited in the voice of other users, rather than in the voice of the brand.

Finally, the number of users who share their Facebook experience after buying a product should be noted. It is important for a brand to arouse in the user a willingness to recommend the product or service with other consumers and to share their shopping and use experience. A great example of online sharing on Facebook are all the digital influencers, who act as partners or sponsored actors, receiving and promoting products as a way to encourage consumers to get the products they show. This is an increasingly dominant phenomenon, where through Facebook and other SNSs, brands rely on large digital influencers, to reach the attention and desire of the consumer. It is a form of communication that transmits trust and credibility to the consumer, since it is the voice of a customer and not the brand itself. But, as in most things, this form of communication also has its problems when the message conveyed is negative, which requires a great deal of attention from the brands. In addition to all these negative comments moving around SNSs, brands are also visible in the other digital networks, even with smaller number of followers, such as the "Reclame Aqui", that are dedicated only to evaluate the brands, through complaints made by consumers. Thus, a brand needs not only to know how to communicate, but also to know how to respond and solve all negative reviews and complaints online, always showing a great esteem for the customer's word.

5 LIMITATIONS

The vast majority of the Portuguese population has a Facebook account. However, the sample of users that answered the questionnaire (108 users) is very small compared to the number of Portuguese citizens users of Facebook (5.8 million). The sample collected was limited due to time constraints within the academic context of the study and the low availability of users to respond. It is therefore important to point out that these conclusions are merely indicative and should be viewed with caution. However, although the study presents data from a small sample of Portuguese Facebook users, the results can suggest some clues about the use of SNSs, namely Facebook in the improvement of digital branding strategies.

6 CONCLUSIONS

SNSs are a phenomenon that is becoming increasingly widespread. In the business sector, SNSs present themselves as a challenge and an opportunity; they are a channel for interaction and distribution of information that contributes to the definition of the reputation of the brands. However, the presence in SNSs requires a constant dynamism and interactivity, so that the appearance of the brand is consolidated and reaches the consumer, who now plays a role in the co-creation of everything from the design of a product to promotional messages. A brand to communicate, needs relevant content, be active, available create emotional involvement and customer follow-up. Adherence to a SNSs as a brand strategy implies a defined structuring of its objectives, in order to approach and relate to its target audience, generating credibility and trust, reducing advertising costs and optimizing results. The investment of time and personal energy in the defence of a brand, is at the moment facilitated by belonging to an online community, which allows the sharing of the ideas and feelings aroused by the brand.

Good SNSs management creates retention mechanisms and consumer loyalty, making it possible to identify the major influencers and market leaders. Recent studies by PSE (Statistical Products and Services) state that Facebook is the preferred communication channel in terms of access to content, representing one of the most influential and interacting media. About 40% of these user accesses to the network are even at work hours. It is also added an analysis of the reach and interactivity of some brands from several sectors, where the retail sector is presented as the smallest promoter of reactions, which leads to the need for fashion brands to start betting more on innovative and motivational branding strategies digital (Consumption, 2018). Facebook, being the most socially representative network, proves to be a good tool to increase a brand's reputation. And in Portugal, it represents an excellent tool for communicating a brand, to reach more than half the national population, without ever ignoring the behaviour and preferences of the Portuguese user and consumer. Thus, SNSs should be an integral part of a brand's communication strategy, not only for its contribution to value creation, but also for its availability of essential tools in the process of brand loyalty.

REFERENCES

Afonso, C., & Borges, L. 2013. *Social Targets.* Rio de Janeiro:Topbooks

Araújo, M. R. 2012. *Marcas de Luxo versus Contrafação: Motivações de Escolha*, Master Dissertation. Universidade Católica Portuguesa, Porto.

Boyd, D. M., & Ellison, N. B. 2007. SNSs sites: Definition, history, and scholarship. *Journal of computer-mediated communication*, *13*(1), 210–230.

Bretzke, M. 2000. *Marketing de Relacionamento e Competição em Tempo Real, com CRM*. Rio de Janeiro: Editora Atlas.

Carter B. 2013. Ganhar Com o Facebook. Lisboa: Marcador Editora.

Ciaco, J. B. 2008. *Blogar é preciso. Navegar não é preciso*. Webinsider (on-line), São Paulo, UOL.

Consumo, G. 2018. *O que "conversam" os Portugueses com as Marcas nas Redes Sociais.* Available at: http://grandeconsumo.com/noticia/20831/o-que-conversam-os-portugueses-com-as-marcas-nas-redes-sociais?utm_campaign=GC+News+10%2F04%2F2018&utm_source=Grande+Consumo&utm_medium=email [Acessed April 2018].

Dionísio, P., Rodrigues, J. V., Faria, H., Canhoto, R., & Nunes, R. 2009. *B-Mercator*, Blended Marketing. Lisboa: Editora D. Quixote.

Dutta, S. 2010. *Managing Yourself]: What's Your personal Social Media Strategy?* Harvard Business Review, available at https://hbr.org/2010/11/managing-yourself-whats-your-personal-social-media-strategy [Acessed Jan. 2018].

Evans, D. 2012. *Social Media Marketing: An Hour a Day*. 2nd Edition. New Jersey: John Wiley & Sons.

Ferreira, R. 2013. *A percepção do valor das marcas – Estudo de caso: O valor da marca DOVE*. Master dissertation. ISG—Instituto Superior de Gestão.

Hei-Man, T. 2008. An Ethnography of SNSs in Cyberspace: The Facebook Phenomenon. *The Hong Kong Anthropologist*, *2*, 53–77.

Hernandez, J. M. D. C. 2002. How to build leading brands / brand asset management / build your own garage, *RAE Eletrônica*, *1*(1).

Kapferer, J. N., & Bastien, V. 2012. *The luxury strategy: Break the rules of marketing to build luxury brands*. London: Kogan page publishers.

Kaplan, A. M., & Haenlein, M. 2010. Users of the world, unite! The challenges and opportunities of Social Media. *Business Horizons, 53*(1), 59–68.

Keller, K. L. 2013. *Strategic brand management: Building, measuring, and managing brand equity* (4th ed.). Harlow UK: Pearson Education Limited.

Kim, Y. A., & Srivastava, J. 2007. Impact of social influence in E-Commerce Decision Making. *Proceedings of the 9th International Conference on Electronic Commerce*, Minneapolis, 19–22 August 2007, 293–302.

Las Casas, A. L. 2010. *Marketing Interativo—A Utilização de Ferramentas e Mídias Digitais*. São Paulo: Saint Paul.

Machado, M., & Keller, K. L. 2006. *Gestão Estratégica de Marcas*. New Jersey: Prentice Hall

Marketest. 2017. "Os Portugueses e as Redes Sociais", Marketest Consulting. Available at: www.marktest.com/wap/a/grp/p~96.aspx [Accessed Fev. 2018].

Palmer, A., & Koenig-Lewis, N. 2009. An experimental, SNSs-based approach to direct marketing. *Direct Marketing: An Internacional Journal, 3*(3), 162–176.

Perrota, K., & Toledo, L. 2006. O posicionamento da marca sob a ótica da internet, REA-Revista Eletrônica de Administração 5 (2),1–17.

Pires, G. D., Stanton, J., & Rita, P. 2006. The internet, consumer empowerment and marketing strategies. *European Journal of Marketing, 40*(9/10), 936–949.

Plascak, N. 2009. A Marca Pós-Moderna: Poder e Fragilidade da Marca na Sociedade Contemporânea. *Revista Signos Do Consumo, 1*(1), 132–136.

Pordata. 2018. Recenseados: total, por nacionalidade e por residência. Available at: https://www.pordata.pt/ [Accessed April 2018].

Ray, A., Riley, E., Elliott, N., Corcoran, S., Greene, M., Parrish, M., & Wise, J. 2011. Now Social Media Marketing Gets Tough. *Forrester*, 1–6. Available at: https://go.forrester.com/ [Acessed Dec.2017].

Silveira, R. C., & Soares, T. O. R. 2011. A influência das redes sociais no comportamento do consumidor: um estudo sobre as decisões de compra do consumidor nas comunidades virtuais. *Congresso nacional de excelência em gestão*, Rio de Janeiro: CNEG, 2011.

Statista. 2017. Facebook—Statistics & Facts, Available at: https://www.statista.com/topics/751/facebook/ [accessed April 2018].

Tybout, A. M. 2006. *Branding: Fundamentos, Estratégias e Alavancagem de Marcas, Implementação, Modelagem Checklists, Experiências de Líderes*. Brasil: Editora Atlas.

Vaz, C. A. 2009. *Google Marketing: O Guia Definitivo de Marketing Digital*. Brasil: Novateca.

Weber, L. 2011. *Marketing to The Social Web: How Digital Customer Communities Build Your Business*. New Jersey: John Wiley & Sons.

Wolton, D. 2010. *Informar não é comunicar. Brasil: Editora* Sulina.

Teaching and education

Teaching design in real contexts to preserve local identity and memory

A.H. Grácio & C. Rijo
CIAUD – Lisbon School of Architecture, Portugal

ABSTRACT: This paper introduces a collaborative project based on the promotion of interdisciplinary dialogue on formal, informal, and culturally plural learning, considering educational practices and design alternatives for the standard non-participatory model, in the classroom. Centered in the field of teaching and research in design, this research develops efforts towards the analysis of the modus operandi and the results obtained by adopting the methodology of the design process in the educational field, in particular in design education. To accomplish these goals, we propose a methodology centered and driven by practical action, as much as possible in a real context, structured according to the different phases of a design project. Signed in collaborative processes, we will characterize the action developed by the interdisciplinary dialogue in formal, informal, generational, and culturally plural contexts. The research assesses the importance of applying less conventional methodologies in design teaching, as well as the perception of how an immersive learning context can be a facilitating tool for learning, fostering creative thinking, and generating new ideas and future paths to the problem in question. Forged in the parallelism with the methodologies used in the real context and adapting them to the teaching-learning process, this project is based on pedagogical tools drawn from the constructive reflections elaborated either from the action of teachers and students in a real context, from surveys of students and key players in each of the proposed projects, or from the moments of formative evaluation of students and teachers. The experience of design teaching that it intends to propose is characterized by the multidisciplinary of its actors. The intention is to integrate practice as an instrument of research and action design considering both practice and reflection on the results of practice, the artifact created as a source of new knowledge and research, or a search capable of leading new knowledge or new practical action. In the end, we assess the importance of applying the new methodologies and how this immersive learning context is a facilitator of learning fostering creativity by providing new ideas and future paths to the problem in question.

1 INTRODUCTION

This paper introduces a collaborative project approach in design teaching, based in the co-creation processes, social design construction and its consequent interaction between artisans and designers.

The knowledge of a certain type of artisanal technique is usually seen as implicit by the craftsman, where specialized skills are personified in artisans and/or its local community. This knowledge, is often technically based, but is also an intangible knowledge, since it consists of a know-how transmitted from generation to generation in an empirical way and is not rationally organized. The project approach, in this particular case was made to Nisa quartz inlaid pottery technique, nowadays only produced in Nisa a small village, in the Alto Alentejo region, Portugal, where through the establishment of new relationships and collaborative processes, it is intended that the crafts knowledge, could be transposed to new models and approaches aiming to result in new products and/or services with renewed communications competences, meanings and symbolic values for the contemporary market, without, however, losing their original identity matrix. This platform of action, which is intended as a regular, not as a singular action, passes not only for the traditional relationship of master/apprentice but also for the capacity to build new working relationships based on a system of co-creation. The processes of co-creation and consequent interaction between artisans and designers have the capacity to nurture a collective and collective creativity resulting from an interactive (and non-interventionist) approach, thus being processes capable of nurturing local handicraft development and his knowledge of a know-how, usually confined to the traditional master/learner transmission. In the case of a concerted approach, rather than isolated acts, in a medium/long-term strategic vision, we will be able to build not only a fabric that contributes to the safeguarding and transmission of the still existing knowledge and which carries serious risks of extinguishing, as contributing to the gathering of new interested in seizing this technique and local heritage.

2 OBJECTIVES

The reflection and process presented, as its consequent interaction between artisans and designers have the capacity to nurture a collective and collective creativity resulting from an interactive (and non-interventionist) approach, thus being processes capable of nurturing local handicraft development and his knowledge of a know-how, usually confined to the traditional master/learner transmission. In the case of a concerted approach, rather than isolated acts, in a medium/long-term strategic vision, we will be able to build not only a fabric that contributes to the safeguarding and transmission of the still existing knowledge and which carries serious risks of extinguishing, as contributing to the gathering of new interested in seizing this technique and local heritage.

There is increased effort to place ecological meaning to material objects and things through the use of social theory. Materials are commonly used to fashion cultural ideals, social structures, social values, human bodies and global networks, demonstrating the social perspective in the design of objects. Material culture and object ecologies have caused renewed definition of the different modes of social communication, social affiliation networks, and the mundane experiences that occur daily in life. Social theorists strive to draw some connection between imagination, emotion, subconscious thought and materiality. It is common to see material agents being used to express symbols, discourses, and ideals.

In the case of the quartz-inlaid earthenware crafts technique, the legacy has been characterized by increased changes in the cultural approach in the initial approaches in delivering the art (Barthes, 1970). With the advent of new science technology and mechanical machines for some aspects of the work, there is increased risk of the art losing its original cultural heritage (Law, 1991). The original art involved men sifting through the clay and kneading it with their own hands. Presently, such work can be done easily with the help of mechanical machines (Cuche, 2010). Mechanical centrifuges powered by motors are also used increasingly nowadays to improve the process of preparing the earthenware products, leading inevitably to a change in the original approach in which the practice was done. The original tradition also involved only women handling the work of selecting quartz materials and inserting in the engraved motifs of the already prepared earthenware. This is likely to change, with more opportunities being presented for men to undertake the work for purposes of producing the quartz in-laid earthenware in commercial proportions.

A close look at the sociological traditions shows that the original cultural heritage held by the objects of quartz inlaid earthenware is increasingly changing. Material culture can be used to explore the mutual relationship that exists between objects and people. The specific uses that a particular community confers to an object help to shape its inherent material culture. Recent trends in the theoretical explanations of person-object relations have changed the traditional approach of explaining such relations. The 'consumer culture' is increasingly being explored in a different perspective due to technological innovations in the contemporary world. Most descriptions are now focused on a consumer culture that values intrinsic materiality at the expense of face value for distinct objects with a cultural value.

Material culture, on the hand, insists on the perception of objects as the visible part of a culture. Such a perception gives the object an increased role in evoking feelings, coordinating socio-economic networks, and animating the cultural practice for a particular community (Douglas, 2002). Objects have the power to seduce, enroll or attract subjectivity and human action. Humans also actively seek out objects out of their innate social action. Objects project a certain level of hold on the people. They eventually draw people to subjectivity as manifest in the patterns existing in the relationships between objects and people. A reliable model that can be used to explain the materiality is to relate the embodied material dimensions for objects with their mythic and narrative characteristics. Object mobility in personal and social territories should also be considered when analyzing materiality of objects within a culture. Such diverse approaches would help one to gain leverage in exploring the material or immaterial changes in the cultural heritage of a community as expressed by their human-object relationships.

By proposing this approach, we expect as a final goal, to foster a platform for students to interact with artisans and its local community, with particular attention to: individual student learning, as well as the structure and process created and defined, that is, what the role of students in the processes of representation, management and feedback received from a local context; identity, and the consequent creation of a sense of belonging in the development of the project, in order to involve all parties and what role, the construction of new symbolic values of identity, in the production of material culture.

3 METHODOLOGY

In his seminal book dubbed The Coming of Post-Industrial Society, Bell (1974) explains how man's society has gradually and consistently transited from a pre-industrial era, to an industrial era, and now to a post-industrial society. In fact, in an article dubbed Sustainable Design and Postindustrial Society: Our Ethical and Aesthetic Crossroads, Bachman (2016) goes at great lengths to succinctly demonstrate how as a society, we have transited from an industrial-product society to a postindustrial knowledge economy and information society. According to Bachman (2016), the post-industrial society is defined by service economy, cybernetic reasoning, and knowledge workers. Unlike its predecessor (i.e. the industrial society), which was characterized by human-machine relationships, Bachman (2016) observes that the post-industrial society is defined largely by human-human collaborations, a phenomenon that makes social design a relevant and essential concept in a post-industrial society. But what is meant by social design in the context of the current paper? Before answering this question, perhaps it is imperative to explore the concept of industrial design (ID).

The Industrial Designers Society of America (IDSA) describes ID as the art of creating products and systems that are primarily aimed at optimizing function, value, and appearance, with an ultimate goal of enhancing mutual benefits for both the manufacturer and the end user. In order to meet that goal, an industrial designer strives to create/develop such products and systems by collecting, analyzing, and synthesizing available, relevant data, taking cognizance of the special requirements and desires of the manufacturer and the ultimate user. Therefore, from a post-industrial society standpoint, the concept of industrial design provides a valuable platform for generating strategic problem-solving processes that promote innovation, enhance business success, while ensuring better quality of life for both the manufacturer and the end user (Veneris, 1990). From the foregoing, it can be argued that the primary objective of industrial design is to improve the livelihood(s) and wellbeing of man.

This insightful background of the concept of industrial design thus provides us with the requisite platform for construing the concept of social design. In the world of design, social design is often construed as the design process that is aimed at improving man's livelihoods and wellbeing. The primary aim of a social designer is often to create new products and processes that profitably develop both human and social capital (Chen et al., 2015). Despite the existence of a universally agreed definition of the term "social capital" (Robison, Schmid, & Siles, 2002), based on various existing attempts to define it (Portes, 1998; Lin, 1982, 2002;

Christopoulos & da Rocha, 2015), social capital entails man's capacity to conveniently access and adequate utilize resources at his disposal. In this regard, collaboration and continued human-human interactions are of the essence in a functional social design system. Therefore, Bachman (2016) notes that in a post-industrial society, social design emerges as an important tool for correcting the large-scale isolation associated with the previous eras, in terms of promoting macro human-human interactions, encouraging systematic reasoning, making cybernetic decisions and dynamic assumptions, encouraging scenario planning and value motives, promoting team attribution, as well as embracing reasoned principle, wisdom, and theory.

Over the last decade, the concept of social design has generated significant interest and gained considerable attention among researchers and practitioners. A number of theories underpinning this apparent surge in interest in social design in the post-industrial society have been advanced by different authors. For instance, Chen et al. (2015) posit that the 2008 global financial crisis coupled with the traditional approach to design (i.e. market-focused) has created a scenario of diminishing job opportunities, especially for young designers. Consequently, the unfolding scenario has compelled designers—mostly upcoming designers, to seek new markets (created by emerging complex societal challenges) through creativity and innovation. In this regard, the post-industrial designers have found it opportune to work, not just with things, but also with abstract social entities such as services and communities to create value for both the manufacturer and the end user. While this proposition is debatable, what is not in doubt is the apparent paradigm shift from the designs of the 1990s to the current designs, which are characterized by a strong sense of modernity through the interweaving of social constructs with things and services, with a view to creating value. Elsewhere, Bachman (2016) attributes the surge in the concept of social design to its perceived relevance in the post-industrial society, where we can envision a future with more sustainability addressing to "the built environment and the complex, collaborative, evidence-based, as well as the cybernetic processes" (Bachman, 2016, p. 30) that are associated with the concept of social design. What this implies in the opinion of this paper, is that besides the traditional design elements of economics, environment, legal frameworks, as well as political dynamics, social design strives to incorporate social and cultural attributes of society into the design process.

In a 2014 British report, Armstrong et al. (2014) classified social design into three main categories namely: (i) design activism; (ii) social entrepreneurship; and (iii) socially responsible design. Nonetheless, Markussen (2015) added social movements to the classification by Armstrong et al. (2014). With the apparent surge in social design coupled with continued scholarly discourses regarding its forms and capacities, it is justifiable to assert that in the context of the post-industrial society, the concept of social design has significantly stretched the whole idea of design beyond its traditional objective and scope. This has made it almost harder for any reasonable researcher or practitioner to succinctly determine the limits of social design and accurately project how best to extend such limits. In this regard, taking stock of the various developments associated with social design in a post-industrial society is of the essence.

This also suggests that the activity of mass production is in itself a luddite approach to implementing design and thereby retaining cultural memory, with a focus on holding on to what works unless either societal needs or changes in design abilities influence a progression in its strategies. According to Douglas (2002), this is also indicative of an over-dependence on the role of engineering in the production of items of material culture, whereby the influence of artistic input is minimized in favor of outputs that have already established themselves in contemporary cultural environments. This highlights the underlying contexts of the globalization of culture as perpetuated in the standardization and mass production of products and services in contemporary society.

When analyzing the influence of communities in the retention of their material culture, it is also imperative to understand the role that their designers play in perpetuating the skills and information that comprise their expertise. Modern approaches have an evidently mass-production emphasis, whereby there are less variations between the ways through which professionals impart their skills upon subsequent generations of designers and artisans (Wood, 2006). However, this approach is questionable considering the aforementioned inhibitions that mass production places on creative or stylistic input in the development process, which also risks applying comparable restrictions on the ability to divert from normative practice in the learning process. Buxton (2010) also highlights the role that the transmission of information plays in the production process, whereby a dissonance between the artist and the other parties involved in the activity can result in a loss of vital elements of the visual vocabulary intended by the designer. This also means that standardized interpretation techniques are not suitable for the dissemination of craft techniques tailored for passing on skills related to the retention or portrayal of cultural identity and memory in production.

When analyzing the communal contributions to the retention of cultural identity and dissemination of items of shared cultural memory, it is also imperative to acknowledge the role that these items play in grounding the targeted cultures in their environments. For instance, Providência et al. (2005) argue that the power of modern consumerism also features among the factors influencing the erosion of cultural memory by emphasizing conformity and creating an alienating culture for outliers. As a result, affected communities have to develop a balance between their needs as individuals living in a modernized society and thereby accommodate new products as symbols of their membership in the globalized world. Nonetheless, Fiske (2002) also calls to question the influence that this dissemination of popular culture has on the willingness of individuals to associate with their local cultures and identify as members of unique communities. Consequently, even as efforts are made to retain craft techniques as symbols of individual communities, there remains a struggle between the outputs of these activities and their ability to fit into the modernizing social landscape and its focus on less individualism in the construction of a global identity.

A deeper review of the role that production approaches play in the construction of modern realities using the base of commoditization reveals the standardized approaches to production as a hindrance to the portrayal of cultural identity in these items. Margolin (2007) points out that the outcomes of design are the product of artists' intentions for the material world, which is an inherently imaginative exercise. To curtail this creativity by requiring conformity to particular design or production approaches is thereby an archaic way to carry on the inclusion of multiple perspectives in a multicultural albeit globalized world. Designers are thereby required to water down their artistic inputs in favor of what mass production elements can output, which risks limiting the ability to create effective cultural reflections in the subsequent products (McDonough & Braungart, 2010). Kopytoff (1986) notes that this arguably allows for their inclusion in contemporary socioeconomic dynamics by improving their conformity to the commoditization process. The willingness to incur this loss should, however, not be taken lightly considering the role that it plays in reducing eroding these objects' ability to communicate the identities of their individual communities.

One of the key considerations in the applicability of traditional craft techniques in the modern world is that they encounter a consumer space that already places a high standard on the items in consideration for inclusion in the collective culture. As a result, the desire to persist with traditional approaches to imparting identity in the form and function of the products that they develop is a key challenge for practitioners of these traditional craft techniques (Walker, 1998). This is also a hindrance to their ability to produce unique artifacts through the inclusion of technological input in design and production phases. However, Pevsner (2005) also highlights this conformity as a key element in the dissemination of cultural products across what is now a globalized society. This brings to mind the influence that cultural inputs have had on social progression, such as in changing global approaches to transportation through the adoption of a mass-market approach to developing the associated products. Therefore, the challenge still persists in the development of an acceptable balance between cultural and functional needs in the creation of products intended to either sustain existing cultural elements or derive from them to create new variants.

4 CONCLUSIONS

The profession of designer has transformed itself drastically in the last decades, in fact only in the last decade it has evolved much since a stage in which it was considered as an activity that should only be concerned with criteria of resolution of problems presented by the industry, where it would have to take into account innovative solutions capable of solving real, man-designed, aesthetically pleasing and useful, mass-consumption, and profitable problems.

Today the design design practice is a very well-established process, where the methodologies are well defined, that is, it is presumed that there is always a previous in-depth research of the contexts of the problems presented to the designer, the in-depth research and data exploration is Always present, as well as a continuous exploration of the improvement in the approach to the problems that arise during the project process, in order to minimize the hypotheses of error. What the market demands today are not only objective products, with the best selection of the raw material to be used, nor the best color, and only the best functionality. What the market is looking for is this and more. Today the designer can and should design products that establish ties and communicate with people. Design is today a much more formalized profession since the corpus of theoretical-critical written knowledge has been developing on a large scale.

The conclusion we can reach is that compliance with criteria such as those mentioned is not a guarantee of success, since if we study and listen carefully to the market, we can perceive that the success of a particular product and/or brand with the consumer, depends (And much) of the identification that the consumer makes with the brand, product or service that he buys, that is, what he apprehends and appropriates. In light of this perspective, we can say that a good design project is defined as a final product that belongs to a system, which in turn belongs to a service, integrated into a concept capable of generating an experience of value, Positive time, in the one who consumes it and appropriates it. That is, design is not just about products but a whole context that is created around them.

For Ricardo Blanco, the "design" object develops over time and space, that is, socially it is a collective creation that stems from the course of history, even stating that all objects always come from other objects, As inserted into an evolutionary continuum in a particular context or social group; According to Blanco the concept of original and the degree of novelty that we gauge to an object, refers to the fact that we recognize a matrix of predecessor object, that is, when we recognize its spatial or cultural origin, because without this reference no We can truly declare originality. In the design act, the designer proposes not only new discourses of cultural identity but also traces of objectified identity; Taking into account that it is in the material culture that the identity characteristics are manifested, that is when the designer's speech and the solution he offers to the problematic proposed to him must operate. An object of daily use acquires its value when it integrates the culture of the project, the visual culture, the production and the culture of use (Blanco, 2007: 18).

It is in this context that the designer has to operate, keeping in mind that, rather than playing the role of the designer, he plays a mediating role with the contexts and realities with which he relates, both from the point of view of the problem, From the point of view of the person who projects, directs and offers the solution to the problem that has been proposed.

In reality, when a designer designs a product, he can not only think of the product as something isolated; It's not just about the end product, but about everything it takes to get to it. This means that in a methodological process it is the responsibility of the designer to search and define the entire course of an object from the point of projection and production (industrially or not), how the envelope will be, how it will reach the market (e.g. via retail or via Online), and how it will be packed and transported to your points of sale. This means that the designer has to project and reflect on the whole universe and the final apprehension of the product when he creates it, that is, he has to think of a whole system and management in which the final product will be integrated, which system is in turn part of a service, and it is the business structure where this service is inserted that establishes relationship between the brand and/or company, and the final consumer.

What the consumer gets when buying an appliance or a service is not only the product that fulfills a particular utilitarian function, but all the story, experience, and surroundings that the product carries with it. When a consumer acquires a watch, or a piece of handicraft, we are actually acquiring the image associated with that watch, i.e. a particular lifestyle, or an ancestral story with a particular meaning. We do not just buy a watch that tells us the time and has two hands, nor do we buy just one clay pot just to carry a liquid. We buy its context, its symbolic value and the bonds we establish with it. In reality, if the meaning of the final product is the result of a positive experience and able to leave a trace and if we do not remain indifferent, we also pass on a positive and valuable image to others. And this is in fact the true ability to propagate objects.

More than creating, completely original and innovative products, it is important to create significant and important products for those who consume them. And it is in the role of the designer that this responsibility lies. The responsibility to draw a context, a system, a service, then a link to your consumer and your universe. In the post-industrial era this became the touchstone, that is, the focus of design became—in a way—more comprehensive, allowing a more precise and precise purpose.

As has been outlined in the foregoing paragraphs, the success of social design relies on effective collaboration between the social designer(s) and existing social entities as well as among the entities themselves. This implies that an effective social design must take into account the various community processes and strive to understand the concept of community building through trust building (Yee & White, 2016). Designers should therefore strive to build up community capacities and align the leadership of the community through action research and organizational transformation. The legacy of action research and the concept of organizational transformation thus provide important design platforms on which to enhance interaction in the scenarios created by social design. Another important platform of action for building the necessary interaction is the amalgamation of the concepts of participatory design and social innovation (Manzini, 2015).

Manzini (2015) points out that designers in the post-industrial society should embrace the methodic approaches of co-creation, participatory design, and social innovation in their design processes. Within the frameworks of the three methodic approaches to design, Binder et al. (2011) reminds designers the necessity to work on small scale scenarios as to use different of design approaches in their practice. In a co-creation scenario, designers are enabled to work in workshops, which are equipped with a wide range of design elements at tools, while participatory approach to design enables the designer to integrate field work with the actual users (Chen et al., 2015). Social innovation on the other hand capitalizes on the laboratory metaphor, which is however, connected to real life situations, and constantly push designers out of the studio scenario (Chen et al., 2015).

The erosion of cultural identity in an effort to improve contemporary society's conformity to a standardized global culture is a threat to the cultural memory of individual communities around the world. Primarily, the originality of culturally-inspired items as well as the design and production processes required places their sustainability at loggerheads with the consumerist agenda of modern socioeconomic environments. Favoring the adoption of technological advancements in an effort to increase the conformity of these cultural items to modern production approaches risks losing the individualism of these products as they become commodities. Moreover, a strict adherence to traditional methods of developing items for purposes of retaining cultural identity is also unsustainable when compared to the ability of the mass-market strategy to create and disseminate standardized items more rapidly. In the impasse that follows, the desire to retain the value of these items as foundations for identity construction means that concessions have to be made through a reduced focus on the craft techniques implemented. Therefore, the retention of cultural memory in material culture today must find a balance between the dissemination of visual markers of cultural identity and the preservation of craft techniques as the source of the cultural value ascribed to the items produced.

In conclusion therefore, from a post-industrial era perspective, the concept of social design seems to be an idea whose time has come and ignoring it or doing nothing is likely to be perilous for the contemporary designer, especially given the knowledge economy and information society we live in today. Therefore, taking stock of the various elements of the concept of social design is of the essence in the post-industrial era. In this regard, designers have an obligation to identify and understand the various existing collaborative process associated with social design. In particular, designers need to embrace the platforms of action research and organizational transformation. In addition, it is imperative for the designers embrace the methodic approaches of co-creation, participatory design, and social innovation in their design processes, since it is not possible to form future designers without these notions, it is also imperative that in the process of design teaching, students can be confronted with established relations in real-life context scenarios, with designer partners, such as artisans and its local communities.

ACNOWLEDGEMENTS

This work would not have been possible without the support of the Superior School of Education from Lisbon Polytechnic, as for the degree in Visual Arts and technologies Students namely from the ones that took the curricular units of Product and Interaction Design from 2015 to 2018. We are also grateful to be research members of CIAUD from Lisbon School of Architecture, who financially supported and allowed participation in CIMODE 2018, as to those we have had the pleasure to work during this project, in particular we would like to mention our colleague Prof. Sandra Antunes and the Craftsmen involved, Mrs. and Mr. Pequito as to Nisa local municipality.

REFERENCES

Armstrong, L., Bailey, J., Julier, G., & Kimbell, L. (2014). *Social design futures*. Available at: http://mappingsocialdesign.org/2014/10/09/social-design-futures-report [Accessed 6th February, 2017].
Bachman, L. 2016. *Sustainable Design and Postindustrial Society: Our Ethical and Aesthetic Crossroads*. The ARCC Journal, 13(2), 30–38.
Barthes, R. 1970. *L'empire des signes*. Skira. Les sentiers de la création, Paris.
Bell, D. 1979. *The social framework of the information society*. The information society, 3.
Buxton, B. 2010. *Sketching user experiences: getting the design right and the right design*. Morgan Kaufmann.
Chen, D.-S., Cheng, L.-L., Hummels, C., & Koskinen, I. 2015. *Social design: An introduction*. International Journal of Design, 10(1), 1–5.
Cuche, D. 2010. Pages de fin. Repères, 4, 148–160. Featherstone, M. 1995. *Undoing culture: Globalization, postmodernism and identity* (Vol. 39). Sage.
Douglas, M. 200). *The world of goods: Towards an anthropology of consumption* (Vol. 6). Psychology Press. [1979].
Fiske, J. 2011. *Reading the popular*. Routledge.
Kopytoff, I. 1986 – *The social life of things: Commodities in cultural perspective*. Cambridge: Cambridge University Press. 2 – The cultural biography of things: Commoditization as process.
Law, J. 1991. *Introduction: Monsters. Machines and Sociotechnical Relations* In: LAW, J.(ed.) A.
Manzini, E. (2015). *Design, when everybody designs. An introduction to design for social innovation*. Cambridge, MA: MIT Press.
Margolin, V. 2007. *Design, the future and the human spirit*. Design Issues, 23(3), 4–15.
McDonough, W., & Braungart, M. 2010. *Cradle to cradle: Remaking the way we make things*. MacMillan.
Pevsner, N. 2005. *Pioneers of modern design: from William Morris to Walter Gropius*. Yale University Press.
Portes, A. 1998. *Social capital: Its origins and applications in modern sociology*. Annual Review of Sociology, 24(1), 1–24.
Providência, B., Guedes, M. D. G., & Cunha, J. 2005. *Cultural and communication common grounds in art craft and design*.
Robison, L. J., Schmid, A. A., & Siles, M. E. 2002. *Is social capital really capital? Review of Social Economy*, 6(1), 1–24.
Veneris, Y. 1990. *Modeling the transition from the Industrial to the Informational Revolution*. Environment and Planning, 22(3), 399–416.
Walker, N. E. 1998. *The design analysis handbook: A practical guide to design validation*. Newnes.
Wood, N. 2006. *Transmitting craft knowledge: designing interactive media to support tacit skills learning* (Doctoral dissertation, Sheffield Hallam University).

Immersive business simulation in footwear design education: A bridge for reality

A.M. Terroso, A. Moreno, N. Amorim, H. Palmares & J. Sampaio
Polytechnic Institute of Cávado and Ave, Design School, Barcelos, Portugal
Institute for Research in Design, Media and Culture ID +, Portugal

ABSTRACT: The training of designers for the footwear industry is an area in higher education still with little expression in the universe of sub areas that today specialize design. Access to literature that describes, prescribes and analyses teaching experiences in this area is therefore also not expressive or even non-existent. In this context, this paper aims to contribute to all the players involved in the educational process of footwear designers, as well as providing information so that this process is effective in the evolution of students and prepare them for their respective professional reality. It describes a process of design and development of footwear collections for the fashion industry, compound by seven phases and that with its real application through an immersive business experience simulation allow students to develop technical, emotional, aesthetic, functional, productive, commercial and social skills.

1 INTRODUCTION

Since ancient times people have protected their feet from the irregularities of the terrain and the variations of the climate. Today, footwear goes far beyond protection and aesthetics and appearance plays a fundamental role in choosing a particular type of footwear; fashion influences choice, comfort and protection are no longer decisive in the purchase option. Some of the footwear companies in Portugal have undergone a transformation in their business model in recent years, the entrepreneurs have acquired experience working for international clients in private label model and realized that with the know how acquired over the years and, the way to add value in Portuguese footwear would be the bet on brands with differentiated products. With this change the sector encountered a problem, to put in the competitive global market a differentiated brand is necessary a design driven business policy. The lack of designers in Portugal with a higher education in footwear was large, being only made up of professionals from other creative areas (eg: fashion and product designers, architects) but without technical and sensory training in the footwear area. The Design School of the Polytechnic Institute of Cávado and Ave (IPCA), in order to meet the need for footwear designers in the current context of the sector, opened in 2015 the first Portuguese higher education course in the area, with a mission to train footwear designers for the Portuguese footwear industry. Characterized by having an essentially practical training component focused on the fashion market. From the first to the third semester students are confronted with a project-based learning course and where technical and theoretical disciplines merge into a practical goal consisting of the design and development of seasonal footwear collections for the fast and competitive fashion market. The 4th and last semester reinforces the skills of the students with an internship in the main Portuguese footwear brands and companies.

One of the key imperatives of the European Union's education and training strategy by 2020 is the promotion of creativity, innovation and student entrepreneurship (European Commission—Steregic framework—Education & Training 2020). Higher education institutions have several local and global imperatives in structuring design courses. These imperatives include industry-related curricula, student-driven attributes and skills, learning-centric approaches to teaching; credit requirements and the need for graduate students to be ready for professional work and lifelong learning skills (Holdsworth & Hegarty 2016; Lasaouskiene & Rauduvaite et al. 2015). Research shows that the application of design-based teaching methods improves the quality of teaching and learning and contributes to cognitive development involving students in solving problems through more innovative solutions and teaches planning and communication procedures (Pietila & Virkkula, 2011). It stimulates work in interdisciplinary teams so that students acquire social skills through the experimentation of real life situations where the teachers assume the role of coaching in the creative direction of the project in order to provide the experience of leading teams and projects (Heinis et al. 2016). Methods based on the constructive knowledge centred on project development as a tool of knowledge catalyst and discovery of solutions to real problems (Arcidiacono et al. 2016). According to the author, there are five fundamental characteristics to the application of methods based on constructive knowledge: (1) to begin with a question that alludes to a problem; (2) students explore the question and apply functional and emotive skills to solve the problem; (3) students and facilitators progress with collaborative activities in search of solutions; (4) students throughout the process have the opportunity to use techniques that help them participate in activities outside their assigned competencies; (5) creates a set of tangible products that answer the question and are shared with the rest of the class. In order to bring these characteristics closer to reality, a sixth has been added that allows challenges and projects in an academic context to be applied to reality and provide a more reliable simulation experience.

1.1 *Objectives and methodology*

Based on the context described, the question was asked: how can we bring the industry closer to the teaching of footwear design in order to provide an experience closer to the business reality of the sector? In order to find the best answer to the question posed, a working method consisting of seven operational phases was outlined, in Figure 1 we can see a summary version with the main phases of the project and the main iterations. Prior to the start of the works, a cooperation agreement was made between the two parties involved. In this agreement, the project outline was described, as well as its timing with tasks and deliverables to be achieved and the duties and rights of each part, safeguarding the intellectual property and commercial rights of both parts. In order to take advantage of the technical skills that the course provides, a project orientation and mentoring team was set up. This team was led by the footwear design project II class, which has the responsibility of making the creative direction and management of the whole project. The classes of digital illustration, technologies and manufacturing processes and drawing and composition work as technical support to the different stages of the process (Fig. 1).

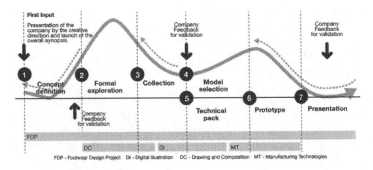

Figure 1. Scheme with the summary of the design process for the development of the project, in the figure we can see the different stages, as well as the iterations realized, the advances and retreats and the participation in the process of the different classes involved.

2 APPLYING THE DESIGN PROCESS

This chapter intends to describe the application of the design process referring to all phases, always accompanied by opinions of improvement of the process that were detected by the players during their accomplishment.

2.1 *The first input*

The first step was the presentation to the students and teachers of the company by the creative and marketing direction and culminated in the launching of the general synopsis turned into a briefing. The presentation focused on a business plan summary with a special focus on market positioning as a concept label that seeks to balance a niche sector with high levels of quality at an affordable price, ie branded products should convey a sense of luxury that is tangible to the masses ensuring innovation through balanced design and price; the culture of the brand is associated with a group of young and ambitious people who come to fashion as part of the personality of each one and where charisma and success are part of personal and social enrichment, manifesting this way of being in an emerging social movement which they call *New Ladies and Gentlemen*.

2.1.1 *Target audience and lifestyle*

The target audience defined by the brand consists of ambitious, confident men and women with a mind-set focused on personal and professional success, Aprille reported that this type of audience fits the generation known as the *millennials* with an age group of 25 to 45 years of a middle and upper class. Aprille sees this audience as *Game Changers*, smart people who are not afraid to risk and contribute to innovation with a disruptive way of thinking and who want to make an impact through their actions and interventions and who mirror that way of being in their way to dress. Aprille's women and man may be CEOs of a publisher, a consultant, or lawyers but they have a special taste for fashion. They are independent and love to reinvent themselves, participate in cultural events and value nature as a way to escape from their professional life. In short, they seek the perfect balance between professional, personal and family life (Fig. 2).

2.1.2 *Fall/winter collection plan*

After describing the target audience and their lifestyle Aprille defined a plan for the fall/winter collection, the objective of the plan was to make a synthesis of all the elements essential to the success of the collection and so that the models developed could meet the public without compromising creativity and innovation. In a first phase, attributes were defined so that the shoes should contain the desired lifestyle, consisted of models with a clean and classy expression where simplicity should be the main motto of the collection, the ability of the models to be sculptural and sophisticated and to allow the models to have a function that goes beyond their merely functional function, considered a living piece and reflect the personality of the wearer, as well as hold their own personality as a sculptural piece.

Figure 2. Panel illustrating Aprille's woman and man profile based on celebrities of the fashion universe.

2.1.3 *Setting: Footwear design gym*

The importance of a space adapted to the context of the project and sufficiently prepared to foster collaboration, creativity and experimentation is in our view one of the preponderant factors for success. The project was developed in Design Gym, a space designed to exercise creativity in three components: the manual, digital component and the physical production of prototypes. These spaces, which we call laboratories for manual creation, digital creation and footwear production (Fig. 3), function as areas dedicated to the design and development of footwear and provide the physical conditions and equipment necessary for the success of the process and where students can have contact with a work environment as similar as possible to a footwear industry with access to design, development and production areas.

2.2 *Formal exploration*

The formal exploration was the second phase of the project and the goal was the definition of the concept of each working group, in this way all groups defined a theme based on several subjects but framed in the target audience and lifestyle intended by Aprille (ch. 2.1.1), the choice of themes went through several typologies, from an action (e.g. the search for a unique attitude based on elegance and sophistication); in an emotion (e.g. formal and minimal purism that provides focus on the essential and in small details); in a function (e.g., seams applied to high quality products such as the leather upholstery technique); in a certain aspect of nature (e.g. the organic geometry proper from a river); in a certain form (e.g. the traditional Portuguese tiles); among others. The defined concept should be synthesized in a single image illustrative of the chosen theme and later transformed into a mood board where the main elements of the concept were identified and once again illustrated (Fig. 4). Through the mood board each group should be able to identify shapes, textures, colours, silhouettes, or patterns that would be inspiring to create the first lines of the models to develop.

Figure 3. Overview of the project setting, laboratories of manual, digital and shoe production.

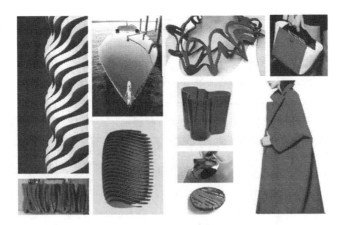

Figure 4. A mood board example made during the formal exploration phase, this concept, based on the organic lines of the Araguaia River in Brazil, gave rise to a survey of a series of inspiring formal elements based on architectural elements, furniture, clothing and fashion accessories and where the curves and overlays are predominant and highlight the colours red, black and blue.

The next step in the formal exploration was the choice of the model to be developed and where each group should choose a type for women's shoes (Pump, stiletto, boot, loafers) and men's shoes (Oxford and Chelsea boots). Respecting the models established in the collection plan defined by Aprille, the students of each group, the footwear design project teacher and the shoe last company Carvalhinhos S.A, carried out the process of selecting the models. The inclusion of a company specialized in the modelling and manufacturing of footwear lasts with more than 90 years of activity has allowed a greater proximity by students to the national reality of the industry. Each group had access to a couple of pairs of physical (female and male) lasts in 3D digital format (Fig. 5) for later modelling the models to be developed, either in a paper-lined form for drawing directly on top of the last or for the process for three-dimensional digital modelling.

2.3 *Collection*

After the phase of formal exploration and selection of women's and men's footwear models the next step consisted in the development of the collection for the concepts pre-established by the working groups. The development of the collection began with manual profile drawings on the chosen lasts and where several alternatives of models were created (Fig. 6). After designing the general lines of the different models alternatives, developed through manual drawings, lateral silhouettes were drawn from the last through vector drawings, this step made the transition from the manual component (manual drawings) to the digital component (digital drawings). Figure 7 shows some digital designs, this digital models allows a greater rigor in the proportions and dimensions of the exterior and interior parts, jumps and soles, accessories as the buckles and cords. In this phase of the project the classes of *Drawing and Composition* and *Digital Illustration* were introduced in the process, taking advantage of the skills of each area. In the case of *Drawing and Composition*, the groups were encouraged to apply the manual drawing techniques taught in the course and apply practical exercises to the project, without repetition of contents and ensuring that students understood, in a

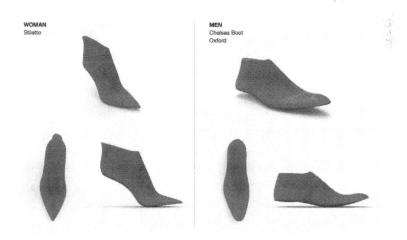

Figure 5. Example of lasts selected for the female stiletto model and for the oxford male model, in the latter case the oxford shape also serves the Chelsea boot model.

Figure 6. Manual profile drawings expressing the first ideas for the models that makes up the collection.

Figure 7. Some of the vector designs drawing for women's models and where the rigor of the lines, the choice of materials, stitching, laces, clasps, buckles and desired colours are simulated.

practical and real way, the importance of manual drawing in the design and development of collections. The same happened with the class of *Digital Illustration*, in this case the students realized the usefulness of the models created manually and that served as basis for the digital drawings. They also assimilated the difference of techniques and purposes in the use of manual designs (faster and more conceptual) and digital drawings (more rigorous and with more attention to detail).

2.4 *Model selection and technical pack*

After the creation and development of the collection, each group prepared a brief presentation to Aprille, in order to select the models to prototype, this step allowed the students to have real and professional feedback of the client for each collection, being at this stage still possible to correct some detail that allows the collection to meet the desires and expectations. To make the situation as real as possible the groups chose a male and a female model in each collection and with which they modeled in 3D (Fig. 8) using a three-dimensional solid modeling software specifically for footwear (Autodesk Shoe Maker). The models allowed renders to be seeing from different perspectives that allow a more real visualization of the models, as well as the simulation of materials and textures. A packaging proposal for the models and graphic image were also designed (Figure 8).

After the validation of the collection with Aprille the students proceeded to execute the technical sheets of the selected models. This sheet consists in the technical information required for its pattern modelling and construction. It consists in the last reference, cutting observations, applications, types of seams, type of assembly and finishing, as well as references to materials. The execution of this sheet allowed students to understand the importance of the technical feasibility of the models in the preparation for industrialization (Fig. 9).

2.5 *Pattern modeling and prototype*

Footwear pattern modeling is one of the most delicate stages in the development process of a collection, this stage is the planning of the different parts and components that make up the shoe, and that will later be used to produce prototypes for validation of the lines, feeting and possibility of industrialization. The students groups, with the support of the technical pack, elaborated the paper and digital modeling of the selected models (Fig. 10). Each group had to make female and male models, this task provides them contact with the modeling techniques for the different models and the constraints associated to both the lines of the models and the behavior of the materials in the different stages of construction. In this phase, the class of *Manufacturing Process* provided support in the execution of the modeling, as well as advice in the application of the construction and assembly techniques most appropriate to each model, in the choice of materials, types of seams and finishes. The class of *Digital Illustration* also played an important role in the execution of the digital modeling, after the planning of the components in paper, each group had to digitize the plans and later realize the CAD drawings of each component with the purpose of making an infographic of the model attached to the technical pack, as we can see in Figure 10.

Figure 8. Two 3D models examples with the digital simulation of the packages.

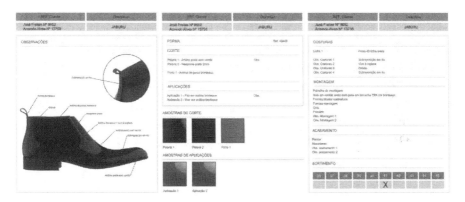

Figure 9. Example of a technical pack for prototype.

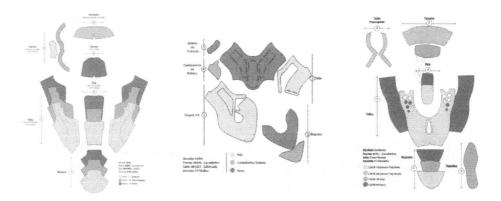

Figure 10. Infographics of the digital pattern modelling for prototyping.

After pattern modeling, in cardboard and digital, the production process of the prototypes was started, where all the stages of a shoe production line were tried, from the cut (using the cutting knife and also laser cut); the preparation of materials for sewing, with the facing and equalization of materials; the sewing of the different components, the assembly and the finishing. This process allowed students a very real bond to what they can find in a footwear production industry, giving them production technical knowledge of all the steps, materials, and components needed to build a shoe. It also allows footwear design students to understand and assimilate the principles of each type of assembly that are part of the footwear industry and link these principles with the models developed, having direct contact with the constraints associated with each technique and being able to adapt their models to be prototyped and later industrialized, Figure 11 shows photographs of some of the prototypes.

Figure 11. Examples of prototyped models, Chelsea boot for man and pumps for woman.

Figure 12. Overview of the showcase "Shoes & Books" that took place in the city of Guimarães.

2.6 *Presentation and promotion of the collection*

The last step of the process was the final presentation of the collection to Aprille and a strategy of dissemination and promotion. The presentation was carried out with all the groups and teachers, the groups made a synthesis of the projects with the options taken in each phase, the feeting test of the prototyped models was carried out to identify details of improvement, either in the lines of the models or in comfort. It was decided that it was necessary to test the models with the target audience, given the classic nature of the models, a concept capable of agglomerating the collections of all groups in a single collection, the *"Shoes & Books—A Classic Combination"* concept was defined. A showcase was held in the city of Guimarães, one of the most classic Portuguese cities, this showcase consisted in a shoe store installation (Fig. 12) that simulates the environment of a selling point and where the models are exposed for the final user validate them. To promote the showcase was written a press release and sent to the major partners in the footwear industry, from producer companies, national brands, business associations, national designers and media.

3 CONCLUSIONS

The approach to the professional reality of the designer during his higher education, based on immersive business experiences and project simulation, has been a very debated theme in different areas of design, however, in footwear design education has not been the same. This experience allows students to have real contact with knowledge and application of hard and soft skills, very useful in the development of professional pathways in this area. Teamwork, time management, planning and scheduling of projects, the need for autonomy in the search of materials and components, the improvement of interpersonal communication skills, among others are some of the soft skills provided. Regarding the footwear design technical skills, the experience has provided a great appreciation and evolution of the students in the development of a footwear collection, starting from the client's briefing and finishing in the presentation and validation of the results by the defined target audience. This process allows students to learn by doing, with objectives and where the sequence of goals helps in the materialization of a meaning for each technical, emotional, functional, productive and commercial option.

ACKNOWLEDGMENT

This work is financed by national funds through the FCT—Foundation for Science and Technology, I.P., under the project «UID/DES/4057/2016».

REFERENCES

Arcidiacono, G.; Yang, K.; Trewn, J.; Bucciarelli, L. 2016. Application of Axiomatic Design for Project—Based Learning Methodology. *Procedia CIRP* 53, 166–172.
European Commission—Strategic Framework—Education & Training 2020. Accessed in 10th October 2017 in: http://ec.europa.eu/education/policy/strategic-framework_en.
Heinis, T.B.; Goller, I.; Meboldt, M. 2016. Multilevel Design Education for Innovation Competencies. *26th CIRP Design Conference* 50, 759–764.
Holdsworth, S. & Hegarty, K. 2016. From praxis to delivery: a Higher Education Learning Design Frame work (HELD). *Journal of Cleaner Production* 122, 176–185.
Lasaouskiene, J. & Rauduvaite, A. 2015. Project-Based Learning at University: Teaching Experiences of Lecturers. *Procedia—Social and Behavioral Sciences* 197, 788–792.
Pietila, M. & Virkkula, E., 2011. Integrating Therapy and Practice According to PBL—Based Project Designs in Secondary Vocational Education of Engineering and Music. *Proc. from the 3rd International Research Symposium on PBL, Coventry University. Aalborg University Press*, 54–63.

Bounding fields of knowledge: Co-designing regional pottery of the Cávado sub-region as a strategic potential of museological differentiation and product development

J. Sampaio, A. Baganha, F. Gomes & A.M. Terroso
Polytechnic Institute of Cávado and Ave, Design School, Barcelos, Portugal
Institute for Research in Design, Media and Culture ID +, Portugal

ABSTRACT: The knowledge and cultural heritage appropriation to other forms of production and expression emerge from interpretation by its contemporaries. This article reports a procedural approach for two utilitarian tableware proposals develop in collaboration with the Pottery Museum of Barcelos (Portugal) and MA students. It describes several stages of an ongoing process that aims to bound theoretical, technical and applied knowledge; promoting direct contact with artisans, based on a co-creative approach among stakeholders; grounded on pottery production know-how and it cultural heritage from production Centre's of Barcelos Region; and adapting it to contemporary uses. By this approach, we aim to develop in student's research, conceptual and productive skills; to build positive partnerships between the MA investigation and the society; and culturally acting and enhancing product quality and economic recognition to the sector, joining diverse fields of knowledge for development of proposals in terracotta and black clay.

1 INTRODUCTION

Barcelos is a Portuguese town in the District of Braga, in the Northern region and Cávado sub-region, with approximately 20 625 inhabitants. It's a municipality with 378,9 km^2 area and proximally 120 400 inhabitants, subdivided in 61 parishes council. After textile manufacturing and clothing making industries, the manufacture of other non-metallic mineral products—where we find Manufacture of Other Non-Refractory Porcelain and Ceramic Products—that emerges as the third most representative activity sector in this region. According to an online companies' directory (Info Empresas, n.d.) it is possible to identify 335 companies or producers of household ceramic goods in Barcelos Region. This sector encountered several problems, like quantity-based business models, low prices and several companies dependent on one or two major clients.

In terms of recognition, the municipality acknowledge regional pottery and ceramic sector as an economic, cultural and tourism asset that is presented in the city in sculptures, museum, tourist office and in municipality's online platforms. As a sign of respect to its cultural heritage, the city demonstrates is tribute promoting the work of several artisans like Rosa Ramalho (a bigger artist), Domingos Côto (creator of Barcelos Rooster) and Rosa Côta, by building big scales models of this artisans in several city roundabouts. Apart of this authorship form of expression, the creation of a website with "figurado" (figure/puppetry) and pottery routes as a touristic asset, that identify active pottery artisans and workshops, along the Pottery Museum these are several examples of promotion of it regional culture.

The Pottery Museum appears in 1963, in the beginning by the name "Regional Museum of Ceramics", it was created following the donation of a very valuable collection gathered by the ethnographer Joaquim Sellés Paes Villas Boas. Initially linked to regional pottery but by the continuous grow of it collection it became a compilation of the national pottery; the institution was renamed as the "Pottery Museum". It collection holds an asset of nearly 9000 objects. It represents national pottery as well as other Portuguese-speaking countries pottery traditions (Museu Olaria, 2014). Following the touristic boom of Portuguese north region in the last years, the museum starts to feel the need to update itself for the growing number and diversity of visitor's nationalities. The idea of creating a museum store/cafe start to build up and became logic as another way to accomplish it mission and objectives of cultural dissemination and preservation; to value pottery as evidence and document of a tradition, society and culture; and support and cooperate in the safeguard, study and spread of the national pottery heritage.

The Design School of the Polytechnic Institute of Cávado and Ave (IPCA), aims to instigate the proximity between Master of Design and Product Development (MDPD) students and the industry or the community by creating a partnership with Museu de Olaria (Pottery Museum). By meeting the needs of several parties, it was possible to preliminarily define a project proposal for utilitarian pottery to be integrated as museum merchandising. This project involves two MDPD students that are on an ongoing dissertation process. This association, able us to support a more complete literacy course, that is mandatory to a scientific education like a master's degree but also to link theory to a practical case study that helps the students to explore the complementary relationship between these two lines of investigation that are often separated.

1.1 *Goals and methodological approach*

According to the previous scenario, we start to focus on how we can involve several parties stimulating a sense of belonging and co-creative approach of proposals, that could act as a strategic approach to add value to regional and national pottery heritage and it contemporary use. In that scenario we define a two axes approach. The first axis defined by

Figure 1. Pottery Museum of Barcelos, permanent exhibition of North Portugal pottery.

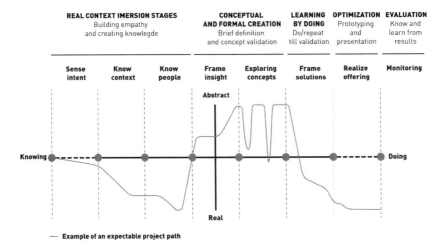

Figure 2. Methodological approach.

what is (real) and consequent *deconstruction* (abstract), followed by exploration and construction of new scenarios of *what could be*. The second axis falls on the gap between analysis and synthesis, acting as bridge from *the problem* (knowing) to *the solution* (doing). In this parameter, the importance of materialization was identified as a way of acting on both axes (*real/abstract* and *knowing/doing*) (Sampaio, 2016). This conceptual framework is shown in Figure 2 and parallels with Vijay Kumar (2012) procedural proposal. Finally, we adapt a *positive core approach* (Cooperrider & Whitney, 2007), as an initial starting point, aiming to promote opportunity identification rather than problem identification. Along with this framework, we identify three major fields of knowledge and action that were related to several actors: historic and theoretical knowledge (held by the museum and Polytechnic tutors); practical and technical knowledge (held by the artisans and producers); and the last conceptual and applied knowledge (held by the examples of professional designers practice in this area).

2 THREE FIELDS OF KNOWLEDGE

In this ongoing investigation were define several tasks to further development:

- Theoretical and historic knowledge: build up knowledge, based on literature review held by regular and weekly presence of students in the museum, with guidance and supervision of the museum staff, having access to the museum archives (publications and collection). By this approach we aim to student build knowledge and ability to analyze and interpret information, to consolidate cultural, social, symbolic and technical knowledge of pottery evolution. Main actors: museum staff and tutoring team;
- Practical and technical knowledge: know people and context, based on field investigation and contact with artisans and productive facilities to identify productive opportunities and restrictions. By this approach, we aim to student build contact network to further develop technical knowledge and skills, interacting with artisans during design process and model making. Main actors: artisans and tutoring team;
- Conceptual and applied knowledge: know approaches and proposals made to a contemporary society and audience. Based on case studies, students must analyze several projects in term of their conceptual, technical and strategic design stance. By this approach we aim to students build critical posture in order to help them to build conceptual framework that enhance student's proposals in order to add value to its products but also for them to act as a cultural and economic actor. Main actors: professional designers and tutoring team.

All this three fields will be the ground stone to build and create their personal brief and support it conceptual and productive framework. This approach must be articulated it with the parties, being the main focus the products creation that intent to add value to regional pottery and be a representative product of the mission of Pottery Museum of Barcelos.

2.1 *Theoretical and historic knowledge: Historic Barcelos' pottery production*

Minho is a privileged region of pottery production, a proof of it, is the commercialization to whole country and Galicia region since back in the days. In this context, stands out the extinct municipality of Prado whose area corresponded to some parishes council of the municipality of Barcelos, Braga and Vila Verde. This area has been referred to as a relevant ceramic productive centre since the 13th century (Remelgado, 2005). Due to the abundance of raw materials, the centre of Prado was one of the most important in Portugal (Araújo, 1998). They produced terracotta/red clay, black-clay and glazed pottery. Although black pottery production was specific to the Paradas de Gatim and S. Mamede de Escariz parishes council, both belonging to the municipality of Vila Verde, Braga municipality. The last Prado potters (3rd quarter of the 20th century) sold their dishes exclusively in the fortnightly fair at Ponte de Lima. The dishes were transported in a cart of oxen, bedded in wheat straw, or carried on the head by women (Fernandes, 2012). Many sales were made to intermediaries who bought dishes that were transported to Lanheses and Viana through fluvial, and then sold them in places like Paredes de Coura, Braga, Guimarães, Porto, among other regions (Fernandes, 2012). Alongside the Prado Pottery Centre, other places have also produced black crockery, mainly in the north and center of Portugal. But over the years, the pottery industry started a slowly path to decline. In the 1940s, there were few localities where clay pieces continued to predominate, which contributed to the extinction of many production centres (Fernandes, 2013). Given the scarcity of supply and the loss of functionality to much of pottery containers, the pieces produced by the current potters no longer meet the utilitarian needs of current consumers (Fernandes, 2013). Historically, pottery production is associated with extreme poverty, a stratified family business that occupied various elements of the family. The learning process of this art started in the early years of life passed by parent to children, being rare to have someone outside the pottery community forums (Fernandes, 2012). In modern times, the tasks were divided by non-specialized workers such as mothers, children, daughters, minors and apprentices, who were engaged in the heavy tasks such as: water transportation, ceramic paste preparation, wood transportation, firing and pottery sale. In the specialized work, we had the master potter on potter's wheel, as well as the painter who executed decorations in the artefacts (Remelgado, 2005).

2.1.1 Regional pottery

Focusing on the regional artefact that were produce, it is possible to identify several proposals of typology organization for utilitarian artefacts using terracotta and black clay. Although potters from county of Barcelos distinguish it in ten different sizes — *cozinheiro, entremeia, canada, meia canada, quartilho, risco, meio quartilho, quatro vinténs/meio risco, quarteirão and meio quarteirão* — only for reasons of production and references to height and diameter, but it is also possible to identify artifact cataloging system based on their function (Remelgado, 2005). If we focus on utilitarian terracotta production, we can identify 9 contexts of use, identifying 40 typologies of artefacts that present in Table 1.

If we focus on black clay production, we identify 9 different contexts of use, counting 28 typologies of artefacts (Fernandes, 1997) that are present in Table 2.

2.1.2 Practical and technical knowledge: Current local production

In terms of black clay production in Barcelos Municipality, the only artisan is Mr. Júlio Alonso with 91 years old, as shown in Figure 3, is the last black clay master, barely active and mainly with the help of his son João Alonso. Although in November 2016, due to an

Table 1. Contexts of use and typology utilitarian terracotta objects of Barcelos region.

Context of use	Objects name*	Obs.
Illumination	*Palmatória; castiçal.*	Candle holder
Hygiene	*Penico; bacia lava-mãos.*	Face and hand cleaning
Savings	*Mealheiro.*	Coins saver
Food storage	*Talha; boião.*	Storing lard, meat and olives,
Liquid storage	*Moringa; Cabaça; caneca; cântaro; poço; vinagreira.*	Storing wine, water and olive oil
Serving food	*Ladeira; saladeira; tijela; rente; frigideira; terrina; patelo; travessa; torteira.*	Serving food, salad, soup and sauce
Serving liquids	*Infusa; caneca; caneca do segredo; galheteiro.*	Serving water, wine, oil and vinegar
Food preparation	*Assadeira* or *pingadeira; canão, frigideira de rabo; alguidar torto; terrina; sopeira; coador; lisboa; coimbra; alguidar; chocolateira; panela.*	Preparing and cooking meat, eggs, rice, coffee, tea and baking
Transport of food	*Confeiteira.*	Transporting food to the countryside

*Portuguese nomenclature.

Table 2. Contexts of use and typology utilitarian black clay objects of Barcelos region.

Context of use	Objects name*	Obs.
Toys	*Cantarinha.*	Small water transporter
Health and care	*Botija; barbeira.*	Hot water bottle and barber rack stand
Agriculture/floral	*Vaso; vaso de portão; cântaro de sulfato; bebedouro; coluna.*	Planting, sulfate, food and water recipient for small animals
House tools	*Cano; telha.*	Water pipe and roof tile
Food storage	*Panelão.*	Storing olives
Liquid storage	*Cântaro; vinagreira.*	Storage water, wine and vinegar
Serving liquids	*Burreta; caneca; infusa.*	Serving water and wine
Food preparation	*Caldeira; terrelo; caçoila de rabo; caçoila* or *caçola; burrito de uma asa; burrito de duas asas; pote; assadeira de uma asa; assadeira de duas asas; panela; chocolateira.*	Cooking soap, rice, chestnuts, tea and coffee
Transport of liquids	*Cantil.*	Transport fresh water

*Portuguese nomenclature.

Figure 3. (left) Visit to Júlio Alonso (91 years-old), the last master of black clay in Barcelo Region, 2017; (right) Julia Ramalho and son António Ramalho, guardians of the family heritage of Rosa Ramalho (grandmother and great-grandmother), 2017.

initiative of Vila Real Municipality the black pottery production process from Bisalhães was recognized as Intangible Cultural Heritage by the UNESCO (Lusa, 2016). This scenario is similar to other north regions of Portugal and demonstrates the need for action in order to preserve this knowledge.

In terms of terracotta clay and ceramic production in Barcelos, we can consider it to be more significant. Although b the above-mentioned companies' directory we identify 335 companies, it was possible to validate 103, that are divided in three sectors: clay pottery (93); manufacture of articles of porcelain or china (13); and manufacture of house ware articles of porcelain, china and fine stoneware (7). To these data we can add 53 artisans, divide in two groups: one that produces "figurado" (35), and the other linked to Barcelos' Roosters and Ceramic painting (18), that are present in the municipality touristic route. As shown in Figure 4, it's possible to understand producer's dispersity, being the Galegos de Santa Maria (32 producers) and Galegos de São Martinho (34 producers), the two parishes councils with a higher producers' density of and productive variety. In terms of territorial occupation, the east side of the municipality has a higher rate. After this geographic research, the students are starting to pay field visit to some distinct producers in order to complement the theoretical knowledge of productive technics and evaluate it adequacy to contemporary production needs, as shown in Figure 3.

2.1.3 *Conceptual and applied knowledge: Case studies*

This field of knowledge, aims to create a better understanding of contemporary use and appropriation made by designer of cultural symbols, regional aesthetics and language as well the production technique. In this task, students are organizing case studies following a similar structure between them, that focus on four major items:

- Identification: Clear project identification, author, promoter, year and references;
- Introduction: Brief description of project scope, its promoters and framework.
- Description: Systematic description bases on conceptual approach, techniques and production technics, markets stance and relation between designer and producers/artisans.
- Conclusion: Overall conclusion about project approach impact and relevance.

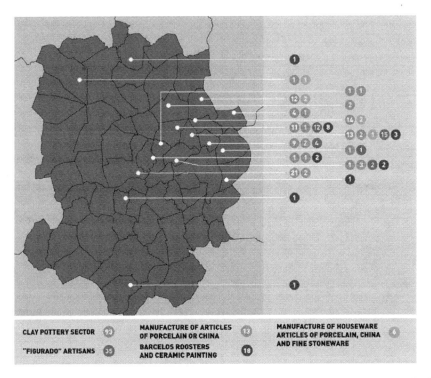

Figure 4. Geographic dispersity of pottery and ceramic producers/artisan in Barcelos' municipality.

Figure 5. Examples of national and international pottery project for case studies.

It's not mandatory that case studies are linked to a museum merchandising subject, although these approaches are important to the investigation scope. By this moment, the selection and analysis of case studies is being made and if it's possible, we stimulate direct contact with project promoter/author. At this stage, in this field of knowledge, it's possible to identify several approaches in national territory like: brand creation (like Bisaro and Tasa projects), independent conceptual/small objects editions (Lateira and Da Panela projects) and industry relation as shown in Figure 5. Also shown in Figure 5, international approaches from Jaime Hayon or from Martin Azua prove to be relevant to generate and overall analysis of approaches and interpretation by different countries and mindsets. This task will bring to surface not just a methodological approach but also a conceptual design and market orientation that will be consider in the next stage of insight framing.

3 NEXT STAGES AND TASKS

After the conclusion of this tasks the students will continue the project base on the methodological approach presented in Figure 1, focusing on completing several tasks:

- Frame insight: Opportunity identification, definition of conceptual framework and brief that emerge from the previous stages. This program will support project guidelines, task schedule, market strategies, stakeholders network and will be created by the students among the tutor and the museum staff, in order to enhance the collaboration between all parties;
- Exploring concepts: Brainstorming, concept generation, mood board, organizing and sorting concepts, communicating concepts (sketching, mock-ups, visualization) in order to be understand and validated by the several parties. Repeat this process until it founds a valid concept;
- Frame solutions: To explore 2D and 3D model's construction, to build a critical approach to shape and concept validation in order to build a proposal. This stage will stimulate direct contact with the producers/artisans to create win-win relation for the several parties;
- Realize offering: Prototype building, defining strategies to marketing, launching and presentation, overall integration in the museum;
- Monitoring: After project launching we need to understand public is reacting to the products. Act in need of changes in productive, communication and branding.

4 CONCLUSION

This procedural approach tries to bring to master research a practical and all-around approach to the subject of investigation. Although this is an ongoing investigation, the collaboration with the Pottery Museum of Barcelos is proving to be a major asset that reinforce and stimulates the students. By this direct contact, the students' immersion in the subject is more effective, acting as a recourse for theoretical, technical and cultural knowledge. These parameters are essential for student to learn and know how to act in a social, cultural and economic level through their design practice; the conceptual and applied knowledge is also a complement analysis that among the previous parameters will enrich the artifact development. By this method we try to reinforce the design practice, bounding theoretical and practical knowledge as well upgrade the conceptual approach in the investigation, that must emerge from a well-grounded brief, based on real knowledge. For next stages, we will bring co-creative approaches to reinforce collaboration between several actors, for them to act as Barcelos' ceramic promoters and contributors, bringing attention to it prominence and diversity. On one side, the process of co-creation will be essential to create visual concepts and models in terms to facilitate the communication and act on both methodology axes (*real/abstract* and *knowing doing*). On the other hand, it allows us to introduce and bring potters to design process and among students they can add their experience and insights to the proposals. By this it will not be so focus on controlling volumes and aesthetic assets but enhancing the symbolic and emotional and cultural outcomes of this partnership. With this project, we hope to contribute in a social, cultural and economic level, adding value to pottery and production Centre's of the region as well to the educational quality of the master, promoting an applied investigation and possibility to act in real life contexts.

ACKNOWLEDGEMENTS

The authors thanks to the staff from the Pottery Museum of Barcelos and to all pottery artisans and producers that are involve and contribute to this investigation.

This work is financed by national funds through the FCT—Fundação para a Ciência e Tecnologia, I.P., under the project «UID/DES/4057/2016».

REFERENCES

Araújo, A. 1998. A louça preta do Prado. *Olaria: Estudos arqueológicos, históricos e etnológicos.* N.º 2 (1) Dezembro, pp. 29–50.
Azúa, M. 2013. *Neo-Rebotijo.* [online] Available at <http://www.martinazua.com/product/neo-rebotijo/> [Accessed 14 Nov. 2017].
Bissaro 2017. *Coleção.* [online] Available at <http://bisarro-ceramics-4h7 g.squarespace.com/colecao/> [Accessed 13 Jan. 2018].
Bourbon, I. 2013. *Da Panela.* [online] Available at <http://cargocollective.com/isabelbourbon/da-panel> [Accessed 2 Dec. 2017].
Cm-Barcelos 2012. *Rota da Olaria.* [online] Available at <http://www.cm-barcelos.pt/visitar-barcelos/artesanato/rotas-de-artesanato> [Accessed 16 Sep. 2017].
Cooperrider, D. 2007. Appreciative Inquiry: A Positive Revolution. In: Holman, P., Devane, T. and Cady, S. (eds.) *The Change Handbook: The Definitive Resource on Today's Best Methods for Engaging Whole Systems.* San Francisco: Berrett-Koehler, pp.73–88.
Fernandes, I. M. 1997. *A louça preta de Prado.* Porto: CM Barcelos, Museu de Olaria.
Fernandes, I. M. 2012. *As Mais Antigas Colecções de Olaria Norte de Portugal.* Barcelos: CM Barcelos, Museu de Olaria.
Fernandes, I. M. 2013. *A louça preta em Portugal: estudo histórico, modos de fazer e de usar.* Ph.D Thesis. Universidade do Minho.
Hayon, J. 2013. *Gardenias for BD Barcelona.* [online] Available at <http://www.hayonstudio.com/design/gardenias-for-bd-barcelona/> [Accessed 10 Nov. 2017].
Info Empresas (n.d.). *Fabricação de outros produtos de porcelana e cerâmicos não refractários.* [online] Available at <http://www.infoempresas.com.pt/C234_FABRICACAO-OUTROS-PRODUTOS-PORCELANA-CERAMICOS-NAO-REFRACTARIOS.html> [Accessed 19 Oct. 2017].
Kumar, V. 2012. *101 Design Methods: A Structured Approach for Driving Innovation in Your Organization.* New Jersey: John Wiley & Sons, Inc. pp. 8–13.
Lusa 2016. *Património—Barro preto de Bisalhães é património da Unesco.* [online] Available at <https://www.dn.pt/artes/interior/barro-preto-de-bisalhaes-e-patrimonio-da-unesco-5525750.html/> [Accessed 16 Nov. 2016].
Museu Olaria 2014. *História* [online] Available at <http://www.museuolaria.pt/?page_id=6> [Accessed 18 Sep. 2017].
Palma, G. 2015. *Catálogo—Projecto TASA.* [online] Available at <http://projectotasa.com/produtos/catalogo/> [Accessed 5 Nov. 2017].
Pereira, R. 2018. *Lateira.* [online] Available At: http://www.rui-pereira.com/index.php?/news1/lateira/> [Accessed 3 Dec. 2017].
Remelgado P. 2005. *A Louça de Barcelos: A Louça Vidrada.* Colecções do Museu 4. Barcelos: CM Barcelos, Museu de Olaria.
Sampaio, J. 2016. *Narrativa e visualização: estratégias para a eficácia na prática do design colaborativo.* PhD Thesis. Universidade de Aveiro, pp. 67–69.

A case study of didactic laboratory approach in fashion design education

X. Lin & A. Dell'Acqua
Department of Design, Politecnico di Milano, Milan, Italy

ABSTRACT: Globalization has created a new normal that affected all aspects of our daily lives. Under the circumstances of this phenomenon, fashion design education is in a great need for adaptation to the shifting paradigm. The traditional approach is not equipped to deal with the new requirements for fashion design curriculum, emerging pedagogies, such as didactic laboratory has been highlighted recently in many design research activities. However, the relevant literature is still limited regard the discipline of fashion design. Therefore, in this paper, we would like to discuss the developing design practice within the didactic laboratory approach of teaching and learning by illustrating a didactic laboratory practice example in fashion design education. We hope that this study could provide a basis for reflection and critical discussions for the future works.

1 INTRODUCTION

Globalization has created a new normal that affected all aspects of our daily lives. It can be compared to the industrial and technological revolution that transformed humanity during the 19th century (Mahgoub, 2007). Under this circumstances, the discipline of fashion; its practice and education, is in a great need for adaptation to shifting paradigm in the age of globalization.

The purpose of fashion design education in this era should be educating fashion designers who can meet the challenges of globalization, work with international partners and clients, apply the latest technology and new materials in their design projects. They are also required to communicate and interact with individuals from other disciplinary and cultural backgrounds. As Ann-Sofie Johansson, the creative advisor of H&M head office, once said, "I believe that teamwork and collaborations today are crucial for everyone, it is not just about 'me' any longer, even though maybe it is still like that in some design schools. I think the future will be even more about collaborations; you need to share knowledge and ideas with others" (Lin and Dell' Acqua, 2017).

However, the traditional approach that follows principles and practices developed in the past is not equipped to deal with the practical realities of contemporary society (Salama, 2006), there is now a paradigm shift in the design education, and new requirements to the already loaded fashion design curriculum should be added.

Therefore, in this paper, we would like to discuss the developing design practice within the didactic laboratory approach of teaching and learning, which, even though in the recent period, has been highlighted by many design research activities, the relevant literature is still insufficient regard the discipline of fashion design. With this in mind, in the second part of the paper, we would like to illustrate a didactic laboratory practice example in fashion design education. We hope that this study could provide a basis for reflection and critical discussions for the future works.

2 THE APPROACH OF DIDACTIC LABORATORY

The educational innovation of the so-called "didactic laboratory" (also referred as design studio) emerged at the end of the seventies in Germany. Those who supported this innovation considered learning not as the product of teaching process, but a process in which direct experience activates learner. The concept of the didactic laboratory is associated with a place where various kinds of experiments (chemical, biological, physical...) are conducted. However, in the last few years, the idea of an educational practice based on a laboratory approach has been asserted. It does not necessarily coincide with the common conception of a laboratory, but it takes the form of a research and learning environment whose focal point is making the most of the methodological aspects (Fasano and Casella, 2001).

The approach of didactic laboratory takes the shape of a workshop organized into projects, from the formulation of the project idea (on what we intend to work) to the definition of the objectives to achieve. This approach is based on an inter subjective exchange between students and teachers through a parallel mode of work and collaboration, conjugating both the teachers' professional and teaching skills, which allows not only transmitting knowledge but, very often, opening new paths towards knowledge. In each didactic activity, the teacher must aim to motivate students toward an active study. This motivation cannot be passed through a selection of contents, [...] but should be close to the young person's mindset, (it should) be able to make them think and especially to get them involved (Landi, 2006).

In the didactic laboratory, students would be able to organize and reorganize their knowledge in a continuous way; it is also supported by the emotional involvement, by the enjoyment and the curiosity of the individual and of the group about the process of discovery that brings to the achievement of a shared objective. The didactic laboratory is a place for building and practicing both theoretical knowledge and practical know-how; to concretize the educational dimension of teaching and learning: students can develop the projects individually and collectively according to their competencies and attitudes.

In order to better understand the approach of the didactic laboratory, we will illustrate an example of a studio course in the fashion design master program from Politecnico di Milano.

3 FASHION ADVANCED DESIGN STUDIO

This Fashion Advanced Design Studio aims to give students through short workshops the most extensive knowledge of the fashion industry. Beginning in September 2017, the entire design studio lasted a semester and was composed by four intensive workshops, followed and participated by a multidisciplinary team of tutors, professionals, designers, and students from different design disciplines and cultural backgrounds.

All the projects we illustrated below were completed in groups, where different competencies and nationalities were mixed. Thanks to the internationalization policy and strategy of our university, in this class we have students come from more than ten different countries, with various disciplinary and institutional backgrounds. We believed that this opportunity to work with others, rather than on one's own, could provide students distinct benefits in their long run.

1.1 *Workshop one—A+MANI*

This workshop introduced students new approaches to the different cultures of the fashion project in Italy. The primary purpose is to connect big companies, small artisans, young students and FabLab to hand down craftsmanship, artisanal knowledge, digital manufacturing and bring it into firms with a contemporary 4.0 way.

During this workshop, students were able to prototype their projects in a process which merges artisanal knowledge, contemporary trends together with creative and technological innovation. Each group can choose from one of the following briefs to design and prototype their capsule collection.

- Brief 1: In collaboration with Italian fashion brand TWINSET.
    1.1 Advanced technological knitting interlace for the design of new TWINSET footwear/bag/knitwear.

- Brief 2: In collaboration with digital agency Kettydo+.
  2.1 Smart fabric: Design session on the hand-weaving loom, hosting wearable technologies.
  2.2 Smart knitwear design. Design table shared with brand TWINSET.
  2.3 Smart luxury knitwear design. Design table shared with brand Luciano Benelli.
- Brief 3: In collaboration with Italian fashion brand Luciano Benelli.
  3.1 The quality and luxury of knitwear vs. 3D printing technology.
  3.2 The quality and the luxury of the design process vs. the wearable tech of smart knitwear.
  3.3 New lifestyles, new technologies, new markets, new concepts of luxury knitwear.
- Brief 4: In collaboration with Italian company Tessitura Monti (company of fabrics for shirts).
  4.1 Design the new custom-made shirt. Shirt as a witness of the values and contents of Italian tailoring contains sartorial DNA.
  4.2 Design the shirt as a new uniform. Shirt as a dress, costume, uniform that defines a role, authority, competence, and plays an essential role in the history of costume and fashion.
  4.3 Design a new iconic shirt. Related to the part of the shirt in the history of art, photography, and cinema.

1.2 Workshop two—instant couture

Draping technique, in this workshop, we refer it to instant couture, is a pattern cutting technique adopted by many designers, where garments are created directly on the mannequin, and then transferred onto paper to make patterns.

This experimental and intuitive method gives designers freedom to manipulate fabrics and adjust details in a lively and quickly way. It also helps our students to develop their conference in experimentation, idea generation, and creativity.

For this workshop, we invited a very experienced Milanese couturier Annagemma Lascari, who owns her brand and atelier in Milan, and have dressed many celebrities for their weddings. During the workshop, she shared the theory behind the practice and took students step-by-step through the fundamental techniques of Instant Couture.

The first day is about to give students the basic idea of what the instant couture is. Students were asked to observe and deconstruct four garments brought by the couturier and design the technical drawings of these garments. The second day is focused on exploring intuitively draping. Students got introduced to this technique by exploring different types of fabrics—synthetic taffetas, cady, and lining, working on a dummy as well as experimenting volumes on objects. The third and fourth day, each group was asked to complete a capsule collection with a theme they chose, and to present it professionally to the couturier.

1.3 Workshop three—denim project

The aim of this workshop, led by Sabrina Mandelli, who is the womenswear head designer for fashion brands Off-White and SSHEENA, is to create a denim collection in a very limited time.

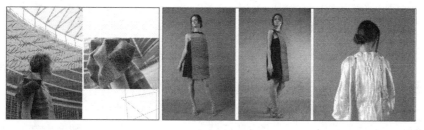

Figure 1. Some of the students' works.

Figure 2. Professionals and artisans were explaining various technologies for students.

Figure 3. Couturier was explaining the technique instruction.

Figure 4. Final work of this workshop.

During the workshop, students went through all the main steps of building a fashion collection, from initial research to final catwalk, showing outfits and running order. Each week we went through different stages of the design process. Week one is about launching the project, developing the mood boards and sketches. Week two is field visits to the denim suppliers. Week three develops the prototypes and simulates the price. Then, at last, the final lesson is to delivery the whole collection and making presentations.

The fashion industry is well known for its glamour and excitement. However, behind all the glamorous, it demands professionalism and excellence from its employees. Through this workshop, we would like our students to experiment an actual and professional environment, shadowing professionals who work for the industry, getting to know all the elements and steps involved and learn about the responsibilities associated with fashion jobs, which would be a great help for building their future careers.

1.4   *Workshop four—a new classic of men's tailoring*

One of the leading arts associated with Naples today is the art of tailoring. The historical and present-day tailoring talent in the city of Naples has found such success that has earned its classification: Neapolitan tailoring.

Though undeniably modern, the roots of today's Neapolitan tailoring stretch back almost 700 years, to the founding in the 14th century of Italy's oldest tailoring association, Confraternita dell'arte dei Giubbonai e Cositori (the Brotherhood of the Jacket Makers and the Tailors).

Figure 5. Designers were reviewing students' work.

Figure 6. Working progress and final collection shooting.

Its members were among the first to create ready-to-wear men's garments, made in Naples and transported to royals, noblemen and the affluent across the continent (Barker, 2017). The reference model for the jackets was English but was wholly adapted to the climate and with an inventive whole Neapolitan. Skilled artisans carried out a revolution vestimentary, which resulted in the elaboration of a new classic: Neapolitan tailoring. The cut becomes free from any line and constriction, and the dress is draped directly on the body, not to limit the spontaneous gestures.

To give students a different experience towards New Classic of Neapolitan men's tailoring, in this workshop, we invited a professor from one Neapolitan university. The workshop aims to design a capsule menswear collection. The macro scenario is the concept "Vintage generation," each group needs to select one to two briefs to create their micro scenario and work out their collection.

– Brief 1: From the classic tie to a new tie.
– Brief 2: Design of graphic tie.
– Brief 3: A capsule collection for a new tailoring shirt.
– Brief 4: From the classic man's jacket to a new classic shirt.
– Brief 5: From the classic man's shoe to a new classic shoe.

4 CONCLUSION

In this paper, we have started by introducing the didactic laboratory approach in teaching and learning. We then illustrated an example of a fashion design studio course by adopting this approach. In this part, we would like to share some conclusions drawn from the feedback and observation from our students and instructors.

For the studio content arrangement:
To complete four projects parallelly and professionally within one semester, was a great challenge for students, which requires a substantial capability to cooperate within the group by

dividing up the work. For most of the time, students had to work under pressure. However, just as Lorenzo Serafini, the creative director of Philosophy said, "Working in fashion industry usually means working overtime during nights and weekends, one minute ago you just finished the fashion show of this season, next minute you need to prepare for the new collection." It is indeed a very intensive industry. Therefore, we have to prepare our students well before they start the real career.

For the teaching and learning approach:
To lead an internationalized class and a cross-discipline studio like this was never an easy job. With students coming from various cultural, disciplinary, and institutional background means everyone has different interpretation and performance of the same concept. In this case, we suggest that particular attention should be paid to the group division phase, to ensure the interculturality and cross-disciplinarity of each group; to maintain a complementary comparative advantage.

As we mentioned at the beginning of this paper, even though in the recent period, the didactic laboratory approach in teaching and learning has been highlighted by many design research activities, the relevant literature is still every limited regard the discipline of fashion design. We hope that this paper could provide a basis for further discussion and action research in the community of researchers, concerning new practices of this didactic approach in supporting students for their future career.

ACKNOWLEDGEMENTS

The authors of the study thank colleagues from Politecnico di Milano—Department of Design, designers of SSHEENA, couturier of Atelier Lascari, colleagues from Università degli Studi della Campania Luigi Vanvitelli, for supervising and supporting this design studio course. Thank also to all our students for their precious works.

REFERENCES

Barker, C. 2017. The History and Anatomy of Neapolitan Tailoring. [Blog] *The Rake*. Available at: https://therake.com/stories/craft/the-history-and-anatomy-of-neapolitan-tailoring/[Accessed 12 Feb. 2018].
Fasano, M. & Casella, F. 2001. The didactic laboratory as a place to experiment models for the interdisciplinary research. In *international conference developing formal thinking in physics*. Udine: Editrice Universitaria Udinese: 396–400.
Landi, L. 2006. The Didactic Laboratory in Classics at the SSIS Veneto, Italy. *The Classical Journal*, 101(3): 319–327.
Lin, X. & Dell' Acqua, A. 2017. The Approach of Didactic Laboratory in Fashion Design Education: A Comparative Case Study. In *Design Management Academy Conference*. 5: 1563–1576. London: Design Management Academy.
Lin, X. & Dell' Acqua, A. 2016. Cross-Cultural Perspective and Politecnical Approach for Fashion Education: A Case Study of an Italo-Chinese International Program. In *Academic Design Management Conference*: 2083–2098. Boston: The Design Management Institute.
Mahgoub, Y. 2007. Design studio pedagogy: from core to capstone, in *Design studio pedagogy: horizons for the future*: 193–200. Gateshead: Urban international press.
Polanyi, M. 1966. *The Tacit Dimension*. New York: Doubleday & Company.
Salama, A. 2006. Committed educators are shaping studio pedagogy, editorial, *Open House International*, 31(4): pp. 1–6.
Smal, D. & Lavelle, C. 2011. Developing a discourse in fashion design: what is research for fashion design? In *Sixth International DEFSA Conference*: 192–198. Pretoria: Design Education Forum of Southern Africa.
Spizzica, M. 1997. Cultural Differences Within "Western" and "Eastern" Education. In Z. Golebiowski & H. Borland (eds.) *Academic Communication across Disciplines and Cultures*: 248–257. Melbourne: Victoria University of Technology.
Woo, J., Clayton, M., Johnson, R. Flores, B. & Ellis, C. 2004. Dynamic Knowledge Map: reusing experts' tacit knowledge in the AEC industry. *Automation in Construction*, 13: 203–207.

Feminine magazines and the development of the female labor force in Valparaiso: Analysis of the graphic system of the Rosita magazine (1947–1972) as a didactic strategy for the informal learning of sewing

F. Gonzalez, U. Bravo & C. Ruiz
Universidad del Desarrollo, Chile

ABSTRACT: The dressmakers and seamstresses formed an important workforce in Chile in the first half of the twentieth century, mainly in Valparaiso, where the textile industry developed at the edge of the port activity. In this context, the publishing industry of women's magazines detected the need of those women who sought to develop these trades from home. Magazines such as Rosita began to incorporate cutting and confection lessons. Rosita developed a true graphic system composed of different sewing columns. Each one oriented to a different stage of the process of making a tailor-made garment. From the perspective of Graphic Design, it is worth asking: How does a graphic information system work in a women's magazine? What visual variables contribute to the understanding of a manufacturing process?. This project seeks to analyze the graphic components of Rosita, in an attempt to determine what extent a graphic system constitutes an informal learning tool.

1 ROSITA: THE PRACTICAL AND SIMPLE FASHION MAGAZINE

This research focuses on the Chilean magazine of the Zig Zag publishing house: Rosita, a feminine publication that had as objective to conform like a manual of dress making for dressmakers and seamstresses, that builds one of the most popular work forces in the country, mainly in Valparaiso, where the textile and tailor-made clothing industry achieved great development.

Being a manual of clothing, it defines itself as the publication of practical and simple fashion, at this point it is important to highlight the term fashion as:

> *A reference of the historical moment that we are reporting. His presence exposes like no other element, the influences, the economic states, the aspirations, the customs, the ideals. This allows us to have a global vision of the context (...)* (Brugmann Study, 2016).

Rosita is an icon of fashion that reflects the society of the time and the trades that are configured around fashion in cities like Valparaiso. Being a reflection of the local culture and forming part of the collective memory, it can be considered part of the Chilean graphic heritage, which is considered as:

> *Our visual heritage, those images that have filled our heads for generations (..) Read the graphic as a reflection of the changes experienced in Chile in the last two centuries and rescue the value of the everyday. Clearly there is a Chilean graphic heritage. Lack of research and dissemination* (Alvarez, 2009).

At the same time, an analysis is made of the didactic graphic system of the garment columns that were responsible for delivering knowledge in a clear and useful way, which transforms Rosita as a design product that empowers the woman of the time.

2 OBJECTIVES AND METHODOLOGY OF THE RESEARCH

How does a graphic information system in a women's magazine work? What visual variables contribute to the understanding of a clothing process? These are the questions that are posed in the development of the investigation, to answer them, objectives are set that guide the investigation:

- Analyze the productive sector of fashion in Valparaiso.
  o Analyze the historical context of the journal in the region.
  o Analyze the publishing sector.
  o Describe Rosita's relationship with the informal training of seamstresses and dressmakers.
- Analyze Rosita's didactic chart.
  o Describe a didactic graphic system.
  o Identify the learning system in the columns.
  o Describe the graphic and learning strategies of the didactics.

2.1 Types of methodology

To achieve the objectives, an exploratory, qualitative and interpretative research is proposed, which includes a mixed methodology composed of:

- Documentary research on the fundamental aspects of the context of women's magazines in Chile from 1910 to 1960, a time when there is a boom in tailor-made clothing. Literature review on the textile industry, the importance of sewing crafts through the censuses of the time the publishing industry and didactics in Graphic Design.
- Ethnographic Research: To describe the way of work of the seamstresses and dressmakers in Valparaiso, the processes of tailor made clothing, also to analyze the formation of these trades and the importance of fashion and women's magazines in the training process informal. Interviews were conducted with women who sewed in the region.
- Formal analysis of the graphic resources of the sewing columns, including iconic variables (types of illustration), typographical variables (typographic variables: Regular, Serif, San Serif, Bold, Italic, Calligraphic, Uppercase and Lowercase), compositional (layout). This through the construction of a tab to systematize the variables of the columns.
- The analysis of the cultural, social, patrimonial and didactic importance of Rosita is made by triangulating the contents of the historical and social context, the interviews and the formal analysis of the columns.

2.2 Types of illustration

The types of images present in the journal are classified according to the differentiation made by Frida Diaz and Gerardo Hernandez (2002) in his book Teaching strategies for meaningful learning. These typologies are represented in Rosita's illustrations and define the types of columns that are made during the signing.

- Descriptive Illustration: It shows how an object is physically, the central and defining characteristics of an object. In the magazine Rosita is reflected in the illustrations that focus on the wardrobe to show a finish or a detail.
- Expressive Illustration: It shows the attitudinal and value aspects of the represented object. In Rosita the figurines were the main expressive illustrations since they show the attitude of the dress put.
- Constructional Illustration: Structural aspects of the object or system represented. In Rosita the supplement that came with each number, the mold or clothing pattern, is the representation of this type of illustration.
- Functional Illustration: Visually describe the different interpellations or functions existing between the parts of a system so that it enters into operation. In Rosita they are presented in the illustrations that showed the process of making step by step.

3 VALPARAISO, THE PORT OF FASHION OF TGE TWENTIETH CENTURY

The Valparaiso region experienced a period of splendor that transformed it into a city with high economic power in the country during the late nineteenth and first half of the twentieth century. At this time, it was recognized as "the first port of the Republic and the most important of the South Pacific" Herrera, Toro (2013), since before 1914 there was no Panama Canal, so it was easier to enter the Strait of Magellan, the port of Valparaiso was a key point in the journey. There came the raw materials for the textile industry that favored the development of the textile industry and custom clothing manufacturing.

This port boom went hand in hand with the entry of foreign immigrants, mostly English, French and Italians, who brought the latest fashion in Europe, establishing the first workshops and making the port a cradle of great tailors, dressmakers and seamstresses; which allowed the development of trades linked to fashion and favored economic development.

The high trade of fabrics, the arrival of European fashion and the development of these crafts makes Valparaiso the main port of fashion, being evident in the squares and parks of the city, true social centers where women went to show through his clothing his social status. Victory Square, Plaza Echaurren, Plaza O'Higgins and Parque Italia were the main places of exhibition of the elite where it was socialized through clothing. Herrera, Toro (2013).

3.1 *The textile industry*

Before 1930, the Chilean textile industry had not developed proportionally to the continuous growth of garment making and with the reduction of imports that caused the First World War, the national industry entered into a crisis.

From 1930 to 1973 the national economy became protectionist and sought to encourage "growth inward" supported by the creation of the Corporation for the Promotion of Production (CORFO) in 1932. Textile factories founded by foreign immigrants (for example: Yarur Chilean Manufactures of Cotton, Cafarena, etc).

As a result of the crisis, women began to make their own clothes ("Do it yourself" time), and even if they were not seamstresses or dressmakers, they had to learn to make clothes. In this context, Rosita began to publish the year 1947, as a tool for those women who did not have access to a formal education.

3.2 *The trades of fashion*

At this time in Valparaiso the demand for clothing by the Valparaiso elite generated an increase in sewing workshops, while the Chilean State, through the Sofofa, created plans to educate this workforce. So, these trades, which are configured around fashion, are valued, seamstresses and become one of the most exercised by women. Between 1940 to 1960 the period of greatest boom of the industrial and tailor-made clothing takes place.

To deepen into the value of these occupations you must differentiate between them and show their roles:

- Dressmaker: She has finished knowledge to obtain patterns, she has knowledge of the volume of the body, of its dimensions, for what she designs, patroness, creates prototypes, chooses fabrics, cuts, adjusts, and sews. They were educated in technical schools, vocational schools or private institutes, although many dressmakers also began as apprentices to a dressmaker and then went to study.
- Seamstress: It is who sews the garments, unlike the tailor or dressmaker, the seamstress does not design or patron. Receive the pieces already cut and your job is to sew them according to the indications you receive. He only has knowledge of the machinery and not of the measurement of the body. It is characterized by learning through "learning by doing" and observation.

These occupations, despite being linked to the domestic environment, was the opportunity that the woman had to acquire a knowledge that would later allow her to generate income, so she gave her a freedom that at that time was not easy get for the woman.

The impact of the labor force can be seen in the trade unions that fought for women's rights in the country and mainly in Valparaiso, a region that was a precursor in this issue. The historian Emilio Toro analyzes the impact of these trades and the scope of these labor unions:

> *The seamstress Micaela Caceres created the "Worker's Society Number 1", with the aim of defending a partner whom they wanted to dismiss for missing work due to the illness of one of their children. Years later, the Feminine Union of Chile, created in 1927, obtained the municipal vote* (Toro, 2013).

Rosita magazine fulfills the function of giving visibility to these trades, contrasting how valued this handbook was that informally educated women who were part of this labor force, versus census figures that did not reflect the increase in these occupations.

3.3 *Fashion magazines: The publishing quarter*

In the twentieth century on par with the development of clothing and the economic development of the country, there is the rise of women's magazines in Chile, these constitute a true publishing district made up of these fashion publications.

The magazines that were fundamental in Creole fashion were: Eva (1942), aimed at a woman of the upper middle class, conservative and was the magazine with the most technical resources. Margarita (1943), focused on the most popular class that followed fashion, included recipes for cooking, crafts, work, among other household columns. And Rosita (1947–1972), specialized in costume making, was more economical and was designed for seamstresses and dressmakers who worked from home (Salinas, 2014).

4 ROSITA: DIDACTIC GRAPHIC SYSTEM

In this research the sewing columns are a graphic message, an intentional and technical unit: that is, they are the union of what to communicate and how to communicate it. In the case of Rosita, it is necessary to teach dressmaking and tailoring, its objective is to transmit stable and usable knowledge for seamstresses and dressmakers.

For the message to achieve this, an important component is the didactic, this term means that the message gives these women a knowledge through concepts or values and that these become part of their life and culture, in this sense the didactic it is linked to the active learning of the receiver. This refers to the need for an interaction with the seamstresses and dressmakers:

> *The didactic chart strives to make comprehensible (understand and learn at the same time) complex or abstract things. This images class stimulates the active participation (active interest) of the codifying individual, who extracts from it the useful knowledge, concepts and values that will be part of his being and will be elements of his personal culture* (Costa, 1991).

That is to say that through the columns of the magazine, women experienced an interactive bidirectional communication, were involved in an induction-deduction mechanism in which they stopped being a passive reader of fashion magazines, to become an active receiver that he had to react to the didactic stimuli of the columns to learn a knowledge and put it into practice.

5 TYPES OF COLUMNS

From the analysis of the results obtained from the signing, it is possible to characterize the types of columns that are present in each published issue of the journal and that are part of the teaching method used by Rosita.

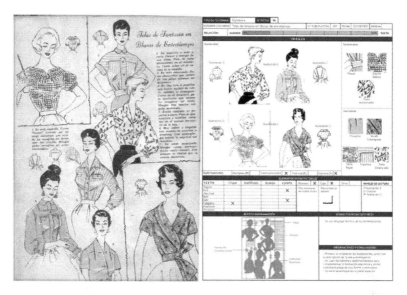

Figure 1. Expressive Column, of Rosita magazine N°461 and a tab made from this.

5.1 *Expressive column: 80% image 20% text*

It is the one with greater incidence in the magazine, the one that is used more and the one that shows more quantity of dresses in a same page. The most important tool is the image, specifically the use of the fashion figurine and complementing this image, a small figurine that shows the structure of the dress.

In this column the teaching of the physical aspect of the dress predominates, mainly the attitude of the dress put, the "as it should be". From this it is deduced that it is a column focused on the needs of the dressmakers in the first stage of making when the client selects the model of the dress, they already know about the construction of the pattern, so they do not need a detailed explanation of construction, they need to have a reference of what is expected from their work.

For this ornamental typography is used that makes the content attractive and there is a lower amount of reading levels since the content must be delivered visually and quickly.

5.2 *Constructional column: 40% image 60% text*

The fundamental characteristic in these columns is that the text is more important than the image, because the content needs more explanation The didactic of this column is to show the planimetry of the pattern through a methodical explanation and is based on the continuity of the learning, in the sense that they are columns that deliver the fragmented content through the published numbers, that is, the column continues in the next number.

It is a learning of the technique of clothing, teaching the basis of the trade of seamstresses and dressmakers, that is why they are fundamental in the global content. From this it is deduced that it is a column focused on the needs of the seamstresses, because they do not have the knowledge to create a pattern from scratch, which corresponds to an intermediate stage in the manufacturing process.

5.3 *Functional column: 50% image 50% text*

They are characterized by a balance between the image and the text. The didactic in these columns is based on the simultaneous reading of these two variables. The receiver when reading the verbal explanation then visualizes the image to understand the process, they are used in the process of making the garment itself.

This type of columns is usually found in the Rosita supplement, in the mold, although it is also found in the magazine, but unlike the mold, in this the content that is communicated is more basic, that is, generally a functional column. The interior of the magazine is used to explain the composure of clothes, which is one of the most basic procedures of clothing.

On the other hand, in the mold it is a more complex knowledge and it is also more complete, since it uses the functional image showing the process of the confection, supported by constructive and expressive images, and from the back it shows the planimetry of the mold.

Another important aspect of the didactic of this column, is the manipulation of the piece, the receiver uses this to cut it and create the pattern. From this it is deduced that it is a column

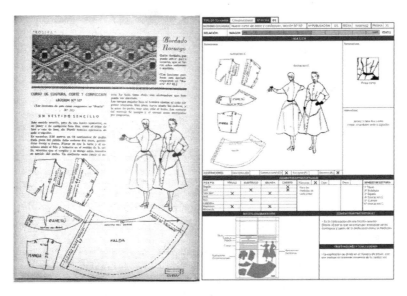

Figure 2. Constructional Column, of Rosita magazine N°167 and a tab made from this.

Figure 3. Functional Column, of Rosita magazine N°991 and a tab made from this.

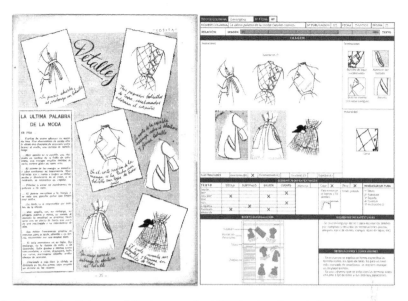

Figure 4. Descriptive Column, of Rosita magazine N°175 and a tab made from this.

focused on the needs of the seamstresses, because they do not have the knowledge to create a pattern from scratch.

They are columns that have a large number of reading levels since the content has to be explained in a detailed and clear way without discarding basic information.

5.4 *Descriptive column: 60% image 40% text*

Its function is to show in detail a finish, for example a fold, a drape, a clip, different types of collars, sleeves or pockets, so they are used in the final stage of the manufacturing process, when the finishes of the garment.

The didactic of these columns is based on the graphic expression of the details through the use of lines, textures, the use of lights and shadows and the angles of these. The text is a complement to what is expressed visually, it is a brief explanation.

From this it is deduced that it is a column focused on the needs of the seamstresses and dressmakers, they are processes that can be carried out by both since it is not of greater technical complexity. Because of this, a large number of reading levels are not used, the content is easy to understand, and the image shows the attractiveness of the termination.

6 CONCLUSIONS

It can be concluded that Rosita delivers knowledge in an accessible way, it is also complete since it delivers a system composed of 4 columns, in which each one is based on a part of the process of making a garment. At the same time, the success of the magazine is based mainly on the fact that Rosita detects the need of these women and takes care of their shortcomings and the things that formal training did not achieve.

The magazine includes its user and provides continuity in terms of content and the time at which it publishes each issue. It is observed that the magazine takes care of the little time that these women had, having to take care of the home in parallel and giving it a fragmented content.

In the 4 types of columns the layout helps to hierarchize the content and balance the visual weights in the column so that the learning is fluid, the image helps to distinguish the contents, that is, the constructive image communicates something different than the expressive, what according to the function of the column are distributed according to that and has different visual weights, but they complement each other. It is important to understand that the variables that intervene in the didactic, are not isolated elements between them, on the contrary they work together to deliver knowledge and that can be understood.

Finally, to conclude, Rosita's graphical information system works through the schematization of content, hierarchization, union of variables, also through the division of the topics and the approach given by each column to a need and a specific user.

REFERENCES

Álvarez, P. 2009. Diseño chileno, de la adolescencia a la adultez. *Revista Patrimonio cultural* (n°50): 16.
Costa, J. Moles, A. 1991. *La imagen didáctica.* España: Enciclopedia del Diseño.
Díaz, F. Hernández, G. 2002. *Estrategias docentes para un aprendizaje significativo* (N°2). México: McGraw-Hill Companies.
Herrera, A. Toro E. 2013. *Costureras, sastres y modistas: Contexto manufacturero en Valparaíso, 1890–1960.* Valparaíso: Centro de Estudios y Conservación del Patrimonio Cultural de Valparaíso.
Hinostroza, T. 2016. Estudio Brügmann: rescatando el patrimonio a través del arte y la moda. *Modacl* Available at: <http://www.modacl.com/index.php/2016/08/29/estudio-brugmann-rescatando-el-patrimo-nio-a-traves-del-arte-y-la-moda/> [Accessed 27 November 2017].
Salinas, J. 2014. *Linda, regia, estupenda: Historia de la moda y la mujer en Chile.* Santiago: El Mercurio Aguilar.
Toro, E. 2013. Escuelas de corte y confección: la desaparición de un ocio. *El mercurio*: 20.

Bê a Bá of sewing: Our online friend

L.C. de M. Tavares & J.P.C. de Brito
Instituto Federal do Piauí—Campus Teresina Zona Sul, Piauí, Brazil

ABSTRACT: This article aims to show the importance of the site "*Bê a Bá da Costura*", an online vehicle that teaches almost everything about sewing. Those responsible, two fashion designers from Teresina/Piauí/Brazil. Students from different courses participated in the research: Apparel Technician of the Federal Institute of Piauí-IFPI, Campus Teresina Zona Sul; Superior Course in Fashion Designer of the same institution and the Superior Course in Fashion and Styling of the Federal University of Piauí-UFPI. We used focus groups and unstructured interviews with UFPI students. We base the research on theorists such as Freire (2001); Castro (1991); Azevedo & Gonçalves (2006); Libânio (2009); Thiollent (1985); Vaillant (2012); Schön (2008), among others. The answers obtained lead us to believe that the site is really an "online friend" of these students, of all who access the tool, consisting of an important pedagogical tool in the teaching-learning process.

1 INTRODUCTION

Between pupil and teacher, it is necessary to have a very close proximity, not as many think, only physical, but is needed a knowledge, to understand the social group in which you are professionally inserted, having also, a concern with the social, feel responsible to it and have feelings of belonging, wanting the best for the students. Finally, the essay of the progressive teacher Freire:

> *Conscious of the limits of her practice, the progressive teacher knows that the question that is posed to her is not to expect radical transformations to take place so that she can act. You know, on the contrary, that you have much to do to help your own radical transformation. It is there, knowing that there is a lot to do, that it is not condemned to fatalistic immobility, immobilism that can not comprehend the dialectic between infra and supra structure, that the problem of limits to its practice is put to him or her. It is precisely at this critical level that, by refusing the naive view of education as a lever for transformation, it also refuses contempt for it, as if education were to be done only "after" radical change in society* (Freire 2001, p. 28).

And it was in 2013, imbued with this feeling, that "*Bê a Bá da Costura*" was created, a working tool whose initial purpose was to supply the lack of sewing teachers in the laboratory during the exercise hours, in particular, it was created for those students who had the most difficulty, what consisted mainly in not having a sewing machine in their homes. We noticed it during the sewing classes, as the students filmed our classes and we, after, realized that these students had a better performance. It was there that we tried to find out why everyone did not have the same attitude and we concluded, after observing their socioeconomic records, that not all of them had a financial situation that would allow them to buy a cell phone with camcorder and much less a sewing machine.

After much discussion we decided to create a blog, which was later transformed into a website, where the students could go to school and attend classes as many times as necessary. We were finding a new and current way of healing the difficulties of the classroom. We were using something called didactic, which in Castro's words is nothing more than "bringing together under this rubric the knowledge that every age values about the teaching process"

(Castro, 1991: 34). We live in the age of the internet, we will take advantage of everything that is at our disposal to better meet the needs of our students.

At the beginning, were only shown the lessons in which students had greater difficulty, later we verified that not only the students were viewing the blog, but the difficulties came much bigger and the requests were much more. Was solved, then, to do the course the way it's done in most clothing's courses, first the notions about sewing machine, then those of how to pass the thread, lessons on how to pedal control, on synchronizing the movements of hands and feet. At the moment, the basic course of sewing is finished, all the lessons related to the first module (Sewing Technology I) have been developed, including the portfolio that is required for all the students in a regular course; the constructions were finished with the special fabrics—striped fabric, plaid fabric, lace, fabric with foot, missing only the construction of a small collection of pieces but, at the request of the followers, one, now, begins the course of modeling to soon start the construction work of current parts that will be, for sure, more exciting. We intended to build, weekly, some current piece or give tips of construction, draw attention to important details in the construction of some pieces; information on how to use certain materials such as gussets, among others. Finally, create a channel for exchanging experiences in the construction of garments. In the meantime, we had to move away for the master's degree, we are back and we intend to finish the modeling course and put into practice our ideas described above.

It's five years of work with a two-year stoppage as stated above, the results are more than one million six hundred thousand views and more than twenty-seven thousand registered on youtube's channel.

The dynamism of the internet has provoked a different relationship between subjects and knowledge, the first begin to realize the need to learn to learn and recognize the constant challenges of digital culture. Law 5622, which deals with distance education in its Chapter I Article 1 states that:

> *[...] is characterized as distance education an educational modality in which the didactic-pedagogical mediation in teaching and learning processes occurs through the use of media and technologies for the information and communication, with students and teachers developing educational activities in different places or times* (Brasil, 2005).

It was in a practically unthinkable way that *"Bê a Ba da Costura"* enters the ranks of distance education, none of its participants has knowledge of how to teach in such a differentiated way, but despite this technical deficiency, the group has tried to overcome with individual studies. They are reading and collecting information, aiming to satisfy the "netizen's needs", as they say in the popular language, but as Azevedo stated,

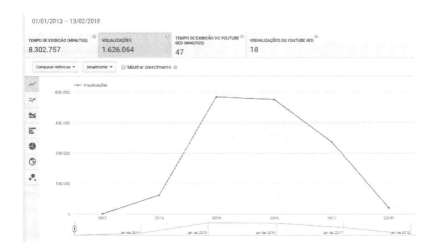

Figure 1. Annual YouTube views.

This interactive process is characterized by the model in which "(...) everyone teaches and everyone learns; the student is fundamentally an agent for the construction of his/ her knowledge and the teacher is the mediator, responsible for facilitating the transformation of information into knowledge" (Azevedo & Gonçalves, 2006, p. 67).

The experience of these five years of *"Bê a Ba da Costura"* leads the group to agree with the above statement. The followers are the most interested and they are who compel the group to produce more and more. It is their requests what leads the work and it is always a pleasure to attend the requests of the online students. And, it is these followers who are asking us to return as quickly as possible; From this we try to do a small research with students of the institution where we work and with students of the Federal University of *Piauí* – UFPI of the course of Fashion Design and Styling, courses that deal directly with the construction of clothing, what gives great importance to the website. Although the sample is small, it was clear how important this work is. We work with focus groups and unstructured interviews and the results of this research is what we want to be seen and analyzed.

2 DEVELOPMENT

One of the political roles of the school is to promote social justice, therefore, it is the responsibility of the student's cultural and scientific formation (LIBÂNIO, 2009 p.1), hence the importance of having in their professional formation these ideals. In the pedagogical process, students and teachers must take their roles consciously, these are not only subjects of learning and teaching, they are human beings with individual histories, each one loaded with unique, different feelings, being part of a society enveloped in many problems. It is up to each one to try to know the other as much as possible. In this dialogical process there is a constant exchange: at the same time it teaches the educator to learn and at the same time to learn the student teaches. The teacher is the mediator between the student and the object of study and establishes with him a social relation that materializes in the classroom, be it physical or virtual, or in the interactions that occur in school or outside.

It is necessary to disclose good examples of learning practices and learning outcomes to sensitize the government, the school community and society about the importance of using technology in schools. The present work is about a successful experiment involving new technologies in this educational area, an educational blog that has become a site and has already surpassed one million six hundred thousand views. A work that initially had as its objective only to subsidize the extra class activities of two classes of a course of Clothing. This work has led those responsible to enter in new (to them) situation—Distance education, forcing these teachers to qualify more and more.

During these five years of *"Bê a Ba da Costura"*, its leaders were often instigated to study more and more. They have already come into contact with readings about distance education, reflective work of teachers, action research, among other subjects in the area. They are very overwhelmed with activities, but very happy because they know that as teachers are fulfilling their social mission. Work is an action research because it is an interference in a given context where there is a problem, with the intention of solving it.

In Thiollent's words, a type of empirical research that is conceived and carried out in close association with an action, or with the resolution of a collective problem, where all researchers and participants are involved in a cooperative and participative way (Thiollent, 1985).

The role of the teacher as a consequence of the processes of social change has been transformed in recent times, the teacher is forced to assume a greater number of responsibilities and consequently increases the number of requirements to which it is submitted. The contemporary literature speaks of a series of characteristics, sometimes contradictory, when it comes to considering what qualifies a good teacher. Many refer to the knowledge of the content to be transmitted, some refer to the methods used to convey this knowledge, others call for the latest advances in information technology. The truth is that teachers have to acquire communicative skills that allow them to interact with students (Vaillant, 2012). Didactics becomes necessary and, as Libânio affirms: "It is the nature of didactics to investigate the links and relations between the act of teaching and the act of learning" (Libânio, 2009, p. 3). This subject should guide the work of every teacher. The teaching and learning process is not the easiest task.

In other words, the teacher has to know what he will teach, how he will teach, who to teach and under what conditions (LIBAN, 1999). This knowledge required by the teacher, requires even a constant recycling, he/she is a professional who has to be constantly in contact with books, is an eternal apprentice. During these four years of *"Bê a Ba da Costura"* many were the moments in which questions emerged, consultations with expert theorists in the subject were needed, discussions with co-workers too and, often we redone all the work trying to find a different way of saying or doing something different in order to better communicate with the students. Distance education reaches a much larger number of students, on the other hand, it requires a better student teacher communication. And this reflection on our work should be a constant.

The team of *"Bê a Ba da Costura"* is aware that a job that constantly reflects on the process of teaching/learning is the most appropriate and shares the ideas above of Schön,

> *Professional schools should rethink both the epistemology of practice and the pedagogical assumptions on which their curricula are based and must adapt their institutions to accommodate reflexive practical teaching as a key element of professional education* (Schön, 2008, p. 25).

Even knowing how imperfect the work still is, the team of *"Bê a Ba da Costura"* cannot fail to celebrate the success achieved. At no point did the staff shy away from bending over the books and trying to improve the work. Always attentive to requests from students and followers and students in the classroom sought and seeks, to give as much attention as possible to all who come in contact with the group. The experience gained in these five years, this reflection on the ways of teaching, this constant search to improve performance as an educator makes him share with the ideas of Castro that states:

> *I do not suppose that a single theory can, for the moment, dominate the field of Didactics. The promising theories are interdisciplinary, which is neither strange nor pejorative, since teaching (already said how and why) is basically complex, and can only be understood from other elements of the human and social sciences. Thus, we might have to integrate models from different areas, provided that they were not contradictory. This is the basic question of interdisciplinarity: the perception of common relations, coherent explanations, use of the same methods of thought* (Castro, 1991, p. 23).

Aware of all these problems we went to the fight, we did our best not to say that Paulo Freire was a progressive teacher, committed to education and to the society of which we are a part. The following is an explanation of how it happened, or rather, how this experience is so important for our work, both as educators and as elements of a society that is going through troubled times, especially in the political and educational area. Impossible to punctuate what is happening with all the visitors of the site, we are here with a small sample that we consider important because they are students of technical and higher courses focused on the area of construction of clothing.

2.1 Class 490, 2017/2

Monday, February 5, 2018, early evening, in the city of *Teresina*, capital of *Piauí* state, northeastern Brazil. Sewing laboratory of Teresina Campus South Zone, clothing course. We asked the students to give their opinion about *"Bê a Ba da Costura"*, we wanted the positive and also negative opinions. Only a positive student affirms: Only positive, everything was positive. We are aware that there are failures, but the need for jobs like this at zero price is so great that the students only see the positive. Another student says that *"Bê a Bá"* is all good. Another note that he learned a lot from *"Bê a Bá"*, who attended classes only paying attention because he knew that he did not need to make almost no notes because he had *"Bê a Bá"* to review the lesson. Another states that *"Bê a Bá"* is her sewing dictionary, everything she wants to know is there. Once again we say that not everything is there, we are far from being such a safe port, but we strive hard to pass on the maximum for our students and our netizens. We are pleased with the feedback given by these students and we are committed to returning with full force to the activities of the site. It should be noted here that we are in February 2018 but we are still finishing the 2017 school year, so that we are giving this group the name mentioned above.

2.2 *IFPI top fashion design course*

We are in the same school, the same day February 5, except, now, in the class of our senior fashion course. Most of the students were not our high school students, some already know and use *Bê a Ba da Costura*, others only know of its existence and the vast majority knows practically nothing of the process of construction of a piece; however the testimonials were very interesting for this work. One of the first students to speak has already given us a very interesting feedback, she saw some classes and found important the feedback given by the site, she states:

> *I watched two videos only, but I found it interesting because it shows the whole process you are doing, explaining, giving tips of materials that could be used and still shows the result of the finished piece and the mannequin dressed.*

Other settings:

> *It only concludes his final work due to Bê a Ba da Costura. "During classes we are sometimes inhibited from asking more than once, making the teacher repeat what has already been taught, but often it is so many information that something goes unnoticed. There in Bê a Bá no, we can repeat as many times as we want that the teacher will speak with the same tone of voice, without bothering"* (Kleison Teixeira Silva).

Very interesting the placement of the student because the teacher always thinks that the student is there in the classroom with all the physical and emotional conditions to learn and that the teacher is being very well understood, that learning is happening a thousand wonders. After repeating once or twice, he thinks the teacher is the limit and if we stop to analyze it is the same limit. Few classes, crowded classes, no time for an informal conversation with any student. We know practically nothing of what is happening with the personal life of each one. Even wanting with all our strength is very difficult to be a progressive teacher, this research points very hard for this.

A former student of the clothing course points out:

> *I watched all the videos and found it really interesting because I only saw the hands, the person never showed up. But it is very good because there is explained detail by detail.* (Deuzalina Alves de Sousa).

Student Antônia Maria de Carvalho, already a seamstress, spoke about the importance of having help at our disposal whenever it is necessary for both professionals and students who are starting up. Another, Ana Delza da Luz Ibiapina spoke of the objectivity of the classes, of how important it is to heal a doubt quickly. I want to know how this is done so I go to class on that subject and that's it, it's settled. It is our duty to point out that they are classes without any interference, saving time and words so that the video is not cumbersome and confusing, however it takes many resources that are not within our reach, we believe that soon we will have there better quality, more powerful machines, special recording rooms etc. We left the class satisfied and their heads bursting with new ideas and dreams.

2.3 *Class 390, 2017/2*

At the moment we are working with this group, although they (the class only have women) were not our students in the first module and not everyone knows the tool, but two who already had the habit of using *Bê a Ba da Costura* said the following:

The image is very clear and the explanations are very good (Jardeane).

Very good, you have the right step. I saw the step by step of the built-in pocket and found it very practical (Sônia).

2.4 UFPI students

We were informed by a professor of the Fashion Design and Styling course of the Federal University of *Piauí*, that the students of that institution and of the aforementioned course used *Bê a Ba da Costura* a lot. As they are still on vacation, we decided to contact by phone and even some students personally. Which was not to our surprise, below some opinions about our work:

Bê a Ba da Costura is an educational tool closer to the reality we live in, it makes learning easier for both the fashion student and a student who is not from the area but who enjoys the sewing practices (Damarys TF Carneiro).

I love it. (Taiza Fernanda).

I always accompany Bê a Bá and I use the techniques, especially the finishing touches passed by the teacher (Tarla Goethe).

I'm not much of a fan of sewing (laughs) but the site helped a lot when we had practices at UFPI (Taiana).

It is a fantastic learning tool because the idea we have is that only people who have a special gift are able to sew and Bê a Bá came to demystify this, the website showed that with technique, with a little practice, with that is the strength of the channel, it is possible to learn a great deal. The channel shows very clearly, very easy, everyone should attend Bê a Ba classes to lose the fear of sewing. And best of all it's a free tool, that's what's missing currently, sharing knowledge for free (Sara Virginia).

3 CONCLUSIONS

We know the importance of the so-called factory floor for clothing and fashion construction, as well as for the country's economy. What is heard of the owners of large and small factories is the shortage of skilled labor, and this was one of the reasons that led to the creation of several clothing courses in federal institutes all over Brazil, in which one tries to qualify the best students.

The experience described was applied in a professional education course at the Federal Institute of *Piauí*, in *Teresina*, at the *Teresina Zona Sul* campus, and it spread through social networks, becoming an experience of distance education which does not make it unfeasible in any educational segment, mainly because the professional education demands much more of the student, therefore, besides the intellectual performance he has to prepare the body for the performance of certain functions which demands time and intensive training. The technological courses are fast and a tool like this is of extreme importance for students and teachers, that is what this research points out.

The work does not end here, the parts referring to the first and second modules are finished, which makes it possible to make garments with a certain level of difficulty, due to numerous requests, a modeling course was started in order to better prepare students, it was stopped for some time but we intend to return as soon as possible. We know that in this area of sewing has much still to be worked, just by way of example we have clothes for men, women, children, bed, table, bath and even clothing for animals.

At first it was not the intention of the group to go this far, but only to give support to the students who were looking for the sewing laboratory to do their extra class activities, a goal that was fully achieved. We did not count on the influence of the social networks that contributed to leverage to a million six hundred thousand visualizations and the work became news in the written and televising media, was praised by colleagues of other institutions, what leaves the staff very flattered, but if we get here, it is our duty to jump higher. Let's get this job done.

REFERENCES

Azevedo, A.B.; Gonçalves, E.M. 2005. The Importance of evaluation in the implementation of a differentiated pedagogical practice. *In: Communication and Society Magazine* Year 27, N. 44. São Bernardo do Campo: São Paulo UMESP.
Brazil, December 19, 2005. Decree-Law no. 5,622. *Distance education is characterized as an educational modality in which the didactic-pedagogical mediation in teaching and learning processes.*
Castro, D. 1991. *The historical trajectory of didactics.* Ideas Series. N. 11. São Paulo: FDE.
Freire, P. Politics and education: essays/Paulo Freire. – 5. ed—São Paulo, Cortez, 2001.
Libânio, J.C. October 2009. Conference text presented at *III EDIPE—National Meeting of Didactics and Teaching Practice.* Anápolis-GO.
Schön, D. 2008. *Educating the reflective professional: a new design for teaching and learning.* Translated by Roberto Cataldo Costa. Porto Alegre: Artmed.
Thiollent, M. 1985. *Methodology of action research.* São Paulo: Cortez.
Vaillant, D. 2012. *Teaching to teach: the four stages of a learning*/Denise Vaillant, Carlos Marcelo – 1st ed. Curitiba: Ed. UTFRP.

Professional education and the formation of garments factory in Brazil

V. Feldman & M.S.B. de Held
Escola de Artes, Ciências e Humanidades da Universidade de São Paulo—EACH—USP, São Paulo, Brazil

ABSTRACT: This article presents a historical panorama about education in Brazil from the colony to the emergence of the first schools of professional education and goes through the formation of the crafts related to clothing and the teaching of drawing in schools and high schools. It was developed from an exploratory bibliographical research, observes the contextual necessity in the creation of means to specialize the workforce of the sector, through the first professional education schools, that will generate specialized labor, necessary for the textile industry and to the garments factory that developed in the period.

1 INTRODUCTION

From the historical panorama about the formation of clothing and fashion in Brazil, from the point of view of education, this article has as main objective to understand the context that led to the establishment of professional and technical education, necessary for the development of the textile industry and clothing for the training of qualified professionals.

To do so, it seeks to outline a brief course of education in Brazil from the colony to the emergence of the first vocational education schools, through the training of crafts related to clothing and the teaching of drawing in schools and high schools. It also verifies the emergence and expansion of the textile industry and the first garments factory in Brazil, which established the favorable scenario to the emergence of these schools.

This exploratory study was developed based on bibliographical research, and is an integral part of a larger research, which is in the preliminary phase, in the proposal to clarify and bring more information to the sector through academic research.

2 HISTORICAL OVERVIEW

Under the context of the history of education, the teaching of women in colonial Brazil, as well as preparing them for household duties, varied according to their ethnic-social condition —as well as men's—, which was preponderant in determining how and where their education would take place. Women from wealthier families were typically more idle and more religious educated than the working classes, middle class or even slaves. The Indian girls, in the other hand, were initiated into Christian dogmas by the Jesuits in their own villages, where they were catechized and sang prayers (Veiga, 2017).

Regarding educational practices, Veiga (2007: 72) points out that:

> Educational practices involving free women are better documented, although the emphasis is on wealthy whites and relates to convents and gatherings. Browns and mulattoes, daughters of white men and poor white women, had alternative education options to religious seclusion, such as learning crafts.

In a record made on female professions existing in Vila Rica (MG) in 1804, and in São Pedro (BA) in 1775, Silva[1] (2004 apud Veiga, 2007, page 72) reports that in Vila Rica the main ones are those of seamstress, nurse, sparkler, spinner, laundress, baker and grocer and saleswoman. In São Pedro, stand out those of seamstress, cook, candy maker, dresser, *ganhadeira*[2], laundress, baker, lacemaker and saleswoman. Veiga (2007: 72) concludes that:

Although some activities required no more than mere imitation, others assumed a systematic learning—salesperson, nurse, seamstress, or lacemaker for example. Finding out how and where this learning took place is still a challenge for historical research.

In addition, important information is provided by the tutors' report to the orphan judge, which included expenses for "sewing masters" and materials for their practice, indicating that in addition to a specific education, women were directed to become master professionals (Veiga, 2007).

With the learning of crafts, even in less complex learning activities, it was possible to acquire knowledge to survive, in view of the large number of women who worked to keep up, since only a small number of them had the means to dedicate themselves only to religion or home.

It is informed through Serafim Leite[3] (Leite apud Veiga, 2007, pp. 71–72) in a report from Father Vieira, regarding the learning of the crafts, about the existence of 24 spinners in the workshops of the colleges of Maranhão and Pará, Indians who learned this activity as a regular practice of work (Veiga, 2007).

In that same period, Rio de Janeiro, in 1808, had around 50 to 60 thousand inhabitants and, despite all the city's problems, such as lack of potable water and sewage, insects, mold, among others, the city was in a privileged place, in the midst of a paradisiacal nature (Italiano; Viana, 2015).

According to Italiano and Viana (2015), the coming of Dom João VI to Rio brought new impetus to the cultural life of the city and in 1808 began the printing of a newspaper edited here in the colony. Theaters, libraries, literary and scientific academies were inaugurated to serve not only the Portuguese but also a rapidly expanding urban population. Fausto[4] (2007 apud Italiano, Viana, 2015, p. 39) states that "many of the new inhabitants were immigrants, not only Portuguese but Spaniards, French and English who would become a middle class of skilled professionals and craftsmen".

Around 1814, with the end of the Napoleonic War (French Revolution), social life in Brazil flourished even more, and new craftsmen and technicians arrived in Brazil—English tailors, French seamstresses. In this period, the maidens Josefine and Durocher, the most famous dressmakers of the first reign, who received orders from Dom Pedro I for his mistress Domitila, the Marchioness of Santos, stand out (Italiano; Viana, 2015).

The social life that was formed, therefore, demanded a new standard of clothing, that was being provided by these new foreign inhabitants, specialized in the art of the dressing.

It should be noted that Madame Durocher, as she was called, has several aspects of exceptionality, such as being the first educated woman in Brazil to have published texts with her own name in the area of medicine (1884); the only one to be admitted as a partner in the Imperial Academy of Medicine (1871), was a midwife, as well as a seamstress and dressmaker. Her life is close to that of other urban women of the nineteenth century who survived various trades and who crossed the Atlantic to make themselves in America (Mott, 1994).

---

1. Silva, Rosa Virgínia Mattos e. *Ensaios para uma sócio história do Português Brasileiro*. São Paulo: Parábola, 2004.
2. Ganhadeira, according to Soares (1996), was the term used to the women who realized the work of gain in Brazil and occupied position of prominence in the urban life, selling mainly food in the streets. There were both black liberties and slaves, and the liberated women experienced a situation different from that of the slaves, for their work did not interfere with their masters; the slaves who were to be marketed in the streets, were obliged to give their masters a pre-established sum, and what exceeded the sum, could be added for their emancipation and that of their relatives or for their family support.
3. Jesuíta Serafim Leite, between 1938–1950, organized an important historical documentation, in *História da Companhia de Jesus no Brasil*.
4. Fausto, Boris. *História do Brasil*. São Paulo: Edusp, 2007.

Durocher had an excellent point of commerce, to the Street of the Ouvidor, so called this to have been settled, since the eighteenth century goldsmiths and jewelers. She sold fabrics, fine haberdashery, dresses and toiletries, like gloves, hats, and feather flowers. Her establishment was close to Ouvidor street, concentration of the chic commerce and of the famous French dressmakers, which are attributed to the time, a fundamental paper for the merchandise disposal (Mott, 1994).

It is important to emphasize that, in this period, practically everything that was consumed in Brazil came from the outside, from fabrics to soap. On this, Braga and Prado (2011, p. 38) point out that:

> *Domestic production (industrial or craft) did not develop, suffocated by imports. Internal textile production occurred on a small scale, producing cotton, linen and jute fabrics for "rustic clothing," lace, nets, mosquito nets, ropes, cords and sacks made from spindles, rocks and hand looms operated by artisans of their own farms or by spinners and weavers in small independent workshops in cities (...).*

Despite the luxury and French appearance of the shops and the dressmakers, they did not dispense the local free labor African slave, like other merchants. Free blacks, more educated and intelligent, sought to enter as workers, per year or per day, in a dressmaker shop or French seamstress, which allowed them later to find work in Brazilian homes (Mott, 1994).

In the second reign, between 1840 and 1889, under the regency of Dom Pedro II, in the years of 1844 it is built the Factory of All Saints, a fabric factory. And from 1850, the first textile industries of Brazil were implemented, outlining the first attempts to establish a labor market (1994).

According to Braga and Prado (2011: 42), "from 1866[5] to 1885 Bahia had 12 textile factories. Across the country, the total had risen in 1882 to 48, 33 of which were in the Southeast (Rio de Janeiro, São Paulo and Minas Gerais)". At the end of the 19th century, São Paulo assumed the national leadership, with the emergence of several major textile enterprises (Braga, Prado, 2011).

Fashion tends to expand. Women of the more affluent class, who were previously little seen on the streets, due to road conditions, began to be seen in the company of relatives and servants, due to the improvements with gas lighting and the regularity of trams.

In this context, with the division of tasks established, in a society quite patriarchal like the Brazilian, the men's clothing in this period is simplified—in the English way and, on the contrary, the feminine—in the French way is elaborate, since the woman, with the bourgeoisie ascention without having to work, returns to take the home and play a more decorative role.

In this way, many intelligent women, who returned to home duties, ended up occupying their free time with activities such as reading, dancing, among other things and sometimes standing out in social commitments for their elegance and refinement. Great influencer women of the fashion arise from this context, coming from the upper middle class which sought refinement, and women arrived in Brazil from the aristocracy.

However, slave labor and the lack of industrial activities, still under development, contributed to the devaluation of the professions, besides studies and manual and technical activities.

According to Barbosa (2012), although art was accepted as a creation and was somewhat welcomed by the wealthier classes, as a symbol of refinement and leisure, the prejudice about manual activity at this time has its roots in slavery, by the Portuguese habit to live from slaves.

With the abolition of slavery, which coincides with the first Brazilian Industrial Revolution, in the substitution of the physical labor by the mechanic one, the respectability of manual labor is established. The Fine Arts, which until that moment were viewed with social appreciation by the wealthier classes, as a gift, luxury, or even as an idle pastime, are beginning to be despised by uselessness, while the Arts applied to industry and linked to technical skills begin to be valued, since a working class with newly freedmen needs to strengthen the country's economic situation, since at that time a quarter of Brazil's population was made up of slaves (Barbosa, 2012).

---

5. In order to establish a parameter of the Brazilian industrial delay, in relation to the same time, England had more than 3 thousand factories of fabrics, counting on 11,250,000 spindles and employing approximately 600 thousand workers.

After Dom Joao VI's arrival in Brazil, a new posture regarding the teaching of drawing was established, initially in a methodological renewal—because manual activities were rejected by the Jesuits in the teaching dedicated to "free men", leaving these functions for when it tried to teach Indians and slaves. Later, it took on a new configuration, when in the Episcopal Seminary of Olinda (founded in 1800), which replaced the Royal College of the Jesuits, included Drawing in the curriculum, indicating a new educational approach with its adoption (Barbosa, 2012).

Barbosa (2012, p. 25), in regard to drawing, states that "Technical Drawing courses were created in 1818 in Rio de Janeiro and in 1817 in Vila Rica and Bahia, which were not very successful".

Part of this is due to the mentality established by the humanistic and abstract system of colonial education, settled by the Jesuits, in which technical education did not provoke great transformations, since the horror of "manual labor" related to the slave system contributed to the devaluation of the professions and studies related to manual and technical activities.

Informality in the professional technical training was widespread, and knowledge was passed from professional to professional, or from father to son, in the day-to-day professional practices.

Some of the few options for technical schools, still scarce in Brazil, were the lyceums of arts and crafts that exist in most capitals. In 1857 and 1873, respectively, were created the one in Rio de Janeiro and Sao Paulo and formed masters or "officers." (Braga, Prado, 2011).

The Brazilian who wanted to learn more about the subject, or the self-educated one who desired improvement, should travel abroad, where not only the first craftsmen came from, but also where materials, methods, technics and technology continued to come from, which reinforced the custom of dependency (Pires, 2002).

However, the concern with technical training becomes more and more necessary with the acceleration of the process of industrial expansion and modernization. Also, in 1894, the Polytechnic School of São Paulo was created to train engineers in various areas of industry and civil construction.

Since the second half of the 19th century, small confections have emerged from sewing ateliers or the initiative of textile manufacturers. According to Braga and Prado (2011, p. 51, 52), some advertising published by newspapers of the 1910s, 1920s and 1930s of Rio de Janeiro and São Paulo confirm that:

> *The making of ready-made clothes did not, during the period, cover the whole set of garments, being restricted to less complex and cost-effective execution of serial production, secondary to the demarcation of fashion styles. Predominantly underwear (called white clothes or underwear), men's shirts, children's clothing, shawls, robes, swimwear. There were also various accessories such as shoes, socks, ties, cuffs and collars, marketed at various points.*

The shops provided these various items of clothing to the prompt delivery in a certain amount, among them, some clothes produced in series, but still in a handmade way.

Moreover, the new concepts of urbanism that were established, provided cities with the development of fashion—families began to attend more public spaces and expose their social status through clothing—as they included shopping centers that facilitated access to several consumer goods, in a segmented way, to serve the different social strata.

This scenario demanded structural changes in clothes, adapting them to urban and technological innovations, and in the establishment of increasingly practical clothing.

These innovations were assimilated, thanks to the increase in the number of women's magazines bringing drawings and molds but, still, by the lack of technical training, many professionals dismantled pieces of clothing to copy them. Over time, this practice provided some security to introduce modifications to the original mold (Braga, Prado, 2011).

The products became available in a more organized way, especially those that imitated the fashions of the higher classes and they provoked an accelerated process in the exchange of fashions, because the "new fashions" were a way of distinguishing itself socially. Specialty stores and department stores appeared (Braga, Prado, 2011).

In these large stores, the sale of ready-made, series-made clothes, even if still of dubious quality, strengthens the nascent concept of *prêt-à-porter* or ready-to-wear, terms respectively of French and American origin to designate ready-to-wear clothing.

The emergence of pioneer garments factory in Brazil was largely the result of small sewing ateliers and family tailors, many of which were founded by the immigrants who arrived here. Small companies had advantages in the production and diversification of pieces that this new scenario demanded, since they had administrative simplicity and agility in the adaptation to the changes of the fashion (Braga, Prado, 2011).

Garments factory begin to acquire considerable sizes by unraveling the process of mass production, many producing on demand for specialty stores and department stores. However, they did not produce their own fashion by making models copied from the Paris couturiers, or from the US garments, which in turn had already adapted French models to industrial production.

The need for specialization becomes latent towards modernity, which would allow companies to continuously improve their procedures, products and services and, above all, to promote the synergy between the textile industry and the emerging clothing industry. For this new generation of machines, processes and products, professionalization was needed (Teixeira, 2007).

3 FINAL CONSIDERATIONS

With the arrival of new craftsmen and technicians, around 1814, a range of clothing-related trades began to be established in Brazil, which until then were scarce. The crafts that were formed in this period were constituted by immigrants coming from Europe, in addition to the few already existing. Informality in technical training was a common occurrence in this period and knowledge was passed from professional to professional or from parent to child by daily practice. In addition, almost all components related to dressing came from abroad, especially from France.

But with the increase in the productive demand, the expansion of the textile industry and the beginning of fashion making, driven by the emergence of series production, the need for skilled workers in their trade becomes essential, which culminates in the emergence of professional and technical education, indispensable to the scenario that was being formed.

REFERENCES

Barbosa, Ana Mae. *Arte-Educação no Brasil*. São Paulo: Perspectiva, 2012. 7ed.
Boucher, François. *História do Vestuário no Ocidente: das origens aos nossos dias*. São Paulo: Cosac Naify, 2010. 480 p.,1054 ils.
Braga, João & Prado, Luís André do. *História da Moda no Brasil: Das influências às autorreferências*. São Paulo: Pyxis Editorial, 2011.
Italiano, Isabel & Viana, Fausto. *Para vestir a cena contemporânea: moldes e moda no Brasil do século XIX*. Isabel Italiano, Fausto Viana; coords. Desirée Bastos, Luciano Araújo. São Paulo: Estação das Letras e Cores, 2015. 320 p.
Mott, Maria Lucia de Barros. Madame Durocher, modista e parteira. *Revista Estudos Feministas,* ano 2, 1º semestre, 1994. p.101–116.
Pires, Dorotéia Baduy. A história dos cursos de design de moda no Brasil. *Revista Nexos: Estudos em Comunicação e Educação. Especial Moda/Universidade Anhembi Morumbi* VI, nº 9 (2002) – São Paulo: Ed. Anhembi Morumbi, 112 p. ISNN 1415–3610.
Soares, Cecília M. As ganhadeiras: mulher e resistência negra em Salvador no século XIX. *Revista Afro-Ásia – Centro de Estudos Afro Orientais (CEAO)*, nº 17 (1996) – Bahia. ISSN: 1981–1411 versão online.
Teixeira, Francisco. *A história da indústria têxtil paulista*. São Paulo: Artemeios, 2007.
Veiga, Cynthia Greive. *História da Educação*. São Paulo: Ática, 2007. 328 p.

Complex geometry and patterns: Colonizing the surrounding space through the costume

M.J. Climent Mondéjar
Catolic University of Murcia, UCAM, Spain

M.P. Moreno Moreno
University of Cartagena, UPCT, Spain

A. Cano Redondo
Superior School of Design of Murcia, Spain

ABSTRACT: 'Complex Geometries and Patterns' is a new elective course (taught for the first time in the first semester of this academic year 2017–18) proposed by the Projects Department of Superior School of Design of Murcia, and designed to bring together students from various specialties (fashion, product and interiors) with the purpose of complementing respectives learning at the different study programmes.

In this article are exposed the teaching experience, its structure, the adopted methodology and the relation between content and format to be developed, taking into account that the ultimate goal was to generate among all (students and teacher) a team capable of designing a prototype of an extendable and foldable suit through the use of recycled textiles (taking materials more common in Construction and Interior Design than in Fashion Design). The proposal consisted in generating a model able of colonizing the adjacent space and qualifying the different intermediate spaces: body-textile and textile-space.

The objective of this exhibition is to assess the academic viability of this model, making critical reflections as an epilogue, which serve to assess this experience, recently concluded.

1 INTRODUCTION

The content of this article is framed in the thematic area 'Teaching and Education', proposed by CIMODE' scientific committee and, specifically, within the context 'Design teaching methodologies'.

The structure of the Studies of Degree in Design in ESD Murcia establishes four specialties: Fashion, Product, Interiors and Graphic, which are regulated by Article 58 of the Organic Law 2/2006 of May 3, Education and Article 23 of Royal Decree 806/2006 of June 30. Since 2010–11 academic year these studies have been progressively implemented.

The study programme contemplates, in its 240 ECTS, three types of different courses:

- Courses of basic training (62 ECTS), common to the four specialties and developed in 1st and 2nd year;
- Compulsory and specialty courses (124 ECTS), typical of each of them and taught in 1st, 2nd, 3rd and 4th grades;
- and Elective courses (29 ECTS), to be choosen between 3rd and 4th.

Each specialty proposes a catalog of elective courses that is approved in the corresponding Department. But this does not mean that the courses have to be designed for each specialty in an airtight way, but can be offered for students of some other specialty or of all of them. This is the case of the subject 'Complex Geometries and Patterns' which, despite having been designed in the Department of Interior Specialty Projects, was also offered to students of Fashion and Product.

The small number of students enrolled, in this case only girls (three from Fashion, one from Product and one from Interiors), makes rethink the methodology of the course for its development during 2017–2018 academic year.

Considering those premises established in the initial planning, in terms of objectives and evaluation criteria, it is decided just to change the format. Individual activities are respected, but the Final Project, designed to develop as a group, is now considered as something similar to a professional assignment to be developed jointly by all the students in the class, under the teaching direction.

To emphasize the imaginary nature as something that could be a professional assignment, it gets really important the active participation of every student, through the prototype that will be designed in the classroom, as an artistic piece that will be held The Night of the Museums in the Museum of Santa Clara (Murcia). It is agreed with the person in charge of Promotion of the ESD and with the director of the Museum that, through the works developed in two elective subjects, 'Complex Geometries and Patterns' and 'Interactive Spaces', a work will be generated that will represent the participation of the students of the ESD in said event.

2 DESIGN OF THE ELECTIVE COURSE

The elective courses develop content whose purpose is to update, complete or extend the training of students, in accordance with the provisions of section 1.c. of article 7 of Royal Decree 633/2010, of May 14, which regulates the basic content of the higher artistic education of Design established in the Organic Law 2/2006, of May 3, of Education.

'Complex Geometries and Patterns' is justified through the constant updating of the daily Design processes related to Interior Design. The main objective is to be able of creating new universes within existing spaces and the implantation of extensible artefacts (portable or not), responding to the current functional requirements of eventual and ephemeral situations or contexts. The mixture of students from different specialties, to work on team proposals, aims to complement the respective learnings of each specialty.

The assignment of competences is done as shown in the following table:

The organizing content of the subject 'Complex Geometries and Patterns' is structured through a theoretical part (referred to the acquisition of knowledge about concepts, spatial organizations, functional typologies, plastic, technical, organizational and economic aspects of the projects related to the geometric design of textile structures and costumes), but first of all it has a marked procedural character, which refers to the acquisition of a method of work, modes, processes and ways of analyzing and understanding the project. Students must be able to design spaces of different categories, bounded by fabrics of different textures and qualities, as well as proposing typological, technical and conceptual innovations responding to this type of requirements.

The final materialization of the reflection will consist of creating a prototype 'EXTENSIBLE SUIT' on a real scale, capable of colonizing the adjacent space.

Table 1. General data of the course 'Complex Geometries and Patterns'.

General data	
Type of subject	Elective
ECTS	3
Academic year	4th
Annual/Semi-annual	First semester
Teaching hours (annual computations)	1/84 hours of workload
Department	Projects
Professors	Interior Design's professors

Table 2. Assignment o the different types of competences in "Complex Geometries and Patterns".

Cross-disciplinary competences	CT1 Organize and plan the work in an efficient and motivating way.
	CT8 Develop reasoned and critical ideas and arguments.
	CT15 Work autonomously and assess the importance of initiative and entrepreneurship in the exercise.
General competences	CG1 To conceive, plan and develop design projects in accordance with the requirements and technical, functional, aesthetic and communicative conditions.
	CG3 Establish relationships between formal language, symbolic language and specific functionality.
	CG8 Proposing research and innovation strategies to resolve expectations wich are focused on functions, needs and materials.
	CG13 Know the economic, social and cultural context in which the design takes place.
	CG19 Demonstrate critical capacity and know how to propose research strategies.
Specific competences interior design	CEI4 Analyzing, interpreting, adapting and producing, information related to the materialization of every projects.
	CEI5 To solve aesthetic, functional, technical and constructive problems that arise during the development and execution of the project.
	CEI6 Interrelate formal and symbolic languages with specific functionality.
	CEI9 Adapt the methodology and proposals to the technological and industrial evolution of the sector.
	CEI15 Reflect on the positive social influence of design, assess its impact on improving the quality of life and the environment and its capacity to generate identity, innovation and quality in production.

3 SYMBIOSIS BETWEEN ARCHITECTURE AND SUIT

We get close to something very like man-as-a-bat where the skin of the enclosure is dependent upon a system of vertebrae that respond very directly to the nervous system of the person within. (Sadler, 2005, p. 129).

How much an human body's artifact have to be separated from the flesh so that it ceases to be considered 'suit' and becomes to be considered 'architecture'? Which is the length to which its envelope must be related to our skin?

The conceptualization of these issues refers to anthropological studies that address the different materialisations about the spatial creation of the human habitat over centuries. In this sense, social psychologists recognize the initiative of clothing as an important precursor of every change related to furniture and, by extension, to interiors architecture. The way of dressing influences sensory perception of human beings about the architecture that surrounds them and that is why men and women has transformed it throughout history. In this sense, it should not been forget Le Corbusier's reflections about disadvantage of men's clothing compared to that of the woman of the 30s, who *"earned a living"* thanks to the reform of their primary daily construction, wich means their clothing linked to the concept of freedom (Le Corbusier, 1999, p. 128).

Suits can go taking more slack not being fit, or even, go completely attached to the body. And, on the other hand, architecture can be delimiting and generating new spaces and its envelope does not necessarily have to be defining the boundary between exterior and interior. Architecture can generate spaces within spaces and thus redefine the initial or pre-existing qualities of a particular site.

Architecture and fashion are able to redefine our second skin (or third, fourth,...) in very different ways: well as something totally material and finite; well as something more conceptual that responds to the space, framed and articulated around a structure; or as something totally immaterial and without form.

As an example of this first category—spaces within defined spaces with a finite envelope, practically impermeable to sight—one could cite the Design Museum dedicated to Apoxyomenos, in Croatia. Apoxyomenos (Figure 1) is a bronze statue representing a young athlete; it was found not too long ago, in 1996, in a city on the island of Lošinj and, after the costly maintenance and restoration work to which it was subjected, as well as being exhibited in some of the most important museums of the world, it returned to Mali Lošinj, where it was built its own exhibition hall inside an existing building, the Kvarmer Palace.

As examples of the second category (intermittent enclosures, defined around a skeleton wich delimits its surroundings conceptually) one can allude to the 'Formwork' (Figure 2) design by Karla Escalona, realized as the final work in the Postgraduate Course of Architecture *Moda y Diseño* (Universidad Politécnica de Madrid) which was exhibited, along with the rest of the work done in AMD, in the ETSAM lobby. The design of the prosthesis (Figure 2) that artist Thomas Thwaites (2016) used to become a goat man is also a good example for this category, or even the metal ribs painted in pink or gray that defined COS pop-up store's commercial space designed by Snackarchitecture, set up inside Austere in Los Angeles.

The third category is that of those envelopes without form, defined in the most abstract way. In architecture it is easy to identify this concept with the Blur Building pavilion, that Diller + Scofidio designed for the Swiss Exhibition 2002 in Yverdon-les Bains (Figure 2). It was a floating pavilion on the Neuchatel Lake all defined by water vapor.

The concept of suit is defined by the dictionary through confusing meanings,[1] but there is no doubt that it can be understood as the closest to the body portable architecture we can find.

Historical proposals in architecture, such as designs by the Archigram Group, designs by Buckminster Fuller, or by Toyo Ito, already made us reflect on this question: when the suit begins to be architecture or when architecture becomes a suit.

Figure 1. 'Apoxyomenos' Tribute Museum, located inside the Kvarmer Palace, Mali Lošinj (Croatia, 2009). Source: Designboom.com.

Figure 2. Karla Escalona's design; prosthesis used by Thomas Thwaites to become Goatman; and Diller + Scofidio's Blur Building, al the Expo 2002 in Yverdon, Switzerland.
Sources: AMD ETSAM; Thwaites; Diller Scofidio + Renfro.

---

1. m. Full dress of a person; 2. m. Peculiar dress of a class of people or from the natives of a country; 3. m. Set of jacket, trousers and, sometimes, vest, made of the same fabric; 4. m. One piece female dress. (RAE, 2018).

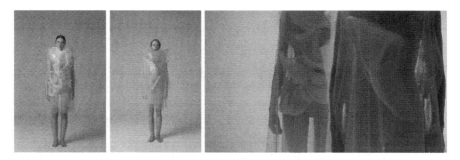

Figure 3. 'Possible Tomorrows', interactive costumes with fingerprint recognition, which enter into movement when an unknown person approaches.
Source: Yellowtrace.

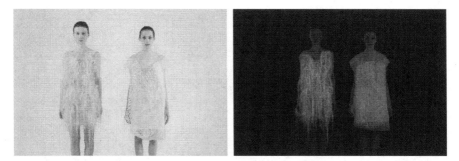

Figure 4. '(No)where (Now)here' collection. Costumes start moving when the person wearing them surprises a spectator who is looking at the garment.
Source: Yellowtrace.

4 YING GAO'S WORK OF AS A REFERENCE

The designer Ying Gao, professor at UQAM (Université du Québec à Montréal) questions through her creations the way we perceive fashion by combining art, architecture, urban design, product design and anthropological studies. She tries to investigate the construction of clothes, being inspired by mathematical patterns and daily transformations at urban environments. Her projects are more inspired by technological progress than by standard textile industry. *"Design is the medium, located in the technological field more than in the textile: sensorial technologies allow garments to become more poetic and interactive"*. (Acance, 2017, December 5th 2017)

Using a combination of atypical materials (organza, optical fiber, PVDF[2] y electronics devices) she has made interesting collections that are shown at ESD's classroom as a theoretical session.

5 WORKSHOP

The students make different designs in order to discuss how the final prototype should be. Projects are mentioned that allude, from more architectural references ('Cushicle' by Archigram, 'Techno Geisha' by Andrés Jaque) to historical references of the world of fashion (Irving Penn and Issey Miyake).

---

2. Polifluoruro de vinilideno, a thermoplastic fluoride that works very well in the manufacture of pipes and installations subjected to high temperatures and which is almost chemically inert.

Figure 5. Sketches and references taken from Issey Miyake and Irving Penn.
Sources: Student self made; Irving Penn.

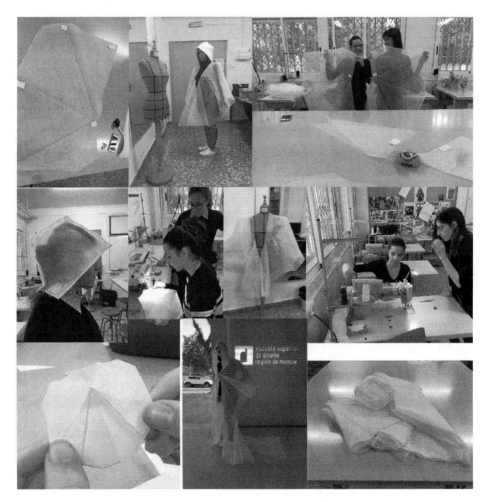

Figure 6. Photos of the work process in the workshop.
Source: Self made.

After making a prototype on a small scale, wich combines ideas of the whole team, we proceed to draw the patterns in CAD to print them on a real scale and start sewing. The textile chosen for the model is a PVC mesh, which is usually used as an anti-insect barrier, installed on stretcher frames at the hollows of conventional facades.

The sewing task becomes interactive, so that the model initially proposed is modified and modeled. The prototype is adjusted to the performatic movements of the process of deployment/extension of the suit. In order to facilitate this dynamic, Velcro is incorporated along the seams, so that not only does it interfere with the perception of its geometry, but it also emphasizes it through the rigidity provided by this material.

Figure 7. Study of the dynamics of the suit's folds, which allow its extension, generating a fusion between the suit space and the architectural space.
Source: Self made.

Figure 8. Photographs taken at Murcia Science's Park (Fernando de Retes, 2013).
Source: Self made.

Once completed, the possibilities of incorporating sensors and lighting are studied and the possibilities of the gesture of shortening or extension of the suit through the mechanics of its folds are analyzed.

Finally, an extensive photographic session is carried out trying to document the possibilities of the prototype.

6 CONCLUSIONS

The methodology of this exercise has combined the theoretical resources explained in the classroom with the systematization of pattern making as a practical mechanism of textile manipulation, and finally, it has incorporated reflections on the visual perception of the object for its understanding and dissemination. These three phases have required the dialectical effort between practical teaching, intuitive learning and observation of referents.

The suit is understood as a third skin that adds new functions to the new spaces it delimits. It has been proved that its materiality makes it impermeable to rainwater, while its porosity makes it breathable.

The final images of the suit in the body of the models, at the environment of an architectural scenario selected for the occasion, are part of the process of manipulating the sequential movement in its expansion into space. Body and suit positions, and its emissive and reflective lighting, have collaborated in the iconographic result that has supposed the corollary of the objectives originally planned, exceeding any forecast. Definitely, as Emilio Lledó (2009) said: *"(…) living in space means accommodating the body and sight limits to the nature that overflows and expands the contour of our skin (…)"*, so this exercise invites us to reflect on the spatial colonization that a system, linked to the body, can generate in the closest space extending exponentially to architecture.

REFERENCES

Acance, F. 2017. *Interactive clothing with fingerprint recognition technology by Ying Gao* [online]. Avaliable from http://ellowtrace.com.au [30 April 2018].
Cook, P. 1999. *Archigram.* Londres: New York Princeton Architectural Press.
Crompton, D. & Schöning, P. 2004. *The Archigram Files.* New York: Monacelli Press.
Herman, A.E. 2017. *Visual intelligence: sharpen your perception, change your life.* Boston: Mariner Books.
Nakamichi, T. 2012. *Pattern Magic.* Barcelona: Gustavo Gili.
Lachmayer, H, Schoenig, P. and Crompton D. 1994. *A guide to Archigram: 1961–74.* London: Wiley-Academy.
Le Corbusier. 1999. *Precisiones, respecto a un estado actual de la arquitectura y el urbanismo.* Barcelona: Apóstrofe.
Lledó, E. 2009. *El marco de la belleza y el desierto de la arquitectura.* Madrid: Biblioteca Nueva.
Penn, I. & Miyake, I. 1988. *Irving Penn regards the work of Issey Miyake.* Paris: Pont Royal.
Perinat Forteza, M. 1997. *Tecnología de la confección textil.* León: EDYM.
Thwaites, T. 2016. *GoatMan: How I Took a Holiday from Being Human.* San Francisco: Chronicle Books.

Project thinking in fashion design: Strategies for facilitating cognitive and metacognitive processes

M.C.F. Sanches
State University of Londrina (UEL), Londrina, Brazil
Brazilian Association of Studies and Research in Fashion (Abepem), São Paulo, Brazil

M.A.R. Silva
State University of Londrina (UEL), Londrina, Brazil

ABSTRACT: This paper results from the overlapping of two researches conducted in the Fashion Design course at the State University of Londrina. One, a doctoral investigation on methodological strategies for fashion design projects from an educational perspective, and two, a master degree research, which aimed to promote self-regulation in learning through cognitive and metacognitive strategies in project teaching. From the shared purpose of improving the project thinking experience in fashion designers, the study was carried by an exploratory approach, resulting in ideas for strategies to connect the multiple phases within the design thinking process. The results presented in this paper reinforce the importance of graphical/visual schemes, of a hands-on approach in education, and of previous knowledge as means to facilitate the teaching and learning of project processes.

1 INTRODUCTION

A design project is always intertwined with the sociocultural context, for its primary purpose is to build interfaces between humans and the environment. The project emerges from a contextual request (that is, a request made by the sociocultural scenario), and ultimately dives, once again, in the same context that created it. Thus, understanding the circumstances around the situation requires great effort. According to Sanches et al. (2015), the design process is a flexible system in constant transformation due to the interaction with the sociocultural context. During this process, activities connect to each other in a dynamic network, embodying new information and transforming itself in order to find the most suitable road to success.

During the education process of designers, it is important to develop integrative abilities to encourage them to connect information in this system, through strategies that transform such connections in possible and coherent solutions. Clearly, this requires the understanding that the project itself is a management process of information/knowledge construction, which demands self-motivation for scientific investigation and learning. Sanches (2017) argues that each new project context requires the ability to explore new worlds, raise questions, set objectives, synthesize concepts, experiment with possibilities, and communicate solutions. Training such abilities in project education depends on strategies that facilitate the dynamic connection of information and the simplification of objectives and concepts.

In order to frame a methodological guidance for the fashion design education, and to assist the construction of strategies, this study reports on the overlapping of two researches conducted in an educational context. One, a doctoral investigation on methodological strategies for fashion design education, and two, a master degree research focused on the promotion of self-regulation in learning, through cognitive and metacognitive strategies in project teaching. The first was the result of a partnership between three universities: the Faculty of Architecture and Urban Planning at the University of São Paulo (FAU-USP), the Universitat Politècnica de València (UPV), and the State University of Londrina. The later took place within the Master Program in Education at the State University of Londrina (PPEdu UEL) and the Fashion Design course in the same institution.

The overlapping of the two researches provided multidisciplinary theoretical foundation and field experience to support this study, which presents a global, dynamic view over the project thinking, focused on improving the learning experience. The results presented in this paper reinforce the advantages of graphical/visual strategies and of active educational approaches in the cognitive and metacognitive processes within the teaching and learning of project thinking.

2 THEORETICAL BACKGROUND

2.1 *The phases of project thinking within the fashion design educational context*

According to Coelho (2008), the concept of "project" is intrinsically related to the idea of planning and organizing towards the future, aiming to accomplish something. Although this definition suits a variety of knowledge fields and human praxis as a whole, it is inextricably linked to the design process. The design project requires a flexible attitude towards the context, in order to capture the big picture and plan on how to deal with it. The overall understanding of the situation provides insights about the problem by integrating multiple dimensions, such as usability concerns, productive aspects and environmental issues, among others. Through a multidisciplinary and transversal approach the problem becomes clear, allowing the definition of strategies to reach solutions.

In face of this scenario where multiple perspectives interact, the path between the contextual request (problem) and the answer (solution) is obviously marked by transformation and the need for numerous and varied thoughts, reflections and knowledge. Therefore, to the aim of studying strategies to facilitate the process, it is necessary to consider the cognitive arrangements applied along the way.

Lawson (2011) explains that the project thinking includes progress, regression and displacement cycles. In the same train of thought, Sanches et al. (2015) explain how this interactive and fluid cycle requires continuous data collection, constant feedback analysis, and that, frequently, parallel lines of thinking may arise. According to Sanches (2016), this process reveals a multidimensional structure embodying a flexible core, which expands to collect new information and shrinks as it connects such data and decides the next step towards the answer—that is, it represents a flexible system adaptable to the changes in context. In this sense, the project management requires a continuous decision-making process, and the ability to deal with constant cycles of analysis, synthesis and evaluation—as underlined by Jones (1992) and confirmed by Lawson (2011).

Theoretical discussions indicate that the design process addresses logical reasoning and heuristic investigations at the same time. According to Lessa (2008), this combination is essential to the analysis, synthesis and evaluation cycles. Studies such as Lawson's (2011), Cross's (2000), Jones's (1992), and Lessa's (2011), highlight that the ability to move between divergent and convergent thinking is desirable in every step of the project. In the divergent mode, thinking moves in multiple and simultaneous dimensions, broadening possibilities, aiming to find multiple possible solutions for the problem. On the other hand, convergent is the unifocal type of thinking—that is, it employs the deductive and interpolative approach to reach an answer that is unique and final. Thus, the selection of a path to follow is always moving from divergent, multiple, wide-open possibilities to convergent, singular, specific answers.

In summary, managing the design process, one must act with proper and efficient attitude in face of contextual requests, which requires a correct arrangement of knowledge, morality and decision-making. As the project progresses, every single action is charged with thinking operations that will establish which procedures and strategies to follow. Thus, in the educational context, the challenge is to set the environment so that students are entitled with an active participation in the strategies used to accomplish the goal. Also, the project teaching-learning environment must encourage reflections on the project practice, where students are prompted to decide on the methodological strategies they use in each step, according to the scope of the project and their previous knowledge.

In an attempt to incorporate such arrangements in the fashion design education and adjust in-class project practice, Sanches (2016; 2017) defines four groups of available procedures: a) procedures used to contextualize and synthesize project requirements (concepts); b) procedures to explore ideas and experiment with possibilities and connections; c) procedures to assess the feasibility of the possibilities; d) procedures to test the technical-productive viability of the ideas.

The author argues that those groups are non-linear and definitely not independent from each other. In fact, they occur in simultaneous and continuous cycles of interaction and feedback. In any case, the comprehension of the sociocultural context is always the starting point for a project, where the relationships between actions should emphasize three focal points: explore the context and identify relationships and requirements (OUTLINE); combine the requirements and design multiple possibilities (GENERATE); and select and organize possibilities into a feasible solution (ASSESS/CONSOLIDATE).

Although the focal points listed represent "phases" of the project evolution, they are definitely non-linear in practice. "Focusing" on the actions depends on repeating cycles of analysis, synthesis and evaluation, in which divergent and convergent processes take place. For instance: while assessing a specific idea in order to select possibilities (convergence process), new development cycles may arise to question or consolidate the original idea (divergence process). This means the aforementioned classification is not a formula, but rather an educational approach to relate the activities and phases within the project development, aiming to encourage reflections on the adjustment of methods as we progress from one step to another.

The methodological approach must allow the understanding that the project is a complex system, in which the variables interact simultaneously. "The project is not a straight line between the problem and its solution, but a spiral where each new segment is thicker than the previous" (Bomfim, 2014, p. 67). The project spiral is structured from the connections of information, the synthesis of concepts, and the transformation of these concepts in meaningful knowledge towards a coherent solution. In order to do so, the methodologies must be holistic and flexible, so that the multifocal dimensions of the design investigation can be clearly identified, as mentioned previously in this topic.

The dynamic investigation of the project context—whose primary concern is to outline the project boundaries—was explained by Sanches (2016) as a "systemic management of information", referring to the exploration of the context that leads to a "questioning" attitude and the definition of relationships between variables to synthesize concepts. Proper management of the variables is essential to improve the project experience. The understanding of such relationships not only promotes the design of multiple possibilities (GENERATE), but also establishes criteria to ASSESS and CONSOLIDATE ideas. However, prior to designing efficient strategies for the educational context, it is important to analyze the cycles of thinking from a cognitive perspective, that is, highlight the most important aspects for the project teaching-learning process.

2.2 *Cognitive and metacognitive processes in project teaching-learning practice*

A project advances through structured, non-linear actions marked by progression and retrocession stages involving the cognitive activity, cognitive organization, and metacognitive processes. This process requires educational efforts in adapting self-regulation strategies along the teaching-learning practice, aiming to promote the organization of knowledge in meaningful structures. Educational Psychology has underlined the students' self-regulation skills as means to consolidate the meaningful learning. The self-regulation is defined as a process in which the individual deliberately controls their internal (cognitive) resources, while managing external resources to self-manage their actions towards the objectives. The main characteristic of self-regulated learning is the deliberate use of strategies.

In order to understand the processes involved in the self-regulated learning, it is necessary to dive in the Cognitive Psychology. This field of knowledge has greatly contributed with the educational context by providing insights on the internal processes involved in knowledge acquisition, retention and recovery, as well as its mental schemes (Rodrigo & Correa, 2004; Sternberg, 2010). Some of the cognitive processes are: perception, learning, memory, reasoning, and problem-solving (Eysenck & Keane, 2017). The individual actively applies these mental processes when receiving and restructuring new information and previous knowledge. In this sense, proposing new and coherent pedagogical practices for design education requires the overall comprehension of mental processes.

The information flows through specific processing stages, beginning with the data being transmitted across the environment and recorded in the sensory memory (echoic and iconic memory). Later, the information moves to the short-term memory (also known as working memory), in order to be codified and passed to the long-term memory, where it will be stored.

The learning process is intrinsically related to memory. According to Sternberg (2010), Eysenck & Keane (2017), this process involves dynamic structures related to the information maintenance and recovery, through codification, storage and recovery mechanisms. Once sensory data is codified, it is transformed into mental schemes, whilst the codified data is stored in memory for later recovery.

Efficiently consolidating information in the long-term memory depends on how we manage and deal with complex data. Meaningful connections must be established to arrange interrelationships between new knowledge and previous experience (also called previous knowledge) in the student's cognitive structure. In order to integrate new information and existing mental structures, one must meaningfully relate it with previous knowledge, thus transforming both recently acquired information and topics learned before (Rodrigo & Correa, 2004). This means that recovering information depends on the degree of meaningfulness established in the codification process.

Processing and organizing knowledge is an individual effort that requires awareness and motivation from the student. The ability to self-regulate the learning process—through the awareness, monitoring and controlling of one's own cognitive processes—was named *metacognition* by the Cognitive Psychology. Waltz et al. (2016) argue that the metacognitive capacity is related to becoming aware of the cognitive skills: on one hand, monitoring refers to the assessment of the cognitive activity and the performance of cognitive tasks; on the other hand, controlling refers to decision-making, as a means to regulate or review the process, based on information collected through the cognitive monitoring.

The project process requires mental labor in order to overcome obstacles and reach goals, involving actions such as: designing, identifying, defining strategies, organizing information, monitoring, and assessing. Designing solutions demands that the student understand the nature of the problem (as well as the contextual requests), identify abilities and limitations, organize and monitor time and deadlines. The metacognitive processes appear before, during, and after the task performance. When the project context is presented to the students, they grow expectations regarding future outcomes. As the project progresses, it is important to measure performance indicators and adapt the planning to reach specific goals.

The cognitive and metacognitive strategies help promoting critical thinking, time management, reach for help when necessary, and the management of complexity in content-context, thus leading to a better organization of information that will then be processed. Within this context, a major concern is to provide students with opportunities to apply strategies during the project process, allowing them to establish meaningful connections and reach success.

2.3 *Towards facilitation strategies*

The two researches presented in this paper uncovered essential principles used to facilitate project practice in educational context. Such strategies not only allow divergent thinking to flow—broadening ideas and improving rapid cognition—but also welcome approaches that sustain and promote convergence in order to synthesize concepts. The strategies provide a global perspective over the context and the ability to synthesize, favoring reflective analysis and encouraging metacognitive reasoning at the same time.

Three key points are essential in accessing and processing information in a systemic fashion, so that transversal associations can emerge. The strategies should: facilitate the documentation and perception over the connections of information extracted from the project context; raise the interest in new knowledge; and create an environment where flexibility and self-determination is promoted. In this sense, the graphical-visual techniques and the transforming aspects of Meaningful Learning are on top of the list.

With the purpose of facilitating the documentation and perception over information connections, Eppler & Burkhard's (2004) study indicates several benefits of visual schemes in transferring and producing knowledge—described in their investigation by the name "knowledge visualization". The authors state that employing knowledge visualization in creative activities can foster the possibility of fluid rearrangements and the connection of collective ideas. Eppler (2013) argues that the majority of mental activities manage the processing and analysis of visual schemes. This idea is supported by many empirical studies that show greater efficiency of nonverbal stimuli (in comparison with verbal ones) in helping performing a variety of activities, since the stimuli channel broadens when prompted with visual inputs.

In addition to that, Roam (2011) indicates that the visual, spatial, synthetic thinking, when combined with the verbal, linear, analytic reasoning, creates graphical organizations that facilitate the expression and the sharing of ideas. This interactive approach to articulate imagery and words—called "vivid thinking" – suggests that a vivid idea emerges from a mental disposition to materialize the idea and to expand the comprehension experience. Within the "vivid grammar", the author stands for the use of maps to represent relationships, and the use of diagrams to synthesize cause and effect interactions in multiple directions.

This approach highlights the benefits of graphical-visual schemes as means to construct collaborative knowledge and as strategy to organize and connect previous knowledge to new ideas, since the global perspective on the dynamic relationships leads to a systemic management of information.

Regarding the second key point—raise interest in new knowledge –, it clearly indicates a need for self-motivation in learning. Therefore, a transforming experience will be easier to embrace as new knowledge if the student is willing to learn. Supportive of this argument is Ausubel's Meaningful Learning theory (2003), whose primary focus is to connect the object of study to previous knowledge in the student's cognitive structure, taking advantage of existing resources as input channel to the new. According to Sanches et al. (2015), this approach helps the promotion of autonomous reasoning and interconnections, since the student is constantly being prompted to organize their cognitive structures to accept mental stimuli and construct new knowledge.

Besides, visual strategies are more useful in constructing meaningful knowledge when used in a positive pedagogical environment, in accordance with the third key point. Both researches used in this paper embraced the pedagogical recommendation of promoting reflective practices and collaborative behavior between students and professors. The stimulation of a proactive and investigative attitude results in a relationship of otherness, so that transformative actions can emerge. In order to manage the project variables, students must exercise their integrative abilities, according to the previous resources available in their cognitive structure.

In the same train of thought, Anastaciou (2012) suggests that the learning process depends on the sharing of possibilities between students and professor. The author argues that the professor's mediating role and the student's self-activity should work together to connect the student and the object of study. Moreira (2000) and Novak (2010) agree with the relevance of affection in the learning process, and explain that, in order to foster motivation, the student must be considered as a whole: human beings with thoughts, feelings, and attitudes.

In this context, it is necessary (and urgent) to have the students deciding on the strategies used to manage their own project process and to stimulate in them a meta-project overview. The ultimate purpose is to encourage a reflective and autonomous attitude, promoting a meaningful and solid educational context.

3 METHODOLOGICAL APPROACH

Since this paper reflects the overlapping of two researches, the procedures adopted cover a broad methodological background we summarize here to clarify the sequence of this specific piece of work. The investigation on design methodologies was carried in a qualitative approach, through bibliographic survey, documentary analysis, and participant observation in hands-on investigations. These methods resulted in a novel graphical-visual scheme (the REC Diagram and the Expressive Categories Map), which was tested and validated in practice during the Project Methodology course in Fashion Design at the State University of Londrina, from 2013 to 2016, involving 77 undergraduate students and 8 faculty members.

In turn, the study on the self-regulation of learning in project education was a semi-experimental research also conducted with undergraduate students in the Project Methodology course in Fashion Design at the State University of Londrina, from 2016 to 2017. Procedures were divided in three phases: pre-testing; pedagogical intervention through curricular integration; and post-testing. The results of the intervention were analyzed by comparison, through a control group of non-participant students.

In line with the same qualitative approach and connecting scientific investigation to educational practice, this study has an exploratory nature. Results from both researches were reviewed and further investigated, with the purpose of producing resources that could serve to assist extended practices. Finally, relationships between facilitation strategies and knowledge construction in project process were synthesized in a diagram.

4 RESULTS AND DISCUSSION: FACILITATION STRATEGIES DIAGRAM

Considering that both researches took place in the same educational context, it is important to mention the pedagogical approach of the course where the study occurred: it encourages autonomous attitude and promotes interdisciplinary skills. Thus, collaborative actions and the new methodological procedures were promptly embraced, which contributed to a transversal and receptive environment for the implementation.

From the analysis of the methods applied—participant observation protocols, focus groups testimonials (collected from students and professors), learning strategies metrics, project notebooks and learning diaries—it is possible to draw some conclusions. First, graphical-visual resources are proven to be efficient in facilitating cognitive organization, the perception of simultaneous relationships, the convergence of overlapping information, and the group communication. Second, the pedagogical orientation provided by Meaningful Learning promoted a collaborative environment, where professor and students could take responsibility for the in-class processes. Third, students reported to have enjoyed the freedom to make decisions and the co-responsibility for the actions—in order to accomplish goals. This confirms the benefits of promoting autonomous attitude and reflections over their particular *modus operandi* in experiencing the process.

In this sense, the graphical-visual strategies allowed to bridge cognitive gaps between previous and new knowledge and to increase the motivation to learn meaningfully. The multiple, perceptive experiences evoked by the tools helped simplify the perception and the integration of the project goals. During the practical application moments, students showed immediate inclination to explore possibilities in the real world, thus leading to the incorporation of tridimensional modeling as strategy to experiment and connect expressive and functional requirements. Based on the results, a diagram on the relationships between facilitation strategies was designed. This diagram can serve as a multifocal guide to preparing the environment for project education.

These approaches, when combined, have proven to be useful facilitation strategies. Figure 1 synthesizes the three categories of approaches: previous knowledge, graphical-visual schemes, and real-world experimentation.

The diagram represents the educational environment (inner circle) embodied by a project context, in which the relationships between the variables (information) energize the design

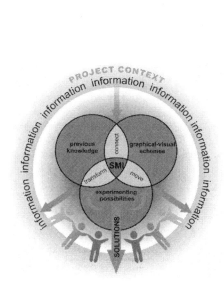

Figure 1. Diagram of relationships between facilitation strategies.
Source: The authors.

process. In turn, the graphical-visual schemes allow bridging the gap between new information and previous knowledge. The simultaneous nature of this approach (combined with the global perception of the process) encourages the connection of new information with previous elements drawn from the context. As the general associations start to narrow (in divergent and convergent cycles), the real-world experimentation transforms abstract ideas in material, feasible possibilities and, at the same time, promotes further understanding on the project context—thus creating new information that will, once again, feed the system and transform the previous knowledge.

As the cycles repeat several times, the solution becomes clear and tangible, also enabling students to construct new knowledge in their cognitive structure. The strategies are reinforced by the collaborative approach—represented by the human figures at the bottom of the diagram—ultimately leading to the intersection of all strategies and the systemic management of information (green area).

5 CONCLUSIONS

The development of mood boards is largely diffused in fashion design—visual metaphors are especially applied in representing the symbolic elements of artifacts. However, as means to promote expression and the cognitive organization, graphical-visual schemes can expand far beyond the aesthetical synthesis. In this study, graphical-visual strategies have proven to facilitate the flow of interactions within the project system and provide a means to materialize abstract ideas, encourage the synthesis of concepts, and improve group communication. Considering that multiple perceptive schemes can occur at the same time, this approach improves the agility in integrating divergent and convergent cycles, while helping students becoming aware of the overall process and of their own attitude towards it.

When visual information is organized, it allows the participants to acquire a global perspective and helps identifying connections between matured ideas and new elements, expanding the overall comprehension over the ongoing process. This approach—when applied in an educational context that encourages proactivity and values previous experience—helps promoting reflective thinking and the independence to make decisions, thus creating a receptive environment to metacognitive processes and self-regulation.

In the educational context presented in this study, positive effects were observed: the approach provided students with motivation to intentionally use cognitive and metacognitive learning strategies, such as the development of graphical-visual schemes (in order to understand the relationships between elements); autonomous comprehension and documentation of their own project attitude; time management; and the handling of emotional and motivational aspects. Such strategies favor learning, since they encourage highly informative mental associations, and facilitate the processing of new knowledge by establishing meaningful connections and the organization of information.

These resources may help unveiling novel approaches towards the facilitation of learning and the development of strategic thinking throughout the project process. The results achieved with this study also promote the discussion over pedagogical practices in the fashion design educational context.

The researches presented in this paper highlights the importance of the faculty role in encouraging the use of facilitation strategies in the educational context, ultimately leading to an autonomous attitude and the reflective thinking over the knowledge construction within the teaching-learning process.

REFERENCES

Anastaciou, L.G., 2012. Ensinar, aprender, apreender e processos de ensinagem. In: L.G Anastaciou e L.P. Alves, (Org.). 2012. *Processos de Ensinagem na universidade*. Joinville: Univille. pp. 17–44.
Ausubel, D.P., 2003. *Aquisição e Retenção do Conhecimento: uma perspectiva cognitiva*. Lisboa: Plátano Ed. Técnicas.
Bomfim, G.A. 1999. Algumas considerações sobre teoria e pedagogia do design. In R.M Couto, J. Farbiarz e L. Novaes, (Org.). 2014. *Gustavo Amarante Bomfim: uma coletânea*. Rio de Janeiro: Rio Books, 2014 c. pp. 57–72.
Coelho, L.A., 2008. *Conceitos-chave em design*. Rio de Janeiro: PUC-Rio e Novas Ideias.
Cross, N., 2000. *Engineering Design Methods: strategies for product design*. 3rd ed. Chichester: Jonh Wily & Sons.
Eppler, M. What Is an effective knowledge visualization? Insights from a review of seminal concepts. In: F. Marchese, E. Banissi, (Org.). 2013. *Knowledge visualization currents*. New York: Springer. pp. 3–12.
Eppler, M., Burkhard, R., 2004. Knowledge visualization: Towards a new discipline and its fields of applications. *ICA Working Paper*, v. 2, pp. 3–25.
Eysenck, M.W., Keane, M.T. 2017. *Manual de Psicologia Cognitiva*. 7rd ed. Porto Alegre: Artmed Editora.
Jones, J.C. 1992. *Design methods*. New York: John Wiley & Sons.
Lawson, B., 2011. *Como arquitetos e designers pensam*. São Paulo: Oficina de Textos.
Lessa, W., 2011. Objetivos, desenvolvimento e síntese do projeto de design: a consciência do método. In D. Westin e L. A. Coelho, (Orgs.). 2011. *Estudo e prática de metodologia em design nos cursos de pós-graduação*. Rio de Janeiro: Novas Ideias. pp. 18–54.
Moreira, M. 2000. Aprendizagem significativa crítica. In: J.A. Valadares et al. (Orgs.). 2000. *Teoria da Aprendizagem Significativa*: Contributos do III Encontro Internacional sobre Aprendizagem Significativa. Peniche: pp. 47–65.
Novak, J. 2010. Learning, Creating, and Using Knowledge: concept maps as facilitative tools in schools and corporations. 2º ed. New York: Routledge.
Roam, D., 2011. *Bla, bla, blh: qué hacer cuando las palabras no funcionan*. Barcelona: Planeta.
Rodrigo, M.J., Correa, N., 2004. Representação e processos cognitivos: esquemas e modelos mentais. In. C. Coll, J. Palácios e A. Marchesi, (Org.). 2004. *Desenvolvimento psicológico e educação: psicologia da educação escolar*. v2, 2rd ed. Porto Alegre: Artmed. pp. 145–160.
Sanches, M.C., 2016. O *projeto do intangível na formação de designers de moda: repensando estratégias metodológicas para a sintaxe da forma na prática projetual*. Ph.D. FAU-USP/Universitat Politècnica de València.
Sanches, M.C., 2017. *Moda e Projeto: estratégias metodológicas em design*. São Paulo: Estação da Letras e Cores.
Sanches, M.C., Barbosa, T.M., Hernandis, B. e Martins, S.M., 2015. Bases para o ensino/aprendizagem de projeto no design de moda: conectando diretrizes didáticas e estratégias metodológicas. *Revista Moda Palavra*, [revista eletrônica] 9 (17), pp. 119–144. http://dx.doi.org/10.5965/1982615×09172016119.
Sternberg, R.J. 2010. *Psicologia Cognitiva*. 5rd ed. São Paulo: Cengage Learning.
Waltz, S.P., Tiziotto, D.L., Akemi, F.M., Boruchovitch, E., Rabello, L.F., M.D.F. 2016. Avaliação do monitoramento metacognitivo: análise da produção científica. *Avaliação Psicológica*, n. 15. pp. 57–65.

Pattern making books: An analysis of documentary research

P.A.A. Spaine, D.M. Brito, L.M. Pereira, N. Pinheiro & R.R. Andrade
UTFPR – Federal University of Parana, Brazil

ABSTRACT: The purpose of the present study is to identify the works utilized as quotation sources and references, by authors in articles related to two-dimensional and three-dimensional pattern making, that were published in the Fashion Colloquium annals from 2011 to 2015. Such analysis aims to support the gathering of means of study that address the garment's pattern making theme. Based on this information, a relation and analysis were made on the utilized works as a source of quotation for the process of theoretical teaching, of elaboration, of execution and of learning about pattern making.

1 FIELD RESEARCH: DOCUMENTARY

According to Tozoni-Reis (2009, p. 30), on documentary research, the source of data collection are documents, such as books, articles, magazines, courses curriculum, laws, normatives, among others that carry necessary information to the study that is being conducted.

On the present study, the documentary research was developed to identify and analyze the bibliographic references utilized and quoted in articles related to two-dimensional and three-dimensional pattern making, that were published in the Fashion Colloquium annals.

For this purpose, it was conducted a gathering on bibliographic references quoted in scientific articles related to pattern making, published in the Fashion Colloquium annals. The investigation occured on five years of published annals of the articles in the congress, that were: 2011 (7° Fashion Colloquium), 2012 (8° Fashion Colloquium), 2013 (9° Fashion Colloquium), 2014 (10° Fashion Colloquium) e 2015 (11° Fashion Colloquium). The annals are available in digital version on the following link: http://www.coloquiomoda.com.br/anais/.

Therefore, it was conducted a gathering of bibliographic references quoted in scientific articles presented in Oral Communication (OC) and in Work Groups (WG), regarding pattern making, published in the analls of the Fashion Colloquium, the biggest scientific fashion congress in Brazil, that presents itself as a place of academic exchange among students, researchers, professors and professionals in the fashion industry.

According to the event's website, the Work Groups (WGs) receive concluded research articles, in the Master's and Doctorate's level, Doctorate in progress and Post-Doctorate Researches or from research groups that are part of the Directory of Research Groups in Brazil (CNPq) or similar, and from international participants, according to the research group's theme expressed in its syllabus.

The Oral Communications present researches concluded by: postgraduate students and postgraduates (academic or professional *lato or strictu sensu*); professors members of research groups that are part of the Directory of Research Groups in Brazil (CNPq) or similar; by international participants; or even by professionals related to teaching.

The objective of this analysis was to identify the utilized works, as a source of citation and reference, by the articles' authors, to support the given research, as a way to gather the main means of study that address the garment's pattern making theme. Based on this information, it was conducted a connection on the utilized works as a source of citation for the process of theoretical pattern making instruction, of elaboration, of execution and of the learning process.

Such analysis supported this research, in regard to the teaching methods presented, in the main books about pattern making utilized in Brazil. Furthermore, the way these information are presented in these issues, constitute an important contribution for the new researchers, regarding the bibliography used in the production of the scientific knowledge.

1.1 *Presentation of the results of the documentary research*

Altogether, 52 (fifty-two) scientific articles were analyzed, that, over the course of five years, discussed, directly or indirectly, garment's two-dimensional or three-dimensional pattern making. For the elaboration of the analysis only the quoted books were considered, therefore, articles, dissertations and theses were dismissed, or were not considered for the proposed research.

The choice for books was determined by The Ministry of Education (MEC, 2016) on which books are considered basic bibliography as a research source in teaching graduation disciplines.

The result of the gathering allowed the completion of a list of 14 (fourteen) most quoted bibliography during 5 (years) of the Colloquium. The objective of the list was to present a relation of books that discuss the content that addresses pattern making as a subject, in order to support the gathering of the issues discussed on those works. The titles of the books are: *Modelagem: tecnologia em produção de vestuário*; *A modelagem sob a ótica da ergonomia*; *Design de Moda: Olhares diversos*; *Fundamentos do design de moda: construção do vestuário*; *Modelagem e técnicas de interpretação para confecção industrial*; *Alfaiataria: Modelagem Plana Masculina*; *Modelagem plana feminina*; *Tecnologia do vestuário*; *El cuerpo diseñado: sobre la forma en el proyecto de la vestimenta*; *Modelagem Industrial Brasileira*; *Moulage: arte e técnica no desing de moda*; *Inventando Moda: Planejamento de Coleção*; *Fashion Design: manual do estilista* and *Modelagem: Organização e técnicas de interpretação*.

1.2 *Book analyses*

▷ Book: *Modelagem: tecnologia em produção de vestuário*

The book *Modelagem: tecnologia em produção de vestuário*, by author and organizer Flávio Sabrá, is a theoretical work that presents subjects related to pattern making process. Therefore, it thoroughlly addresses the Basic theory of pattern making, discussing topics such as:

- Garment's history
- Current definition of pattern making
- Types of pattern making and what basic pattern is
- The professional's profile/the professional's workplace
- Work tools: different types of materials used in the execution of pattern making
- Normalization and standardization: the importance to standardize measurements for the elaboration of pattern making;
- Stages of development of a pattern making;
- Types of pattern making: three-dimensional, flat, tailoring, pattern making for stretch fabric and CAD/CAM system
- Industrial pattern making process
- Pattern making construction system: addressing how the large scale pattern making is developed to fit the clients;
- Process of setting the patterns on the fabric, in order to avoid waste.

The work presents a study on Ergonomics and its relation to pattern making and exposes different types of ergonomics: physic, cognitive and of awareness, and their levels of operation, in other words, conception, correction and awareness Ergonomics.

As for the Anthropometry, it presents a study on: body measurements; types of measuring process: static and dynamic; and body types: ectomorph, mesomorphic and endomorph. It also analyses the process of measurements' normalization and standardization.

For that matter, the work exposes a superficial study of the Body applied to the pattern making and emphasizes that it is crucial to "envision" the body before elaborating any garment's proposal. The book does not address the following issues: types of teaching methods on pattern making contruction, grading and mathematics.

▷ Book: *A modelagem sob a ótica da ergonomia*

The book *A modelagem sob a ótica da ergonomia*, by Maria de Fátima Grave, consists in a theoretical gathering that addresses, thoroughly, a study of the Body, exposing the following themes: human anatomy and its systems; prominent considerations of the body; body movements; space dimensions; textile surfaces; and gravity. As for Ergonomics, it researches morphology and discusses its relevance in the elaboration of pattern making.

The work also presents a chapter that analyzes garments and inclusion, from the following themes: the problems of garments that do not consider a detailed study of the body and of ergonomics in its conception; the physical disability and its particularities in the elaboration of products. Such considerations are relevant because they interfere in pattern making; and also includes the particularities of pattern making for people with disabilities. The book does not address the following issues: basic theory of pattern making, types of teaching methods of pattern making construction; grading, anthropometry and mathematics.

▷ Book: *Design de Moda: Olhares diversos*

The book *Design de Moda: Olhares diversos*, by author and organizer Dorotéia Pires, consists in a theoretical gathering that is based on three different fronts of fashion design:

- Part 1: concepts and interfaces of design;
- Part 2: culture of projects in fashion design;
- Part 3: configuration of fashion products.

Therefore, in part 3, which addresses the Basic theory of pattern making, there is a chapter that discusses draping, formal innovation and new body architecture. As for Ergonomics, there is a chapter that analyzes the applicability of ergonomics in fashion and presents a study on the second skin, or clothing, as well as creates a reflection on the importance of body awareness in the garment's conception.

Regarding Anthropometry, there is a chapter that thoroughly presents its definition, its relation to the garment's project and its applicability on pattern making elaboration. The book does not address the following issues: types of teaching methods for pattern making construction, grading and mathematics.

▷ Book: *Fundamentos do design de moda: construção do vestuário*

The book *Fundamentos do design de moda: construção do vestuário*, by Anette Fischer, is a theoretical and practical work that addresses issues related to the process of garment's construction, from the pattern making to the sewing and finishing processes.

As for the content on Basic theory of pattern making, it presents a definition of flat and three-dimensional pattern making, the tools and equipments for pattern making contruction, the basic patterns and the interpreted patterns, the shapes and the silhouettes. As for Grading, it presents its definition, an analysis of measurements and its application and importance for the clothing industry.

Regarding Pattern making types: teaching method, it presents the construction process of flat and three-dimensional pattern making. When discussing the flat pattern making, it analyzes and exemplifies the manipulation process of darts, cuttings, extensions, sleeves, collars, pockets, bias and fittings. When it comes to three-dimensional pattern making, it discusses and exemplifies the construction process of the *toile*, the geometric shapes and the interpretation process through draping. The book does not present the following subjects: anthropometry, ergonomics, study of the body and mathematics applied to pattern making.

▷ Book: *Modelagem e técnicas de interpretação para confecção industrial*

The book *Modelagem e técnicas de interpretação para confecção industrial*, by Daiane Pletsch Heinrich, consists on a theoretical and practical work that addresses the process of execution and interpretation of basic pattern making, adapted to the industrial pattern making.

Regarding the content of Basic theory of pattern making, it presents: a definition of industrial, flat two-dimensional and computerized two-dimensional pattern making; the principle and techniques of basic pattern trace and of pattern interpretation; the grading; the planning and fitting of the patterns; and the measurement process.

As for Grading, it addresses its definition, applicability and the methods of enlargement and reduction. With regard to Anthropometry, it addresses the measurement process for pattern making. On the body study, it presents a wearability and fitting table. As for Mathematics, the book addresses, through images, the basic notions that should be utilized in the patterns elaboration, although, without any type of explanation. The book does not address the following content: ergonomics applied to pattern making.

▷ Book: *Alfaiataria: Modelagem Plana Masculina*

The book *Alfaiataria: Modelagem Plana Masculina*, by Stefania Rosa, is a theoretical and practical work that addresses issues related to the elaboration process of men's wear patterns. With regard to the Basic theory of pattern making, it presents the following contents:

- The advent of pattern making, definition and types;
- The importance of pattern making in the garment's construction;
- Stages for developing flat pattern making;
- Basic information related to the pattern;
- The types of material that are used in the elaboration of flat pattern making.

With regard to the Types of pattern making: teaching method, it presents the teaching process of the pattern making construction of men's wear flat tailoring. As for Grading, it analyzes its definition and applicability and creates guidelines on how to make Grading, in men's wear tailoring, through light colored pictures.

As for Anthropometry, it presents its definition, application and the measurement process for the construction of the garment's pattern.

With regard to Mathematics applied to pattern making, this is the only book on pattern making instruction that contains a small analysis on fundamental mathematical concepts for the construction of a pattern. It presents images and simple explanations about the following issues: stitch, line, parallels, angles, perpendiculars and geometric shapes. The book does not address contents related to ergonomics and to the study of the body applied to pattern making.

▷ Book: *Modelagem plana feminina*

The book *Modelagem plana feminina*, by Paulo de Tarso Fulco e Rosa Lúcia de Almeida, is a work that presents the process of construction of women's wear flat pattern making.

With regard to the Types of pattern making: teaching method, it presents the measurement tables and the elaboration stages of the women's wear pattern making basis: skirt, body, sleeve, pants and collars. The book does not address the contents related to: basic theory of pattern making, grading, ergonomics, anthropometry, study of the body and mathematics applied to pattern making.

▷ Book: *Tecnologia do vestuário*

The book *Tecnologia do vestuário*, by Mário de Araújo, is a theoretical work that makes a detailed analysis of the garment's production process. With regard to the Basic theory of pattern making, it presents: the definition of pattern making, the definition of two-dimensional pattern making, the existence of the computerized pattern making and the materials utilized on the execution of the flat pattern making. Besides that, it analyses the process of Grading through its definition and examples on how to grade patterns, however, only theoretically, without illustrative images.

As for Ergonomics, it superficially analyzes its relation with garment's technology. The anthropometry is addressed only through a presentation of measurement tables of the human body, without any definition or detailed study on the subject. The book addresses contents related to teaching methods, to the study of the body and to mathematics applied to pattern making.

▸ Book: *El cuerpo diseñado: sobre la forma en el proyecto de la vestimenta*

The book *El cuerpo diseñado: sobre la forma em el proyecto de la vestimenta*, by Andrea Saltzman, is a theoretical work that discusses issues related to the body, the garment and to the relation of the body with the garment. In that way, relating to the body, it develops a study on: skin, anatomy, joints, proportions, real body and ideal body, morphology, support, textile anatomy, constructive thinking, flat articulations and silhouettes.

The book does not address the contents related to the basic theory of pattern making and to the types of pattern making: teaching method, Grading, Ergonomics applied to pattern making, Anthropometry applied to pattern making and Mathematics applied to pattern making.

▸ Book: *Modelagem Industrial Brasileira*

The book *Modelagem Industrial Brasileira*, by Sonia Duarte e Sylvia Saggese, is a work with a practical purpose that addresses the process of construction of women's and kid's wear flat pattern making. With regard to the Basic theory of pattern making, it presents the types of materials for the construction of a pattern.

With regard to the Types of pattern making: teaching method, it presents the measurement charts and the stages of execution of women's wear pattern making, that are: body basis, darts, cleavages, collars, pleats, types of sleeves, skirt basis and its variations, pants basis and its variations and stretch fabrics pattern making.

With regard to kid's wear flat pattern making, it presents the measurement table and the construction process of: body basis, sleeve basis, skirt basis, pants basis, basic dress, sleeved dress, collar and jumpsuits. The book does not address the contents related to: basic theory of pattern making, grading, ergonomics, anthropometry, study of the body and mathematics applied to pattern making.

▸ Book: *Moulage: arte e técnica no desing de moda*

The book *Moulage: arte e técnica no design de moda*, by Annete Duburg e Rixt van der Tol, which approach is theoretical and practical, discusses the process of execution of garment's three-dimensional pattern making.

With regard to the Basic theory of pattern making, it makes a brief analysis on the garment's history and on draping in present days, introduces the definition of three-dimensional pattern making and the materials utilized in the execution of draping, and explains the mannequin's preparation, the measurement table and the detailed process of constructing the *toile*.

As for the Types of pattern making: teaching methods, it addresses the process of execution of the three-dimensional pattern making on the following patterns: skirts, body, dresses, sleeves, collars, coats and *blazers*, pants and drapes. The book does not address the following subjects: grading, ergonomics, anthropometry, the study of the body and mathematics applied to pattern making.

▸ Book: *Inventando Moda: Planejamento de Coleção*

The book *Inventando Moda: Planejamento de Coleção*, by Doris Treptow, is a theoretical work that presents a gathering of issues related to the development process of garment's product.

With regard to the Basic theory of pattern making, it presents, in a superficial way, the importance of pattern making and explains, without images, the construction steps of pattern making and the existing pattern making types: flat, three-dimensional and computerized. As for Grading, it analyzes its definition, however, without showing the application in a pattern. The book does not address contents related to the types of pattern making: teaching method, ergonomics, anthropometry, study of the body and mathematics applied to pattern making.

▸ Book: *Fashion Design: manual do estilista*

The book *Fashion Design: manual do estilista*, by Sue Jenkyn Jones, is a gathering of topics related to the fashion product. With regard to the basic theory of pattern making, it presents a definition of pattern making and the materials used for the development of a pattern and explains both flat and three-dimensional pattern making.

The book does not address the following topics: types of patern making: teaching methods, ergonomics, grading, anthropometry, study of the body and mathematics applied to pattern making.

➤ Book: Modelagem: *Organização e técnicas de interpretação*

The book *Modelagem: Organização e técnicas de interpretação*, by Ligia Osório, is a theoretical and practical work that addresses the subjects related to the process of execution of garment's flat pattern making.

In that way, it exposes the Basic theory of pattern making in a detailed and thorough way, aside from presenting the pattern making definition, the types of pattern making, the existing types of patterns, the pattern's interpretation, the principles of pattern's construction, the necessary rules to the pattern making construction, the factors that must be analyzed in a pattern's construction, the darts processes and the constructive resources utilized in the pattern making execution.

As for the Types of pattern making: teaching method, it addresses the women's wear pattern making when explaining the execution process of the following patterns: base of the body, dresses, sleeves and fists, buttons and collars. The book does not address the following issues: grading, ergonomics, anthropometry, study of the body, mathematics applied to pattern making.

1.3 *General considerations on the books*

The analysis of the fourteen selected works, identified as theoretical, practical or theoretical and practical, highlighted that none of them contains, in a detailed form, all the subjects identified on the study of the present research, that are: basic theory of pattern making; types of pattern making: teaching method; Grading; Ergonomics; Anthropometry; Body; and Mathematics applied to pattern making.

The most significant points of the gathered work, relating to each of the themes, are:

- Basic theory of pattern making: despite the fact that it is fundamental in the learning process of pattern making, only 10 (ten) books addressed this subject, in either a superficial or a thoroughly way.
- Types of pattern making: teaching method: the gathering allowed to indentify that, from the researched books, only 7 (seven) taught some kind of method of pattern making execution.
- Grading: this is a relevant issue on the learning process of pattern making, however, it was addressed only in 4 (four) of the analyzed bibliography, and, in some cases, it was explained in a superficial manner, without exemplification of the technique's application.
- Ergonomics applied to pattern making: meaningful content for the execution of garments and for the learning process of pattern confection, however, it was only addressed in 4 (four) of the analyzed books.
- Anthropometry applied to pattern making: as well as in the ergonomics case, the anthropometry content is a necessary knowledge for the pattern making construction, but it was also only addressed in 4 (four) of the analyzed books.
- Study of the body applied to pattern making: this is a crucial knowledge for the construction of garments, but, in the same way, it was only presented in 4 (four) books.
- Mathematics applied to pattern making: it consists in a basic and fundamental subject for the comprehension of the execution of garment's patterns, however, it was only addressed in 1 (one) of the researched books.

Therefore, the collected data in the documentary field research allowed the identification that there is a lack of books on pattern making and teaching methods that address, in a thorough manner, the pattern making subject, in order to facilitate the learning process of students and the execution of the garment's patterns.

In addition, it enabled the achievement of the object of study, in other words, to identify and analyze the bibliographic references utilized and quoted, in articles on elaboration/execution/learning process/teaching of two-dimensional and three-dimensional pattern making, published in the annals of the main fashion congress in the country.

REFERENCES

Araújo, M. 1996. *Tecnologia do vestuário*. Lisboa: Fundação Calouste Gulbenkian. COLÓQUIO DE MODA. Disponível em: http://www.coloquiomoda.com.br/anais/.

Duarte, S. & Saggese, S. 2002. *Modelagem industrial brasileira*. Rio de Janeiro: Guarda-Roupa.

Duburg, A. & Tol, R.v.d. 2012. *Moulage: arte e técnica no design de moda*. Tradução Bruna Pacheco. Porto Alegre: Bookman.

Fischer, A. 2010. *Fundamentos do Design de Moda: construção de vestuário/* Anette Fischer; Tradução Camila Bisol. Brum Sherer. Porto Alegre: Bookmen.

Fulco, P. de T. & Silva, R. L. de A. 2010. *Modelagem plana feminina*. Rio de Janeiro: Senac Nacional.

Grave, M. F. 2004. *A modelagem sob a ótica da ergonomia*. São Paulo: Zennex.

Jones, S. J. 2011. *Fashion design: manual do estilista*.Trad. Iara Biderman. São Paulo: Cosac Naify.

MEC. Ministério da Educação. *Instrumento de Avaliação de Cursos de Graduação presencial e a distância.* Disponível em: <http://download.inep.gov.br/educacao_superior/avaliacao_cursos_graduacao/instrumetos tos/2012/instrumento_com_alteracoes_maio_12.pdf> Acesso em 20 de out. 2016.

Osório, L. 2007. Modelagem: organização e técnicas de interpretação. Caxias do Sul: Edusc.

Pires, D. B. (Org.). 2008. *Design de Moda: olhares diversos*. São Paulo: Estação das Letras e Cores, p. 319–336.

Rosa, S. 2009. *Alfaiataria: modelagem plana masculina*. Brasília: SENAC-DF.

SABRÁ, F. (Org). 2009. *Modelagem: tecnologia em produção de vestuário*. São Paulo: Estação das letras e Cores.

Saltzman, A. 2004. *El cuerpo diseñado: sobre la forma en el proyecto de la vestimenta*. Buenos Aires: Paidós.

Spaine, P.A.A. 2016. *Diretrizes para o ensino e construção da modelagem: um processo híbrido*. Bauru, 2016, 18`. Tese (Doutorado em Design) – UNESP—Universidade Estadual Paulista "Júlio de Mesquita Filho", FAAC—Faculdade de Arquitetura, Artes e Comunicação.

Tozoni-Reis, M.F.C. 2009. *Metodologia de Pesquis*a. Curitiba-PR: IDESDE Brasil S.A.

Treptow, D. 2013. *Inventando moda: planejamento de coleção*. 5. ed. São Paulo-SP: Doris Treptow.

Interdisciplinarity: Creation of looks and techniques used by UFPE CAA students

A. Camargo & R. Alves
Federal University of Pernambuco (UFPE), Caruaru, Pernambuco, Brasil

ABSTRACT: This article sets out to demonstrate an action of interdisciplinarity developed between the subjects of Planning a Collection and of Assembling and Piloting Women's Clothing, of the Design course, at the Federal University of Pernambuco (UFPE), Academic Center of the Agreste region (ACA). The final products, four thematic looks, were generated from the creative process, based on the concept of mental maps and founded on the theoretical content of Fashion Design. The action led to the materialization of creation, by confectioning garments. It also provided students with an interdisciplinary experience and their becoming closer to the labor market.

1 INTRODUCTION

For students of the course in Design, with an emphasis on Fashion, the confection of a garment is, in principle, the result of their combining knowledge acquired during their experience at university. At the same time, a good number of the students enter Higher Education after having obtained practical knowledge of sewing, especially in Caruaru, one of the towns that belongs to the garment hub in the Agreste region of Pernambuco, a state in Northeast Brazil. In this town, as children, they become familiar with this craft on account of the influence of the activity of their family in the garment industry, in factories and "factions"[1] (*facções* in Portuguese).

Many of these factories operate clandestinely i.e. in the so-called informal economy, without the guidance of qualified professionals. However, the consumer market, in general, demands a quality standard in the pieces made. These are based on market trends, the customer's needs and are managed by the creation of the designer.

In the act of his/her creation, the designer draws up, implements, tests and models. Everything is done meticulously at his/her pace and with considerable degrees of precision and exhaustive effort. What characterizes a designer is precisely the systemic work from which a conjunction of activities emanates in order to compose a product.

More specifically, Niemeyer (1998, p. 25) stresses that design is the "simultaneous equationing of social, anthropological, ecological, ergonomic, technological and economic factors when conceiving of elements and material systems necessary for the life, well-being and the culture of human beings". The interdisciplinary nature of design is encompassed in this affirmation. This integration also contributes to the moment of disconnection, that is, the immersion of the designer in his/her work, thereby corroborating the act of creating.

Interdisciplinarity is the scope of this paper, which sets out to study the confection of looks that arise from the theme linked to Brazilian art, and which are generated by students who were linked to the subjects of Planning a Collection and of Assembling and Piloting Women's Clothing, in the second semester of 2017.

The relevance of this study is based on the underpinning sense of interdisciplinarity which resides in the coexistence of at least two subjects and in the presence of an action of reciprocity. Lenoir (1998) points out that in the pedagogical field, the classroom becomes a laboratory, in

---

1. Faction is the term for small businesses that provide partial services, for example, those that do not cut cloth, but only sew it.

which interdisciplinary studies and actions are experimented with in concrete didactic situations. In this process, several other variables (related to the object of study) are considered, which should be contextualized and during which links between scientific disciplines should be established. In this case, each proposed look is the intersection between the two (or more) subjects.

2 DESCRIPTION OF THE ACADEMIC SUBJECTS INVOLVED IN THE ACTION

2.1 *Planning a collection*

The objective of the academic subject in question is to insert the student into an apparent consumer market environment, where what, among other matters, must be considered is the spirit of the age (*zeitgeist*), which can be measured from the current behavioral macro-tendencies. These are founded on themes that weave their way through the maxims described, starting with what is signaled in politics, economics, technology, health, and sports, thereby reverberating the relationship of human beings with the environment and their own bodies.

The final design of the subject is individual or in a group, and creation is the time at which students can release/develop their creative skills. This subject stimulates students' sensorial bias, since sensibility governs the pleasure of creating. Satisfaction is about generating and giving shape to the creation, that is, about making an idea tangible that is based on producing a look that has been confectioned in the subject of Assembling and Piloting Garments.

The creation methodology that we use in this subject is the one implemented by Treptow (2013), in his book "Inventing Fashion" and based on foundations discussed in the book "Fashion Design", by Jones (2005). Treptow (2013) maps the way towards conceiving of a collection that achieves satisfactory goals in academic terms and/or in a company. The reason for choosing it as a guiding tool is due to its phased arrangement, which, in a didactic way, facilitates the student's understanding when undertaking research.

The steps used in the simplified method for planning and developing collections are given below:

1. Behavior research: at this stage, what must be found out are the consumption habits of the target audience, what their current interests are, places frequented, idols, books, music, i.e., narrowing the niche with a view to reaching a full understanding of for what the collection will be designated.
2. Market research: This board is linked to the relationship with the other competitors, thereby situating "styles and prices practiced by competitors, parallel products aimed at the same target audience, new brands, future competitors" (Jones, 2005, 78).
3. Research on trends: this board can be conceived from two strands, namely: macro—tendencies and micro-tendencies. With regard to macro-tendencies, Clério (2015) discusses that these are related to behavioral tendencies which have a planetary, humanist and social breadth, where a more continuous relationship with the pattern of human behavior is observed. In general, they have been directed for several seasons and transfer content to the design of the micro-tendencies. These are of an aesthetic nature and from which the style elements of the collection are extracted. Treptow (2013) points out that the visual information that appears regularly in the research of tendencies, such as the length of trousers, a type of collar, a pattern, represents these elements of style.
4. Technological research that "accompanies launches of techniques and machinery that can be applied to the confection. For example, stamping techniques, technological fabrics, modeling and cutting software, alternative systems" (JONES, 2005, page 78).
5. Research on the theme of a Collection/moodboard.

In this study, we will concentrate on the moodboard, where we explain that this deals with visual references that can speak of a story, an idea, a concept. This can be based on an image, or a pictorial composition.

A fashion designer committed to his/her work needs to undertake a very great deal of research and to develop a theme or concept that will serve as inspiration for his/her creation. This theme will guide the designer in his/her design action, thereby fostering a cohesive job of work, besides defining certain limits when using elements of style, which is something that will ensure that the collection has a visual unity.

2.2 Assembling and piloting garments

This academic subject deals with the importance of the process of assembling and prototyping industrial and artisanal processes. It takes place in the phase of bringing the creation into concrete reality as an integral part of the activities that concern fashion design.

We emphasize that fashion is strongly related to cultural, social, economic and technological events. In this perspective Moura (2008, 73) stresses that its "development and expression occur as a result of the interrelationships between creation, culture and technology."

The expression of fashion, especially the fashion of clothing, is approached by Moura (2008, p. 37) as a result of several factors that converge when a designer draws up and develops products, both in industry in the midst of technological advances, as well as by artisanal means that use manufacturing processes. Thus, the fashion designer is seen as the professional: "[...] responsible for the creation and design of the product, for its development and for monitoring its production."

This happens in a way that, in addition to having to have knowledge about creative processes, the fashion designer is also required to have technical knowledge regarding the design part and its development, while keeping the focus on the end-consumer. Such a condition implies having knowledge of the available resources (equipment and machinery) in order to draw up and carry out projects.

Regarding the technological resources available, Moura (2008) comments: when the great creators launch their collections, these are revisited, recreated and diluted to suit the materials available in the market and the techniques or technologies available so that the productive and commercial process can be made viable. This is a factor that, in our view, does not impede making other solutions viable. However, to do so what is necessary is to be familiar with techniques of modeling, cutting and sewing and arising from this to move towards the creation of new arrangements.

In these terms, our understanding of fashion design is that it the process which is described by Moura (2008), namely that of conceiving and designing a model. The conception referred to by Moura is about undertaking research, seeking cultural and aesthetic references, as well as the chart that displays the trend of the season for which the conception is being created. While the project comprises:

> *The fabric chosen, the modeling, the detailing of trimmings and, as the case may be, [...] the complements. In the modeling phase, the project can [...] undergo changes and adaptations [...]. After this comes the confection and production of the prototype piece, the sewing phase, in which the project can still undergo changes and adaptations.*
>
> *Then, the test phase is entered into in a model appropriate to the size chosen for the prototype. After finishing this first part of the fashion design process, a return is made to modeling in order to develop the grid of sizes that will be followed by the processes of cutting and producing in scale.* (Moura, 2008, p. 70).

Having understood these elements (choice of fabric, modeling, confection of the prototype, development of the grid of sizes, industrial processes of cutting and production), as they are integral parts of the project in fashion design, the next step in the subject of Assembling and Piloting Garments is to learn the assembly techniques. Sewing is the main technique used to enable the assembly of the prototype garment.

The prototype is considered, says Araújo (1996, p. 88), an element of fundamental importance within a company, which is developed in the sampling sector, and needs to be "analyzed, tested and modified in order to improve both manufacturing times and methods as well as the use of raw material and accessories".

3 ACT OF CREATING

Art has been a source of inspiration for fashion designers, given the partnerships that there have been between designers and artists, since the early decades of the 20th century. Pride of place must go to the creations of Poiret who, in the 1920s, used prints in his creations which were designed by Fauvist artists. Following this premise, there figures Schiaparelli who maintained a friendship with Dadaist and surrealist artists, and who actively participated in his creations, special mention being made of Salvador Dalí in the 1930s. Yves Saint Laurent, in the 1960s, brings Mondrian's paintings to fashion. This trend is also evident nowadays on considering the immersion of the creative director of Osklen in the work of the modernist artist Tarsila do Amaral, for the creation of the spring/summer 2017/2018 collection of the label.

It is not the purpose of this study to make a complete retrospective of the designers who sought support from art to conceive their creations. Thus, the very brief retrospective given above was to situate the motivational bias that led to designating the theme of the project as being linked to Brazilian art. In this case, the following brief account that refers to Brazilianness is a macro-tendency that was present in the period between sports events, namely the World Cup and Olympics, which were events held in Brazil in 2014 and 2016, respectively.

To create is to give shape to the imagination. And it is on this premise that students begin the creative process. This is based on brainstorming carried out in the classroom and supported by adapting the concept of mental maps, presented by Mazzotti and Broega (2012, p.4), who indicate that it is a "technique for visualizing thought, which is used to generate, organize and structure ideas". The authors of this paper stress that in order to draw up a mental map, words, drawings, colors and symbols can be used to contribute visually to breaking down the theme explored.

The guiding thread of the exercise is established from the central theme (Brazilian art), where the students, starting from a network structure, try to generate ramifications with the objective of exploring the basic idea, as well as unravelling it, thereby generating the concept which will guide the creation.

The schema (Figure 1) followed by the pupils for generating the mental map is given below. The basic idea was Brazilian art followed by painting. At this point, there was the unfolding into modern art and contemporary art, followed by the region of birth of the artist chosen. The visual references are based on the generative concept of a moodboard.

In the first look shown, titled Cangaço, the mental map was guided by the work of Aldemir Martins, a modernist artist from the state of Ceará, associated with the work of Espedito Seleiro, a craftsman from Ceará who makes use of the image of the cangaço period to decorate his creations of bags, shoes, saddles; all made of leather. The moodboard (Figure 2) was generated from the shapes and colors of the cangaço, thereby strengthening this concept. The adoption of the warm color, yellow, to represent the Northeastern sun and the tones of the earth, symbolizing the earth, clay, and leather, the latter being the raw material used by Seleiro in his works (Figure 3).

The cangaço refers to a period in the late 19th century to around 1930, when there were bands of outlaws in the semi-arid (which is called the Sertão) of Northeast Brazil, the most infamous of whom was called Lampião. Popular culture still reflects this period especially in the festivities of the month of June when members of sertanejo music bands wear the clothes of the cangaço (see photo in Fig. 2).

The mental map of look 2 (Figure 4), called Witch of the Sea, is linked to the modern art of Vera Sabino, an artist from the southern region of the state of Santa Catarina. The theme of her works deals with the sacred and/or the profane, and there is a pertinence in the allusion to the imaged concept of witches, a motif chosen by the student to add meaning to her concept. The feathers/scales, portrayed in her pictures, are the style elements of the look. This, starting from the imagery of feathers linked to flight, freedom and, scales, that refer to the fish that inhabit the sea around the island of Florianópolis, the hometown of the artist. The cape, the reference to witches, comes in the color blue so as to make a parallel with the artist's bias to the sacred. In this instance, the blue is seen in the image of the Virgin in the top right-hand corner of Figure 4 and this contrasts with purple, a color that belongs to the

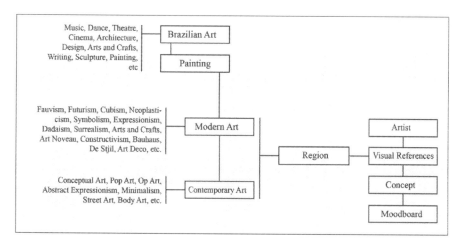

Figure 1. Diagram of the mental map used in this project. Source: by a student, 2017.

Figure 2. Moodboard Look 1. Numbers 1 and 2 represent the forms which generate the elements of style. Source: from the student, 2017.

Figure 3. Look 1. Source: from the student, 2017.

Figure 4. Moodboard Look 2. Number 1 represents the form that generated the style element. And, Number 2 signals the idea of the cape in the look. Source: from the student, 2017.

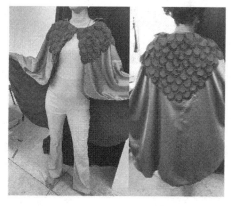

Figure 5. Look 2. Source: from the student, 2017.

symbiology of witches, but purple is also found in sacred vestments, thus corroborating the dichotomy that the artist explores between the sacred and the profane (Figure 5).

From Look 3, called Asymmetric, we can discuss if it is a synthesis of elements found in the work of Tarsila do Amaral. The mental map passed along the paths of modernist painting until it formed the concept of the look that was based on the asymmetry that there are in the pipes of the culverts, depicted in the picture of the factory workers, 1933 (Figure 6, item 2). The trapezoidal shape of the roofs of houses synthesized by Amaral (Figure 6, items labelled 1), as well as the use of the earthy color in tune with that used in her works, inspired the style element set out on the sides of the look (Figure 7).

The work of Eli Heil, who is from Santa Catarina, inspired the creation of look 4, called Maternity. The mental map followed the path described until arriving at the work of the artist, the main inspiration of which is its circular form. The "eyed hands" painting (Figure 8, item 1) was the basis for the conception of this idea. Interpreted by the student as the abstract simulation of a childbirth (Figure 9).

Figure 6. Moodboard Look 3. Numbers 1 and 2 represent the forms which generated the style element. Source: from the student, 2017.

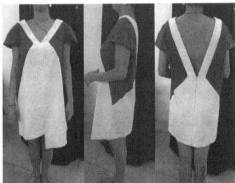

Figure 7. Look 3. Source: from the student, 2017.

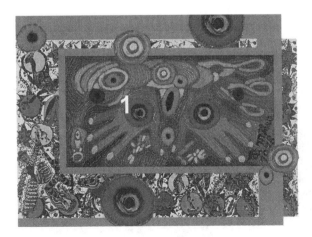

Figure 8. Moodboard Look 4. Number 1 represents the form which generated the style element. Source: from the student, 2017.

Figure 9. Look 4. Source: from the student, 2017.

4 PRODUCTION OF THE PIECES

In order to make the creative act concrete, in compliance with the design process, according to the definition of fashion design attributed by Moura (2008), the students of the subject Assembling and Piloting Garments, configured the looks that they created into four main phases: 1) Defining Materials; 2) Modeling; 3) Prototyping; and 4) Making the final piece.

4.1 Look 1, Cangaço, inspired by the works of the artist Aldemir Martins

In the Cangaço Look, the fabrics selected to represent the rusticity of part of the culture of the Northeastern region of Brazil were: yellow Brim (100% cotton), Softened Suede in a golden color (100% polyester) and Eco Leather (10% PVC with 100% polyester lining).

For the configuration of the pierce, a combination of flat and *moulage* modeling techniques was used, simultaneously. The next stage consisted of defining the cut plane, the distribution of the molds on the fabric and the cut. Then, the order of assembly was established with a description of the procedures used, an indication of the appropriate machine and the time that it takes to do. These procedures were applied to the prototype and then transposed to make the final piece.

The final look was configured in the form of a garment that is an *evasé* skirt in *apliqué* brown leather. The upper part was made with the golden fabric on the sides and the leather in the central part, thereby forming a plunging V neckline, front and back (Figure 3).

4.2 Look 2, Witch of the sea, inspired by the works of artist Vera Sabino

The fabrics selected to represent the malleability and fluidity of the sea that bathes the South region of Brazil were: Two Way with elastane in blue (80% polyester and 20% elastane) for catsuit; Gloss Silk in a greyish blue (97% polyester and 3% elastane) for the outside part of the cape; Chiffon in navy blue color (100% polyester) for the inner part of the cape; Chiffon in wine, magenta and green colors (100% polyester) for the application at the upper part of the cape.

To give shape to the project, the flat modeling technique was used. In the following step, the piece was prototyped, using the procedures of cut, assembly and proofing. The effect of laser cutting was also tested and evaluated on Chiffon fabrics, used in the application of the upper part of the cape. These served as the basis for confectioning the final look.

The composition of the final look consisted of a celestial blue body-hugging catsuit. overlaid with a flowing cape made with a double layer of fabric and an *apliqué* finish on the upper part of the shoulders. (Figure 5)

4.3 Look 3, Asymmetric, inspired by the works of the artist Tarsila do Amaral

The fabrics selected to represent the rural and urban relationship as part of the culture of the Southeastern region of Brazil were: Mixed Linen in off-white (50% linen and 50% cotton); Mixed Linen in the color Marsala (50% linen and 50% polyester).

To configure the piece, the flat modeling technique was used. The next stage consisted of studying the cut plane, so as to match the vertices of the different geometric shapes of each of the molds. Then, the order and assembly procedures were established. These procedures were prototyped and served as a reference for confectioning the final piece.

The final piece, the representative of the geometry inserted in the image of the architecture of the rural space, present in the works of Tarsila, was configured in the form of an asymmetrical tubite garment with a V neckline at the front and at the back (Figure 7).

4.4 Maternity look inspired by the works of the artist, Eli Heil

The fabrics selected to represent the softness of feminine curves during gestation, present in part of Eli Heil's works, were: Bubble Crepe in a pink tea color (100% polyester) for the look; Chiffon in wine, dark blue and light blue colors (100% polyester) for the application of polka-dots on the hemline of the pantaloons.

The piece was configured using the flat modeling technique, followed by prototyping—cutting and assembling of the piece in a handicraft way, laser cutting and manual sewing of the polka-dots. Subsequently, the same procedures and techniques were transferred in order to make the final piece.

The final look was composed of a crop-top blouse that was finished in a hemline decorated with polka-dots, in the color wine. The second piece, pantaloons with maxi colored polka-dots on the hemline and finish in the hem with trimmings of polka-dots in the color wine (Figure 9).

5 FINAL REMARKS

This paper demonstrated the result of a practical action of interdisciplinarity developed between the subject of Planning a Collection and the subject of Assembling and Piloting Garments, of the course of Design, at UFPE CAA.

The creative process was guided by the content taught in the subject of Planning a Collection, based on the theoretical bases of mental maps, with subsequent creation of moodboards, which were able to translate the aesthetics chosen for the looks.

The confection of the pieces was developed from the techniques and procedures taught in the subject of Assembling and Piloting Garments, such as: modeling, cutting plane, cutting, prototyping, proofing and confection of the final pieces.

In our view, interdisciplinarity enabled the project to be fully implemented. It articulated the creative processes sketched on paper and made concrete in the form of a clothing product. The result of the materialization of the creation gave great value to the action in the students' eyes because it enabled them to draw closer to the labor market.

However, some limitations can be highlighted, such as the difficulty of finding local materials on the market that meet the idea that the students conceived. Re-adaptations had to be made to the material available.

For future papers, from this same perspective, we suggest that a greater articulation be made with the community external to the university, both through partnerships, which aim at acquiring materials for the confection of looks, as well as at gaining publicity in the media, such as an exhibition and/or fashion show.

REFERENCES

Araújo, M., 1996. *Tecnologia do vestuário*. Lisboa: Fundação Calouste Gulbenkian.
Clério, L. 2015. *Curso desenvolvimento de coleções de Moda*. Senac Caruaru.
Jones, S. J., 2005. *Fashion design—manual do estilista*. São Paulo: Cosacnaify.
Lenoir, Y., 1998. Didática e interdisciplinaridade: uma complementariedade necessária e incontornável. In: *Fazenda, I. C. A. (Org.) Didática e interdisciplinaridade*. Campinas: Papirus.
Mazzotti, K., Broega, A. C., 2012. O desenvolvimento da criatividade, através da utilização de técnicas de mapas mentais e analogias, no processo de criação em moda. Available at: <https://repositorium.sdum.uminho.pt/bitstream/1822/21905/1/Global_Fashion%20Madrid_2012.pdf> [Accessed on 20 Feb 2018].
Moura, M., 2008. A moda entre a arte e o design. In: *Pires, D. B. (Org.). Design de moda: olhares diversos*. Baureri, SP: Estação das Letras e Cores.
Niemeyer, M., 1998. *Design no Brasil: origens e instalação*. Rio de Janeiro: 2AB.
Treptow, D., 2013. *Inventando moda—planejamento de coleção*. 5th.ed. São Paulo: Edição da Autora.

Active applied methodologies on fashion design teaching: Possibilities and challenges

L.E.F.S. Rebello, G.D. Marques & M. Mansur
Universidade Estácio de Sa, Rio de Janeiro, Brazil

ABSTRACT: Globalization, together with the increasing use of social medias on the last few decades have expanded the discussion about the role of a college degree education over the formation of citizens who are able to be part of the life in society in an integrated and effective way. As knowledge, although essential, the information offered, when only retained or stored, becomes a component of mere reproduction, and existing maintenance, setting learners as limited expectors of the world. Regarding the professor, he or she can contribute to the promotion of pupils' autonomy as well as to the maintenance of the behavioral controls over them. In Brazil, the college fashion courses have already been subjected to changes due to the new approach brought from fashion being seem as design. Nowadays, a project methodology is present inside the speech of these courses and it appears like it belonged there since the beginning. However, the increasingly urge to generate more autonomous professionals, made the implementation of teaching through active methodologies reached also the Fashion Design courses classrooms. In face of the experience as Brazilian fashion design professors, who are implementing the active methodology on hybrid college subjects since 2017, the main purpose of this article is to point out possibilities and challenges faced by this teaching model inside the fashion design courses, in Brazil. The study is justified since it highlights pedagogical paths to endeavor to overcome the challenges that the formation of autonomous fashion designers who can face an increasingly demanding market, especially regarding the formation of more autonomous students who will reach the market prompt to offer viable solutions to problems, to work in teams, to elaborate innovative projects as well as to meet the global needs of sustainability.

1 INTRODUCTION

Globalization, together with the increasing use of social medias on the last few decades have expanded the discussion about the role of a college degree education over the formation of citizens who are able to be part of the life in society in an integrated and effective way. Considering college degree education, even today it persists the idea that in order to be considered a good university professor, would suffice possess a broad knowledge about the field of the discipline taught along with a good oratory. Actually, this reality conflicts with the upper level student profile, especially when it regards the Millennium generation student. These young people who were born in a world controlled by the internet arrive with their personalities stablished along with a huge baggage of knowledge, both reflections from a globalized and informative society.

Nevertheless, in opposite to what is expected, the higher education system remains reproducing a traditional teaching model. Furthermore, perhaps for convenience or either for lack of perspective, students also settle and follow reproducing texts, practices and limited content, focusing on evaluations and the diploma, disregarding the quality of their own education. In addition to that there was, quite often, a common speech from these students, they point out the blame on their educational institution or some of their teachers; also, they do not consider themselves responsible because they do not recognize their responsibility in the process of knowledge acquisition.

With this in mind, we have a huge challenge to overcome: to contribute to the formation of more students that will reach to the labor market being able to offer viable solutions to problems, work as a team and develop innovative projects that meet the global needs for sustainability.

We are using the term autonomy in the sense of the University governing by itself; the rights or faculty to govern by its own laws and rules; freedom or moral or intellectual independence. In spite of this concept have a nation as focus, different areas of human activity take hold on that philosophy. Looking at our practice with a critical look, can we affirm that we are fulfilling this role?

It is possible to say that as emphasis is placed on learning, the teacher predominant role ceases from being exclusively teaching, to help the student to learn. However, the emphasis on learning also has spawned serious misconceptions, once many teachers simply exempt themselves from the obligation to teach. Therefore, what would be the "breakeven point"?

Nowadays, there is a great urge that teachers from Universities develop vocational skills which must prepare students in a critical and argumentative social formation. It is necessary, therefore, to replace the traditional teaching standards for active learning methods, which can be used as a teaching resource in daily teaching practice (Borges & Alencar, 2014).

Finally, setting as a starting point the experience as professor in a Brazilian fashion design course, who is trying to implement the active methodology in hybrids disciplines, since 2017, this article aims to highlight the possibilities and challenges faced by this model in higher education fashion design courses, in Brazil. In other words, the proposition is to discuss the potentialities and objections for the application of active methodologies on the fashion design education.

2 ACTIVE METHODOLOGIES IN THE HIGHER EDUCATION SYSTEM

Active methodologies are based on ways to develop the process of learning, using actual or simulated experiences, aiming to solve successfully the challenges arising out of the core activities of social practice, in different contexts (Berbel, 2011).

Incidentally, due to the Internet and open disclosure of many courses and materials, it is now possible to learn anywhere, anytime along with many different people. This may be complex, necessary and a little scary, because we do not possess prior successful models of flexible learning in a highly connected society. Moreover, it can be affirmed that formal education is increasingly *blended* as well as hybrid, since it does not only happen in the physical space of a classroom, but inside multiple spaces of everyday life, including the digital ones.

In despite of the weaknesses and structural problems in the Brazilian education system, there is currently a search for alternatives on important educational sectors, such public and private. The changes arise two ways: the progressive and the profound ones.

Following the path of progressive changes, institutions remain the predominant curriculum model—discipline—but prioritize an increased student participation; with active methodologies such as teaching towards more interdisciplinary projects, hybrid or blended learning and the "flipped classroom".

The institutions that follow the deeper changes model propose more innovative and disruptive models, with no disciplines; which may reconfigure projects, physical spaces and methodologies, based on activities, challenges, problems, and games where each student learn at his own pace and needs and also learn from each other in groups and projects, with supervision from advisors' teachers.

The purpose of this article is to discuss the process of progressive changes, once we deem that it is more feasible in the context of higher education in which we are inserted as teachers. We believe that active methodologies are starting points on the progress for more advanced processes of reflection, cognitive integration, generalization and reworking of new practices.

Theorists such as Dewey (1950), Freire (2009), Rogers (1973), Novack (1999), among others, have been emphasizing for a long time, the importance on overcoming the traditional bank education and focus on student learning, by involving, motivating and dialoguing with him.

Even schools without so much technology can modify the concept of room and space, once they apply advanced pedagogical projects.

In higher education, healthcare has pioneered work with troubleshooting, already in the 60's by McMaster University in Canada. (Cyrino & Nunes-Pereira, 2004) and today there is a consolidation of these active methodologies, as well as some Engineering courses develops the methodology of projects. Nowadays, the mobile technologies are beginning to be introduced in order to arrange the one-on-one meeting in a more flexible way. The hybrid model is very important for those who work with problems and projects. Institutions like the Unisal, in Lorraine, develop active methodologies such as peer Learning (Peer Instruction), Team Learnig among others. Other methods used are PBL – Project Based Learning (learning through projects or problems); TBL – Team-based Learning (learning by teams), WAC – Writing Across the Curriculum (written based on the disciplines) and Case Study.

The Peer Instruction is one of these breakthrough methodologies applied by teachers in different courses. This is a method that has as main goal which is to make classes more interactive, diverging from the traditional model where the student takes a passive stance in the classroom. Also, it proposes that the students interact with each other in classrooms, seeking to explain for each other, the concepts used and apply them on the solution for conceptual problems proposed.

The PBL method – Project Based Learning (learning through projects or problems), is characterized for using real-life practice questions for awakening critical thinking and developing behavioral skills, which are needed to the formation of professional who are contextualized with the industry. This type of approach was originally used by the school of Medicine of McMaster, Canada, during the 60's decade, based on the method of case study of Harvard Law School, USA, from de 20's decade; and also based on the model of the school of medicine at the University of Case Wester No, USA, from the fifties. The idea was to modify the framework of dissatisfaction of the McMaster medical students, due to the large volume of fragmented, irrelevant and without context subjects, which lacked on real questioning by the high and low loads of concepts in diagnosis and handling with patients. Nowadays, their applications are virtual in a diverse group of areas of knowledge, medicine and nursing, education, social and political sciences, business administration, financial and economic areas as well as in engineering. Moreover, on the exact sciences field the method became more known as *Project Based Learning* or project-based learning, since the complexity of the cases just taking students to work for long periods, developing multidisciplinary projects.

The TBL method – Team-based Learning (learning by teams) is an instructional strategy developed for Administration courses in the 1970's by Larry Michaelsen, made specially for large classes of students. It looked forward to creating opportunities and obtain the benefits of working in small learning groups, so that they would form teams from 5 to 7 students who were going to work in the same physical space (Burguess & McGregor, 2014).

The method WAC – Writing Across the Curriculum (written based on the disciplines) aims to lead the students to read and write in several disciplines, exploring the underlying issues regarding the similarities and differences, in terms of constructing scientific knowledge in different areas).

The Study Case method is one of the most used methods for researching, although there are some criticism about it, mainly concerning the impossibility to give a scientific rigor due to the possibility of distortion of results by the researcher. According to Yin (2001) the case study is just one of many ways to do a research on social sciences.

3 FLIPPED CLASSROOM AND THE LEARNING AUTONOMY 'NOVELTY'

First of all, a reflection regarding the flipped classroom concept may not be initiated without highlighting a recurring situation in many courses, including the fashion design. Our twelve years' experience in teaching Brazilian fashion design courses, allowed to perceive that, once fashion is a field where image and competition stand out, many teachers tend to see themselves as experts in the discipline they instruct, assuming a "catwalk" stance. Their egos are feed by the idolatry from some students who assume a non-reflexive and mimetic approach. It is a fact that some classes, practical or theoretic, instructed by these professors, at most of the time reproduce the processes which they went through their formation, centered only in their point of view, skills and knowledge. These teachers perceive themselves as an expert in a given field of knowledge and care for your content to be known by students (Walnut & Oliveira, 2011).

Consequently, the student, lacking from role models and idols, seems to put this teacher on a pedestal; almost like irrepleaceable gods, who must be imitated. Meanwhile, every word this professor says is almost a dogma, leaving no room for reflections or questioning and restricting the student role as a submissive learner, withour any autonomy. Notwhithstanding, they believe it is the best path, since they intend to be likewise this professor in the future.

Paulo Freire (2007) tends to make clear his beliefs that the educators should not turn the educational experience into something mechanical. The educator not only has the duty of respecting, but also should vail the students experience and associate to the taught discipline along with a discussion with the students about the reality to which it must be associated, so they can have a clear mindset.

In this context, the autonomy on the process of knowledge acquisition has been a subject of reflection in different spaces as breaking news. Would it be?

Paulo Freire, in Pedagogy of Autonomy (1990), your last published work in life, presents propositions for needed pedagogical practices to the educational system as a way of building autonomy for students, valuing and respecting their culture and their collection of empirical knowledge along with their individuality.

It is a fact that students who perceive themselves autonomous in their academic interactions reach positive results regarding motivation by showing intrinsic enthusiasm, the perception of competence, sense of belonging, curiosity, internalization of values; the engagement, with positive emotions, persistence, presence in class, seeking not fail and reducing evasion; regarding development, self-esteem, self-worth, preference for great challenges, creativity; regarding learning, better conceptual understanding, deep processing of information, use of self-regulated strategies; concerning improved grades performance, during the activities in class, on the results from standardized tests; and psychological state, featuring wellness indicators, satisfaction with life, vitality (Berbel, 2011).

The approached for the reversed classroom, or *flipped classroom*, emerges as a strategy which aims to change the paradigms of classroom teaching, changing its traditional organizational logic.

Generally speaking, the main objective of this approach is that the student would have a prior access to course material—printed or online—and could discuss the content with the teacher and other colleagues. In this perspective, the classroom becomes a dynamic and interactive space, allowing the realization of group activities, encouraging debates and discussions, as well as improving the student learning by exposing him to many points of view. In an effort to achieve that, it is necessary that the student assume his role in the knowledge acquisition process, in addition the teacher must be open and ready for a constant update on their knowledge together with a non-ethnocentric vision about cyberculture.

The application of this approach, as may be noticed, presupposes that the teacher prepares the material beforehand and make it available to students through some online platform (videos, audios, games, text, etc.) or physically (printed texts), in order to make the one-on-one debate more qualified due to prior students' reflection on the subject that will be addressed. Therefore, the traditional teaching model suffers a reversion. In this context, the room becomes a rich knowledge environment, due to the adoption of exercises, group activities and discussions. In addition to that, the vertical relationship—where the professor transmits information and students absorb—is replaced by an exchange of points of views, in which the teacher assumes a role as an education conductor, solving doubts, deepening the theme discussed and stimulating debate, in order to provide the student with a wider and fuller learning experience.

Finally, it may be no surprise that for many teachers, is a challenging process that leaves them unarmed and exposed to the supposed opponent. There is no room for improvisation in a not prepared class, even more less for the *slides* repeated from one year to the next one. Each class has its own particularities, as it works in fashion, the speed of time and the volume of information make that the novelty from yesterday will be outdated today. It is worth mentioning that the choice of an active methodology should be made consciously, well-thought and, above all, set not to take the joy of teaching away from the professor. It is also important to highlight the institutional role for the University on giving this teacher and this student tools that will assist them along this process.

4 APPLYING WAC, TBL, PBL AND CASE STUDY METHODOLOGY IN FASHION DESIGN TEACHING

In Brazil, graduation courses in fashion design follow the same guidelines for Design course[1]. In art. 3 from the guidelines it is possible to read that:

The graduation program in Design should lead, as a desired graduated student profile, capacitation for the appropriation on reflective thinking and artistic sensitivity, so that the designer will be able to produce projects involving visual, artistic, aesthetical cultural and technological information systems, including an historical adjustment, cultural traits and development of communities as well as the characteristics of their users alongside with their socio-economic and cultural context.

To begin with, if the main point is to make the students pro-actives, it is needed to embrace methodologies in which students are engaged in increasingly complex activities where they need to make decisions and to evaluate the results, using the support of appropriate materials. If creativity is the goal, they need to try countless new possibilities to show their initiative.

Furthermore, a proposition for applying active teaching methodologies in graduation courses in fashion design has already been released; looking for alternatives of action, in order to surpass the Brazilian fashion design courses weakness which were identified during the exercise of the teaching work.

As a matter of fact, some fashion design courses in Rio de Janeiro are already working with active methodologies. However, the proposal has not yet been very well assimilated. Hence, higher education institutions and teachers seem to have a limited vision of active methodologies which ends up causing resistance to the proposal.

Although hybrid teaching and the Flipped Classroom systems are in focus on the methodological approaches that put the student as your main training agent, the use of active methodologies, as pointed out earlier, is not restricted to these actions.

Another very important point that must be discussed talks about the fact that for many courses related to the creativity industry -architecture, design and fashion design, among others—project methodology is the active teaching methodology. Often, this leads to a fusion of the two proposals, which may be distinct paths but also correlated.

With this in mind, the PBL methodology believes that only previous theoretical knowledge is not enough to solve problems. The difference is that in PBL the resolution is given and prepared actively as a result of discussions and teamwork. After these discussions, new knowledge is acquired for so that they can be applied in that particular problem on the agenda. In other words, knowledge is generated on demand needs.

In the fashion design scenario, the relation between theory and practice is explored on projects of products creation, which in general, are individual. The suggestion questioning on PBL methodology differs from thematic proposition widely used by fashion design courses. Deliberate or work on each individual project involves a much larger analysis, when it comes about fashion design. The student needs to be stimulated to reflect over the product which is about to be created. Moreover, he needs to go beyond the creative insight to fetch the viability of his product as much as the execution and market incorporation.

---

1. Diretrizes Curriculares Nacionais para Cursos de Graduação em Design. Available at: http://portal.mec.gov.br/cne/arquivos/pdf/rces05_04.pdf. Accessed: 02/15/2018.

In regard of the development of the TBL method in fashion design courses, opportunities need to be created so the student may acquire and apply knowledge through a sequence of activities that include prior steps from the meeting with the teacher and those which he accompanies.

In a first moment, starting with a problematization or project, the student should be encouraged to prepare himself for the teamwork. It is necessary that he brings to the classroom the result from previous readings or other activities which were defined by the teacher in advance, such as watching a movie, perform an internet search, bibliographic or field research, among others. In TBL this preparation is called individual preparation (pre-class). The preparation of individual pre-classes activity is a critical step. If students do not complete pre-classes tasks individually, they will not be able to contribute to the performance of their team. The lack of this preparation delays the development of group cohesion and results in resentment of the students that have been prepared, as they realize the overload caused by their colleagues who are less willing and/or less capable. Which leads to the question, how to ensure that students complete this step?

The answer is in the second stage of the TBL—staging-warranty evaluation that must be performed individually—the teacher can demand that each student introduces what they have researched on a pre-elaborate form, propose a test of awareness on the subject, or even an individual inspection academic model or field diary. After the individual part, the students gather in groups (teams) where they may deepen the discussions and improve their knowledge about a particular subject or a theoretical development of a design practice. Once again, the teacher would evaluate the process while maintaining the sense that at this stage, the developed activities seek to ensure that the student is prepared and ready to solve problems individually, to contribute with your team and apply knowledge in the following stage of the TBL methodology.

It is worth mentioning that the word team is not used as a synonym for group. Assembling a team in this process means to apply a market simulator. The team should be assembled by skills and abilities and not simply by affinity and friendship. We must remind the students that in the fashion market the error from one of the team members—Modeler, seamstress, designer, etc.—jeopardizes the entire final production. It might seem that this form of grouping is exclusionary. However, many students settle at groups leaving others to do the work for him. In this model there is no room for accommodation.

The last stage of work based on the TBL methodology is the application of knowledge (concepts) acquired by the teams' resolution on problem scenarios. This step should take up the most load of time for the job. It is a critical step that occurs in the classroom. The teacher should provide students, gathered in their teams, the opportunity to apply knowledge to solving issues presented in the form of problem scenarios which are relevant and present during daily professional practice. In other words, students should be challenged to interpret, infer, analyze or synthetized, and create projects.

However, for the application of PBL and active methodologies the TBL may have the desired effect, another active methodology must be stimulated. On the weakness found in our students concerning written reading, the strategy of driving the undergraduates on reading and writing in various disciplines, exploring the underlying issues, the similarities and differences, in terms of construction of scientific knowledge in several areas, is a path to be navigated with perseverance.

The experience with classes from first and second semester proved to be much more efficient than with more advanced classes, which led to a belief that the deployment of active methodologies must happen slowly and gradually, respecting the required adaptation from teachers and students to the new process.

The imposition of change in more advanced classes had a more negative result due to protectionist previously deployed models. The pre-class suggestion that causes the student to create a study routine, going far beyond the implementation of practical tasks, requiring research and reflection as the basis for the creative process in the classroom, were understood as if the teacher were delegating his or hers responsibility of teaching. On the other hand, the classes of first and second period, saw the proposition as something innovative and necessary to their academic autonomy and training.

Alongside the project it has been observed, the urge to prepare teachers for this change. Also, it has been noticed a very large opposition on the idea of student autonomy, especially when it involves the creative process. These teachers feel themselves often threatened by the possibility of been seem without an answer to the 'problem' brought by this new type of student.

As a result, it is possible to affirm that the biggest challenge found so far is the fear of students and teachers over the novelties, which can be confused by the academic fragility of both sides, something permissible in a fashion course that until yesterday had its own teaching base divided between the practice of professional market with little academic training and academics with little or no training in fashion.

As a possibility, let it be noticed the growth and recognition that fashion has been reaching both in the creative industry and in the academic area. Also, the ongoing dialogue between researchers of fashion with other areas, strengthening the solid construction of knowledge and sharing experiences.

Finally, we believe that we are on the right track, however there is a long way until the first classes that have passed through the methodological change which applied the active methodology in the fashion scenario reach the market. In the specific case of this article experience the marc is 2020. Until then we will continue exchanging experiences, correcting errors and registering the hits.

5 FINAL CONSIDERATIONS

The challenge is set. We know that every proposed innovation always begins with a struggle. In this case, where the active methodologies are a proposition for students and professor to get out from their comfort zone, the struggle and opposition may be even greater. Is the price you pay for getting out of the usual mindset, breaking the amateur status or the informal work in order to achieve academic excellence.

Hence that, many educators who work on the fashion field but do not possess an academic view may say that virtually this proposition does not work; that this is not the way fashion works. However, as we can notice, the fashion market has changed. Fashion and design are strictly related, if not the same thing, since they incorporate all the market demands for professional who are willing to look beyond the walls of their vanity. The "outside world" is a battlefield who may only be overcame by the strongest and best prepared to survive.

REFERENCES

Berbel, N.A.N. 2011. As metodologias ativas e a promoção da autonomia de estudantes. Ciências Sociais e Humanas, Londrina, v. 32, n. 1, p. 25–40, jan./jun. 2011. Available at http://www.proiac.uff.br/sites/default/files/documentos/berbel_2011.pdf. Accessed: 02/14/2018.

Borges, T.S. & Alencar, G. 2014. Metodologias ativas na promoção da formação crítica do estudante: o uso das metodologias ativas como recurso didático na formação crítica do estudante do ensino superior. Cairu em Revista. Jul/Ago 2014, Ano 03, n° 04, p. 1 19–143. Available at: https://ufsj.edu.br/portal2-repositorio/File/napecco/Metodologias/Metodologias%20Ativas%20na%20Promocao%20da%20Formacao.pdf. Accessed: 02/14/2018

Burguess, A.W., McGregor D.M. & Mellis C.M. 2014. Applying established Guidelines to team-based learning programs in medical schools: A systematic review. *Acad. Med.* 2014 Feb; 19; 1–11.

Freire, P. 2007. Pedagogia da Autonomia. São Paulo, Editora Paz e Terra.

Megido, V.F. (org.). 2016. A Revolução do Design: Conexões para o século XXI. SP: Editora Gente.

Moran, J. 2015. Mudando a educação com metodologias ativas. Available at: http://www2.eca.usp.br/moran/wp-content/uploads/2013/12/mudando_moran.pdf. Accessed: 02/18/2018.

Silva, R.N. da & Borba, E.O. 2011. A importância da Didática no Ensino Superior. Available at http://www.ice.edu.br/TNX/storage/webdisco/2011/11/10/outros/75a110bfebd8a88954e5f511ca9bdf8c.pdf. Accessed: 02/02/2018

Yin, R.K. 2001. Estudo de caso: Estudo de caso planejamento e métodos. 2. ed. Porto Alegre: Bookman.

Sustainability in fashion and design

Fashion that cares for the future

E. Araujo, I. Castro, I. Pinto & I. Araujo
Universidade do Minho, Braga, Portugal

ABSTRACT: This paper has two main objectives: first, to provide empirical evidence about the extent to which consumers are informed and willing to develop practices that help pave the way to sustainability in fashion; second, to clarify the type of methods and practices that can be implemented, with a view to fostering public perception about several types of behaviour they can undertake, or possible practices they can develop to achieve increasingly sustainable practices when consuming fashion products, thereby contributing to societal and cultural changes in terms of measures that will ensure greater care for the future.

1 INTRODUCTION

The fashion industry has been under increasing pressure to reduce waste in terms of consumption of materials and natural resources such as water and reduce the use of chemical products such as pesticides. It has also been obliged to encourage recycling, reduce pollution, and improve factory conditions, in particular after the collapse of the Rana Plaza complex in Bangladesh (Boscio, 2015). This paper discusses the importance of establishing methods to foster greater citizen knowledge about sustainable consumption of fashion products. It is assumed that, despite major theoretical development associated to this topic, citizens are still poorly informed about what sustainable fashion means, or which behaviours they are expected to change, in order to ensure a better future for society.

The movement towards fostering greater sustainability in fashion is nothing new (Parker, 2009, Gwilt, 2010; Shen et al., 2013). Nevertheless, in face of present-day societal challenges imposed by climate change and environmental degradation, sustainability in fashion increasingly lies at the heart of political and scientific debates. The UN's millennium challenges include the commitment to ensure that capitalist interests meet natural needs, by empowering people to act responsibly as citizens and consumers, and by changing industries—leading them to act more conscientiously. In fact, sustainability means the use of less toxic chemicals, the use of less land and water, and the reduction of greenhouses gases (Kaikobad et al., 2015). That is why sustainability is linked to ethics, which according to the ethical fashion forum (2018): "represents an approach to the design, sourcing and manufacture of clothing which maximizes benefits to people and communities while minimizing the impact on the environment". But this debate goes beyond environmental issues. Social and cultural questions, such as the relevance of working conditions, workers' wellbeing, and health (Henninger, 2016) are also important. In fact, ethical fashion proposes a new form of production, as well as a new form of consumption, based on values and principles (Cohen, 2015, p. 8). Over time, an increasing number of consumers value and understand the importance of conscious consumption (Marcondes, 2016), adopting ethical consumption practices and search of products whose characteristics reflect these principles (Henninger, 2016). Ethical fashion has not always been consensual. Greater industrialisation has provided several means of accelerating the fashion industry's production processes (fast fashion) and this has led to various different modes of aggressive capitalism (Marcondes, 2016), with a greater impact on the environment, concomitant with higher natural resource exploration and exponential growth of waste generation.

The fact is that sustainability is now increasingly adopted, and much work has been done to improve sustainability in different contexts (Annamma, 2012). Unresolved issues include the ways that citizens are responding to these changes and trends, and the extent to which they are receiving and consuming these new products. At this point it is relevant to conceptualise sustainability in the fashion industry with social studies of science and technology, since the latter explore the concept of public perception of science and scientific culture. In fact, the adoption of sustainable fashion products implies restructuring lifestyles and becoming effectively aware of various aspects that condition the choice of sustainable products, beyond price considerations. This paper explores this main idea, clarifying the need to develop more strategies to inform consumers about what sustainable fashion means and how it can be integrated within citizens' lives. It is developed on the basis of exploratory research conducted in Portugal involving university students to whom several questions were posed regarding their knowledge about sustainability choices in the fashion industry. The text is therefore divided into five sections. The first section is devoted to the theoretical framework. It highlights the core ideas underpinning the main concepts to be used. The second section clarifies the method to be used. The third and fourth sections describe and discuss the main results of the exploratory research conducted in Portugal involving higher education students about the issue of sustainable consumption. Finally, the fifth section presents the guidelines that can help fashion innovators to implement a set of measures designed to combine economic interests with societal and cultural interests, thus actively contributing to caring for the future.

2 THEORETICAL FRAMEWORK

As Sachs (2015) explains, sustainability is a sociocultural matter, that involves the network of relations and interdependences between the natural and social worlds (Adam and Groves 2007, Smith 2014; Gwilt and Rissanen 2010; Fletcher and Tham 2015). This challenge requires greater research and innovation into materials and products, and also requires changes to production equipment and vending methods. It also demands changes in consumer behaviour, given that consumers are dependent on the amount and type of information they have regarding products' composition and other effects (Smith 2014; Gwilt and Rissanen 2010; Fletcher and Tham 2015).

For example, the brand Eileen Fisher is perceived as being committed to several aspects that maintain and increase the overall sustainability of its production lines. The fibres used by the company are primarily vegetal fibres e.g. cotton and linen, and the animal fibres, such as wool, are "from sheep that are humanely raised" (Fisher, undated), as well as for being committed to using recycled synthetic fibres, such as polyester. The brand has also been working with bluesign technologies since 2009 "aiming to shift their global dye houses toward responsible chemical, water and energy usage" (Fisher, undated). The brand is also investing in alternative energy sources as well as cutting reliance on air shipping. By 2020, the company hopes that its US operations won't just be carbon neutral, they will be carbon positive (Fisher, undated). In that way, the brand starts by designing each item of clothing to last, but when people are "done with it" they take it back to resell and the pieces that they can't resell are transformed into "tomorrow's raw material, to be reborn as new textiles or refashioned as new clothes" (Fisher, undated). In addition to Vision 2020, Eileen Fisher makes donations to numerous social organizations.

Mango is a Spanish fast-fashion brand which has made a step towards including sustainability as an important component of its business plan, through the launch of "Mango Committed". The product line consists of a collection manufactured with fabrics that respect the environment. According to the brand's communications director, Guillermo Corominas, in an interview with Vogue Online, "the sustainable fabrics used for this collection, such as organic cotton and recycled polyester, have international certificates such as GOTS (Global Organic Textile Standard), OCS (Organic Cotton Standard) or GRS (Global Recycled Standard) among others" (Conlon, 2017). Besides the materials used, the items are also dyed with low environmental impact dyes in neutral colours (Network, 2017). This product line is a step forward in the development of "sustainable fast-fashion", however, one may question its true efficiency when it comes to how "environment friendly" it can be, if it maintains itself as a fast fashion brand whose business models are based on fast sales turnaround and large volumes, while ignoring the consequences. Like Mango, other fast-fashion brands have invested in the inclusion of sustainable product lines, such as H&M with "h&m conscious" and Zara with "join life". The Swedish brand H&M, Hennes & Maurtitz, has adapted itself to help preserve the environment. In 2011, H&M launched "Conscious Exclusive", a product line that combines several other brand partnerships, with organizations such as the ETI (Ethical Trading Initiative) and WaterAid. The collection has captured the attention of several celebrities and been a pioneer in the incorporation of environmental issues in the fast-fashion business. The capsule collection is made from organic and recycled materials, as well as innovative materials and, according to the brand, this is done in order "to prove that there's no contradiction in loving beautiful clothing and living a sustainable life" (Magazine, 2016). Zara is one of the biggest and most successful fast-fashion brands. The "join life" product line observes Zara's standards for garment design but has greater awareness of the choice of materials for garment production. Zara also implemented the "clothes collecting" initiative focuses on the development of communities, promoting education and offering support in emergencies such as natural disasters or armed conflicts (Life, undated).

Despite the interest of the fashion industry in promoting sustainable modes of production and market innovation, several paradoxes remain when considering the ways to achieve less harmful capitalism, that cares for the future on several fronts. The essential questions are somehow identical: is the fashion industry changing consumers' beliefs and behaviour towards protection of the natural and social world, or is it taking advantage of sustainability as a fad? Are consumers informed and willing to make a step towards fuller engagement with sustainably challenges? Are industries really interested in stopping climate change, exploitation, and inequalities, or are they jus interested in using sustainability as a flagship for gaining market competition? Therefore, the main issue to be addressed in this paper is therefore the extent to which people are informed about what is happening in science sustainability innovation and how they use that knowledge in their lives (Jasanoff 2010; Maasen and Weingart 2008). Matters such as the intersections between scientific knowledge and common-sense knowledge are of great importance in studies that defend the need for science and technology to communicate better with citizens, also giving them the ability to participate in decisions about scientific processes and goals.

In this line of thought, the strategies to foster greater public understanding of science comprise a wide range of methods that evaluate, first and foremost, what people know about a certain issue; what factors can explain those ideas, and finally propose modes of changing their perceptions—to become more critical about science and technology uses and more skilled and responsible when making choices. In Portugal, studies about citizens' scientific culture, public perception of science, or public participation in science have documented the general low level of knowledge or awareness of these matters. Despite the importance of social class, age and academic qualifications, there is widespread consensus that citizens do not care about many issues related to the scientific knowledge that underpins many devices, processes, or services. Additionally, participation in science is often regarded as lengthy and difficult to implement in the context of high pressure to innovate.

The following table shows the main points addressed by studies devoted to the public perception of science and to public participation in science of relevance to the fashion world.

3 METHOD

This paper presents several arguments in favour of a strategy towards greater engagement in education for sustainable consumption and increasing innovation to ensure a sustainable fashion industry. The arguments are sustained by a short exploratory research that we have developed in Portugal with university students, concerning the way that they understand and experience green-based innovation, both in terms of fashion clothing products, and cars—under

Table 1. Surveyed students by age and sex.

Sex	Age		Total
	0–19	20–40	
Female	34	81	115
Male	8	26	34
Total	42	107	149

Source: Online survey with university students.

the theme of a larger project entitled "Care for the future". The study cited in this paper involved an online survey of university students. After planning and testing the questionnaire in function of the dimensions presented below, a message was sent to students clarifying the objectives, and guaranteeing anonymity. The University in which the study was developed currently has about 20,000 students. We managed to get answers from 149 students, ending up with a non-probabilistic sample. Due to the inexistence of studies into "green" fashion, this survey acted as a first step to provide consistency to some hypotheses about the need to improve science for societal strategies in fashion-related areas. In this sense, we only use descriptive statistics to show the answers about trends concerning the major dimensions in analysis. In subsequent phases of this study it would be possible to make a more broken-down data analysis to achieve a fuller understanding of the variability of the data, in terms of variables such as age, gender, or social class.

3.1 Dimensions in study

- Importance of sustainable practices
- Factors influencing the purchase decision
- Actual knowledge about sustainability-related issues in fashion products
- Knowledge about the *Zara Join life* product line
- Perception about the combination of economic and socio-cultural interests and goals

4 FINDINGS

4.1 Importance attributed to sustainable practices

When asked about the extent to which sustainable practices are important for choosing fashion products or services, a significant percentage of people stated that it is "very important". However, the biggest number of respondents stated that it is only "somewhat important".

4.2 Factors influencing the purchase decision

Consumers admit they barely attribute relevance to sustainability-allusive indications in the product, or even consider it when deciding to purchase a product. In accordance, when asked about what factors they consider to be more important to purchase a certain type of fashion product, the clear majority assumes the cost of the product as the most influential factor in the purchase decision, followed by design and style. Nonetheless, sustainability makes a difference when combined with the price and the style (42,3%). Only a very small percentage of respondents declare choosing products in function of the price and sustainability features (2%).

In accordance, when we asked students about whether the absence of sustainability proprieties in the product would mean that they wouldn't purchase it, they answered negatively, stating that it would make no difference to their decision, as shown in the following table.

4.3 Actual knowledge about sustainability-related issues in fashion products

We asked people to say whether they knew about the type of negative effects of polyester fibres on the environment (do you know that polyester fibre is not bio-degradable and that after a simple wash, masses of particles are released into the ocean?). Not surprisingly, most of the respondents said they did not know about this.

Cotton is a type of fibre that needs enormous quantities of water to be produced, as well as to be dyed. When dyed, it also generates several environmental and health effects, because

Table 2. Importance of sustainability (Source: Online survey with university students).

Importance	Freq.	Perc.
Very important	36	24,2
Somewhat important	97	65,1
Indifferent	5	3,4
Of little importance	10	6,7
Unimportant	1	0,7
Total	149	100,0

Table 3. Factors influencing the purchase decision.

Factors	Freq.	Perc.
Price, style and sustainably	63	42,3
Only price	5	3,4
Only style	2	1,3
Price and sustainability	3	2,0
Price and style	75	50,3
Style and sustainability	1	0,7
Total	149	100,0

Table 4. The extent to with the absence of sustainability features could be a reason for not purchasing the product. (Source: Online survey with university students).

Agreement	Freq.	Perc.
Agree	7	4,7
Disagree	142	95,3
Total	149	100,0

Table 5. Knowledge about the effects of polyester fibres (Source: Online survey with university students).

Knowledge about the negative effects of the use of polyester fibres	Freq.	Perc.
Yes	42	28,2
No	107	71,8
Total	149	100,0

Table 6. Level of agreement with the idea that cotton is a sustainable fibre (Source: Online survey with university students).

Agreement	N	Perc.
Agree	105	70,5
Not Agree	44	29,5
Total	149	100,0

Table 7. Knowledge about Zara's sustainability orientated fashion products Source: Online survey with university students.

Questions	Yes Freq.	Perc.	No Freq.	Perc.	Total Freq.	Perc.
Do you know about the Join Life product line?	30	20,1	119	79,9	149	100
Do you know about Zara's programme to collect used clothes?	32	21,5	117	78,5	149	100
Have you ever participated in a Join Life programme?	4	2,7	145	97,3	149	100

Table 8. Level of agreement with the possibility of matching economic interests with social/cultural interests and goals.

Agreement	Freq.	Perc.
Yes	36	24,2
Yes, probably	73	49,0
No opinion	22	14,8
Probably no	17	11,4
No	1	0,7
Total	149	100,0

large quantities of dye are released, in some way, into the environment. Nevertheless, a strong indicator of the general lack of knowledge about the best choices to take in favour of the environment and care for the future is the high percentage of people who consider that the consumption of 100% cotton products is a green and smart option.

4.4 Knowledge about the Zara Join life line

Further to these questions, we also asked students about their knowledge about a specific branch of fashion products that try to be sustainability orientated.

As the information in the table indicates, the clear majority of respondents neither knew about this branch of products (even regular consumers), nor knew about Zara's programme to collect used clothes. Despite a high percentage declaring that they would be open to using it, the vast majority answered negatively (97,3%).

4.5 Perception about matching economic interests with socio-cultural goals

Despite considering that conducting quantitative research about this matter is not an easy endeavour, since it favours politically correct answers, we can anticipate that this group of people is rather optimistic, concerning the chances to make economic interests match social and cultural interests and goals. Only 20% believe that this is not possible.

The clear majority assert that it is possible to combine these apparently dualistic interests, as shown in the table.

5 DISCUSSION

Awareness of the uncertainty associated to the classification of any material or process is crucial in order to understand the need to perceive sustainability as a key topic of work in any of the stages involving the fashion industry which is, per se, dominated by ephemeral and short-term cycles. In its own context, sustainability is a highly complex matter, since it

Table 9. Recommended practises.

Schools and Universities	Firms and local authorities	Governmental and regulatory /other agencies
Enhancing education for consumption, with a focus on the limits and effects of consumerism.	Promote ecological and sustainable innovation in engineering and design fashion industries.	Improve studies about public perception of science considering all the actors involved.
Enhancing children's and young people's knowledge about fashion and environmental effects of the use of chemicals, and other potentially health damaging materials or products.	Enhancing reflexivity of local authorities and firms for environmental and social effects of the fashion industry.	Promote the creation of self-assessment standards orientated to the different types of fashion-related industries.
Improve ecological and sustainable innovation in engineering and design fashion industries.	Promote intensive and regular surveys and research about public perception of fashion industry-based science.	Promote integrated participative measures to be jointly implemented by different stakeholders with distinct publics
Make public participation effective though the involvement of students in projects integrated in formal educational programmes.	Promote public participation and make it effective though the use of various methods, including the media and social media.	Promote scientific communication and dissemination programmes.

involves more than capturing consumer interest to purchase certain types of products. At its core, our study astonishingly indicates that young people, who often purchase items in shopping malls, tend to attribute low importance to sustainability issues when choosing fashion products, and are unaware of the major effects of the fashion industry on society, throughout the production chain, which are not all strictly environmentally-related. In fact, we also know that the fashion industry's production processes are highly based on exploitation of more vulnerable populations and countries.

Despite being educated, the survey respondents revealed an absence of elementary knowledge about sustainability practices, and the extent to which such knowledge might change their behaviour as consumers. In this regard, the data is consistent with other studies cited herein, that document the general lack of scientific culture in Portugal, as well as the absence of studies about the public perception of science. The data also indicate the need for actors in the fashion industry (including researchers and innovators) to achieve a fuller consideration of sustainability issues as an integral part of the innovation chain, paying careful attention to product costs, to avoid labelling sustainability in the fashion industry with social class.

6 CONCLUDING REMARKS

The data collected does not allow us to explore the contribution of other variables that have a strong sociological weight when thinking about sustainability, e.g. gender. However, data indicate that one of the present-day challenges is to better inform young people, since they are also one of the key mass consumers. Therefore, it becomes clear that there are cultural variables that shape people understandings about choices in consume and which are of strong importance for rethinking sustainability in fashion, in multiple perspectives. In countries such as Portugal, it becomes evident that sustainability in fashion is still a matter of expert knowledge, that is, of people working in fashion science and innovation, or fashion industry. Hence, there is still a long way to accomplish in order citizens improve their knowledge on the better choices to make when buying and using fashion products. In a synthetic manner, we are going to provide a look into some practical implications conveyed by the results with potential to be enacted in industry, and public institutions that can contribute to change that scenario and improve the "care for future" perspective in fashion industry, assuming that there is a chance to combine economic with social and cultural interests, in a manner that capitalist logic becomes more aware of its consequences and more conscious about its responsibilities in the time ahead.

REFERENCES

Adam, Barbara & Groves, Chris. (2007). *The future matters*. London, Brill.
Adam, Barbara (n.d). *Towards a new sociology of the future*. [Online] Available from: http://www.cardiff.ac.uk/socsi/futures/newsociologyofthefuture.pdf [Accessed 6th February 2018].
Boscio, Chere Di. (2015) *Eluxe Magazine*. [Online] Available from: https://eluxemagazine.com/fashion/hm-conscious-collection-2015/ [Accessed 17th April 2018].
Cohen, Nathalie. (2015). *Moda ética: A Consciencialização Através da Comunicação*. Guimarães: Universidade do Minho.
Conlon, Scarlett. (2017). *Vogue*. [Online] Available from: http://www.vogue.co.uk/gallery/mango-launches-new-committed-sustainable-collection- [Accessed 17th April 2018].
Eileen Fisher. (n.d.). *Eileen Fisher – Vision 2020*. [Online] Available from: https://www.eileenfisher.com/vision-2020/ [Accessed 17th April 2018].
Fletcher, Kate; Tham, Mathilda (eds.). (2015). *Routledge Handbook of Sustainability and Fashion*. London, Routledge.
Gwilt, Alison & Rissanen, Timo (eds.). (2010). *Shaping sustainable fashion: changing the way we make and use clothes*. [Online] London, Earthscan. Available from: https://www.scribd.com/document/353512029/alison-gwilt-timo-rissanen-shaping-sustainable-fashion-changing-the-way-we-make-and-use-clothes-routledge-2011 [Accessed 6th February 2018]
H&M Magazine. (2016). H&M *Magazine: Everything you need to know about conscious exclusive*. [Online] Available from: https://www.hm.com/au/magazine/culture/h-m-inside/2016/04/everything-you-need-to-know-about-conscious-exclusive [Accessed 17th April 2018].
Henninger, Claudia; Alevizou, Panayiota, Oates, Caroline. (2016). What is sustainable fashion? *Journal of Fashion Marketing and Management*, 20(4), 400–416.
Jasanoff, Sheila (ed.). (2010). *States of Knowledge: The Co-Production of Science and the Social Order*. International Library of Sociology. London, Routledge.
Kaikobad, Najmul; Bhuiyan, Md. Zafar; Zobaida, Helena & Daizy, Afroza. (2015). Sustainable and ethical fashion: the environmental and morality issues. *Journal of humanities and social science*. 20(8), 17–22.
Maasen, Sabine & Weingart, Peter (eds). (2008). *Democratization of Expertise? Exploring Novel Forms of Scientific Advice in Political Decision-Making. Sociology of the Sciences: A Yearbook*. Dordrecht, London, Springer.
Marcondes, Dal. (2016). *Envolverde: A economia do fim do mundo*. [Online] Available from: http://envolverde.cartacapital.com.br/a-economia-do-fim-do-mundo/ [Accessed 17th April 2018].
Parker, Lis. (2009). Sustainable fashion: A handbook for educators. United Kingdom, Labour Behind the Label.
Sachs, Jefrey. (2015). *The age of sustainable development*. Columbia University: Columbia University Press.
Shen, Dong., Richards, Joseph.; Liu, Feng. (2013). Consumers' awareness of sustainable fashion. *Marketing Management Journal*, 6(9), 134–147.
Smith, Joe. (2015). The world in a wardrobe: expressing notions of care in the economy and everyday life. In: Fletcher, Kate and Tham, Mathilda (eds.) *Routledge Handbook of Sustainability and Fashion*. London, Routledge. pp. 139–146.
The Fashion Network. (2017). *The Fashion Network: Mango lança Mango Committed, uma coleção de moda sustentável*. [Online] Available from: http://pt.fashionnetwork.com/news/Mango-lanca-Mango-Committed-uma-colecao-de-moda-sustentavel,792938.html#.WpnSdedUmUl [Accessed 17th April 2018].
Zara Join Life. (n.d.). *Zara Join Life*. [Online] Available from: https://www.zara.com/uk/en/sustainability-11449.html?v1=967743 [Accessed 17th April 2018].

Design, tradition and culture—past and future united to develop depressed regions

I. Oliveira & M.G. Guedes
Universidade do Minho, Guimarães, Portugal

ABSTRACT: Small economies need to adopt new practices and concepts to survive in an economic context driven by constant technological advances. Design, tradition, and culture can be linked to handcrafts, reshaping the traditional methods, and transforming them as a mean of self-production in a tendency of revivalism and value.

This essay intends to highlight the benefits of a possible partnership between design and handcrafts, as a contribution to the economic and social development of depressed regions. In a bibliographic search, it is shown the point of view of several authors about cultural value and the tradition of "forgotten" regions, underlining the critical role of inclusion and at the same time turning evident their local value and entrepreneurship.

The conclusion we intend to achieve is the understanding of how social design and handcrafts can be a synonym of regional innovation, and through the creation of new products, how they may contribute to the economic and social growth of underdeveloped regions.

1 INTRODUCTION

As a result of globalization, innovation has won an enormous focus in almost every area. Nowadays, we are looking for sustainable initiatives, which in the long term, enhance and promote regional development; have the ability to retain and capture new inhabitants; boost the tourism and, consequently, boost the investment and attract/encourage companies to stay. Every day there is a need of changing sustainable lifestyles and the search for new processes and new ways of doing something appear not only with technological change but also on the economic and social levels (Krabbendam, 2013).

It is in this context that concepts such as regional innovation arise. They combine planning and urban interventions, the projection of infrastructures and promote cultural events (Abreu, 2017), because a region must be seen as a civic place for cooperation and inclusion, must create connectivity and conditions for different lifestyles where one can grow intellectually, where there are purchasing power, jobs, and leisure opportunities.

Innovation stands for a process of creativity, transformation or experimentation and, once allied to entrepreneurship, can explore opportunities that improve social, economic and environmental surroundings. Plonski (2017) considers that a process of regional innovation is a set of structured actions, and is a multidisciplinary field applied to knowledge and the practice of administration, economics, and engineering among others.

A few decades ago, innovation was understood as the specific practice of development and launch of new industrial products, through modern design, new processes or new technologies.

Nowadays, innovation allied to design is considered the primary goal by several economic sectors, regional development, and other social segments, highlighting the search for change in society in a context of the exploration of new resources and new opportunities which emphasize positive results in the process of creating new realities.

The contribution of social design to the regional development is still feeble. However, modern economies face a global recession, which inevitably strikes all countries (low-income taxes), present a fundamental characteristic and might work as leverage: substantial regional asymmetries. Even in the most developed countries, it is possible to find less developed regions, and despite the efforts to support their sustainable development the asymmetries still maintain.

Economic, social and environmental crises caused the emergence of new paradigms, and markets opened new paths and created new guidelines to development, which changed the use of resources, the structure of social systems related to work and social security, as well as the global production, and distribution of goods and services. The less developed countries position themselves strive for internal innovation and design skills growth. According to Canclini (2008), the culture of a region is an essential asset in the transformation of the social system, since it can function as a process capable of improving ways of living, ideas, and values.

Thus, through the view of several authors, it will be relevant to see if a partnership work between designers and artisans will be enough to create regional innovation and if through the development of new products, they will be able to contribute to the economic and social growth of a less developed region.

2 THE INFLUENCE OF SOCIAL, ECONOMIC AND ENVIRONMENTAL FACTORS ON THE SUSTAINABLE REGIONAL DEVELOPMENT

According to Sachs (1993), a process of local development should be a process of socio-structural changes, characterized by its autonomy in the use of available resources, by searching bio, cultural and economic developed sites. The changes to implement must be sustainable at local, regional and global levels. However, regional asymmetries remain, and though there was a little change, the persistence of a few unbalanced aspects represents an issue to have in mind when facing regional and developing growth, whenever possible sustainably and socially.

Since the establishment of the European Union, and particularly in recent years, a lot of considerations have emerged around the convergence and cohesion processes concerning regional development (Costa et al., 2005). Needs of adaptability began to arise in the market, invest in regional sustainable development, apply "eco-efficient" concepts trying to profit from natural and local resources with the purpose of rendering some value to a product. The regional development based on sustainability, besides revealing a concept of regional development and influencing the society's lifestyles, also reveals entrepreneurship and the performance of a community (Sachs, 1993). It is also capable of generating positive economic, social and environmental significant results.

Fonseca (2004) stands that the idea and importance of human capital start getting more and more visibility, so from an infrastructure policy, we should move to a service policy, with human capital reinforcement. Gonzalez claims (2005), from a regret economy, we need to move to a business economy. Therefore, this new way of creating regional innovation starting by profiting from resources in benefit of the local economy and cultural value and tradition might be a functional, sustainable and lucrative solution and a criterion of differentiation of the market.

One of the principles to sustainable and regional development is to maintain a right balance between production and the available regional resources (natural and local). Sustainable and regional development are supported by the implementation of new businesses and activities that bring more added value income to the local economy, boosting cultural, economic and social change processes. However, to achieve this effect, it is necessary to establish long-term regional goals and to motivate the participation of the local community in the process. It is also required to develop social and economic methodologies to follow up results at quantitative and qualitative levels.

According to Barbosa (2008), sustainable development turns into the combination of three components (economic, social and environmental quality). The concept of sustainability present in the economy and in the social sphere enables the development and choice of

Figure 1. Parameters to achieve sustainable development (adapted from Barbosa, 2008).

sustainable strategies that can contribute to improving a population's quality of life (image 1). The goal will be to develop gradually the value of goods and resources (natural and local) that can be transformed and promote their insertion into planning and the daily economy. There is no economy if natural resources disappear so that the ideal would be linking these two areas (Alves, 2010).

To Alves (2010), three interconnected elements compose the regional development: the cultural ability to think about oneself and to innovate; the political-administrative ability to make autonomous decisions and organize their implementation; the ability of production, ensuring their reproduction according to the social objectives established collectively.

Economic and sustainable management serves to contribute to the evolution of a new way of thinking from the social point of view to benefit the development of depressed regions. This process must highlight human, material and cultural resources, changing the local social context to allow the population to solve the problems of their region effectively. Regional development, based on sustainability, shows ethics, and at the same time appeals to marketing, reaching public opinion rapidly.

In conclusion, a development on regional level must provide the decrease of inequalities preserve the environment and stimulate sustainable development within a local system to a global one.

3 INNOVATION, DESIGN AND TRADITION – A PATH TO ENTREPRENEURSHIP

When we talk about innovation or innovate change, it can mean that change will lead to novelty, challenge or risk of instability. To introduce new products and services in the market a company (or a production unit) needs to be creative, paradigmatic, experimentalist and interdisciplinary; the principle of innovation has been a key element of changing not only on a technological level but also on political, economic, social and cultural levels as well. Innovation is also frequently linked to design.

Casagrande (2004) stands that innovation and design can be linked, as they both try to build processes and products that represent a dialog between technology and society, based on human creative process. However, this change has only been seen from a global perspective, putting aside depressed regions which cannot reach the global market due to their social and economic fragilities.

In times where global market appeals, more and more, to consumption and contributes to an environmental decline there is a need of changing, not only a technological changing but also a social one (Krabbendam, 2013). From this moment, new concepts become fundamental when we talk about regional innovation, where the use of natural and local resources contributes to the improvement and wealth of that same region.

At present, is frequent to come across with situations in which the initiative and the ability of production existing in less developed regions is frequently is overvalued. New entrepreneurs tend to adopt a vision that places the local resources as core-assets to create new products and create innovative solutions to achieve sustainable production. They aim to contribute to economic growth and stimulate social behaviours that conquer prestige and quality positioning to local production. Thus, a new perspective of design (social and inclusive) arises allied to handcrafts, as a new method of creation inspired by local culture, tradition, and resources.

To Krucken (2009), the intervention of design will contribute to overvalue new products, to promote local manufacture processes, establish links between producers and consumers, support the development of a sustainable value chain, and to encourage the creation of new micro or small companies. Shea (2012) states that design is changing and is helping to build the interconnection between tradition and innovation. This focus is placing design right in the centre of local communities economic development.

The design focus in social contexts brings to products an aura of genuineness, of traditional, of cultural rooted, and original. Consumers in general, and conscious consumers especially, are actively attracted to those values when integrated into new, and exquisite products. Brands structured over local cultural and traditional values as market differentiators, offer consumers not just a product use satisfaction: bring to their life a cultural narrative which reinforces the emotional links with the brand.

Handcrafts reveal the identity and the culture of a particular region, and an interdisciplinary approach may enhance the development and reinvention of products. Maintaining its cultural essence allied to the innovative thought of design, several techniques, and traditional materials can be combined into new processes of production as an answer to the needs of the market (Delgado, 2015).

The design allied to tradition (handmade processes) may promote a cooperative and inclusive work. In this context, design assumes the position of an interface of communication between an inherited past (handcrafts) and the desired future (innovation) (Albino, 2011).

4 SOCIAL DESIGN – THE NEW SCENARIO FOR THE DEVELOPMENT OF DEPRESSED REGIONS

Setting strategies of entrepreneurship and creating regional products to the development of economic growth of depressed regions has been a worry in the past few years. The consumer begins developing ideas of goods and services within its territory; the development of sustainable solutions, as a synonym of entrepreneurship and regional innovation, start en8gendering a new approach to consumption. Fagianni (2006) argues that, after the creation of a globalized market and fast industrial development, the issue quality is no longer a differential factor, the innovation of a product must appeal to emotion.

Thus, integrated into the contemporary world, social design emerges as the focus of the relationship between economy and culture (Gomez, 2010). It should be used to aggregate value to local production and create competitive and symbolic differentiation in the market, to maintain the traditional cultures alive (Hall, 2006).

To promote this type of solutions, we need the contribution of several skills, and social design can reinforce this dynamic and interactivity with a logical collaborative and inclusive innovation, opening the way to the intervention of society. More than art, design is a cultural phenomenon (Schneider, 2010).

Design operates as a way of creating products that fulfill their function while remaining "communicable" and symbolic (Ono, 2004). According to this new perspective, when associated with its cultural and regional origin, a product can be placed competitively in the market and become an over-valued asset. Barroso (1999 *sit. by* Pichler, 2012) claims that developing new products with cultural references means reporting the product to its place of origin. The design must "be aware of" the creation of new products, through the use of raw materials or typical techniques of production in the region or by the use of symbolic elements which can recognize the producers. Pichler (2012) also stands these products represent the cultural material of a particular place and the behaviour of its population.

The innovation towards sustainability requires social participation (Krucken, 2009), and the responsibility of the designer is to instill new consumption criteria in the society based on quality rather than quantity. Krucken (2009) declares it is essential to communicate the quality and the social and environmental contents of a product, presenting information that really shows sustainability to stimulate the development of the producer—consumer—region relationship.

When design collaborates in the development of a product in favour of its origin, it must contribute as well as to the benefit of its producers and its consumers. Social design arises as an ally, promoting cooperativism and inclusion, valuing land capital, making regional recourses advantageous in favour of development, creating entrepreneurship and rising the skills offer/answer in the global market (Krucken, 2009). So, the possibility of partnership work between designers, artisans or any other producers in the region may generate regional innovation through the development of new products and like this contribute to economic and social growth in these underdeveloped regions.

5 CONCLUSION

In recent decades, there has been an increasing interest in sustainability issues and entrepreneurship dynamic. Increasingly, it competes for high economic performance, and it requires a certain level of creativity, betting on regional innovation as a competitive advantage in the conquest of the global market.

After the research carried out on strategies of economic and social development regarding entrepreneurship and regional innovation with the intervention and influence of design, we concluded that social and economic development of cooperation and inclusion would be fostered, respecting culture and tradition of each region to be explored.

The present study initiates the approach to new researches on the subject and represents a contribution to the reflection on the relationship between entrepreneurship, regional innovation and the creation of competitive advantage in depressed territorial areas based on social knowledge.

It is evident throughout the article that several authors consider this new aspect of social design can enhance economic and social development respecting culture and tradition in each region which is intended to explore while promoting their products. The union between design and the region, an identity, allows the union of different elements in the development of new products that, despite a contemporary approach, may appeal to emotion by connecting them to their cultural roots.

Design as a creative and innovative activity can use local characteristics that transfer to the products new details and compositions, uniqueness and symbolism of the region where it was conceived. Here, the designer must assume the challenge of translating and interpreting the local culture as a differentiation and competitive factor. The communication of historical elements, associated to a product, allows knowing the "history of the product", and the influence of design might be the answer to the growth and the enrichment of the territory.

We can concluded, social design and handcrafts can be essential when it comes to regional value, innovation and the development of a value chain through the creation of new products and it can contribute to the economic and social growth of less developed regions. It is supported in many levels, that to have regional innovation, an opportunity for entrepreneurship and the creation of a competitive and creative advantage in the global market, there must be sustainable regional development, both economically and socially. It is important to understand what the favourable elements are presented in a particular region to be able to exploit it adequately, favour and stimulate economic and social relationships, to increase a sustainable economy.

ACKNOWLEDGMENTS

This work is financed by FEDER funds through the Competitivity Factors Operational Programme—COMPETE and by national funds through FCT—Foundation for Science and Technology within the scope of the project POCI-01-0145-FEDER-007136".

REFERENCES

Abreu, J., 2017. *Success Full – Casos de Sucesso nos Municípios Portugueses*, 1º ed. Santo Tirso: Idioteque.
Albino, C., Roda, R. & Providência, F. 2011, Reinterpretationof tradition values, in Minho territory: Handcraft – a Reading key. *Strategic Design Research Journal*, 4(2): 99–105.
Alves J., 2010. O Desenvolvimento regional sob a óptica da sustentabilidade: uma reflexão sobre a economia e o meio ambiente. *Ágora – Revista de Divulgação Científica*, vol. 17, nº 2: 13–23.
Barbosa, M.G., 2008, O desafio do desenvolvimento sustentável. *Revista Visões*, 4ª edição, nº 4, vol. 1.
Canclini, N. 2008. *Culturas Híbridas*, 4ª ed. São Paulo: Edição Universidade de São Paulo.
Casagrande, E. 2004. Inovação tecnológica e sustentabilidade: possíveis ferramentas para uma necessária interface. *Revista Educação & Tecnologia*.
Delgado M.J. & Albuquerque, M.H. 2015. The Contribution of Regional Costume in Fashion. *Procedia Manufacturing* 3: 6380–6387.
Fagianni K., 2006. *O poder do design*. Brasília: Thesaurus.
Fonseca, M. 2004. A política regional da União Europeia: uma utopia viável?. *E-topia: Revista Electrónica de Estudos sobre a Utopia*, nº 2.
Barrera, Y.N. 2010. La Cultura del Diseno, estrategia para la generacion de valor e innovacion en la PyMe del Area Metropolitana del Centro Occidente. *Cuardenos del Centro de Estudios en Diseño y Comunicación*, nº 34: 109–209.
Gonsalez, A., 2005, Regional divergence in the euro area, *International Conference on the Role of Government in Regional Economic Development* in https://www.ecb.europa.eu/press/key/date/2005/html/sp050919.en.html, retrieved 03 Mars 2018.
Hall S. 1992. Identidade cultural na pós-modernidade, 11ª ed. Rio de Janeiro: DP&A Editora, 2006.
Krucken, L. 2009. *Design e território – Valorização de identidade e produtos locais*. S. Paulo: Studio Noble.
Ono M. 2004. Design, cultura e identidade, no contexto da globalização. *Revista Design em Foco*, Vol. I, Nº 01: 53–66.
Pichler, R. & Mello, C. 2012. O Design e a Valorização da Identidade Local. *Design e Tecnologia*, Vol. 2, nº 4: 1–9.
Plonski G., 2017, Inovação em transformação, Estudos Avançados nº 31, nº 90.
Sachs I., 1993. *Estratégias de Transição para do século XXI: Desenvolvimento e Meio Ambiente*. S. Paulo: Studio Nobel.
Schwarz, M. & Krabbendam D. 2013. *Sustainist Design Guide: How Sharing, Localism, Connectedness and Proportionality Are Creating a New Agenda for Scial Design*, 2ª ed. Amesterdão: BisPublishers.
Shea, A. 2012, *Designing for Social Change: Strategies for Community—Based Graphic Deisgn (Design Briefs)*, 1st ed. New York: Princeton Architectural Press.
Schneider B., 2005. *Design – uma introdução: O design no contexto social, cultural e econômico*. São Paulo: Blucher, 2010.

Circular economy: An approach for the fashion industry

E. Pinheiro
State University of Maringá (UEM), Maringá, Brazil
Federal University of Technology – Paraná (UTFPR), Ponta Grossa, Paraná, Brazil

S.M.B.D. Barcelos
State University of Maringá (UEM), Maringá, Brazil
University of Minho (UM), Cianorte, Paraná, Brazil

A.C. de Francisco
Federal University of Technology – Paraná (UTFPR), Ponta Grossa, Paraná, Brazil

ABSTRACT: The sustainability issue has been widely discussed by researchers and companies from various sectors, and the circular economy has the potential to provide new production and consumption standards to help society achieve greater sustainability by reducing environmental, social and economic impacts. Given this context, this article aims to analyse the circular economy studies approach targeted at the fashion industry. This study is justified by the need to implement sustainable practices in the sector and the circular economy to present itself as a concept that can contribute to such actions. This is a bibliographic research whose data analysis used a qualitative approach. The proposed study based on a literature review addressing the circular economy and the fashion industry has the corroboration of Methodi Ordinatio and analysis in VOSviewer Software. The main result discusses the opportunities that the circular economy approach provides for the minimisation of environmental impacts and scope for recasting collaborative processes.

1 INTRODUCTION

The markets require increasingly more sustainable products and services as well as information about their impact on the environment (Buxel et al., 2015). In this scenario, the circular economy (CE) is seen as a new business model that will lead to sustainable development and harmony between economy, environment and society (Ellen Macarthur Foundation, 2012; Guiszelini et al., 2016).

CE promotes a more appropriate and environmentally healthy use of resources for the implementation of a greener economy, characterised by a new business model and job opportunities (Ellen MacArthur Foundation, 2012). The improvement of the welfare and environment impacts are also evident on equity within and between generations in terms of use and access to resources.

In the face of the changes and others still necessary for the transformation of industrial processes, the garment industry which is dominated by just-in-time and is subject to rapidly changing fashion trends making uninteresting textile products to consumers at seasonal intervals, production faster and of lower quality, which consequently generates use of materials and quantities of waste pre and post-consumer (Niinimäki & Hassi, 2011; Clancy et al., 2015), perceives the need for changes to reduce environmental impact.

This study was conducted with the aim of analyzing circular economy studies with targeted approach for the fashion industry and opportunities that the CE provides for minimizing environmental impacts and recast collaboratively processes.

2 CIRCULAR ECONOMY

The implementation of the circular economy (CE) in the world still seems to be in the early stages as well as their concepts are very recent and still require an additional improvement as to how they can affect the population carrying capacity, employment, international trade, the role of institutions, etc. (Ghisellini et al., 2016; Goyal et al., 2016).

The changes began to occur to a circular economy model from 2005, mainly due to growing evidence of misfits and future supply and demand achievements in relation to the finite nature of resources (Goyal et al., 2016). Circular economy has the potential to understand and implement radically new standards and help society achieve greater sustainability and well-being at low or no cost of material, energy and environmental. For this implies the adoption of cleaner production standards at company level, an increase in responsibility and awareness of producers and consumers, the use of technologies and renewable materials (where possible) as well as the adoption of policies and tools appropriate, clear and stable (Ghisellini et al., 2016).

The circular economy is defined in line with the vision of Ellen MacArthur Foundation (2012) as a system designed to be restorative and regenerative. Goyal et al., (2016) have defined the circular economy as a closed loop system comprising three aspects: reducing the consumption of non-renewable raw materials and toxic; reuse of products and services for superior design of offers, processes, systems and business models; Recycling waste into new resources for use and consumption. The circular economy is a widely used approach in different countries, but its implementation in developing countries like Brazil, is in very early stage in general.

2.1 *Fashion industry*

The fashion industry has undergone a remarkable expansion in recent years, especially with the consolidation of the fast fashion approach, which emphasizes one chain model of clothes supplies that responds quickly to the latest fashion trends often updating the clothing available in stores (Fletcher, 2013; Zamani et al., 2015).

To meet the trend of fast fashion and the pace of production, the industries of fashion uses cheaper labor, excess chemicals and water. The industry waste seems obvious when considering the environmental and social footprint of clothing production and before this face, there are new rules, norms and beliefs in the sector by placing the circular economy and the question of clothing waste on the agenda (Stål & Corvellec, 2018).

3 MATERIALS AND METHODS

The study uses a bibliometrical secondary data analysis, according to the methodology Methodi Ordinatio (Pagani et al., 2015) consisting of 9 phases (Table 1).

The keywords were set after Phase 2. Since them, "circular economy" AND "clothing industry"; "circular economy" AND "fashion industry"; "Industrial ecology" AND "clothing industry"; "Industrial ecology" AND "fashion industry". The survey was conducted during the last week of November 2017.

In Phase 3 the information was retrieved from the selected databases for searching, namely: Elsevier (http://www.sciencedirect.com); Scopus (https://www.scopus.com); Taylor & Francis (www.tandf.co.uk/journals); Springer (http://www.springerlink.com); Emerald (www.emeraldinsight.com); Web of Science (webofknowledge.com); Hindawi (http://www.hindawi.com); Wiley (http://www.wiley.com); Informs (http://journals.informs.org); Inderscience (www.inderscience.com); IEEE (ieeexplore.ieee.org/xpl/periodicals.jsp) and JSTORE (http: //www.jstor.org). For presenting publications with the keywords searched in a large quantity and greater availability of access to published materials. We opted for the realization of search for articles published from 01/01/2005 to 01/2018, targeting a wider range of jobs.

Table 1. Phases methodology methodi ordinatio.

Phase identification	Phase description	Description of methodological procedures
Phase 1	Establishing the intention of research	Research on the topic circular fashion and economy.
Phase 2	Preliminary exploratory search of keywords in data bases	Keywords were tested "circular economy" and "fashion industry" in considered relevant databases in the area of production engineering research, sustainability and fashion.
Phase 3	Definition and combination of keywords and data bases	Articles were searched for and found with a combination of keywords.
Phase 4	Final search in the data bases	The final search for the data resulted in a gross total of 252 results.
Phase 5	Filtering procedures	All sought work in all databases have been gathered.
Phase 6	Identifying impact factor, year and number of citations	The articles were organized in a spreadsheet column in the following order: title (title); impact factor (IF); year (year), and; number of quotes (CI).
Phase 7	Ranking the papers using the InOrdinatio	The procedure was the application of InOrdinatio equation.
Phase 8	Finding the full papers	Finding the full papers.
Phase 9	Reading and systematic analysis of the papers	Reading and analysis.

During Phase 5 jobs were eliminated in duplicate; work whose title, abstract or keywords were not related to the topic researched; papers presented at conferences, and; book chapters. This resulted in a total of 91 papers, especially articles 28 containing the word "fashion" or "textiles" or "apparel" in the title. Note that some of the bases did not allow more rigorous filters, which turned out to collect many articles whose themes were not related to the topic researched, generating this number of elimination, or collecting book chapters, that was not the point at this time. The VOSviewer tool was used in the analysis. VOSviewer is a free computer program information technology developed by Waltman and Van Eck (2012) for the construction and visualization of bibliometric maps.

4 RESULTS

Found 252 articles after elimination of duplicates and filter materials that were not close proximity to the subject articles 91 remaining. To find such a result all stages of Methodi Ordinatio were strictly followed, as described in Section 3. The results of the method showed the sequence of relevant articles for this study considering the impact factor of the journal (IF), publication year (Year), number of quotes (CI) and finally INordinatio with results ranging from >2730, 006 and <410, 0041.

Selected articles were available in forty-nine (49) different journals. The Fig. 1 shows twelve (12) Journals that had two or more publications. Journal Cleaner Production (5.715 IF) having the largest number of articles published on the subject followed by the Journal of Industrial Ecology (4.123 IF).

Figure 2 shows the map of the authors that stand out in this study and equality of publications between them and a connection between Daphne Comfort and Peter Jones, and; Nancy M. P. Bocken and Jonathan M. Cullen can be clearly appreciated.

Table 2 shows the 10 with more than two authors published studies a total of 237 identified researchers. The Hirsch index, or h-index is included for each author, and is a measure of the quality of professional authors, depending on the number of times their scientific production has been cited (Schreiber, 2015).

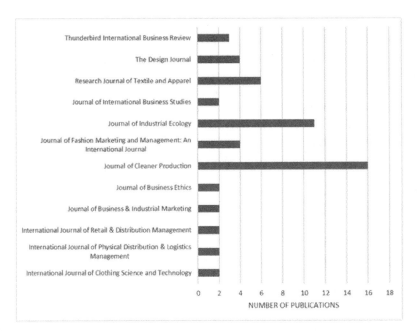

Figure 1. Journals with research theme publications.
Source: Authors.

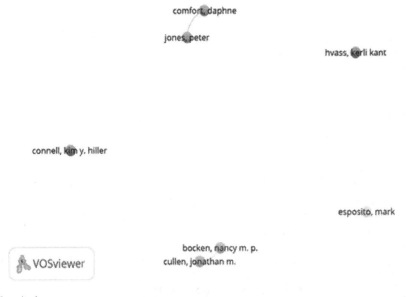

Figure 2. Authors.
Source: VOSviewer—Preparation.

Figure 2 and Table 2 show the author's developed the studies in the United Kingdom (3 authors) and the other in the United States, Denmark and The Netherlands. These authors, two authors have developed studies in fashion: Kim Y. Hiller Connell and KK Hvass.

According to the purpose of the study and to better understanding of the issue have been raised with the help of tool VOSviewer the occurrence of keywords and cluster. The map was

Table 2. Authors.

Authors	Affiliation	City	Country	Number of publications	Number of citation	H-index
Jones, Peter	University of Gloucestershire	Cheltenham	United Kingdom	151	1337	18
Comfor, Daphne	University of Gloucestershire	Cheltenham	United Kingdom	96	1101	17
Cullen, Jonathan M.	University of Cambridge	Cambridge	United Kingdom	20	557	11
Esposit, Mark	Harvard University	Cambridge	U.S.	21	106	4
Bocken, Nancy MP	Delft University of TechnologY	Landbergstraat, Delft	Netherlands	12	91	6
Connell, Kim Y. Hiller	Kansas State University	Manhattan, Kansas	U.S.	8	60	4
Hvass, K. Kant	Copenhagen Business School	Copenhagen	Denmark	2	8	2

Source: Authors.

Table 3. Cluster of words.

Cluster 1	Cluster 2
Industrial ecology	Circular economy
Recycling	Economics
Sustainability	Sustainable development
Waste management	–

Source: VOSviewer—Preparation.

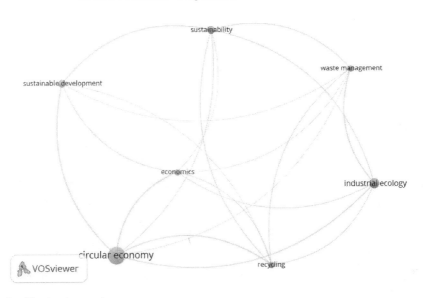

Figure 3. Viewing keywords.
Source: Vosviewer—Preparation.

created based on co-occurrence of keywords complete count of words. The keywords identified in the studies are: circular economy, economics, recycling, industrial ecology, sustainable development, sustainability and waste management. Table 3 shows the formation of two clusters of words that shows the clusters of terms.

Figure 3 highlights the CE and themes that seem to link with each other. The CE has connection with all items and "recycling, economics and sustainable development" the connection is stronger because the discussion in the studies are more consistent in this cluster.

5 DISCUSSION

The literature review and the use of Methodi Ordinatio contacted that there are relevant academic studies published and that they contribute to improve the basis for the CE implementation of discussion in various sectors, industrial parks and cities in developed and developing countries. It is emphasized that the present discussion is based solely on the results obtained from the application of the method.

The Methodi Ordinatio consider the Impact Factor (IF) of the journal, year of publication and the number of citations. Thus, the greater number Inordinatio was 2730.006 (IF: 5.715; Year: 2016; CI: 264). In this study the authors make a comprehensive review of the global CE (Ghisellini et al., 2016).

One of the problems presented and discussed widely in the literature, among business and society is the generation and waste management. Waste increase when a society develops and the problem aggravated by globalization. In urban centers, municipal solid waste is mainly disposed of in landfills, recycled or recovered. Due to rising environmental problems and restrictions of landfills, waste prevention is gaining more attention (Ghisellini et al., 2016). For fashion industry, zero waste is a tool that shows effectiveness to be based on the notion of minimization of waste materials in the production of clothing. This generates innovation for sustainability, reducing the use of raw materials through the development and adoption of new production processes and more efficient (Todeschini et al., 2017).

Thus, the circular economy can be used to inspire long-term systematics and changes in society and the fast fashion industry for sustainable operations. Linking the concept of circular economy to fast fashion allows the recovery of materials and energy, creating recycling flows in a closed circuit in the treatment of clothing waste, and promoting "shared economy" to extend the life of a part in the use phase (Zamani et al., 2015).

The circular economy model highlights three characteristics, and the first aspect involves adoption of the reduced paradigm around the resource consumption during the production and consumption processes (Goyal et al., 2016). Upcycling approach incorporates industrial waste of textile and apparel manufactures and post-consumer into clothing projects to generate new goods of equal or greater value, usefulness and/or quality than the original products (Goworek et al., 2012; Dissanayake & Sinha, 2015).

The second feature involves adoption of reuse paradigm, making the foundations of consumption (Goyal et al., 2016). Reuse is an example of "shared economy" or "collaborative consumption", extending the life of clothing (Zamani et al., 2015).

The third feature involves the adoption of recycled paradigm. On the one hand, this allows the recycling of resources, while on the other hand, creates a new industry, resulting in job creation and socio-economic well being enabling recovery of materials and energy, creating circuit in recycling flows in closed waste treatment clothing (Zamani et al., 2015; Goyal et al., 2016). As recycling often involves high energy processes, it is considered the last choice between the 3Rs (reduce, reuse, recycle), although it is an important alternative for the CE to implement principles to reduce the need to use new materials and resource consumption (Todeschini et al., 2017).

In addition to the features mentioned, sustainability reports are also important channels to display professional standards about what is involved and/or suggests sustainability (Stål & Corvellec, 2018). Another relevant aspect is the use of sustainable raw materials that include the development and adoption of different types of ecological raw materials such as organic cotton, hemp, bamboo, lyocell and recycled fibers, which consequently make the most durable clothes if compared with standard materials of clothes (Goworek et al., 2012; Todeschini et al., 2017).

6 CONCLUSIONS

The analysis showed that the circular economy is an important issue, relatively new, not yet known by the general public. It was verified the progress of the research on CE, from a broad perspective, to a more specific form of analysis, showing that the circular economy can help the fashion industry for waste reduction, reducing the negative environmental impact, etc. It is a search field and incipient applicability, which needs further research on its real benefits and potential impacts, especially in sectors characterized by the ephemeral nature of its products, as the fashion industry.

Still, it is evident that the circular economy concept can be used to inspire a long-term systematic changes in the fast fashion industry to sustainable operations, for example, the use of sustainable materials in the manufacture of products and the reporting of sustainability and hence can minimize problems such as management and waste management.

Finally, the circular economy model highlights three characteristics: reduction of resource consumption, reuse and recycling. It involves an adoption of the reduced paradigm around the consumption of resources during the processes of production and consumption. Also occurs the incorporation of concepts such as upcycling and zero waste which allows to link the concept of circular economy to fast fashion with the recovery of materials and energy, creating circuit in recycling streams closed in the treatment of clothing waste.

It is noteworthy that studies are needed to show boundaries, barriers, limitations to the applicability and enforcement of the CE in various sectors as well as further research on minimizing environmental risks. The studies are scarce in the area, especially when you mention the fashion industry and its production processes. Studies on disposal of post-consumer apparel products, including the life cycle of these also require comprehensive studies.

It should be stressed that studies are needed to show limits, barriers, limitations for the applicability and execution of CE in the various sectors, as well as further investigations on the minimization of environmental risks. Studies are limited in the area, especially when mention is made of the fashion industry and its production processes. Studies concerning the destination of post-consumer clothing products, including the life cycle of these also require comprehensive studies.

ACKNOWLEDGEMENTS

The authors gratefully acknowledge the "National Council for Scientific and Technological Development—CNPq" and the State University of Maringá (UEM).

REFERENCES

Bocken, N.M.P., Pauw, I. de., Bakker, C., & van der Grinten, B. 2016. Product design and business model strategies for a circular economy, *J of Industrial and Production Engineering*, 33(5), 308–320, DOI: 10.1080/21681015.2016.1172124.

Buxel, H., Esenduran, G., & Griffin, S. 2015. Strategic sustainability: Creating business value with life cycle analysis. *Business Horizons*, 58(1), 109–122.

Clancy, G., Fröling, M., & Peters, G. 2015. Ecolabels as drivers of clothing design. *Journal of Cleaner Production*, 99, 345–353.

Dissanayake, G., & Sinha, P. 2015. An examination of the product development process for fashion remanufacturing. *Resources, Conservation and Recycling*, 104, 94–102.

Ellen Macarthur Foundation, 2012. Towards the Circular Economy. Available in: http://www.ellenmacarthurfoundation.org/business/reports. (accessed 20 Jan 2018).

Fletcher, K. 2013. *Moda sustentável e têxteis: viagens de design*. Routledge.

Ghisellini, P., Cialani, C., & Ulgiati, S. 2016. A review on circular economy: the expected transition to a balanced interplay of environmental and economic systems. *Journal of Cleaner Production*, 114, 11–32.

Goworek, H., Fisher, T., Cooper, T., Woodward, S., & Hiller, A. 2012. The sustainable clothing market: an evaluation of potential strategies for UK retailers. *International Journal of Retail & Distribution Management*, 40(12), 935–955.

Goyal, S., Esposito, M., & Kapoor, A. 2016. Circular Economy Business Models in Developing Economies: Lessons from India on Reduce, Recycle, and Reuse Paradigms. *Thunderbird International Business Review*.

Niinimäki, K., & Hassi, L. 2011. Emerging design strategies in sustainable production and consumption of textiles and clothing. *Journal of Cleaner Production*, *19*(16), 1876–1883.

Pagani, R.N., Kovaleski, J.L., & Resende, L.M. 2015. Methodi Ordinatio: a proposed methodology to select and rank relevant scientific papers encompassing the impact factor, number of citation, and year of publication. *Scientometrics*, 105 (3), 2109–2135.

Schreiber, M. 2015. Restricting the h-index to a publication and citation time window: A case study of a timed Hirsch index. *Journal of Informetrics*, 9(1), 150–155.

Stål, H.I., & Corvellec, H. 2018. A decoupling perspective on circular business model implementation: Illustrations from Swedish apparel. *Journal of Cleaner Production*, *171*, 630–643.

Todeschini, B.V., Cortimiglia, M.N., Callegaro-de-Menezes, D., & Ghezzi, A. 2017. Innovative and sustainable business models in the fashion industry: Entrepreneurial drivers, opportunities, and challenges. *Business Horizons*, *60*(6), 759–770.

Van Eck, N., Waltman, L., Dekker, R., & Van den Berg, J. 2010. A comparison of two techniques for bibliometric mapping: Multidimensional scaling and VOS. *Journal of the American Society for Information Science and Technology*, 61, 2405–2416.

Waltman, L., & Van Eck, N. 2012. A new methodology for constructing a publication level classification system of science. *Journal of the American Society for Information Science and Technology*, 63, 2378–2392.

Zamani, B., Svanström, M., Peters, G., & Rydberg, T. 2015. A carbon footprint of textile recycling: A case study in Sweden. *Journal of industrial ecology*, *19*(4), 676–687.

Reversal logistics: Case study in the franchises O Boticário in the city of Itabuna-BA

E.N. Velanes
UNIFACS, Salvador, Bahia, Brazil

ABSTRACT: The objective of this article is to analyze some actions and benefits in O Boticário franchises in the city of Itabuna—BA, considering that the company has responsibility towards the society through the program Recycling of Reverse Logistics, bringing with it a conceptual proposal of O Boticário "sustainable beauty" products, thus covering the economic, environmental and social dimensions. The research uses as a methodology the case study, conducted through an individual interview with the five managers of the franchises located in the city of Itabuna—BA, with the main objective of explaining the importance of how to guide consumers to a clearer awareness about the correct disposal of waste products after consumption. Within this context, O Boticário franchises in the city of Itabuna—BA, has been effectively contributing to the Packaging Recycling Program with the intention of reducing environmental impacts, improving the quality of life of present and future generations.

1 INTRODUCTION

Society is extremely broad, in a constant evolution of change and transformation, always looking for the new, the innovator, the desire for a healthier life, where the contemporary consumer is increasingly aware of his goals, seeking pleasure and satisfaction at the time of purchase or consumption.

According to the Brazilian Institute of Public Opinion and Statistics (Ibope), currently 70% of Brazilians seek to purchase products aimed at sustainability, even if they have to pay more for it. Given this scenario, it is necessary to understand what sustainability is, aiming as origin and evolution, associating business with social responsibility as a way to contribute to a better world. Willard (2014).

The company plays a vital role towards society, aiming at social responsibility as a foundation in the concept of sustainable development, thus creating prospects for a better world. In this consensus, O Boticário franchises in the city of Itabuna—BA was used as a reference for the case study as a responsible company, since it proposes society as one of the pillars of the Strategic Sustainability Plan, with actions in short, medium and long terms/deadlines by 2024. Thus the brand analyzed has been providing new opportunities for strategies, becoming stronger, so that the franchise networks have a solid management in the competitive market of sustainable business. Through the Reverse Logistics, post-consumer is to raise awareness of the importance of the correct disposal of empty packaging, enabling customers to return empty products at any time of the year to any of Boticário's stores. We can thus associate business and sustainable beauty as movements of change and responsibility, aiming at social and environmental benefits.

2 SUSTAINABILITY IN BUSINESS

Sustainability represents a new way of looking at life as a whole, aiming at a healthier world in the face of an ever-growing market, associating business and sustainability.

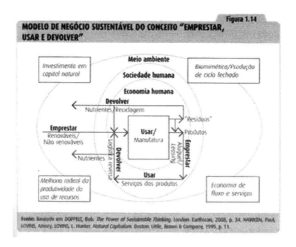

Figure 1. Sustainable business model of the "lend, use and return" concept.

"The Brazilian consumer has been increasingly concerned about the acquisition and use of socio-environmentally sustainable products and services". Ventura (2010).

Thus, sustainable business aims at sustaining the financial economic viability of enterprises while preserving the environmental integrity of current and future generations, building fairer relationships in society, resulting in a positive and solid reputation.

Society lives in constant change, where a large part is concerned with ethical issues when consuming their products, so the market has been adapting to these new conscious demands, seeking to meet the needs and desires of consumers. Given this market that is aware of the value and conservation of the environment focused on the quality of life, many entrepreneurs, associations and cooperatives have created their actions for strategies for business development with responsibility, economic, social and environmental.

According to Willard (2014), a sustainable business model "lend, use and return" was established as the best for the environment, for society and for the company, as shown in Figure 1.

It is understood that these five characteristics of a sustainable and cyclical business model of the concept of lending, using and returning, would be a process where everyone would win, both society, company and also the environment. Willard (2014).

Nowadays, both by people and by companies and organizations, sustainability is a major concern in the world.

In fact, sustainability is the search to find forms of development that meet the demands of the present without compromising the ability of future generations to meet their own needs. Cmmad (1991).

Given this current scenario, there is a great concern of consumers regarding sustainability (social, ecological and economic), where the positioning of the consumer and the company is increasingly providing possibilities of a conscious consumption, acting from the creative process of product development, their distribution and their arrival to the final consumer.

3 SUSTAINABLE BEAUTY O BOTICÁRIO

According to the O Boticário brand, people are increasingly seeking alternative ways to contribute to a more sustainable, non-nature-damaging world. It is perceived that the same believe that the products with more meanings, somehow associate stories in their creations manifesting their desire.

Due to this context, a Sustainable Beauty Packaging Recycling Program was created, aiming to provide an adequate destination for empty packaging.

But why is this important? Firstly, because it reduces the environmental impact caused by the inadequate disposal of these packages (when discarded in landfills or landfills, it takes many years to decompose in the environment). And secondly, because it ensures benefits for the whole society—in particular, the income generation that the sale of these materials makes possible for the cooperatives of scavengers partners of the Boticário Group. Krigsner (2015).

3.1 *Reverse logistics packaging recycling program*

Reverse Logistics deals with the planning process, as well as the efficient and effective implementation and control of the waste that returns to the logistics cycle, which can be post-consumer goods, with no further use to the first consumer, and after-sales goods. are characterized by goods little or nothing consumed by the customer. According to Leite (2003), the post-consumption category classifies its goods as: disposable, durable and semi-durable. Disposables are characterized by the shelf-life, which lasts only a few weeks, typical of packaging. The durable ones have years of life and decades and the semi-durables a few years. Meanwhile, the O Boticário franchise has been working on this sustainability model through Reverse Logistics, aiming at encouraging consumer participation, making available at all collection points the proper destination of the post consumer containers, as shown in Figure 2.

In this context, empty containers that are returned in one of the O Boticário stores will be sent to the cooperatives, or to the places where the recycling processes are carried out, according to the company, "trained employees transform the waste into raw material for new productive cycles".

These containers are not reused by the brand, when they are in the cooperative, are characterized, separated according to the type of materials such as paper, glass, plastics and others, are reinserted in new production processes. O Boticário (2017).

As far as the product life cycle is concerned, prospecting and technologies such as materials, design and methods that improve the environmental, social and economic performance of the products, from the extraction of the raw material to the disposal of the packaging, as shown in the Figure 3.

3.2 *Sustainability management in franchising network O Boticário*

The company has its position in the market associating business and sustainable beauty. The Post-consumer Recycling Program was created in 2009 by GrupoBoticário, bringing another awareness tool to a better world, where it had as partners all the franchises in Brazil, making

Figure 2. Cycle of Reverse Logistics.
Source: Adapted by the authors of https://www.mundodomarketing.com.br/.../o-boticario-cria-iniciati.

Figure 3. Packaging recycling program.
Source: Adapted from https://www.mundodomarketing.com.br/.../o-boticario-cria-iniciati.

Figure 4. Packing collector.
Source: Adapted by the authors of https://www.mundodomarketing.com.br/.../o-boticario-cria-iniciati.

this a reality in practice, currently a collector in each O Boticário store, where empty containers are deposited, as shown in Figure 4.

> *According to the founder of the brand O Boticário, it is 40 years transforming people's lives through beauty, currently the brand is the leader in national perfumery and one of the most beloved in Brazil. In 2016, the Care for Good line was reformulated with packaging made from vegetable plastic, produced from sugarcane. As a result, the company has become more sustainable the day after in the perfumery and cosmetics sector, with approximately 900 franchisees, which total more than 3,700 stores in 1,700 Brazilian cities, with the generation of 22 thousand direct and indirect jobs. As a business partner, t for the commitment to good management practice assumed by the group, becoming an agent in the region, directing and guiding all those involved in the awareness process, acting and transforming the franchise network into a large source of inspiration for people and businesses in their relationship and the environment. Krigsner (2015).*

3.3 Incentive campaign: O Boticário exchanges brand empty containers for new makeup

According to the Boticário Group, working in partnerships with the public of its value chain is fundamental to multiply initiatives that contribute to the development of society as: Employees, suppliers, franchisees, other retail audiences and the community. It understands that gender equality is the way to a more balanced world. Krigsner (2015).

Since 2015, the symbology that identifies the materials that comprise them has been applied in all packages, besides being a further incentive for the consumer, it also allows the correct destination of each package.

The Boticário Recicla campaign was created with the purpose of orienting the customer and the consumer on the correct disposal of empty post consumption packages, since they took place from May 22 to 26, 2017. According to the brand, the consumer would have to register in the site (botirecicla.com.br), so that it could make the exchange of its empty packaging by a product of the Intense line of makeup. This registration also guaranteed a voucher of 20% discount on the purchase of Intense Line products, as shown in Picture 1.

It is understood that the brand has attitudes of respect to the planet and people, taking into consideration that the analysis of the case study was only made in the O Boticário franchise of the city of Itabuna (BA), referring to the Program Recycling of Packaging named Sustainable Beauty.

The company states that it participates as an entire life trajectory during 38 years of franchising, always seeking to inspire positive attitudes and to raise awareness of the right to diversity, following closely and participating with its employees of all the campaigns that are directed by the Boticário Group.

Through the individually analyzed interviews with the five store managers O Boticário of Itabuna (BA), regarding the behavior of the consumer before the incentive campaign, they state that there was a great flow of people who went to the stores to make the exchange of empty containers for the targeted product, but the managers also say that there is still a lot to do, since many consumers do not have the same knowledge of the campaigns offered by O Boticário, and not even acknowledge the existence of the post-consumer collector in all O Boticário do Brasil stores.

In this consensus, O Boticário believes that the campaigns that are created are yet another incentive tool for the consumer to go to the stores and return the empty packaging, but it is aware that there is still a lot to do so that it can appeal to more and more consumers to the stores, bringing to this sustainable reality.

Picture 1. Intense make up line.
Source: Adapted from https://www.mundodomarketing.com.br/.../o-boticario-cria-iniciati.

The company affirms that it has been creating an internal incentive campaign with its employees and all those involved, in a proposal to raise awareness of the importance of post-consumption, thus, they collect more and more empty containers in stores, because in exchange they win prizes. The store that collects the most empty containers, where they are packaged and directed to the cooperative responsible for each region, is rewarded with awards: trip packages and others, bringing the team awareness of the importance of correct post-consumer waste and adding the satisfaction of all involved.

> *If each person wakes up in the morning with the intention of going through the day having positive attitudes, we will be building a better world, one day after another. GrupoBoticário has been working in this spirit of sustainability with all partners and throughout society.* Krigsner (2015).

3.4 Results of the analysis of the reports on the actions and benefits of reverse logistics of the "package recycling program" GrupoBoticário

The results listed here are from the "O Boticário" Sustainability Report, positive attitudes.

According to the current president of O Boticário Arturn Grynbaum (2015), nine more cooperatives were accredited totaling 21 partnerships and more than 900 cooperatives are benefited directly in this program.

Also in 2015, a survey carried out by O Boticário revealed that 30% of the public that consumes their products associate the company with some initiative of recycling, collecting or reusing packaging. Grynbaum (2015).

It also concluded that, the challenge would be to transform that interest in practice, where we count on the strength of the capillarity of the points of sales.

In 2016, the analysis report on sustainability shows us that the company has been executing the best practices of the packaging recycling program in operation in the sales channels, transforming reality in search of a better future in partnership with employees, franchisees, suppliers consumers. Grynbaum (2016).

In 2016, 1,200 waste pickers were benefited from the "Reverse Logistics" action, from the packaging recycling program, creating an opportunity for professionalization. Grynbaum (2016) Incentive campaigns "Boti Recicla", which took place from 22 to 26 May 2017, to exchange empty packaging for the new product of the line of makeup Intense, was a success in the stores. Artur Grynbaum (2017).

According to the current president Artur Grynbaum (2017), more than half a million people who participated in this action. In some stores, the stock for the exchange of products took only two days, exhausting the coupons available for the exchange of the product. He also states that this engagement of our customers fills us with joy and pride, because it shows that people understood the relevance of initiative to contribute to sustainability and the environment.

Artur Grynbaum also adds that (2017) the company was given a certainty that we are on the right track with our packaging collection program, which has existed since 2006.

O Boticário's permanent packaging collection program, besides contributing to the benefit and profit of the environment, the brand and the consumers together, makes a difference in the life of 2 thousand people and 34 recycling cooperatives throughout Brazil. Grynbaum (2017).

4 FINAL CONSIDERATIONS

From the information collected through the individual interviews with the five managers of O Boticário's store in the city of Itabuna (BA), it is possible to make some considerations. First, it is of great value to the social responsibility of the brand O Boticário to society, and this analysis showed the importance of guiding consumers to a clearer awareness of the correct disposal of waste products after consumption.

Therefore incentive campaigns on the exchange of empty packaging for a new product had a satisfactory result. The environment, the consumers and all those involved in this recycling process definitely benefited from it.

However, it is also clear that despite the satisfactory outcome of the incentive campaign and the increase in consumers who have returned to change their empty packaging for new products, much remains to be done as many consumers are not aware of the incentive campaigns created by GrupoBoticário. Some are not even aware of the post-consumer collectors that are located in all O Boticário stores. The Post-consumer Recycling Program is another tool for raising awareness of a better world, besides benefiting the lives of 2,000 people and 34 recycling cooperatives throughout Brazil. O Boticário (2017).

The brand "O Boticário" believes in the possibility of achieving economic growth with respect to society and the environment, transforming the reality of its publics with which they relate. The belief expressed in this Sustainability Analysis Report through analysis of strategies and results expressed here makes clear how to look at the future, inspiring positive attitudes for the construction of a more beautiful world.

In this perspective of analysis, consumers are increasingly demanding, where they live in constant changes, always in search of the new, innovation and companies that care about sustainability. It is perceived that, the personality of each consumer influences their behavior when buying or choosing the product.

It is necessary to understand the market and its trends in different ways, since consumer choices associate preferences by places, brands, products, among others. It is believed that sustainability is a new way of seeing life, associating business and sustainable beauty.

It is known that it is not easy to keep a company 100 percent sustainable, but there is a need for this concern for re-education, both by franchisees, employees, suppliers and consumers.

At the same time it is possible, to contribute to a better world, making it necessary to comply with all steps of the correct disposal of the packaging after consumption, ensuring respect for the environment, so we can build a healthier and more conscious society.

REFERENCES

Cmmad. 1991. *Nosso futuro comum*. Rio de Janeiro FGV.

DeCarli, Ana Merysehbe. Manfredini, Mercedes. 2010. *Moda em sintonia*. Caxias do Sul, RS: Educs;

Grynbaum, Artur. 2016. Relatório de Sustentabilidade, page 5 O Boticário, São Paulo de 2017.

Grynbaum, Artur. 2017. *O Boticário troca embalagens vazias da marca por maquiagens*. In https://www.embalagemmarca.com.br/2017/05/boticario-troca-embalagens-vazias-da-marca-por-maquiagens-novas>.

Krigsner, Miguel.2016.. Beleza Sustentável: Reciclagem das nossas embalagens de produto. Vida Linda. In http://vivalinda.boticario.com.br/estilo-de-vida/beleza-sustentavel-reciclagem-das-nossas-embalagens-de-produtos>.

Leite, P. R. 2003. *Logística reversa: meio ambiente e competitividade*. São Paulo: Pearson Prentice Hall.

Oliveira, Priscilla. 2017. *Mundo do Marketing*, in <https://www.mundodomarketing.com.br/ultimas-noticias/37365/o-boticario-cria-iniciativa-para-programa-de-reciclagem.html>.

Thofehn, Guilherme. 2016. *Redação Viva Linda com o Boticário*. in http://vivalinda.boticario.com.br/estilo-de-vida/reciclagem-de-embalagens-pos-consumo-de-produtos-de-beleza*.

Willard, Bob. 2014.*Como fazer a empresa lucrar com sustentabilidade*. São Paulo: 1ª Edição Saraiva.

Sustainable fashion. Strategies for sustainability and new forms of value creation in fashion and textile

M.A. Sbordone
Università degli Studi della Campania "Luigi Vanvitelli", Naple, Italy

L. Di Lucchio
Università La Sapienza Roma, Rome, Italy

R.A. Sanches
University of São Paulo, São Paulo, Brazil

ABSTRACT: In contemporary society, strongly characterized by new forms of cultural exchange, the Fashion sector is consistently evolving, capable of producing product and process innovation. In fact, consistently with the social changes taking place, consumers demand ever more performative goods which, in addition to expressing intangible cultural and symbolic values, are the result of a new value creation more sustainable. Thus emerges the urgent paradigm of understanding, developing and promoting a different way of doing fashion. In this paradigm the key word is "sustainability" acknowledged in social and economic terms, as well as environmental. The paper refers about the achievements in progress of a research process that sees the convergence of three different approaches related to Design for Fashion. The first approach aims to recognize and to establish heterogeneous supply chains of Fashion System' small and medium companies that hold a high quality level in terms of product and capabilities, to be drove towards new design-oriented values. The second approach aims to experiment the theme of environmental sustainability in the field of textiles. The last approach is aimed to the strategic development of new processes between design-production-consumption. Three phases were highlighted in the paper: a) the construction of widespread heterogeneous production chains, which set small entities with a distinctive cultural, technical and social characterizing contribution, b) the taking in charge of environmentally sustainable solutions for the production of fabrics and, finally, c) the development of products that transfer not only tangible values but also all intangible values to final consumers as well. These, in short, are the conditions for the creation and diffusion of a new form of mix value for the Fashion sector, identified in the Sustainable Fashion. At this stage of the research the paper presents only the results of the first phase.

1 INTRODUCTION

The present contribution intends to open a reflection on the ability of design to set up sustainable strategies in the production, diffusion and creation of goods in the fashion and textile sector. Fashion as a commodity sector, benefits from the relationships it establishes in very diversified areas—productive, social and cultural—thus demonstrating an attention to the processes of value creation that go through material production, creation processes and on modalities of communication and dissemination of tangible and intangible values. In fact, in the Fashion sector, goods draw on the sphere of sensibility and aesthetics, culturally orienting the markets. In the fashion processes, more than in others, the contextual links that Granovetter (1973) defines as 'weak or peripheral' occur: which, characterized by relational nature, are able to actualize the value mix and gradually adapt the goods to the new social and consumption contens. In this sense, Fashion is called upon to rethink these links and reconstruct them according to the emergence of new values. Among these, there is certainly the urgent and critical environmental sustainability of production, consumption and distribution processes. In contemporary society, public engagement, with local communities interested in environmental issues and sustainable development, is growing, and there are many emergencies that require reflection for the development of innovative and advanced solutions. The fashion industry is responsible for engaging in the adoption of measures to mitigate and combat phenomena such as: pollution, deforestation, desertification, water consumption, natural resources and the use of human resources, in addition to indiscriminate use of agrochemical products, with consequences on ecosystems. Obviously the theme of sustainability is not recent (Papanek, 1985), but thanks to the political and cultural debate today it is accepted as essential in all the productive sectors, even if perceived differently depending on the peculiarities of the system.

Even in the industrial segments of textiles and clothing the sustainability factor has been implemented and the aspects related to the impact in economic terms are carefully assessed.

The constant search for efficiency in the use of resources, the need to increase industrial competitiveness, the transition to more effective production models, are the incentives topics for innovation in the scientific and technological development in terms of sustainability.

Faced with this commitment, the textile industry must attempt valid alternatives to meet the expectations of more sustainable productions and the Fashion system will have to do the same by influencing the communication mechanisms to encourage more ethical forms of consumption.

But precisely because of the 'weak and peripheral' links inherent in Fashion's value contribution, the concept of sustainability assumes a peculiar character and finds a specific convergence between the more specifically technical aspects of environmental sustainability—which focuses on research, development and experimentation with textiles—with the cultural ones of a more social and economic sustainability—which instead seeks to understand and build new values for the fashion product. Research on new textile materials is a complex process, currently concentrated in four interconnected areas: use of renewable resources, systems to reduce energy consumption, water and chemicals, recycling and biodegradability.

2 SUSTAINABLE FASHION: FOR A NEW PARADIGMA

The fashion supply chain starts in agriculture (natural fibers) or in the chemical industry (artificial or synthetic fibers), through spinning, weaving and finishing, up to the final packaging of garments and their marketing (Rech, 2012).

According to Sanches (2011), the Fashion production system can be divided into six phases: raw materials, spinning, weaving and textile finishing, followed by the packaging of fashion items and the consumer market. Rech (2012) still divides this chain into three macro-sectors, which are:

- the production of the raw material, through the transformation of fibers and filaments;
- the processing industry, with the phases of spinning, weaving, finishing (preparation, dyeing and finishing) and manufacture of textile articles;
- the consumer market, with the actions relating to communication and to the distribution and marketing channels of the final product.

In the fashion industry, as in every sector, sustainable development is a necessity and has stimulated changes or even changes in management models, use of materials, product creation and process development.

According to Uniethos (2013), incorporating sustainable development into the production chain means re-evaluating or discovering new ways of balancing economic, environmental and social aspects.

2.1 Guidelines on materials

Textile fibers are classified into chemical and natural (Kadolph; Langford, 2016). Those of chemical origin are divided into artificial and synthetic fibers. Artificial fibers are obtained from natural polymers transformed by the action of chemical reagents and synthetic fibers from polymers obtained from chemical synthesis. Those of natural origin are produced by nature in a form that makes them suitable for textile processing.

This first synthetic classification makes us get how the interaction between textile materials and the environment is complex (Horrocks, Anand, 2000), it is no casually, in fact, that scientific research on new textile materials moves on four interconnected areas.

The first concerns the search for materials from renewable sources, where the focus is on the importance of considering not only the capacity for renewal of materials, but also the conditions of their production process, which determine impacts on air, on water and on workers.

The second includes research on materials that entail a reduced level of input in terms of resources (water, energy and chemicals) with direct consequences on CO_2 emissions and therefore with effects on climate change.

The third concerns research on fibers produced under better working conditions for farmers and producers, not least the action against child labor and low wages which, as we know, cause, among other criticisms, a great disparity in the markets competition with the consequent slow extinction of small producers in mature capitalist countries.

The last area concerns research aimed at obtaining materials produced with less waste, such as biodegradable and recyclable fibers coming from the recycling of industrial and consumer waste flows (Fletcher, Grose, 2011).

So, as mentioned, it is a wide and complex field of experimentation that cannot always be faced by all the operators in the Fashion sector, as well as for qualitative reasons both in terms of economic resources, skills and infrastructures suitable for research.

2.2 Premises on social and economic sustainability in fashion

As mentioned, it is clear that in the Fashion sector sustainability cannot draw exclusively to the environmental aspects but must enhance the social and economic ones.

The attention settled to social and economic sustainability rises in the field of fashion from a reflection and a rethinking of the processes of consumption and establish value.

In fact, in the last century in all sectors, as in Fashion, the "mass production" has given way to a "mass" of identity goods overturning the paradigm of "industrial capitalism" to become the aesthetic paradigm of "capitalism artist". Lipovetsky and Serroy (2013) identify their characteristics by connoting the change in "creative transesthesia". It is an evolutionary stage of capitalism that proposes goods that satisfy evolved needs, intercepting values that pertain to the sphere of sensibility, of affection and which induce to always desire the new.

A mechanism that requires a continuous renewal of the purchasing experience that needs to be implemented, of continuous updates on the level of the search for the 'sense' of goods; sensible values typical of aesthetics understood as a term for comparing the degree of involvement and individual emotional interaction. Fashion has long been intercepting this dynamic and, by adopting the demands of continuous renewal, has imposed what is increasingly identified as the 'Fashion model'. A decentralized system that uses the tools of the cognitive economy or artist whose levers for the creation of value are based on imagination, on intuitive knowledge (tacit-knowledge) and on the supply of identity goods that are renewed at a fast pace. It is clear that talking about social and economic sustainability in the field of fashion means in particular constructing relational systems that establish heterogeneous mixes of values along the whole process that goes from production to design to consumption.

2.3 *The research adoption*

Research in Design for Fashion has for a while been focused on production chains that contribute to forming self-organized "clusters" in a social, productive and economic relational fabric. The purpose of these research is to identify and develop ecosystems of relationships that draw from the contexts, from the original places of the productions of excellence, and are characterized by those that are called tacit knowledge (tacit knowledge) not formalized. One of the essential factors in making these ecosystems valid and efficient is the transfer of knowledge and therefore of innovation that inevitably resides in the relational model that underlies it. It is on this relational model that the "sustainability" is based on the social and economic responsibility of the Fashion sector. In fact, this must be as decentralized as possible due to the configuration of production units, the diffusion of services, the distribution over several territories, the operative disposition of the groups of research, design and production activities.

The decentralized and "sustainable" model of Fashion must, therefore, be characterized by the constant acquisition of "innovative knowledge" (Rullani, 2010) that multiply creative connections, accelerating the "new utilities" that become pivotal in the value creation processes. Knowledge, cooperation, peer-to-peer interactions, networking, are indispensable tools in the process of assigning the value mix, and the 'innovative knowledge' (Rullani, 2010) represent the driver to establish, in existing productions and in new, the degree of sustainability expected.

In this sense, training, industry and craftsmanship are at the center of this strategic reshape of activities according to a defined approach of Research Adoption. The latter stimulates the productive and capacity ecosystem by activating new forms and modalities of the creative ecosystem in the direct comparison with a given territorial productive environment; expression of manufacturing skill that expands in goods, and whose commercial and cultural value requires knowledge-based innovation processes. Figures 1, 2 and 3 show the workshop at Lanificio Leo.

2.4 *Sustainable Fashion: The case study of the workshop "Lanificio Leo meets Cesare Attolini and Kiton" (in progress)*

In order to assess whether the assumptions done can guide operational practices, precisely in the perspective described by the Research Adoption, several workshops were organized in collaboration between companies, design schools and professionals.

Figure 1, 2, 3. Lanificio Leo—The historical enterprise museum (Sbordone et al., 2018).

Figure 4, 5, 6. Workshop at the companies Cesare Attolini and Kiton (Sbordone et al., 2018).

In particular, these assumptions were the basis of the workshop held at the historic Lanificio Leo company. (2)

Figures 4, 5 and 6 show the phases of the production process of the companies Cesare Attolini and Kiton.

The aim of the workshop was to develop new fabrics designed and made ad hoc for some excellent manufacturers in the Campania Region: "Kiton" and "Cesare Attolini". (3)

Starting from the typical processing methods of the company, the evidence of the production characteristics and the textile materials, in an environmental sustainability perspective, we tried to expand the ecosystem of relations between the manufacturing companies and the fabric supplier company.

The methodology adopted was that of Listening Design, in the approach to product design. A team of professionals and students of the Fashion courses have "listened" and understood the peculiarities of the production systems involved.

Subsequently, the characters that implied both environmental sustainability (we chose to work on fabrics made of extra-virgin wool from local productions) were identified, as well as those related to social and economic sustainability in order to increase volume and diversity of the supply chain relationships, thanks to the introduction of "Lanificio Leo" as a new stakeholder in the supplier circuit.

In fact, the adoption of the woolen fabrics of Lanificio Leo for the manufactures of "Kiton" and "Cesare Attolini", represented:

- on the one hand, the recovery and enhancement in productive terms of fabrics with good environmental sustainability and economic sustainability, by virtue of local production, raw materials and semi-finished products;
- on the other hand, the identification of a model of direct dialogue between different productive territorial contexts, which would respond to the innovation needs of the supply chain (cluster).

In terms of product innovation, apparel product lines have been developed and produced, highlighting the characteristics of the manual skills that extend from the processing of the fabric to the manufacture typical of high craftsmanship in the dialogue between different components of the actors' ecosystem.

So the design-driven innovation approach that has crossed the Research Adoption, has allowed to obtain concrete results in terms of product innovation and product-service. Figures 7, 8, 9, 10, 11, 12 and 13 show new samples of fabrics and developed by the company "Lanificio Leo" during a workshop with the students employed in the creation of collections by the luxury manufacturing companies Kiton and Attolini.

Figures 7, 8, 9, 10, 11, 12, 13.   Workshop at Lanificio Leo and test of new fabrics (Sbordone et al., 2018).

3 DISCUSSION

Today, it is clear that the hypertrophy of goods feeds an excessive "overdeveloped" consumption (Lipovetsky, 2013), whose one-dimensional effects induce the dynamics of diffused aesthetics of the daily that creates economic value from the aesthetic and emotional value.

The main characters are to be ascribed to phenomena concerning: the growing weight taken by the nature of the goods that relies on the symbolic-communicative and affective factors; the Design Process in the sense of new configurations of value that, through the design of tangible and intangible goods, moves from the logic of design-driven, the result of the connection of economic, socio-cultural and aesthetic know-how; the capillary aestheticization of the places of commerce, distribution, private life, work and entertainment; the constant "de-differentiation" (Lipovetsky, 2013) of the economic and aesthetic spheres, with the consequent hybridization of operating methods and practices.

The Fashion system, as one of the typical sectors that derives its value from the phenomena of the phenomena described above, needs to be increasingly configured as a true cognitive supply chain, based on heterodox ways of creating value.

Therefore, Fashion, proposing goods with a clear intangible and relational matrix, in advance of the other productive realities, requires a different application of the sustainability component. Having long abandoned the traditional approach at the base of the modern paradigm that bases external relations to companies exclusively on the exchange of material goods, in the Fashion only environmental sustainability can be a sterile approach if not accompanied by a full assimilation of processes of social and economic sustainability.

It is precisely by broadening its perspective of sustainability, incorporating complex symbolic values into productive activities, Fashion is at the center of the configuration of truly sustainable cognitive and productive networks.

4 CONCLUSION

In conclusion, the research and experimentation described up to now, even if still under way, suggests in these first steps a possible path towards the realization of the Sustainable Fashion paradigm. A content theory path that goes beyond the simple application or adoption of some eco-design reccomendations, proposing a different reconfiguration of Fashion sector, which profoundly affects economic, production and social processes, both locally and in terms of global level.

In this first phase of research, the construction of widespread production chains was experimented, bringing together small and medium-sized enterprises with a high cultural, technical and social content characterizing and exclusive.

The second phase will focus on the introduction of environmentally sustainable solutions, relative to the production of fabrics with low environmental impact fibers.

Finally, the third phase foresees the development of a set of criteria for the construction of an approach to the creation and development of new products that transfer not only tangible values but also and above all intangible values to the final consumers.

Notes

1. The authors of this paper have been carrying out experimental and didactic research on the three approaches described for some time. The paper reports the results of a work in progress of exchange of knowledge between the three authors.
2. Lanificio Leo is today one of the most significant cases of a company-museum and a brand with an international vocation whose identity values, transmitted also thanks to ethical communication, are focused on enhancing the local manufacture of wool weaving. http://www.lanificioleo.it
3. Cesare Attolini and Kiton are among the great brands of traditional Neapolitan men's tailoring. Cesare Attolini is one of the inventors of the Neapolitan style and especially of the jacket. The manufacturers Kiton and Attolini realize in their production centers, completely handmade items. It takes 30 to 33 hours to make a dress, from 130 to 300 tailors at work every day, each dedicated to the execution of a single step. http://www.kiton.it/ http://cesareattolini.com.

REFERENCES

Benkler, Y. 2006. *The Wealth of Networks: How Social Production Transforms Markets and Freedom.* Yale University Press, London.
Fletcher, K. & Grose, L. 2011. *Moda e Sustentabilidade, design para mudança*. Editora Senac. São Paulo.
Freeman, I. 2006. *The Development of Social Network Analysis*. Empirical Press, Vancouver.
Granovetter, M. 1973. The Strength of Weak Ties. *American Journal of Sociology.* V. 78, N°. 6, pp. 1360–1380.
Guido P. 1995. *Il marketing del valore.* Sperling & Kupfer, Milano.
Hildreth, P. & Kimble, C. 2004. *Knowledge Networks: Innovaton Through Communities of Practice.* Idea Group, London.
Horrocks, A.R. & Anand, S.C. 2000. *Handbook of Technical Textiles – 19*. The Textile Institute. Woodhead Publishing. Cambridge.
Kadolph, S.J., Langford, A.L. 2016. *Textiles.* 12th ed. Ed. Prentice Hall. New Jersey.
Lipovetsky, G. & Serroy, J. 2013. *L'esthétisation du monde: Vivre à l'âge du capitalisme artiste.* Gallimard, Paris.
Milan, G.S., Vittorazzi, C. & Reis, Z.C. 2010. *A Redução de Resíduos Têxteis e de Impactos Ambientais: Um Estudo Desenvolvido em uma Indústria de Confecções do Vestuário.* XIII Seminários em Administração. São Paulo, 17p. Available <http://sistema.semead.com.br/13 semead/resultado/trabalhosPDF/282.pdf.> [Accessed 15 February 2016].
Papanek, V. 2012. *Design for the Real World: Human Ecology and Social Change*, 2nd ed. Chicago Review Press, Chicago.

Rech, S.R. 2012. Conceito de Produto de Moda. *Actas de Diseño*, Buenos Aires, V. 6, Nº. 13, pp. 187–191.
Rullani, E. & Plechero, M. 2007. *Innovare. Reinventare il Made in Italy*. Egea, Milano.
Rullani, E. 2011. Reti di imprese, un nuovo percorso per crescere e competere. *Confindustria*, Vicenza.
Sanches, R.À. 2011. *Utilização de fibras sustentáveis na fabricação de vestuário*. São Paulo, Escola de Artes, Ciências e Humanidades, Universidade de São Paulo. Tese de Livre-docência, São Paulo.
UNIETHOS. *Sustentabilidade e Competitividade na Cadeia da Moda*. 2013. Available <http://www.abit.org.br/conteudo/links/estudo_sustentabilidade_uniethos.pdf.> [Accessed 10 June 2015].
Verganti, R. 2009. *Design-Driven Innovaton, Cambiare le regole della competizione innovando radicalmente il significato dei prodotti e dei servizi,* Rizzoli Etas, Milano.

Generation Y's sustainability attitude-behaviour gap

J.P. Bernardes, F. Ferreira & A.D. Marques
Universidade do Minho, 2C2T, Guimarães, Portugal

M. Nogueira
Universidade Europeia, IPAM Lab, Porto, Portugal

ABSTRACT: This paper intends to study the gap between the attitude Millennial consumers (or Generation Y) have towards sustainability and sustainable purchasing habits and what they actually do in terms of their green purchasing behaviour. In order to understand this gap, some internal and external influencing factors were obtained from the literature review which can positively or negatively influence the green purchasing behaviour of the Millennial consumers. They are: Awareness, Financial availability, Personal benefits, Subjective norm and Trust, and were analyzed in the form of a questionnaire that was applied to Millennials across Portugal. To further this research, some of the influencing factors were selected in order to obtain a better understanding of the attitude-behaviour gap regarding the purchase of sustainable footwear. With this paper it is possible to conclude that although Gen Y's attitude towards sustainability is very positive, they still don't have a green purchasing behaviour.

1 INTRODUCTION

These days, the survival of the human being is not so much a matter of how many we are, but how much space that each of us takes up on planet Earth, how does it become necessary to produce to satisfy the individual consumption, the amount that each one of us consumes energy, or, the waste produced by each person in a certain space of time. All the factors mentioned above form the ecological footprint (Kiperstok A., 2005).

According to Canciglieri JR. et al. (2011) it is necessary to use the natural resources of our planet with enough awareness, so that future generations are not affected by the current lack of sustainable planning.

The fashion industry is evolving at a fast pace manner and witnessing situations of contradictions: On one hand, consumers are becoming more conscious of the impact of their purchasing behaviour, and willing to support a more sustainable fashion industry but on the other hand are complying with unsustainable business models where production is achieved to the lowest price in the shortest time possible (Morgan L. & Birtwistle G., 2009).

Consumers are an important part of the fashion system and can create a valuable influence in the pursuit of sustainability in the fashion industry. There are several aspects to consider when analyzing fashion consumers and sustainability: The consumer knowledge about sustainability, consumer behaviour and consumption habits and feelings associated with sustainable consumption. In fact, the attitude and the behaviour of consumers are in line with the rapid production. Encouraged by low prices and heavily influenced by marketing campaigns and constant changes of trends, consumers tend to speed up their fashion consumption (Birtwistle G. & Moore C. M., 2006).

According to Zemke R. (2001), the new consumers, or the generation Y (Millennials) are defined as a group of people born more or less between 1980 and 2000 and their core values include confidence, loyalty to civic duty, sociability, morality, intelligence and diversity.

Millennials maintain a positive attitude in relation to sustainability in general. Young consumers are very aware of the opportunities that companies have to help the environment and reduce the ecological footprint. Also, this is a generation that thinks that it is the companies' duty to invest in a better environment and society by producing products in a sustainable way. However, there is a clear contradiction between how the Millennial generation thinks about sustainability and what they effectively do when it comes to a sustainable consumption (Schweitzer L. & Lyons S. T., 2010).

That way, for young consumers, green attitudes are not in any case predictions of behaviour (McDougle L. et al., 2011 and Paladino & Serena, 2012). Hume M. (2010) stresses that there is a clear pattern of contradiction between what Generation Y knows and thinks with what they actually do in regards of purchasing 'green'.

In sum, the main goal of this paper is to understand the gap between what Portuguese Millennials think about sustainability and what they actually do in their life to promote a more sustainable lifestyle.

2 STATE OF ART

2.1 *Sustainable design*

Based on the desire of preserving the environment without sacrificing economic growth and social development, emerges the concept of sustainable development. This concept has evolved to a sustainability perspective that brings together three vital aspects: environmental, economic and social. Nowadays design should incorporate the principles of sustainability in order to work with the natural world and not impoverish the remaining resources on the planet. A sustainable design has several names, such as "Eco-Design", "Green", "Eco-conscious", and its meaning depends on the context, the situation and the social, cultural, financial, and of course, environmental impacts (Brundtland G.H., 1991). A sustainable design aims to create value by finding the triple bottom line: economic, environmental and social benefits. (Charter M. & Tischner U., 2001). As stated by the Brundtland report, Our Common Future (WCED 1987, p. 41), sustainability is defined as the development that meets the needs of the present with yourself compromising the ability of future generations to meet their own needs.

When added to the design, Vezzoli C. & Manzini E. (2008) agree that the environmental requirements must be considered from the first phase of the design process, as well as cost, performance, legal, cultural and aesthetic requirements. Niinimäki K. (2011) stresses that the fashion industry has focused more on technical and financial aspects, and not so much on sustainable issues.

A successful and sustainable design follows and complies with strict criteria and considers in response to customers, users, participants, people, marketing, company, brand, channel, culture and environment. A sustainable design is created for and about customers as people and not just as consumers. So, the design has a healthy result for the ecosystem (Shedroff N. & Lovins L. H., 2009).

It is imperative that designers first understand the overall situation, before beginning the creative process. So, the sustainable design becomes a process that addresses a problem and a specific need, combining different fronts, not only through research, but also focusing on detailed issues. Designers have a responsibility to be involved in the entire process of creating product, through the adoption of life-cycle solutions, which reduces the environmental impact. Informed decisions about where, how, when and who will produce the product comply with the principles of sustainable design. The real challenge is not creating just another green and eco-friendly product, but to create somethings that adds real value for the consumer (Shedroff N. & Lovins L. H., 2009).

2.2 *Millennials as 'green' consumers*

A report from the University of Borås (2012) has shown that there is a gap between the concern for the environment of consumers and their daily actions. This is seen as an indication that consumers don't need more information about the negative environmental aspects of consumption, but argue that consumers need multiple measures to enable them to act more environmentally friendly.

Isenhour C. (2010) states that no matter how aware, reflective, concerned with sustainability and committed to do less environmental impact individuals are, but we're influenced by the societies in which we live. It also states that people are social creatures whose attitudes, behaviours and actions not only reflect their own personal values but who are also formed and highly influenced by a consumer culture complex.

Isenhour C. (2010) adds that corporate leadership should complement the responsibility of the consumer and public policies and programmes should be implemented to encourage the process and that mutual cooperation is the only way to carry out a significant change in the long run on the planet. Simply providing information to consumers about the nature of ethical purchases, or even appeal to their moral values to try to invoke a change of behaviour, probably will not create a anti-consumerism movement of unethical or irresponsible environmentally brands (Eckhardt M. et al., 2010).

3 RESEARCH METHODS

Since only studying the green attitudes will not produce conclusive results on the behaviour of young consumers to buy green products, it is interesting to know what factors influence the attitude-behaviour relationship. Several studies suggest that there is a difference of attitude of green purchasing behaviour due to the complex nature of personal and situational influences (Csutora M., 2012; Kolkailah S.K., et al., 2012).

Personal factors are internal influences. The attitude is a personal factor important in the purchase of sustainable products but doesn't explain why the young consumers don't buy this type of products. Other examples of personal factors are awareness, trust, priorities, emotion and control (to the extent that a consumer believes to have control over the events that affect) (Csutora M., 2012).

Situational influences are the external influences that a person cannot control, but which affect the relation attitude-behaviour. Examples of situational factors: time, opportunity, money or the ability to perform the desired behaviour (Ajzen I., 1991).

Csutora M. (2012) also states that to strengthen the relationship between sustainable attitudes and buying sustainable products, is of great importance that the influence factors serve as support. Even consumers with a sustainable negative attitude tend to buy green when the factors of influence are highly favorable, while consumers with a sustainable positive attitude may be discouraged from buying green products when the factors of influence are highly restrictive.

The following aspects present the possible factors influencing the attitude-behaviour relationship, identified in the literature review and will be explained it the results and discussion of this paper. These factors can positively or negatively influence the green purchasing behaviour of the Millennial consumers, and are: Awareness, Financial availability, Personal benefits, Subjective norm and Trust. That way, each of the influencing factors stated above will now be analyzed allowing a better understanding of the Generation Y's sustainability perceptions and consumption habits in Portugal. A questionnaire was built in order to obtain answers to these particular factors and was shared with Millennials from all across Portugal. In total, 635 answers were obtained. All statements regarding the analysis of the factors were constructed using a Likert scale (1 – Strongly disagree; 2 – Disagree; 3 – Neither agree nor disagree; 4 – Agree; 5 – Strongly agree). After a more general approach regarding Millennial's consumption habits, some of the influencing factors were selected in order to obtain a better understanding of the attitude-behaviour gap regarding the purchase of sustainable fashion, in this case, footwear.

4 RESULTS AND DISCUSSION

In this section, each of the influencing factors is going to be analyzed according to the results of the questionnaire that was applied to Portuguese Millennials. From the 635 responses,

67,7% were female and 32,3% male. Regarding their age, 48,2% are between 17 and 20, 31,2% between 21 and 25, 11,8% for the 26 to 30 years and 8,8% for the Millennials with ages between 31 and 37.

4.1 *Awareness*

Consumer awareness is defined as the amount of time that was spent in the processing of information about green and sustainable products (Baker W. et al., 1986). It is important that consumers are aware of the fact that buying green can be a criterion of purchase (Jones S. & Eden C., 1981).

Regarding the questionnaire, to the statement "I'm aware that I should consume green", 471 Millennials answered positively, (34,2% for "Agree" and 40% for "Strongly agree"), however when they were asked "For me it is important to consume green" their responses, although positive, weren't as positive as the previous statement, having 258 answers for the "Agree" (174) and "Strongly agree" (84). Out of the 635, 141 answers were negative, with 94 Millennials answering "Disagree" and 47 "Strongly disagree". Also, it is interesting to note that, when asked "I believe that by buying green, I can contribute positively to the environment", a smashing majority (75,5%) answered "Agree" (38%) and "Strongly agree" (37,5%).

4.2 *Financial availability*

The financial availability to purchase green products has a strong negative influence on the behaviour of buying. According to Kollmuss A. & Agyeman J. (2006) a necessary condition to buy products of this type is that the price and quality of these products must be comparable to the regular products that a consumer would normally buy. However, the sustainable products are often more expensive than the regular and consumers are unwilling to incur any additional cost (Kolkailah S.K. et al., 2012).

However, the results from the questionnaire are not totally in line with these conclusions. To the statement "I don't buy green because it is expensive", only 5,2% answered "Strongly agree" and 17,8% "Agree". This way, 47,6% (302) of Millennials seem to believe that the higher price for green products isn't a barrier for a possible purchase. But what is interesting in these results, is that, even though the majority feel that the higher price isn't a barrier, 199 Millennials are not willing to pay extra, and 174 are. So opinions are divided and the some more barriers are still to identify.

4.3 *Personal benefits*

Personal benefits, personal concern or personal interest, all refer to the efforts that a person has to do to satisfy their needs and desires. Consumers will assess whether the expected behaviour (for example, what to do) corresponds with their own personal concerns (Davis I., 2012). There is the tendency to decide in favor of their own interests, which is thus related to the fact that consumers often fail to estimate the actual impact of buying green in their lives (Rokka J. & Uusitalo J., 2008). A consumer will buy a green product, when in addition to the environmental benefit, they realize some benefit individual direct in acquiring the product (Nottage A., 2008).

To the statement "I find personal benefits in consuming green", "Strongly Agree" and "Agree" obtained 123 and 225 responses, respectively. This way, 54,8% of all the Millennials believe that consuming green is something that will add value to their life and that is beneficial. However, despite these results, the gap between attitude and behaviour is becoming more noticeable. To the statement "I buy green products", only 3,8% answered "Strongly agree" and 12,8% "Agree". On the negative spectrum of the answers, 17% answered "Strongly Disagree" and 27,4% "Disagree". Even though the majority of Millennials agree that there are personal benefits in consuming green, only a small part actually does buy and consume green.

4.4 Subjective norm

The subjective norm is the assessment of feelings in relation to the perception of what the people most important to them think (friends, family, influencers, teachers), when a decision to purchase green is being considered. The normative influences strongly the intention that an individual has to act when it comes to the purchase of green products (Ajzen I., 2012). Still, the social pressure from their friends and family becomes an explanatory variable to their purchase intentions, states Vermeir I. & Verbeke W. (2006).

In order to understand if other opinions have an important role in the green purchasing decision moment, Millennials answered to the following statement: "I value my friends and family's opinion regarding the products I buy and consume" and also to "If my friends and family buy green it is most likely that I do the same". To the first statement Millennials seem not to care about what their loved ones have to say about what they consume, having "Strongly Disagree" a total of 178 answers and "Disagree" 180 answers, both representing a majority with 56,3% combined. The answers to the second statement shows a balance between the negative answers ("Strongly Disagree" with 16,4% and "Disagree" with 22,4%) and the positive ones ("Strongly Agree" with 7,7% and "Agree" with 23,8%). This shows that the majority of Millennials don't care about what other think about their consumption habits, however they can be influenced in to consuming green if their major influencers do it.

4.5 Trust

The trust factor or skepticism is concerning the motivation of an organization to engage in green products and is often mentioned in the literature as a being a factor of negative influence on attitude-behaviour relationship. Research suggests that the perceptions that consumers have regarding the companies' green attitude results in a conclusion on the company's efforts (Drumwright M.E., 1996).

Regarding the questionnaire, in order to understand if Millennials trust companies' efforts in being green they were asked to answer, "I trust companies that claim to be green". Only 9,4% answered "Strongly Agree" and 26,8% "Agree". "Neither Agree nor Disagree" had an impressive 40% and the remaining answers were negative (6,1% to "Strongly Disagree" and 17,5% to "Disagree"). Low levels of confidence concerning the motivation of an organization to engage in green products would impact particularly Generation Y, since this generation is known for being more skeptical regarding commercial messages than the generations previous (Schmeltz L., 2012).

But how do Millennials think about buying 'green fashion'? Some of the influencing factors "Awareness", "Financial Availability" and "Trust" were now selected in order to obtain a better understanding of the attitude-behaviour gap regarding the purchase of footwear:

Regarding the "Awareness" factor, Millennials were asked "Are you aware of any sustainable footwear brand?", 93,1% said "No", which means that out of 635 only 44 knew about a sustainable footwear brand. Also, to the statement "I search about the footwear materials before buying them" only 23,4% answered positively ("Agree" and "Strongly Agree").

The "Financial availability" was also applied to understand the purchasing behaviour of sustainable footwear, and therefore Millennials had to answer to que statement "I'm willing to pay extra for sustainable footwear". As a result, "Neither Agree nor Disagree" was the answer with the higher percentage, with 40,3%. "Strongly Agree" was chosen by 8,3% and "Strongly Disagree" by 6,6%.

In order to understand the "Trust" Millennials had in sustainable footwear brands, they had to answer the following statement: "I believe that sustainable footwear has higher quality". Only 42 answered "Strongly Agree", and 159 "Agree". In the negative spectrum of the answers, 22 answered "Strongly Disagree" and 40 "Disagree". Both positive and negatives answers were not chosen by the majority of Millennials as "Neither Agree nor Disagree" was selected by a smashing 342 people (53,9%).

5 CONCLUSIONS

After selecting and analysing the influencing factors that can have a positive or negatively impact the gap between Millennials' attitude and behaviour concerning sustainability and sustainable fashion, some interesting conclusions are now possible to obtain.

Millennials are a generation that are fully aware of their role in society and therefore know that they should consume green and that by doing so they are contributing positively to the environment and improving their lives as well. This generation believes that being green adds value to their life and that is extremely beneficial. In what comes to fashion, in particular, footwear, only a very small percentage knew about a sustainable footwear brand, and only a quarter of the sample claimed to search on the internet about footwear materials before they buy.

When the price aspect is considered their beliefs seem to not be as important. It is a known fact that sustainable products have a higher price and Portuguese Millennials don't seem to see be willing to incur in extra costs to consume green, including buying sustainable footwear.

Also, their decisions about what to consume can be influenced by their close ones (family, friends) but in the end they will only choose to consume green if they really see a benefit for themselves.

Trust wise, it is also possible to note that low levels of trust negatively influence the attitude-behaviour in purchasing green. The majority of Millennials don't seem to trust companies that claim to be sustainable and therefore their decisions about what to consume don't seem to take the companies statements in consideration. They need to see to believe. Also, this generation still thinks that sustainable footwear doesn't have the same quality as 'regular' ones. This may be due to their lack of research and interest in these types of products.

In sum, it is possible to conclude that even though Portuguese Millennials have a very positive attitude towards sustainability and green products, this attitude is not reflected in actual behaviour. They are aware that they should consume green, but the price is still a very big barrier and it overweighs their personal benefits in living a more sustainable lifestyle. Also, Millennials don't spend too much time researching sustainable alternatives and have very little knowledge in the matter.

The main limitation of this study is that the results from the analyzed sample are 100% Portuguese, therefore it is not possible to extrapolate these results to other countries. Even though it is the same generation worldwide, each culture has its own impacts on the Millennials education, and therefore, consumption habits.

As future research it would be very interesting to add 'Marketing Efforts' as an influencing factor by analyzing the marketing strategies done by sustainable footwear brands and how they engage with the Millennial generation. Also, it would be interesting to study different ways to increase consumer awareness in what comes to buying green, having the 'Marketing Efforts' factor as the starting point.

ACKNOWLEDGMENTS

This work is financed by FEDER funds through the Competitivity Factors Operational Programme – COMPETE and by national funds through FCT – Foundation for Science and Technology within the scope of the project POCI-01-0145-FEDER-007136.

REFERENCES

Ajzen I. 1991. The theory of planned behavior. *Organizational Behavior and Planned Decision Processes*, 50, p 179–211.

Ajzen, I. 2012. The theory of planned behavior. In P. A. M. Lange, A. W. Kruglanski and E. T. Higgins (Eds.), *Handbook of theories of social psychology*, 1, p. 438–459. London, UK: Sage.

Baker, W., Hutchinson, J. W., Moore, D., and Nedungadi, P. 1986. Brand familiarity and advertising: Effects on the evoked set and brand preference. *Advances in Consumer Research*, 13(1), p. 637–642.

Birtwistle, G. and Moore, C. M. 2006. Fashion adoption in the uk: a replication study. Paper presented at the Anzmac Conference Brisbane CA.

Brundtland, G. H. 1991. Nosso Futuro Comum. Rio de Janeiro, Editora de Fundação Getúlio Vargas.

Canciglieri, JR., O.; Cardoso, R.; Pereira, S. 2011. Uma Visão Tecnológica sobre o Desenvolvimento de Produtos e a Sustentabilidade. Congresso Brasileiro de Gestão de Desenvolvimento de Produto, Porto Alegre.

Charter, M. and Tischner, U. 2001. Sustainable solutions: developing products and services for the future, Sheffield, U.K., Greenleaf Pub.

Csutora, M. 2012. One more awareness gap? The behaviour–impact gap problem. *Journal of consumer policy*, p. 1–19.

Davis, I. 2012. How (not) to market socially responsible products: A critical research evaluation. *Journal of Marketing Communications*, p. 1–15. doi: 10.1080/13527266.2012.696076.

Drumwright, M. E. 1996. Company advertising with a social dimension: The role of noneconomic criteria. *The Journal of Marketing*, p. 71–87.

Eckhardt, M., Russell, B. and Timothy, D. 2010. Why Don't Consumers Consume Ethically?. *Journal of Consumer Behavior*, 9(6), p. 426–436.

Hume, M. 2010. Compassion without action: Examining the young consumers consumption and attitude to sustainable consumption. *Journal of World Business*, 45, p. 385–394.

Isenhour, C. 2010. On Conflicted Swedish Consumers, the Effort to Stop Shopping and Neoliberal Environmental Governance. *Journal of Consumer Behaviour*, 9(6), p. 454–469.

Jones, S., and Eden, C. 1981. OR in the community. *Journal of the Operational Research Society*, p. 335–345.

Kiperstok, A. 2005. Sustentabilidade ambiental: produção e consumo; I congresso internacional de cooperação universidade-indústria.

Kolkailah, S.K., Aish, E.A., and Bassiouny, N. 2012. The impact of corporate social responsibility initiatives on consumers' behavioral intentions. *International Journal of Consumer Studies*, 2, p. 19–35.

Kollmuss, A., and Agyeman, J. 2002. Mind the gap: why do people act environmentally and what are the barriers to pro-environmental behavior?. *Environmental education research*, 8(3), p. 239–260.

McDougle, L., Greenspan, I., and Handy, F. 2011. Generation green: Understanding the motivations and mechanisms influencing young adults' environmental volunteering. *International Journal of Nonprofit and Voluntary Sector Marketing*, 16, p. 325–341.

Morgan, L. and Birtwistle G. 2009. An investigation of young fashion consumers' disposal habits. International. *Journal of Consumer Studies*, 33, p. 190–198.

Niinimäki, K. 2011. From Disposable to Sustainable. The Complex Interplay between Design and Consumption of Textiles and Clothing. Doctoral Degree Dissertation.

Nottage, A. 2008. The green agenda gets personal. Marketing, July, 33–5.

Paladino, A., and Serena, N. 2012. An examination of the influences on 'green' mobile phone purchases among young business students: An empirical analysis. *Environmental Education Research,* doi:10.10 80/13504622.2012.687044.

Rokka, J., and Uusitalo, J. 2008. Preference for green packaging in consumer product choices—do consumers care?. *International Journal of Consumer Studies*, 32, p 516–25.

Schmeltz, L. 2012. Consumer-oriented CSR communication: Focusing on ability or morality? Corporate Communications: *An International Journal*, 17(1), 29–49.

Schweitzer, L. and Lyons, S. T. 2010. New generation, great expectations: A field study of the millennial generation. *Journal of business and psychology*, 25 (2), p. 281–292.

Shedroff, N. and Lovins, L. H. 2009. Design is the problem: the future of design must be sustainable, Brooklyn, N.Y., Rosenfeld Media.

University of Borås. 2012. Mot en mer Hållbar Konsumtion. En Studie om Konsumenters Anskaffning och Avyttring av Kläder. Report nr. 20.

Vermeir, I., and Verbeke, W. 2006. Sustainable food consumption: exploring the consumer "attitude–behavioral intention" gap. *Journal of Agricultural and Environmental Ethics*, 19(2), p. 169–194.

Vezzoli, C. and ManzinI, E. 2008. Design for environmental sustainability, Berlin; London, Springer.

Zemke R. 2001. Here come the millennials. *Training Magazine*, 38 (7), p. 44–49.

Collaborative economy: Case study of new business models

S.R. Fernandes, J.M. Lucas & M.J. Madeira
University of Beira Interior, Covilhã, Beira Baixa, Portugal

A.I.C. Barreiros
Polytechnic Institute of Castelo Branco, Castelo Branco, Beira Baixa, Portugal

I.D. Honório
State University of Londrina, Londrina, Paraná, Brazil

ABSTRACT: Sustainable development is directly related to the economic model of production, distribution and consumption. Collaborative consumption has gained worldwide strength through new business technology models, offering shared use products and services. The purpose of this paper is to explore a multiple case study Collaborative Consumption in the fashion industry.

It is based on a bibliographical survey of several business models and classification in the concept of collaborative consumption. It is concluded that business models of collaborative consumption in the area of fashion are growing and there are several alternatives to be explored, both for the user and for the investor.

Collaborative consumption can be considered more sustainable than in0064ividual consumption, since it increases the use of fashion products, not requiring the need to produce new. The article, which deals with collaborative economics, with a focus on collaborative consumption, is a scientific contribution to the recent economic phenomenon.

1 INTRODUCTION

The linear industrial economy, based on the production and sale of the new products, generated the depletion of natural resources. The fashion products, in this traditional process, is among the most consumed: people buy more than they use and often discard products with the label.

> *Large amounts of non-renewable resources are extracted to produce clothes that are often used for only a short time, after which the materials are mostly sent to landfill or incinerated. More than USD 500 billion of value is lost every year due to clothing underutilization and the lack of recycling. Furthermore, this take-make dispose model has numerous negative environmental and societal impacts* (EMF, 2017)

This has led to new concepts and procedures regarding the production, commercialization and consumption of goods, and among them the collaborative economy. The collaborative economy, through new shared business models and the redistribution of fashion products, can increase the period of their use. It intercepts the early discard of the fashion product and reduces the pressure on the use of natural resources, reducing negative impacts throughout the production chain.

From the beginning of the century the collaborative economy has been a more present object of study in the scientific productions and it is perceived an advance in both its understanding and its application. In 2001, Rifkin published the book "The Age of Access", in which he presented the future transitions of economic activities based on access to goods and services through shared use. The term collaborative consumption was consecrated in 2010 with Rachel Botsman and Roo Rogers (2011) in the book "What's mine is yours". In the same year, Lisa Gansky created the term Mesh, which refers to interconnectivity through digital technology, used to give to people access to goods and services (Gansky, 2010). OuiShare Academy is a global project with a very ambitious goal: to make the economy more collaborative, with impacts and potentials understandable and accessible to all:

> *OuiShare is a global community empowering citizens, public institutions and companies to build a society based on openness, collaboration and sharing, (…) OuiShare Fest is an international event that gathers entrepreneurs and social innovators, non-profit and business leaders, grassroots activists and public officials.* (Ouishare, 2018)

Already in 2011, Time Magazine pointed to Collaborative Economy as one of the 10 ideas that will change the world (Walsh, 2011).

> *These kinds of phenomena are going to continue to spread in the years ahead, and as hundreds of millions of people shift large part of their economic activity to the sharing economy, they will change the course of economic history.* (Rifkin, 2014)

In 2016 the European Commission launched "The future of the EU collaborative economy —Using scenarios to explore future implications for employment". It recognizes the potential of the collaborative economy in creating new sources of income and benefits for consumers through new models of businesses for the temporary use of goods and services, such as lodging, car sharing, bicycles, televisions, lawnmowers, drills, among others (Bock et al., 2016).

For Manzini and Vezzoli (2008), shared uses are ways of optimizing the useful life of the products because they reduce the need for new products, resulting in less environmental impact. Tukker (2004) complements, in saying that this reduces the amount of products discarded, being able to reduce in 50% the environmental impact, when compared with the individual use.

The collaborative economy opens the way for new, more sustainable business models. For the Europe Network Circular (CED, p. 9), Collaborative Economy "refers to optimizing the use (or even production) of a product or service through the use of sharing. Among other things, collaborative economics questions the effectiveness of needs and the indispensability of satisfying them through the use of material resources". It is based on the hierarchy of multi-R's: Re-thinking, Re-drawing, Re-reduce, Reuse, Redistribute, Repair, with a view to sharing and preserving the social, economic and environmental dimensions.

Collaborative economics is comprehensive and adaptive in several areas. It includes: Collaborative Consumption; Collaborative Production; Open Knowledge, Collaborative Financing, Open Government and Horizontal Organizations, Exchange of Value Systems – (Stokes et al., 2014).

This article aims to present new business models that fit the collaborative consumption. (Botsman and Rogers, 2011) divide collaborative consumption into three systems: Product Service Systems (SSP), Redistribution Markets, and Collaborative Lifestyles.

2 COLLABORATIVE CONSUMPTION, PRODUCT SERVICES SYSTEMS (SSP)

The Products and Services System (PSS) allows sharing the products of a company. Users pay for the use of the product without needing to have it.

Businesses with the shared wardrobe concept Fashion Library are increasingly earning followers in several countries. In English is called Fashion Library, in Spain uses the name Ropateca, in Brazil it is named *Modateca, Roupateca*. These business models are expanding to the whole of Brazil. In Portugal, to date, there is no record of a business model with these characteristics.

These new business models work through a signature system similar to Netflix and allow customers to pay a monthly service fee. There are companies that offer various plans so that one can access a fixed number of garments on loan basis at any time. These models are an attractive offer for users who want frequent changes of clothes.

2.1 Brief operating history

In Germany, in Hamburg, Kleiderei members acquire a monthly ticket of 14 euros. They are entitled to one coordinate per month. The clothing should then be returned clean after seven days. The store's motto is: "lend it to a friend". Members can try out new styles and wear quality clothes with low prices (Kleiderei, 2018).

In Amsterdam, the Netherlands, one of the most well-known, Fashion Library, is the Lena store, in which members make a signature worth € 19.95 a month. Moreover, they have at their disposal high quality clothes, vintage, famous brands and ecological brands. The slogan is "Collect moments, not things." The clothes should be returned clean. The store receives donations of clothes in good condition (Morgan, 2015).

In Copenhagen, the Resecond fashion library opened in 2012, has about 200 members who can change their own dresses, with a stock of around 300 pieces. The members pay a fee, for six months. When they become members they bring clothes to contribute to the stock of the library.

In Helsinki, Finland, the fashion library was founded in 2011; account for 750 pieces most of them from Finnish fashion designers and has about 100 members. The fashion library of Helsinki offers three types of associations, ranging from 160 to 460 euros, for a period of six months (Pedersen, et al., 2015).

In Barcelona, Spain, Ropateca, with the same concept, has 400 pieces of different clothes and accessories. It has vintage and contemporary styles, for party-wear and casual wear for all seasons. Customers are associated and pay a value of 15 euros. They can carry three pieces of clothing per month. In this case, they must also deliver clothes clean. However, they have the option of paying 20 euros and do not need to return them clean. Another alternative is to pay 5 euros and rent one piece per week (Ropateca, 2018).

In Italy, Milan and Brescia, the company EGO, Ahoroupa Ecológico Organizzato, also follows the concept of shared fashion, (Vezzoli, 2010). The EGO offers 365 different outfits per year with eight styles, ranging from classic to casual, romantic, diva, dandy, sporty, ethnic and technical. There are two collections per year: one fall/winter and one spring/summer, each one with 120 pieces. Registered customers go to EGO once a week and are entitled to a look for each day of the week. The monthly amount is 86 euros per month plus 89 euros annual subscription. Customers do not need to wash or even ironing to return them (Salvina, 2009).

In New York, United States, Gwynnie Bee offers a signature clothing service for the day-to-day in the 'plus size' female segment. It offers customers more than 4,000 styles and various subscription plans. Members can keep items for as long as they want, then just send them back to get other items of interest. The brand was founded in 2011 and decided to invest in this niche market due to the fact that 75% of the US adult female population have size 44—or above—and that 67% of the total population wears size 48 or above. The hygiene of the clothes is made by the brand itself (Gwynnie Bee, 2018).

Rent the Runway Unlimited Company charges a monthly membership fee. The membership renews automatically and is charged on the same date every month, and the member can cancel at his own discretion. Members have more than 350 high-end designer items for $139 a month and they can rent three designer dresses, blouses, skirts or accessories at a time, and keep them for as long as they like. RTR College Rep Program is a specific program for students. The hygiene of the clothes is made by the brand itself (Rent the runway, 2018).

YCloset, founded in 2015, is a Chinese clothing-sharing platform where users pay a monthly subscription fee to rent clothing and accessories. YCloset charges a monthly membership fee of 499 yuan, which allows users to rent unlimited clothing and accessories through their mobile application. According to YCloset CEO Liu Mengyuan, the clothing-sharing business will become commonplace on e-commerce platforms in the future due to the low value of membership and the huge variety of styles offered by the brand. YCloset also collaborated with the luxury brand of Kenzo and Acne Studios. The Laundromat service chain also is a strategic partner.

3 COLLABORATIVE LIFESTYLES

Collaborative lifestyles refer to business models with shared systems of leisure, work and learning spaces. People with common interests seek to divide and exchange tangible and intangible assets.

In, São Paulo, Brazil, the House of All is an example of a modular business model with a wide range of products and services within the collaborative lifestyle. Modes: Bubbles Lab—House of Bubbles is a space for sharing fashion, design, innovation and technology. It works as an experimental system, prior to purchase. There is also the House of Work, which is an office equipped with the following modalities: House of Learning and House food, for the chefs, where they rent the facilities per day; Co tattooing Space is a framework for the tattoo artist to create and carry out their work. House of Gaming is a gaming station. House of All members can also enjoy Hot Tub and Bar (House of All, 2018).

The House of Bubbles follows the same concept of clothing library, which they preferred to call *Roupateca*. The space has more than 100 clothes and accessories. There are Brazilian brands such as Animale, Cris Barros, Le Lis Blanc, Osklen and Reinaldo Lourenço, and international brands such as Cristian Dior, Comptoir des Cotonniers and Zadig & Voltaire. It works by subscription, in the amount of 100, 200 or 300 reais (approximately, 23, 46 and 67 euros); customers can withdraw from one to six pieces at a time and these must be returned cleaned within 10 days. The space also has a self-service laundry (Braun, 2017).

In 2017 the House of Bubbles started a system of parts concession and remuneration. When a lease part is rented, the owner receives a value. The values are set according to the part value. The clothes can be picked up in São Paulo and delivered to House of Bubbles in Rio de Janeiro. The Brand also intends to operate with an online booking system and members can receive the clothes at home for little additional value. Mini-Bubbles LAB is a collection for babies and children with self-service laundry. Business is expanding. There are already Houses in Rio de Janeiro, Belo Horizonte and the owner of this brand intends to open new units in Curitiba and Ribeirão Preto and, in the future, in France, Germany, United States and South Korea.

4 REDISTRIBUTION MARKETS, P2P PLATFORM

In the redistribution markets, second-hand goods of unused private property are redistributed through online platforms, social networks or physical stores. In this case, there is possession of the goods that can be permanent or temporary. The transaction can be financial or direct exchange of product. According to EMF (2017) the resale market for second-hand products online has been growing more than four times faster than the traditional second-hand store market (35% a year versus 8% a year).

4.1 *Child segment*

Mybabystock, located in Mahón, Balearic Islands, Spain is an online child care platform for families concerned with financial and environmental sustainability. It is based on the model of reuse and redistribution of clothes and accessories for children, as well as rent of articles such as baby strollers, cribs, safety seats, appliances (sterilizers, bottle warmers, kitchen robots). Among the brands offered are Baby Comfort, Bebedue, Buggypod, Casual Play, Chicco, Firstwheels, Fuli & C, Jané, Micralite, OK, Baby and Quinny. Mybabystock connects the source and demand, guaranteeing the quality of the products. In this case study the child segment was highlighted because it is one of the sectors with the highest activity to be explored within the collaborative consumption. The market opportunity can begin with redistribution through selling the rental of maternity clothing and maternity kit unfolding in all phases of the child's growth (Mybabystock, 2018).

5 FINAL CONSIDERATIONS

The innovation of new sustainable business models presents many possibilities for exploration. The redistribution market has great potential because it has low risk and presents good rewards for investors. It intensifies the use of fashion products without the need to produce new products.

> *The starting point for a sustainability innovation is to try to decouple the success of a business from the relentless expansion of material consumption by seeking to minimize resource depletion, pollution and associated effects such as climate change* (Fletcher & Grose 2011, p. 100).

There are many success models of redistribution. EBay, Amazon, Gumtree and Craigslist are all examples, especially for P2P transactions that connects consumer to consumer, being able to sell or buy, rent or exchange. Collaborative consumption on P2P platforms works with reputation mechanisms that can build trust among users. To sell a product it may be charged a membership fee to the platform; if the product is sold, a fee will be charged on the final value. These platforms that have no physical inventory and in which users own their clothing have great potential for scalability with exponential growth. They can be implemented anywhere in the world, at reduced cost for deployment. The great investment is made in technology, insurance and partnership with companies providing services such as washing, arrangements, delivery and others.

OLX and Custo Justo are P2P platforms that act as classified ads online. These connect seller and user, without charge. They do not offer evaluation system and are not responsible for fraud. The PSS model per signature system, although the client has the possibility to cancel whenever he wants, generates user loyalty and develops a lasting relationship based on loyalty.

Companies can also exploit innovation through eco-efficient systems. Rent the Runway Unlimited is an example of using non-toxic product hygiene, delivery and return reusable packaging, and protective clothing plastic cover. All are returned for recycling.

There are many possibilities to be explored in the collaborative lifestyle with hotels in tourist locations, such as ski resorts and national parks, among others, that can offer sports clothing and other equipment.

Regardless of the business model, collaborative consumption can represent a significant change in behaviors to increase the usage time and lifespan of the fashion product. It is worth mentioning that the business models presented in this article all value the quality of the products.

ACKNOWLEDGEMENTS

The authors thanks to Santander-Totta for the IDB/ICI-FE/Santander—Totta-UBI/2017 doctorate grant, to FibEnTech R&D Unit, Covilhã, Portugal, and to Design, Sustainability and Innovation Research Group (Desin), UEL Design Department.

REFERENCES

Bock, A. & Bontoux, L. & Figueiredo, S. & Szczepanikova, A. 2016. *The Future of the Eu Collaborative Economy: Using Scenarios to Explore Future Implications for Employment*. EUR 28051 EN: Publications Office of the European Union.

Botsman, R. & Rogers, R. 2011. *O que é meu é seu: como o consumo colaborativo vai mudar o nosso mundo*. Porto Alegre: Bookman.

Braun, S. 2017. *"Guarda-roupa" compartilhado será inaugurado em Pinheiros*. Available at https://vejasp.abril.com.br/blog/liquidacao-cia/8220-guarda-roupa-8221-compartilhado-sera-inaugurado-em-pinheiros/ [Accessed 11 January 2018].

CEN, Circular Europe Network (SD). 2015. *Orientação gerais para a implementação de Estratégias Integradas de Economia Circular Nível Regional.* Brussels: ACR.

EMF, Ellen MacArthur Foundation. 2017. *A New Textile Economy: Redesigning Fashion's Future.* http://www.ellenmacarthurfoundation.org/publications.

Fletcher, K. & Grose, L. 2011. *Moda & Sustentabilidade: Design para Mudança.* São Paulo: SENAC.

Gansky, L. 2011. Mesh: *Por que o Futuro dos Negócios é Compartilhar.* Rio de Janeiro: Atla Books.

Gwynnie Bee. *How it works.* Available at https://closet.gwynniebee.com/pages/how-it-works [Accessed 18 January 2018].

House of All. Available at www.houseofall.co [Accessed 11 January 2018].

Kleiderei: Hamburg's Lending Library for Clothes. Available at http://www.young-germany.de/topic/play/art-fashion/the-kleiderei-hamburgs-lending-library-for-clothes. [Accessed 15 January 2018].

Manzini, E. & Vezzoli, C. 2008. *O desenvolvimento de produtos sustentáveis: os requisitos ambientais dos produtos industriais.* São Paulo: Ed USP. (original: Lo sviluppo di prodotti sostenibili. Rimini: Maggioli Editore, 1998).

Morgan, H. 2015. *This Fashion Library Lets You Check Out Clothing—Literally. Ecouterre.* Available at http://www.ecouterre.com/amsterdams-lena-fashion-library-lets-you-check-out-clothing-literally/lena-fashion-library. [Accessed 15 January 2018].

Mybabystock. Available at http://mybabystock.com/content/11-alquiler-de-material-y-carritos-de-bebe [Accessed 21 January 2018].

Ouishare. Ouishare Academy. Available at http://academy.ouishare.net/ [Accessed 20 January 2018].

Pedersen, E. & Netter, S. 2015. Collaborative consumption: business model opportunities and barriers for fashion libraries. *Journal of Fashion Marketing and Management,* 19 (3), 258–273.

Rent the runway. Available at https://www.renttherunway.com/pages/about / [Accessed 23 January 2018].

Rifkin, J. 2016 *Sociedade com custo marginal zero.* São Paulo, M.Books do Brasil Editora Ltda.

Ropateca. Available at http://smoda.elpais.com/moda/bibliotecas-de-ropa/ [Accessed January 2018].

Ropateca. Available at http://www.lavanguardia.com/local/barcelona/20150518/54431247422/bibliotecas-ropa-llegan-barcelona.html [Accessed 15 January 2018].

Salvina. 2009. EGO, *Ecologico Guardaroba Organizzato.* Available at http://www.guadagnorisparmiando.com/curiosita/ego-ecologico-guardaroba-organizzato. [Accessed 10 January 2018].

Stokes, K. & Clarence E. & Anderson L. & Rinne A. 2014. *Making Sense of the UK Collaborative Economy. Collaborative Lab.* England: Nesta.

Tukker, A. 2004. Eight types of product-service system: eight ways to sustainability? Experiences from Suspronet. *Business Strategy and the Environment.* v.13, p. 246–260.

Vezzoli, C. 2010. Design de Sistema para Sustentabilidade: *teoria, métodos e ferramentas para o design sustentável de "sistema de satisfação".* Salvador: Editora EDUFBA.

Walsh, B. 2011. *Ideas that will change the world: today's smart choice: don't own Share.* Time. http://content.time.com/time/specials/packages/article/0,28804,2059521_2059717_2059710,00.html [Accessed 20 November 2017].

Ycloset. Available at http://www.scmp.com/business/china-business/article/2110128/china-clothes-sharing-start-ycloset-secures-us50 m-new-funds [Accessed 15 January 2018].

Imbrications and distances between the creative economy and the sustainability of small fashion brands

U.S.T. Barbosa, H.A. Dieb, G.M.J. Sales, L.U. Dantas & A.N. Targino
Centro Universitário de João Pessoa, João Pessoa, Paraíba, Brazil

ABSTRACT: This paper addresses sustainability and creative economy as part of the fashion system, and comments possible relations between both. We use a recent bibliographic of a decade, with authors and researchers linked to the contemporary fashion and its imbrications with initiatives related to the cultural and creative industry and sustainable practices. It is also part of a recent methodological resource, named netnography, to research fashion brands in the Northeast of Brazil to understand how these are inserted in their businesses, considered as new management models based on such concepts. The choice of brands was made based on the authors' statements, such as the use of creativity as a factor of innovation in the market, valorization of local production, and sustainable factors that are associated with the principles of the Creative Economy and Sustainability, observed in texts and images of the brands, transmitted in the sites and their social networks.

1 INTRODUCTION

Fashion lives a paradigm shift. Experts, such as trend researcher Lidewij Edelkoort, say fashion as we know it is in its last days. In her Anti Fashion Manifesto (available in http://www.edelkoort.com/2015/09/anti_fashion-manifesto/), the researcher says that the speed of production caused by the Fast Fashion system, the economic demands of luxury brands that quickly change their creators that do not drive sales; the obsolescence of fashion products, are killing creativity and sacrificing fashion education that is increasingly focused on the market and less on authoring. In this way, it is necessary to rethink how to produce, consume and discard the products generated by this industry.

To do so, thinking about the existing approaches between Creative Economy and Sustainability is the order of the day to promote the changes expected so that fashion returns to its prominent place as a creative and innovative area. Indeed, fashion is embedded in what is called the creative industry, since it is at the confluence between being an industry with a strong consumer appeal and requiring professionals to create products with aesthetic, creative and authorial appeal. An easy task in the world where speed requires cutting creative processes to shorten delivery time to the market. Cietta (2017b) recalls that fashion, whether in the creative stage or in the industrial stage, must think about time, not just the chronological, but the time of maturity, for its acceptance or failure. This is crucial to the moment when fashion lives. How much of the old speech will keep companies/brands that want to stay in their old business models, focused on linearity of production and only on tangible goods for consumption? And how many will be able to adapt and create committed and purposeful discourses? Are small brands more easily making that transition or creating their speeches with transparency?

Based on these questions, this article proposes to give a brief presentation of the concepts of Creative Economy and Sustainability and how some local brands in the cities of João Pessoa (PB) and Recife (PE) in Brazil are inserting in their businesses new models from such concepts. It also tries to verify the greater difficulties perceived by being small brands still in the process of adaptation to a new market that still discusses these concepts, that even reached a definitive consensus. And finally, it comes to the considerations with the perceptions of the possible relations between the two areas and of the challenges faced and will face. For the analysis of the data of this research, we used the method of systematic bibliographic review, following the guidelines pointed out by Lorgus et al. (2001) and analysis of brand discourses on websites and social networks, called netnography (Kozinets, 2014). For the choice of brands, we used criteria that are minimally adapted to the concepts of Creative Economy and Sustainability.

2 SUSTAINABILITY

The goal of better environmental quality is a global theme. The concern and the attempt to alleviate, or even remedy, the damage caused by a devastating production, has been studied for some time, but there is still much to be discussed. In 1972, the United Nations Conference on the Human Environment took place in Stockholm, Sweden, organized by the United Nations Organization, where, according to Denardin (2012), the major problems arising from air pollution generated by industries were discussed. Since then, a number of actions have been implemented to save the planet's resources. In the 1980s, the term sustainability has become directly linked to the way we understand the world, "leaving aside a vision in which the economy, society and the environment are interdependent systems [...] and adopting a vision of integrated systems" (Salcedo, 2014, p. 16). According to De Carli (2012), the principle of sustainability aims to achieve three pillars: the social, directly linked to fair work; the environment, which seeks to find a balance of ecosystems; and finally, the economic one, which is concerned with the efficiency of the productive processes, taking into account the tendencies of greater generation of benefits, with the low use of resources.

According to Martins et al. (2012) the fashion industry, presents a series of environmental impacts, starting from the great consumption of raw material, the use of energy generated, to the huge disposal of textile waste. Besides this fact, the rampant consumption of fashionable products and, consequently, their discards, considerably increase the impact on the environment. The reduction of misuse and wastage resulting from the aforementioned processes, together with the reuse of waste and materials, as well as other specific resources and methods are a possible way forward, in an attempt to produce a less aggressive and more sustainable fashion.

The current fashion market model, called Fast Fashion is in its fastest possible form, and what is new today, becomes old and obsolete tomorrow in a very quick manner. Fast Fashion is linked to the production model in which the industry produces on a large scale, usually with inferior raw materials, with cheap labor and, in turn, the ever lower price, which causes a high production and therefore an extremely exacerbated consumption. Another aggravating factor of Fast Fashion is the exploitation of the labor involved in the production process of these industries. According to Salcedo (2014), the working conditions found in these industries are terrible and depressing, in which the main problems in these working environments are: miserable wages, endless working hours, unsanitary conditions, insecurity and repression concerning union decisions, and difficulties for collective bargaining. However, in parallel to the Fast Fashion market, the fashion scene has been building paths in an attempt to produce a more ethical, less impactful and less aggressive production, which also values the valorization of the work and worker involved. Slow Fashion arose not in opposition or to fight against the Fast Fashion model, but rather as another approach to design and to the fashion market.

Berlim (2012) points out that Slow Fashion proposals may have been born from the Slow Food movement created by Carlo Petrini in Italy in 1986. Like Slow Food, Slow Fashion and Slow Design favor the opposition of standardization and speed of production, seeking creation, invention and innovation with pleasure to consume. In addition, in the case of Slow Fashion, its consumers value information about the origin of products and are concerned about impacts generated. They seek the pleasure in the consciousness of being promoting an ethical productive chain. In this model system, fashion designers, creators, producers, suppliers, dressmakers, artisans and consumers are more aware and concerned about the impact that the fashion industry has on the environment, thus narrowing the relationship among these professionals, establishing a relation between the parties involved. Once created, these bonds are important values in an interrelational chain, which will contribute to the loyalty between the product and the consumer and, consequently, the consolidation of the brand in the fashion market. And in this new model, capitalist aspirations are no longer the main objective, and there are other market relations,

> [...] just as we need to take care of nature (because we are connected to it) so that we continue to have the resources to produce and survive, we need to take care of the community around us, which is the network that produces and consumes what we do (Carvalhal, 2016, p. 169).

These new production ideas, focused on promoting social and environmental well-being, may be related to what is known as the Creative Economy, which brings together all these initiatives together with the appreciation of knowledge and culture, to promote new business models.

3 CREATIVE ECONOMY

It is still a difficult task to come up with a more assertive definition of what the Creative Economy is. For Greffe (2015), for example, it emerges as a new way of perceiving economic development, being a new way of making the economy. An economy that does not focus primarily on the monetary value of goods and products, but which is concerned with other factors, such as social and environmental, based on creativity. Lala Deheinzelin is a great advocate of this economy, believing that this should be the ideal path for all nations. For her, it is a sector that brings together activities that have culture and creativity as its main raw material. This being a vast field, it states that it is possible to include varied business models within the Creative Economy, from the individual who works with education to a luxury car brand (Deheinzelin, 2008).

We see that, initially, the term creative economy was treated as Cultural Industry or Creative Industry, being found such meanings in many authors who study the area, defining them and trying to explain the difference between these perceptions. Both terms tend to have the same meaning, with some different points, and their interpretation will depend on the region, according to the meaning of the words for each language/culture. The Culture Industry in many cases refers to sectors that have culture or creativity as the central factor of economic development and are uniquely creative. The Creative Industry integrates other factors of production, considered non-creative, as well as creativity, but it is only part of the process (Cietta, 2017b).

The term Creative Industry emerged in 1994 in Australia through the publication of the Creative nation: commonwealth cultural policy report, but it was in the United Kingdom that the shift in vision was most prominently observed during the term of Prime Minister Tony Blair in the same decade (Cietta, 2017b). In Australia, several studies are conducted, in addition to policies and programs for the Creative Economy, mainly conducted jointly with Queensland University of Technology and the Department of Communications, Information Technology and the Arts (DCITA). In the UK, the Department for Culture, Media and Sport (DCMS) is responsible for the implementation of policies in the sports, cultural, tourism and creative industries, aiming to analyze and map the existing creative industries, measuring their contribution to the economy, thus identifying measures to foster greater economic development (Cietta, 2017b).

Greffe (2015) says that these countries realized that artistic and cultural activities could be a mean of re-creating jobs, at a time when the industrial production model was beginning to collapse. For Howkins (2013) the developmental impact that the creative economy causes are not only restricted to the creative sectors, because the skills and how they are used, impact on other areas as well. One can conclude that it would be an incentive to new markets and new ways of doing business, that is, when a venture shows signs of being able to evolve, others mirror and try to follow the same path.

In Brazil, the discussion on Creative Economy began to gain momentum following the 11th United Nations Conference on Trade and Development held in June 2004 in the city of São Paulo and continued with the participation in the General Conference of the United Nations Educational, Scientific and Cultural Organization held in Paris (Cietta, 2017b). These contributions resulted in the creation of the Creative Economy Secretariat, with federal scope, which started to direct the sector through the elaboration of a plan, with the policies, guidelines and actions for the sector (Friques, 2013). The Plan of the Secretariat of the Creative Economy defines the Creative Economy as being:

> *The creative sectors are all those whose productive activities have as main process a creative act generating a product, good or service, whose symbolic dimension is determinant of its value, resulting in the production of cultural, economic and social wealth* (Ministério, 2011, p. 22).

It is clear that the main characteristic of the Creative Economy is always to seek innovation in products and services, which is a differential factor for business. Howkins (2013) believes that creativity must be widely valued in many productive sectors, because for him "creativity is present in science, technology and engineering and indeed in all new and innovative products and services" (p. 17). By including creativity as the main engine of business, Reis (2008) says that creativity is renewed with "competition between creative agents" (p. 15), rather than exhausting itself, such as tangible resources. The stimulus to this type of economy can present several benefits for society,

> *These and other characteristics make the creative economy an opportunity to rescue the citizen (inserting them socially) and the consumer (including them economically), through an asset emanating from his own formation, culture and roots. This framework of coexistence between the symbolic universe and the concrete world is what transmutes creativity into a catalyst of economic value* (Reis, 2008, p. 15).

The same author continues the theme by saying that the sustainability of this type of business, which values and has its contribution from the cultural production, needs to be made on the basis of talent training, so that the cultural producer can survive its production; that this cultural production circulates, guaranteeing a constant renewal; and that access to such production is guaranteed to all (Reis, 2008). This is another notable feature of the businesses involved in the Creative Economy, the appreciation of local production. There is a lot of talk in the term 'the place for the global', that is, the stimulus to local production should be a priority, while production should be recognized not only locally.

4 POSSIBLE RELATIONSHIPS—CREATIVE ECONOMICS AS A SOLUTION TO ENVIRONMENTAL CONCERNS

The Creative Economy encourages other means of production than the traditional industrial model, in which mass production is valued and there is an excess of raw material extraction (here considered as tangible resources). Instead of relying on the tangible resources typical of the traditional economy, it considers what is capable of multiplying infinitely, which are the intangible aspects, such as knowledge, human values, culture, creativity, experiences and attributes, that together, form the values of a brand and actually make it sustainable, in the broader term of the word. Deheinzelin (2008) says that work done based on creativity and culture operates in four dimensions: economic, social, environmental and symbolic, adding the latter as a fourth aspect to what several authors present about sustainability, such as De Carli (2012).

In her various publications, which can be verified on her website (http://laladeheinzelin.com.br/), Lala Deheinzelin says that Creative Economy can be the solution to a sustainable future, since it focuses on values other than money for products and services. Cietta (2017a) complements this fact, stating that Creative Economy and sustainability are related by the fact that the Creative Economy conveys meanings in its products or services, and not necessarily because the creative product is sustainable. Concerning these meanings, the author cites a chain that sees creative products as a text, or we can also associate them with an image, since both require varied interpretations, results of encounters, of individual or group visions on their meanings. Manzini et al. (2011) already said that thinking about environmental sustainability challenges the model of development based on capital accumulation. They believe that it is necessary to change the way society lives, "in which economic well-being and health, which are now measured in terms of growth in production and consumption of raw materials, to a society in which it is possible to live better consuming less and developing the economy by reducing the production of material products" (p. 31).

When we think about sustainability, we must think more broadly, not just think about consuming something with greater durability. We must think about whether we really need to consume it. A new economy, or a new way of doing business, can provide the consumer with a new experience of consumption, going beyond just consumption. Carvalhal (2016) presents several factors that have been perceived in recent times, believing that this new way of doing business is more conscious and human, focused on collaborative, is more "distributed, sustainable, social and ethical" (p. 241). These new businesses, based on adding value to their products/services beyond money, are emerging along with a new consumer who is also more concerned with issues beyond their basic needs. "The output is not to consume less, but to make the vehicle of consumption more sustainable" (Cietta, 2017a, p. 162). Carvalhal (2016, p. 137) adds that, in this sense,

> *The younger generations do not want to have a car, 'this is Grandpa's thing'. Millennials want access, and not ownership. They are really beginning to see themselves as part of a large human family. The so-called 'empathic civilization' brings a mentality no longer suited to capitalism, but to the economy of sharing. A vision that conceives humanity as a single family and the planet or biosphere as the shared community.*

There are various moves in this direction to provide these new experiences. And business must find its way. Carvalhal (2016) continues his thinking by saying that the way the new consumer is acting goes against traditional forms of "consumption, business, communication and life itself" (p.144), and it is necessary to differentiate to reach the objectives, it is necessary to seek an economy "more connected with personal premises and values" (p.144). This thinking converges with what Cietta says (2017a), which suggests that trade in sustainable products is still a niche market, and that these characteristics can be a great differentiating factor for those who intend to follow this path.

This fact brings us another point, that of communicating. The positioning of a brand can be modifying, not in the sense of changing the direction of the economy, having environmental and social consequences, as to changing the habits of its consumers. When a business has ethical attitudes, it can change the way people consume (and not think) the people/consumers who follow them. Cietta (2017a) complements our thinking by saying that conscious consumption is a consequence and not an objective. For him it is necessary to change the way of producing, with a more sustainable production and directly related to the content of the message.

An example that demonstrates that this relationship is possible is the brand 'Insecta Shoes' (https://www.insectashoes.com/), a company that produces shoes and accessories that are considered ecological and vegan. We can consider that it is a brand that fits within the principles of the Creative Economy by having innovation and creativity as the main sales vehicle. Its products are differentiated and already have a characteristic of their own, creating an identity with their consumer. In addition, it is a company that says (and shows itself on its website and social networks) to be sustainable. In fact, we can prove this speech by browsing its online page, because they can put into practice the pillars of sustainability. They carry out various actions with different communities, training less favored people and providing income to NGOs (Non-Governmental Organizations) with the purpose of promoting and supporting their actions, contributing in the social and economic development of several sectors involved. The environmental aspect of the brand is also quite clear, since it acts in different ways. It reuses fabrics that would be thrown in the garbage in the development of its products, being this the main raw material. In addition, it employs the policy called 'Close the cycle', which consists of returning defective or worn products, being repaired or recycled, and when resold, has its values passed on to some environmental recovery project.

5 IS IT POSSIBLE TO RELATE? LOCAL BRANDS AND THEIR BUSINESS MODELS

Following the approximation between the Creative Economy and the concept of sustainability, five brands were selected in the capitals of two states of Northeastern Brazil—Paraiba and Pernambuco—to exemplify/verify the capacity of inclusion in a market dominated by major brands, with a new management model. The choice of brands was made based on some criteria extracted from the authors' statements associated to the principles of Creative Economy and Sustainability, such as the use of creativity as a factor of innovation in the market, the valorization of local production, and sustainable factors, already discussed a lot, observed in texts and images of the marks, transmitted in the sites and social networks of the same ones. Next, these brands will be presented briefly and discussed the relations detected with the terms that are object of study in this work.

The first brand analyzed was 'Mayrles Emille', which is originated in the state of Paraiba. It manufactures and sells decorations made with fibers and colored cotton fabrics. In their profile on Instagram (@mayrlesemille), these pieces are presented as "contemporary and sustainable handmade jewelry". This approach already indicates where the value of the piece is deposited, indicating small-scale mode of production and the principles it holds (sustainability, environmental and social). This is confirmed in the images that it conveys in the media where one observes the presence of handicraft items complementing their pieces. There is also an appreciation of environmental aspects when displaying a necklace made of leaves, stones and shells 'collected' on the beach, pieces of wood, patchwork of fabric and fabric.

Another brand analyzed was 'Camila Demori', which is also originated in Paraiba. It produces fashion accessories "with sustainability and sophistication". The pieces, in their majority, have organic formats, remembering leaves, algae and marine gravels. It is made exclusively and makes use of the reuse of PET bottles blended with paper and natural fibers. The production is small-scale, given the originality of each piece, created and executed by the designer. Communicates with clients through social networks, such as WhatsApp and Instagram (@camilaeco_joias).

'Calma Monga!' is a brand located in the state of Pernambuco, which produces bags, backpacks and accessories, working with original and sustainable parts. It uses the "Fair Trade, Local and Vegan" banner. In its profile on the social network Instagram (@calmamonga), the team defines itself as happy and happy with what it does. Certainly, the information has to do with the connection with the culture of the state notably present in the colors of the popular festivals, elements of the pop culture, and in the engagement with social causes.

'Jailson Marcos' is the creator of the Pernambuco brand of the same name. It produces sandals and shoes in a unique style, explained to visitors as an 'aesthetic dialogue between the current Design and the Brazilian craft tradition'. His first references came from the sandals of the countryside legends, like Lampião and his team, and mixed with the concepts of architecture, aerodynamics and sportswear, realizing the existence of a strong cultural appeal in its products (http://www.jailsonmarcos.com/).

The 'Studio LAMA' (@studio_lama_), based in Recife, Pernambuco, produces contemporary jewelry. They are pieces of paper—necklaces and earrings—that mix materials and processes, such as acrylic and silver and make use of laser cutting, milling cutters and innovative use of 3D printers, with hand—made elements in goldsmithing techniques. Participate in creative markets and markets.

It can be noticed that the word that appears most in the discourse of the analyzed brands (in 4 of the 6 companies), identified in their social networks, is 'being sustainable'. The question is whether this concept is understood in its full complexity. To achieve that, it is not enough to use biodegradable materials that do not pollute the environment. It does not mean imprinting a trace of culture or generating income fairly. Being sustainable requires a broad field of action. It is a set of pillars of equal stature. If one falls, the others cannot keep their balance.

To relate the approximations between these brands and this management model, there were difficulties, precisely because the information, or nonexistent, or what is analyzed does not correspond to the complexity of the terms used. Thinking about the social aspect, the analysis points out that all brands have their production in the place of origin and use local labor. In two, production is the responsibility of the owner. We deduce that fair work is a premise among all. But only 'Calma Monga!' refers, in his speeches, to fair work in a clear way. The cultural/symbolic value is defined as the use of knowledge and appreciation of the human and social values in which the marks are inserted. At this point, it is noticed that some of the brands clearly invest in the use of these resources, such as 'Jailson Marcos'. This is the essential factor of the Creative Economy and associated with Sustainability can be said that cultural valorization is fundamental, to rescue communities and traditions, including social and cultural aspects.

Speaking of the environmental aspect, most brands report using natural or environmentally friendly materials. The doubt, which always remains, is whether materials of animal and vegetable origin such as leather and wood are considered natural. As for the production processes, these are mostly handcrafted, including some operations that use electromechanical equipment. Only the LAMA Studio brand innovated in the production process, including digital manufacturing and laser cutting, bringing the new technology resources to its advantage, thus seeking a production with reduction of waste, for example. When we think about the economic aspect, all brands are entitled to be part of the Slow Fashion system, which is based on a small-scale production with fair and sustainable work, even if they reach markets beyond their place of business, due to the fact that almost exclusively over the internet. As Carvalhal (2016) says, it is necessary to "act locally, and think globally", both in relation to the environment and the processes adopted in the business, basic premises of both terms analyzed here. In all indications, these brands suggest that they are treading a path where these characteristics or prerequisites of the Creative Economy, though still immature, are present.

6 FINAL CONSIDERATIONS

Faced with the theories studied and data collected, it is possible to get to certain considerations. It is a fact that these brands that are conforming to these new principles and finding adept consumers, follow their path, parallel to the conventional models of the big brands. It may seem somewhat utopian, but there are confluences of various factors that announce the success of these little novelties, in the face of the large torrent of the industrial economy, which is sick, exhausting the population of the planet.

The agents of the conventional models of production, still strongly in force, can be understood as part of the mechanism of production much more than workers. These distances themselves from the target of the system—the consumer—complacently manipulated by a vicious system. What is at stake is the profit, which is distributed among a few. However, the damage seems to be being seen by a new, more questioning generation, which finds no solution but change.

It no longer makes sense so many inequalities and obscure logics. The production-consumption relationship will not cease to exist, but equilibrium is possible if we humanize the processes. It can be difficult to persuade the inattentive consumer and the greedy producer for profits to change their perspective and envision a new ethic that already pulsates. But upon seeing it, they certainly will not hesitate to be a part of it.

It is possible to perceive in all the research carried out and presented here that there are relations between the two areas. In fact, we believe that one should not exist without the other. The Creative Economy in all its history is a method, that the small producers are inserted in the market, being that a social aspect, of insertion, and at the same time economic, because it generates income for these small producers, which are fundamental to sustainable development. And, as a local market is valued, which has a small-scale production, based on the cultural aspect, the environmental factor is being considered, since consumption becomes differentiated and conscious, and in most cases, it is thought with more care in the raw materials and labor used.

REFERENCES

Berlim, L. 2012. *Moda e sustentabilidade: uma reflexão necessária*. São Paulo: Estação das Letras e Cores.
CarvalhaL, A. 2016. *Moda com propósito: manifesto pela grande virada*. São Paulo: Editora Paralela.
Cietta, E. 2017a. A moda como produto cultural e a economia criativa: entrevista com Enrico Cietta. Entrevistado por: Maria de Fátima da S. Costa de Mattos. In: *dObras*, v. 10, n. 21, Maio.
Cietta, E. 2017b. *A Economia da Moda – Porque hoje um bom modelo de negócios vale mais do que uma boa coleção*. São Paulo: Estação das Letras e Cores.
De Carli, A.M.S. 2012. Moda uma prática de múltiplas economias. In: De Carli, A.M.S; Venson, B.L.S. (Org.). *Moda Sustentabilidade e Emergências*. Caxias do Sul, RS: Educs—Editora da Universidade Caxias do Sul. pp. 85–102.
Deheinzelin, L. 2008. Economia criativa é a estratégia de desenvolvimento do século. In: *Revista Dealer*, mai./jun. 2008. Available: http://laladeheinzelin.com.br/wp-content/uploads/2010/07/Economia-Criativa-%C3%A9-a-estrategia-de-desenvolvimento-do-S%C3%A9cuclo-Revista-Dealer-junho–2008.pdf [Accessed June 2016].
Denardin, K.S. 2012. Sustentabilidade na Moda: casos de reaproveitamento e economia solidária. In: De Carli, A.M.S; Venson, B.L.S. (Org.). *Moda Sustentabilidade e Emergências*. Caxias do Sul, RS: Educs—Editora da Universidade Caxias do Sul. pp. 157–179.
Friques, M.S. 2013. O escopo da Economia Criativa no contexto brasileiro. *REDIGE*, V. 4, N. 01, abr. 2013. Available: http://www2.cetiqt.senai.br/ead/redige/index.php/redige/article/viewFile/179/233 [Accessed June 2016].
Greffe, X. 2015. *A economia artisticamente criativa*. São Paulo: Iluminuras: Itaú Cultural.
Howkins, J. 2013. *Economia Criativa—Como ganhar dinheiro com ideias criativas*. São Paulo: M. Books do Brasil Editora Ltda.
Kozinets, R.V. 2014. *Netnografia: realizando pesquisa netnográfica online*. [Edição Online]. Porto Alegre: Penso.
Lorgus, A.L.; Odebrecht, C. 2011. *Metodologia de pesquisa aplicada ao design*. Blumenau: Edifurb.
Manzini, E.; Vezzoli, C. 2011. *O Desenvolvimento de Produtos Sustentáveis, os requisitos ambientais dos produtos industriais*. São Paulo: Editora da Universidade de São Paulo.
Martins, S.B.; Daher, M.A.Z.; Pinheiro, N. 2012. A Moda, sustentabilidade e inclusão: retraços que tecem histórias. Resultado de uma Oficina. In: *VIII Coloquio de Moda*, 2012, Rio de Janerio. En Moda Escola de empreendedores. Rio de Janeiro: Senai—CETIQt.
Ministério da Cultura. 2011. *Plano da secretaria da economia criativa: políticas, diretrizes e ações 2011 a 2014*. Brasília: Ministério da Cultura.
Reis, A.C.F. (Org.). 2008. *Economia criativa como estratégia de desenvolvimento: uma visão dos países em desenvolvimento*. São Paulo: Itáu Cultural: Garimpo Soluções.
Salcedo, E. 2014. *Moda ética para um futuro sustentável*. Barcelona: Editorial Gustavo Gili.

New sustainable fashion business models: "the *coworking*" case study

L.F. Barcellos & A.C. Broega
Universidade do Minho, Guimarães, Portugal

ABSTRACT: The current fashion scenario presents unsustainable practices and business models, which are focused on the mass market, reducing the costs and stimulating hyperconsumption. In opposition to this, new sustainable fashion business models are emerging with high ethical attitudes and more conscious with the future of the planet. Thus, it becomes relevant to explore sustainable fashion business models. In this context, the main goal of this article is to identify new existent models and the concepts that underlie them. The method used was the exploratory research, through case analysis, to stimulate the understanding of the theme. The results embrace coworking spaces, which are identified by the concepts they involve and how a new business model can contribute to a more sustainable fashion system.

1 INTRODUCTION

The environment where the actual society lives have been constructed by a culture based on a way of projecting, producing, distributing and consuming which is originated during the Industrial Revolution, especially after the 80 s of the 20th century (Malaguti, 2009).

The desire for power and ambition to grow were preponderant for the industrial revolution to take place. These values guided people's lives and developed a capitalist society governed by the pursuit of money. From this, the society comes to experience an environment immersed in deep poverty, with a scarcity of resources and economic vulnerability (Carvalhal, 2016).

In this context, according to Malaguti (2009), many environmental problems were generated. The necessity to reconsider the relationship between man and nature is required to establish limits on production, consumption and in the concept of environmental responsibility of the man's relationship with the objects and the environment. Based on Carvalhal's (2016) rational, the actual world is in a tremendous economic crisis as result of the internal crisis of values, conscience, environmental, market, trust, political, affective and even spiritual. The current socioeconomic system goes against human and social development, and it is necessary to find a new way of thinking about economics.

From this conjuncture and searching for complementarity, segmentation, agreements, and partnerships, a new way of human relations practices is being experienced in the usage of territories or even in social relationships. According to those as mentioned earlier, the practices and actions in consonance with the concept of sustainable environmental, social and cultural development increasingly need to emerge in the sectors of the economy.

According to Freire & Araújo (2016), the fashion sector, specifically, has been criticized throughout history for undertaking unsustainable practices in which business models are focused on the mass market, cost reduction and stimulus to exacerbated consumption, building fast fashion thinking or disposable fashion mentality.

The fast fashion movement contributes to the subhuman work conditions in countries where labour laws are inefficient and secondary. Besides, the environmental costs are enormous to the detriment of a possible lower cost of production. Some brands outsource their production, and with that, they also outsource their responsibility, fomenting slave labor. Another issue is that resources are being used on a large scale, as well as a significant amount of waste is being generated in production. This way, the short concerning with the workers and with the environment, where everything is produced, is evident.

Contrary to this movement, the slow fashion has been growing, stimulating the quality, creativity, ethics, the valorisation of the product and its impact on the environment. Additionally, this movement values the variety and cultural importance, as well as the positive impact generated by the society (Carvalhal, 2016).

It should be noted that the "slow" in question refers not only to slowing down production and consumption but also a different worldview based on a new way of thinking, acting and consuming with awareness and ethics. This suggests a disruption with the old values, in favor of a shared growth (Carvalhal, 2016).

Fashion is the second most polluting industry, but it is also the second most productive job industry in the world (Carvalhal, 2016). With this, it is understood the relevance of the study in question, mainly for dealing with issues of the textile/fashion sector. The reflexion about new sustainable fashion business models is essential for the construction of a new era in the industry, making this sector a transformative agent and broadcasting new ideas.

2 SUSTAINABILITY

Considering that the present study refers to sustainable business models in fashion, it is essential the theoretical understanding about concepts related to sustainability.

According to Berlin (2012), in the 1960s, after disastrous events and ecosystem imbalances, researchers and environmentalists began discussions about the environmental crisis, perceived as a global problem. Years later, in 1972, the Union Nation Conference on the Human Environment happened in Stockholm, Sweden. This first conference focused on environmental issues. After more than ten years of this Conference, the concept of sustainable development emerged, considered complex and questioned since their origin. The definition of the term involved the following aspects:

– The dissonance between the partners of industrial production and consumption.
– The possibility of a future for the next generations, with natural resources available.
– The decrease, if not the eradication of misery in the world.

Thus, the concept of sustainability was delimited as a new paradigm of the relationship between the human being and the environment, not ignoring the economic question.

In the last two centuries, the evolution of the industrial sector has generated gains and damages to humanity. The harms were the degradation of the environment, the biodiversity loss, the climate change, the increased greenhouse effect, the acid rain, the soil deterioration, the waste and misuse of natural resources, the excessive garbage growth, and in particular, the hunger and misery (Berlim, 2012).

In pursuit of increased profit, the environment has been deteriorated, and the dignity of much of society has been renounced, compromising the future of human life. Still according to Berlin (2012), currently the debates are about the necessity of reconciling economic growth, job creation, access to health and education, and improvement in the quality of life. For this to be possible, it is essential to preserve the natural environment, biodiversity, and natural resources.

As already mentioned, the term sustainability has multiple definitions. However, it is not the intention of this article to expose the wide variety of meanings attributed to the term. According to Berlin (2012), the concept of Sachs relates sustainability with eight criteria, such as social, cultural, environmental, territorial, economic, national and international political. In addition, Sachs emphasizes the importance of the rational and ecologically sustainable use of nature.

The relationship between sustainability and fashion, first association in this work, has many different facets, however, that complement each other. Fashion and Sustainability can mean much more than generating ecologically correct products. It can reveal the generation of new transforming tendencies, promoting the re-signification of the logic of consumption.

2.1 New sustainable thinking in the fashion market: Slow fashion (concept)

According to Irokawa et al. (2017), in this current scenario of the fashion industry, characterized by *hyperconsumption*, accelerated production, a high degree of competition and the search for low cost of production, fashion brands, appealingly, bring news at all times, based in ephemeral tendencies, sustaining an exaggerated consumption. In this dynamic, the fashion sector is becoming increasingly unsustainable, a reason for reflection and questioning by professionals and researchers in the area.

In response to this new perspective, a new sustainable way of thinking in the market emerged, called slow fashion. This line of thinking proposes reflections about current production and consumption values, believing in more ethical attitudes, consumer awareness and, above all, consumer choices. The knowledge of the production process, the perception of quality and the durability of the pieces begin to be considered in the consumption of a fashion product. These new conditions imposed by consumers take some fashion brands redefine their values and seek more transparency in their business (Irokawa et al. 2017).

The idea of "slow fashion" was created by British designer Kate Fletcher, a consultant, and professor of sustainable design at the Center for Sustainable Fashion in 2007, through the concepts of the slow food movement, designed by the Italian journalist Carlos Petrini, in 1986. The principles were set in opposition to the conceptions of the fast fashion movement. The slow fashion values the diversity as opposed to mass production; global-locality against globalization; self-consciousness rather than image; symbiosis to the detriment of parasitism; confection and maintenance for a long life cycle and not the constant news; the real price adding the social and ecological costs as opposed to the cost based on labour and materials; the search for strengthening social capital and valorisation of the territory in denial to the weakening of social capital, small and medium production scale as opposed to large scale, sustainable rather than unsustainable (Irokawa et al. 2017).

According to Carvalhal (2016), slow fashion is part of a transformative process in the fashion market, which stimulates the quality, creativity, ethics, valorisation of the product and the impact on the environment. Similarly, Irokawa et al. (2017), believes that the idea of slow fashion is not only restricted to the concern about the origin of the raw material and the use of resources in a conscious way, but, above all, it is proposes to make consumers aware of the production chain, valuing each stage of production.

2.2 New sustainable forms of structuring in the fashion market

In this context already mentioned, new sustainable fashion business models are arising to meet the needs of current's world. New ways of "making" fashion, building, working and educating become preponderant, having as central axis practices in consonance with the theme environmental, social and cultural sustainability. The textile/fashion sector needs to reinvent itself, to rethink the whole process to build a future that protects the planet and improves people's lives.

After research on new sustainable fashion business models, the coworking spaces were found. The researchers of this article were interested in knowing this business model, their characteristics and which concepts are involved when professionals of the fashion area use these spaces.

2.3 The coworking model

The *coworking* spaces emerged as a new way of structuring in the fashion sector, presenting an innovative character concerning business model. Therefore, it becomes crucial to understand how it is established in the fashion area and for that, first it is necessary to conceptualize it in a general way, so that later to be characterized, in particular, in the fashion market.

According to Rus & Orel (2015), long before the 2008 economic crisis, new forms of work have been created instead of traditional jobs in large hierarchically organized companies, outside the big corporations that these new forms of work and support are being explored.

With considerable growth in recent years, *coworking* has emerged, as a new type of workspace and organizational arrangement appropriate to the creative class, with the concept of community as the main characteristic.

The first *coworking* space was Spiral Muse Coworking, created by Brad Neuberg in San Francisco in 2005. They have achieved success and have moved to a larger space, an industrial loft, where an old hat factory operated. This initiative stimulated the creation of two *coworking* spaces, Citizen Space and Indy Hall, in 2006, also in the United States. The founders were proponents of the Open Source Movement, a rebellion among computer programmers against large corporations that used proprietary rights to profit by restricting the flow of information, stifling innovation and learning of grassroots growth. The focus of the movement was taken to the *coworking* movement, a global organization, with the vision and values of collaboration, openness, community, sustainability, and accessibility (Rus & Orel, 2015).

Still, according to Rus & Orel (2015), independent professionals sought this type of space to escape the isolation of their homes and as a way of uniting the workplace with the community, thus making them more productive. These sites usually provide minimal structures that allow users to use space part-time or full-time, deciding whether or not to participate in activities, thus maintaining their independence.

However, according to a study by Spinuzzi, there are two types or configurations of *coworking* spaces, one is the "The Good Partners" model, which emphasizes collaboration among the members in common projects, and the other is the "The Good Neighbors" model, in which the participants work alone but collaborate to the community within the space. In the first case, the community arises from the collaborative work between the participants, and in the second, the collaboration arises from an awareness of joining efforts among the participants to build a community (Rus & Orel, 2015).

Some authors consider that these work environments are opportunities for contact, collaboration, expansion of networks and exchange of information. So Rus & Orel (2015), mentions that Kenline (2012), defined *coworking* as a socio-cultural community ecosystem of exchanges, in which a network of people are connected by shared social networks and resources. This definition points that the *coworking* model is not limited to the physical spaces used by individuals in their field of work, but is defined as a system in which social networks and resources are connected, forming a collaborative economy.

3 METHODOLOGY

This article results from in-depth research on new sustainable fashion business models during the accomplishment of the dissertation work in Design and Marketing. The research methodology was based on an exploratory study of the theme studied, whose main objective of this article is to identify and understand the new sustainable fashion business models to promote knowledge about the theme.

According to Gerhardt & Silveira (2009), the exploratory research aims to promote an approximation about a theme, in order to make it more explicit or to construct hypotheses. The method presented here is by definition exploratory since there is an analysis of examples that stimulate the understanding of the study in question. Thus, searches were done on the internet about spaces of *coworking* and the respective interpretation and analysis of the information found. The analysis was based on the identification of the concepts involved in these new business models and all the interpretation of the facts made through the vision of the researchers of the article.

4 PRESENTATION AND CONSIDERATION ON RESULTS

Established general concepts regarding *coworking*, the characterization of this new way of structuring the business, specifically in the fashion sector, will be done. After research, eleven *coworking* spaces were found focused on the fashion area and located in several countries, such as Portugal, United States, Germany, Spain, England, and Brazil. These spaces have common elements in terms of central idea and some other aspects that differentiate them from each other. In an attempt to understand these already created models, in this article will be presented each *coworking* space found in the fashion area and will be exposed components that resemble them, common characteristics found and also the specificities that differentiate them, in a general way.

The *coworking* space, Sew Stitches, located in Nevada, United States, was created for fashion professionals to share the workspace, knowledge, extend their networks of contact and develop a community. For this, they offer a place for project development, which can be rented by the hour.

The Nadelwald, a *coworking* space, created in Berlin, Germany, is a project that has as its principle the creation of a shared workspace, from the provision of flexible hours, the use of specialized machines by all members, sewing workshops, and a creative exchange. In addition, there is a physical and online store for the commercialization of products created by members of the community. There are several types of flexible fares, varying according to the hours, days or months that the professionals want to work in the place.

In Brazil, in Pernambuco, there is the Creativity Warehouse, a *coworking* space, belonging to the Porto Digital project (one of the main technological parks and innovation environments in Brazil). The place offers structure and services to support the local economy, providing technology to artisans and fashion designers and stimulating the creativity and entrepreneurship of professionals.

The Sewing Café project was developed in Leicestershire, England, as a *coworking* space, with the central idea of making everyone accessible to sewing practice. With this, they offer a space for manufactory (social sewing), and offer machines that can be rented for hours by the members.

The *coworking* space, Fábrica Santo Tirso, located in the city of Santo Tirso, Portugal, was created as a workspace and business in the fashion area, for experimentation and promotion of cultural events. The idea is to become the city and the North of Portugal a cultural, commercial and tourist axis in the scope of Fashion and Design. There are incubation projects, training courses and various activities in the fashion area.

The Teté Café Costura, a *coworking* space, located in Madrid, Spain, was developed with the concept of social sewing, a space to acquire knowledge in sewing, where it can rent sewing machines. The idea is based on developing a place where creativity is present so that sewing and fashion communicate in a natural way. Periodically, fashion professionals are invited to transmit and share their knowledge through workshops. The idea is to transform the environment into a relaxed place so that the exchange of knowledge can happen in the cafes and collective spaces. Besides, the idea is to make this space, a meeting place for creativity in the area of sewing and fashion.

The space of *coworking*, Space 530, designed and implemented in a luxury building in the center of Manhattan, New York, United States, accept several fashion brands on the rise and companies that seek to use technology in the fashion market. Members can use office space (private suites or collective space), showroom space to experiment new collections with product launches and fashion exhibits. Currently, it unites more than 40 fashion brands, which becomes the space valuable for making contacts. It has some flexible membership options, which do not require a long-term commitment, and clients can enter into contracts with minimum terms, without a large deposit. In addition, there are dependencies such as cafe, lounges, pantries, reception with staff and equipped presentation/conference rooms.

The CRU Cowork, located in the city of Porto, Portugal, is a *coworking* space developed with the intention of implementing a regime of sharing and collaboration for creative activities. The motivation for the creation of the project was to provide optimized space management and resources that imply low-cost jobs. The space is shared by independent professionals, who exhibit and promote their work in a shop and an art gallery. Focused on creative activities, not exclusive to fashion, fashion professionals are locals and pay a monthly income that allows getting a work environment. They have facilities such as meeting rooms available for up to 6 people, spaces for workshops, for practical training or theoretical expositions with up to 16 participants, screen printing manufactory, a photography studio with available material and photography laboratory.

In São Paulo, Brazil, the Lab Fashion was implemented as a *coworking* space that offers a physical structure for fashion designers to create, develop, undertake, stimulate creativity, expand their contact networks and develop new forms of production. With a privileged location, there are meeting rooms, specialized service, sewing room with sewing machines and specific materials available, room for individual attendance, rented by hour/period, with a fitting room, tables removable and mirrors. Also, they offer training courses, lectures, workshops and consulting with multidisciplinary teams on business in the fashion field. It focuses on the interaction between members and aims to reduce costs for those who use, compared to the costs of an individual structure.

The *coworking* space Nemona was idealized as an incubator and network of fashion and sewing in Berlin, Germany. It started with an equipped manufactory with specialized machines, cutting tables and sewing items for tailors, designers and dressmakers to produce their pieces. Currently, about 150 designers and fashion producers work there. The main objective is to promote a network of fashion design in Berlin to strengthen sales, increase local production and generate high-quality jobs in the textile sector. They promote workshops and courses, repair and commercialize sewing machines, as well as providing a technical team to support sewing. This project has generated jobs and stimulated new business models in Germany, such as online portal "Sourcebook.eu". It is a B2B platform, e-commerce for business-to-business transactions, developed for textile traders, producers and service providers of the Berlin fashion industry. Today, Nemona is working in professionalization with a focus on sales, with the support of the Neukölln Fashion Economy subproject. The focus of the work is on education, economics, and integration and the aim is improve the economic situation of Neukölln fashion workers in the long term, with the generation of jobs in the city and making the district a place of reference in the fashion area. The project includes the development of projects in partnership, the presence in several events in the area of fashion and art and the business fair and the creation of temporary stores. Nemona also performs as an intermediary platform in the industry, promoting fairs, communicating requests for cooperation and information from industry partners. Besides, professionals in the area of business management and fashion production organize the workshops. One relevant aspect of this *coworking* space refers to the projects idealized and implemented in the field of sustainability, some of them being winners of the Werkstatt N Label.

Lastly, the Malha project was identified, considered more than a space of *coworking*. It is a collaborative network platform for the fashion ecosystem, which connects creators, entrepreneurs, producers, suppliers, and consumers by building a collaborative, local and independent fashion. Implemented in July 2016, in the city of Rio de Janeiro, idealized by influential professionals in the fashion sector, the project was elaborated in the garden of Templo, the first space of creative *coworking* in Rio, where great brands, designers, entrepreneurs, and educators talked about challenges and needs in the industry. This collective idealized to work with a new fashion, thinking about the future of the sector through a new business model that could better meet the needs of today's world.

Space, machinery, and equipment (technology), photo studio and content generation are shared, encompassing all of the production chain. The community is formed by brands that have values focused on sustainability or interest in developing from this perspective. The concept of sustainability, in all axes, environmental, social and cultural are very well founded in the structure and actions developed in the project.

The Malha was created with the intention of realizing fair trade, an essential premise considered for the collective economy. This means realizing commercial partnerships based on dialogue, transparency, better conditions of exchange and guarantee of rights for those who produce and work in the fashion market. The space offers free events for the benefit of the local community as a way to develop local cultural movements, receiving cultural producers, content mobilizers, activators of independent movements, who are interested in taking their ideas, projects and be part of a movement for a fashion more collaborative. This way of organizing facilitates the interaction of people through the actions and joint activities of associates and the discovery of affinities among them, which are fundamental for the establishment of fast partnerships.

5 FINAL CONCLUSIONS

Sustainability in the fashion sector has been an ever-present theme, and with that, new sustainable fashion business models are emerging in the market. With this new thinking, focused on practices in line with environmental, social and cultural sustainability, fashion is beginning to take a path with a greater commitment to the environment and greater awareness of the future of the planet.

With the research done, it is understood that most of the spaces of *coworking* spaces in the fashion area have as a central idea, the creation of a workplace for professionals of the sector, with the intention of establishing a system for sharing, collaboration, and integration among the members, allowing the coexistence of different ideas and projects in the same space.

The sharing of space, knowledge, the possibility of extending the network of contacts, developing a community are also elements that characterize, in most cases, this business model. Many of these spaces are created to support the local economy, providing access for those who work and seek to work in fashion, increasing local production and generating high-quality jobs in the textile sector.

Besides, in *coworking* spaces, the management of space and resources occurs in an optimized way, which results in jobs of low cost.

The physical structure is provided for professionals to create, develop, undertake, stimulate creativity, and even develops new forms of production and can be used by independent professionals or fashion brands.

Some of these spaces have an area for workshops for practical training, theoretical expositions, experimentation of new collections with product launches, and fashion exhibitions and photography studio.

After research, which appeared as a point of greater relevance, when it comes to a more sustainable future for the fashion sector, was the fact that some *coworking* spaces were created with the intention of developing the local community, allowing access to a workplace for fashion professionals to perform their functions.

There is *coworking* space, such as the Malha, which is considered more than that, it is a new business model more complex and broad. A collaborative network platform that from a collaborative space and interaction between professionals creates the possibility of a connection between the different parts of the system that involve the whole process of producing a fashion product, providing significant gains in scale and benefits along the chain.

In this way, it is understood that spaces of *coworking* are emerging as business models of sustainable fashion, even if gradually, but that already show that there is a movement in the fashion sector traveling in more ethical and sustainable ways. These business models emerge with a new approach, which distances from the perverse logic of *hyperconsumption* and financial success as priorities and they approach a way of thinking with more awareness about the reflexes of business in society and on the planet.

ACKNOWLEDGEMENTS

"This work is supported by FEDER funds through the Competitiveness Factors Operational Programme – COMPETE and by national funds through FCT – Foundation for Science and Technology within the scope of the project POCI-01-0145-FEDER-007136".

REFERENCES

Armazém da Criatividade. [Online] Available at: http://www.armazemdacriatividade.org/#home [Acesso em 26 09 2017].
Betahaus. [Online] Available at: https://www.betahaus.com/magazine/nadelwald-from-coworking-to-cosewing [Acesso em 26 09 2017].
Berlim, L., 2012. *Moda e Sustentabilidade. Uma Reflexão Necessária.* São Paulo: Estação das Letras e Cores.
Carvalhal, A. 2016. *Moda com propósito.* 1ª ed. São Paulo: Paralela.
CRU. [Online] Available at: https://cru-cowork.com/[Acesso em 26 09 2017].
Deskmag. [Online] Available at: http://www.deskmag.com/en/cosewing-spaces-coworking-for-diy-fashionistas-626/2[Acesso em 26 09 2017].
Fabrica Santo Thyrso. [Online] Available at: http://www.fabricasantothyrso.com/[Acesso em 28 02 2018].
Freire, K. & Araújo, R. 2016. *Design estratégico e modelos de negócio para moda sustentável: o estudo do caso vuelo.* IN: Encontro de Sustentabilidade em Projeto. Anais ENSUS 2016 – IV Encontro de Sustentabilidade: 646–657. Florianópolis.
Gerhardt, T. & Silveira, D. 2009. *Métodos de Pesquisa.* 1ª ed. Porto Alegre: Editora da UFRGS.
Irokawa, E., Maia, S. & Câmara, J. 2017. *Slow fashion: Possíveis caminhos para a indústria da moda contemporânea.* Bauru: 13º Colóquio de Moda.
Lab Fashion. [Online] Available at: http://labfashion.com.br/[Acesso em 26 09 2017].
Malaguti, C. 2009. *Design e valores materializados: cultura, ética e sustentabilidade. Estudos avançados em design: Sustentabilidade I: 55.* In: De moraes, D. Kruken, L. (Org.). Barbacena: Editora da Universidade do Estado de Minas Gerais. Available at http://www.tcdesign.uemg.br/pdf/Sustentabilidade_I.pdf.
Malha. [Online] Available at: https://www.malha.cc/[Acesso em 28 11 2017].
Nemona. [Online] Available at: http://www.nemona.de[Acesso em 26 09 2017].
Rus, A. & Orel, M., 2015. *Coworking: A Community of Work.* Teorija in Praksa, 52(6), pp. 1017–1038, 1244.
Sew Stitches Cafe. [Online] Available at: http://www.sewstitchescafe.com/[Acesso em 26 09 2017].
Space 530. [Online] Available at: https://www.space530.com/[Acesso em 26 09 2017].
Teté Café Costura. [Online] Available at: https://www.tetecafecostura.com/[Acesso em 16 01 2018].
The Sewing Cafe. [Online] Available at: http://www.thesewingcafe.co.uk/[Acesso em 26 09 2017].

Sustainability: The reconfiguration of fashion, the textile industry and the role of the designer in a new social scenario

M. de L. Iracet
FADU – Universidad de Buenos Aires, Argentina

ABSTRACT: Since prehistory, we have developed a material culture that has varied over the centuries. We have gone from a community culture to an increasingly individualistic one that understands objects (textiles and clothing) as disposable products. Since the Industrial Revolution, manufacturing processes have accompanied modern thinking, technological innovations and new textile materials. At present, SUSTAINABILITY challenges the Fashion and Textile Industry. Sustainability presents us with a new way of thinking and practicing design. This re-configuration raises an integral vision of the design process, intervening it and combining it with social, economic, political and environmental disciplines. It promotes a state of awareness of the design process and the impacts it causes on resource flows, on workers, on communities and on ecosystems. In this framework, the designer of clothing and textiles must be trained in his new role of participatory/active designer, functional to his society; designer/communicator who notifies through various information channels; designer/educator, who seeks feasible modifications within an academic environment.

1 INTRODUCTION

With the advent of the 21st century, the fashion industry, and more specifically, the textile industry began to wonder what would be the fields of work in the textile production and manufacturing process, how the textile industry will impact and affect the environment, how the fashion industry will be re-signified in a new social context and what will be the role of the designer of clothing and textiles in this new century? The 21st century proposes a renewed look at the world where social, cultural, technological and environmental transformations, together with global and local policies are the keys to thinking about a reconfiguration in the fashion and textile industry.

With reference to the textile sector, groups of designers, engineers, entrepreneurs and manufacturers are currently thinking about the future of the textile industry in two areas: the first of them works with specific textiles, which are made from a miniaturization technology (technological process, nanotechnology, by which attempts are made to reduce the size of electronic devices). These workspaces, dedicated to the research of intelligent textiles, promote changes based on the physics and mechanics of the fibers. The second axis begins, from the reflections on the maelstrom by the technological consumption, the eagerness to overcome all the limits and the obsession by the economic growth (own of the thought of the postmodern man). This movement is inspired by Slow Food.

Slow Food is a philosophy founded in Italy in 1986 by Carlo Petrini that relates the conscience and the responsible nature of its production with the pleasure of food which aims to preserve the culinary traditions of the region, culture and agricultural diversity of slow food. Therefore, it promotes the culture and values of the slowness in fashion. This concept not only defines speed, but has a different vision of the world, requires a change of infrastructure and a lower production of articles. It questions the obsession with the growth of the world of fashion, mass production, and the globalized style, and becomes the protector of diversity. It promotes a state of awareness of the design process and the impact it causes on resource flows, on workers, on communities and on ecosystems. These sectors work on new paradigms contemplating economic, social, environmental, and fair-trade policies, the latter being what is known as Fairtrade. Fairtrade was created in 2005 to ensure that cotton farmers received a minimum price for the raw product and a premium for community investment. These elements are contained in the concept of sustainability. This construction forms a new field of research, reflection and analysis, since it questions the fashion and textile industry regarding its complex production, commercialization and consumption system. For this reason, this paper proposes a selection and analysis of authors whose view on sustainability reveals different approaches: social, environmental, business and those concerning the field of the fashion and textile industry. With this new multidisciplinary dynamic, we will verify that the reconfiguration proposed by sustainability is possible, as well as we will indicate in which stages it is located.

2 THE LOOK OF THE AUTHORS

The authors that we will mention give us their perspective, about what is written, with reference to the concept of sustainability developed in their texts.

Therefore, the diversity of the concept "sustainability", in full development, research and experimentation, proposes the analysis and reflection of the production in the fashion and textile sector. It presents an integral vision of the design process, intervening it and combining it with social, economic, political and environmental disciplines.

The World Commission on Environment and Development (WCED, 1987) also called the Brundtland Report (IB) defines the concept of sustainable development as "one that meets the needs of the present without compromising the ability of future generations to meet their own needs". This definition has taken different aspects: economic, social and environmental policies, according to the field of research.

For authors such as Carl Frenkel, Ruggero Schleicher-Tappeser, Stuart Walker and John Ehrenfeld; It could be the starting point of the interpretation of the concept of sustainability that is still in full development today. *In Earth's Company* (1998) Carl Frenkel analyzes the evolution of the concept and understands that the harmony of three elements – such as economics, the environment and social equality – is the fundamental characteristic of sustainability, while Ruggero Schleicher – Tappeser, in *Assessing Sustainable Development in the European Union – The Sustainable Quality Management Approach in the Context of Structural Funds, Greenleaf Publishing. Grenner Management International Journal* (2001) states that we are facing a new paradigm, that sustainable development is not only a new concept, but a concept that invites us to change priorities and objectives. This process of change and transformation is open since it cannot be reached openly. However, the author Stuart Walker in *Sustainable by Design* (2010) understands that there are differences between sustainable development and sustainability. The first is understood as a broad design process, in which social, economic and environmental aspects intervene, while sustainability is a style, in which these aspects – social, environmental and economic – permeate various ventures.

Finally, a more philosophical (and ethical) view of sustainability is proposed by John Ehrenfeld (1999, p. 230) in *Cultural Structure and the Challenge of Sustainability*, defining it as such:

> *As a way of life or being in which people, companies, governments, and other institutions are responsible for looking after the future, since it belongs to them now, to share equitably the ecological resources on which the survival of human beings and other species and to ensure that all inhabitants of today and of the future can and will meet their human needs and aspirations.*

If until now we have found conflicts and dilemmas around the definition of sustainability it is still more experimental this concept in the field of design, which lives in a constant state of exploration and transformation. Authors and researchers such as Kate Fletcher, Linda Grose and Dilys Williams add other concepts of sustainability but referring to the design of fashion and textiles.

The authors Fletcher and Grose in their book *Gestionar la sostenibilidad en la moda. Diseñar para cambiar* (2012, p. 25) propose a reconfiguration in this field and in the fashion system. They reflect on the transformations that the textile sector will have to undergo in relation to the entire life cycle of the product (from the choice of fiber until it is no longer in use) and in this regard they state the following:

> *(…) something as apparently simple as choosing one fiber and not another is really intimately related to global issues and personal values; as if we prefer a slow and deep change instead of rapid innovations, or in what way and at what scale we think it is necessary to intervene to achieve sustainability.*

They add, in addition, that "the process of sustainability forces to change the system of the fashion …" (Fletcher and Grose, 2012, p. 10), reason why the practices of design, known until now are in question. Also, this reconfiguration to which they refer is in constant transformation and the analysis and conclusion of these will depend on the field (or geographical location) of experimentation. For this reason, this experimental global-local model reveals that it is not possible (at least until now) to implement a model of universal work in fashion and in the textile sector because the social, environmental and political variables are specific to it to each State. On the other hand, with respect to the new paradigm of sustainability Fletcher and Grose (2012, back cover) add the following:

> *Sustainability is possibly the fundamental issue of the 21st century, an issue that raises to the fashion industry such diverse issues as labor abuses, the use of toxic chemical substances and unbridled consumption (…).*

Likewise, Dilys Williams reflects on *Sustainable Apparel* (2015, p. 163) with the same approach as the previous authors on the methods used (until now) to produce textile objects and on the fashion system. Understands sustainability in relation to fashion sense: its processes, relationships, textile materials and resources that are available in it. But this adds to the social resources, understanding that these "social resources" are what affect nature and our own future"

The Fashion Design and Sustainability chapter of the same text focuses the debate within the practices of fashion design. Observe the role of the designer—designer as determiner, the creator of contour objects; facilitator, designer as co-creator and designer as a condition of creator, the creator of learning devices—and the transformations that this promotes in relation to the function they fulfill within their cultural space.

On these foundations of social design, and participatory, authors like Henry Sanoff and Ezio Manzini have spoken. The first understands that participatory design contemplates the work of the designer with and according to the community to which it belongs. Establishing the idea of participatory/active design seeking the welfare of our society. While Manzini in his book *Cuando todos diseñan. Una introducción al diseño para la innovación social* (2015) means social innovation as ideas (new products, services and models) that satisfy social needs and create new relationships or forms of collaboration. Improve the capacity of society to function, as a driver of change to sustainability. In addition, he explains that society has suffered, in recent years, profound and vertiginous changes in technical and social systems, with the designer being a bridge between them. It understands social innovation as an open process, susceptible to modifications, and in this panorama distinguishes two types of design: diffuse design, which anyone can perform and expert design, that is, professionals trained in the discipline. Both can jointly drive social change and Manzini (2015, p. 37) points out that:

> *Social innovation has the possibility of changing the world, but for this a new culture and a new way of proceeding are needed. To achieve this, design must be transformed into a broad activity that permeates the multiple nodes of the socio-technical networks in which we live and work, unprecedented in history.*

The author proposes a dynamic design attitude motivating the designer to propose new meanings and solve social and technical problems.

In addition to these academic publications at the World Summit of the Earth in Rio de Janeiro, Brazil (1992) the attending politicians, NGOs and participating business leaders recognized that none of the three problems of sustainable development (economic growth, social equity and care of the environment) could be solved separately. The WBCSD (World Bussiness for Sustainable Development), which incorporated the concept of sustainable development—under the term corporate sustainability (business strategies that are trying to incorporate social value/environmental and its external "actors" while adding value for shareholders) – in the business world. This incipient application of sustainability in the fashion industry and in the textile sector has given way to new business models, new forms of marketing and communication, which are still under observation.

In this collection and selection of views of some authors, the importance of sustainability in the 21st century is evident. Its emergence does not only apply to a new terminology but, on the contrary, it is understood as a new paradigm in the process of development and experimentation. This new construction, which presents diverse views and approaches—social, economic, philosophical and business –, accounts for the complexity with which sustainability is presented. But more difficult is its implementation in the field of the fashion and textile industry, since sustainability proposes a reconfiguration of the entire life cycle of the object, be it a garment or a textile piece. It is for this reason that an interdisciplinary look and joint work is promoted to address it.

3 THE ROLE OF THE DESIGNER

In this frame, we appreciate a change in the role of the textile designer, reshaping it not only in its terminology but also in its functions as a surface maker.

We must point out that the only sustainable textiles (in relation to the environmental variable) are compatible with nature in ALL their life cycle, from the choice and cultivation of the fiber until the garment is disused. This reveals the complexity of the production of a textile. That is why the textile designer's first dilemma is the selection of the type of fiber with which he will work, manufacture and then discard (when the textile is disused), and how it will impact society and the environment. Knowing the types of fibers, their origins (natural, synthetic or artificial), their properties, applications of use and disintegration is key to a first approach to textile design. Then, the production and manufacturing, marketing and discarding processes will be developed, and in all these aspects, sustainability must be involved. It must be borne in mind that there is not and cannot be a universal resolution for all cases. Each place will meet the specific regional needs.

This re-signification also proposes to us, to think of the textile designer as a designer/ activist, functional to his society; a designer/communicator, who notifies through various information channels these transformations; and a designer/educator, who seeks feasible modifications within an academic environment. But for this to be possible the designer will develop a new participatory role. When the textile designer is involved with sustainability, one of the roles that must be fulfilled is that of activist or generator of objects gestated with community participation. This concept is based on the principle of participatory democracy and, in reference to this, sustainable textile design works on the concept of a community participation plan. The professionalization of the design led to the fact that in teaching the importance was in the competitiveness of the product, in this way the object reflected the mentality behind its design process. Currently the designer must depart from the role he had, in the past, as a designer, director and supervisor, to get involved in a riskier and more committed field. It must interact with the performers of each stage, know each process, learn about work methodologies, know the culture, its inhabitants and ensure the welfare of each worker throughout the life cycle of the product. It must also provide knowledge about manufacturing processes, supervise the textile supply chain, control fair trade for workers in all areas and other business issues to improve industrial, social and ecological behavior.

Knowing all these variables means that the current designer acquires a global view of the product it generates.

Sustainability, like other currents and ideologies, offers a new way of understanding the world and the place that man occupies in it. This construction reinforces the concept of "objects as bearers of ideas" and must be gestated and executed for society, introducing the new designer at the heart of it, according to Ariel Guersenzvaig in *Los mitos del diseño* (1975), on concepts of Victor Papanek, not with an egocentric or egomaniac look but as a generator of solutions to specific needs (design for the need) and with a social function. Because beyond the aesthetic in the genetics of design we find its objective which is to improve the quality of life of people. The designer is a connector between technical-technological and social innovation.

These concepts of participatory design are fundamental in the writings of Ezio Manzini who responds to a view of the designer as a facilitator or solicitor of consumer objects and service innovations with a social orientation. He urges the designer to consider himself part of the community with which he collaborates by working as equals with other members. The designer is and must be considered part of the community to which he belongs. Under this paradigm Manzini affirms that, in the 21st century, social innovation will be intertwined with design as it happened with design and technique in the 20th century. Social innovation will drive design to another stage, this concept of "design for social innovation" being an alternative to achieve sustainability. The author understands that "design has the necessary potential to play a prominent role as a trigger and support for social change and, therefore, to become a design for social innovation", this is not a new design discipline, but "one of the ways in which the current design materializes" (Manzini, 2015, p. 71).

Undoubtedly, the role of the textile designer today has changed. While designing is creating, building, innovating, inventing, discovering, generating, among so many other verbs that we could use to define there is a new use in the terminology associated with the designer that manifests itself in the discourse of the authors studied: active, participative, dynamic, facilitator, communicator, educator; all these terms that indicate an action, a doing. It is not the objective of this paper to make a semiotic analysis of this terminology, but it can be taken into account that this new concept (or action) of doing is intended depending on another. Do for and with the other. Complementing knowledge, working in a linked way, grouping knowledge to improve the quality of life of the society to which we belong. This new participatory role of the designer also involves getting involved in a new scenario in which multiple actors (social, environmental, economic, cultural, business, among others) participate, with whom he will have to live and share for centuries. Therefore, the multidisciplinary and interlaced work (which weft and warp threads) of variables could be the key in this reconfiguration or new role of the designer. Each individual, from the function or place he occupies in his community, is functional to it.

4 SOME REFLECTIONS AND QUESTIONS

This new social, technological, sustainable and changing roles scenario poses challenges, questions and contradictions: will sustainability be a construction of thought in the design disciplines? Is it possible, in the near future, to design fully sustainable textile products? Will these textile objects be functional to the community? Can the designer integrate "others" (non-designers) in various processes of the fashion industry and the textile industry? Will sustainable textiles be consumer or luxury products? Will sustainable objects be available to everyone? Will sustainable fashion or sustainability be a fashion? Will the new models of social innovation be those applied in the fashion industry? Although all these questions cannot be answered at present, they are triggers of future investigations, case studies and reflections that allow us to reveal these unknowns.

Some cases of study in Argentina, whose analysis is the subject of another writing, show that, until now, the reconfiguration in fashion and textiles proposed by sustainability has nodes that have permeated and have been installed in the cities of large capital cities. In general terms, some luxury brands implement, at some stage of the life cycle of an object, this reconfiguration. While in the less urbanized, almost rural areas, the life cycle of a garment or piece of textiles is presented in a more sustainable way, almost intuitively, due to the care of the environment, the due use of resources and the collective work of the communities. The reconfiguration proposed by sustainability is like a ball that is woven and unwoven at the same time. It remains to be seen how it is applied and installed in the field of the fashion and textile industry.

REFERENCES

Bauman Z. 2011. Daños Colaterales. Desigualdades sociales en la era global. Buenos Aires. Fondo de Cultura Económica, 1ª ED.
Ehrenfeld J. 1999. Cultural Structure and Challenge of Sustainability. In Sexton, K., Marcus, A.A., Easter, K.W. and Burkhardt, T.D. Better Environment Decision. Strategies for Government, Businesses and Communities. Washington. Island Press.
Fletcher K. & Grose L. 2012. Gestionar la sostenibilidad en la moda. Diseñar para cambiar. Barcelona. Blume.
Frankel C. 1998. In Earth's Company. New Society Publishers.
Guersenzvaig A. 1975. Los mitos del diseño (Edugrafología). Iconographic Nº 9 (Croydon, Inglaterra) Recopilado en Fundamentos del diseño gráfico. Fuente: Interacciones.org.
Manzini E. 2015. Cuando todos diseñan. Una introducción al diseño para la innovación social. Madrid. Experimenta Theoria.
Manzini E., Jerou F. & Corporate Services 2012. Diseño e innovación social para la sostenibilidad. Parte primera, A Fondo, Newsletter Experimenta 63, 10/10/2012. Fuente: artículos mártir del diseño, agosto 2005. Extracto de un artículo de Enzio Manzini publicado en Designmatters 10. Design to Improve Life. Danish Design Centre. Summer 2005. WWW. ddc.dk/ Designmatters _ 10_dansk.
Sanoff H. 2008. El diseño participativo. In ITESO, 04/12/2008.
Schleicher Tappeser R. 2001. Assessing Sustainable Development in the European Union – The Sustainable Quality Management Approach in the Context of Structural Funds. European Conference. Seville.
Walker S. 2006. Sustainable by Design: Explorations in Theory and Practice. London. Earthcan.
WCDE, World Commission on Environment and Development 1987. Our Common Future. Oxford: University Press.
Williams D. 2015. Sustainable Apparel: *Fashion Design and Sustainability.* Cambridge. Woodhead Publishing.

Tucum fiber: Reflections about Amazonian biodiversity, traditional knowledge and sustainable fashion

L. Pennas & J. Baruque-Ramos
School of Arts Science and Humanities of the University of São Paulo (EACH-USP), São Paulo, Brazil

ABSTRACT: Tucum fiber is originated from the leaf of *Astrocaryum chambira* Burret, an Amazonian palm tree species. The process, which begins with the removal of the fiber until the production of a handmade yarn known as "Linha do Tucum", is part of the knowledge of "Vila Ecologica Ceu do Jurua" community located in the municipality of Ipixuna (Amazonas, Brazil). This study aims to reflect on cultural and environmental aspects related to the Tucum fiber, addressing the concepts of Amazonian biodiversity, traditional knowledge and sustainable fashion found in this scenario. It may be concluded that, being a fiber removed from a palm native from Amazon region and, because the processes of extraction, artisan spinning and natural dyeing are part of the knowledge of a community, the Tucum involves aspects of Amazonian biodiversity and traditional knowledge and is regarded as a promising possibility of providing raw materials for sustainable fashion.

1 INTRODUCTION

The fiber of the Tucum is a vegetal fiber, extracted from the leaves of the palm tree *Astrocaryum chambira* Burret, which is employed by people from the Amazon region for the manufacture of handicrafts and other utensils (Ferreira 2005).

In the community of "Vila Ecologica Ceu do Jurua" (located in the municipality of Ipixuna, southwest of Amazonas, near the border with Acre state, Brazil), the fibers of the tucum palm are removed, and through traditional knowledge maintained from generation to generation, used for the artisanal production of a yarn known regionally as "Linha do Tucum", with which the members of the community produce handicrafts (Abreu & Nunes 2012).

This process of harmonious coexistence with the forest, left by the ancient populations, together with the current popular knowledge are true cultural heritage, and so must be valued and preserved by the present generations, to ensure that future generations can also enjoy these valuable Amazon resources and benefits (Sousa et al. 2015).

According to Brown and Freitas (2002), the Jurua Valley Extractivist Reserve, southwest of the Amazon Region, is considered one of the richest regions in biodiversity around the world and the interventions carried out by the traditional communities of the Jurua Valley contribute to its conservation. In this way, according the current discussion on the role of communities in environmental conservation, which is concentrated on the integration of traditional knowledge with scientific knowledge (Hanazaki et al. 2010; Bohensky & Maru 2011).

The term "sustainable fashion" emerged in the early 2000s as a new way of producing and consuming fashion that would be environmentally, socially and economically responsible (Thomas 2008). Terms such as "Green Fashion" and "Eco Fashion" were previously used in the 1990s, but the term "Sustainable Fashion" is associated with the concept of sustainable development (Hethorn and Ulasewicz 2008) in order to ensure that the needs of the current generation are addressed without compromising the needs of future generations (WCED 1987).

Thus, the present study aims to reflect on cultural and environmental aspects related to the Tucum fiber, addressing the concepts of Amazonian biodiversity, traditional knowledge and sustainable fashion within this scenario.

2 BIBLIOGRAPHICAL REVIEW

2.1 *Amazonian biodiversity*

Amazon Forest is one of the richest areas of biodiversity on the planet. With an area of approximately 5.5 million square kilometers, 3.6 million of which are in Brazil, this region offers an incalculable wealth of diversity of organisms, as well as the largest concentration of fresh water on the planet (Marinelli et al. 2008).

As Brazil is a megadiverse country, it has an enormous potential for the exploitation of biodiversity to obtain natural products, pharmaceuticals, textiles, cosmetics, food, among others, allied to environmental conservation, especially considering how little is known about this biodiversity (Ferro et al. 2006).

This extraordinary richness, with abundant natural resources, favorable climatic conditions and fertile soil, enables a great production of vegetable fibers in Brazil, such as: cotton, banana fiber, coconut fiber, sisal, jute, mallow, piassaba, buriti, pineapple, curaua, etc. (Santos 2013).

2.2 *Traditional knowledge*

Taking in account the global environmental crisis and its innumerable possibilities that have somehow stimulated the recognition of the social diversity which conforms the Amazon region, it has become recurrent the need to rethink the use of biodiversity, adding value to extractive products and holding in high regard the traditional knowledge of forest peoples (Porto-Gonçalves 2001).

Knowledge of Amazonian biodiversity and natural resources is part of the culture of the forest peoples. This knowledge is transmitted through the generations and is one of the main sources of information about this region (Abreu & Nunes 2012).

The direct dependence of traditional communities on the environment constitutes a modality of the relationship between society and nature (Diegues & Arruda 2001; Toledo & Barrera-Bassols 2009). It is a relation considered of low environmental impact (Cunha & Almeida 2000), which can, therefore, to propitiate the conservation of the biological diversity (Shen et al. 2012; Silva & Fraxe 2013). In this sense, faced with the environmental crisis, the way of life and the knowledge of these populations have been listed as important instruments for environmental conservation (Pereira & Diegues 2010).

2.3 *Sustainable fashion*

Sustainable fashion does not have an exact definition; rather, it is an umbrella term for a variety of pro-activities and practices adopted to achieve each specific situation in which sustainability issues appear. Some alternatives give more attention to environmental practices, others to social practices, but everything is interconnected. As the raw material is the base element of the fashion industry, innovations related to this field are the most common approaches to sustainability in fashion (Aakko & Koskennurmi-Sivonen 2013).

However, naming fibers as "green" or "eco-friendly" is not enough because that is not analyzed the whole context, apart from the fact that none fiber is considered 100% sustainable. All of them present their strengths and their weaknesses (Chen & Burns 2006).

According to Berlim (2009), what pertains to sustainability in Fashion, involves both research related to the development of products from ecological materials, as the choice of fibers, and the manufacturing process. However, it also covers social, environmental, economic, political and cultural aspects, analyzing everything from the production process to the consumption and disposal of these products.

Slow fashion arises as a vital alternative to the current fashion system, emphasizing aspects such as quality, durability and diversity of styles, which helps to build an appreciation for the pieces, reducing the rhythm of consumption. In this way, a balance is sought in the economic,

Figure 1. (a) Tucum palm (*Astrocaryum chambira* Burret).
Source: fogoepalmeiras.wordpress.com (b) Tucum Palm (*Astrocaryum chambira* Burret).
Source: jb.com.br.

social and environmental spheres, promoting small-scale production, handicrafts and traditional techniques, consequently offering a starting point that differs from the current fashion system (Fletcher, 2010).

2.4 Tucum

2.4.1 Tucum palm (Astrocaryum chambira Burret)

The scientific name of the tucum palm is *Astrocaryum chambira* Burret and its scientific classification is as follows: Kingdom – Plantae; Division – Magnoliophyta; Class: Liliopsida; Order: Arecales; Family – Arecaceae; Genus – Astrcaryum; Species – *A. chambira*; Binomial Name: *Astrcaryum chambira* Burret (Kahn 2008).

The *A. chambira* palm (Figure 1(a) and (b)) is botanically characterized by having a solitary, erect stipe, up to 15 m long and 30 cm in diameter, internodes densely covered by black spines up to 20 cm long (Galeano & Bernal 2010).

It can be found along the Amazon region of Colombia (Amazonas, Caqueta, Guaviare, Meta, Putumayo and Vaupes), Venezuela (Amazonas), Ecuador (Morona-Santiago, Napo), Peru (Amazonas, Loreto) and Brazil (Acre, Amazonas), in wetlands at an altitude of 100–500 m (Henderson et al. 1995; Galeano & Bernal 2010).

2.4.2 "Vila Ecologica Ceu do Jurua" community

The Jurua Ecological Village community is located in the region formerly known as Estorroes, one of the productive units of the "Seringal Adelia" (hevea or rubber plantation), in the Jurua Basin, southwest of the state of Amazonas in the municipality of Ipixuna, GPS 6° 50' and 6°55'S; and 71°15' and 71°10' W, near the border with Acre.

The process of manufacturing of the yarn known as "Linha do Tucum" begins from the fibers extracted from the palm leaf of *A. chambira*. It is traditional knowledge passed from generation to generation. It involves craftsmanship and can be divided into four stages: collecting the leaves, removing the fiber (which is referred as "linen"), combing and pulling the "linen", spinning and warping the yarn (Nunes 2009; Abreu & Nunes 2012).

3 MATERIALS AND METHODS

According to the conceptualization of Gil (2008), the present study is a qualitative approach, with exploratory and descriptive objectives. The employed procedures were the bibliographical research developed from literature already elaborated, consisting mainly of books and scientific articles.

3.1 *Cultural and environmental aspects related to the Tucum fiber*

The information about the "Vila Ecologica Ceu do Jurua" and about fiber management were obtained from the project carried out with this community, between 2008 and 2009, which resulted in the elaboration of the book "Linha do Tucum: artesanato da Amazônia" (Froes et al. 2010). This project was executed by a multidisciplinary team in partnership with "IECAM – Instituto de Estudos da Cultura Amazonica" (Institute of Studies of Amazonian Culture). The scientific article "Weaving Tradition and Valuing the Traditional Knowledge of the Amazon: the case of the tucum yarn" by Abreu and Nunes (2012) and the documentary "The Tucum yarn, the yarn of loyalty", directed by Noilton Nunes (2009) also were consulted.

4 RESULTS AND DISCUSSION

4.1 *Environmental and cultural aspects related to the Tucum fiber*

Belonging to the domains of the Western Amazon Rainforest, the Middle Jurua region is considered one of the biological megadiversity areas, i.e., presenting extremely high biological importance (Brown & Freitas 2002).

In the mythical conception of traditional societies, there is a symbiosis between man and nature, both in the field of activities, techniques and production, as in the symbolic field (Diegues 2008). Marques (2001) demonstrates that there is a relationship of respect, gratitude, fear and complicity with nature, which is a direct cause of the environmental preservation of the localities the traditional populations inhabit.

For the people of the Jurua Ecological Village community, the yarn that is spun with the fibers of tucum, received as traditional knowledge, is also known as "Linha da Lealdade" ("loyalty yarn"), because it is considered by them of great utility and importance for the life in the forest, having enabled, for many years, the making of important daily life objects as nets, casting nets, lines for hooking, rope to tie canoes, among others. The art of Tucum fiber spinning is ancestral, comes from the indigenous Kulina and Katukina peoples and, through cultural exchanges, this knowledge was passed to the rubber tappers who came to inhabit the region (Froes et al. 2010).

As mentioned before, the yarn production process can be divided into four stages: collecting the leaves; removing the fiber; combing to pull the "linen"; spinning and warping the yarn. All the steps are performed by the community residents themselves (Nunes 2009; Abreu & Nunes 2012).

The natural dyeing is a step apart that employs many species of the region to color the Tucum yarn. Among the many coloring materials found in the forest, jatoba (*Hymenaea courbaril*), urucum (*Bixa orellana*), *Hibiscus acetosella*, Brazil wood (*Paubrasilia echinate*) and acai (*Euterpe oleracea*) can be mentioned as some species used by the local people in dyeing,

Figure 2. Tucum fiber. Source: Author's Collection.

Figure 3. Tucum artisanal Piece.
Source: Nunes, 2009.

rescuing cultural practices and valuing the final product. The Tucum fiber is used only in the production of local handicrafts, for example, bags and necklaces made with macrame technique with the Tucum yarn and seeds of native species (Froes et al., 2010).

As an example of a Tucum fiber employment as an artisanal product, the Figure 3 shows a craft piece made by the people who lives there in the Jurua community. In this case, the craftsmen employ macrame, technique to weave yarns that does not use any type of machinery or tool. In the macrame the work is made with the fingers to cross the yarns and to tie it with knots, creating fringes, geometric crosses and other shapes (Redeasta 2014).

4.2 *Reflections on sustainable fashion, Amazonian biodiversity and traditional knowledge*

As mentioned before, the term sustainable fashion involves practices, activities and choices that encompass both social aspects and fairer modes of production; economic features of the fashion market; as well as environmental aspects, such as processes and raw materials that are less damaging to the environment as a whole (Fletcher & Grose 2012).

When analyzing the environmental aspects of fibers, it is usually taken into consideration the amount of employed energy and the pollution caused by the production and processing of such fibers, whether it is a renewable source fiber or not, as well as its biodegradability and the recycling possibilities (Chen & Burns 2006). On the other hand, even with current methods of life-cycle analysis, it is difficult to tell which fiber is better. Natural fibers cannot be treated as the best environmentally friendly option, simply because they grow in nature (Grose 2009).

Sarmento (2014) states that it is inefficient to think only of the product life cycle without understanding the socio-environmental connections existing in each material but believes that most designers still pay little attention to activities that could favor the maintenance of biodiversity and aggregate services, working with materials often without even questioning their origin, where they come from, who extracts them and how the extraction is made.

The concept of socio-biodiversity seeks to integrate nature conservation with the cultural approach, encompassing products, knowledge, habits and traditions proper to a particular place or territory (MDA, 2012 apud Sarmento 2014).

The knowledge of traditional populations, according to Arruda and Diegues (2001), is composed of the "set of knowledges and know-hows regarding the natural and supernatural world, transmitted orally, from generation to generation." Schmidt (2001) adds that the construction of this knowledge has a broad relation with the physical and social environment inhabited by these populations.

It is understood that the valorization of traditional knowledge can be perceived as a way of conserving local biodiversity—that is to say, the biological diversity of the areas in which these populations are found depends on the continuity of resource management (Pereira & Diegues 2010).

The current discussion on the role of communities in environmental conservation is based on the integration of traditional knowledge with scientific knowledge (Hanazaki et al. 2010; Bohensky & Maru 2011) and the inclusion of the inhabitants in the management of natural resources for conservation and sustainable use (Charnley et al. 2007; Potter-Bolland et al. 2012). Knowledge dialogues between academia and society can be seen as a two-way street: the academy has a lot to learn from peoples who, in turn, have something to learn from academics (Barros et al. 2015).

The social value associated with the traditional textile arts of handmade spinning, manual weaving, natural dyeing, among others, together with the resurgence of these values and interests, make the ideals of sustainable fashion a reality with social, economic and cultural impacts (Trejo 2014).

Thus, according to all the questions pondered, it can be inferred that the process of production of the Tucum yarn, linked to the traditional knowledge passed from generation to generation, associated to the preservation of the biodiversity present in its surroundings, promoted by the community that produces it, may be related to aspects of cultural, social and economic sustainability. This could provide controlled production of raw materials (fibers, yarns, fabrics, natural dyes, etc.) that might turn out very interesting, from both a material and aesthetic point of view, to sustainable fashion. Of course, this study intends to give a first qualitative perspective on this possibility, about which more in-depth qualitative and quantitative studies should be carried out in order to make this promising prospect feasible.

5 CONCLUSIONS

Sustainable fashion comes as a movement, which slows down the pace and the way fashion is currently being made and seeks an environmental, economic and social balance.

The search for new textile raw materials, alternatives to those commonly used, with less environmental impact, represent one of the bases that "making sustainable fashion" can lean on. Ancestral artisan knowledge, such as spinning and hand weaving, natural dyeing, among others, is being rescued by those who want to make fashion conferring greater emotional value to the handmade and, therefore, is also an example of ways that sustainable fashion can follow.

In addition to that, it was possible to understand from the study that the recognition of worth of a traditional knowledge can indeed be a form of conservation of the local biodiversity, since the relation that these people have with nature involves respect, dependence and gratitude, which means that their management of the natural resources can be considered of low environmental impact. Valuing a traditional knowledge of a population may be the same as keeping a species alive.

It may be concluded that the Tucum, because it is a fiber removed from a palm native to the Amazon region and its processes of extraction, artisan spinning and natural dyeing are part of the knowledge of a community, involves aspects of Amazonian biodiversity, conservation and traditional knowledge, and constitutes a promising possibility of providing raw materials for sustainable fashion.

REFERENCES

Aakko, M. Koskennurmi-Sivonen, R. 2013. Designing Sustainable Fashion: Possibilities and Challenges. *Research Journal of Textile and Apparel* 17:13–22.

Abreu, R. Nunes, N.L. 2012. Tecendo a tradicao e valorizando o conhecimento tradicional na Amazonia: o caso da" linha do tucum". *Horizontes Antropologicos* 18(38): 15–43.

Barros, F.B. Porro, N.S.M. Moreira, E. Matos, K. 2015. Dossie "Direitos e Conhecimentos de Povos e Comunidades Tradicionais". *Fragmentos de Cultura* 25(2): 139–142.

Berlim, L. 2009. *Moda, a possibilidade da leveza sustentavel: tendencias, surgimento de mercados justos e criadores responsaveis*. Masters Dissertation. Universidade Federal Fluminense.

Bohensky, E.L. Maru, Y. 2011. Indigenous knowledge, science and resilience: what have we learned from a decade of international literature on "integration"? *Ecology and Society* 16(4): 6.

Brown Júnior, K. Freitas, A.V.L. 2002. Diversidade biologica no Alto Jurua: avaliacao, causas e manutencao. In: Cunha, M.C.; Almeida, M.B. (Org.). *Enciclopedia da Floresta. O Alto Jurua: praticas e conhecimentos das populacoes*. Sao Paulo: Companhia das Letras.

Charnley, S. Fischer, A.P. Jones, E.T. 2007. Integrating traditional and local ecological knowledge into forest biodiversity conservation in the Pacific Northwest. *Forest Ecology and Management* 246:14–28.

Chen, H-L. Burns, L.D. 2006. Environmental Analysis of Textile Products. *Clothing and Textiles Research Journal* 24: 248–261.

Clark, H. 2015. Slow + Fashion – an Oxymoron or a promise for the future? *Fashion Theory* 12:427–446.

Cunha, M.C. Almeida, M.W.B. 2000. Indigenous people, traditional people, and conservation in the Amazon. *Daedalus* 29(2): 315–338.

Diegues, A.C. Arruda, R.S.V. 2001. *Saberes tradicionais e biodiversidade no Brasil*. Sao Paulo: MMA/COBIO/NUPAUB/USP.

Diegues, A.C. 2008. *O mito moderno da natureza intocada*. 6 ed. Sao Paulo: Hucitec.

Ferreira, E.L. 2005. *Manual das palmeiras do Acre, Brasil*. Rio Branco: Instituto Nacional de Pesquisas/Universidade Federal do Acre.

Ferro, A.F.P. Bonacelli, M.B.M. Assad, A.L.D. 2006. Oportunidades Tecnologicas e estrategias concorrenciais de gestao ambiental: o uso sustentavel da biodiversidade brasileira. *Gestao e Producao* 13(3): 489–501.

Fletcher, K. 2010. Slow Fashion: An Invitation for Systems Change. *Fashion Practice* 2: 259–266.

Fletcher, K. Grose, L. *Fashion & Sustainability: Design for Change*. London: Laurence King Publishing.

Fogo e Palmeiras. 2013. *Ecologia de fogo e impacto nas palmeiras*. [Online] Available at: https://fogoepalmeiras.wordpress.com/palmeiras-do-alto-jurua/imagens-palmeiras/#jp-carousel-463 [Accessed 22 Mar. 2018].

Froes, V. Domingues, G. Ferreira, E. Quinet, A. Abreu, R. Abreu, N.L. Katz, N. 2010. *Linha do Tucum: Artesanato da Amazônia*, Rio de Janeiro: Instituto de Estudos da Cultura Amazonica.

Galeano, G. Bernal, R. 2010. *Palmas de Colombia: Guia de campo*. Bogota: Universidad Nacional de Colombia, Facultad de Ciencias, Instituto de Ciencias Naturales.

Gil, A.C. 2008. *Como elaborar projetos de pesquisa*. Sao Paulo: Atlas.

Grose, L. 2009 'Sustainable cotton production', In: Blackburn, R.S. *Sustainable textiles: Life cycle and environmental impact*. Oxford: Woodhead Publishing.

Henderson, A. Galeano, G. Bernal, R. 1995. *Field guide to the palms of the Americas*. Princeton (New Jersey): Princeton University Press.

Hethorn, J. Ulasewicz, C. 2008. *Sustainable fashion: Why now? A conversation about issues, practices, and possibilities*. New York: Bloomsbury Academic.

Jornal do Brasil. 2011. *Palmeiras Tucum no Rio de Janeiro? Tem sim senhor!* [Online] Available at: http://jb.com.br/ciencia-e-tecnologia/noticias/2011/08/11/palmeiras-tucum-no-rio-de-janeiro-tem-sim-senhor/ [Accessed 04 Apr. 2018].

Kahn, F. 2008 El genero Astrocaryum (Arecaceae). *Revista Peruana de Biologia* 15: 31–48.

Kawamura, Y. 2005 *Fashion-ology: An Introduction to Fashion Studies*. Nova York: Bloomsbury Academic.

Lira, S.R.B. Silva, M.L.M. Pinto, R.S. 2009. Desigualdade e heterogeneidade no desenvolvimento da Amazonia no seculo XXI. *Revista Nova Economia* 9(1): 153–184.

Marinelli, A.L. Monteiro, M.R. Ambrosio, J.D Branciforti, M.C. Kobayashi, M. Nobre A.D. 2008. Desenvolvimento de compositos polimericos com fibras vegetais naturais da biodiversidade: uma contribuicao para a sustentabilidade amazonica. *Polimeros: Ciencia e Tecnologia* 18(2): 92–99.

Marques, J.G. 2001 *Pescando pescadores: ciencia e etnociencia em uma perspectiva ecologica*. Sao Paulo: Nucleo de Apoio a Pesquisa sobre Populacoes Humanas e Areas Umidas Brasileiras.

Ministerio do Desenvolvimento Agrario. [Online]. Available: <http://www.mda.gov.br>. [Accessed 20 Mar. 2018].

Nunes, N. 2009. *Linha do Tucum: a linha da lealdade*. [Online] Imagine Filmes. Available: <http://youtu.be/RpbJH8lgzJ0>. [Accessed 12 Jan. 2018].

Pereira, B.E.; Diegues, A.C. 2010. Conhecimento de populacoes tradicionais como possibilidade de conservacao da natureza: uma reflexao sobre a perspectiva da etnoconservacao. *Desenvolvimento e Meio ambiente* 2(22): 37–50.

Porto-Gonçalves, C.W. 2001. *Amazonia, Amazonias*. Sao Paulo: Contexto.

Potter-Bolland, L. Elis, E.A. Guariguata, M.R. Ruiz-Mallen, I. Negrete-Yankelevich, S. Reyes-Garcia, V. 2012 Community managed forests and forest protected areas: an assessment of the conservation effectiveness across the tropics. *Forest Ecology and Management* 268: 6–17.

Redeasta 2014. *Acessorios em Tucum, beleza em um toque.* [Online] Avaiable at: http://redeasta.blogspot.com.br/2014/05/acessorios-em-tucum-beleza-com-um-toque.html [Accessed 22 Mar. 2018].

Santos, F.R.S. 2013 *Desenvolvimento e aplicacao de compositos a base de matriz polimerica reforçado com fibras de curaua (Ananas erectifolius) e residuos de madeiras amazonicas.* Masters Dissertation. Universidade Federal do Amazonas.

Sarmento, F. 2014. *Design para sociobiodiversidade: Perspectivas para o uso sustentavel da borracha na Floresta Nacional dos Tapajos.* PhD final thesis. Faculdade de Arquitetura e Urbanismo da Universidade de Sao Paulo.

Schmidt, M.V.C. 2001. *Etnosilvicultura kaiabi no parque indigena do Xingu: subsidios ao manejo de recursos florestais. Masters Dissertation.* Universidade de São Carlos.

Shen, X. Li, S. Chen, N. Li, S. McShea, W.J. Lu, Zhi. 2012. Does science replace traditions? Correlates between traditional Tibetan culture and local bird diversity and Southwest China. *Biological Conservation* 145: 160–170.

Silva, F.P. Fraxe, T.J. 2013. Saberes De Populacoes Tradicionais: Etnociencia em Processos de Bioconservacao. *Contribuciones a las Ciencias Sociales*, 2013(08).

Sousa, S.G.A. Araújo, M.I. Wandelli, E.V. 2015. Saberes tradicionais dos povos amazonicos no contexto do processo de transicao agroecologica. *EA Saberes Tradicionais/Alternativos – Ambientalmente sustentable* 2(20): 1699–1717.

Thomas, S. 2008 From 'green blur' to ecofashion: Fashioning an eco-lexicon. Fashion Theory – *The Journal of Dress, Body and Culture* 12(4): 525–539.

Toledo, V.M. Barrera-Bassols, N. 2009. A Etnoecologia: uma ciencia pos-normal que estuda as sabedorias tradicionais. *Desenvolvimento e Meio Ambiente* 20: 31–45.

Trejo, H.X. 2014 *Exploring the New York slow fashion value chain: Local Animals, Fibers, and Knitwear.* Masters Dissertation. Cornell University.

WCED. *Our Common Future.* World Commission on Environment and Development. 1987. [Online] Available: < http://www.un-documents.net/our-common-future.pdf> [Accessed 06 Oct. de 2017].

Silk: Protocols for the verification of socioeconomic impacts in the production of cocoons

S.M.B.D. Barcelos
State University of Maringá (UEM), Maringá, Brazil
University of Minho (UM), Cianorte, Paraná, Brazil

M.G. Guedes
University of Minho (UM), Azurém, Guimarães, Portugal

E. Pinheiro
State University of Maringá (UEM), Maringá, Brazil
Federal University of Technology—Paraná (UTFPR), Ponta Grossa, Paraná, Brazil

A.C. de Francisco
Federal University of Technology—Paraná (UTFPR), Ponta Grossa, Paraná, Brazil

ABSTRACT: Paraná is the largest Western producer of silk yarn, considered the best quality yarn in the world. The present study focuses on the living and work conditions of producers of silk in Paraná. The primary goal is to map the socio-economic impacts resulting from silk production activity and propose an intervention model to enhance producers revenue and personal and social satisfaction. The research focused topics such as sustainability in their socio-economic aspects, sustainable fashion, the use of silk in fashion products, the importance to silk producers of SIA and EIA methodologies. To perform this study, it was necessary to gather in-depth information on the primary causal factors of these impacts, and the method selected was the design and application of three questionnaires addressed to silkworm producers. The results demonstrated we developed a useful research tool to gather social and economic information among this specific population.

1 INTRODUCTION

In the sericulture sector, Brazil is the fifth largest producer of raw silk yarn in the word and the only producer on a commercial scale in the West (Abit, 2017). Paraná is responsible for 83% of the production of silkworm cocoons (Bombyx mori). The production is concentrated in rural areas and depends on family labor. It is a well-established activity in the region and generates the income the families need for their survival. In fact, as silk production is concerned, the municipalities of the Northwestern State of Paraná are the most prominent producers (Cirio, 2017).

However, although the activity is vital to the families and the region income, the situation is not stable due to a considerable amount of factors. The lack of investment in infrastructures, and technologies, as well as the low education level, and the lack of professional training of producers create severe difficulties in developing production quantity and quality. One possible solution to reverse this situation, and increase the local added value across the social environment, could be the adoption of fair trade practices linked with solidarity economy principles, and development of political economy (Pereira et al., 2009).

The intervention over local silk productions and producers requires the collection of accurate information what is not simple nor easy due to social and cultural contexts surrounding researchers and the observed population. In consequence, we aimed to develop research protocols based on SIA (Social Impact Assessment) and EIA (Economic Impacts Assessment) methods, oriented to collect data on the social and economic impacts present in the production of silk cocoons in the context of the rural family economy.

1.1 Aspects of social and economic sustainability

Sustainable development in its various aspects has often prioritized in the discussions, the environmental sphere, as it relates to the production and consumption have the speed equivalent to the reimbursement of the raw materials used, thus preserving the ecosystem. In the economic sphere, it is assumed the expansion of the efficiency of production and consumption, with a considerable economy of natural resources and vulnerable resources or poorly distributed (Nascimento, 2012).

Economic development occurs in the institutional internal means, through productive activities, commercial transactions, negotiations, relationships and supply chains among other aspects, that make institutions allied and determining for sustainable development (Munck & Souza, 2009).

In the social sphere, sustainability encourages equality among the citizens, which can benefit equally from improvements achieved in the region, and there is no absorption of goods, natural resources, and energy that are harmful to a population part, caused by the conflict of interest. Thus, cultural factors influence profoundly economic sustainability since it is mostly related to changes in lifestyle, values, and habits. Local consumption patterns much move toward the prioritization of quality and enjoyment values, instead of the volume of goods in quantity (Nascimento, 2012).

1.2 Silk as a fashion product

Silk and products derived from silk have particular properties not observed in any other fiber. Silk presents low density and excellent insulation properties which makes it comfortably light, and warm in cold weather and fresh in hot weather. Silk presents high resistance to deformation, and is the most durable natural fiber, what makes it an excellent material to produce fashion items that are durable and keep its good shape longer. Silk also dyes easily, offers delightful sensations to the touch and presents a soft and pleasing shine. This material is renewable, and its production requires low capital investment, and depends mostly on workers know how and the capability to cultivate mulberry to feed the larvae. Silk is a noble material, and fashion always used it in the most exquisite fashion items. Nowadays world production of silk is higher than ever what makes it affordable to traditional and technic applications, and it is used to manufacture yarns blended with other fibers enhancing yarn quality, without rising prices excessively. When fashion made sustainability a brand value materials, quality and performance are critical. Good quality fibers such as silk are now used not just by high-end fashion but also by most good quality fashion designers. The increased demand is satisfied mainly by China and India that produce 84% and 14% of world silk. Brazil is the 5th world silk producer and the most significant western producer. Brazil silk production is most important since the country is also investing in fashion markets. Brazilian designers are conquering growing prestige in world fashion, and fashion brands begin to follow sustainability trends. Silk production is essential not just to answer world silk demand but also to support Brazilian designers demand in competitive and differentiated conditions. Thus, the association between the local fashion industry and the silk industry is essential (Inserco, "n.d.").

The IFM—Institut Français de la Mode developed a survey to find out what was world consumers' perception of silk. The study showed that consumers with high cultural, high-income, and high formal education levels are the ones that appreciated silk the most. Silk and its marketed products are synonymous with status, luxury and power mainly in Europe, USA, Japan and Canada (Garcia, Oliveira & Pereira, 2011).

The comfort and enjoyment of products that exploit the softness, smoothness, and sensuality are rated Relax Silk. Already the Cultural Silk is related to social and environmental values of some cultures, wherein some products is inserted local identity, and its consumption is in the form of social action linked to strong emotional capital that integrates silk. The Japanese kimono is an example of a traditional product made of silk, where the mystical side returned to the Cultural Fashion Silk (Garcia, Oliveira & Pereira, 2011).

Products such as sweaters, sweatshirts, scarves, shirts, dresses, wallets, clutches, ties and socks are some of the products that can be made with silk or silk blends (Inserco, "n.d.").

1.3 *The socioeconomic importance of silk in Paraná*

The world textile and clothing value chain is exceptionally dynamic, and although is changing towards a new rhythm, and a slow design approach, world fashion market are continuously launching new products, and services. The annual fiber consumption is of 89.1 million tons, 70% of chemical fibers and 30% natural fibers. This consumption resulted in 84 million tons of yarn, fabric, mesh and made. However, Brazil has a small share of the global market, occupying the 23rd position in the ranking of world exporters, despite being a major producer and consumer of textile and clothing industries (Abit, "n.d.").

In Brazil, textile and clothing industries (TCI) represent 5.7% of the total industrial production, accounting for $58.2 billion a year. From a social point of view, they represent 16.4% of total industrial jobs, employing 1.6 million workers. TCI has a substantial social impact in Brazil, and the regions more specialized are even more dependent on these industries performance (Iemi, 2013).

Focusing on the textile production for sericulture, Brazil produced in the last harvest 2.99 tons of green cocoons, and the State of Paraná accounts for 83% of this production, which is 2.46 tons of green cocoons. Paraná produces 100% of Brazil's silk exports (82% of exports are of silk yarn, and 18% is of non-carded silk waste, not combed). The green cocoons harvest of 2016/2017 counted with the production of 161 Paraná municipalities that occupied 1,867 sericulture workers, 2,017 sheds, and productivity of 622 kg/ha of mulberry. The success of silk production (or the need of populations to ensure more family's revenue) brought a significant growth on sericulture exploitations, with 30% more new silkworm producers. The new producers reduced the impact of old workers retirement exit, and brought a beneficial effect on product quality as the average price per kg of first quality green cocoons in the last harvest was R$ 17.46 and US $ 5.54 correspondingly, representing a growth of 3.4% in the average price for kg in the previous year (Cirio, 2017).

The textile industry in Brazil presents an odd ratio total textile/total silk exports: Brazil occupies the fifth position among the largest textile producers in the world with 89.1 million tons of fibers, and occupies the same position among the largest world silk producers with only 2.99 tons.

National textile and clothing producers can now enjoy available local resources due to sustainable regional development. As far as silk is concerned, in the northwest of Paraná, small businesses networks had been formed to stimulate cooperation in the context of Silk Valley Project, which seeks the concentration of the silk production chain (Garcia, Oliveira & Pereira, 2011).

It was also established the partnership with the Technological Incubator of Maringá supported by teachers and students of the State University of Maringá. The cooperation with the technological Incubator has two goals: one is to develop new products that can be market as Silk Valley original products, and the other is to define silk production processes that improve the quality of cocoons, and consequently, raise the income of the local families. (Technological Incubator of Maringá, 2016).

Sustainability is allied to Fairtrade, which aims to benefit small producers, directing them to the optimization of production and placing them on the market. Several communities can be encouraged to expand output by the government and private institutions as well, to rally to the development of policies and the solidarity economy on fair trade. Consequently, the income of these local producers is expanded, as can be seen in the rural town of New Hope, with the Silk Fair project and the creation of the Cooperative Artisans Brazil (Pereira et al., 2009).

Sustainable projects in which producers can survive and progress independently are essential in disadvantaged regions where resources are scarce. Different sectors united in favor of these social and environmental communities can help human resources to acquire the necessary competencies to raise the regional income (Pereira et al., 2009).

The project "Silk Fair" stands out as an alternative for reuse of low quality and non-marketed cocoons. Those rejected cocoons are processed by a company in Maringá, which extract the low-quality silk, spin, and dye it. The resulting silk yard, low priced, is then distributed among older people living in the rural village of New Hope to produce scarves. This artisans work granted those older villagers an income supplement and enhance their quality of life (Shiohara & Sellos, 2012).

Projects that bring new economic activities to poor and dependent communities are most needed since they generate extra income and opportunities for sustainable growth by stimulating the social entrepreneurship around welfare and the preservation of natural environment (Pereira et al., 2009).

Fairtrade encourages sustainable development by helping small local producers to obtain better work conditions and fair prices. Commercial activities thrive when the integration of social and environmental costs is reflected in prices, which can happen if intercultural communication is used to inform and educates consumers about their influence on building sustainable societies (Pereira et al., 2009).

However, before new sustainable projects can be successfully implemented, is necessary to determine how these partnerships and/or projects affect the people involved, directly or indirectly. The Social Impact Assessment (SIA), and the Economic of the Impacts Assessment (EIA) methods can be used to assess the conditions of production, the benefits and the limitations silk production present, and well as it is possible to guarantee that of the local community needs are fulfilled.

2 METHODOLOGY

The present study is applied research, qualitative, exploratory and descriptive. It was carried out a previous literature review on published studies on the topics discussed, and the development of specific questionnaires focused on gathering information on economic and social aspects related with family life and work conditions of silkworm producers. They are the key players in Paraná silk production, and collect information among them present specific problems since they are not used to share opinions, and expose themselves to outsiders and they are dispersed over a vast territory.

The questionnaires developed for the SIA and EIA were based on Guidance for Assessing and Managing the Social Impacts of Projects (Vanclay et al., 2015); and EVALSED—Assessment of the Socioeconomic Development (European Commission, 2013). After the pre-testing the questionnaires, the research was conducted using focus group technic during the period of delivery of the cocoons produced in the year 2016/2017. The researcher contacted the producers present in the selling area and booked the meeting with them. Once they gather, they received information about the research goals, and the type of questions they would be asked to answer. After this explanation, they manifested if they accepted or not to participate. After that, the researcher presented the questions and allowed the participants to discuss them, freely. All information was analyzed and interpreted at the end of each session.

2.1 Social impact assessment

Social Impact Assessment (SIA) is a process of identifying and managing the social difficulties that affected communities; the process effectively involves the individuals of those communities in participatory processes for identifying, assessing and managing of social impacts of projects. Although, SIA is a method to prevent adverse consequences of projects and a tool-making in the process to evaluate the social contributions and manage the social issues throughout a project development cycle, from creation to extinction (Vanclay et al., 2015).

2.2 Economic impacts assessment

The assessment of the economic impacts (EIA) is based on different theories that drive the interest of several stakeholders (workers, employers, consumers, producers, territories), by comparing scenarios, taking into account the importance of economic activities. Likewise, we evaluate the extent of the impact quantitatively, whether direct or indirect (European Commission, 2013).

The information is acquired through secondary information or documents were drawn up by experts, and field information gathered by surveys. You can draw up a list of the effects, direct or indirect consequences and what the expected changes are (European Commission, 2013). Two questionnaires were developed based on these guidelines and directed to silkworm producers to assess the social and economic impacts in the production of silk cocoons.

2.3 Analysis and results

The two questionnaires were designed to assess the social and economic impacts of the production of silk cocoons over the silkworm producers and their families.

After the development of the research protocols based on SIA and EIA methods, the pre-test was carried out in some rural properties as well as in a silk spinning company. Adjustments were made until the questionnaires were considered entirely viable, allowing the identification of some critical data.

The initial results showed the social characteristics of the target population: the producers working in sericulture are rural people, aged between 25 and 63 years, having as qualifications the elementary school. In general aspects, the place of production is owned by the producers or is granted, 60% work eight (08) to twelve (12) hours per day in the production. 01 (one) to 03 (three) family members also work in the silk production, and 70% of these family members are dependent on the average monthly income, which is approximately three (03) to 05 (five) minimum wages (based on the minimum wage of R$ 937,00).

Among the negatives aspects, some of the most important ones relate to health: 90% of silkworm producers report respiratory allergy to cocoons, that some actions in the production entail in pain or discomfort, such as backaches, physical fatigue, pain in the arms (shoulders), and legs during the activities such as the treatment of silkworm, cut the mulberry trees and cleaning the beds. Also, 90% of silkworm producers have contact with herbicides and/or chemicals like formaldehyde and chlorine. Only 70% reported use PPE (Personal Protective Equipment) as masks, boots, gloves, overalls, and cap.

As leisure activities, 70% watch TV and 60% listen to the radio every day, but rarely or never read newspapers, magazines or books. Besides working in sericulture, 80% say they do not engage in any physical activity.

The questioned silkworm producers mostly demonstrate to be satisfied with their lives and believe that the income earned in the silk production is sufficient to meet the basic needs of the family.

In the economic field, it was possible to ascertain that 100% of the construction of sheds are made autonomously. The silk farms have on average 1.3 to 6 hectares of production of silkworm silk cocoon and mulberry trees plantation, with an average output of 2.4 tons of cocoons per year.

The assistance 100% of the producers have, is labor protection and technical support from the spinning company for the sericulture activity. The company that is the only buyer of silk cocoons production supports producers and their families with food aid and food baskets during the period between harvest and the granting of mulberry seedlings in the case of new planting and/or extension of the planting mulberry. However, repairs are performed autonomously in 77% of the cases. 77% of the producers supplement their income carrying out activities other than sericulture, such as coffee, and soybeans plantation, horticulture, milk production and mason services.

Both questionnaires addressed the main aspects of the social and economic life and work conditions of silk producers. However, since the method of gathering information from this specific population has been tested, it is possible to extend the scope of the issues to consider in the research, such as questions about the political aspect of the social context to assess the level of democratization and participation of the producers in politic decisions that directly affect their lives, as well as measure the level of satisfaction regarding the quality of life they have.

3 CONCLUSIONS

The sericulture is an essential activity with significant economic and social impact in the State of Paraná. Producers and the spinning industry work together and are winning international prestige to the quality of their silk production. In parallel with this economic impact, silk is rooted in the local culture as a traditional activity that generates profitability for small farmers and the existent silk spinning company. However, the preliminary research brought to light a set of negative impacts possibly unknown or underestimated directly linked with production activities. The method used demonstrated to be efficient to gather reliable information and thus allow local policymakers to develop actions to correct negative social and economic impacts of silk production.

Finally, the designed questionnaires proved to be effective tools to researchers and can be extended to cover a broader range of social and economic issues in the sericulture. They can, as well, be applied in other regions that practice this activity.

ACKNOWLEDGEMENTS

The authors gratefully acknowledge the University of Minho, State University of Maringá—UEM and Federal Technological University of Paraná—UTFPR. This work is financed by funds from the ERDF through the Operational Program for Competitiveness Factors—COMPETE and national funds by FCT—Foundation for Science and Technology under the POCI-01-0145-FEDER-007136 project.

REFERENCES

Abit - Associação Brasileira da Indústria Têxtil e de Confecção. 2017. *Perfil do Setor* [Online]. (Updated December 2017). Available at: http://www.abit.org.br/cont/perfil-do-setor. [Accessed 26 December 2017].

Abit - Associação Brasileira da Indústria Têxtil e de Confecção. "n.d". *O poder da moda* [Online]. Available at: http://www.abit.org.br/conteudo/links/Poder_moda-cartilhabx.pdf. [Accessed 26 December 2017].

Cirio G.M. 2017. *Sericicultura no Estado do Paraná*. Safra 2016/2017 - Relatório TAK II [Online]. Available at: http://www.agricultura.pr.gov.br/arquivos/File/deral/Prognosticos/2018/Sericicultura_2016_17.pdf. [Accessed 27 December 2017].

European Commission. 2013. *EVALSED: the resource for the evaluation of Social and Economic Development* [Online]. Available at: http://ec.europa.eu/regional_policy/sources/docgener/evaluation/guide/guide_evalsed.pdf. [Accessed 10 July 2017].

Garcia, J. B. Jr., Oliveira, P. A. & Pereira, M. F. 2011. *Organização de ativos locais para a estruturação e adensamento da cadeia da seda na região de Maringá* [Online]. Available at: http://www.coloquiomoda.com.br/anais/anais/ColóquiodeModa_2011/GT14/ComunicacaoOral/CO_88468_Organizacao_de_ativos_locais_para_a_estruturacao_e_adensamento_da.pdf. [Accessed 30 November 2017].

Iemi - Instituto de Estudos e Marketing Industrial. 2013. *Relatório setorial da indústria têxtil brasileira* [Online]. (Updated November 2014). Available at: http://sites.uem.br/pettextil/relatorio-setorial-da-industria-textil-brasileira. [Accessed 24 December 2017].

Incubadora Tecnológica de Maringá. 2016. *Responsabilidade social: projeto Artisans Brasil – seda justa* [Online]. Available at: http://www.incubadoramaringa.org.br/responsabilidade-social. [Accessed 25 July 2017].

Inserco - International Sericultural Commission. "n.d.". *Statistics* [Online]. Available at: http://inserco.org/en/?q=statistics. [Accessed 7 December, 2017].

Munck, L. & Souza, R. B. 2009. A relevância do ser humano no contexto de institucionalização e legitimação do paradigma da sustentabilidade. *Revista de Gestão USP* [Online], v. 16, no. 3, p. 1–14.

Nascimento, E. P. 2012. Trajetória da sustentabilidade: do ambiental ao social, do social ao econômico. *Estud. av.* [Online], 2012, vol. 26, no. 74, p. 51–64. Available at: http://www.scielo.br/scielo.php?script=sci_arttext&pid=S0103-40142012000100005. [Accessed 20 January, 2016].

Pereira, M. F., Galeti, N. A., Uchida, K. K. & Garcia, J. B. 2009. Criação de sustentabilidade via princípios de comércio justo: o caso da Artisans Brasil. *A Economia em Revista – AERE* [Online], v. 17, no. 2. Available at: http://periodicos.uem.br/laboratorio/ojs/index.php/EconRev/article/view/12945/8024. [Accessed 06 December 2015].

Shiohara, M. & Sellos, V. 2012. O poder público e a coletividade na promoção de políticas públicas socioambientais municipais: o projeto "Seda Justa" no município de Nova Esperança. *Revista Jurídica* [Online], v.1, n.28. Available at: http://revista.unicuritiba.edu.br/index.php/RevJur/article/view/425. [Accessed 03 December 2015].

Vanclay, F., Esteves, A. M., Aucamp, I. & Franks, M.D. 2015. *Social Impact Assessment: guidance for Assessing and managing the social impacts of projects* [Online]. Available at: http://www.iaia.org/uploads/pdf/SIA_Guidance_Document_IAIA.pdf. [Accessed 20 July 2016].

A perspective towards a circular fashion design system

Beatrice Melo & Ana Cristina Broega
University of Minho, Guimarães, Portugal

ABSTRACT: The current fashion design, production and consumption system, known as 'fast fashion', is characterized by the manufacturing of low quality garments in a short period of time carried out in developing countries. In parallel with the deficits in social responsibility and human rights, the prevailing 'take-make-dispose' system in the fashion industry is one of the main causes of environmental pollution that concerns the climate change, scarcity of natural resources and health problems for the living beings. Due to these facts, discussions on Circular Design strategies—for waste reduction, components recycling and materials reuse—became increasingly relevant through the globe. This paper's aspiration is to outline a perspective towards a Circular Fashion. The concept of the Design for Sustainability requires a holistic view throughout designing strategies as well as the establishment of cyclical systems for the production site. Notwithstanding, the efficient integration of social and cultural dimensions are vital for the Sustainable Fashion's triumph.

1 INTRODUCTION

Globalization has provided to large companies the feasibility of manufacturing goods in developing countries. Circumstances such as the leverage of poverty and social inequality together with non-compliance with human rights, the waste accumulation and incineration followed by the rise of greenhouse gas emissions are the reflection of a 'fast fashion' industry. While many people worldwide have their basic needs deprived by the State, others are willing to satisfy their most unnecessary desires imposed by the market.

In parallel, a wave of concern about social and ecological responsibility spreads globally. The visualization of common interests strengthens the union of communities fighting for great transformations, affecting both the consumption decisions and the production structure. On that account, it is up to the fashion designer to assume the important role of rethinking the way clothing is been developed, so it could be disassembled, recycled and reused in the future. The perception of the whole, rather than the segmented processes, values the creation of sustainably efficient models. As a consequence, researches in Circular Design strategies targeting the achievement of a Sustainable Fashion where the materials can be segregated either for biological or technical cycles along with the development of alternative economic systems are being intensified.

The paper is outlined according to the following: contextualization of the 'fast fashion' industry, introduction of the principle of sustainability in the garments industry, overview of the latest applications of Circular Design and cyclical systems in the fashion value chain, opportunities that are being created by the Circular Economy (CE) and challenges faced while implementing sustainable practices through the industrial processes.

2 METHODOLOGY

In order to contextualize the emerging concept of "Circular Fashion" and its implementation throughout the fashion industry, a literature review was conducted. The timeframe comprised papers published arising out of 2012 until November 2017. In addition, the search included the most recent action plans pronounced by the European Commission as well as reports drawn up by eminent global institutions and consultancy groups working on sustainable strategies. The documents were restricted by their subject areas: Circular Economy application (opportunities and challenges), textile waste's data and alternative models for industrial processes management targeting the shift of the contemporary 'take-make-dispose' system into a Circular Fashion system. The review was performed by means of the following keywords: Sustainability in the fashion production chain, Circular Economy in fashion design, Circular Design, Sustainable Fashion.

3 CONTEXTUAL FACTORS: AN INTRODUCTION TO SUSTAINABILITY IN FASHION

According to Fashion United (2016), the apparel industry currently has a global economic value of three trillion dollars, generating two percent of the world's Gross Domestic Product (GDP), acknowledging its essentialness globally. In parallel, in consonance with the UNEP Year Book 2009, organized by the United Nations Environment Programme, thereabout two tonnes of garbage produced each year worldwide. The values are the consequence of a society obsessed with excessive consumption connected with the culture of rapid disposal of products, as Fletcher (2008) states, a linear result of a 'fast fashion' industry. As reported by Stahel (2010), the mass production and consumption systems are driven by transnational companies[1], intensifying the socio-environmental problems faced by the fashion industry; the products are manufactured at low cost, low quality and large scale in failed states[2]. Thus, conforming to Ahmed and Peerlings (2010), the State's economic growth and the generation of new employment opportunities are allied with unsatisfactory conditions of work, education, health and housing in developing countries.

As reported by Ellen MacArthur Foundation (2017) on the subject of Circular Economy in the textile and fashion field, "total greenhouse gas emissions from textiles production, at 1.2 billion tonnes annually, are more than those of all international flights and maritime shipping combined" (2017, 'Executive Summary', para.2). Besides that, the report presents an analysis of the Global Material Flows for Clothing, in which the steps of the fashion's value chain are set—starting with the fibers production until the closed-loop recycling—along with the quantity of materials available in each of them. The index also demonstrates that circa 48 million tonnes of clothes are disposed per year after their use; among them, 35 million tonnes are landfilled or incinerated.

Introduced through the system of mass production and mass consumption, these data reflects the establishment of the fashion industry as one of the main causes of environmental pollution. The lacks of enhanced planning of production processes and the large disposal in a 'twinkling' after consumption are responsible over and above the alarming climate changes, scarcity of natural resources, global waste production and health problems for living beings.

Notwithstanding the mentioned situations, new educational methodologies, design techniques and economic models are emerging throughout a systematic change in the fashion industry which meets the definition of Sustainable Development stated by the World Commission on Environment and Development (1987, 'IV. Conclusion', para.1): "development that meets the needs of the present without compromising the ability of the future".

---

1. Transnational Company: "The continuation of a de facto open immigration policy is prompted by employers' demand for fresh sources of low-wage labor in the advanced countries, while the relocation of production facilities abroad is motivated by a similar demand by certain industrial sectors" (Portes, 2000, p. 257).
2. Failed States: "Even in the narrowest interpretation, 'failed states' are identified by the failure to provide security for the population, to guarantee rights at home or abroad, or to maintain (not merely formal) democratic institutions. The concept must surely also cover 'outlaw states' that dismiss with the contempt the rules of international order and its institutions, carefully constructed overwhelmingly under US initiative" (Chomsky, 2007, pp. 109–110).

Therefore, the concept of Sustainable Fashion, according to Fletcher and Grose (2011), Gwilt (2014) and Salcedo (2014), is linked to the innovative outline for the creation, manufacturing and consumption of fashion products, regarding the social, environmental and economic impact. Combined with this idea, alternative approaches define socially and environmentally responsible practices for rethinking the manner that fashion has being designed, such as: ethical fashion, eco fashion, conscious fashion, socially responsible fashion, fairtrade in fashion, slow fashion, cradle-to-cradle design, circular design, zero-waste design, upcycled design, product as a service, closed-loop system, take-back system, etc.

Also associated with sustainable initiatives, rooted in the principle of 3Rs—reduction, reuse and recycling—visions such as the Sound Material-Cycle Society (SMC)[3] in the Eastern world and the Circular Economy in the Western are conceiving models that aim to extend products' service cycle and minimize waste production. The Ministry of Environment Government of Japan (2008) and the Ellen MacArthur Foundation (2013) state these circular models are inspired by ancient cultural practices integrated into the new technologies. To build them it was taken into account the 'biological cycles' as well as the 'technical cycles' – the first one considers waste as nutrients for the soil, while the second, the recycled components are reused to manufacture brand-new goods.

4 CIRCULAR DESIGN APPLIED INTO THE FASHION VALUE CHAIN

Mcdonough and Braungart (2009) argue that the designer can decide between: implementing a design strategy for destruction or change. "To eliminate the concept of waste means to design things—products, packaging, and systems—from the very beginning on the understanding that waste does not exist" (McDonough & Braungart, 2009, p. 104). These authors have established the concept of cradle-to-cradle design, or Circular Design, stating that products need to be designed to track two distinct metabolisms: biological and technical. Materials and products intended to follow the biological cycle can be safely decomposed and integrated into the natural environment. In the technical metabolism, the valuable nutrients are 'upcycled' rather than recycled—to retain their high quality in a closed-loop industrial cycle" (McDonough & Braungart, 2009, p. 110).

Ellen MacArthur Foundation considers the Circular Design as one of the main aspects for the creation of CE, defining it as "improvements in material selection and product design (standardisation/modularisation of components, purer material flows, and design for easier disassembly)" (2013, p. 9).

The principle of Circular Design is directly allied to cradle-to-cradle design. Thence, the recognition of the importance of the design during the creative process has a repercussion over the production system, culminating to the feasibility of the products and materials' circularity. According to Moreno et al. (2016), the relation between the design and the environment started during the 70's decade. At that time, the ecological aspects were the main standards for a Green Design. In 1990, the Sustainable Design included the social dimension, concerning justice in labor. In consonance with the authors, the current concept is termed Design for Sustainability, contemplating, innovative designing strategies are being developed aiming for the wellness of the next generation.

According to the Ellen MacArthur Foundation, the current economic model can be labeled as "take-make-dispose" – portrayed by the easy extraction of raw materials, quick manufacturing, use and discard of products. Moreover, the Foundation has been investigating cyclical structures, concomitantly with a market behavior analysis and interviews with specialists in the field for the purpose of transforming the present system. "Recently, many companies have also begun to notice that this linear system increases their exposure to risks, most notably

---

3. Sound Material-Cycle Society (SMS): "[...] based on reduced consumption of natural resources and lower environmental burdens. Such a society can be achieved by stepping up efforts toward sustainability and integrating those efforts into actions to create a low-carbon society and a society in harmony with nature" (MOEJ, 2008, intro.)

higher resource prices and supply disruptions" (Ellen MacArthur Foundation, 2013, p. 6), for this reason, the establishment of a Circular Economy in the industrial system enables "the elimination of waste through the superior design of materials, products, systems, and, within this, business models" (Ellen MacArthur Foundation, 2013, p. 7).

The CE system, developed by Ellen MacArthur Foundation, adheres to the following principles:

- The absence of waste—since the components, categorized by biological and technical, can return wither to the biosphere or to the construction of new materials and products;
- The design development targeting either the technical or biological cycle and the design for disassembly and revitalization;
- The construction of business resilience deriving out of alternative systems;
- Commitment to the usage of renewable energies;
- Ideas focused on new systematic structures.

The application of a circular system (Figure 1) aims to benefit the global economy while assisting the wellness of businesses and consumers. Due to this fact, the global imbalance—considering dilemmas such as: population growth, infrastructure problems, political risks, market globalization, climate change—connected with the pressure for disruptive solutions is vital to turn the concept 'viral'.

The urge to apply a circular system in the textile and fashion industry has encouraged the European Union to conceive additional policies. The European Commission (EC) launched in 2016 the action plan: Closing the loop: new Circular Economy package. The package intended to adopt the CE as a guide in which the waste generation can be scaled down through value creation of materials and products. Through this strategy, the European Commission's report (2016) outlined opportunities and challenges in implementing this circular approach.

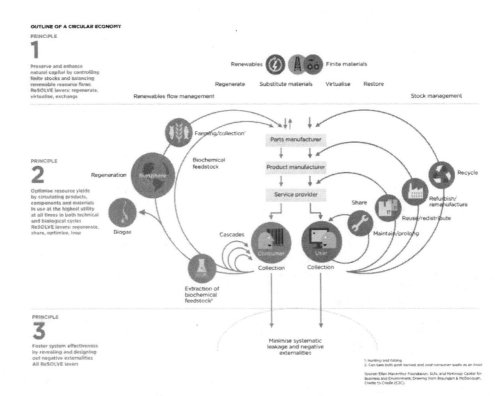

Figure 1. Outline of a circular economy. "Circular economy system diagram". Ellen MacArthur Foundation (2015).

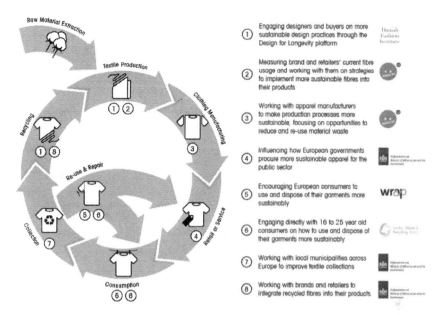

Figure 2. Delivery partners and key action areas. "European Clothing Action Plan (ECAP)". MADE-BY (s.d.b).

The circumstances favoring the Circular Economy effect addressed by the EC were: (a) Reduction of the use of resources during the manufacturing process; (b) The guarantee of raw materials for an efficient production, enabling businesses competitiveness; (c) "Need to redesign materials and products for circular use" (European Commission, 2016, p. 4); (d) Reinforcement of the global economy through job creation. Concurrently, the barriers encountered by the European Commission for an effective performance were: (a) Transitions to investing in new technologies and improved infrastructure; (b) Incentive for the usage of secondary raw materials over the value chain; (c) Demand of specific knowledge and technical skills during design, production, recycling and reuse processes; (d) Changes in consumer and business behavior; (e) Involvement of the public and private sector as well as the participation of the civil society.

As represented in Figure 2, strategic partnerships were drawn together with socio-environmental organizations to achieve the progression towards a circular system in textile and fashion manufacturing. Among them, the selected institutions were: Danish Fashion Institute[4], MADE-BY[5], Rijkswaterstaat Ministry of Infrastructure and the Environment[6], WRAP[7] e London Waste and Recycling Board[8].

---

4. Danish Fashion Institute (DFI): "Danish Fashion Institute (DAFI) is a network organization established by the Danish fashion industry in 2005. The purpose of DAFI is to develop, strengthen and promote Copenhagen and Denmark as a destination for responsible fashion and societal solutions" (Danish Fashion Institute, s.d.).
5. MADE-BY: "an award-winning not-for-profit organisation with a mission to 'make sustainable fashion common practice'. Through targeted consultancy, partnerships and stakeholder engagement we work with well over 100 brands and retailers" (MADE-BY, s.d.a).
6. Rijkswaterstaat Ministry of Infrastructure and the Environment: "Rijkswaterstaat is responsible for the design, construction, management and maintenance of the main infrastructure facilities in the Netherlands. This includes the main road network, the main waterway network and watersystems" (Rijkswaterstaat Ministry of Infrastructure and the Environment, s.d.).
7. WRAP (Waste and Resources Action Programme): "WRAP's vision is a world in which resources are used sustainably. WRAP works with governments, businesses and communities to deliver practical solutions to improve resource efficiency" (WRAP, 2017).
8. London Waste and Recycling Board: "London Waste and Recycling Board is a statutory Board established by the GLA Act 2007 to provide a strategic approach to waste management in London. The Board is chaired by the Mayor of London (or his representative)" (LWARB, s.d.).

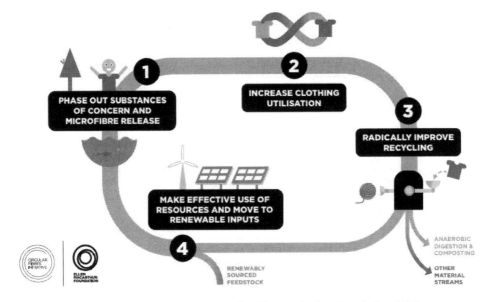

Figure 3. "Ambitions for a new textile economy". (Ellen MacArthur Foundation, 2017).

The latest report released by the Ellen MacArthur Foundation in late November 2017, The New Textiles Economy: Redesigning Fashion's Future, focuses on solving environmental issues through the redesign of the materials and products along with the use of renewable energy. The report considers the following aspects for the development of an innovative Textile Economy (Figure 3):

1. Analyse the components of the current materials and the creation goods con sidering the living beings' safety;
2. Shift the linear system to a circular one, taking the design, usability and marketing structures in mind;
3. Improve textile recycling through clothing design;
4. Adopt the usage of sustainable resources and renewable inputs

Socio-cultural development shall be included for an efficient transition to a Sustainable Fashion. According to Parker (2011), the implementation of sustainability in fashion requires the industry's commitment with: "environmental protection, social justice, economic fairness and cultural validity" (p. 4). However, due to the complexity of these dimensions, they have been treated as a background picture by the Circular Economy. Disregarding the relevance of the social and cultural aspects as pillars of the principle of sustainability, they are visualized by the CE as consequences of environmental and economic positive initiatives.

5 CONCLUSIONS

The decision of integrating a Circular Economy in manufacturing processes enables the progress of the traditional economy throughout the waste reduction, minimization of greenhouse gas emissions and the diminishment of the use of toxic components on textile production. This ecologically responsible conduct applied to the fashion industry foments the environmental protection and, therefore, society wellness.

Acting in a sustainable manner has become a new trend. The contemporary concept of Sustainable Fashion dictates a 'style' beyond the urge of following the seasons. In virtue of that, the scarcity of resources' dilemma attempts to be solved throughout the determination of residues as nutrients to the soil or waste as a new resource for manufacturing future products. Innovative practices allowing the extension of materials 'use-cycle' and products' 'service-cycle' are being developed. In such a way, the garments must be created for its disassembly, 'upcycling' and reuse, supporting the incorporation of a Circular Fashion.

Unfortunately, there is still not enough technology for components segregation of mixed fibers. However, the vision for improving businesses performance through the implementation of circular systems and the demand for specific sustainable knowledge—conceptual and instrumental—for the fashion value chain generate new job opportunities.

Nonetheless, the cultural and social dimensions shall be included while constructing alternative fashion systems. The interconnection of these fundamental aspects with the economic and environmental values will provide a powerful basis to Design for Sustainability. The transition towards a Circular Fashion will also depend on the risks that the public sector, corporations and civil society decide to take. The collaboration of all the institutions involved, in conjunction with community participation, strengthens the demand for innovative sustainable solutions. Forasmuch as each action, smallest as it can seem, will have a global impact, to improve the current fashion system is required the application of a holistic approach—in which the environmental, social, cultural, economic and political aspects are considered for creation, manufacturing, use, reuse and upcycling of apparel products.

ACKNOWLEDGEMENTS

This work is supported by FEDER funds through the Competitiveness Factors Operational Programme—COMPETE and by national funds through FCT – Foundation for Science and Technology within the scope of the project POCI-01-0145-FEDER-007136.

REFERENCES

Ahmed, N., & Peerlings, J.H.M. 2009. Addressing Workers' Rights in the Textile and Apparel Industries: Consequences for the Bangladesh Economy, *World Development*, 37(3): 661–675.

Chomsky, N. 2007. *Failed States: The Abuse of Power and the Assault on Democracy* (ed. 2007). New York: Owl books.

Danish Fashion Institute (s.d.). *About the organization.* [Online] Ava http://danishfashioninstitute.dk/en/node/719.

Ellen MacArthur Foundation 2017. *A New Textiles Economy: Redesigning Fashion's Future.* [Online] Available at https://www.ellenmacarthurfoundation.org/assets/downloads/publications/A-New-Textiles-Economy_Full-Report.pdf.

Ellen McArthur Foundation 2013. *Towards the Circular Economy: Economic and business rationale for an accelerated transition.* [Online] Available at https://www.ellenmacarthurfoundation.org/assets/downloads/publications/Ellen-MacArthur-Foundation-Towards-the-Circular-Economy-vol.1.pdf.

Ellen McArthur Foundation 2015. *Circular Economy System Diagram.* [Online] Available at https://www.ellenmacarthurfoundation.org/circular-economy/interactive-diagram.

European Commission 2016. *Closing the loop: new Circular Economy package.* [Online] Available at http://www.europarl.europa.eu/RegData/etudes/BRIE/2016/573899/EPRS_BRI(2016)573899_EN.pdf.

Fashion United 2016. *Global Fashion Industry Statistics – International Apparel. Available at* https://fashionunited.com/global-fashion-industry-statistics.

Fletcher, K. 2008. *Sustainable Fashion & Textiles: Design Journeys.* Oxford: Earthscan.

Fletcher, K., & Grose, L. 2011. *Moda & sustentabilidade: design para mudança* (Brazilian edition). São Paulo: Editora Senac São Paulo.

Gwilt, A. 2014. *Moda sustentável: um guia prático* (Brazilian edition). São Paulo: Editora G. Gili Ltda.

LWARD (s.d.). [Online] Available at http://www.lwarb.gov.uk/who-we-are/.

MADE-BY (s.d.a). [Online] Available at http://www.made-by.org/about/.

MADE-BY (s.d.b). *European Clothing Action Plan.* [Online] Available at http://www.made-by.org/projects/ecap/.

McDonough, W. & Braungart, M. 2009. *Cradle to Cradle: Re-making the way we make things.* London: Vintage.

MOEJ 2008. *A Sound Material-Cycle Society through the Eyes of Hokusai.* [Online] Available at https://www.env.go.jp/recycle/3r/approach/hokusai_en.pdf.

Moreno, M., Rios, C., Rowe, Z. & Charnley, F. 2016. A Conceptual Framework for Circular Design. *Sustainability.* 8, 937.

Parker, E. 2011. *Steps towards sustainability in fashion: snapshot Bangladesh.* [Online] Available at http://ualresearchonline.arts.ac.uk/2754/.

Portes, A., 2000. Globalization from Below The Rise of Transnational Communities. In Kalb, D., van der Land, M., Staring, R., van Steenbergen, B. & Wilterdink, N. (Eds), *The Ends of Globalization: bringing society back* in (pp. 253–272). Lanham: Rowman & Littlefield Publishers Inc.

Rijkswaterstaat Ministry of Infrastructure and the Environment (s.d). [Online] Available at https://www.rijkswaterstaat.nl/english/index.aspx.

Salcedo, E. 2014. *Moda ética para um futuro sustentável* (Brazilian edition). São Paulo: Editora G. Gili Ltda.

Stahel, W. 2010. Durability, Function and Performance, In T. Cooper (ed.) *Longer Lasting Products: Alternatives to the Throwaway* Society (pp. 77–106). Farnham: Gower.

World Commission on Environment and Development 1987. *Our Common Future.* Oxford: Oxford University Press.

WRAP 2017. *Our vision.* [Online] Available at http://www.wrap.org.uk/about-us/about.

Fashion and sustainability's valences: Exposing gaps

R. Puppim, C. Jordão, L.M. Arruda, D.P. Beduschi & A.C. Broega
Universidade do Minho, Guimarães, Portugal

ABSTRACT: The present article is a quantitative and qualitative study on Fashion and Sustainability, in selected scientific publications. First, it presents a proposal of subareas and criteria divided into five categories on contributions of relevant authors. In the sequence, it exposes the adoption of strategies for statistical verification of researchers presented at *Colóquio de Moda* (Brazil) and the International Congress of Fashion and Design—CIMODE (Europe). Then, it debates the results of this survey, equalizing, qualitatively, the quantitative presented. Finally, it explores interpretations of the analysis, pointing out gaps for future and further research on Fashion and Sustainability.

1 INTRODUCTION

The terms "Eco", "Ecological", "Sustainable", among others, receive more and more attention in the current daily consumption. However, it is rare to find a clear explanation of the origin of these terms and reasons why they are in vogue. Veiga (2010) estimates that the beginning of the use of the term Sustainability is in the late 1970s, when scientists assessed the ability of an ecosystem to be impacted through human action, evaluating its resilience. The author also points out that, only in the 1990s, the word spread out and was used by the general studies. The movements and congresses on this theme, such as the *Rio 92*, the Kyoto Protocol and *Agenda 2*1, helped to increase the concerns and, consequently, the researchers on several areas of knowledge related to the theme.

On the Fashion area, this is not a different trend. As readily seen today, more and more brands, companies and designers are emerging and developing their works under the label of a sustainable product. Nevertheless, it is necessary to caution and question about the legitimacy of a product, brand or work to be, as a matter fact, sustainable. Flusser (1999) already exposed that the power to doubt is essential to the human being think because to doubt can trigger a stimulating process of investigation and inquiry.

Thus, it can promptly come across a fashion products panorama full of fake intentions and, in fact, do not possess any requirement or criterion to be characterized as sustainable. This is because the term sustainability has the power to add intangible value to products and services, and call the attention of the consumer interested (even if not much) on resolutions of the planet's environmental problems. This attempt to add the adjective "sustainable" on fashion, all over the world, is named, by researchers, as Greenwash, Manzini & Vezzoli (2011). Which reveals, that sophisms are also possible in products that present themselves as eco friends.

Understanding this issue is crucial to comprehend the means of distinguishing fashion products and services that effectively seek to reduce the environmental damage, and others, which merely aim in the use of the term an easy marketing strategy to sale. A way to perceive this difference, perhaps the most important, is to study the research and scientific investigation published on the subject Fashion and Sustainability.

Although, it is difficult to conceptualize the term Sustainable Fashion since that Lipovetsky (2009) already defended that Fashion is a system, emerged from a modern context and that, consistently, instigates to consume and to believe that it must be continually renewed, to be socially part of the system. Through this complexity of fashion, as a system, seems to be arduous and stigmatized to failure on any attempt to blend the term to the adjective "sustainable", which, in opposition, proposes different concepts of fashion, such as conscious consumption, Boff (2012). So, it leads to the uncomfortable question: How to propose a reconciliation of the two terms, if they are averse and distinct, by their essences?

The author Brown (2010) deeply explored these relations, proposing that the term to be adopted is Eco-Fashion since it presents a better understanding of how to conjure Fashion and Sustainability. Therefore, authors and researchers of Sustainability in Fashion do generally not use the term Sustainable Fashion, because it is a paradox of difficult assimilation.

In this scenario, investigations and projects can be evidenced on scientific events, such as the *Colóquio de Moda* (a reference for Brazil and South America) and the International Congress of Fashion and Design—CIMODE – (a reference for Iberian Europe). On these databases, the surveys show that the consolidation of the Sustainability branch, as part of the Fashion Field, is increasingly a focus of research. However, because it is a segment that has not yet been, entirely, consolidated, due to the recent interest of researchers, it is clear that there are still gaps to be explored. The selection of these congresses is due to that in Latin America, and Southern Europe (more on Iberian peninsula) studies and projects have not yet been consolidated and implemented compared to Germany, Denmark or Nordic countries, (Berlim, 2012).

One of the main points when studying Eco Fashion is the requirement of sustainable materials, that is, ecologically correct, economically viable, socially just and culturally diverse, as portrayed by Berlim (2012). By structuring a project with raw materials that are sustainable, it will enable the entire production cycle to be more sustainable, as it is the initial phase of the development cycle. These materials can be used to recycle textile waste or can even be organic products, and/or with maximum use and efficiency of water, energy, and discards. In both cases (recycling or clean production), it is understood that there is a need to thoroughly know the technical complexity of the materials, which is not just a Greenwash, as already mentioned.

This article explores the state of the art of sustainability on fashion, from the scientific publications of the presented databases. It is structuring as follows: the present Introduction; followed by a Prelude which explores concepts and concepts of Sustainability. To legitimize, the reasoning of this perception of the word, suggesting possible subdivisions for studies in Fashion and Sustainability (or Eco-Fashion). To the methodology presented, where the procedures for the collection of relevant quantitative data of the surveys are presented. It follows the Analysis and Discussion of the results, where the data is treated, it analyzes the results of the proposed strategies, critical reflections on the subject, the conclusions are drawn. Finally, Final considerations express the mainly gaps importance for future and further studies.

2 PRELUDE

Initially, it is explored the bibliography of authors that concerns on Sustainability, as an area of study. This is important since the necessity of understanding the language techniques, of recent popularization, as already presented. To comprehend this methodology adoption, as an allegorical comparison, it is necessary to understand how walls are built before stating which, structural masonry or the wood, is the best (or worst) type of material to build them. On this sense, it is revisited renowned authors of this segment (Veiga, 2010; Boff, 2012; Sachs, 2009; Van Bellen, 2006).

In Van Bellen (2006), sustainability indicator systems are presented, especially the *Ecological Footprint,* the *Dashboard of Sustainability* and the *Barometer of Sustainability*, noting the nuances and relevance of each method, which provides the Sustainability Management bias. In Sachs (2009), considered by many scholars as one of the great authors of Sustainability, it is presented the criteria for sustainable development, especially Social, Cultural, Ecological, Environmental and Economic, aiming at the Social Sciences for Sustainability. While in Veiga (2010), the emergence and consolidation of the pure term Sustainability are analyzed and instigated to reflect on its real applicability, taking the perspective of Economy to Sustainability. Finally, in Boff (2012), criticisms are rationalized on how the world is naturalized and practices unsustainable, mainly focused on development, education, and even the individual, revealing Sustainability from the point of view of Philosophy.

After elaborating these complicated scenarios, it can, therefore, move to an intermediate level, of another understanding about Sustainability: it is dialogic with Design (a study area where the Fashion can be framed/part of), mediating the trajectory to our final research convergence. In this context, Manzini and Vezzoli (2011) are recapitulated as pioneers and consecrated authors of this tonic, and it is introduced to Queiroz (2014), for the fundamentally critical scope brought to Sustainability in Design. The firsts deals with an introduction to the possibilities of making industrial products (and, consequently, Design) sustainable, focusing on the developed by the Designer, in which the reference point (such as the environment, society, and policies) is emphasized and possible tools to evaluate the effectiveness of what has been proposed and/or developed. On the other hand, the second author, due to the profile of the critical theory of Design, sees rigorous standards for design sustainability, considering a "utopia" the combination of Sustainable Design, naming transgressions to the miscellaneous that occurs between the terms (Sustainability and Design).

Most of the author mentioned above address with what has already been put forward in this article: the incongruous paradox (or intent) of creating an adjectival noun, being it Sustainable Fashion or Sustainable Design. Another aspect that stands out and causes mental discomfort is an (apparently) visible contestation: by not following all aspects/criteria of sustainability, a product/project should not be qualified as sustainable. Also, also that products/projects that are "sustainable" only in one aspect do not contribute in any way to Sustainability.

For being so overwhelming, the authors drive back to Flusser (1999) – as already presented –, and, thus, confronts Queiroz's discourse with the other authors of Sustainability. In particular, Boff (2012) warns that even little actions, in favor of a more sustainable result than the common one, bring added value to human interactions and the development of critical thinking to sustainability, even in daily life. The author believes that significant changes cannot be required to occur in a short period.

In this confrontation, it contemplates and embraces more from Boff's view than Queiroz's, but it is possible to be taken to a higher level of criticality. Assimilating the proposals of that as a potential object of analysis and investigation of this research, but not ruthlessly judging proposals (in articles, by example), from other authors, on how it could be done.

The combination of Sustainability and/with/on Fashion are presented because of the vast number of publications that express on Fashion and Sustainability, based on the criterion of accessibility/reachability of this type of literature and relevance in contributions to the area. Thus, a significant number of authors who study this theme are re-evaluated, namely Brown, (2010); Berlim, (2012); Lee, (2009); Fletcher, (2011); Thompson & Thompson, (2013); Gwilt, (2014); Salcedo, (2014); Styles, (2014); Schulte, (2015); etc.

After the analysis of the publications, the authors direct the convergences of the approach to Sustainability in Fashion, as scope. From this angle, it is synthesized as a subdivision proposal that can be shown in Table 1, to make the analysis easier to understand, suggesting sub-areas from the readings. In the first column, it is listed the authors mentioned, filled in the following columns with their theories (in concepts or chapters presented) for the different valences/dimensions proposed.

Thus, we deduce that the proposal of five sub-areas for studies in Fashion and Sustainability —being those presented in the first line of Table 1 – harmonize with many of the concepts and proposed divisions of the authors studied. They are:

- Raw material: corresponding to the materials that make the garments, textiles or non-textiles;
- Processes: representing the approach in the stages of the production cycle of clothing and accessories, such as creation, modeling, prototyping, dyeing, sewing or even other operations;
- Consumption: referring to consumers' use, durability and purchasing habits—concerning clothing—such as durability analysis, the frequency of new purchases, or even the users understanding the origin of the product;

Table 1. Selected authors and their approaches to sustainability in fashion.

Sub-areas author	Raw material	Processes	Consumption	End of cycle	Transparency
Lee (2009)	Chapters 3 and 6	Chapter 1	Chapter 10	Chapter 2	Chapter 4, 7 and 9
Brown (2010)	–	*Slow design*	*Reuse New models*	*Reuse, Redesign & recycle*	*Fairtrade*
Fletcher (2011)	Chapter 1 *Materials*	Chapter 2 *Processes*	Chapter 4 *Consumer Care* Chapter 7 *Optimized lifetime* Chapter. 9 *Services and shares*	Chapter 5 *Discarded*	Chapter 14 *Engagement*
Berlim (2012)	Chapter 3 *Materials*	Chapter 1.1	Chapters 1.2 and 1.3	Chapter 3.4	Chapter 1.4
Thompson & Thompson (2013)	Part 1 *Materials*	Part 2 *Processes*	–	Part 3 *Life Cycle*	–
Gwilt (2014)	–	Chapter 4 *Produção* Chapter 4 *Produção*	Chapter 6 *Use*	Chapter 7 *End of life*	–
Salcedo (2014)	Chapter 4 *Raw material*	Chapter 5 *The manufacturing processes*	*The durability of the garment Role of the user Product life-cycle*	*Recycling Waste manage*ment *design without waste* Chapter 7 *Management of the end of t*	*Bhe life cycle Social well-being*
Styles (2014)	*Textiles*	*Slow fashion*	*Retail revolution*	*Recycling*	–
Schulte (2015)	–	Chapter 1.4	Chapter 1.5	–	Chapter 2.1

Source: Created by the authors of this article.

- End of the cycle: concerning to the disposal section of products and materials, where it encourages reuse and recycling;
- Transparency: responding to the view of validation and suitability of the *sustainable* adjectives on fashion articles, to prospect environmental certifications and sustainability label.

It must be emphasized that, in general, the authors also contribute to the theoretical and critical questions of Sustainability on Fashion, which was not a sub-area proposal, since it is a scientific basis for investigating Fashion and Sustainability. Another relevant aspect is that even authors who do not adopt specific themes or concepts about one (or another) item of the sub-area proposal, formally, hardly fails to, at least, reference the topic/theme, collaborating to consolidate the suggestion of this article.

3 METHODOLOGY

To answer the objectives of this research, on analyzing how Fashion and Sustainability have been approached by the researchers of the area, a methodology of secondary data collection was used through literature review, resulting on the following criteria:

- Type of publication – Scientific Articles;
- Database typology – Proceedings of Scientific Congresses on Latin American and Iberian – European origin, namely: *Colóquio de Moda* and *Congresso Internacional de Moda e Design*;
- Written language – Portuguese and English;
- Publication Period – from 2012 to 2017;
- Key-words
  - In Portuguese: – Moda/Roupa/Vestimenta/Indumentária/Traje + Eco/Sustentável/ Sustentabilidade/Ecológica(o);
  - In English – Fashion/Clothes/Clothing/Garment/Dress/Costume/Apparel/Wear/Dressing/Attire + Eco/Sustainable/Sustainability/Ecologic(al).

On this scenario, the choice, about the type of publication of projects and research in progress or concluded that have systematization of the material, was made in order to notice which focuses have been chosen at this theme. On the databases typology it was chose two scientific congresses: The *Colóquio de Moda*, which has 13 editions (by 2017), the oldest in Brazil, being a reference in South America, with significant international participation; and CIMODE, corresponding to the 4th edition by 2018, unifying studies in the Fashion and Design segment, with a strong presence of European authors. Regarding the determination of the writing language, Portuguese is the preferred one to contemplate more than 90% of the articles presented, followed by English. Afterwards, it was defined the period of publication between 2012 and 2017, since these are more up-to-date diffusion surveys. Finally, it was delineate a set of keywords, with which it is intended to reach different incidences of the combined terms.

Thus, reflecting on the proposed convergence of sub-areas/criteria (expressed in the previous section), it is possible to, statistically study the amounts of scientific publications under the theme of Sustainability in Fashion, resulting in Tables 2 and 3, where the *Colóquio de Moda* surveys are presented (initially based on the Silva and Giuliano's study in 2017), and concerning CIMODE respectively.

It is emphasized, here, that some of the published articles portrayed more than one of the proposed criteria. Texts were evoking other approaches, other than those suggested, of Fashion and Sustainability, in particular (for recidivism), the theme of studies and sustainable business and branding proposals.

Table 2. Articles on sustainability at the *Colóquio de Moda*, according to established criteria.

Edition/ Areas	Total articles	Articles on sustainability	Raw material	Processes	Consumption	End of cycle	Transparency
2012 *Rio Janeiro/RJ*	236	16,1% (38 articles)	2	6	11	5	1
2013 *Fortaleza/CE*	228	16,7% (38 articles)	3	9	10	6	2
2014 *CDSul/RS*	164	18,3% (30 articles)	4	11	12	8	–
2015 *Curitiba/PR*	234	19,2% (45 articles)	5	9	14	10	1
2016 *JPessoa/PB*	224	24,1% (54 articles)	8	11	19	9	–
2017 *Bauru/SP*	242	12,8% (31 articles)	5	11	9	5	2
Total			13,1 (%)	26,7 (%)	36,4 (%)	20,9 (%)	2,9 (%)

Source: Created by the authors of this article, based on Silva and Giuliano (2017) and *Colóquio de Moda* proceedings 2012; 2013; 2014; 2015; 2015; 2017.

Table 3. Articles on sustainability in CIMODE, according to established criteria.

Edition/areas	Total articles	Articles on sustainability	Raw material	Processes	Consumption	End of cycle	Transparency
2012 Guimarães, PT	228	8,3% (19 articles)	3	8	4	1	–
2014 Milan, IT	175	7,4% (13 articles)	5	4	4	3	1
2016 Buenos Aires, AR	246	13,0% (32 articles)	2	9	7	7	1
		Total	16,9 (%)	35,6 (%)	25,4 (%)	18,6 (%)	3,4 (%)

Source: Created by the authors of this article, based on the CIMODE proceedings 2012; 2014; 2016.

4 ANALYSIS AND DISCUSSION

From the numerical results presented, it is possible to explore the formal aspects of this exposition (quantitatively and qualitatively). Initially, it is possible to notice that the number of *Colóquio de Moda* events (annual) is double in relation to the CIMODE event (biannual), and the percentage of articles presented on average is smaller in CIMODE (13% its greater value for CIMODE versus a maximum of 24.1% and a minimum of 12.8% for *Colóquio de Moda*). However, these results express the consolidation of Sustainability in Fashion, as a subject addressed and investigated at present.

In both cases, the oldest editions of the events presented few specific articles related to the subject, and currently, both have their own theme sections: at the *Colóquio de Moda* with the Working Group 10 – Fashion and Sustainability and the Axe 8 – Sustainability (for oral presentation modalities). The Congress CIMODE stands out on the Sustainability in Fashion and Design track (presented on its last two editions). From this research, it was noticed that even with the introduction of these specific tracks to the Scientific Areas of the Congress, the articles that explore this issue, are explores throughout other axes, such as Consumer & Marketing, Productive Systems and Processes, Education & Culture, among others.

Regarding the sub-areas (criteria) proposed for Sustainability, in the results, the "Transparency" criterion has a lower incidence of publications, with 2.9% and 3.4%, in CIMODE and *Colóquio de Moda*, respectively. This result may be linked to the complexity of a product's likelihood of being sustainable. Considering that it refers only to international label and certifications, since that only within these, it can merge the suitability of the proposal upon in the criteria of sustainability. The articles presented here mostly portray the views surrounding this question, but do not present such certification samples.

On the "End of Cycle" criterion, where there are numerous projects, and proposals for recycling of textile waste are found, has 29.9% for the *Colóquio de Moda* and 18.6% for CIMODE. The ecological and environmental perspectives are permanent on most texts of this segment. Specifically, this area is distinguished as the most condescending of the others, although its articles are also linked to the other's proposals. Therefore, many approaches report new raw materials, consumption and/or productive processes on this segment.

Concerning the "Consumption" criterion, it is possible to find the largest volume of publications for the *Colóquio de Moda*, with 36.4%, and the second largest for CIMODE, with 25.4%. It is credited to be the most radical/strict segment, in contrast to the fundamental concepts of Fashion, considered almost antagonistic. These are samples of collective projects, where the term "ethical fashion" is also recurrent.

On the other hand, the criterion *Processes* also shows statistical relevance, the first in the CIMODE event (35.6%) and the second in the *Colóquio de Moda* (26.7%). However, these values can be understood by the different tracks in which it is distributed, counting on different types of tasks and activities of the whole productive system of Fashion. Manual processes (handcrafts) and the valuation of human labor are the most recurrent aspects. A question pointed a few times in the publications studied instigates to reflect about the sustainable level of the process' proposals, which can advance for a non-sustainable raw material, or that does not stimulate the conscious consumption of (possible) purchasers, or even, that it does not allow recycling of material or produces in a large quantity of waste/discard. This leads to the understanding of this criterion as an intercessor to the others, sometimes improving, sometimes lowering the sustainability of products.

Finally, on the "Raw Material" criterion, textile materials are the majority, but there are also non-textiles, such as leather and similar surfaces, or even, more daring proposals, suggesting non-traditional inputs. On that topic, it can be seen that the great ally of studies is technology, in most cases, blending itself with sustainability. Although this percentage is lower than the three previous criteria (with 13.1% in the *Colóquio de Moda* and 16.9% in CIMODE), this sub-area expresses factual ease, since it makes (potentially) the rest of the entire production chain (in the processes, in the consumption and the end of cycle) more effective and efficient to the applications of sustainability.

5 FINAL CONSIDERATIONS

It is noticed that Fashion and Sustainability are gradually managing to interact with each other. Although it should not be said that "Fashion and Sustainability" are already at a significant level of study, knowledge and research, but it is assertive and safe to say and recognize the *status quo* that has been known, as demonstrated here, by numbers. Consecutively, it is possible to identify a pattern, in order to group themes within Fashion and Sustainability. This agglutination proposed in sub-areas/criteria represents the desires pointed out by the authors who research about this subject. Thus, researchers can be assimilated to the sub-areas, depending on their specialty, familiarity and/or research aspiration, mostly, wishing to contribute to their environmental concerns.

Given the data collected, it could be spontaneous said that the criterion of "Transparency", which reports on the suitability of sustainable, through the tables presented, would be the least studied. Alternatively, under the same argument, that too much was explored in "Processes" and "Consumption". However, it is necessary to make consideration and balanced analysis of the numbers in this perspective. It is concluded that none of these sub-areas is filled, with no gaps. Once again, it is important to remind that, since Fashion and Sustainability studies (and even more when "Fashion and Sustainability" as an association) are not yet fully defined, they are entirely new/recent and uncontested knowledge.

Despite the low volume presented in the "Transparency" segment, this is an important issue and needs to be further explored, debated and investigated, since it is verified that most authors do not dominate the subject, not understanding precisely the concepts of Fair-trade and practices for ISO 14000 (International Organization for Standardization), for example. It is not yet verified the demonstration by cases, projects, brands and/or products addressed in researches that understand the depth and importance of certification. This gap allows the opening of space in the market for the practices, previously exposed, of Greenswashes.

However, even though it is not the most relevant area, it is essential to highlight the need for studying the "Raw Materials" issue, which investigates aspects such as the organic potential of natural fibers (such as cotton and linen). The exploration of wild silk that does not affect the natural life cycle of other organisms (*Bombix Mori*); the reduction of water and energy use for raw material production; the potential of the reprocessing of the textile materials to turn them back to fiber, when discarded, among others.

Most of the authors of Fashion and Sustainability consulted (Berlim, 2012; Lee, 2009; Fletcher, 2011; Salcedo, 2014; Styles 2014), and especially Thompson and Thompson (2013), argue for the importance of research and development more sustainable input materials, boosting the rest of the production chain, thus choosing the raw material, as the most important for this segment of studies. Moreover, in this sense, it is encouraged to expand and engage, realizing that Fashion and/or Design can make complementary and supplementary contributions in the field of Fashion production.

However, the core of this type of research is exploratory, trying to understand details in a resolute way to respond with possible solutions. In other words, the mechanism for better functioning of this machine (here understood as research) is in the Textile Engineering area. After all, both Fashion and Sustainability, as areas of knowledge, could be benefited and consolidated from contributions from other fields of study by Thompson & Thompson (2013).

At the moment, it is believed that Engineering (in particular, Textile) along with Fashion must surrender to the principles for/and Sustainability, with the purpose of developing and implementing new materials and raw materials that reach the environmental, functional, efficient, intelligent and contemporary requirements.

ACKNOWLEDGEMENTS

This work is supported by FEDER funds through the Competitiveness Factors Operational Programme – COMPETE and by national funds through FCT – Foundation for Science and Technology within the scope of the project POCI-01-0145-FEDER-007136.

REFERENCES

Berlim, L. 2012. *Moda e Sustentabilidade: uma reflexão necessária*. São Paulo: Estação das Letras e Cores.
Boff, L. 2012. *Sustentabilidade: o que é – o que não é*. Petrópolis: Editora Vozes.
Brown, S. 2010. *Eco Fashion*. London: Laurence King Publishing.
Colóquio de Moda 2018. *Anais – Edições Anteriores*. Available at: http://www.coloquiomoda.com.br/anais/anais/edicoes/index.php (Accessed 04 February 2018).
Colóquio de Moda de 2018. *Anais*. Available at: http://www.coloquiomoda.com.br/anais/(Accessed 04 February 2018).
Congresso Internacional de Moda e Design 2018. *Anais – Edições Anteriores*. Available at: http://www.design.uminho.pt/cimode2016/pt-PT/(Accessed 09 February 2018).
Fletcher, K. 2011. *Moda e sustentabilidade: design para a mudança*. São Paulo: SENAC-SP.
Flusser, V. 1999. *A dúvida*. Rio de Janeiro: RelumeDumará.
Gwilt, A. 2014. *Moda sustentável: um guia prático*. São Paulo: Editora Gustavo Gili.
Lee, M. 2009. *Eco Chic: o guia de moda ética para a consumidora consciente*. São Paulo: Larrousse do Brasil.
Lipovetsky, G. 2009. *O império do efêmero: a moda e seus destinos na sociedade moderna*. São Paulo: Companhia das Letras.
Manzini, E. & Vezzoli, E. 2011. *O desenvolvimento de produtos sustentáveis: os requisitos ambientais dos produtos industriais*. São Paulo: EDUSP.
Queiroz, L.L. 2014. *Utopia da sustentabilidade e transgressões no design*. Rio de Janeiro: 7 Letras.
Sachs, I. 2009. *Caminhos para o desenvolvimento sustentável*. Rio de Janeiro: Garamond.
Salcedo, E. 2014. *Moda ética para um futuro sustentável*. São Paulo: Editora Gustavo Gili.
Schulte, N.K. 2015. *Reflexões sobre moda ética*. Florianópolis: Editora UDESC.
Silva, C.V. & Giuliano, C.P. 2017. Sustentabilidade e Moda: um estudo bibliométrico dos Anais do colóquio de moda. *Revista Conhecimento Online* 9 (2): 92–104.
Styles, R. 2014. *Ecologist guide to Fashion*. London: Thames & Hudson Publishing.
Thompson, R. & Thompson, M. 2013. *The manufacturing guides: sustainable materials, processes and production*. London: Thames & Hudson Publishing.
Van Bellen, H.M. 2006. *Indicadores de sustentabilidade: uma análise comparativa*. Rio de Janeiro: Editora FGV.
Veiga, J.E. 2010. *Sustentabilidade: a legitimação de um valor*. São Paulo: SENAC-SP.

Design and social innovation: Methodological principles for community qualification

J. Oenning
Federal University of Technology, Paraná, Brazil

J.B. Garcia Jr.
Instituto Vale da Seda, Brazil

J. Cunha
2C2T Research Unit, University of Minho, Guimarães, Portugal

ABSTRACT: The analysis of activities and processes configured by qualitative methods has shown efficient results within the field of Social Innovation, in the sense of building collaborative work and community involvement for sustainable social development. This paper reports the first experimental part of the doctorate research project, where these methods, geared toward social problems, help to elucidate questions necessary for design and contribute to the structuring of new forms of production of goods and services. The results obtained in this stage become tools used to create and structure a collaboration network, and thus propose efficient and sustainable solutions that benefit social groups and generate new opportunities for work and income.

1 INTRODUCTION

Social Innovation is an innovative solution to social problems of a given community, aiming at improving people's living conditions and, therefore, their well-being. It focuses on several factors such as physical and emotional self-esteem, level of competence of individuals, family relationship (belonging to a community), social infrastructure and environmental conditions (Kocziszky and Somosi, 2016; Ayob and Teasdale and Fagan, 2016). For Mulgan (2006), social innovation refers to the activities of innovative services that aim to meet social needs and that stand out in the field of organizations, which purposes are responses to social needs.

In the area of design for social innovation, the projects team is challenged to rethink its role in the projects and in which perspectives it wishes to develop its ideas, offering new solutions to social problems and collaborating for a more assertive vision of sustainable futures. The network of actors, who participate directly or indirectly in the development of solutions, starts from an identified social problem and gathering of tools to explore the skills and knowledge of each one of them, aiming for social construction (Santana, 2013; Morelli, 2007).

These perspectives establish new routines that seek the improvement of the structures through the qualification of the individuals acting in the process of social innovation. It is hoped to expand the diversity of products and services offered by that group, as well as to generate empowerment and integration of excluded groups through the appropriation of knowledge, reassessing the role of work in their lives and generating collective benefits (Moulaert and MacCallum and Hillier, 2013).

2 SOCIAL EXPERIMENTAL PROJECT IN SOCIAL INNOVATION

In order to develop a more action oriented research methodology, it is extremely important to analyse the relationships between all the actors involved with Social Innovation and the specific needs and main objectives of each one. The central focus for the definition of the methodology is based on the Social Innovation precepts, which define that the work must be developed in a collaborative way among all the actors, highlighting the relations between researcher-actors and the social world in which they will work. Decision-making is shared, the development of the work itself is inclusive and collaborative (Oenning and Cunha and Garcia Jr., 2017). This analysis can take advantage of a holistic methodology that allows the theoretical perspectives to dialogue with the empirical observations from an ethical position (Moulaert and MacCallum and Hillier, 2013). Some applied qualitative research techniques were: Semistructured Interview and Semistructured Observation.

2.1 *Structuring of the methodology and data collection—semistructured interview*

In the initial phase of a project, a freer and more exploratory process helps to understand the most important issues. This type of interview consists of organizing the questions by guiding topics, in which the interviewer introduces the subject of the research and leaves the interviewee free to talk about it, having the opportunity to talk about their experiences in a free and richer way. The interviewer can make small punctual interferences to deepen aspects that may be relevant to the subject of study, maintaining an openness in the communication process (Bogdan and Biklen, 1994; Fraser and Goudim, 2004). Therefore, for the gathering of information, guidelines were drawn which formed the basis of the semi-structured interview.

1. Understand the life context of women working in the NGO, how they organize themselves for work and how they are paid by it.

The NGO, called "Christian Association for Women: House of Talents", where the research was applied was founded in 2004 in the neighbourhood "Cinco Semblages" in the city of Londrina, in the state of Paraná, in southern Brazil. The objective is to make it possible for women who are not able to enter the formal labour market, for the most diverse reasons, or who are in a situation of social vulnerability, to have the opportunity for qualification and income generation in the areas of clothing manufacture and development of handicrafts. In addition, it aims to offer an environment of human development and social interaction, where the users can exchange personal and professional experiences. Is currently formatted in three distinct groups: women engaged in handicrafts such as canvas painting, kitchen cloths, knitting, crocheting and macramé; the women working in cutting and sewing operations and courses; and women working in the beauty salon, as well as the group of students who attend classes of all the specialties developed there. Their income comes from the sale of made-up products (cushions, rugs and towels), repairs and adjustments of pieces that the people of the community bring, and also 50% of the cutting and sewing and handcraft students' tuition fees. The other 50% is destined for maintenance of the machines and workplace and for purchase of consumables. Another source of income is an annex, the thrift store (store that sells second hand pieces), which has a collection of clothes donated by welfare programs. All female employees have incomplete secondary education as a level of education, and are not included in the conventional labour market.

2. Knowing the skills and abilities already established in the NGO's handicraft, cutting and sewing unit.

The collaborators are available to the community during business hours to meet the demands that arrive daily. In the atelier's work, the main activities developed are described in Table 1:

3. Presenting the concept of the research project and the actors involved.

If for Manzini (2008, p. 61) Social Innovation "refers to changes in the way individuals or communities act to solve their problems and create new opportunities," and these innovations"

Table 1. Atelier services provided by the NGO.

Dresses and skirts	Pants, shorts and jeans	Shirts and blouses	Knitwear, crochet and macramé
Circumference, handle, sheath and zipper closures	Waistband, bar, zipper and side seams	Buttons, cuffs, sleeve and collar	Kitchen cloth bars developed with the techniques to decorate and add value to the piece

are guided more by behavioral changes than by technological or market changes" it was understood the importance not only to include the NGO collaborator women in the context of the new work they were about to develop, but also to explain the central motivations guiding those perspectives. The purpose of this explanation was to locate them in the larger context of the research project they would be part of and sensitize them to these changes and opportunities that could help to be built with mutual benefits. An explanation was given concerning the research in Design and Social Innovation, developed in partnership between the Vale da Seda Institute, volunteer designers and the NGO, with the general objective of changing or promoting an improvement in the way each "actor" works in the moment they are connected to the network, in order to create new common and individual opportunities for all.

4. Discussing and building together, new possibilities and work opportunities within the context of the Research Project.

The search for results in Social Innovation, in this network context, when several actors will act in a project for a common good, goes through a discontinuity of the systems in force both in the required results and in the means to achieve them. The project model for new solutions proposed by Manzini (2008), was applied as an aid, at this stage of the research, to help elucidate this paradigm break in the relationship between the work that the NGO already does, and on which it is supported, and the new possibilities and opportunities that are emerging. Change the perspective: change the center of interest of things to the results, focusing on the activities to be performed (Manzini, 2008). Currently, the center of interest for the NGO's sustainability and the profitability of the sewing workshop collaborators are the classes, and the repairs and adjustments of clothes. However, the installed productive capacity creates the opportunity of an organization for provision of services, and concerning the clothing area, the new perspectives are drawn from new business models for which, apparently, the NGO has a structure, to develop and generate income that goes beyond the area of knitting and crocheting. These models present themselves as a strong alternative to the manual work employed in the designer's work, deepening the co-creative and collaborative relationship that Social Innovation advocates. To imagine alternative solutions: to plan different possible combinations of products, services, organizational skills and roles played by the actors involved (Manzini, 2008). Considering that the installed capacity of the NGO and the low productive demand can turn to a new work model with the idea of producing small-scale parts for local brands or for the NGO itself. For Manzini (2008), it is possible to offer support so that solutions in social innovation come about and are replicated without losing their original qualities. This support is understood as a capacity building stage for the preparation of this Institution, so that it is understood which areas can act and what resources will be needed in each process. Evaluate and compare alternative solutions: use an appropriate set of criteria to evaluate the effective economic, social and environmental suitability of the identified alternatives (Manzini, 2008). In the economic sphere, a new look at actions that increase the value and quality of the products and services already developed, considering the collaborative practices in their place of origin, generates a greater and equal economic development. In the social, opportunities are horizontal, the development of capacities and abilities offers the improvement of the living conditions of the individuals and groups involved; the participatory relations that are established in community stimulate the concern with the well-being and the respect with the workers and consumers (Silva, 2014). For the environment situation, the amount of material used by the NGO from donations, such as fabric cuts from the textile industry that are discarded because they no longer meet the demands of large-scale production, indicates many possibilities of use within a sustainable perspective. This is characterized by upcycling, focusing on its improvement, involving refurbishment or remodeling of products (Kim, 2015).

Considering that Social Innovation is seen as a new dimension of management science and has defended collective action among all actors (Moulaert and MacCallum and Hillier, 2013), it is important at this moment that the researcher assumes his or her role as an observer, in which he or she proposes to establish relationships with the group in the field work, with deep and active participation in the whole research routine, experiencing everyday situations (Lima et al., 1999). After the interview, a holistic observation was made, highlighting the most relevant questions for the research. It was found that the work structure, although simple, meets the many requirements of a small operational clothing work unit. It is a wide space with four tall tables approximately 3 meters long by 1.5 meters wide for the development of flat modeling, and with a table with office chairs for the development of handicraft activities such as knitting, crochet and macramé, there are four straight industrial machines and two overlock machines. The NGO had no technical mannequin for the development of three-dimensional modeling, which was acquired and donated, to carry out future training activities. Noting the motivation for qualification and learning, it is important to emphasize that Social Innovation is "the result of knowledge applied to social needs through the participation and cooperation of all the actors involved, generating new and lasting solutions for social groups, communities or for the society in general "(Silva, 2014 apud Bignetti, 2011, p. 4). Thus, the willingness to learn new techniques to apply them in a collaborative business that generated income for all the women involved was perceived. Aware of the installed capacity, the available raw material and the number of women who have an interest in generating income within the NGO, they signaled an interest in developing and commercializing products (creation, modeling, clothes production and sale) to a specific public with great demand identified by them, aiming to meet their aesthetic, physical and psychological needs.

2.2 Outline of the training course

Mulgan (2014) makes an analysis of how design methods applied to social innovation have contributed to successful project design development by proposing creative solutions beyond the reach of conventional structures and methods. Within this analysis, the four main elements that are classified as effective in the work of designers in social innovation services are highlighted. They are:

– Understanding users' experiences: it means deepening and reaching the roots of the problem that needs to be solved, mapping the users' real experiences, to elucidate new points of view that can help in the desired transformations.
– Ideation: tools for a diagnosis of ideas, which can have a great impact on creativity and help teams to develop a range of wider options. Some of these tools, exemplified below, are easy to use and promote the emergence of more radical ideas:

  1. Inversion (Ex: peasants become bankers, patients become doctors);

  2. Integration (Ex: personal counselors, one-stop counter, portals, accelerated flow);

  3. Addition (Ex: obtaining GPs to do a new test, libraries with speech therapy);

  4. Creative extremism (Ex: pushing ideas and methods to their furthest limits).

– Quick Prototypes: As new ideas begin to appear, they can be tested quickly, allowing for quick learning with practice rather than with very detailed planning. The creation of new systems and services becomes fast and collaborative;
– Systems: designers adopt ideas before a systemic thinking, that is, the focus of the questions occurs in the right direction, considering all the possibilities that could be the real reasons for the problem to exist, thus increasing the chances of solving it (Mulgan, 2014).

Figure 1 shows the design elements for the stage of ideation proposed by Mulgan (2014) in the development of projects in social innovation described under the perspective of the organizational reality of the NGO.

These notes allow the visualization of key situations that are interconnected with the design concepts and help in choosing the tools to be worked out in the Action Plan for Qualification.

2.2.1 Action plan for qualification

The Young Foundation (2010) clarifies that the process of Social Innovation is established not only by its measurable results, but also by the means adopted to reach them. Therefore, new ideas can be classified in products, but also as services or models that meet social needs and create new relationships and collaborations. The basis for building a strategic plan is based on the principle, according to Manzini (2013), of providing motivating concepts capable of providing new knowledge, respecting and supporting individual capacities, in order to make solutions accessible to a greater number of people. In addition, it aims to develop service models as stimulating as compatible with the economic and cultural interests of the participants. Therefore, the steps described here were designed not only to respect the facilities and limitations of the women involved, but also to promote knowledge of the areas of greatest interest, always aiming at improving work opportunities and income generation. The three steps below were intended to expand the existing capacities of the women collaborators of the NGO through an experimental work, in which the theory could be learned in practice, with the aim of providing new skills and articulating other possible solutions for the productive and collaborative work. For this, it was understood that the practical development of creation and assembly of whole pieces (and not only the practice of adjustments as it had been developed up to then) could mean an advance in terms of knowledge and new approaches in future works.

Step 1 Sewing: for a better performance in the development of the pieces, it was shown some knowledge about two techniques of seams normally used in flat fabrics, to which the collaborators of the NGO had no knowledge: French and English sewing. Then, small prototypes of the possibilities of seams were made, with fabric flaps, developed in the straight industrial machine.

Step 2 *Moulage*: The next step consisted of an explanation of the *moulage* technique, with a brief theoretical explanation followed by practice in the development of a blouse base, a skirt base and an interpretation of each, as well as the planning of all the molds.

Step 3 Construction of garments: after presenting these two new know-hows, French and English sewing techniques and *moulage* technique, the training phase was continued, with the objective of deepening this knowledge in a practical way. The purpose of this phase was to create the opportunity for the construction of a systemic thought closer to the reality of the development of fashion products within a production unit. A platform with steps that facilitate this process was created (Figure 2) and later applied in the development of garments.

This platform was applied in the design and development of garments that, theoretically, had different degrees of difficulty in their execution. Figure 3 demonstrates this

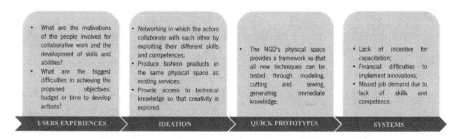

Figure 1. Effective design elements for social innovation related to the research project.

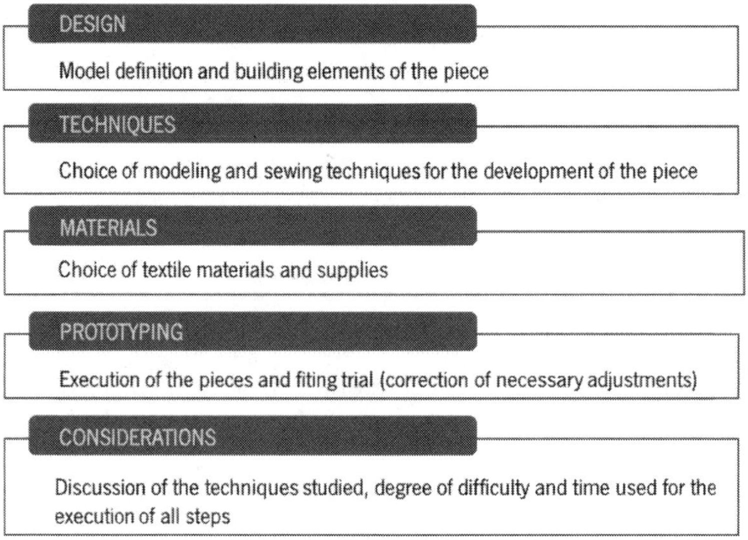

DESIGN
Model definition and building elements of the piece

TECHNIQUES
Choice of modeling and sewing techniques for the development of the piece

MATERIALS
Choice of textile materials and supplies

PROTOTYPING
Execution of the pieces and fiting trial (correction of necessary adjustments)

CONSIDERATIONS
Discussion of the techniques studied, degree of difficulty and time used for the execution of all steps

Figure 2. Platform of steps for the development of garments.

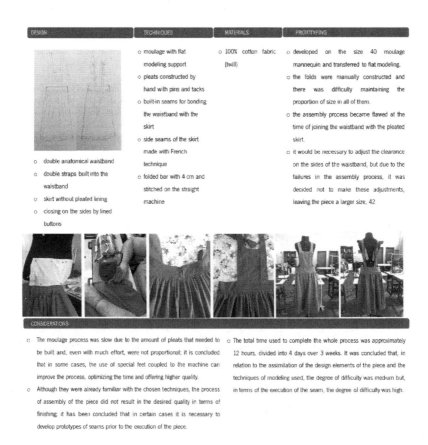

Figure 3. Skirt with anatomical and pleated waistband: Development process through the applied platform.

process applied in one of these pieces and in addition to the description of these steps, photos of the development process and the final result of the pieces were included for better visualization.

3 CONCLUSIONS

After completing this first stage of the Action Plan for Capacitation, it is important to analyze the impact achieved in terms of knowledge, but also to understand the meaning that women gave to their actions.

The work of developing and executing prototypes was essential to achieve what Hillgren and Seravalli and Emilson (2011) understand as ways of creating teams and generating greater capabilities among the members, where each one performs a task and everyone shares the final result. It was noticed that the exchange of information, discussion and decision-making, as well as the difficulties faced, made the involved group more organized and safer to carry out new projects.

Prototyping provided the possibility of visualization at all stages, allowed the transfer of knowledge among all, enabling continuous responses as well as adaptation to the obstacles that have arisen and, finally, generating innovation.

The result achieved came from a complementation between existing knowledge and skills and new possibilities never before experienced. The assembly of the pieces, for example, has rescued the experiences that women have lived through life, often reported verbally through life stories told, since they all learned to sew with their mothers when they were younger. This recovered knowledge was used in conjunction with the new ones, and not only new skills were acquired, but also some weaknesses were recognized, due to the failure of some strategies adopted that did not obtain the best expected result. This fact generated interest in the improvement of the assembly process of pieces and in a better finishing for future works.

For Santana (2014), social innovation is socially constructed as individuals collectively engage in purposeful actions and reflexively monitor the outcome of their actions, and the involvement with the new techniques reflects this once they are incorporated in the work routine of the NGO through, for example, the transfer of the moulage technique to the students of the cutting and sewing course, as well as the sewing techniques, which were inserted in the work of the adjustments of pieces, improving the quality of the service offered.

The authors Moulaert, MacCallum and Hillier (2013) argue that the project in social innovation is closely linked to the sociocultural context of the people and entities involved and allows empirical observations to be made throughout the process, ensuring an ethical position in the actions. This involves getting involved in an open way, respecting some characteristics or even limits that may arise. Manzini (2013) warns that interventions that are to be made in organized communities should not put their balance at risk while retaining their original qualities. These assertions help to realize that the long time spent in the development of garments was due to the fact that the research project activities were fit on idle days when the demand for labor is lower. For the security of women and to guarantee community service, the process could not be accelerated, using the time available for the development of activities that already generate income, to the detriment of training activities.

ACKNOWLEDGEMENTS

This work is supported by FEDER funding on the Programa Operacional Factores de Competitividade-COMPETE and by national funds through FCT—Foundation for Science and Technology within the scope of the project POCI-01-0145-FEDER-007136 and UID/CTM/00264. The authors would like to thank for the support of the Federal University of Technology—Paraná, especially the Fashion Design Department and the Postgraduate Board.

REFERENCES

Ayob, N., Teasdale, S., Fagan, K. 2016. Evolution of Social Innovation Concept. Cambridge University Press, Jnl Soc. Pol. 45, 4, 635–653.

Bogdan R.C., Biklen, S.K. 1994. *Investigação Qualitativa em Educação: Uma Introdução à Teoria e aos Métodos*. Porto, Editora Porto.

Fraser, M.T.D., Goudim, S.M.G., 2004. Da fala do outro ao texto negociado: discussões sobre a entrevista na pesquisa qualitativa. Paidéia, 14 (28), 139–152.

Hillgren P. & Seravalli A. & Emilson A. 2011. Prototyping and Infrastructuring in design for social innovation. Co-Design Vol. 7, Nos. 3–4, September–December 2011, 169–183.

Kim, H.E. 2015. A Study on the Characteristics and Trends of Sustainable Fashion through Esthetica at London Fashion Week. In: Fashion & Text. Res. J. Vol. 17, No. 2, 168–177.

Kocziszky, G., Somosi, M.V., 2016. Generating Social with knowledge engineering. Procedia – Social and Behavioral Sciences 223, 167–174.

Lima, M.A.D.S., Almeida, M.C.P., Lima, C.C. 1999. The utilization of participant observation and semi-structure interview in the research in nursing. *Revista Gaucha Enfermagem*. Porto Alegre, V.20 n. Esp., pp. 130–142.

Manzini, E. 2013. Making Things Happen: Social Innovation and Design. *Design Issues*, MIT Press, 30: 57–66.

Manzini, E. 2008. *Design para a inovação social e sustentabilidade: Comunidades criativas, organizações colaborativas e novas redes projetuais*. Rio de Janeiro: Editora E-papers.

Moulaert, F., MacCallum, D., Hillier, J. 2013. Social innovation: intuition, precept, concept, theory and practice. In: Moulaert, F. & MacCallum, D. & Mehmood, A. & Hamdouch, A (eds), *The International handbook on Social Innovation collective action, Social learning and Transdisciplinary research*. 13–24.

Mulgan, G. 2006. The Process of Social. *Innovation. Innovations: Technology, Governance, Globalization*, volume 1, issue 2: 145–162.

Oenning, J., Cunha, J., Garcia Jr., B. 2017. The relevance of different players on the design project: garment, identity, motivation and social innovation. *In: 1rd International Textile Design Conference, Lisboa 2017*.

Santana, G.C., 2014. Social innovation: Moving the field forward. A conceptual framework. Technological Forecasting & Social Change 82, 42–51.

Silva, J.S.G 2014. Estratégias em design orientadas para a inovação social com enfoque no desenvolvimento local. *PhD Thesis Programa de Pós-Graduação em Design do Departamento de Artes & Design da PUC-Rio*. Rio de Janeiro, Abril de 2014.

Young Foundation 2010. Study on Social Innovation. In: Bureau of European Plicy Advisors. Brussels: EC.

Author index

Abreu, M.J. 287
Adsuara, M. 447
Adverse, A. 37
Albuquerque, P.B. 389
Alcoceba, B. 341, 447
Almeida, M.D. 263
Álvarez Caselli, P. 129
Alves, R. 567
Amaro, M. 45
Amorim, N. 497
Andrade, R.R. 407, 559
Araujo, E. 585
Araujo, I. 585
Arruda, L.M. 683
Azevedo, A. 437
Azevedo, S. 87, 399

Baganha, A. 507
Barata, J.A.B. 399
Barbosa, U.S.T. 95, 637
Barcellos, L.F. 645
Barcelos, S.M.B.D. 599, 667
Barreiros, A.I.C. 631
Barreto, M. 479
Barrocas, L. 241
Baruque-Ramos, J. 659
Beduschi, D.P. 683
Bentz, I. 3
Bernardes, J.P. 623
Bezerra, G. 241
Bona, S.F. 79
Braga, I. 287
Brandão, R. 389
Bravo, U. 521
Brito, D.M. 407, 559
Broega, A.C. 263, 645, 675, 683

Camargo, A. 567
Campos, R. 415
Cano Redondo, A. 169, 543
Cardoso Braga, J. 365
Carvalho, M.A. 415
Castro, I. 585

Castro, S. 303
Cidreira, R.P. 13
Climent Mondéjar, M.J. 543
Conti, G.M. 331
Cruchinho, A. 231, 279
Cruz Bermeo, W. 121
Cunha, J. 103, 303, 691

da Silva, L.R. 349
Dantas, L.U. 95, 637
de Brito, J.P.C. 529
de Carvalho, F.A.H. 71
de Carvalho, M.H. 71
de Francisco, A.C. 599, 667
de Held, M.S.B. 537
de L. Iracet, M. 653
de las Heras, D. 447
de M. Tavares, L.C. 177, 529
de Miranda, A.P. 455
de S. Araújo, V.P. 177
Delgado, M.J. 429
Dell'Acqua, A. 515
Di Lucchio, L. 615
Dieb, H.A. 95, 637
Domingues, I. 455
Duarte, A.Y.S. 373, 381

Effe, R. 63

Faria, A.P. 103
Feldman, V. 537
Fernandes, C.E. 463
Fernandes, S.R. 631
Ferreira, F. 623

Gaddi, R. 161
Garcia Jr., J.B. 691
García, R. 447
García-Ergüín, M. 447
Gennaioli, D. 137
Glück, M. 191
Gomes, F. 507

Gomes, S.H.A. 373, 381
Gomez, L.S. 469
Gonzalez, F. 521
Grácio, A.H. 489
Guedes, M.G. 593, 667

Honório, I.D. 631
Horta, A. 37

Italiano, I.C. 421

Jerónimo, N.A. 87
Jordão, C. 683

Karam Jr., D. 27

Leal, D.V. 27
Lenzi, G.P. 199
Lima, M.S. 349
Lin, X. 515
Lobo, T. 255
Lucas, J.M. 231, 631
Luceño, L. 295

Machado, B. 53
Madeira, M.J. 463, 631
Mansur, M. 575
Marcelo, A.S. 279
Marques, A.D. 623
Marques, G.D. 575
Marques, R.R. 95
Martínez-Medina, A. 169
Melo, B. 675
Mentone, D.A.N. 373
Mesquita, C. 215
Miguel, R. 87, 231, 399
Montagna, G. 429
Morais, C. 303
Moreira da Silva, A. 21
Moreira da Silva, F. 365
Moreira, B. 437
Moreira, S. 231
Moreno Moreno, M.P. 543
Moreno, A. 497

Moura, A. 279
Moura, M. 263

Nogueira, M. 623
Norogrando, R. 271
Noronha, R.G. 349

Oenning, J. 691
Okasaki, A. 207
Oliveira, I. 593
Oliveira, M.M. 287
Osava, R.H. 373
Otani, M.M. 421

Palmares, H. 497
Pan, Y. 223
Pennas, L. 659
Pereira, L.M. 407, 559
Pereira, M. 87, 231, 463
Peres, P. 279
Peroba, A.R.V. 215
Pinheiro, E. 599, 667
Pinheiro, N. 407, 559
Pino, B. 145

Pinto, I. 585
Pires, B.F. 13
Providência, B. 103, 389
Puppim, R. 683

Rabàdan, A. 3, 323
Rebello, L.E.F.S. 575
Reis, B.M. 87
Ribeiro, L. 463
Ribeiro, L. 479
Ribeiro, L.W. 469
Ribeiro, R. 37
Rijo, C. 489
Rocha, M.A.V. 241
Rodrigues, M.C.P. 27
Rojas, I. 145
Ruiz, C. 521

Sales, G.M.J. 95, 637
Sampaio, J. 497, 507
Sanches, M.C.F. 551
Sanches, R.A. 373, 381, 615
Santos, A.F.B. 247
Sbordone, M.A. 615

Scoz, M. 53, 79
Silva, A.A. 37
Silva, M.A.R. 551
Silva, M.S. 463
Silva, P.P. 421
Souza, L.N. 381
Souza, N. 455
Souza, P.M. 421
Spaine, P.A.A. 407, 559
Spry, F. 313

Targino, A.N. 637
Terroso, A.M. 497, 507

Vallejos, F.P. 357
Velanes, E.N. 607
Viana, T.C. 37
Vicentini, C.R.G. 381
Vidal Miranda, J. 129
Vieira, G.R.R. 183
Vignes, L. 113
Vilela, A.P.X. 247

Watson, L. 153